"From the perspect
book has very littl
contains 20th-centu
the religious faith o
have been aware of such a concept. … Controversy may assist book sales, as happened in the case of The Satanic Verses, but it would be a cynical and questionable strategy to publish a book simply because it arouses the wrath of many people. Simply to publish this work as it is basically gives this religious group a platform to express their distinctive theology, which is highly polemical and dismissive of other perspectives…" – **An anonymous Sunni Scholar who was described by the editor of Palgrave-Macmillan as "a very well-established professor."**

"A bold and beautiful translation that serves as a timely reminder to all affirmers that the Qur'an is not a static scripture, but a living, breathing, ever-evolving text whose sacred words are as applicable today as when they were first uttered by the Prophet Muhammed fourteen centuries ago." - **Reza Aslan, PhD., former CBS and CNN News Consultant; Author, *No god but God: The Origins, Evolution, and Future of Islam*. Zealot: The Life and Times of Jesus of Nazareth.**

"A testament to the fact that faith need not suffocate reason. This is bound to be among the smartest of 'smart bombs' in the battle of ideas within Islam." **- Irshad Manji, Fellow, Yale University, and author, *The Trouble with Islam Today: A Muslim's Call for Reform in Her Faith.***

"Every conversation begins with a single voice. This Reformist Translation of the Quran and its ancillary materials should begin many conversations, between and among Muslims and non-Muslims alike. In many parts of the Muslim world this is a dangerous discussion, and sometimes that danger can reach well into the West, as evidenced by the 1990 fatwa-inspired murder of Rashad Khalifa in Tucson, Arizona. It is an important discussion, however, and the editors of this book have assumed this risk to argue for a perspective that sets violence aside both in discourse and living. One can imagine that a broader adoption of their perspective across the Muslim world would reduce strife and invite greater examination of Islam by non-Muslims as something other than a threat. It would expand the conversation." **- Mark V Sykes Ph.D. J.D. Director, Planetary Science Institute.**

"I completely agree with you in your rejection of the right of any group to arrogate to themselves the sole interpretation of the Quran. The Quran, being a book containing divine knowledge and wisdom, can only be understood progressively. It has to be interpreted anew by every generation and through a scientific methodology.... Your effort is praiseworthy. Well done. Keep it up." **- Kassim Ahmad, former president of Malaysian Socialist Party and head of Malaysian Quranic Society who was declared "apostate" by religious authorities for his work.**

"A timely and stimulating contribution to scholarship on Islam that offers cogent testimony to the diversity of views within the Islamic community. This new translation challenges those in East and West alike who see Islam as irreconcilably opposed to the scientific and democratic impulses of modernity." - **Germaine A. Hoston, Ph.D, Professor, University of California. San Diego, former Director of the Center for Democratization and Economic Development.**

"This translation is the best tool for those who want to understand the uncorrupted Message of Islam - justice and peace. This translation shows that the Quran is but the confirmation and continuation of God's system memorialized through Abraham, demonstrated in Torah through numerous prophets, and in the Hebrew Gospel through Yeshu'a/Jesus, the righteous of God. This translation is a message of peace, justice and judgment. I pray that the Reformist translation of the Quran will replace all others not only because it is the best but also because it is the closest to the original Arabic text." **- Gershom Kibrisli, theologian and communal leader, The Karaim of the Early Hebrew Scriptures, Holy Land & Benelux.**

"With its lucid language, brilliant theological and philosophic arguments, Edip Yüksel removes the smoke of distortions and ignorance generated by clergymen that have concealed the light of the Quran from masses. Pulling our attention to numerous scientific evidence supporting the authenticity of the divine nature of the Quran, the Reformist Translation is destined to create a Copernican Revolution in the realm of religions. I highly recommend it to Agnostics, Skeptics, Christians, Hindus, Buddhists, or anyone who seeks truth about God without suppressing or compromising their brains." – **Caner Taslaman, PhDs in Theology and Political Science; Post doc fellow at Harvard and Oxford Universities; Prof of Philosophy at Yildiz Tech Uni; author, *The Quran: Unchallengeable Miracle; The Big Bang and God; The Theory of Evolution and God;* and *The Invented Religion*; www.quranic.org.**

"Allah has gifted humanity with many signs for guidance. The Qur'an not only directs us towards these signs all over creation, but also in itself is a most miraculous sign. We can never know the full meaning of the Qur'an even as we exercise our minds and peacefully surrender our hearts to Allah so that we become able to read the signs and grow to know more. A Reformist Translation directs us to this miracle by offering an intense and challenging addition to the practice of sincere reading for knowing. I cannot accept its Qur'an only perspective, even as I support the efforts of these translators to engage, as they have, in reading and growing with knowledge while relaying to others some new possibilities of meaning for the sake of reflection and peacefully surrender. I hope many will examine their efforts to gain benefit and challenge." ***- Amina Wadud, PhD, Author: Inside the Gender Jihad: Reform in Islam.***

"*Quran: A Reformist Translation* is distinct from other translations of the Qur'an in several important ways. First, to the best of my knowledge, it is only the second English translation of the Qur'an produced by *Qur'anists*--advocates of the concept of the Qur'an as the sole legitimate scriptural source of religious law and guidance in Islam. As *Qur'anists*, Yuksel and his colleagues reject the *Hadith* as sources of religious law and guidance and do not rely on them in this translation and commentary. The first *Qur'anist* English translation was done by the late Rashad Khalifa, a seminal figure in the late twentieth-century *Qur'anist* movement who directly influenced both Yuksel and Shulte-Nafeh. *Quran: A Reformist Translation* is also unique because it is the product of collaboration between two key figures in the present-day *Qur'anist* movement: Edip Yuksel and Layth Saleh al-Shaiban. The *Qur'anist* approach offers religious rather than secularist challenges to traditional understandings of Islam, whether Sunni, Shia, or academic, on a number of critical issues; so this translation and commentary have the potential to spark extreme controversy among Muslims and non-Muslims." - **Aisha Y. Musa, PhD, Assistant Professor of Islamic Studies, Florida International University; author of *An Examination of Early and Contemporary Muslim Attitudes toward Hadith as Scripture* (Doctoral Dissertation, Harvard University, 2004).**

"Very Interesting and Timely" ***- Riffat Hassan, Ph.D. Professor of Religious Studies and Humanities at the University of Louisville, Kentucky. A pioneer of feminist theology in the context of the Islamic tradition.***

"I have to say that this translation --following in the same spirit of Dr. Khalifa's *Quran: The Final Testament* -- not only is daring, brave and non-traditional in its approach, it will open a lot of eyes that the Quran is dynamic in nature and relevant for all times. The spirit of *using the Quran to explain the Quran*, namely trying to understand a certain word by searching for the meaning of the same word in many different contexts within the Quran, is very evident in this translation. This is a book that encourages the reader to use his/her intelligence faculty in order to understand the message, true to the key message of the Quran itself." **- Gatut Adisoma, PhD, Indonesia.**

"I strongly suggest that this English representation of the Qur'an can only be fully appreciated by slowly absorbing it, cover to cover. As well, I suggest that the most integrated understanding of the Qur'an can only be realized by synthesizing the full message in one's heart first, as a single experience. With this in mind, this Reformist Version does an unusually fine job in clarifying those elements (such as gender imbalance) which have been perceived as dissonant within the whole message in the "standard" translations. This version, which is not revisionistic, presents an integrated consistency rarely found in other translations and it elucidates issues not commonly grasped by modern readers (in any language). Those with an open mind and heart, who only understand modern Arabic and not the dialect in which it was originally revealed, have the opportunity to experience comfort and inner peace by absorbing this clean, Reformist translation. With this in mind, this version can only be judged following a thoughtful read of the entire volume…" **- Jeff Garrison, MD, Colorado.**

"THANK YOU, DEAR GOD THANK YOU, I'm coming back because of your work. Edip, your work has freed me from year of condemnation, cruelty, and misinterpretation of Islam by my ex-husband. Your work has freed me from the pain I've carried for so long and gave me back basic self-esteem that was mine from God but was slowly eroded by misogyny. Your work gave me the wings I had lost. Again, thank you." **Martina D., a reader.**

*****

For a sample reader response please see the last Appendix.

EDIP YUKSEL is a Kurdish-Turkish-American author, lawyer, philosophy professor, inventor and a progressive activist who spent four years in Turkish prisons in the 1980's for his political writings and activities promoting Sunni revolution in Turkey. He experienced a paradigm change in 1986 transforming him from a Sunni Muslim leader into a rational monotheist, a peacemaker. Edip Yuksel has written more than thirty books and hundreds of articles on religion, politics, philosophy, and law in Turkish and English. After receiving his B.A. from the University of Arizona in Philosophy and Near Eastern Studies, Edip received a J.D. degree from the same university. Edip teaches Philosophy and Logic at Pima Community College. He is inventor and co-founder at Beltways. He is fluent in Turkish, English and Classic Arabic, proficient in Persian, and barely conversant in Kurdish, his mother tongue. Books include; Quran: a Reformist Translation, *Manifesto for Islamic Reform; NINETEEN: God's Signature in Nature and Scripture; Running Like Zebras, 10 Questions for Atheists; Peacemaker's Guide to Warmongers. Test Your Quranic Knowledge.*

LAYTH SALEH AL-SHAIBAN is an author of various books and articles on Islam, founder of Progressive Muslims, and co-founder of Islamic Reform. Layth works as a financial adviser, and lives in Saudi Arabia.

MARTHA SCHULTE-NAFEH is a retired Professor of Arabic language at the University of Texas at Austin. Before, she worked as Assistant Professor at the University of Arizona. Martha received her B.S. from Wharton School, University of Pennsylvania in Economics, received her M.A., in Linguistics from the University of Arizona in 1990, and her Ph.D. from the same university in NES - Arabic and Linguistics 2004. She also taught English as a Foreign Language at American University in Cairo, Egypt.

LESLIE TEREBESSY joined our team in 2022, providing initial editorial contribution for the special edition of Oxford Institute for British Islam. Leslie taught at the Islamic Science University of Malaysia (2002-2008). He was a Research Fellow at IAIS (2008-2017). He earned an M.A. in political philosophy from the University of Toronto, (1999); an M.Ed. (1986) at the University of Toronto's Ontario Institute for Studies in Education (OISE) and a B.A. (Honors) from the University of Guelph (1976). His interests include Islam, education, literature, exegesis, and finance.

Edip Yuksel &
Layth Saleh al-Shaiban:
Translated the main text of the Reformist Translation of the Quran.
Edip Yuksel:
Authored the annotations, subtitles, endnotes, introductory materials, and appendices.
Martha Schulte-Nafeh:
Provided linguistic consultation and feedback.
Leslie Terebessy:
Joined the team in 2022. Provided substantial editorial support.

Cover design: Mustafa Nazif

A message for those who prefer reason over blind faith, for those who seek ultimate peace and freedom by submitting themselves to the Truth alone.

- 19.org
- quranix.org
- pima.academia.edu/EdipYuksel
- twitter.com/edipyuksel
- twitter.com/19org
- youtube.com/user/edipyuksel

For Kindle or printed version, visit:

- amazon.com/s?k=Edip+Yuksel

This is the revised 2022 edition of QRT, in which keywords such as *Muslim* is translated as Peacemaker, *Mumin* is translated as Affirmer, *Hajj* as Debate, *Malak* as Controllers, *Deen* as System. This is work in progress. God willing, there will be a major revision to improve the language and style of the translation.

You may contribute to the improvement of this translation by visiting our site:

- brainbowbooks.com

QURAN
A Reformist Translation

Translated and Annotated by
Edip Yuksel
Layth Saleh al-Shaiban
Martha Schulte-Nafeh
Leslie Terebessy

- Quran is the most read, the least understood and the most betrayed book in the world. However, the light of the Quran is shining again.
- The Reformist Translation of the Quran offers a non-sexist and non-sectarian understanding of the divine text; it is the result of collaboration between three translators, two men and a woman.
- It explicitly rejects the authority of the clergy to determine the meaning of disputed passages.
- It uses logic and the language of the Quran itself as the ultimate authority in determining meanings, rather than ancient scholarly interpretations rooted in patriarchal hierarchies.
- It offers extensive cross-referencing to the Bible and provides arguments on various philosophical and scientific issues.
- It is God's message for those who wish to progress, those who prefer reason over blind faith, those who seek peace and ultimate freedom by submitting themselves to the Truth alone.

brainbowpress
Hundred Fourteen Books
United States of America

A few reminders

- Do not accept information on faith; critically evaluate it with your reason and senses (17:36: 10;100)
- Ask the experts if you do not know (16:43)
- Use intelligence, reason, and historical precedents to understand and carry out God's commands (7:179; 8:22; 10:100; 12:111; 3:137)
- Do not dogmatically follow the status-quo and tradition; be open to new ideas (22:1; 26:5; 38:7)
- Be open-minded and promote freedom of expression; listen to all views and follow the best (39:18)
- Do not follow conjecture (10:36,66; 53:28)
- Study God's creation in the heavens and land; explore the beginning of creation (2:164; 3:190; 29:20)
- Attain knowledge, since it is the most valuable thing in your appreciation of God (3:18; 13:13; 29:43,49)
- Be a free person, do not follow crowds, and don't be impressed by crowds (2:112; 5:54,69; 10:62; 39:36; 46:13)
- Do not follow blindly the religion of your parents or your nation (6:116; 12:103,112)
- Do not make profit from sharing God's Message with others (6:90; 36:21; 26:109-180)
- Read to know, and read critically (96:1-5; 55:1-4)
- Do not ignore divine revelation and signs (25:73)
- Do not be pedantic; do not miss the main point by indulging in small and inconsequential details (2:67-71; 5:101-102; 22:67)
- Hold your judgment if you do not have sufficient information; do not rush into taking a position (20:114)
- Know that God, the Gracious, is the teacher of the Quran (55:1) and unlike dead prophets and saints, God hears you.
- The Quran should be translated to all world languages (26:198-200; 41:44). Monotheists will get the clear divine message despite errors in translations since they study comparatively by using their God-given critical thinking skills. (17:36; 39:18) The teacher of the Quran is God Himself (11:1; 55:1-2).
- Quran is the most read, the least understood and the most betrayed book in the world. (See 25:30)
- In year 1974, exactly 19x74 lunar years after the revelation of the Quran, by revealing the secret of the number 19 prophesied in Chapter 74, through a messenger (3:81; 72:19-28; 40:28-78)) whose name mentioned 19 times in the Quran and the creature prophesized in 27:8), God has started a new era. The liberating and progressive light of rational monotheism has started shining. (74:30-37).

Following a Fatwa-Review of Establishment, Palgrave/Macmillan Abandoned the Publication of this Book

In 2004, my colleagues and I signed a contract with Palgrave/Macmillan publishing house for the publication of Quran: a Reformist Translation. The editor and other staff of the publishing house were very encouraging and enthusiastic, and during the summer of 2006, I was personally introduced to the director of the publishing company at its New York headquarters. Palgrave even published an announcement about the upcoming Reformist Translation in their 2006 Fall/Winter Catalogue, which was later postponed to the summer of 2007. The publishing house posted information about the Reformist Translation for pre-orders at Amazon.com and other online bookstores. However, in December 2006, the editor informed me that the board had determined that my manuscript was not acceptable for publication.

Apparently, they were convinced or intimidated by a review (more accurately, a fatwa) of "a very well-established professor," who misleadingly likened our annotated translation of the Quran to Salman Rushdie's Satanic Verses. This was akin to a medieval publishing house turning down Martin Luther's 95 Theses after consulting "a very well-established" Catholic Bishop! It is telling that Palgrave's "very well-established scholar" in his several-page review, had only one substantive criticism, which consisted of our usage of a word, yes a single word in the translation: *progressive*.

I believe that without hearing my defense against this Sunni version of excommunication in the guise of a "scholarly review," the publishing house committed an injustice against my person and our work. I called the publishing house and asked them to give me the chance to respond to the reviewer and defend myself and work against his disparagement and distortions; I was told he remain anonymous.

We were not surprised to hear negative remarks, insults, or false associations from a reviewer who considers a rejection of backward and bankrupt sectarian dogmas "heresy." However, we were surprised to learn that the board of the publishing house cancelled the publication of a potentially controversial yet crucial book that would introduce the message of the Quran—the message of peace, justice, reason, and progress—without the distortion of sectarian teachings. Any scholar who can see beyond his or her office can see the growing reform movement, open or clandestine, particularly in Turkey, Malaysia, Iran, Egypt and Kazakhstan where people take great risks to question popular dogmas.

You may visit the following websites for the full text of the letter of the Sunni scholar whose advice was taken at face value by Palgrave/Macmillan, and for our response to the letter. You may also find in the following websites, recent updates, reactions, and feedback from reviewers, our responses, and the activities of the global reformist movement:

www.19.org
www.islamicreform.org

Let the world hear the message. Let the West hear the voice of monotheism, the voice of reason, peace, justice and progress. Let the East and the Middle East hear the clear message of the book that they have abandoned for centuries.

9:32 They want to extinguish **God**'s light with their mouths, but **God** refuses such and lets His light continue, even if the ingrates hate it.

9:33 He is the One who sent His messenger with guidance and the system of truth, to make it manifest above all other systems, even if those who set up partners hate it.

Edip Yuksel
Tucson, Arizona
July 2007

ACKNOWLEDGEMENT: We thank the following individuals for their feedback, criticism and suggestions: El Mehdi Haddou, M.D., Aisha Musa, Ph.D.; Germaine Hoston, Ph.D.; Mark Sykes, Ph.D., J.D; Chris Moore; Linda Moore; Engin Gurkan, Ruby Amatullah; Oben Candemir, M.D.; Miriam Janna; Daniyar Shekebaev; Arnold Mol, Caner Taslaman, Ph.D.; Timur Irdelp, Abdullah Al-Mamun, Ali Seyed, and the thousand others who shared their knowledge with us through their words, mails, books, and critical reviews. Nima Aghili, J.D., and Ahamed Abdou for their feedback and extensive editorial assistance. We should also mention Muhammed Kodzoez and Tufan Karadere for volunteering their technical skills to design quranix.org and 19.org, particularly a private internet forum that facilitated our early discussion on the methodology and key words. Special thanks to Ogan Timinci and his son Tayfun. Kindle and Electronic versions were made possible by brother Omar AL Zabir of UK

Table of Contents

We decided that it is unnecessary to insert page numbers in the translation section, since each verse is indicated with a number and they are ordered within chapters starting from verse 1:1 ending with verse 114:6. The sum of all verses in the Quran is 6346, including the 112 unnumbered Basmalahs.

Asterisks (*) in the end of verses refer to our notes, comments, cross-references, and discussions in the ENDNOTES section, which follows immediately the respective chapter. To visually separate ENDNOTES from the text of the translation of the Quran, we used a smaller font, six digits for verse numbers, and full justification for the text.

The subtitles in *italics* are not part of the Quran. We inserted subtitles to make the reading easier for those whose attention might be compromised by fast-paced modern life complicated with numerous tasks and technologies

Translators' Introduction:

The Case for a Reformist Translation of the Quran

> "And you shall know the truth, and the truth shall make you free." (Old Testament, Isaiah 61:1; 42:6-7; New Testament, John 8:32).
>
> "And had We sent it down upon some of the foreigners, and he recited it to them, they would not have affirmed it." (26:198-199)
>
> "Had We made it a non-Arabic compilation, they would have said, 'If only its signs were made clear!' Non-Arabic and Arabic? Say, 'For those who affirm, it is a guide and healing. As for those who reject, there is deafness in their ears, and they are blind to it. These will be called from a place far away.'" (41:44)
>
> "A book that We have sent down to you, that is blessed, so that they may reflect upon its signs, and so that those with intelligence will take heed." (38:29)
>
> "The messenger said, 'My Lord, my people have deserted this Quran.'" (25:30)
>
> "Say, 'O people of the book, let us come to a common statement between us and between you; that we do not serve except God, and do not set up anything at all with Him, and that none of us takes each other as lords beside God.' If they turn away, then say, 'Bear witness that we have peacefully surrendered.'" (3:64)
>
> "... and God made a covenant with those who received the Book: 'Proclaim the scripture to the people, and never conceal it.'" (3:187)

Arguments against the practices and teachings of "Orthodox Islam" have created controversies in the past and are likely to continue to do so in the future. But the Holy Quran, the word of God Almighty, demands that its words be proclaimed without misdirection or obstruction (3:187).

This volume is undertaken in obedience to this command. *Quran: A Reformist Translation* is an English version of the Quran that takes an accurate reading of what is in the Quran itself as the standard. It thus abandons the rigid preconceptions of all-male scholarly and political hierarchies that gave rise to the series of writings and teachings known as the "*Hadith & Sunna*", which, according to the Quran itself, carry no authority (9:31; 42:21; 18:110; 98:5; 7:3; 6:114). It is a progressive translation of the final revelation of God to all of humanity – a translation that resonates powerfully with contemporary notions of gender equality, progressivism, and intellectual independence. It is a continuation of the modern monotheist movement that started an era of paradigm change and reform in 1974 with the fulfillment of a Quranic prophecy based on the number 19 mentioned in chapter 74. [1]

---

1 As the fulfillment of a great prophecy, this discovery provided a verifiable and falsifiable powerful evidence for God's existence, His communication with us, and perfect preservation of His message. Besides, it saved monotheists from the contradiction of trusting past generations regarding the Quran, knowing that they have fabricated, narrated, and followed volumes of man-made teachings, demonstrated ignorance, distorted the meaning of the Quranic words, and glorified gullibility and sheepish adherence to the fatwas of clerics. (See Chapter 74 and Appendix 3).

Centuries after the revelation, understanding of the Quran was inevitably influenced by the cultural norms and practices of tribal cultures in ancient Arabia, which were attributed to Prophet Muhammed and his close friends, and introduced as secondary religious sources besides the Quran. As the sample comparisons offered below demonstrate, these norms and practices distorted the perceptions of what is in the text. When translation is liberated from these traditions the Quran conveys clearly a message that proclaims freedom of faith, promotes male and female equality, encourages critical thought and the pursuit of knowledge, calls for accountability and repudiation of false authority, as well as the replacement of political tyranny and oppression through representation in government, Above all, it is God's command for the realization of justice for every man, woman, and child irrespective of ethnic origin or religion.

By presenting the peaceful and unifying message of the Quran, we hope to increase understanding and reduce tensions between the "Muslim World" and people of other religions, especially those whom the Quran calls the People of The Book (Jews and Christians). This translation will also highlight, without apology or distortion, the major differences between our approach and that of orthodox translations and commentaries.

Quran: A Reformist Translation offers a non-sexist and non-sectarian understanding of the divine text; it is the result of collaboration between three translators, two men and a woman. We use logic and the language of the Quran itself as the ultimate authority in determining likely meanings, rather than previous scholarly interpretations. These interpretations, though sometimes useful as historical and scholarly reference resources, are frequently rendered inadequate for a modern understanding and practice of Islam because they were heavily influenced by patriarchal culture, relied heavily on the hearsay teachings falsely attributed to the prophet Muhammed, and were frequently driven by hidden or overt sectarian and political agendas. We therefore explicitly reject the authority of the clergy to determine the likely meaning of disputed passages.

In the Reformist Translation, we also offer extensive cross-referencing to the Bible and attempt to provide scientific and philosophical reasoning to support and justify the translation. We intend for the translation to reflect the original message of the Quran for those who have scholarly or personal curiosity in it, and to provide an alternative perspective, unfettered by the constraints of uncritically accepted interpretations that rely on hearsay accounts.

We argue that any modern commentary on the Quran – and all translations are, by definition, commentaries upon the Arabic text – should not be monolithic but should instead reflect the perspective and critical evaluation of diverse disciplines and populations. We also argue that the voices of women, suppressed for so many centuries by Sunni or Shiite alike, should be taken into account in any interpretation of these extraordinary verses. To correct the egregious historical biases so obvious in previous English translations, we have chosen to take an inclusive approach incorporating input from scholars, lay readers, and even non-Muslims. The final word choices of the actual translation, however, are ours. We alone are responsible for them before God; if we have made an error, we appeal only to God for forgiveness.

Since we reject all man-made religious sources and accept only God's signs in nature and scripture as the guide for eternal salvation, one might wonder about the presence of subtitles and extensive endnotes in this translation. We would like to emphasize that the subtitles and endnotes are not part of God's word and they do not constitute a source or authority. We think that this is the best available English translation and the most accurate in its rendering the meaning of the scripture. Any translation, by the very nature of the deficiency of one-to-one correspondence among human languages, is not identical to the original text and reflects our fallible understanding that is tainted with limited time, limited knowledge and human weaknesses. However, a translation done with a monotheistic paradigm and clear mindset, despite some inevitable loss in rendering the richness and eloquence of the original text and minor translation errors, will convey the intended message of the Quran to the world in any language.

> 41:44 Had We made it a non-Arabic compilation, they would have said, 'If only its signs were made clear!' Non-Arabic and Arabic, say, 'For those who affirm, it is a guide and healing. As for those who reject, there is deafness in their ears, and they are blind to it. These will be called from a place far away.'

One might notice that we did not refer to books of *Hadith and Sunna*, since they are commonly idolized and associated as partners with the Quran. Their perceived value is wrongly based on the sanctified names of the narrators and the authority of the collectors, rather than their substance. Muslims, or peacemakers and submitters to God alone, might read and benefit from studying any book, including books of *Hadith* and sectarian jurisprudence, without considering them infallible authorities or partners with God's word. Similarly, we might benefit from the ideas of philosophers and scientists regardless of their religious position or affiliation.

The Quran puts the search for truth and its affirmation as a top priority for us and uses the word "truth" more than two hundred times. Additionally, the Quran uses the word "the truth" or "Truth" (*Haq*) as an attribute of God. We should affirm and appreciate any truth, be it in the scripture or nature. Search for truth should be continuous since we are imperfect humans. However, the only truth that is relevant for our eternal salvation is the one expressed and implied in the Quran, which is basically the same with the message of revealed books in the past. When we assert that God alone is the authority in defining the system of Islam, we mean that no signature and no authority beside God will be considered as justification for the truth-value of a proposition regarding Islam. Thus, a statement made by this or that scholar or clergymen, by this or that messenger or prophet, has no truth-value just because they said so. Prophets, messengers, and scholars, all are humans like us, and their personal statements or understanding of God's word have no authority in defining Islam, the system exclusively defined by God (9:31; 42:21; 18:110; 98:5; 7:3; 6:114).

Each of us is responsible for our own understanding. We may exchange views and benefit from one another's insights, but we should never value each other's understanding on the basis of rank, diploma, reputation, the number of halos, the size of turban, or the length of the beard (17:36; 10:100; 39:18).

Thus, our understanding, inferences and arguments in this translation should not be accepted just because they are those of scholars in theology, linguistics, or philosophy. The reader is invited to take a critical perspective to this as well as to other translations, thinking carefully about the words and following those that seem the best (39:18), The point is that as each individual is responsible alone to God on the Day of Judgment the reader must not delegate their intellectual responsibility to others but rather search constantly and sincerely for better understanding (20:114), knowing that only the truth can liberate one from falsehood and idolatry.

To assist the reader in this endeavor, we have chosen to include subtitles and endnotes for the following reasons (Endnotes are indicated by asterisks placed in the end of verses):

1. The translation from one language to another can cause the loss of meaning or ambiguity. For instance, translating an Arabic word with multiple alternative meanings into an English word that does not reflect all those meanings, would limit the rich implication of the original text. For instance, the Arabic word *AYAAT* in its plural form is used in the Quran to mean signs, miracles, lessons, and

revelation; and the usage of the word *KaFaRa* has additional contextual meaning. We needed to inform the reader about the various meanings of these words so that they would not be confused regarding different word choices, depending on their context. (See 2:106,152)

2. When it was possible, we preferred a word-by-word literal translation, with which we had to compromise the fluency of the language. However, in the case of language-specific idioms, we preferred to convey the intended meaning. The Quran provides us with guidelines in translation. For instance, one of the purposes of the repetitive mentioning of some events is to teach us how to translate. The Quran translates some historical non-Arabic conversations to Arabic in different words. Please compare 7:12-17 to 15:32-40; 11:78 to 15:67-74; and 20:10-24 to 27:7-12 & 28:29-35. These examples instruct us to focus on conveying the meaning rather than being constrained by the literal translation. We occasionally used the endnotes to inform the reader about why we translated a particular word or phrase the way we did.

3. We also utilized the subtitles and endnotes to alert the reader to orthodox or sectarian distortions.

4. Since the participants of this annotated translation are independent thinkers, and have different backgrounds, education, and experience, naturally there are some differences in our understanding some verses. The endnotes are an attempt to accommodate some of these differences.

5. Through endnotes, we also aimed to provide the readers, especially the Christian readers, some cross-references to the Bible, which shares a common message and numerous events and characters with the Quran.

Space does not permit a detailed discussion in this Introduction of all the methodological and philosophical issues that have arisen and how we have resolved them. For the reader's convenience, we have incorporated some of these into Appendices, as follows:

Appendix 1:	Some Key Words and Concepts
Appendix 2:	The "Holy" Viruses of the Brain
Appendix 3:	"On it is Nineteen"
Appendix 4:	Which One do you See: Hell or Miracle?
Appendix 5:	Manifesto for Islamic Reform
Appendix 6:	Why Trash All the *Hadith*s?
Appendix 7:	A Forsaken God?
Appendix 8:	Eternal Hell and the Merciful God?
Appendix 9:	There is No Contradiction in the Quran
Appendix 10:	*Sala* Prayers According to the Quran
Appendix 11:	Blind Watch-Watchers or Smell the Cheese
Appendix 12:	Sample Comments and Discussions on RTQ

This translation will certainly contain errors. However, we think that its potential errors do not have effect on its message. When working on this translation we tried to remember 20:114; 55:1-2; 42:38; 2:286; 22:2-3; and 11:88 frequently. We would like to extend our gratitude to those who review and evaluate this translation and inform us about its errors. We invite the reader to help us improve this translation by entering your corrections and suggestions at:

brainbowbooks.com

Before presenting you a comparative analysis of several verses, we should introduce you a general description of the Quran and its content.

The Quran

Al-Qur'an (also transliterated as Koran or Kuran), which means The Reading or The Recitation, is a proper name, not a generic common adjective. Thus, it rejects the existence of any other Quran (10:15). Besides, though the Bible and Psalms are also readings, they are never referred to as al-Quran or Quran, but as *Injeel*, *Torah*, or as *Zabur*. Each have their own dictionary meanings, like all other Arabic and Hebrew proper names do. If the Quran was not a proper name of the book revealed to Muhammed but just a generic common word meaning "reading/recitation," it should have also been used for other books. The Arabic words *Kitab* (book/law), or *Zikr* (message), on the other hand, are used as generic or common names, since they are used to not only to refer the Quran, but to the *Torah*, the *Injeel*, and all other books revealed to numerous prophets. Conversely, the Quran has not been used even once for other books.

Al-Quran is used only for a particular book given to Muhammed. So, the Quran is unique. There is no other quran besides the Quran. Some may suggest translating the word Quran as "reading," but the word "reading" is not unique, but the Quran is. Note that the Quran refers to itself more than sixty times, and mostly as a single word. In Arabic, a "reading" is *Qiraa*; Arabic textbooks or reading books are called *al-Qiraa*, not *al-Quran*.

A complete study has shown that the scripture uses the unique name of 'Quran' when speaking about certain aspects of itself or properties that are conveyed:

- Quran is to be used to warn people (6:19).
- Quran has been made easy to learn (54:17).
- Quran is in the hearts of those who are endowed with knowledge (29:49).
- Quran is preserved in a master record (85:21-22).
- Quran is an interesting book (72:1).
- Quran cannot be replicated by humans and jinn (17:88).
- Quran is a numerically structured book (83:9, 20).
- Quran is a book of prophecies (74:1-37; 27:82-84; 44:10:15; 54:1-6; 2:26; 18:9-27; 25:4-6; 10:20; 41:53; 38:1-12; 3:81...).
- Quran can humble a mountain (59:21).
- Quran contains a healing for the hearts (17:82).
- Quran is what the prophet's followers have abandoned (25:30).
- Quran is without doubt revealed by a higher power (27:6).
- Quran is to be paid attention to when read (7:204).
- Quran disturbs the ingrates with its words (10:15).
- Quran gives detail to the divine laws (10:38).
- Quran is related to the seven pairs (15:87).
- Quran places barriers for those who have no intention to understand it (17:46).
- Quran alone makes the rejecters run away in aversion (17:47).
- Quran is recited at dawn (17:78).
- Quran is full of examples cited for mankind (17:89).
- Quran contains the *'Zikr'*/Reminder (38:1).
- Quran disturbs the rejecters; they seek to speak over it (41:26).

Though most of the followers of Sunni and Shiite sects show apparent respect to the physical papers where the 114 chapters of the Quran is recorded or listen to its recitation with utmost respect, for centuries they have adopted sectarian teachings contradictory to the Quran. They read it without understanding; they listen to it without hearing. Even if they understand its message, they prefer following the teachings of their scholars or hearsay narratives falsely attributed to Muhammed.

Since its revelation, tens of thousands of books have been written to analyze, to study, to understand, to distort, to praise, or to criticize the Quran. Here, I would like to quote from Harold Bloom's description of the Quran's literary aspect:

> "The Koran, unlike its parent Scriptures, seems to have no context... Strangely as the other Scriptures are ordered, they seem models of coherence when first contrasted to the Koran. The Koran has one hundred and fourteen chapters or sections (called suras) which have no continuity with one another, and mostly possess no internal continuity either. Their length varies enormously, their order has no chronology, and indeed the only principle of organization appears to be that, except for the first sura, we descend downwards from the longest to the shortest. No other book seems so oddly and arbitrarily arranged as this one, which may be appropriate because the voice that speaks the Koran is God's alone, and who would dare to shape his utterances?
>
> "Sometimes I reflect that the baffling arrangement (or lack of it) of the Koran actually enhances Muhammed's eloquence; the eradication of context, narrative, and formal unity forces the reader to concentrate upon the immediate, overwhelming authority of the voice, which, however molded by the Messenger's lips, has a massive, persuasive authority to it, recalling but expanding upon the direct speeches of God in the Bible." (Harold Bloom, Genius, A Mosaic of One Hundred Exemplary Creative Minds, Warner Books, 2002, pp. 145-146)

The observation above is accurate in its literary aspect. However, when we study the Quran according to its message, semantic web, scientific accuracy, and its mathematical structure, we find it to be the most coherent, the most consistent, the most meticulous, and the most organized book in the world. Thus, the Quran is like a beautiful tree that relaxes the eyes of its spectators with its apparently unorganized branches, leaves and flowers; yet its chemical, biological, and microscopic structure is so marvelously organized it leaves its students in perpetual awe and wonder.

Reaction to the Quran is not all positive. For instance, Prof. Gerd Puin of Germany, who had worked on Sanaa Manuscript, was quoted saying the following:

> Gerd-R. Puin's current thinking about the Koran's history partakes of this contemporary revisionism. "My idea is that the Koran is a kind of cocktail of texts that were not all understood even at the time of Muhammed," he says. "Many of them may even be a hundred years older than Islam itself. Even within the Islamic traditions there is a huge body of contradictory information, including a significant Christian substrate; one can derive a whole Islamic anti-history from them if one wants." ... Puin speaks with disdain about the traditional willingness, on the part of Muslim and Western scholars, to accept the conventional understanding of the Koran.

> "The Koran claims for itself that it is 'mubeen,' or 'clear,'" he says. "But if you look at it, you will notice that every fifth sentence or so simply doesn't make sense. Many Muslims—and Orientalists—will tell you otherwise, of course, but the fact is that a fifth of the Koranic text is just incomprehensible. This is what has caused the traditional anxiety regarding translation. If the Koran is not comprehensible—if it can't even be understood in Arabic—then it's not translatable. People fear that. And since the Koran claims repeatedly to be clear but obviously is not—as even speakers of Arabic will tell you—there is a contradiction. Something else must be going on." (Toby Lester, What is the Quran? Atlantic Monthly, January 1999)

Ironically, all Sunni and Shiite scholars agree with Puin regarding the impossibility to understand the Quran without the help of external sources such as hearsay reports called *hadith*, distorted history called *syrah*, or the so-called *asbab-i nuzul*, the reasons for revelation. As we will demonstrate, this perception is both true and false. It is true since it is a self-fulfilling assessment. Those with prejudice against the Quran will not have access to the real meaning of its verses. See: discussion on verse 3:7 in the Introduction: *How Much of the Quran Can/Should We Understand?*

The Quranic description of the earth, the solar system, the cosmos and the origin of the universe is centuries ahead of the time of its first revelation. The Quran contains many verses related to a diverse range of sciences and not a single assertion has been proven wrong. For instance, the Quran, revealed between 610-632 AC., states or implies that:

- Time is relative (32:5; 70:4; 22:47).
- God created the universe from nothing (2:117).
- The earth and heavenly bodies were once a single point, and they were separated from each other (21:30).
- The universe is continuously expanding (51:47).
- The universe was created in six days (stages) and the conditions that made life possible on earth took place in last four stages (50:38; 41:10).
- The stage before the creation of the earth is described as a gas nebula (41:11).
- Planet earth is floating in an orbit (27:88; 21:33; 36:40).
- The earth is round (10:24; 39:5) and resembles an egg (10:24; 39:5; 79:30).
- Oceans have subsurface wave patterns (24:40).
- Earth's atmosphere acts like a protective shield (21:32).
- Wind also pollinates plants (15:22).
- A bee has multiple stomachs (16:69).
- The workers in honeybee communities are females (16:68-69).
- Creation has an evolutionary system (7:69; 15:28-29; 24:45; 32:7-9; 71:14-17).
- The earliest organisms were incubated inside flexible layers of clay (15:26).
- The stages of human development in the womb are detailed (23:14).
- Our biological life span is coded in our genes (35:11).
- Photosynthesis is a recreation of energy stored through chlorophyll (36:77-81).
- Everything is created in pairs (13:3; 51:49; 36:36).
- The atomic number, atomic weight and isotopes of Iron are specified (57:25).
- Atoms of earthly elements contain a maximum of seven energy layers (65:12).
- There will be new and better transportation vehicles in the future (16:8).

- The sound and vision of water and the action of eating dates (which contain oxytocin) reduce labor pains (19:24-25).
- There is life (not necessarily intelligent) beyond earth (42:29).
- The Quran correctly refers to Egypt's ruler who made Joseph his chief adviser as king (*malik*), not as Pharaoh (*12: 43,50,54,72)*.
- Many of the miracles mentioned in the Quran represent the ultimate goals of science and technology.
- The Quran relates that matter (but not humans) can be transported at the speed of light (27:30-40); that smell can be transported to remote places (12:94); that extensive communication with animals is possible (27:16-17); that sleep, in certain conditions, can slow down metabolism and increase life span (18:25); and that the vision of blind people can be restored (3:49).
- The number of months in a year is stated as 12 and the word Month (*shahr*) is used exactly twelve times.
- The number of days in a year is not stated, but the word Day (*yawm*) is used exactly 365 times.
- The frequency of the word year (*sana*) in its singular form occurs 7 times and plural form 12 times and together 19 times; each number relating to an astronomic event.
- A prophetic mathematical structure based on the number 19 implied in chapter 74 of the Quran was discovered in 1974 by the aid of computer shows that the Quran is embedded with an interlocking extraordinary mathematical system, which was also discovered in original parts of the Old Testament in the 11th century.
- And there's more – much more. Also, see 68:1; 79:30; 74:1-37.

Quran v. Hadith

There is no difference between the teaching of Sunni or Shiite Religions and of Al-Qaeda, ISIS, or Taliban. They all use the same men-made religious sources fabricated by Ibn Filan and Abu Falan after the Quran, such as Hadith, Sunna, Ijma, Syrah, sectarian jurisprudence… Sunni and Shiite scholars distorted the meaning of many Quranic verses, which I have discussed in *Quran: a Reformist Translation*. I have exposed the Sunni/Shiite religions in books such as *Manifesto for Islamic Reform, Peacemaker's Guide to Warmongers*, various articles, speeches, debates & interviews such as: "ISIS follows Sunni Religion".

For a detailed discussion of the table below, see: MANIFESTO for ISLAMIC REFORM

	Quran	**Hadith**	**ISIS**
Religious dogmas, stories should be accepted on faith	No	Yes	Yes
Adulterers are killed by stoning to death	No	Yes	Yes
Slavery & having sex slaves are permitted	No	Yes	Yes
The heads of prisoners of war are chopped	No	Yes	Yes
Apostates and heretics are killed	No	Yes	Yes
Insulting prophet Muhammed is punished	No	Yes	Yes
A 54 y/o man marrying with a 9 y/o girl is ok	No	Yes	Yes
Muslims who do not pray may be beaten and even killed	No	Yes	Yes
Sculpture/picture of animals and humans is prohibited	No	Yes	Yes
Muhammad was illiterate	No	Yes	Yes
Testimony is: La ilaha illa Allah Muhammad Rasulullah	No	Yes	Yes
Muhammad was Better than Jesus	No	Yes	Yes
Shrimp, Lobster… are prohibited	No	Yes	Yes
Christians/Jews should pay extra tax (jizya)	No	Yes	Yes
Women cannot work with men	No	Yes	Yes
Women must cover their hair	No	Yes	Yes
Women are buried alive in sacks; must cover their faces	No	Yes	Yes
Some Quranic verses contradict and abrogate each other	No	Yes	Yes
There was a verse about stoning adulterers to death, but after Muhammad's death, a hungry goat ate it	No	Yes	Yes
The Quran is not detailed; it needs hadith, sunna, fatwas and stories of Ibni Filan and Abu Falan	No	Yes	Yes
Expecting intercession, asking help from dead people, praying at shrines is good	No	Yes	No

Sample Comparisons

On the following pages, you will find several comparisons between our translation and that of traditional orthodox English renditions of the Quran. By the word "tradition," we refer to the works that heavily rely on hearsay reports such as *hadith*, *sunna,* and sectarian jurisprudence.

We chose to compare our work primarily with the translation of Yusuf Ali, Pickthall, and Shakir, since they reflect most of the common errors and distortions, and because they are popular translations among the English-speaking Sunni population.

We use standard reference numbers in referring to specific passages of the Quran: the number preceding the colon is always the chapter number, and the subsequent numbers are always verse numbers.

Should Men Beat Their Wives?

A famous (and controversial) passage in the Quran has brought about a great deal of misunderstanding about Islam. When in 1989, I started translating the Quran to Turkish, verse 4:34 was among a few verses that I noted down on an orange paper for further research. I had problem with my understanding of it and I let its solution to God, in accordance with the instruction of verse 20:114. I shared the story of my discovery of its original meaning with my Turkish readers in "Errors in Turkish Translations of the Quran" (1992). Below are three translations of that verse, reflecting a deformed mindset followed by our translation:

! **Disputed passage**: The traditional rendering is: you may beat them.

Yusuf Ali	Pickthall	Shakir	Reformist
Men are the protectors and maintainers of women, because Allah has given the one more (strength) than the other, and because they support them from their means. Therefore the righteous women are devoutly obedient, and guard in (the husband's) absence what Allah would have them guard. As to those women on whose part ye fear disloyalty and ill-conduct, admonish them (first), (Next), refuse to share their beds, **(And last) beat them (lightly)**; but if they return to obedience, seek not against them Means (of annoyance): For Allah is Most High, great (above you all). (4:34)	**Men are in charge of women**, because Allah hath made the one of them to excel the other, and because they spend of their property (for their support of women). So good women are the obedient, guarding in secret that which Allah hath guarded. **As for those from whom ye fear rebellion**, admonish them and banish them to beds apart**, and scourge them**. Then, if they obey you, seek not a way against them. Lo! Allah is ever High Exalted, Great. (4:34)	Men are the maintainers of women because Allah has made some of them to excel others and because they spend out of their property; the good women are therefore obedient, guarding the unseen as Allah has guarded; and (as to) those on whose part you fear desertion, admonish them, and leave them alone in the sleeping-places **and beat them**; then if they obey you, do not seek a way against them; surely Allah is High, Great. (4:34)	**The men are to support the women** by what God has gifted them over one another and for what they spend of their money. The reformed women are devotees and protectors of privacy what God has protected. **As for those women from whom you fear disloyalty**, then you shall advise them, abandon them in the bedchamber, **and leave them**; if they obey you, then do not seek a way over them; God is High, Great. (4:34)

DISCUSSION OF 4:34

"Verse 4:34 of the Quran orders Muslims to beat their wives; therefore, Islam is a male-dominant religion." Many of us have heard this criticism from Christians, Atheists, Agnostics, and others. Though wife-beating is not a Muslim specialty, and domestic violence is an endemic problem in the West as well as the East, the issue nevertheless is whether it is justified by God. Most people reading conventional translations of 4:34 feel that something is deeply wrong. How could God, the Most Wise order us to beat our women? What kind of solution is that? It contradicts verses in which God describes marriage:

> "Among His signs is that He created for you spouses from among yourselves, in order to have tranquility and contentment with each other. He places in your heart love and care towards your spouses. In this, there are signs for people who think." (30:21)

Besides, the punishment of adultery by married couples who had accepted the jurisdiction of the Quran on their free will is stated in the beginning of chapter 24. The procedure mentioned in those verses leaves no doubt that husband has no right to beat his wife.

Obviously, these mixed messages have bothered many contemporary translators of the Quran. To avoid the moral and intellectual problems, they try to soften the word "beat" when they translate the verse 4:34. For instance, Yusuf Ali uses a merciful parenthesis after "beat," adding the word "(lightly)." This insertion does not appear in the Arabic text; it serves as a kind of apology for his translation of the surrounding material.

Later, Rashad Khalifa, a leading figure in the modern Islamic reform movement, rather than questioning the orthodox translation of the word, demonstrates his discomfort with his own version of orthodox translation by an implausible argument in the footnote and a contradictory subtitle before the verse: "Do not beat your wife." (However, Rashad Khalifa does not duplicate the orthodox distortion of other key words in the verse).

Many orthodox translators have tried to beat around the bush when it comes to explaining this passage, and perhaps just as many have beaten a hasty retreat from those inquiring after the author's intention -- but all have found themselves, in the end, beaten by 4:34.

Now please reread the sentence above. You will see that the word "beat" has been used three times, conveying totally different meanings each time: a verbal phrase meaning "avoid approaching directly" ("beat around the bush"); a verbal phrase meaning "depart quickly" ("beat a hasty retreat") and the status of having been defeated ("beaten"). Interestingly, the Arabic verb traditionally translated by male translators as "beat" or "scourge" -- *iDRiBuhunne* – also has numerous different meanings in Arabic, which is reflected by the Quran.

When I finished the Turkish translation (1991), this verse was on the top of my list to study carefully. Whenever I encounter a problem regarding the understanding of a Quranic verse, I remember 20:114 and pray accordingly: "Most Exalted is God, the only true King. Do not rush into (understanding) the Quran before it is revealed to you, and say, 'My Lord, increase my knowledge.'"

Almost all of the translations have mistranslated the four key words or terms of this particular verse. These are:

- *Qawwamun*;
- *Faddallallahu ba'dahum ala ba'd*;
- *Nushuzahunna*; and
- Fadribuhunna

In one of my books published in Turkey in 1992, "Errors in Turkish Translations," I discussed the real meaning of these words and the motivation and reasons for mistranslating them. Let's first start from the last one.

A Famous Multi-meaning Word

The main problem comes from the word *iDRiBuhunna*, which has traditionally been translated as "beat them." The root of this word is *DaRaBa*. If you look at any Arabic dictionary, you will find a long list of meanings ascribed to this word. In fact, you will find that that list is one of the longest lists in your Arabic dictionary. It can be said that *DaRaBa* is the number-one multi-meaning word in Arabic. It has so many different meanings; we can find numerous different meanings ascribed to it in the Quran.

- To travel, to get out: 3:156; 4:101; 38:44; 73:20; 2:273
- To strike: 2:60,73; 7:160; 8:12; 20:77; 24:31; 26:63; 37:93; 47:4
- To beat: 8:50; 47:27
- To set up: 43:58; 57:13
- To give (examples): 14:24,45; 16:75,76,112; 18:32,45; 24:35; 30:28,58; 36:78; 39:27,29; 43:17; 59:21; 66:10,11
- To take away, to ignore: 43:5
- To condemn: 2:61
- To seal, to draw over: 18:11
- To cover: 24:31
- To explain: 13:17

As you see, in the Quran alone we can attest to the verb *DaRaBa* having at least ten different meanings. *DaRaBa* also has other meanings that are not mentioned in the Quran. For example, in modern Arabic, you do not print money--you *DaRaBa* money. You do not multiply numbers--you *DaRaBa* numbers. You do not cease doing work--you *DaRaBa* doing work. In Turkish, we have many verbs similar to the Arabic DaRaBa, such as Çalmak, which means to play, steal, or strike. In English, we have two verbs that are almost equivalent to *DaRaBa*. These are "strike" and "beat." Consider, for the sake of comparison, that Webster's Dictionary gives fourteen different meanings for the verb "to strike," and eight for the verb "to beat"! (One strikes a match, strikes a deal, strikes an opponent, strikes gold, goes "on strike" against an unfair employer; one beats another team, beats out a rhythm, beats a retreat, and so on.).

The common denominator among all the nuanced usage of the Arabic words derived from the root DRB is this: "Moving from one point to another point."

Finding the Appropriate Meaning

Whenever we encounter a multi-meaning word in the Quran we must select the proper meaning (or meanings) given the context, the Arabic forms, the usage of the same word elsewhere in the Quran, and a certain amount of common sense. For instance, if one were to translate *DaRaBa* in 13:17 as "beat" (as one could conceivably do), the meaning would be ridiculous:

> ."\. . *God thus beats truth and falsehood..." (13:17)*

A more sensitive rendering of the context, however, yields a better translation:

> **"... God thus explains truth and falsehood..."** *(13:17)*

Another example of mistranslation of *DaRaBa* can be found in the translation of 38:44. Almost all the translations inject a rather silly story to justify their rendering of the passage. Here is how Yusuf Ali translates the first portion of this verse, which is about Job:

> "And take in the hand a little grass, and strike therewith: and break not (the oath)." (38:44)

Yusuf Ali, in the footnote, narrates the traditional story: "He (Job) must have said in his haste to the woman that he would beat her: he is asked now to correct her with only a wisp of grass, to show that he was gentle and humble as well as patient and constant".

However, without assuming the existence of this strange, male-viewpoint story (which has no other reference in the Quran), we can translate the verse as:

Yusuf Ali	**Reformist**
And take in thy hand a little grass, and strike therewith: and break not (thy oath)... (38:44)	Take in your hand a bundle and travel with it, and do not break your oath… (38:44)

Another Take on 4:34

In keeping with the translation, we have used in 38:44, we translate the controversial "beating" portion of 4:34 as "leave her" (Literally, the phrase might also be rendered "strike them out," meaning, in essence, "Separate yourselves from such wives.").

Additionally, the word *nushuz*, which is generally translated as "opposition" or "rebellion" in 4:34, has another meaning. If we study 4:34 carefully we will find a clue that leads us to translate that word as embracing a range of related ideas, from "feeling separated" to "flirting" – indeed, any word or words that reflects the range of disloyalty in marriage. The clue is the phrase before *nushuz,* which reads: "… they honor them according to God's commandments, even when alone in their privacy." This phrase emphasizes the importance of loyalty in marriage life and helps us to make better sense of what follows.

Interestingly, the same word, *nushuz,* is used later in the same chapter, in 4:128 – but it is used to describe the misbehavior of husbands, not wives, as it was in 4:34. In our view, the traditional translation of *nushuz*, that is, "opposition" will not fit in both contexts. However, the understanding of *nushuz* as marital disloyalty, in a variety of forms, *is* clearly appropriate for both 4:34 and 4:128. A careful study of other two verses where the word *NaShaZa* is used in different contexts clarifies the meaning. In verse 2:259 bones are separated from each other (*yaNShuZu*), and in verse 58:11 people are ordered to separate or break up (*uNShuZu*) to open space for newcomers.

The fourth key word is *QaNiTat*, which means "devoted to God," and in some verses is used to describe both man and woman (2:116,238; 3:17,43; 16:120; 30:26; 33:31,35; 39:9; 66:5,12). Though this word is mostly translated correctly as "obedient," when read in the context of the above-mentioned distortion it conveys a false message implying women must be "obedient" to their husbands as their inferiors. The word is mentioned as a general description of Muslim women (66:12), and more interestingly as a description of Mary who, according to the Quran, did not even have a husband! (66:12).

A Coherent Understanding

When we read 4:34, we should not understand *iDRiBuhunna* as "beat those women." We should, instead, remember that this word has multiple meanings. God gives us three ways of dealing with marital disloyalty on the part of a wife. In the beginning stage of such misbehavior, the husband should begin to address the problem by giving advice. If this does not work, he should stop sleeping in the same bed and see if this produces a change in behavior. And if there is still no improvement in the situation, the husband has the right to abandon her.

The Quran gives analogous rights to women who must deal with disloyal husbands (4:128); this is in accordance with the principle that women have "similar" rights to men in such situations, as stated clearly in 2:228. These would hardly be "similar" rights if women had to suffer physical beatings for marital disloyalty, and men did not!

Beating women who are cheating and betraying the marriage contract is not an ultimate solution, and it is not consistent with the promise of equitability and comparable rights that appears in 2:228. (This is an important consideration, because the Quran proclaims, and Muslims believe, that it is utterly free from inconsistencies.) But "separating from" the disloyal wives–that is, leaving them alone–before divorcing them, is consistent, and it is the best solution. It is also fair.

A personal note on the translation of 4:34:

I was the first Quran translator who noticed and exposed the historical distortion in 1990 while translating the Quran to Turkish in Masjid Tucson. I was sure that there was a concerted effort by Sunni and Shiite scholars, to distort the meaning of verses about women, especially this one. However, I wanted to support my personal conviction through scholarly work. I was following verse 20:114 and asking the Gracious, the teacher of the Quran, to guide me regarding the true meaning of the problematic word, DRB. God responded my prayer in less than two years. During my visit to Turkey in 1992, while in Ankara, somehow, I wanted to watch a movie in theater. It was a very rare occasion for me. In the movie titled *The Sheltering Sky* (1990), screenplay by Mark Peploe and directed by Bernardo Bertolucci, the first wife of the sheikh is unhappy with the tourist woman and tells her to leave the room by saying: *idrib*, that is "beat it." As it appears, by God's grace, one of the events that provided me external support was that movie. By "coincidence," I came across that movie during my visit to Ankara in 1992. It was one of the few Arabic words uttered in the entire movie. At that moment, I learned that Arabic speakers in North Africa were still using the expression "*idrib*" in the same sense as in verse 4:34. Soon after that, I came across, again by "coincidence" an article published in Cumhuriyet newspaper. Cengiz Ozakinci, a Turkish author promoting secularism, was questioning the translation of 4:34, and asking "why do they translate the word DRB in 4:34 as 'beat' while it has multiple meaning?" After the publication of my Turkish translation of the Quran, Mesaj in early 1990s, I wrote an English article, exposing the four key distortions on verse 4:34, and published it at my website 19.org under the title "Beating Women, or Beating around the Bush, or ..." The article pulled the attention of academics, especially after Dr Kecia Ali heralded it in an article. Since then, thank God, a few Quran translations corrected that major error.

Should Thieves' Hands Be Cut Off?

If non-Muslims "know" anything about Islam, it is that they "know" that the Quran mandates a severe punishment for thieves: the cutting off of their hands. Here are three traditional translations of the famous passage on the left and our translation on the right:

! **Disputed passage**: Traditional translations render the punishment for thieves as "cut off," while the verb has other meanings too.

Yusuf Ali	**Pickthall**	**Shakir**	**Reformist**
As to the thief, male or female, **cut off his or her hands**: a punishment by way of example, from Allah, for their crime: and Allah is Exalted in power. (5:38)	As for the thief, both male and female, **cut off their hands**. It is the reward of their own deeds, an exemplary punishment from Allah. Allah is Mighty, Wise. (5:38)	And (as for) the man who steals and the woman who steals, **cut off their hands** as a punishment for what they have earned, an exemplary punishment from Allah; and Allah is Mighty, Wise. (5:38)	The male thief, and the female thief, you shall **mark or cut off their means** as a punishment for their crime, and to serve as a deterrent from God. God is Noble, Wise. (5:38)

DISCUSSION OF 5:38

The Quran often uses words with more than one applicable and relevant meaning. This leads to verses that mean two, three, or more things at the same time, verses that make the translator's job exquisitely difficult.

We come now to such a verse. The verb form we translated as "mark, cut, or cut off" comes from a root verb -- *QaTa'A* – that occurs in the Quran many times. In almost all of its occurrences in the Quran, this verb means "to sever a relationship" or "to end an act." Only in two instances (12:31 and 12:50) is this verb clearly used to describe a physical cutting; in another instance (69:46), the verb might be interpreted in that way. A related form of this same verb -- one that implies repetition or severity of action -- occurs in the Quran seventeen times. This particular form is used to mean physically cutting off; or as a metaphor for the severing of a relationship; or to describe physically cutting or marking, but not cutting *off*.

Thus, the verse recommending punishment for theft or burglary, in the context of the Quran and its terminology (and not the terminology or interpretation attributed to Muhammed or his followers) provides us with a *single* verb … but one that God has permitted to incorporate a *range* of possible penalties. For instance:

- Cutting or marking the person's hands as a means of public humiliation and identification;
- Physically cutting off the person's hands; or
- Cutting off a person's means and resources to steal and burglarize (presumably through rehabilitation or imprisonment).

The act of imposing any of these penalties, or any of their combinations, would of course depend on the facts of each case, the culpability and mental capacity of the accused, and the ability of the society as a whole to act in accordance with God's other instructions in the

Quran. Note, for instance, that a Muslim society cannot punish a hungry person for stealing food, since letting a member of the society go hungry is a much bigger crime than the act of stealing food. Such a society demonstrates the characteristics of a society of unappreciative people! (See 107:1-7; 89:17-20; and 90:6-20). Considering theft solely as an individual crime and advocating the severest possible interpretation of the Quran in rendering punishment, is neither fair nor consistent with the scripture promoting mercy (7:52). Cutting off organic hands is permanent and in case of error it is an irreversible injustice. Thus, a peacemaker society will not cut off organic hands, but will cut off causal hands of thieves by eliminating causes related to poverty, mental health, education, culture, politics, and economic inequalities.

Interestingly, Chapter 12, which uses the same expression for little cut inflicted on one's hand while peeling fruits (12:31); yes, this very chapter provides us an example of punishment for committing theft: detention and prison! It is also instructive, since this punishment (arrest and temporary prison) is practiced by a Prophet, namely, Joseph. See 12:70-76.

Should Muslims Levy an Extra Tax on Non-Muslims?

Verse 9:29 is mistranslated by almost every translator. Shakir translates the Arabic word *jizya* as "tax," Pickthall as "tribute." Yusuf Ali somehow does not translate the word at all. He leaves the meaning of the word at the mercy of distortions:

! **Disputed passage**: The meaning of the Arabic word *jizya* (reparation/compensation) has been distorted to mean extra tax for non-Muslims.

Yusuf Ali	Pickthall	Shakir	Reformist
"Fight those who believe not in Allah nor the Last Day, nor hold that forbidden which hath been forbidden by Allah and His Messenger, nor affirm the religion of Truth, (even if they are) of the People of the Book, until they pay the **Jizya** with willing submission, and feel themselves subdued. (9:29)	Fight against such of those who have been given the Scripture as believe not in Allah nor the Last Day, and forbid not that which Allah hath forbidden by His messenger, and follow not the Religion of Truth, until they pay the **tribute** readily, being brought low. (9:29)	Fight those who do not believe in Allah, nor in the latte day, nor do they prohibit what Allah and His Messenger have prohibited, nor follow the religion of truth, out of those who have been given the Book, until they pay the **tax** in affirmation of superiority and they are in a state of subjection. (9:29)	Fight those who do not affirm God nor the Last day from among the people who received the book; they do not forbid what God and His messenger have forbidden, and they do not uphold the system of truth; until they pay the **reparation**, in humility. (9:29)

DISCUSSION OF 9:29

We should be reminded that the context of the verse is about the War of *Hunain*, and fighting is allowed only for self-defense. See: 2:190-193, 256; 4:91; and 60:8-9.

Furthermore, note that we suggest REPARATION instead of Arabic word *jizya*. The meaning of *jizya* has been distorted as a tax on non-Muslims, which was invented long after Muhammed to further the imperialistic agenda of Kings. The origin of the word that we translated as Compensation is *JaZaYa*, which simply means compensation or, in the context of war, means war reparations, not tax. Since the enemies of Muslims attacked and aggressed, after the war they are required to compensate for the damage they inflicted on the peaceful community. Various derivatives of this word are used in the Quran frequently, and they are translated as compensation for a particular deed.

Unfortunately, the distortion in the meaning of the verse above and the practice of collecting a special tax from Christians and Jews, contradict the basic principle of the Quran that there should not be compulsion in religion and there should be freedom of belief and expression (2:256; 4:90; 4:137; 10:99; 18:29; 88:21, 22). Since taxation based on religion creates financial duress on people to convert to the privileged religion, it violates this important Quranic principle. Dividing a population that united under a social contract (constitution) into privileged groups based on their religion contradicts many principles of the Quran, including justice, peace, and brotherhood/sisterhood of all humanity.

Some uninformed critics or bigoted enemies of the Quran list verses of the Quran dealing with wars and declare Islam to be a religion of violence. Their favorite verses are: 2:191; 3:28; 3:85; 5:10,34; 9:5; 9:28-29; 9:123; 14:17; 22:9; 25: 52; 47:4 and 66:9. In this article, I refuted their argument against 9:29, and I will discuss each of the verses later.

Some followers of Sunni or Shiite religions abuse 9:5 or 9:29 by taking them out of their immediate and Quranic context. Sunnis and Shiites follow many stories and instructions falsely attributed to Muhammed that justify terror and aggression, which is currently used as a pretext and propaganda tool by imperialist or neocolonialist powers to justify their ongoing terror and aggression against countries with predominantly Muslim population. For instance, in a so-called authentic (or authentically fabricated) *hadith*, after arresting the murderers of his shepherd, the prophet and his companions cut their arms and legs off, gouged their eyes with hot nails and left them dying from thirst in the desert, a contradiction to the portrayal of Muhammed's mission in the Quran (21:107; 3:159). In another authentically fabricated *hadith*, the prophet is claimed to send a gang during night to secretly kill a female poet who criticized him in her poetry, a violation of the teaching of the Quran! (2:256; 4:140; 10:99; 18:29; 88:21-22). Despite these un-Quranic teachings, the aggressive elements among Sunni and Shiite populations have almost always been a minority.

Can One Marry Underage Orphans?

A passage of the Quran has persistently been interpreted as sanctioning marriage to young orphan girls:

! **Disputed passage**: The traditional rendering suggests that the objects of marital intention are the orphans, not the mothers.

Yusuf Ali	Pickthall	Shakir	Reformist
"They ask thy instruction concerning the women say: Allah doth instruct you about them: And (remember) what hath been rehearsed unto you in the Book, **concerning the orphans of women to whom ye give not the portions prescribed, and yet whom ye desire to marry**, as also concerning the children who are weak and oppressed: that ye stand firm for justice to orphans. There is not a good deed which ye do, but Allah is well-acquainted therewith." (4:127)	They consult thee concerning women. Say: Allah giveth you decree concerning them, and the Scripture which hath been recited unto you (giveth decree), **concerning female orphans and those unto whom ye give not that which is ordained for them though ye desire to marry them**, and (concerning) the weak among children, and that ye should deal justly with orphans. Whatever good ye do, lo! Allah is ever Aware of it (4:127)	And they ask you a decision about women. Say: Allah makes known to you His decision concerning them, and that which is recited to you in the Book **concerning female orphans whom you do not give what is appointed for them while you desire to marry them**, and concerning the weak among children, and that you should deal towards orphans with equity; and whatever good you do, Allah surely knows it. (4:127)	They ask you for divine instruction concerning women. Say, "God instructs you regarding them, as has been recited for you in the book **about the rights of orphans whose mothers you want to marry** without giving them their legal rights. You shall observe the rights of helpless children, and your duty to treat orphans with equity. Whatever good you do, God has full knowledge of it. (4:127)

DISCUSSION OF 4:127

Though the Quran permits polygamy for men (4:3), it severely discourages its actual practice by requiring certain significant preconditions: men may marry more than one wife only if the later ones are widows with children, and they should treat each wife equally and fairly. (See 4:19-20; 127-129.). Unfortunately, verse 4:127 has been traditionally misinterpreted and mistranslated in such a way as to suggest that God permits marriage with juvenile orphans. This is clearly not the case.

The Arabic expression *yatama-l nisai-l lati* in 4:127 has been routinely mistranslated as "women orphans, whom..." The expression is also sometimes translated as "orphans of women whom..." This later translation, though accurate, makes the crucial reference of the objective pronoun "whom" ambiguous: Does the phrase after "whom" describe orphans or women?

As it happens, the Arabic plural pronoun in this verse is the female form, *allaty* (not the male form *allazyna*), and it can *only* refer to the women just referenced, not to the orphans. This is because the Arabic word *yatama* (orphans) is male in gender!

All the English translations of the Quran that we have seen have mistranslated this passage. This is remarkable because correct translation requires only an elementary knowledge of Arabic grammar. This error is thus much more than a simple grammatical slip; it is, we would argue, willful misrepresentation. The traditional interpretation of this passage offers an apparent justification for marriage with children, which flatly contradicts the Quran.

Like so many passages in the Quran, 4:127's meaning was severely distorted in order to gain the favor of rich, dominant males. Over the centuries, male scholars with active libidos have used fabricated *hadith* to pervert the meaning of this and other Quranic verses relating to marriage and sexuality. (See the discussion of 66:5, below.)

What are the Characteristics of a Model Muslim Woman?

Verse 66:5 lists some ideal characteristics of an appreciative woman. The last three characteristics, however, have been mistranslated.

! **Disputed passage**: The traditional rendering emphasizes virginity.

Yusuf Ali	Pickthall	Shakir	Reformist
"It may be, if he divorced you (all), that God will give him in exchange Consorts better than you—who submit (their wills), who believe, who are devout, who turn to God in repentance, who worship (in humility), **who travel (for Faith) and fast, previously married or virgin."** (66:5)	It may happen that his Lord, if he divorce you, will give him in your stead wives better than you, submissive (to Allah), believing, pious, penitent, devout, **inclined to fasting, widows and maids.** (66:5)	Maybe, his Lord, if he divorces you, will give him in your place wives better than you, submissive, faithful, obedient, penitent, adorers, **fasters, widows and virgins.** (66:5)	"If he divorces you, his Lord will substitute other wives in your place who are better than you; peacefully surrendering (to God), affirming, devout, repentant, serving, **active in their societies, responsive, and foremost ones**." (66:5)

DISCUSSION OF 66:5

Traditional translations mistranslate the last three adjectives used here to describe Muslim women. They distort their meaning as "fasters, widows and virgins." When the issue is about women, somehow, the meaning of the Quranic words passes through rapid mutations. For instance, we know that the Sunni and Shiite scholars who could not beat cows and examples (since the same verb is used in relation to a heifer as in 2:68, and for citing examples, yet translated differently) found it convenient and fair to beat women (see 4:34). Those of us who have rejected other religious sources besides the Quran are still struggling to clean our minds from these innovations that even have sneaked into the Arabic language long after the revelation of the Quran. There is, in fact, nothing whatsoever about fasting, widows and virgins in this verse. We are rediscovering and re-educating ourselves in matters pertaining to the Quran.

The third word from the end of the verse, *SaYiHat,* which we have translated as "active in their societies" simply means to travel or move around for a cause. About two centuries after the revelation of the Quran, when the rights of women were taken away one by one through the all-male enterprises called *hadith*, *ijtihad* and *tafseer*, Muslim communities found themselves thinking and living like the enemies of Islam in the Days of Ignorance. The misogynistic mind of orthodox commentators and translators simply could not fathom the notion of a Muslim woman traveling around alone to do anything – and so they pretended that the word in question was not *SaYaHa,* but *SsaWM* – fasting! Socially active women were indeed more difficult to control than the women who would fast in their homes; they were even less costly, since they would eat less. For the usage of the verb form of the root, see 9:2. The word *SaYaHa* has nothing to do with fasting; the Quran consistently uses the word *SaWaMa* for fasting (2:183-196; 4:92; 5:89,90; 19:26; 33:35; 58:4).

The second word from the end is *THaYiBat, which* means "those who return" or "those who are responsive". Various derivatives of the same root are used to mean "reward" or "refuge" or "cloths". For instance, see 2:125; 3:195. The Arabic words for widow are *ARMiLa* or *AYaMa*. The Quran uses *AYaMa* for widow or single; see: 24:32.

The last word of this verse, *aBKaR*, which means those who are "young," "early risers" or "foremost," has traditionally, and implausibly, been interpreted as "virgins" in this passage. The resulting distorted meaning of the verse supports a sectarian teaching that justifies a man marrying more than one virgin. The Arabic word for virgin is *BaTuL* or *ADRa*.

Here, I will focus only on the last word.

This false interpretation has become so popular that it is apparently now considered beyond any challenge. We have not seen any published translation (except Edip Yuksel's Turkish translation, Mesaj) that does not duplicate this centuries-old error. It is particularly important, therefore, that we explain exactly *why* we have translated this verse as we have.

The Arabic root of the word we have translated as "foremost" is *BKR*, and it occurs 10 times in the Quran. In seven of these occurrences, the word describes time; in two (including the present verse) it describes women; and in one case it describes a heifer.

Before deciding that *aBKaR* means "virgin," the translator should look closely at how the Quran itself employs the root word. Here are the references in Quranic sequence:

- 2:68 ---> Young (heifer)
- 3:41 ---> Early morning
- 19:11 ---> Early morning
- 19:62 ---> Early morning
- 25:5 ---> Early morning
- 33:42 ---> Early morning
- 48:9 ---> Early morning
- 54:38 ---> Early morning
- 56:36 ---> ??? Woman
- 66:5 ---> ??? Woman
- 76:25 ---> Early morning

When this word is used in reference to women (in 56:36 and 66:5), orthodox scholars, without hesitation, leap to translate it as "virgin." We reject this interpretation and choose instead to translate the word as "foremost."

Reportedly, one of the earliest converts to Islam, and the first elected Islamic leader after the death of Muhammed, was his father-in-law, a man who became known to Muslims as Abu Bakr. Abu means "father." This nickname was not given to Muhammed's father-in-law because he was the "Father of a Virgin"! Being the father of a virgin was not something unique, since every father of a daughter would deserve such a title at one time in their lives. Besides, Abu Bakr's daughter did not remain virgin; she married Muhammed and was called by many as the "mother of Muslims." Some sources relate the name to "young camels." However, the best explanation must be related to his standing among Muslims. Abu Bakr was one of the foremost ones, one of the progressive ones, one of the early risers, one of the first converts to Islam. He was foremost – or, if you prefer, progressive.

We must note, too, the Arabic conjunction *Wa* (and), which appears before the word *aBKaR* (foremost ones). The end of the verse has traditionally been mistranslated, as though the word were actually OR:

> *"... previously married OR virgins".*

The traditional commentators and translators knew full well that a contradiction might present itself had they translated *Wa* accurately:

> *"... previously married AND virgins".*

We have chosen to render this passage as "responsive and foremost," thereby retaining a legitimate alternate translation of the word traditionally translated "previously married" – as well as the impossible-to-evade "and" of the Arabic text.

There is, however, a portrait of the ideal Muslima (female Muslim): submitting, affirming, devoted, repentant, worshipful, active in her society, responsive and foremost.

Was Muhammed Illiterate?

During the month of Ramadan, every evening, after the lengthy congregational prayers millions crowding the mosques ask God to bless the soul of his *Nabbiyy-il Ummy,* meaning, in the orthodox interpretation, "illiterate prophet." "Illiterate" (or "unlettered") is one of the most common titles used by Muslim clerics and imams to praise Muhammed, the deliverer of the Quran.

The Arabic word *ummy*, however, describes people who are not Jewish or Christian. The meaning of this word, which occurs six times in the Quran, has nevertheless been rendered as "one who can neither read nor write." This deliberate manipulation by Muslim scholars has become widely accepted as the true meaning of the word. For example, Yusuf Ali and Pickthall follow this pattern, while Shakir prefers not to translate the Arabic word.

! **Disputed passage:** Orthodox sources distort the meaning of *ummy* to turn Muhammed illiterate.

Yusuf Ali	**Pickthall**	**Shakir**	**Reformist**
Say: "O men! I am sent unto you all, as the Messenger of Allah, to Whom belongeth the dominion of the heavens and the earth: there is no god but He: it is He That giveth both life and death. So believe in Allah and His Messenger, **the Unlettered Prophet**, who believeth in Allah and His words: follow him that (so) ye may be guided." (7:158)	Say (O Muhammed): O mankind! Lo! I am the messenger of Allah to you all - (the messenger of) Him unto Whom belongeth the Sovereignty of the heavens and the earth. There is no Allah save Him. He quickeneth and He giveth death. So believe in Allah and His messenger, **the Prophet who can neither read nor write,** who believeth in Allah and in His Words, and follow him that haply ye may be led aright. (7:158)	Say: O people! surely I am the Messenger of Allah to you all, of Him Whose is the kingdom of the heavens and the earth there is no god but He; He brings to life and causes to die therefore believe in Allah and His messenger, the **Ummi Prophet** who believes in Allah and His words, and follow him so that you may walk in the right way. (7:158)	Say: "O people, I am God's messenger to you all. The One who has the sovereignty of heavens and earth, there is no god but He; He gives life and causes death." So affirm God and His **gentile prophet**, who affirms God and His words; and follow him that you may be guided. (7:158)

DISCUSSION OF 7:158

The Quran itself provides guidance for the true meaning of *ummy*. If we reflect on the verse 3:20 below, we will easily understand that *ummy* does not mean an illiterate person:

> "And say to those who received the scripture, as well as those who did not receive any scripture (*ummyyeen*)..." (3:20)

In this verse, the word *ummy* describes Meccan polytheists. It is obvious that *ummy* does not mean illiterate because it has been used as the counterpart of the people of the scripture. If the verse was " ... And say to those who are literate and illiterate," then the orthodox translation of *ummy* would be correct. According to 3:20, the people of the Arabian Peninsula consisted of two main groups:

- The people of the scripture, i.e., Jews and Christians.
- Gentiles, who were neither Jewish nor Christian.

If the people who were neither Jews nor Christians were called "*ummyeen*" (3:20; 3:75), then the meaning of *ummy* is very clear. As a matter of fact, the verse 3:75 clarifies its meaning as Gentile.

Mecca was the cultural center of the Arabs in the 7th century. Poetry competitions were being held there. It is a historical fact that Meccans were not familiar with the Bible, thus making them Gentiles. So, the verse 62:2 describes Meccan people by the word *ummyeen*:

> "He is the One who sent to the Gentiles (ummyeen) a messenger from among them, to recite to them His revelations, purify them, and teach them the scripture and wisdom. Before this, they had gone far astray." (62:2)

The unappreciative opponents claimed that Muhammed was quoting verses from the Old and New Testaments (25:5; 68:15). The verse below refutes their accusation and gives the answer:

> "You did not read any previous scriptures, nor did you write them with your hand. In that case, the objectors would have had reason to harbor doubts." (29:48)

This verse tells us that Muhammed did not read nor write previous scriptures. The word *min qablihi* (previous) suggests that Muhammed *did* read and write the final scripture.

Muhammed was a literate Gentile (*ummy*)

After this examination of the true meaning of the word *ummy*, here are the reasons and proofs for the fact that Muhammed was a literate Gentile:

To magnify the miraculous aspect of the Quran, religious people thought that the story of illiteracy would be alluring.

The producer(s) of the illiteracy story found it relatively easy to change the meaning of *ummy*. Nevertheless, the word appears throughout the Quran, and consistently means "Gentile" (2:78; 3:20; 3:75; 62:2). In verses 3:20 and 3:75, the Quran uses the word *ummy* as the counterpart to the *ehlil kitab* ("People of the Book," a phrase that in both verses refers to "Jews and Christians").

The Quran describes Meccan people with the word *ummyyeen* (Gentiles) (62:2). According to the orthodox claim, all Meccan people must have been illiterate. Why then were the poems of pre-Islamic Meccan poets hung on the walls of the Ka'ba (the ancient monotheistic shrine of Abraham)?

The Arabs of the 7th century used letters as numbers. This alphabetical numbering system is called "Abjad." The merchants of those days had to know the letters of the alphabet to record their accounts! If Muhammed was a successful international merchant, as is universally accepted, then he most probably knew this numbering system. The Arabs stopped using the "Abjad" system in the 9th century when they took "Arabic numbers" from India.

The different spelling of the word *bism* in the beginning of the Basmalah and in the first verse of chapter 96 supports the literacy of Muhammed. It is not reasonable for an illiterate to dictate two different spellings of the same word which is pronounced the same.

The very first revelation from the Controller Gabriel was, Muslims believe, "Read!" And the first five verses of that revelation encourage reading and writing (96:1-5). The second revelation was "The pen and writing" (68:1). These facts compel some questions that orthodox scholarship would rather avoid. Does God command an illiterate man to "read"? If so, could Muhammed read *after* Gabriel's instruction to do so? The story told in *hadith* books about the first revelation asserting that Muhammed could read only after three trials ending by an angelic "squeeze" contradicts the other stories claiming that Muhammed died as an illiterate!

Traditional history books accept that Muhammed dictated the Quran and controlled its recording. Even if we accept that Muhammed did not know how to read or write before revelation of the Quran, we cannot claim that he preserved this illiteracy during the 23 years while he was dictating the Quran! Let us accept, for the sake of argument, that Muhammed was illiterate before the revelation of the Quran. Why then did he insist on *staying* illiterate for 23 years after the first revelation: "Read!"? Did he not obey his Lord's command? Did he receive another command forbidding him from reading and writing?

Was it so difficult for Muhammed to learn to read and write? If a person still does not learn to read and write after 23 years of careful dictation of a book, what kind of intellect is that?

If Muhammed was encouraging his followers to read and write (which he did when he recited 2:44 to them), then why should he have excluded himself?

Muslim scholars, who disagree on a bewildering array of subjects, somehow have managed to agree on the story of Muhammed's illiteracy. Perhaps the glorification of illiteracy, using it as a positive attribute of a worshipped figure, is one of the causes of the high current level of illiteracy in Muslim communities.

PS: There is another meaning of *ummy*, which does not exclude *Gentile*, is "the one who is the resident of the capital city." Mecca was the capital city of medieval Arabia, and it is referred in the Quran as "Umm ul-Qura" that is "the mother of cities" (42:7).

Do We Need Muhammed to Understand the Quran?

The so-called "Orthodox Islam," by answering the above question affirmatively, has sanctified a collection of medieval hearsay reports and traditions attributed to Muhammed. Unfortunately, this was accomplished through the distortion of the meaning of several Quranic verses regarding the role of the prophet. The following verse is one of several crucial verses that have been used to promote *hadith* and *sunna* as the second source of Islam.

! **Disputed passage:** The traditional rendering implies the need for hearsay.

Yusuf Ali	**Pickthall**	**Shakir**	**Reformist**
(We sent them) with Clear Signs and Books of dark prophecies; and We have sent down unto thee (also) the Message; that thou mayest **explain** clearly to men what is sent for them, and that they may give thought. (16:44)	With clear proofs and writings; and We have revealed unto thee the Remembrance that thou mayst **explain** to mankind that which hath been revealed for them, and that haply they may reflect. (16:44)	With clear arguments and scriptures; and We have revealed to you the Reminder that you may **make clear** to men what has been revealed to them, and that haply they may reflect. (16:44)	With proof and the scriptures. We sent down to you the Reminder to **proclaim** to the people what was sent to them, and perhaps they would think. (*16:44*)

DISCUSSION OF 16:44

Traditionalists have opted for what we consider an inaccurate rendering of the Arabic root word "BYN."

The word *'lituBaYyeNa* is a derivative of "BYN," which is a multi-meaning word. It means:

- To reveal what is concealed; or
- To explain what is vague.

Thus the first meaning is the antonym of "hide," and the second is the antonym of "make vague." We have translated this passage in accordance with the first meaning and understand the passage as relating to God's order to Muhammed to proclaim the revelation which is revealed to him personally. We believe that the Quran is clear, as its text itself insists.

Indeed, "proclaiming" is the whole mission of the messengers of God, as the Quran maintains (16:35). To be sure, prophets sometimes have trouble in proclaiming the revelation (33:37, 20:25). But if the Quran is a profound book written in Arabic so that people may understand (12:2), if it is to be explained by God (75:19), and if it is simple to understand (5:15; 26:195; 11:1; 54:17; 55:1-2), then it is hard to see why or how the prophet is to assume the additional mission of explaining the divine message.

We emphasize, once again, that 75:19 holds that *God* explains the Quran and makes no mention of Muhammed or any other prophet, or indeed any human explanation whatsoever. Thus, the word *lituBaYyeNa* of 16:44, which we have translated as "proclaim," is similar to the one in 3:187. Verse 3:187 tells us that the people who received the revelation should

> "... proclaim the scripture to the people, and never hide it." (3:187)

Do The Verses of The Quran Abrogate Each Other?

There are schools of thought that have committed the travesty of allowing the Quran – regarded as the definitive word of God – to be abrogated (replaced, overridden) by the *hadith* and the *sunna.*

Though in his footnote, Yusuf Ali refers to the other meaning of the word AYAH, and the different interpretation of the verse, by translating the word as "revelations," he affirms the traditional position known as "abrogation" of one revelation by another revelation. A few scholars, as we say, have gone so far as to use *hadith* to abrogate the Quran; a more common view, however, is that certain verses in the Quran either have been removed or some verses cancel each other's judgment out. We reject these contentions.

! **Disputed passage:** The traditional commentaries and translations render 2:106 to justify abrogation in the Quran, thereby justifying the rejection of many Quranic verses.

Yusuf Ali	Pickthall	Shakir	Reformist
None of Our **revelations** do We abrogate or cause to be forgotten, but We substitute something better or similar: Knowest thou not that God hath power over all things? (2:106)	Nothing of our **revelation (even a single verse)** do we abrogate or cause be forgotten, but we bring (in place) one better or the like thereof. Knowest thou not that Allah is Able to do all things? (2:106)	Whatever **communications** We abrogate or cause to be forgotten, We bring one better than it or like it. Do you not know that Allah has power over all things? (2:106)	We do not duplicate (or abrogate) any **sign** or cause it to be forgotten, unless we produce a better, or at least an equal one. Do you not know that God is Omnipotent? (2:106)

DISCUSSION OF 2:106

Ayaat, the plural of *aya*, is used in the Quran to mean both a) signs/miracles and b) verses/revelations of the Quran itself. Since verses of the Quran are miracles/signs, the plural form occasionally conveys both meanings simultaneously.

A single verse of the Quran is not deemed to be a miracle since some short verses of the Quran (for instance: 55:3; 69:1; 74:4; 75:8; 80:28; 81:26) are not unique and can be found in daily conversations of Arabic-speaking people. In fact, the Quran determines the minimum unit of miraculous nature as a chapter (10:38), and the shortest chapters consist of 3 verses (103; 108; 110). Therefore, only the plural form of *aya*, that is *ayaat*, can be used as reference to the verses/revelation of the Quran.

However, the singular form, *aya*, in all its 84 occurrences in the Quran is *always* used to mean sign or miracle. Therefore, we choose to translate the singular form *aya* in verse 2:106 as "sign," rather than as "verse (of the Quran)." Verse 13:38-39 clearly supports this understanding.

By declaring the word of God to be vague and ambiguous, early scholars opened the gate for unlimited abuse and distortion. Furthermore, by distorting the meaning of 2:106, they claimed that many verses of the Quran had been abrogated (amended) by other verses or *hadiths*. By this "abrogation theory," they amended verses which they did not understand, or which did not suit their interests, or which contradicted their *hadiths*. Repeating the same error committed by the Children of Israel (2:85), Muslims fulfilled the prophetic description of their action in 15:91-93. Some of them abrogated 5 Quranic verses, some 20 verses and some 50. Below, you'll find an extreme (and bizarre) example of an abrogation based on the supposed "authority" of 2:106 as interpreted in the orthodox manner.

The verse that was abrogated by a goat!

Some fabricated *hadiths* claim that the prophet Muhammed stoned a particular couple to death for illicit sexual relations. This punishment, we believe, would have been in conflict with 24:2 of the Quran, which sets out a separate penalty for adultery, and makes no mention whatsoever of capital punishment.

Since *hadith*-manufacturers realized that *hadiths* were not enough to abrogate the clear verses of the Quran, they went so far as to fabricate a "verse" supporting stoning and attributed it to God. "*Al-shaykhu wa al-shaykhatu iza zanaya farjumuhuma nakalan bima kasabu...*" They tried to inject the stoning penalty for adulterers into the Quran!

When they failed, they fabricated stories which only the people who are described in 10:100 of the Quran (those who believe without understanding and comprehending) could accept. According to the story, the 'stoning verse' was recorded in the Quran during the time of Muhammed; but just after his death, a goat entered the house of the prophet's wife Aisha and ate the page on which that verse was inscribed. Thus, we are assured, the "stoning verse" had been abrogated physically!

This story can be found in the so-called authentic *hadith* collections, such as Ibn Maja, Nikah, 36/1944 and Ibn Hanbal, 5/131,132,183; 6/269.

How could a verse of a perfect scripture be abrogated by a goat? As an answer to this question, Ibn Qutayba, a proponent of *hadith* and *sunna*, in his classic book entitled "Solving the Contradictions Among *Hadith*s" puts forward the contention that "the goat is a holy animal." And he asked a counter question: "Why not trust in God's power? As He destroyed the people of Aad and Thamud, He is also able to destroy His revelations by using even a goat!"

How Much of the Quran Can/Should We Understand?

Is the Quran meant for everyone to understand? Or are some parts inaccessible to mere human beings?

Verse 3:7 is one of the commonly mistranslated verses; it is extremely important since it deals with these very questions. Different interpretations of this verse can lead to two totally different conceptions of Islam! Both Yusuf Ali and Pickthall reflect the majority's view on behalf of "not understanding," while Shakir prefers the minority view of "understanding."

! **Disputed passage**: The traditional rendering suggests that some Quranic verses can never be understood fully.

Yusuf Ali	Pickthall	Shakir	Reformist
He it is Who has sent down to thee the Book: in it are verses basic or fundamental (of established meaning); they are the foundation of the Book: others are allegorical. But those in whose hearts is perversity follow the part thereof that is allegorical, seeking discord, and searching for its hidden meanings, **but no one knows its hidden meaning except God.** And those who are firmly endowed by knowledge say: 'we believe in the Book; the whole of it is from our Lord:' and none will grasp the Message except men of understanding. (3:7)	He it is Who hath revealed unto thee (Muhammed) the Scripture wherein are clear revelations - they are the substance of the Book - and others (which are) allegorical. But those in whose hearts is doubt pursue, forsooth, that which is allegorical seeking (to cause) dissension by seeking to explain it. **None knoweth its explanation save Allah.** And those who are of sound instruction say: We believe therein; the whole is from our Lord; but only men of understanding really heed. (3:7)	He it is Who has revealed the Book to you; some of its verses are decisive, they are the basis of the Book, and others are allegorical; then as for those in whose hearts there is perversity they follow the part of it which is allegorical, seeking to mislead and seeking to give it (their own) interpretation. **but none knows its interpretation except Allah, and those who are firmly rooted in knowledge say**: We believe in it, it is all from our Lord; and none do mind except those having understanding. (3:7)	He is the One who sent down to you the book, from which there are definite signs; they are the essence of the book; and others, which are multi-meaning. As for those who have disease in their hearts, eager to cause confusion and eager to derive their interpretation, they will follow those with multi-meaning. **But none knows their meaning except God and those who are well founded in knowledge**; they say, "We affirm it, all is from our Lord." None will remember except the people of intellect. (3:7)

DISCUSSION OF 3:7

The Arabic word we have translated as "multi-meanings" is *mutashabihat*. The word comes from *shabaha* ("to became similar").

The word can be confusing for a novice. Verse 39:23, for instance, uses *mutashabihat* for the entire Quran, referring to its overall similarity -- in other words, its consistency. In a narrower sense, however, *mutashabihat* refers to all verses which can be understood in more than one way. The various meanings or implications require some special qualities from the person listening to or reading the Quran: an attentive mind, a positive attitude, contextual perspective, the patience necessary for research, and so forth.

It is one of the intriguing features of the Quran that the verse about "*mutashabih*" verses of the Quran is itself *mutashabih* -- that is, possessing multiple meanings. The word in question, for instance, can mean "similar," as we have seen; it can mean "possessing multiple meanings;" it can also mean "allegorical" (where one single, clearly identifiable element represents another single, clearly identifiable element).

As you may have noticed, interpretation of the last part of 3:7 depends on how one punctuates the verse. (There is no punctuation in the original Arabic text.)

If one stops after the word "God," then one will assume, as centuries of Sunni and Shiite scholars have, that even those who possess deep levels of knowledge will never be able to understand the "*mutashabih*" verses. However, if the sentence does not stop there, the meaning will change to the opposite: Those who possess knowledge *will* be able to understand the meaning of allegorical or multi-meaning verses.

Here are five reasons for why we prefer the second understanding of this verse.

REASON ONE: The passage clearly emphasizes the unhealthy intentions of those who fail to understand multi-meaning verses and choses the meaning that contradicts other verses/signs. With the disease of doubt in their hearts, they try to confuse others by focusing on their own faulty interpretations of these verses. We think that the passage emphasizes this point because the Quran tells us elsewhere that only sincere people possess the qualities necessary to understand the Quran (as emphasized in 17:46; 18:57; and 54:17).

REASON TWO: The Quran tells us repeatedly that it is easy to understand. (It does so at many points, including 5:15; 11:1; 26:195; 54:17; and 55:1-2.). If one punctuates this verse in the traditional way, there is an apparent contradiction – the Quran is, at least in some places, *impossible* for any human being to understand -- and Muslims maintain that the Quran does not contradict itself.

We believe the Quran broadcasts a very clear, coherent message. However, there is sometimes a problem with our receiver. If our receiver does not hear the broadcast or cannot understand it well, then something is wrong with our receiver, and we must check it. If the signal is weak, we need to recharge our batteries, or reset our antennas. If we do not receive a clear message, we need to set our tuning to the right station to get rid of the noises and interference from other sources. We may, of course, ask for some help from knowledgeable people or experts for this task. If the receiver does not work at all, then we should make a sincere effort to fix the broken parts. However, if we believe that the problem is in the broadcast, then nobody can help us.

REASON THREE: It is beyond dispute that the Quran encourages Muslims to study its words with patience. It advises us not to rush into understanding without sufficient knowledge (20:114). Nevertheless, it claims to be easy to understand (see REASON TWO, above). This, however, is *not* a contradictory position!

Experience with the book suggests that both statements are accurate. Although it can be explored for a lifetime without grasping all its subtleties, the Quran is in fact quite easy to understand, revolving as it does around three basic ideas:

- There is only one God.
- This life is a test.
- There will be an accounting for everyone after death.

We know that the Quran really is comprehensible *and* worthy of sustained, careful study, just as it promises. Whoever opens his/her mind and heart as a monotheist and takes the time to study it, will understand it, and that this understanding will be enough for salvation. Such people will also be inspired to explore it deeply and will find ample rewards for doing so.

REASON FOUR: To accept all the verses of the Quran, one does not need to be deeply rooted in knowledge. To be an “affirming person" is a sufficient condition to affirm all the verses. However, one needs to have deep knowledge of the Quran to understand "*mutashabih*" (multi-meaning) verses accurately. Therefore, 3:7 mentions a narrow category (those who are deeply rooted in knowledge) in relation to those multi-meaning verses.

REASON FIVE: If we follow the orthodox punctuation and translation of 3:7, then, we must, by logical extension, establish a clear definition of *what the "mutashabih" verses are* ... to avoid trying vainly to understand them or teaching others based on them. We thus need a definitive list of the "mutashabih" verses to avoid being among those who are condemned in this verse. There is a problem, however: No one has ever been able to compile such a definitive list! What could the criteria for the list possibly be? Surely one person's lack of understanding of a verse should not make a verse "taboo" for all other people. If that were the case, the lowest degree of understanding would be the common denominator for understanding and interpreting the Quran! In this Alice-in-Wonderland school of thought, there would be a perpetual race towards ignorance!

Unless one is committed to determining the truth by majority vote, then one may want to reflect upon the five reasons listed here for interpreting the verse as we have.

(A side note: There are a few Sunni commentators who support our understanding of this verse. For instance, the classic commentary of al-Baydawi prefers this understanding. Please note that Yusuf Ali also affirms this fact in the footnote of 3:7: "One reading, rejected by most Commentators, but accepted by Mujahid and others, would not make a break at the point marked *Waqfa Lazim*, but would run the two sentences together. In that case the construction would run: 'No one knows its hidden meanings except God and those who are firm in knowledge. They say', etc.")

Is the Earth Flat?

The Quranic description of the earth, the solar system, the cosmos and the origin of the universe is centuries ahead of the time of its first revelation. For instance, the Quran, delivered in the seventh century C.E., states or implies that:

- Time is relative (70:4; 22:47).
- God created the universe from nothing (2:117).
- The earth and heavenly bodies were once a single point and they were separated from each other by an explosion (21:30).
- The universe is continuously expanding (51:47).
- The universe was created in six days (stages) and the conditions that made life possible on earth took place in the last four stages (50:38; 41:10).
- The stage before the creation of the earth is described as a gas nebula (41:11).
- Planet earth is floating in an orbit (27:88; 21:33).
- The earth is round like an egg (10:24; 39:5; 79:30).
- Earth's atmosphere acts like a protective shield for the living creatures (21:32).
- Wind also pollinates plants (15:22).
- Creation follows an evolutionary system (15:28-29; 24:45; 32:7-9; 71:14-7).
- The earliest organisms were incubated inside flexible layers of clay (15:26).
- The stages of human development in the womb are detailed (23:14).
- Our biological life span is coded in our genes (35:11).
- Photosynthesis is a recreation of energy stored through chlorophyll (36:77-81).
- The atomic number, atomic weight and isotopes of Iron are specified (57:25).
- Atoms of elements found on earth contain a maximum of 7 energy layers (65:12).
- The sound and vision of water and the action of eating dates (which contain oxytocin) reduce labor pains (19:24-25).
- There is life (not necessarily intelligent) beyond earth (42:29).
- The word Sabbath (seventh day) occurs exactly 7 times.
- The number of months in a year is stated as 12, and the word Month (*shahr*) occurs exactly twelve times.
- The number of days in a year is not stated, but the word Day (*yawm*) occurs exactly 365 times.
- A prophetic mathematical structure based on the number 19 implied in chapter 74 of the Quran was discovered in 1974 by the aid of computer shows that the Quran is embedded with an interlocking extraordinary mathematical system (I have written several Turkish and yet-to-be-published English books on this subject).

And there's more – much more. Many of the signs/miracles mentioned in the Quran, for instance, represent the ultimate goals of science and technology. The Quran relates that matter (but not humans) can be transported at the speed of light (27:30-40); that smell can be transported to remote places (12:94); that extensive communication with animals is possible (27:16-17); that sleep, in certain conditions, can slow down metabolism and increase life span (18:25); and that the vision of blind people can be restored (3:49).

Does it make sense, then, that a book that displays such astonishing knowledge of the physical universe would declare that the earth is flat?

As for the shape of the earth, Yusuf Ali extends, Pickthall spreads, and Shakir expands. Other traditional translations are essentially no different. They extend, expand, or spread the earth.

! **Disputed passage:** Traditional translations extend and spread the egg-shaped earth.

Yusuf Ali	Pickthall	Shakir	Reformist
And the earth, moreover, hath He **extended** (to a wide expanse); (79:30)	And after that He **spread** the earth, (79:30)	And the earth, He **expanded** it after that. (79:30)	And the earth afterwards, he made it **like an egg**. (79:30)

DISCUSSION OF 79:30

Almost all English translations of the Quran mistranslate the word *DaHY*, a word that is still used for "egg" among Arabic-speaking populations in North Africa. Why? The answer can be found in the footnotes of some classic commentaries of the Quran, which were written centuries ago.

Early commentators and translators of the verse were stuck on the word *DaHY* since it means (as we have translated) egg. In the verse, it is used with a transitive verb, with the third person pronoun Ha, which means "made it round like egg" or literally "egged it." But the commentators thought that the earth was flat!

Since they knew God's word could not contain errors or contradiction, they assumed their understanding of the verse must have been wrong. Thus, they tried to interpret the description. They argued that the word *DaHY* (egg) must have implied *maDHY* (nest), and inferred that God meant "nest" by the word "egg." Therefore, the earth is extended like a nest.

Early commentators had an excuse for reaching such a conclusion, since they did not know that the earth was spherical and slightly distended (like an egg); however, modern commentators of the Quran have no excuse to parrot this medieval misunderstanding. They should have known that the verse means what it says: the earth is shaped like an egg.

The external physical appearance of planet earth is like an egg, and its cross-section displays geological layers like an egg. Not only is the earth egg-shaped and resembles an egg regarding its layers, but so is its orbit around the sun. In fact, this is Kepler's famous First Law of Planetary Motion: the orbit of a planet about a star is an ellipse, as opposed to a perfect circle.

Is it Obvious OR
is it Darkening, Scorching, Shriveling, and Burning?

Many verses in chapter 74 have been commonly misunderstood and mistranslated. We have written a lengthy article titled, *Which One Do You See: Hell or Miracle?*, exposing the problems in traditional commentaries and translations. Here we will only pick a sample from the lengthy list.

Verse 74:29 is very interesting and crucial in understanding the rest of the chapter. Though it consists of only two words, *LaWwaHa lil-BaSHaR,* this verse is translated in several different ways. Here are some examples from English translations:

! **Disputed passage**: Many keywords in Chapter 74 have been mistranslated to describe the punishment of Hell, while they describe an intellectual punishment.

Yusuf Ali:	"darkening and changing the color of man"
Pickthall:	"It shrivelleth the man"
Irving:	"as it shrivels human (flesh)."
Shakir:	"It scorches the mortal".
M. Ali:	"It scorches the mortal"
Dawood:	"it burns the skins of men."
M. Asad:	"making (all truth) visible to mortal man."
R. Khalifa:	"obvious to all the people."
Reformist:	"obvious to humankind."

DISCUSSION OF 74:29

The derivatives of the word LWH are used in the Quran to mean a surface used for recording information, board, and flat wood; and nowhere is it used to mean scorch or burn. Before the fulfillment of the prophecy, translators and commentators of the Quran had difficulty in understanding the simple meaning of this word and thus, resorted to external sources and often odd meanings, such as scorch, or burn. In fact, the drive to justify a particular meaning for some "difficult" Quranic words is one of the many reasons for fabricating *hadith.*

Those who do not know Arabic might think that the words are difficult to understand and translate. In fact, the meaning of these two words, *LaWwaHa* and *BaSHaR* is very clear in the Quranic context. The word *LaWwaHa,* which comes from the root *LWH,* is the sister of the word *LaWH* (85:22) and its plural *aLWaH.* The plural form *aLWaH* is used in verses 7:145, 150, 154 for the "tablets" given to Moses, and in verse 54:13 for broad planks used by Noah to build his ark. The medieval commentators, not knowing the mathematical implication of the verses, mostly chose an unusual meaning for the word: scorching, burning, shriveling, etc. Ironically, most of them did affirm the obvious meaning of the word as "open board, tablet" (See Baydawi, Fakhruddin Al-Razi, etc.). Few preferred the "obvious" to the obscure. For instance, Muhammed Asad, who had no idea of the mathematical code, preferred the most obvious meaning. Rashad Khalifa who fulfilled the prophecy and discovered the implication of the entire chapter reflected the same obvious meaning. That "obvious" meaning, was obscured by the smoke of "scorching fire" burning in the imaginations of generations before him.

In 7:145; 7:150; 7:154, the word *aLWaH*, the plural of *LaWHa* is used to depict the tablets on which the Ten Commandments were inscribed. In 54:13 it is used to describe the structure of the Noah's ship made of wood panels. In 85:22 the same word is used for the mathematically protected record of the original version of the Quran. As for the *LaWaHa* of 74:29, it is the amplified noun-adjective derived from the root of the verb LWH, meaning open tablets, succeeding screens, obvious, manifesto, or clearly and perpetually visible. Ironically, the Quran uses different words to describe burning or scorching. For instance, for burning, the derivatives of *HaRaQa* (2:266; 3:181; 7:5; 20:97; 21:68; 22:9; 22:22; 29:24; 75:10), or for scorching the derivatives of *SaLaYa* (4:10; 4:30; 4:56; 4:115; 14:29; 17:18; 19:70; 27:7; 28:29; 29:31; 36:64; 37:163, 38:56; 38:59; 52:16; 56:94; 58:8; 69:31; 74: 26; 82:15; 83:16; 84:12; 87:12; 88:12; 92:15), or *NaDaJa* are used (4:56).

Again, we should note that the understanding of pre-1974 commentators was not without basis. Though their understanding did not rely on the Quranic usage of the words which created some problems (such as explaining the verse 74:31), they had some justifiable excuses to understand the way they understood. The word *LaWaHa* also meant burn and *BaSHaRa* was another rarely used word for skin in Arabic language. As I mentioned above, the multiple meanings of these verses allowed the impatient pre-1974 generations to have an understanding, though a temporary and not primarily intended one. In fact, it was better for them to have patience and not rush to speculate on these verses without knowledge (20:114; 75:16-19). It was the computer generation that was destined to understand their real meaning (10:37-46).

A Portion of the Message or A Fistful of Dirt?

Moses left his people in the charge of his brother Aaron and went to the mountain to receive commandments from God. While at the presence of God, Moses is lightly criticized for leaving his people behind:

> "'And what has caused you to rush ahead of your people O Moses?' He said: 'They are following in my footprints, and I came quickly to you my Lord so you would be pleased.' He said: 'We have tested your people after you left, and the Samarian misguided them.'" (20:83-85).

However, God knew that Moses' people were not following his message. Taking advantage of Moses' absence, the Samarian tried and succeeded to a degree to revert The Children of Israel to the religion of their Egyptian masters. Recognizing that many people respond better to concrete and tangible objects than to an abstract idea of a transcendental God, the Samarian collected jewels and melted them in a pot to form a calf (7:148). He produced a calf statue, which made sounds in the wind due to the craftily designed holes in their bodies, a symbol for intercession. Samarian was proud of his knowledge of human psychology, and he had the audacity to tell Moses the following:

! **Disputed passage**: Ignoring reason, the context of the verse, and the semantics of the Quran, traditional translations and commentaries create a bizarre story through mistranslation of the words *AThaR*, and *NaBaZa*. To accommodate the mistranslation, they also add several non-existing and non-implied words, such as, "Muhammed," "Gabriel," "Dust," and "into calf."

Yusuf Ali	Pickthall	Shakir	Reformist
He replied: "I saw what they saw not: so I took **a handful (of dust) from the footprint of the Messenger, and threw it (into the calf)**: thus did my soul suggest to me." (20:96)	He said: I perceived what they perceive not, so I seized **a handful from the footsteps of the messenger, and then threw it in**. Thus my soul commended to me. (20:96)	He said: I saw **(Jibreel)** what they did not see, so I took **a handful (of the dust) from the footsteps of the messenger, then I threw it in the casting**; thus did my soul commend to me. (20:96)	He said: "I perceived what they did not perceive, so I **took a handful from the teaching of the messenger, and I cast it away**. This is what my soul inspired me to do." (20:96)

DISCUSSION OF 20:96

According to traditional translators and many who were unknowingly influenced by them, Samarian possessed extraordinary powers! According to the story that they all copy, the Samarian secretly followed Moses and somehow was able to see the Controller-messenger of God, *Jibreel* or Gabriel, riding a horse. He thought that the dirt stepped on by the feet of *Jibreel's* horse must have had magical powers. So, the story goes, he collected some dirt and took it back to where his people were dwelling. He mixed the dirt with the melting jewelry and voila, the sound-making calf!

The word *athar,* which is mistranslated as "footprint" or "footsteps," has the following meanings: trace, teaching, relics of knowledge, remains, mark, footprint, memorial, etc. The Quran uses the word *athar* to mean teaching or message (20:84; 37;70; 43:22-23). The meaning of the word should have been clear from several verses before, where in verse

20:84, it is used to mean "teaching." God did not ask Moses why he came alone; He questioned Moses about leaving his people too early. Moses understood God's question—though the majority of translators did not—and responded by saying that his people were following his teaching. God informs him about Samarian's plan to mislead his people back to idolatry.

Should Skeptics Hang Themselves to the Ceiling?

The traditional rendering of verse 22:15 is so bad that it becomes an absurdity, a joke. The amazing thing is that anyone who studies the Quran should easily understand its meaning, since the expressions are used in other verses and contexts. Instead of first looking at the usage of words and expression in other parts of the Quran, the traditional translators look for inspiration from the early commentators who mostly relied heavily on *hadith* hearsay. Regardless of the source, except for a few, such as Muhammed Asad, Muhammed Ali, and Rashad Khalifa, many translations have duplicated the bizarre and absurd traditional rendering.

! **Disputed passage:** Traditional translations insert non-existing words and produce a myriad of absurd and hilarious challenges.

Yusuf Ali	**Pickthall**	**Shakir**	**Reformist**
If any think that Allah will not help him (**His Messenger**) in this world and the Hereafter, **let him stretch out a rope to the ceiling and cut (himself) off**: then let him see whether his plan will remove that which enrages (him)! (22:15).	Whoso is wont to think (through envy) that Allah will not give him (**Muhammed**) victory in the world and the Hereafter (and is enraged at the thought of his victory), **let him stretch a rope up to the roof (of his dwelling), and let him hang himself**. Then let him see whether his strategy dispelleth that whereat he rageth! (22:15).	Whoever thinks that Allah will not assist him in this life and the hereafter, **let him stretch a rope to the ceiling, then let him cut (it) off,** then let him see if his struggle will take away that at which he is enraged. (22:15).	Whosoever thinks that God will not help him in this world and the Hereafter, **let him extend (his request) to the heaven via some means, then let him cut off** (his dependence on anyone else) and see whether this action has removed the cause of his anger. (22:15)

DISCUSSION OF 22:15

Reading the verse within the immediate context of the Quran alone is sufficient to shed light on its meaning. For instance, just four verses before, verse 22:11 reminds us of the importance of unconditional trust in God:

> "And from the people there is he who serves God in alteration. So if good comes to him, he is content with it; and if an ordeal comes to him, he makes an about-face. He has lost this world and the Hereafter. Such is the clear loss." (22:11)

The following verse informs us that those people who oscillate in their service to God depending on the circumstances, associate others as partners to God.

> "He calls upon besides God what will not harm him and what will not benefit him. Such is the far straying." (22:12)

We learn from the Quran that most polytheists are in denial (6:23), though they hope for the intercession of prophets and saints (72:21 and 7:188; 10:49; 13:16; 48:11). The following verse expresses the reality of idolatrous people:

> "He calls on those who harm him more than they benefit him. What a miserable patron, and what a miserable companion." (22:13).

Verse 14 mentions God's blessing on those who do not pollute their affirming of monotheistic message through polytheistic ideas and practices:

> "God admits those who affirm and promote reforms to gardens with rivers flowing beneath them. God does as He wishes." (22:14)

And the following verse, 22:15, shows them a way: try to reach God through prayer, science, charity, and cut off all your dependence and hope from other things besides God. You should cut off your dependence to gods or holy people other than God since you will do so in the hereafter (2:166). In other words, if you can remove all the idolatrous ideas and dedicate yourself to God alone, you will witness God's help and victory. Those who mix monotheism with idolatry are not of those who affirm truth, even if they think so (49:14; 6:23).

Thus, starting from verse 11, when we reach verse 15, the theological relationship among the verses becomes evident, and the meaning of verse 15 shines clearly. The message of 22:15 is identical to the message given in 6:41.

Unfortunately, many translations and commentaries ignore the context of the verse and the usage of certain words in the Quran and copy the traditional false inferences and references inside the parentheses.

After reading these absurd and ridiculous translations and misleading parentheses, if an investigating person read the following verse, 22:16, he or she would be repelled from the Quran, since it asserts that the revelation of the Quran is clear!

- Many add Muhammed's name in the parenthesis, though the verse does NOT mention Muhammed and the context is not about Muhammed, but about God alone.
- Many insert the word "ceiling," though *sama* does not mean "ceiling," but rather means "heaven"; "ceiling" in Arabic is *saqf* (43:33; 21:32).
- Many insert a "rope," though the verse does not mention "rope," which is *habl* (111:5; 3:103).

The insertion, twisting, and distortion are not limited to the three examples above. The prominent Pakistani radical scholar Mawdudi, in his commentary, Tafhim al-Quran (Towards the Understanding of the Quran), lists six alternative views of previous commentators on this verse:

1. Whoever thinks that God would not help Muhammed (pbuh), should hang himself to a ceiling?
2. Whoever thinks that God would not help Muhammed (pbuh), should climb to the sky with a rope and try to stop God's help.
3. Whoever thinks that God would not help Muhammed (pbuh), should ascend to the sky and stop God's revelation.

4. Whoever thinks that God would not help Muhammed (pbuh), should ascend to the sky and stop God's blessings to him.
5. Whoever thinks that God would not him, should hang himself to the ceiling of his home.
6. Whoever thinks that God would not help him, should ascend to sky to ask for help.

Mawdudi finds the first four comments to be meaningless in their context, since he rightly argues that the pronoun "he" cannot be referring to Muhammed, but instead refers to a person who has doubt about God's help. Though Mawdudi concedes that the last two renditions fit the flow, he finds them to be far from reflecting the meaning of the verse. After rejecting all these alternative interpretations, Mawdudi presents his own version:

> "Whosoever fancies that Allah will not help him in this world and in the Hereafter, let him reach out to heaven through a rope, and then make a hole in the sky and see whether his device can avert that which enrages him."

Well, Mawdudi too disappoints; even more than the others. He asks the doubtful person to get a rope, climb to the sky, and somehow open a hole in the sky to peep through! However, Mawdudi's understanding is perhaps better than others who nicely ask the opponents of Muhammed to commit suicide by hanging themselves to the sky and THEN think about their feelings and doubts!

Now it is time to challenge all those scholars, mewlahums, commentators, translators, and their admirers, who missed the obvious and simple meaning of the verse:

If they think that our reformed translation is wrong, then they should extend a rope to the sky and hang themselves, and then think whether this trick of theirs would remove the cause of their anger! If they want, they might just open a hole in the sky and read my translation through their magic peephole!

If the above challenge is a meaningful and wise challenge, then they should take it! No, if it is an absurd and silly challenge, then they should expunge from their translations the Muhammed, the rope, the ceiling, the climbing, the hanging, the committing suicide, the thinking after killing self, and the opening of a peephole in the sky □

Smite at their Neck or Strike the Control Center?

The Quran does not promote war; but encourages us to stand against aggressors on the side of peace and justice. War is permitted only for self-defense (See 2:190,192,193,256; 4:91; 5:32; 8:19; 60:7-9). We are encouraged to work hard to establish peace (47:35; 8:56-61; 2:208). The Quranic precept promoting peace and justice is so fundamental that peace treaty with the enemy is preferred to religious ties (8:72). The Quran orders affirmers not to transgress and ask them to keep the prisoners of wars alive to be released afterwards.

! **Disputed passage:** The translations that is based on hadith/sunnah distort the meaning of the word RiQaB and translate is as "neck."

Yusuf Ali	Pickthall	Shakir	Reformist
Therefore, when ye meet the Unbelievers (in fight), **smite at their necks**; At length, when ye have thoroughly subdued them, bind a bond firmly (on them): thereafter (is the time for) either generosity or ransom: Until the war lays down its burdens. Thus (are ye commanded): but if it had been Allah.s Will, He could certainly have exacted retribution from them (Himself); but (He lets you fight) in order to test you, some with others. But those who are slain in the Way of Allah, He will never let their deeds be lost. (47:4)	Now when ye meet in battle those who disbelieve, then it is **smiting of the necks** until, when ye have routed them, then making fast of bonds; and afterward either grace or ransom till the war lay down its burdens. That (is the ordinance). And if Allah willed He could have punished them (without you) but (thus it is ordained) that He may try some of you by means of others. And those who are slain in the way of Allah, He rendereth not their actions vain. (47:4)	So when you meet in battle those who disbelieve, then **smite the necks** until when you have overcome them, then make (them) prisoners, and afterwards either set them free as a favor or let them ransom (themselves) until the war terminates. That (shall be so); and if Allah had pleased He would certainly have exacted what is due from them, but that He may try some of you by means of others; and (as for) those who are slain in the way of Allah, He will by no means allow their deeds to perish. (47:4)f	So, if you encounter those who have rejected, then **strike the control center** until you overcome them. Then bind them securely. You may either set them free or ransom them, until the war ends. That, and had **God** willed, He alone could have beaten them, but He thus tests you by one another. As for those who get killed in the cause of **God**, He will never let their deeds go to waste. (47:4)

DISCUSSION OF 47:4

The expression "*darb al riqab*" is traditionally translated as "smite their necks." We preferred to translate it as "strike the control center." The Quran uses the word "*unuq*" for neck (17:13,29; 8:12; 34:33; 38:33; 13:5; 26:4; 36:8; 40:71). The root *RaQaBa* means observe, guard, control, respect, wait for (20:94; 28:18; 28:21; 44:10; 44:59; 11:93; 54:27; 5:117; 50:18; 4:1; 33:52; 4:92; 5:89; 58:3; 90:13; 2:177; 9:60; 47:4). "*Riqab*" means slave, prisoner of war, since they are controlled.

Even if one of the meanings of the word *riqab* were neck, we would still reject the traditional translation, for the obvious reason: The verse continues by instructing peacemakers (*muslims*) regarding the capturing of the enemies and the treatment of prisoners of war. If they were supposed to be beheaded, there would not be need for an instruction regarding captives, which is a very humanitarian instruction.

Unfortunately, the Sunni and Shiite terrorists have used the traditional mistranslation and abused it further by beheading hostages in their fight against their counterpart terrorists, Crusaders, and their allied coalition, who torture and kill innocent people even in bigger numbers, yet in a baptized fashion that is somehow depicted non-barbaric by their culture and media.

The Quran gives two options regarding the hostages or prisoners of war before the war ends: (1) set them free; or (2) release them to get a fee for their unjustified aggression. Considering the context of the verse and emphasis on capturing the enemy, we could have translated the segment under discussion as, "aim to take captives."

The Old Testament contains many scenes of beheadings and grotesque massacres. For instance, see: 2 Samuel 4:7-12; 2 Kings 10:7, and 2 Chronicles 25:12.

1:1 In the name of **God**, the Gracious, the Compassionate.*
1:2 Praise is to **God**, Lord of the worlds.
1:3 The Gracious, the Compassionate.
1:4 Master of the day of judgment.*
1:5 You alone we serve; you alone we ask for help.
1:6 Guide us to the straight path;
1:7 the way of those whom you blessed; not of those who received anger, nor of the strayers.*

ENDNOTES

001:001-7 The first Chapter of the Quran is an outline of the Quran. The Quran repeatedly emphasizes God's grace and mercy and asks us to never give up hope. It invites us to always be conscious of the Day of Judgment in which God ALONE will judge our worldly choices and deeds. It underlines the importance of attaining freedom by submitting ourselves only to the Truth, and nothing but the Truth. It reminds us of the spiritual, individual, social, and political duties that will guide us on and to the straight path; additionally, it informs us of God's blessings to ancient communities and their various triumphs and blunders. For the literary aspect and purpose of shift in person, see the footnote of verse 39:53.

001:001 This first verse, shortly known as Basmalah, consists of four Arabic words and has a unique function in the Quran. It crowns every Chapter except Chapter 9. Basmalah is the foundation of a simple to understand, and impossible to imitate, interlocking mathematical structure intertwined in the text of the Quran and the Bible (We would like to remind the reader that this has nothing to do with numerological speculations or the statistically insignificant study published in the so-called Bible Code.) The miraculous function of the number 19 prophesized in Chapter 74 was unveiled in 1974 through a computerized analysis of the Quran. Though, in retrospect, the implication of 19 in Chapter 74 was obvious, it remained a secret for 1406 (19x74) lunar years after the revelation of the Quran. Ironically, the very first word of Chapter 74, The Hidden One, was revealing, yet the code was a divinely guarded gift allocated to the computer generation for they were the ones who would need and appreciate it the most. The following is just the tip of the iceberg: Basmalah consists of 19 letters and each of its words (*Ism*, *Allah*, *Rahman*, and *Rahim*), occurs in the Quran in multiples of 19. Many parameters of the mathematical structure of the Quran are related to this first verse. For instance, the missing Basmalah in the beginning of Chapter 9 is restored 19 chapters later inside Chapter 27. Thus, the frequency of this most repeated verse becomes 114 (19x6). The mathematical structure of the Quran has also resolved many scholarly arguments or mysteries. For instance, now we know why the Basmalah of the first Chapter is numbered while other Basmalahs in the beginning of chapters are not. As we have demonstrated in various books, through hundreds of simple and complex algorithms, we witness the depth and breadth of the mathematical manipulation of Arabic, an arbitrary human language, to be profound and extraordinary. This is indeed a fulfillment of a Quranic challenge (17:88). While the meaning of the Quranic text and its literal excellence were maintained, all its units, from chapters, verses, words to its letters were also assigned universally recognizable roles in creation of mathematical patterns. Since its discovery, the number 19 of the Quran and the Bible has increased the faith of many affirmers, has removed doubts in the minds of many People of the Book, and has caused discord, controversy and chaos among those who have traded the Quran with man-made sectarian teachings. This is indeed a fulfillment of a Quranic prophecy (74:30-31).

We translate the word *ISM* as NAME, but in the Arabic language and Arab culture, names are attributes. In fact, the word *ISM* shares the same root with the Arabic verb for "describe" (22:78; 53:27; 19:7). Like Hebrew, Arabic proper names are descriptive. God has all the beautiful attributes (7:180; 17:110; 20:8; 59:24). Basmalah contains three names or attributes: *Allah* (the god; God), *Rahman* (Gracious), and *Rahim* (Compassionate/Caring/Loving). The Arabic word *Allah* is not a proper name as some might think; it is contraction of *AL* (the) and *ELAH* (god). The word *Allahumma* is a different form and the letter "M" in the end is not an Arabic suffix as a novice might think. The word *Allahumma* may not be considered a divine attribute since it cannot be used as a subject in a sentence or as an attribute of a divine subject. It is always used in supplication and prayers, meaning "o my lord" or "o our lord." *Allah* and *Rahman* are two attributes that are invariably used as names rather than adjectives. Since God sent messengers to all nations (10:47; 16:36; 35:24) in their own language (14:4), they referred to their creator in their own language. See 7:180.

While some tried their hardest, for centuries, to turn the creator of the universe into an Arab God, others too have attempted to transform Him into an Anglo-Saxon male. The former ignored the fact that the languages of many nations who received God's message in their own language did not contain the word *Allah*. The latter ignored the fact that Jesus or (J)esu(s), never uttered the English word 'God,' but referred to his Lord with Hebrew or Aramaic words such as *Eli*, *Eloi*, *Elahi*, or *Ellohim* (Mark 15:34), which are almost identical to corresponding Arabic words.

The Old Testament contains several verses containing the attributes of 'Gracious' and 'Merciful' as used in Basmalah: Exodus 34:6; 2 Ch 30:9; Nehemiah 9:17,31; Psalms 103:8; 116:5; Joel 2:13; Jonah 4:2.

For more information regarding God's attributes, See 7:180 and 59:22-24.

001:004 For more information about the Day of Judgment see 82:15-19. Also, see the account of the New Testament: Matthew 12:36.

001:007 Traditional commentaries attempt to restrict the negatively described groups to Christians and Jews. This self-righteous attitude has led the Muslim masses to ignore their own corruption and deviation from the straight path.

The Quran mentions communities as well as individuals who received retribution such as the People of Noah (26:25-102), the People of Thamud (7:78; 11:61-68), the People of Lot (26:160-175), the People of Madyan (11:84-95), Ayka (26:176-191), Aad (11:59-60; 26:123-140), and Pharaoh (3:11; 11:96-99; 20:78-80).

The Bible provides numerous examples of divine retribution against nations: The old world (Genesis 6:7,17), Sodom (Genesis 19:24), Egypt (Exodus 9:14), Israel (Numbers 14:29, 21:6), People of Ashdod (1Samuel 5:6), People of Bethshemesh (1Samuel 6:19), Amalekites (1 Samuel 15:3). The Bible also gives examples of divine punishment against individuals, such as: Cain (Genesis 4:11-12), Canaan (Genesis 9:25), Korah (Number 16:33-35), Achan (Joshua 7:25), Hophni (1Samuel 2:34), Saul (1Samuel 15:23), Uzzah (2Samuel 6:7), Jeroboam (1Kings 13:4), Ahab (1Kings 22:38), Gehazi (2Kings 5:27), Jezebel (2Kings 9:35), Nebuchadnezzar (Daniel 4:31), Belshazzar (Daniel 5:30), Zacharias (Luke 1:20), Ananias (Acts 5:1-10), Herod (Acts 12:23), Elymas (Acts 13:11).

2:0 In the name of God, the Gracious, the Compassionate.

2:1 A1L30M40*

2:2 This is the book in which there is no doubt, a guide for the conscientious.

Affirmers, Ingrates, and Hypocrites

2:3 Those who affirm the unseen, and observe the contact prayer (*sala*), and from Our provisions to them, they spend.*

2:4 Those who affirm what was sent down to you, and what was sent down before you, and regarding the Hereafter they are certain.*

2:5 These are the ones guided by their Lord, and these are the winners.

2:6 As for those who do not appreciate, whether you warn them or do not warn them, they will not affirm.*

2:7 **God** has sealed their hearts and their ears; and over their eyes are covers. They will incur a great retribution.

2:8 Among the people are those who say, "We affirm **God** and the Last day," but they do not affirm.

2:9 They seek to deceive **God** and the affirmers, but they only deceive themselves without noticing.

2:10 In their hearts is a disease, so **God** increases their disease, and they will have a painful retribution for what they have denied.

2:11 If they are told "Do not make evil in the land," they say, "But we are the reformers!"

2:12 No, they are the evildoers, but they do not perceive.

2:13 If they are told, "affirm as the people have affirmed," they say, "Shall we affirm as the fools have affirmed?" No, they are the fools, but they do not know.

2:14 If they come across the affirmers, they say, "We affirm," and when they are alone with their devils they say, "We are with you, we were only mocking."

2:15 **God** mocks them and leaves them prolonged in their transgression.

2:16 These are those who have exchanged guidance for straying; their trade did not profit them, nor were they guided.

2:17 Their example is like one who lights a fire, so when it illuminates what is around him, **God** takes away his light and leaves him in the darkness not seeing.

2:18 Deaf, dumb, and blind, they will not revert.

2:19 Or like a storm from the sky, in it are darkness, thunder, and lightning. They place their fingers in their ears from stunning noises out of fear of death; and **God** is aware of the ingrates.

2:20 The lightning nearly snatches away their sight, whenever it lights the path, they walk in it, and when it becomes dark for them, they stand. Had **God** willed, He would have taken away their hearing and their sight. **God** is capable of all things.

2:21 O people, serve your Lord who has created you and those before you that you may be conscientious.

2:22 The One who made the land a habitat, and the sky a structure, and He sent down from the sky water with which He brought out fruit as a provision for you. So do not make any equals with **God** while you now know.

Verifiable and Falsifiable Evidence

2:23 If you are in doubt as to what We have sent down to Our servant, then bring a chapter like this, and call upon your witnesses other than **God** if you are truthful.*

Paradise and hell Descriptions are Allegorical

2:24 If you cannot do this -- and you will not be able to do this -- then beware of the fire whose fuel is people and stones; it has been prepared for the ingrates.

2:25 Give good news to those who affirm and promote reforms, that they will have gardens with rivers flowing beneath. Every time they receive of its fruit, they say, "This is what we have been provisioned before," and they are given its likeness/allegories. Moreover, there they will have pure spouses, and there they will abide.*

2:26 **God** does not shy away from citing the example of a mosquito, or anything above it. As for those who affirm, they know that it is the truth from their Lord. As for the unappreciative, they say, "What does **God** want with this example?" He lets many stray by it, and He guides many, but He only lets stray the evildoers.

2:27 The ones who break their pledge to **God** after making the covenant, and they sever what **God** had ordered to be joined. Moreover, they make corruption on earth; these are the losers.

2:28 How can you not appreciate **God** when you were dead and He brought you to life? Then He makes you die, then He brings you to life, then to Him you return.*

2:29 He is the One who created for you all that is in the earth. Moreover, He attended to the universe and made it seven heavens, and He is aware of all things.

Creation

2:30 Your Lord said to the controllers, "I am placing a successor on earth." They said, "Would You place in it he who would corrupt it, and shed blood, while we sing Your glory, and praise You?" He said, "I know what you do not know."*

Innate Ability to Relate and Discriminate

2:31 He taught Adam the description of all things, and then He displayed them to the controllers and said, "Inform Me the descriptions of these things if you are truthful."*

2:32 They said, "Glory be to You, we have no knowledge except what You have taught us, You are the Knowledgeable, the Wise."

2:33 He said, "O Adam, inform them of the descriptions of these." When he informed them of their descriptions, He said, "Did I not tell you that I know the unseen of the heavens and the earth, and that I know what you reveal and what you were hiding?"

Testing Iblis (The one who Despaired) with Humans and Humans with Iblis

2:34 We said to the controllers, "Submit to Adam," so they submitted except for the Despaired, he refused and became arrogant, and became of the ingrates.

2:35 We said, "O Adam, reside you and your mate in the paradise, and eat from it bountifully as you both wish, and do not approach this tree, else you will be of those who did wrong."*

2:36 So, the devil tricked both and he brought both of them out from what they were in, and We said, "Descend, you are all enemies of one another. In the land shall be abode for you and luxury for a while."*

2:37 Adam then received words from His Lord, so He forgave him; He is the Forgiver, the Compassionate.

2:38 We said, "Descend from it all of you, so when the guidance comes from Me, then whoever follows My guidance, they will have nothing to fear, nor will they grieve."

2:39 Those who do not appreciate and deny Our signs, they are the dwellers of fire, in it they will abide.

Pledge with the Children of Israel

2:40 O Children of Israel, remember My blessings that I had blessed you with, and fulfill your pledge to Me that I may fulfill My pledge to you, and revere Me alone.

2:41 affirm what I have sent down, authenticating what is already with you, and do not be the first to reject it! Moreover, do not exchange My signs for a petty gain; and of Me you shall be conscientious.*

2:42 Do not obscure the truth with falsehood, nor keep the truth secret while you know.

2:43 Observe the *Contact prayer*, and contribute towards betterment, and bow with those who bow.*

2:44 Do you order the people to do goodness, but forget yourselves, while you are reciting the book? Do you not reason?

2:45 Seek help through patience, and the *Contact prayer*. It is a difficult thing, but not so for the humble.

2:46 The ones who conceive that they will meet their Lord and that to Him they will return.

2:47 O Children of Israel, remember My blessings that I had blessed you with, and that I had preferred you to all the worlds!

2:48 Beware of a day where no person can avail another person, nor will any intercession be accepted from it, nor will any ransom be taken, nor will they have supporters.*

2:49 We saved you from the people of Pharaoh; they were punishing you with severe retribution, killing your children, and shaming your women. In that was a great trial from your Lord.

2:50 We parted the sea for you, thus We saved you and drowned the people of Pharaoh while you were watching.

2:51 We appointed a meeting time for Moses of forty nights, but then you took the calf after him while you were wicked.*

2:52 Then We forgave you after so that perhaps you would be thankful.

2:53 We gave Moses the book and the criterion so that perhaps you would be guided.

2:54 Moses said to his people, "O my people, you have wronged yourselves by taking the calf, so repent to your Maker, and face yourselves. That is better for you with your Maker, so He would forgive you. He is the Forgiving, the Compassionate."*

Those Who Cannot Withstand the Energy of Lightning Want to See God!

2:55 When you said, "O Moses, we will not affirm you until we see **God** openly!" the lightning bolt took you while you were still staring.*

2:56 Then We resurrected you after your death that you may be appreciative.

Children of Israel in Sinai, yet Unwilling to Pay the Price of Freedom

2:57 We shaded you with clouds, and sent down to you manna and quails, "Eat from the goodness of the provisions We have provided you." They did not wrong Us, but they wronged themselves.

2:58 We said, "Enter this town, and eat from it as plentifully as you wish; and enter the gate humbly, and talk amicably, We will then forgive your mistakes, and We will increase for the good-doers."

2:59 But the wicked altered what was said to them into a different saying, thus We sent down upon the wicked an affliction from the sky because they had transgressed.*

2:60 Moses was seeking water for his people, so We said, "Strike the rock with your staff." Thus, twelve springs burst out of it; each tribe then knew from where to drink. "Eat and drink from **God**'s provisions, and do not roam the earth as corruptors."*

2:61 You said, "O Moses, we will not be patient to one type of food, so call for us your Lord that He may bring forth what the earth grows of its beans, cucumbers, garlic, lentils, and onions." He said, "Would you trade what is lowly with what is good?" Go down to Egypt, there you will find what you want. They were thus stricken with humiliation and disgrace, and they remained under **God**'s wrath for they were not appreciating **God**'s signs and killing the prophets with no justification. This is for what they have disobeyed and transgressed.*

Conditions for Salvation

2:62 Surely those who affirm, and those who are Jewish, and the Nazarenes, and those who follow other religions, all who affirm **God** and the Last day, and promote reforms, they will have their reward with their Lord, with no fear over them, nor will they grieve.*

2:63 We took your covenant, and raised the mount above you, "Take what

We have given you with strength, and remember what is in it that you may be conscientious."

2:64 Then you turned away after this. Had it not been for **God**'s favor upon you and His mercy, you would have been of the losers.

2:65 You have come to know who it was amongst you that transgressed the Sabbath, We said to them, "Be despicable apes!"*

2:66 We did this as a punishment for what they had done on it and afterwards, and a reminder to the righteous.

Seeking Unnecessary Details in Divine Instructions and its Consequences

2:67 Moses said to his people, "**God** orders you to slaughter a heifer." They said, "Do you mock us?" He said, "I seek refuge with **God** lest I be of the ignorant ones."*

2:68 They said, "Call upon your Lord for us that He may clarify which it is." He said, "He says it is a heifer neither too old nor too young, an age between that. So now do as you are commanded."

2:69 They said, "Call upon your Lord for us that He may clarify what color it is." He said, "He says it is a yellow heifer with a bright color, pleasing to those who see it."

2:70 They said, "Call upon your Lord for us that He may clarify which one it is, for the heifers all look alike to us and we will, **God** willing, be guided."

2:71 He said, "He says it is a heifer which was never subjugated to plough the land, or water the crops, free from any blemish." They said, "Now you have come with the truth." They slaughtered it, though they had nearly not done so.

2:72 You had murdered a person, and then disputed in the matter; **God** was to bring out what you were keeping secret.

2:73 We said, "Strike him with a piece of it." It is thus that **God** brings the dead to life, and He shows you His signs that you may reason.*

2:74 Despite this, your hearts hardened; like rocks. Or even harder, since some rocks allow rivers to gush out, others crack and release water, and some rocks fall out of reverence for **God**. **God** is never unaware of what you do.

2:75 Do you expect that they would affirm your thesis; when a group of them had heard **God**'s words and reasoned accordingly, yet after that they altered them knowingly?

2:76 When they come across those who affirm, they say, "We affirm!", and when they are alone with each other they say, "Why do you inform them about what **God** has disclosed to us? Lest they use it as an argument against us concerning your Lord. Do you not reason?"

2:77 Do they not know that **God** knows what they conceal and what they declare?

2:78 Amongst them are Gentiles who do not know the book except by hearsay, and they only conjecture.*

Those Who Attribute Their Religious Fabrications to God

2:79 So woe to those who write the book with their hands then say, "This is from **God**," so that they can seek a cheap gain! Woe to them for what their hands have written, and woe to them for what they have gained.*

2:80 They said, "The fire will not touch us except for a number of days." Say, "Have you taken a pledge with **God**? If so, then **God** will not break His pledge. Or do you say about **God** what you do not know?"

2:81 Indeed, whoever gains a bad deed and his mistakes surround him; those are the people of the fire, there they abide eternally.

2:82 Those who affirm and promote reforms, they are the people of paradise, in it they abide eternally.

Covenant With the Children of Israel

2:83 We took the covenant of the Children of Israel, "You shall not serve except **God**, and regard your parents, and regard the relatives, and the orphans, and the needy, and say kind things to the people, and observe the *Contact prayer*, and contribute towards betterment." Then you turned away, except a few of you; you are stubborn.

2:84 We have taken a covenant with you, "You shall not spill each other's blood, nor drive each other out from your home circles." You agreed to this while bearing witness.

2:85 But here you are killing each other and driving out a group of you from their homes; you act towards them with evil and animosity. If they come to you as prisoners, you ransom them while it was forbidden for you to drive them out! Do you affirm some of the book and reject some? The punishment for those amongst you who do so is humiliation in this worldly life; and on the day of Resurrection, they will be returned to the most severe retribution. **God** is not oblivious of what you all do.

2:86 These are the ones who have purchased this lowly/worldly life in exchange of the Hereafter. The retribution will not be reduced for them, nor will they be supported.

2:87 We gave Moses the book, and after him, We sent the messengers. Later We gave Jesus son of Mary the clear proofs, and We supported Him with the Holy Spirit. Is it that every time a messenger comes to you with what your minds do not desire, you become arrogant? A group of them you deny, and a group of them you fight/kill!

2:88 They said, "Our hearts are uncircumcised/covered!" No, it is **God** who has cursed them for their rejection, for very little do they affirm.*

Authenticating the Previous book

2:89 At a time when they were asking for victory against those who were unappreciative, a book came to them from **God**, authenticating what is with them. But when what they knew came to them, they did not appreciate it! **God**'s curse is upon the ingrates.

2:90 Miserable indeed is with what they traded themselves by not appreciating what **God** has sent down. They did so because of their resentment that **God** would send down from His grace to whom He pleases from among His servants; thus, they have incurred wrath upon wrath. The ingrates will have a humiliating retribution.

2:91 If it is said to them, "affirm what **God** has sent down," they say, "We affirm only what was sent down to us," and they do not appreciate what came after it, while it is the truth authenticating what is with them. Say, "Why then did you fight/kill **God**'s prophets if you were those who affirm?"

2:92 Moses had come to you with clear proofs, then you took the calf after him; you were wicked!*

2:93 We took your covenant, and raised the mount above you, "Take what We have given you with strength, and listen." They said, "We hear and disobey!" and they had consumed the calf inside their hearts by their rejection. Say, "Miserable indeed is what your affirmation instructs you if you are those who affirm!"*

2:94 Say, "If the abode of the Hereafter has been set exclusively for you to the exception of all other people, then you should wish for death if you are truthful!"

2:95 They will never wish for it because of what their hands have done; and **God** is aware of the wicked.

2:96 You will find them the most obsessive people regarding long life, as well as those who have set up partners. Each one of them wishes that he could live a thousand years. It will not change for him the retribution even if he lived that long; **God** is watchful over what you do.

2:97 Say, "Whoever is an enemy to Gabriel," who has sent it down into your heart by **God**'s permission, authenticating what is already present, and a guide and good news for those who affirm.

2:98 "Yes, whoever is an enemy to **God** and His controllers, and His messengers, and Gabriel, and Michael, then so **God** is the enemy of those who do not appreciate."

2:99 We have sent down to you clear signs; only the evil ones would not appreciate them.

2:100 Whenever they make a pledge, does a group of them break it? Alas, most of them do not affirm.

2:101 When a messenger came to them from **God**, authenticating what was with them, a group of those who had already received the book placed **God**'s book behind their backs as if they did not know.

Science of Magic

2:102 They followed what the devils recited regarding Solomon's kingship. Solomon did not reject, but it was the devils that rejected by teaching people magic and what was sent down on the two controllers in Babylon, Haroot and Maroot. They would not teach anyone until they would say, "We are a test, so do not be unappreciative!" Thus, they teach what can separate a person from his mate; but they cannot harm anyone except by **God**'s permission. They learn what harms them and does not benefit them, and they have known that he who purchases such a thing has no place in the Hereafter. Miserable indeed is what they traded themselves with; if only they knew!*

2:103 Had they affirmed and been aware, it would have brought them a reward from **God** which is far better if they knew!

2:104 O you who affirm, do not say, "Shepherd us," but say, "Pay attention to us," and listen. For the ingrates is a painful retribution.*

2:105 Neither do those who have not appreciated among the people of the book, nor from among those who have set up partners, wish that any good comes down to you from your Lord. Nevertheless, **God** chooses with His mercy whom/whomever He wishes; and from **God** is the greatest favor.

Signs and Miracles

2:106 We do not abrogate a sign, or make it forgotten, unless We bring one which is like it or even greater. Did you not know that **God** is capable of all things?*

2:107 Did you not know that to **God** belongs the kingship of the heavens and earth, and that you do not have besides **God** any guardian or supporter?

2:108 Or do you want to ask your messenger as Moses was asked before? Whoever replaces affirmation with rejection, has indeed strayed from the right path.

2:109 Many of the people of the book have wished that they could return you to being unappreciative after your affirmation, out of envy from themselves after the truth was made clear to them. You shall forgive them and overlook it until **God** brings His will. **God** is capable of all things.

2:110 Observe the *Contact prayer*, and contribute towards betterment, and what you bring forth of good for yourselves, you will find it with **God**. **God** sees what you do.

Eternal Salvation is not Exclusive to a Race or Sect

2:111 They said, "None shall enter paradise except those who are Jewish or Nazarenes;" this is what they wish! say, "Bring forth your proof if you are truthful."*

2:112 No, whosoever peacefully surrenders himself to **God**, while being a good-doer; he will have his reward with his Lord. There will be no fear over them, nor will they grieve.

2:113 The Jews said, "The Nazarenes have no basis," and the Nazarenes said, "The Jews have no basis," while they are both reciting the book! Similarly, those who do not know have said the same thing.

God will judge between them on the day of Resurrection in what they disputed.

2:114 Who are more wicked than those who boycott **God**'s temples and seek their destruction because His name is mentioned there? It is they who should not be able to enter them except in fear; they will have humiliation in this world and in the Hereafter a painful retribution.

2:115 To **God** belongs the east and the west, so wherever you turn, there is **God**'s presence. **God** is Encompassing, Knowledgeable.

2:116 They said, "**God** has taken a son!" Be He glorified. To Him is all that is in the heavens and in the earth, all are humbled to Him.

2:117 Initiator of heavens and earth, when He decrees a command, He merely says to it, "Be," and it is.

2:118 Those who do not know said, "If only **God** would speak to us, or a sign would come to us!" The people before them have said similar things; their hearts are so similar! We have clarified the signs for a people who have conviction.*

2:119 We have sent you with the truth as a bearer of good news and as a warner. You will not be questioned about the people of hell.

Bigots

2:120 Neither the Jews nor the Nazarenes will be pleased with you until you follow their creed. Say, "The guidance is **God**'s guidance." If you follow their wishes after the knowledge has come to you, then none can help or protect you against **God**.

2:121 Those whom We have given the book and they recite it truthfully the way it deserves to be recited; they affirm it. As for those who do not appreciate it, they are the losers.

2:122 O Children of Israel, remember My blessings that I have bestowed upon you, and that I have preferred you to all the worlds.

2:123 Beware of a day when no person can avail another person, nor will any change be accepted from it, nor will any intercession help it; they will not be supported.

Abraham, the Model Monotheist and Reformer

2:124 Abraham was tested by commands from His Lord, which he completed. He said, "I will make you a leader for the people." He said, "Also from my progeny?" He said, "My pledge will not encompass the wicked."*

2:125 We have made the sanctuary to be a model for the people and a security. Utilize the place of Abraham to reach out. We had entrusted to Abraham and Ishmael, "You shall purify My sanctuary for those who visit, those who are devoted, and for those who kneel and prostrate."

2:126 Abraham said, "My Lord, make this town secure, and provide for its inhabitants of the fruits for whoever affirms **God** and the Last day." He said, "As for he who does not appreciate, I will let him enjoy for a while, then I will force him to the retribution of the fire, what a miserable destiny!"

2:127 As Abraham raised the foundations for the sanctuary with Ishmael, "Our Lord accept this from us, You are the Hearer, the Knowledgeable."

2:128 "Our Lord, let us peacefully surrender to You and from our progeny a nation peacefully surrendering to You; show us our rites and forgive us; You are the Most Forgiving, the Compassionate."

2:129 "Our Lord, and send amongst them a messenger from among themselves, that he may recite to them Your signs and teach them the book and the wisdom; and purify them. You are the Noble, the Wise."

2:130 Who would abandon the creed of Abraham except one who fools himself? We have selected him in this world, and in the Hereafter, he is of the reformers.

2:131 When his Lord said to him, "Peacefully surrender," he said, "I peacefully surrender to the Lord of the worlds."

2:132 Abraham enjoined his sons and Jacob, "O my sons, **God** has selected this system (peacemaking) for you, so do not die except as ones who have peacefully surrendered."*

2:133 Or were you present when death came to Jacob and he told his sons, "Who shall you serve after I am gone?" They said, "Your god, and the god of your fathers Abraham, and Ishmael, and Isaac; One god and to Him we peacefully surrender."

2:134 That was a nation that has passed away. To them is what they have earned, and to you is what you have earned; and you will not be asked regarding what they did.

Monotheism: The Right Paradigm

2:135 They said, "Be Jews or Nazarenes so that you may be guided!" Say, "No, rather (follow) the creed of Abraham, monotheism; for he was not of those who set up partners."*

2:136 Say, "We affirm **God** and what was sent down to us and what was sent down to Abraham, Ishmael, Isaac, Jacob, the Patriarchs, what was given to Moses and Jesus, and what was given to the prophets from their Lord; we do not make a distinction between any of them and to Him we peacefully surrender."*

2:137 So, if they affirm exactly as you have affirmed, then they are guided; but if they turn away, then they are in opposition, and **God** will suffice you against them; He is the Hearer, the Knower.

2:138 Such is **God**'s color. Whose color is better than of **God**? To Him we serve.

2:139 Say, "Do you debate with us regarding **God**? He is our Lord and your Lord, and we have our work and you have your work, and to Him we are devoted."

2:140 "Or do you say that Abraham and Ishmael and Isaac and Jacob and the Patriarchs were all Jewish or Nazarene?" Say, "Are you more knowledgeable or is **God**?" Who is more wicked than one who conceals a testimony with him from **God**? **God** is not oblivious of what you do.

2:141 That was a nation that passed away; to them is what they earned and to you is what you have earned; you will not be asked about what they did.

Formalistic Ritual is Not the Essence

2:142 The foolish from amongst the people will say, "What has turned them away from the focal point that they were on?" Say, "To **God** is the east and the west, He guides whomsoever He wishes to a straight path."*

2:143 As such, We have made you a balanced/impartial nation so that you may be witnesses over the people, and that the messenger may be witness over you. We did not make the focal point that you came on except so that we know who follows the messenger from those who will turn on their heels. It was a big deal indeed except for those whom **God** had guided; **God** was not to waste your affirmation. **God** is Kind and Compassionate over the people.

Focal Point

2:144 We see the shifting of your face towards the sky; We will thus set for you a focal point that will be pleasing to you, "You shall set yourself towards the Regulated Temple; and wherever you may be, you shall all set yourselves towards it." Those who have been given the book know it is the truth from their Lord. **God** is not oblivious of what you do.

2:145 Even if you come to those who have been given the book with every sign, they will not follow your focal point, nor will you follow their focal point, nor will some of them even follow each other's focal point. If you were to follow their desires after the knowledge that has come to you, then you would be one of the wicked.

2:146 Those to whom We have given the book know it as they know their own children, and a group of them hides the truth while they know.

2:147 The truth is from your Lord; so do not be one of those who doubt.

2:148 To each is a direction that he will take, so you shall race towards good deeds. Wherever you may be, **God** will bring you all together. **God** is capable of all things.

2:149 Regardless from where you start, you shall set yourself towards the Regulated Temple; it is the truth from your Lord; and **God** is not oblivious of what you do.*

2:150 Again, regardless from where you start, you shall set yourself towards the Regulated Temple. Wherever you may be, you shall set yourselves towards it; that the people will have no room for debate with you, except those of them who are wicked. You shall not fear them but fear Me so that I may complete My blessings upon you and that you may be guided.

2:151 As We have sent a messenger to you from amongst yourselves; they recite Our signs to you. Thereby he purifies you, teaches you the book and the wisdom, and teaches you what you did not know.

2:152 So remember Me that I may remember you, and be thankful to Me and do not become unappreciative.*

2:153 O you who have affirmed, seek help through patience and the *Contact prayer*; **God** is with the patient ones.

2:154 Do not say of those who are killed in the sake of **God** that they are dead; no, they are alive but you do not perceive.*

2:155 We will test you with some fear and hunger, and a shortage in money and lives and fruits. Give good news to those who are patient.*

2:156 The ones who, when afflicted with adversity, say, "We belong to **God** and to Him we will return."

2:157 To these will be support/solidarity from their Lord and a mercy; they are the guided ones.

Serenity and Chivalry at Debate Conference

2:158 The serenity and chivalry are amongst **God**'s sentiments. Therefore, whosoever debates the basis, or prepares for it, commits no error by moving between the two states. Whoever volunteers for goodness; indeed, **God** is Appreciative, Knowledgeable.

Hiding the Truth

2:159 Surely those who conceal what We have sent down to them which was clear, and the guidance, after **God** had made it clear in the book; these will be cursed by **God** and be cursed by those who curse;

2:160 Except those who repent, reform and proclaims (the truth); from those I will accept their repentance, for I am the One who accepts repentance, the Compassionate.

2:161 Surely, those who did not appreciate and then died as ingrates; they will be cursed by **God**, and the controllers, and all the people.

2:162 They will abide therein, where the retribution will not be lightened for them, nor will they be reprieved.

God's Signs in Nature

2:163 Your god is but One god, there is no god but He, the Gracious the Compassionate.

2:164 Surely, in the creation of heavens and earth, the succession of night and day, and the ships that sail in the sea for the benefit of people, and what **God** has sent down of water from the sky therewith He brings the earth back to life after it had died, and He sends forth from it every creature, and the movement of the winds and the clouds that have been designated between the earth and the sky are signs for a people that reason.*

Followers versus Leaders

2:165 Among the people are some who take other than **God** as equals to Him, they love them as they love **God**; but those who affirm love **God** more strongly; and when those who were wicked see the retribution, they will see that all power belongs to **God**, and that **God** is severe in retribution.*

2:166 When those who were followed will disown those who followed them, they will see the retribution; all intermediaries have abandoned them.

2:167 Those who followed them said, "If we only could have a chance to disown them as they have disowned us." It is such that **God** will show them their works, which will be regretted by them; they will not leave the fire.

The Followers of the Devil Prohibit Lawful Food

2:168 O people, eat of what is in the earth as good and lawful, and do not follow the steps of the devil; he is to you a clear enemy.

2:169 He only orders you evil and sin, and that you may say about **God** what you do not know.

2:170 If they are told, "Follow what **God** has sent down," they say, "No, we will follow what we found our fathers doing!" What if their fathers did not reason much and were not guided?

Following the Ancestors Blindly

2:171 The example of those who are unappreciative is like the one who repeats what he has heard of calls and shouts; deaf, dumb, and blind; they do not reason.

2:172 O you who affirm, eat of the good things We have provided for you, and be thankful to **God** if it is Him you serve.*

2:173 He has only forbidden for you what is already dead, the blood, the meat of pig, and what was dedicated to other than **God**. Whoever finds himself forced out of need and without disobedience or animosity, there is no sin upon him. **God** is Forgiving, Compassionate.

The Professionals Who Hide the Truth

2:174 Surely, those who conceal what **God** has sent down of the book, and exchange it for a petty gain; they will not eat into their stomachs except the fire, and **God** will not speak to them on the day of Resurrection, nor will He purify them, and they will have a painful retribution.*

2:175 These are the ones who have exchanged straying for guidance, and retribution for forgiveness. What made them steadfast against the fire!

2:176 This is because **God** has sent down the book with truth; and those who have disputed about the book are in far opposition.

Piety is Not Following a Set of Rituals, But

2:177 Piety is not to turn your faces towards the east and the west, but piety is one who affirms **God** and the Last day, and the controllers, and the book, and the prophets, and he gives money out of love to the near relatives, and the orphans, and the needy and the wayfarer, and those who ask, and to free the slaves, and he observes the *Contact prayer*, and contributes towards betterment; and those who keep their pledges when they make a pledge, and those who are patient in the face of good and bad and during persecution. These are the ones who have been truthful, and they are the righteous.

Limitation to Capital Punishment via the Rule of Equivalency and Forgiveness

2:178 O you who affirm, equivalent execution has been decreed for you in the cases of killings: the free for the free, the slave for the slave, and the female for the female. Whoever is forgiven anything by his brother, then it is to be followed with good deeds and kindness towards him; that is alleviation from your Lord, and a mercy. Whoever transgresses after that, he will have a painful retribution.*

2:179 Through equivalent execution, you will be protecting life, O you who possess intelligence, that you may be righteous.

Write a Will

2:180 It is decreed for you that if death should come to any of you, that it is best if he/she leaves a property, then a will should be for parents and relatives, according to recognized norms. This is due right for the righteous.*

2:181 Whoever alters it after having heard, the sin will be upon those who alter it. **God** is Hearer, Omniscient.

2:182 If anyone fears deviation or sin from his beneficiary, then he does not commit sin by reconciling what is between them. **God** is Forgiving, Compassionate.

Fasting

2:183 O you who affirm, fasting is decreed for you as it was decreed for those before you that perhaps you may be righteous.*

2:184 A number of days. Whoever of you is ill or traveling, then the same number from different days; and as for those who can do so but with difficulty, they may redeem by feeding the needy. Whoever does good voluntarily, it is better for him. If you fast, it is better for you if you knew.

2:185 The month of Ramadan, in which the Quran was sent down as a guide to the people and a clarification of the guidance and the criterion. Therefore, those of you who witness the month shall fast it. Whoever is ill or traveling, then the same number from different days. **God** wants to bring you ease and not to bring you hardship; and so that you may complete the count, and glorify **God** because He has guided you, that you may be thankful.

2:186 If My servants ask you about Me, I am near answering the calls of those who call to Me. So let them respond to Me and affirm Me that they may be guided.

2:187 It has been made lawful for you during the night of fasting to approach your women sexually. They are a garment for you and you are a garment for them. **God** knows that you used to betray yourselves so He has accepted your repentance and forgiven you; now you may approach them and seek what **God** has written for you. You may eat and drink until the white thread is distinct from the black thread of dawn; then you shall complete the fast until night; and do not approach them while you are devoted in the temples. These are **God**'s boundaries, so do not transgress them. It is thus that **God** makes His signs clear to the people that they may be righteous.*

Unfair Financial Gains, Bribery and Corruption

2:188 Do not consume your money between you unjustly by bribing the decision-makers so that you may consume a part of the other people's money sinfully while you know!*

2:189 They ask you regarding the new moons, say, "They are a timing mechanism for the people as well as for the Debate Conference." Piety is not that you would enter a home from its back, but piety is whoever is aware and comes to the homes through their main doors. Be conscientious of **God** that you may succeed.*

War as Defense

2:190 Fight in the cause of **God** against those who fight you, but do not transgress, **God** does not like the aggressors.*

2:191 Fight them wherever you meet them, and expel them from where they expelled you, and know that persecution is worse than being killed. Do not fight them at the Restricted Temple unless they fight you in it; if they fight you then fight them. Thus is the reward of those who do not appreciate.

2:192 If they cease, then **God** is Forgiving, Compassionate.

2:193 Fight them so there is no more persecution, and so that the system belongs to **God**. If they cease, then there will be no aggression except against the wicked.*

Follow the War Restrictions

2:194 The Restricted Month is for the Restricted Month. The restrictions are mutual. Whoever attacks you, then you shall attack him the same as he attacked you; and be conscious of **God;** and know that **God** is with the righteous.

2:195 Spend in the cause of **God**, but do not throw your resources to disaster. Do kindness, for **God** loves those who do kindness.

2:196 Complete the Debate Conference and the visit for **God**. But, if you are prevented, then make what is affordable of donation, and do not shave your heads until the donation reaches its destination. Whoever of you is ill or has an affliction to his head, then he may redeem by fasting or giving a charity or a sacrifice. But if you are able, then whoever continues the visit until the Debate Conference, then he shall provide what is affordable of donation. If one cannot find anything, then he must fast for three days during the Debate Conference and seven when he returns; this will make a complete ten; this is for those whose family is not present at the Regulated Temple. Be conscious of **God**, and know that **God** is severe in retribution.*

Debate Conference in Four Months

2:197 The Debate Conference is in the known months. So, whosoever decides to perform the Debate Conference in them, then there shall be no sexual approach, nor vileness, nor quarreling in the Debate Conference. Any good that you do, **God** is aware of it; and bring provisions for yourselves, though the best provision is awareness; and be conscientious of Me, o you who possesses intelligence.*

2:198 There is no blame on you to seek goodness from your Lord. So when you disperse from the stations of identification/recognition, then remember **God** towards the regulated place of sentiments, and remember Him as He has guided you, for you were before that straying.

Those Who Set Partners With God Were Practicing Debate Conference

2:199 Then you shall disperse from where the people dispersed, and seek **God**'s forgiveness; **God** is surely Forgiving, Compassionate.

2:200 When you have completed your rites, then remember **God** as you remember your fathers or even greater. Among the people are those who say, "Our Lord, give us from this world!", but in the Hereafter he has no part.

2:201 Some of them say, "Our Lord, give us good in this world, and good in the Hereafter, and spare us from the retribution of the fire."

2:202 These will have a benefit for what they have gained; and **God** is quick in computing.

2:203 Remember **God** during a number of days. Whoever hurries to two days, there is no sin upon him; and whoever delays, there is no sin upon him if he is being righteous. Be conscious of **God**; and know that it is to Him that you will be gathered.

Do Not be Impressed by Appearance

2:204 Among the people are those whose words you admire in this worldly life, but **God** is witness as to what is in his heart, for he is the worst in opposition.

2:205 If he gains power, he seeks to corrupt the earth, destroy culture and progeny. **God** does not like corruption.

2:206 If he is told, "Be conscious of **God**," his pride leads to more sin. Hell shall be sufficient for him; what a miserable abode!

2:207 Among the people is he who dedicates his person by seeking **God**'s favors; **God** is Kind to His servants.

2:208 O you who affirm, join in peace, all of you, and do not follow the footsteps of the devil. He is to you a clear enemy.

2:209 But if you slip after the proof has come to you, then know that **God** is Noble, Wise.

2:210 Are they waiting for **God** to come to them shadowed in clouds with the controllers? The matter would then be finished! To **God** all matters will return.*

Nearsightedness

2:211 Ask the Children of Israel how many clear signs We gave them. Whoever changes **God**'s favor after it has come to him, then **God** is Mighty in retribution.*

2:212 This lowly/worldly life has been made pleasing to the ingrates, and they mock those who affirm. Those who are aware will be above them on the day of Resurrection; **God** provides for whomever He wishes without counting.

2:213 The people used to be one nation, then **God** sent the prophets as bearers of good news and as warners, and He sent down with them the book with the facts so that they may judge between the people in what they were disputing. But after receiving the proof, the people disputed in it due to animosity between them. **God** guided those who affirmed with His permission regarding what they disputed in of the truth. **God** guides whoever/whomever (He) wishes to a straight path.*

Testing through Adversities

2:214 Or did you calculate that you would enter paradise, while the example of those who were before you came to you; they were stricken with adversity and hardship, and they were shaken until the messenger and those who affirmed with him said, "When is **God**'s victory?" Yes indeed, **God**'s victory is near.

2:215 They ask you what they should spend, say, "What you spend out of goodness should go to your family and the relatives and the orphans, and the needy, and the wayfarer. Any good you do, **God** is fully aware of it."

2:216 Warfare has been decreed for you while you hate it; and perhaps you may hate something while it is good for you, and perhaps you may love something while it is bad for you; **God** knows while you do not know.

2:217 They ask you about fighting in the restricted month. Say, "Fighting in it is great offense," yet repelling from the path of **God** and not appreciating Him and the Restricted Temple, driving its inhabitants out is far greater with **God**. Persecution is worse than being killed." They still will fight you until they turn you back from your system if they are able. Whoever of you turns back from his system, and dies as ingrates, then these have nullified their work in this life and the next; these are the people of the fire; there they will abide eternally!

2:218 The affirmers, and those who have emigrated and strived in the cause of **God**; these are seeking **God**'s compassion, and **God** is Forgiving, Compassionate.

Intoxicants, Gambling, and Charity

2:219 They ask you about intoxicants and gambling. Say, "In them is great harm, and a benefit for people; but their harm is greater than their benefit." They ask you how much are they to give, say, "The excess." Thus, **God** clarifies for you the signs that you may think.*

2:220 In this world and the next. They also ask you regarding the orphans, say, "To reform their situation is best, and if you are to care for them, then they are your brothers." **God** knows the corrupt from the reformer, and had **God** wished He could have made things difficult. **God** is Noble, Wise.

Do not Marry Those Who Set Up Partners With God

2:221 Do not marry the females who set up partners until they affirm. An affirming servant is better than one who sets up partners even if she attracts you. Do not also marry the males who set up partners until they affirm. An affirming servant is better than one who sets up partners even if s/he attracts you. These invite to the fire, while **God** is inviting to paradise and forgiveness by His leave. He clarifies His signs for the people that they may remember.

2:222 They ask you about menstruation? Say, "It is painful; so retire yourselves sexually from the women during the menstruation, and do not approach them until they are cleansed. When they are cleansed, then you may approach them as **God** has commanded you." **God** loves the repenters and He loves the cleansed.*

2:223 Your women are cultivation for you. So, approach your cultivation as you wish towards goodness. Be conscious of **God** and know that you will meet Him; give good news to those who affirm.*

Abusing God's Name

2:224 Do not make **God** the subject of your oaths; to be pious, aware, and reform relations among people; and **God** is Hearer, Knower.

2:225 **God** will not call you to account for your casual oaths, but He will call you to account for what has entered your hearts. **God** is Forgiving, Compassionate.

Divorce

2:226 For those who are discontent with their wives, let them wait for four months. If they reconcile, then **God** is Forgiving, Compassionate.*

2:227 If they insist on the divorce, then **God** is Hearer, Knowledgeable.

2:228 The divorced women shall wait for three menstruation periods; and it is not lawful for them to conceal what **God** has created in their wombs, if they affirm **God** and the Last day. Their estranged husbands are more justified to return to them, if they both wish to reconcile. The women have rights similar to their obligations, according to the recognized norms. But the men will have a degree over them. **God** is Noble, Wise.

2:229 The divorce is allowed twice. So, either remain together equitably, or part ways with kindness. It is not lawful for you to take back anything you have given the women unless you fear that they will not uphold **God**'s limits. So if you fear that they will not uphold **God**'s limits, then there is no sin upon them for what is given back. These are **God**'s limits so do not transgress them. Whoever shall transgress **God**'s limits are the wicked.

2:230 So if he divorces her again, then she will not be lawful for him until she has married another husband. If the other husband divorces her, then they are not blamed for coming back together if they think they will uphold **God**'s limits. These are **God**'s limits; He clarifies them for a people that know.

Do not Leave the Divorced Women on the Street

2:231 If you have divorced the women, and they have reached their required interim period, then either remain together equitably, or part ways equitably. Do not reconcile with them so you can harm them out of animosity. Whoever does so is doing wickedness to his person. Do not take **God**'s signs lightly; remember **God**'s blessings towards you, and what was sent down to you of the book and the wisdom, He warns you with it. Be conscious of **God** and know that **God** is Knowledgeable in all things.

2:232 If you divorce the women, and they have reached their required interim period, then do not prevent them from remarrying their husbands if they amicably agree amongst themselves out of what is best. This is to remind any of you who affirm **God** and the Last day, this is better for you and purer; and **God** knows while you do not know.

Rights and Responsibilities After Divorce

2:233 The birth mothers suckle their children two full years, for those who wish to complete the suckling. The man for whom the child is born is responsible for both their provisions and clothing equitably. A person should not be burdened beyond its means. No mother shall be harmed because of her child, nor shall a father be harmed because of his child. For the guardian is the same requirement. So if they wish to separate out of mutual agreement and council, then there is no blame on them. If you want to hire nursing mothers, then there is no blame on you if you return what you have been given according to the recognized norms. Be conscious of **God**, and know that **God** is watching over what you do.

Widows

2:234 For those of you who pass away and leave widows behind, then their widows will have a required interim period of four months and ten days. When they reach their required interim, then there is no blame on you for what they do to themselves according to the recognized norms. **God** is Ever-aware of what you do.*

2:235 There is no blame upon you if you openly propose marriage to these women, or you keep it between yourselves. **God** knows that you will be thinking of them, but do not meet them secretly, unless you have something righteous to say. Do not consummate the marriage until the required interim is reached in the book. Know that **God** knows what is in your minds, so be conscientious of Him, and know that **God** is Forgiving, Compassionate.

Breaking the Engagement

2:236 There is no blame on you if you divorce the women before having sexual intercourse with them, or before committing to what was agreed for them. Let them have compensation, the rich according to his means, and the poor according to his means. Compensation, which is according to the recognized norms, is a responsibility for good-doers.*

2:237 If you divorce them before having sexual intercourse with them, but you have already agreed to the dowry, then you must give half of what you have agreed, unless they forgive or the guardian over the marriage contract forgives. If you forgive, it is closer to awareness. Do not forget the favor between you; **God** is Seer over what you do.

Observe the Contact prayer (Sala)

2:238 Preserve the *contact prayers*, and the middle *contact prayer*, and stand devoutly for **God**.*

2:239 But if you are in a state of worry, then you may do so while walking or riding. If you become secure, then remember **God** as He has taught you what you did not know.*

Alimony For Widows and Divorcees

2:240 Those of you who pass away and leave widows behind, leave a will for them that they may enjoy for one year without being evicted. If they leave, then there is no blame on you for what they do to themselves out of the recognized norms; and **God** is Noble, Wise.*

2:241 For the divorced women compensation is an obligation upon the conscientious.

2:242 It is such that **God** clarifies to you His signs that you may reason.

Fight Against Injustice and Oppression

2:243 Did you not see those who were evicted from their homes in groups, all the while they were wary of death; so **God** said to them, "Die," then He resurrected them. **God** has great favor over the people, but most people are not thankful.

2:244 Fight in the cause of **God** and know that **God** is the Hearer, the Knowledgeable.*

2:245 Who will lend **God** a loan of goodness that He may multiply it for him? **God** collects and He distributes, and to Him you will return.

Saul and the Ark of the Covenant

2:246 Did you not note the leaders of the Children of Israel after Moses, they said to their prophet, "Send us a king that we may fight in the cause of **God**;" he said, "Are you not concerned that if fighting is decreed upon you, you will then not fight?" They said, "Why should we not fight in the cause of **God** when we have been driven out from our homes with our children." Yet, when fighting was decreed for them, they turned away, except for a few of them! **God** is fully aware of the wicked.*

2:247 Their prophet said to them, "**God** has sent Saul to you as a king." They said, "How can he have the kingship when we are more deserving than him, and he has not been given an abundance of wealth?" He said, "**God** has chosen him over you and increased him in knowledge and stature." **God** grants His sovereignty to whom He chooses; and **God** is Encompassing, Knowledgeable.*

2:248 Their prophet said to them, "The sign of his kingship shall be that he brings to you the ark in which there is tranquility from your Lord and the legacy of what was left behind by the descendants of Moses and the descendants of Aaron being carried by the controllers. In this is a sign for you if you affirm."

2:249 So when Saul set out with the soldiers, he said, "**God** will test you with a river, whoever drinks from it is not with me, and whoever does not taste from it except one scoop with his hand is with me." They all drank from it, except a few of them. So when he and those who affirmed with him crossed it, they said, "We have no power today against Goliath and his soldiers!" But the ones who understood that they would meet **God** said, "How many a time has a small group beaten a large group by **God**'s leave, and **God** is with the patient ones!"*

2:250 When they came forth to Goliath and his soldiers, they said, "Our Lord grant us patience, and make firm our foothold, and grant us victory over the ingrates."

2:251 So they defeated them by **God**'s leave. Thus, David killed Goliath, and **God** gave him the kingship and the wisdom and taught him what He wished. Had it not been for **God** pushing the people to challenge one another, then the earth would have long been corrupted. But **God** has done favor over the worlds.*

2:252 These are **God**'s signs, We recite them to you with truth, and you are of the messengers.

One Message

2:253 Such messengers, We have preferred some to others; some of them talked to **God**, and He raised some of them in ranks, and We gave Jesus son of Mary the proofs and We supported him with the Holy Spirit. Had **God** wished, the people after them would not have fought after the proofs had come to them, but they disputed, some of them affirmed and some of them did not appreciate. Had **God** wished they would not have fought, but **God** does whatever He wishes.*

2:254 O you who affirm, spend from what We have provided for you before a day will come when there is no trade, nor friendship, nor intercession; and the ingrates are the wicked.*

God Alone

2:255 **God**, there is no god but He, the Living, the Sustainer. No slumber or sleep overtakes Him; to Him belongs all that is in the heavens and in the earth. Who will intercede with Him except by His leave? He knows their present and their future, and they do not have any of His knowledge except for what He wishes. His throne encompasses all of the heavens and the earth and it is easy for Him to preserve them. He is the High, the Great.*

Freedom

2:256 There is no compulsion in the system; the proper way has been made clear from the wrong way. Whoever rejects the transgressors, and affirms **God**, has grasped the firm branch that will never break. **God** is Hearer, Knower.*

2:257 **God** is the ally of those who affirm, He brings them out of darkness into the light. As for those who reject, their allies are the transgressors; they bring them out of the light into darkness. These are the people of the fire; there they will abide eternally.

Abraham is Genuinely Seeking Extraordinary Evidence for Extraordinary Ideas

2:258 Did the news come to you of the person who debated with Abraham regarding his Lord, though **God** had given him a kingship? Abraham said, "My Lord is the One who gives life and death," he said, "I bring life and death." Abraham said, "**God** brings the Sun from the east, so you bring it from the west!" The one who did not appreciate was confounded! **God** does not guide the wicked people.*

2:259 Or the one who passed through a town, where all its inhabitants had passed away. He said, "How can **God** possibly resurrect this after its death?" So **God** put him to death for one hundred calendar years, then He resurrected him. He said, "How long have you stayed here?" He said, "I have stayed here a day or part of a day." He said, "No, you have stayed here for one hundred calendar years! Look at your food and drink, they have not changed, but look at your donkey. Thus, We will make you a sign for the people; and look at the bones how We grow them, and then We cover them with flesh." When it was clear to him what happened, he said, "I now know that **God** is capable of all things!"*

2:260 Abraham said, "My Lord, show me how you resurrect the dead." He said, "Do you not already affirm?" He said, "I do, but to assure my heart." He said, "Choose four birds, then cut them, then place parts of the birds on each mountain, then call them to you; they will come racing towards you. Know that **God** is Noble, Wise."*

Sharing the Wealth with Others

2:261 The example of those who spend their money in the cause of **God** is like a seed that sprouts forth seven pods, in each pod there is one hundred seeds; and **God** multiplies for whomever He chooses, and **God** is Encompassing, Knowledgeable.

2:262 Those who spend their money in the cause of **God**, then do not follow what they have spent with either insult or harm; they will have their reward with their Lord, they have nothing to fear nor will they grieve.

2:263 Kind words and forgiveness are far better than charity that is followed by harm. **God** is Rich, Compassionate.*

2:264 O you who affirm, do not nullify your charities with insult and harm; like the one who spends his money in vanity to show the people, and he does not affirm **God** and the Last day. His example is like a stone on which there is dust, then it is subjected to heavy rain, which leaves it bare. They cannot do anything with what they earned; and **God** does not guide the ingrates.

The Quality of Charity

2:265 The example of those who spend their money seeking **God**'s grace, and to save themselves, is like the example of a garden on a high ground which is subjected to a heavy rain, and because of that it produces double its crop! If no heavy rain comes, then light rain is enough. **God** is Seer over all you do.

2:266 Does anyone of you desire that he have a garden with palm trees and grapevines, and rivers flowing beneath it, and in it for him are all kinds of fruits, then he is afflicted with old age and his progeny is weak, and a whirlwind with fire strikes it and it all burns? It is thus that **God** makes clear for you the signs that you may reflect.

2:267 O you who affirm, spend from the good things that you have earned, and from what We have brought forth from the earth. Do not choose the rotten out of it to give, while you would not take it yourselves unless you close your eyes regarding it. Know that **God** is Rich, Praiseworthy.

2:268 The devil promises you poverty and orders you to evil, while **God** promises forgiveness from Him and favor. **God** is Encompassing, Knowledgeable.

Philosophy

2:269 He grants wisdom to whom He chooses, and whoever is granted wisdom, has been given much good. Only those with intelligence will take heed.

2:270 Whatever you spend out of your monies, or whatever you pledge as a promise, certainly, **God** knows it. The wicked have no supporters.

2:271 If you declare your charity, then it is acceptable; but if you conceal it and give it to the poor, then that is better for you. It reduces some of your sins; and **God** is Ever-aware of all that you do.

2:272 You are not responsible for their guidance, but it is **God** who will guide whomever He wishes. Whatever you spend out of goodness is for yourselves. Anything you spend should be in seeking **God**'s presence. Whatever you spend out of goodness will be retuned to you, and you will not be wronged.*

2:273 As for the poor who face hardship in the cause of **God**, and cannot leave the land; the ignorant ones think they are rich from their abstention; you know them by their marks, they do not ask the people repeatedly. What you spend out of goodness, **God** is fully aware of it.

2:274 Those who spend their money in the night and in the day, secretly and openly, they will have their reward at their Lord, there is no fear over them nor will they grieve.

Usury is Evil; Don't Exploit the Needy

2:275 Those who consume usury, do not rise except as the one who is being struck by the devil directly. That is because they have said, "Trade is like usury." Indeed, **God** has made trade lawful, yet He has forbidden usury. Whoever has received understanding from His Lord and ceases, then he will be forgiven for what was before this and his case will be with **God**. But whoever returns, then they are the people of the fire, in it they will abide eternally.*

2:276 **God** wipes out the usury and grants growth to the charities. **God** does not like any ingrate sinner.

2:277 Those who affirm and reform, and observe the *Contact prayer*, and contribute towards betterment, they will have their rewards at their Lord and there is no fear over them nor will they grieve.

2:278 O you who affirm, be conscious of **God** and give up what is left of the usury if you are the affirmers.

2:279 If you will not do this, then take notice of a war from **God** and His messenger. However, if you repent, then you will have back your principal money, you will not be wronged nor will you wrong.

2:280 If the person is facing insolvency, then you shall wait until he becomes able. If you relinquish it as a charity it is better for you if you knew.

2:281 Be conscious of a day when you will be returned to **God** then every person will be paid what it has earned, they will not be wronged.

Record Your Financial Transactions

2:282 O you who affirm, if you borrow debt for a specified period, then you shall record it. Let a scribe of justice record it for you. No scribe should refuse to record as **God** has taught him. Let him record and let the person who is borrowing dictate to him, and let him be conscious of **God**, and let him not reduce from it anything. If the borrower is mentally incapable, weak or cannot dictate himself, then let his guardian dictate with justice; and bring two witnesses from amongst your men. If they are not two men, then a man and two women from whom you will accept their testimony, so that if one of them errs, then one can remind the other. The witnesses should not decline if they are called, and you should not fail to record it no matter how small or large including the time of repayment. That is more just with **God** and better for the testimony, and better that you do not have doubts; except if it is a trade to be done on the spot rolling between you, then there is no blame on you if you do not record it. Have witnesses/evidence if you trade. No scribe shall be harmed nor any witness; for if you do so then it is vileness on your part. Be conscientious of **God** so that **God** teaches you; and **God** is aware of all things.*

2:283 If you are traveling or do not find a scribe, then a pledge of collateral. If you trust each other then let the one who was entrusted deliver his trust, and let him be conscious of **God**, and do not hold back the testimony. Whoever holds it back, then he has sinned in his heart; **God** is aware of what you do.

2:284 To **God** is what is in the heavens and in the earth, and if you declare what is in your minds or hide it, **God** will call you to account for it. He will forgive whom He wishes, and punish whom He wishes, and **God** is capable of all things.

Do not Make Distinction Among God's Messengers

2:285 The messenger affirms what was sent down to him from his Lord and the affirmers. All affirmed **God**, His controllers, His books, and His messengers, "We do not discriminate between any of His messengers;" and they said, "We hear and obey, forgive us O Lord, and to you is our destiny."*

Our Only Lord

2:286 **God** does not impose a person beyond its capacity. For it is what it earns, and against it is what it earns. "Our Lord do not adjudge us if we forget or make mistakes. Our Lord do not place a burden upon us as You have placed upon those before us. Our Lord, do not burden us beyond our power; pardon us, and forgive us, and have compassion on us; You are our patron, help us against the ingrates."*

ENDNOTES

002:001 A1L30M40. The meaning of 14 different combinations of alphabet letters/numbers initializing 29 chapters of the Quran remained a secret for centuries until 1974. Many scholars attempted to understand the meaning of these initial letters with no results. A computerized study that started in 1969 revealed in 1974 a 19-based mathematical design that was prophesied in Chapter 74. The frequency of the 14 alphabet letters in 14 different combinations that initialize 29 chapters are an integral part of this mathematical structure.

Arabs, during the time of revelation, did not use what we now know as Arabic Numerals. A notable book, The Universal History of Numbers by George Ifrah, dedicates Chapters 17-19 to ancient numerical systems, titled: Letters and Numbers, The Invention of Alphabetic Numerals and Other Alphabetic Number-Systems. The book provides extensive information on the Hebrew, Armenian,

								ا 1
ي 10	ط 9	ح 8	ز 7	و 6	ه 5	د 4	ج 3	ب 2
ق 100	ص 90	ف 80	ع 70	س 60	ن 50	م 40	ل 30	ك 20
غ 1000	ظ 900	ض 800	ذ 700	خ 600	ث 500	ت 400	ش 300	ر 200

Phoenician, Greek, Syriac, Arabic, and Ethiopian Alphabetic Numerals. According to historical evidence, during the era of Muhammed, Arabs were using their alphabet as their numbering system. The Arabic alphabet then was arranged differently and was named after its first four letters, *ABJaD*. Each of the 28 letters of the alphabet corresponded to a different number starting from 1 to 9, from 10 to 90, and then from 100 to 1000. When Arabs started adopting Hindu numerals in 760 AC, long after the revelation of the Quran, they abandoned their alphabet/numeral system in favor of a pedagogically arranged alphabet, which is currently in use. Below is the list of the 28 Arabic letters/numbers. When they are put next to each other, their numerical values are added to each other to attain a total number. For instance, the numerical value of ALLaH (God) is 1+30+30+5=66, of ShaHYD (Witness) is 300+5+10+4=319, and of WA**H**iD (One) is 6+1+8+4=19. (The word WAHiD is misspelled as WaHiD in modern manuscripts. But the oldest available manuscripts and numerous oldest fragments uses the word with Alif, as WAHiD)

Thus, the Quran is not only a literary book, but also a book made of numbers. The numerical structure of the Quran has two features: intertwined patterns among the *physical frequency* of its literary units, such as letters, words, phrases, verses, and chapters AND intertwined patterns among the *numerical values* (*ABJaD*) of these literary units. Considering the relevancy of the common meaning of words and letters during the time of revelation, we cannot ignore the fact that the combinations of letters that initialize 29 chapters are primarily numbers. After Arabs commonly adopted the Indian numerals in the 9th century, unfortunately, the numerical system prevalent during the revelation became an antique abused by charlatan psychics, and astrologists. It is ironic that an innovation (today's Arabic numerals) replaced the original (*ABJaD*), but through the passage of time people started considering the original to be the innovation! The allergy of Muslim scholars to *ABJaD* and its use in the Quran, does not allow them to open their eyes to marvels of the numerically structured book (83:7-21).

To verify and witness the mathematical system of the Quran, one does not need to know Arabic, though the knowledge of Arabic might increase the magnitude of its appreciation. Though there are hundreds of examples of the 19-based mathematical system, there are surely many still out there yet to be discovered. Meanwhile, there are also some issues that need to be resolved; for instance, the count of the letter *Alif* remains to be settled. We know the mathematical structure of the Quran verifies both the extraordinary nature of the text and its miraculous preservation from tampering. However, the system still involves the entire Quran and we have yet to discover a miraculous system verifying the divinity and authenticity of each chapter, *independently*.

Observers reacting to this prophetic feature fall into three main groups: 1) unappreciative skeptics or fanatic religionists who reject its existence without sufficient investigation; 2) appreciative seekers of truth who witness it and experience a paradigm change; and 3) gullible people who are impressed by it without fully appreciating the statistical facts and consequently indulge in "discovering" their own "miracles" through arbitrary numerical manipulations and jugglery. Ironically, the third group's exaggerated claims serve as justification for the first group's beliefs. For more information on this issue, please visit, 19.org.

See 74:1-56.

002:003 We translated the word *yuminun* as "they affirm." The verbs and adjectives generated from the root *AMaNa* have been translated with different forms of the word "believe." However, *believe* does not accurately describe the positive meaning ascribed to the root *AMaNa* by the Quran. According to the Quran, guesswork, wishful thinking, following the majority, and believing without investigation and evidence, are not acts of *AMaNa*. Most people, according to the Quran, do not *YuMiNun* (11:17; 12:103-107; 13:1; 40:59). Knowing that the great majority of people do "believe in God" or believe in some form of religious teachings, choosing to translate the appropriate Quranic phrase, say in 12:103, as "the majority of people do not believe" creates a contradiction. Belief in something, according to its common usage, may not need reasoning or affirmation of truth, but affirmation implies an intellectual process to search for and find the truth. *Muminun* are rational monotheists.

No wonder the Quran repeats all the derivatives of the word *AMN* (acknowledge, affirm) 811 times, while repeating all the derivatives of *ALM* (know) 782 times and of *ARF* (recognize) 29 times, totaling the same number of times *AMN* (affirm) is mentioned, that is, 811. The antonym of AaMaNa is KaFaRa (become ingrate, reject) occurs in the Quran 697 times. Interestingly, the difference between the two, that is affirmation and denial is exactly 114, the number of chapters of the Quran.

This verse, evaluated within the context of the Quranic precepts, does not mean that we should believe without evidence. The Quran categorically rejects blind faith or credulity (17:36). Unfortunately, the word belief or faith is commonly used as a euphemism for wishful thinking or joining the closest, the most crowded, the loudest bandwagon. Those who cannot justify their faith through deductive or inductive arguments, and those who cannot provide compelling reasons for why they believe certain dogmas and disbelieve others, are not considered *muminun* (affirmers) according to the definition of the word as used in the Quran. If the author of the scripture is also the creator of nature, and if it is He who is rewarding us with scientific knowledge and technology when we rationally and empirically investigate its laws, then why would He discourage us from using our mind and senses to investigate claims about Him and words attributed to Him? Otherwise, schizophrenia or inconsistency become divine attributes. God is not at the end of a dark tunnel of blind faith; but we can discover and get in touch with Him by tuning our minds and hearts to receive His message broadcast in the frequency of wisdom and knowledge. Therefore, according to the contextual semantics of the Quran, faith or belief denote conviction as a result of reason and compelling evidence. Thus, to distinguish the Quranic terminology of belief and faith from its common usage, we preferred the word "acknowledge" or "affirm."

A philosophical argument regarding the attributes of God is worth discussing. Plotinus, Moses Maimonides, and later Leibniz asserted that as imperfect beings we could not know God. How can limited creatures like us comprehend divine attributes? This is a fair question. Indeed, knowing or comprehending God is far beyond our capabilities. Do we really comprehend the meaning of Infinity, Perfection, Omnipotence, or Omniscience?

Do we really understand existence of a First Cause beyond time and space? It is no easier to describe and comprehend God than to describe and comprehend the Singularity of the Big Bang or multiple universes, or quantum physics. As the Catholic philosopher, St. Augustine, cleverly argued, when we claim we do not comprehend perfection, we show that we understand the meaning of the word, for if we did not understand its meaning and implication, we would not be able to claim that we do not comprehend it. Understanding the meaning of a word or attribute without fully comprehending it is interesting. It might simply mean that we mentally distinguish a set from other sets without knowing all its members.

However, we can get a limited idea about Singularity through its offspring or effects; similarly, we can get an idea about God through His creation. Thus, all the attributes of God that we use are related and limited to our environment and to us. To water molecules, the most important attribute of God might be COHA, the Combiner of Oxygen and Hydrogen Atoms, or the Evaporator. To a whale, it might be the Creator of Planktons. To the stars in the galaxies, it is the Designer of Fission, Creator of Gravity, or Gravity itself. Not only are we limited in comprehending attributes that are related to us, but we are also unable to comprehend an infinite number of attributes, in relation to other creatures, possible universes, or God Himself. Our language is limited to our common experiences through the five senses and by the knowledge we gather through deductive and inductive inferences. Obviously, concepts, beings, and events beyond our experience and inferences cannot be communicated via human language; it would be akin to attempting to describe the colors of the rainbow to a group of congenitally blind people.

Nevertheless, possibility does not mean probability, let alone reality. Just as lack of direct evidence does not prove the existence of an asserted event or being, similarly not all proofs depend on direct evidence. The verse refers to the events and things whose existence and nature can be accepted through indirect evidence or inferences. These evidences and inferences are substantiated by a combination of verifiable and falsifiable objective hypotheses as well as personal experiences in harmony with their objective counterparts (41:53). Furthermore, the Quran affirms the maxim that extraordinary claims need extraordinary evidence and it thus informs us that all of God's prophets and messengers were supported by *ayat*, that is, signs, evidences, or miracles (2:211; 6:4,142; 11:64; 13:38; 29:50-51; 79:20). Though the Quran criticizes the dogmatic and prejudiced challenge of unappreciative people against a messenger to demonstrate miracles, it approves of the requests by the sincere seekers of the truth to witness miracles to remove doubts from their hearts. Moreover, the Quran deems appreciation of miracles as a sign of intelligence and objective thinking (2:118; 7:146).

Not only do our five senses have limited capacity, but their numbers too are also most likely limited; we could have six, seven, or more senses. A person blind from birth can never imagine the nature of color. If all humans were blind, the stars and galaxies, and almost the entire universe would disappear. The rare-sighted person would not be able to prove the existence of stars to the blind majority; the entire universe would thus be considered a metaphysical concept by the all-blind scientific community. In an island inhabited by deaf people, a hearing person might be able to prove his extraordinary ability to detect sound, but he would never be able to make the deaf experience a similar auditory effect. It is possible there are, perhaps, beings that can only be detected by a sixth, seventh, or a Nth sense.

The Arabic word *ghayb* means things, events, and facts that are beyond the reach of our five senses or beyond our access in time and space (31:34). There are known knowns, known unknowns, and unknown unknowns. Hundreds of years ago, bacteria, viruses, radio waves, genes, the sex of a child before birth, black holes, and subatomic particles were beyond our perception and knowledge. Our five senses and our brain have a limited ability to perceive the external world, which we presume to exist; for instance, we cannot hear all sound waves and even dogs have a more acute hearing sense than us. We cannot see most of the spectrum of light; many animals or modern technological devices, such as radio and TV, are much better in their reception. Similarly, our nose cannot detect all the odor molecules; some insects are much better in smelling. Our knowledge of the world helps us invent and improve devices that increase the domain and range of our five senses and these new horizons, in turn, increase our knowledge further and help us invent newer devices. This perpetual process, as a byproduct, rewards us with technology, which we might use for good or bad ends.

A reminder to English-speaking readers: In Arabic, the pronoun "he" or "she" does not necessarily mean male or female, though the use of "she" is more restricted (for instance, see 16:97,111). Like in French, all Arabic words are grammatically feminine or masculine. Though God is referred by Arabic *pronoun HU* (He or It), God is never considered male. Christendom, greatly influenced by St. Paul's story of an incarnated god, has a propensity to envision God as a sort of Superman. No wonder Christian cartoonists depict God as a bearded old man. According to the Quran, God is beyond anything that we can imagine (42:11; 6:103). We entertained the idea of using the pronoun IT rather than HE for God, but we abandoned it since we thought it would create confusion.

002:003 All the necessary details of the *Sala* prayer are provided by the Quran. See, the Appendix titled *Sala Prayer According to the Quran*. You may visit www.islamicreform.org, www.19.org, or www.free-minds.org to read articles or to participate in forum discussions on this issue.

002:004 Previous scriptures still contain the divine message even though they were intentionally or unintentionally tampered by humans during the process of oral transmission, collection, writing, duplication, translations, translations of translations, revisions, and re-revisions of the translations. Both the *Torah* (the Old Testament) and *Injeel* (Gospels) promote monotheism (Deuteronomy 6:4-5, Mark 12:29-30). Theological distortions can be easily identified the light of the Quran or via modern textual criticism, such as the work of Bart Ehrman and Bruce Metzger. See 2:59, 79; 5:13-15; 7:162.

002:006-7 Those who decide to reject God and/or His message will be able to do so in accordance with God's system designed to test the result of one's free will to choose good or evil, to choose freedom by submitting to the Truth or slavery by submitting to falsehood (2:37-38; 57:22-23). As long as individuals follow the religion of their parents or societies on faith, without critical evaluation, they will not be able to witness the divine evidence, nor will they be guided to the truth (6:110). For a brief discussion on the freedom of choice, please see our discussion on 57:22-23.

The Quran delineates three categories of people in the context of their reaction to the truth: affirmers, Rejecters, and Hypocrites. The first and second groups are associated to each other because they are opposites while the third group is associated with the first two since it resembles both. The actions of affirmers associate them with paradise according to cause and effect. According to the cause-and-effect rule of association, the actions of unappreciative rejecters and hypocrites lead them to hell.

We translated the word *KaFeRun* as "ingrates." The word *KaFaRa* means "became unappreciative" of God's blessings, such as reasoning, senses, revelation, messengers, life, health, friends, relatives, trust, liberty, provisions, etc. In only one verse, the word *KaFaRa* is used to describe a positive rejection (14:22). Mostly in early revelations during Meccan years, the verb *KaFaRa* with its various derivatives was used to literally mean "unappreciative." Later, its semantic evolves, transforming to a more specific and a more serious attitude; it becomes a defining characteristic of those who actively oppose the message of the Quran (2:152; 27:40; 31:12; 76:3).

The skeptical philosopher David Hume came up with an ingenious thesis claiming that there are only three ways of making associations between concepts: resemblance; congruity in space or time; and cause-effect. For instance, bees and ants may be associated with each other because of their resemblance of being insects working in a social structure. Bees and flowers may be associated with each other because of their congruity in space. Bees and honey may be associated because of cause and effect since bees make honey. Resemblance may include lack of resemblance or resemblance in quantity. We associate the decimal system with our ten fingers both because of their resemblance in quantity and assumed cause and effect relationship between them.

The Quran, while using the already established associations or relationships in our language, also emphasizes those relationships or creates new ones between concepts by either juxtaposing or contrasting them within a particular context. Furthermore, the Quran is unique in creating a new semantic relationship (we call it Nusemantic from numerical semantic) between concepts through their frequencies, verse numbers or numerical values of their letters. For instance, Jesus and Adam resemble each other because they both lacked two parents. It turns out their names are both mentioned exactly 25 times in the Quran, thereby numerically strengthening their relationship. Another example of nusemantics: the Quran creates a special relationship between the word *ALLAH* and the attribute The One with Great Bounties by making the frequency of the word *ALLAH* equal to the numerical value (not frequency) of the letters comprising that attribute, that is, 2698. Extensive evidence of the numerical structure of the Quran provides many nusemantic relationships. For instance, the frequency of the word *YaWM* (day) in its singular form in the Quran is exactly 365 times, thereby creating a nusemantic relationship between the word DAY and SOLAR YEAR rather than DAY and LUNAR YEAR, thus rejecting the claim of those who wish to impose medieval Arabic culture, complete with the use of a lunar year, onto the Quran. However, the plural word of day *aYyAM* (days) occurring 27 times and the dual form *YaWMayn* (two days) occurring 3 times, relate to the approximate number of days both in a lunar and solar month. Similar nusemantic relationships are extensive and were first observed by an Egyptian scholar Abdurazzaq al-Nawfal in the 1960's and was published in Al-ijaz ul-Adady fil-Quran il-Karym (The Numeric Miracle/Challenge in the Quran). For the prophecy regarding the mathematical structure of the Quran, see 74:30.

002:023-24 These verses primarily refer to the mathematical system of the Quran. The most popular position among Muslim scholars is that the miracle of the Quran is its literary excellence. This is, however, an imaginary or false claim, since there are no objective or universal criteria for comparing literary texts and thereby preferring one over the other. The miraculous nature of the Quran or the Quran's claim of divine authorship and its universal appeal cannot be relegated merely to the literary taste of those who know Arabic, which may change subjectively. It is impossible to argue that the Quran is divine by comparing its literary aspects to those of Al-Mutanabbi, Taha Hussain or any other Arabic literary work. Bukhari and Ibn Hanbal, the two so-called authentic *hadith* collections of Sunnism, tacitly confess the impossibility of distinguishing the Quranic statements from man-made literary works since they narrate a text, which is still uttered by millions of Sunni *mushriks* in "bonus" prayers called *nawafil.* According to their report, Ibn Masud, one of the most prominent comrades of prophet Muhammed, and allegedly one of the most prolific sources of *hadith*, had included that prayer to his Quran as an independent chapter while expunging the last two chapters. The claim of medieval Sunni and Shiite sources regarding the requirement for two witnesses for each verse to authenticate their divine nature, regardless of the truth-value of the claim, is another affirmation of their inability to distinguish a Quranic revelation from fabricated statements via the so-called literary excellence criterion. See 4:82; 10:20; 74:30, and the Appendix titled *On it Nineteen*.

002:025-26 The descriptions of Hell and Paradise are allegorical. See 13:35; 17:60; 47:15; 76:16. The language of 2:25 is similar to 74:31 and may consequently imply a miraculous prophecy that might cause a great controversy in the future. The fact that the verse follows 2:23, which challenges unappreciative people regarding the miraculous nature of the Quran, supports our expectation.

The Bible contains numerous verses referring to Hell. For instance: "Again, the Kingdom of Heaven is like a dragnet, that was cast into the sea, and gathered some fish of every kind, which, when it was filled, they drew up on the beach. They sat down, and gathered the good into containers, but the bad they threw away. So will it be in the end of the world. The controllers will come forth, and separate the wicked from among the righteous, and will cast them into the furnace of fire. There will be the weeping and the gnashing of teeth." (Matthew 13:47-50)

002:028 After creation, we were put to death. Through birth we were given life on this planet; we will die again, and we will be resurrected for the Day of Judgment. Death is simply the stage of unconsciousness that is required by God's system where our mind is forced to travel between universes.

002:030 We may infer that Adam was the name of the new species created by God through evolution and he became the successor of a violent primate dominating the earth.

002:031 The verse reminds us of one of the most important divine gifts: the capacity of creating abstract ideas, categorization, and discrimination. Through Adam, our common ancestor, our genetic program is made able to recognize and distinguish numerous animals, plants, devices, chemical compounds, emotions, ideas, etc.

The Old Testament has a parallel account: "And out of the ground the Lord God formed every beast of the field, and every fowl of the air; and brought them unto Adam to see what he would call them: and whatsoever Adam called every living creature, that was the name thereof. And Adam gave names to all cattle, and to the fowl of the air, and to every beast of the field; but for Adam there was not found any help meet for him" (Genesis 2:19-20).

002:034 Jeffrey Lang, professor of Mathematics at Kansas University, in his remarkable book Losing My Religion: A Call for Help, (Amana Publications, 2004), notes the following about this verse, "If I had any doubt about the Qur'an's position is that the human character is potentially greater than the angelic one, this verse removed it. When Adam succeeds intellectually where the angels failed, God tells them, "Bow down to Adam." They then bow down, demonstrating their affirmation of his superiority. Bowing is also a symbol of subservience and thus the Qur'an seems to be indicating that the angels/controllers will serve mankind in its development on earth. (p.32).

002:035 Serving God is living according to the nature and reason. Human nature requires enjoyment of God's blessings within the limitations put by the Creator. By choosing to act against the advice of our own Creator we acted unreasonably. See: 51:56.

002:036 Though the Quran holds both Adam and his spouse equally responsible for their sinful act, the Old Testament's related verses have been distorted by all-male Jewish clerics, and this misogynistic distortion was later exploited further by St. Paul to promote male hegemony.

The dubious St. Paul justifies his misogynistic ideas with the following rationale: "Whereunto I am ordained a preacher, and an apostle, (I speak the truth in Christ, and lie not) a teacher of the Gentiles in faith and verity. I will therefore that men pray everywhere, lifting up holy hands, without wrath and doubting. In like manner also, that women adorn themselves in modest apparel, with shamefacedness and sobriety; not with braided hair, or gold, or pearls, or costly array; But (which becomes women professing godliness) with good works. Let the woman learn in silence with all subjection. But I suffer not a woman to teach, nor to usurp authority over the man, but to be in silence. For Adam was first formed, then Eve. And Adam was not deceived, but the woman being deceived was in the transgression. Notwithstanding she shall be saved in childbearing, if they continue in faith and charity and holiness with sobriety" (1 Timothy 2:7-15). "Let your women keep silence in the churches: for it is not permitted unto them to speak; but they are commanded to be under obedience, as also says the law. And if they will learn anything, let them ask their husbands at home: for it is a shame for women to speak in the church" (1 Corinthians 14:34-35). "Likewise, ye husbands, dwell with them according to knowledge, giving honor unto the wife, as unto the weaker vessel, and as being heirs together of the grace of life; that your prayers be not hindered" (1 Peter 3:7).

Ironically, the Old Testament contradicts St. Paul's attempt to deprive women from freedom of expression: The Old Testament mentions several women as prophets: Miriam (Exodus 15:20), Deborah (Judges 4:4-5), Huldah (2Kings 22:14), Noadiah (Nehemiah 6:14). For the Old Testament account of this event, see Genesis 3:1-24. See the Quran 4:1; 4:34; 49:13; 60:12. Also see 7:19-25; 20:115-123.

002:041 The Quran mentions Jews frequently; sometimes for their good deeds and accomplishments, and sometimes for their blunders and failures. God knew that the followers of Muhammed would fall for the same trap centuries later. The Quran narrates the history so that we do not repeat the same errors. Unfortunately, those who followed the hearsay stories about prophet Muhammed committed the same sins. For instance, compare these verses with the beliefs and actions of today's Sunni and Shiite *mushriks*: 2:48; 2:67-71; 2:80; 3:24; 9:31. The Quran was not revealed to dead people, but to the living (36:70). Thus, we should be warned and enlightened by the divine criticisms levied to the previous generations. Some people have problem with the expression "do not exchange my signs for a petty gain" or "do not trade it with a cheap price" since they commit the fallacy called Accent. The accent should not be on "cheap gain" or "cheap price" rather should be on "do not exchange…" or "do not trade it…"

002:043 For *zaka* (purification/betterment through sharing the blessings) see 7:146.

002:048 Belief in intercession is a mythology common in many religions. Satan, via religious clergymen, infected the faith of many people with the virus called intercession. Intercession generates false hope that promotes human-worship. The living religious leaders distinctly discover the power of intercession for their political and economic exploitation: if their followers believe that "holy dead humans" could bestow them eternal salvation and save them from God's justice, then they would be more susceptible to follow their semi-holy leaders blindly.

None has the power of saving criminals from God's judgment. The Quran considers the faith in intercession to be *shirk* or polytheism. If there *is* any intercession, it will be in the form of testimony for the truth (2:48,123,254; 6:70,94; 7:53; 10:3; 20:109; 34:23; 39:44; 43:86; 74:48; 78:38). Ironically, the Quran informs us that Muhammed will complain about his people deserting the Quran, not "his *sunna*" as they claim (25:30). If there were any intercession by Muhammed, this would be the one. Muhammedans are so ignorant and arrogant, like their ancestors, that they too are in denial of their associating partnership with God through attributing the power of intercession, or other false powers, to God's servants (6:23-26; 16:35; 17:57; 39:3, 38; 19:81-82). Those who affirm the Quran do not favor one messenger over another (2:285), since all the messengers belong to the same community (21:92; 23:51-53).

The Quran gives examples of many idolized concepts and objects. For instance, children (7:90), religious leaders and scholars (9:31), money and wealth (18:42), angels/controllers, dead saints, messengers and prophets (16:20, 21; 35:14; 46:5, 6; 53:23), and ego/wishful thinking (25:43, 45:23) all can be idolized.

In order to infect the human mind with the most dangerous disease called *shirk* (associating partners with God, or polytheism), Satan infects the unappreciative minds with a virus that destroys the faculty of recognition and self-criticism. As a result, the faulty and defective recognition program hands the mind over to the Satan's control. Therefore, most of those who associate partners with God in various ways do not recognize their polytheism (6:23). Polytheists show all the symptoms of hypnosis; their master hypnotist is Satan.

We are instructed to glorify and praise God (3:41; 3:191; 33:42; 73:8; 76:25; 4:103), not His messengers, who are only human beings like us. We are instructed by the Gracious and Loving God to utter the name of messengers by their first names, without glorifying them, and Muhammed is no different from other messengers (2:136; 2:285; 3:144). Muhammed was a human being like us (18:110; 41:6), and his name is mentioned in the Quran as *Muhammed*, similar to how other people are mentioned in the Quran (3:144; 33:40; 47:2; 48:29).

Uttering expressions containing *salli ala* after Muhammed's name, as is commonly done by Sunni and Shiites alike, is based on a distortion of the meaning of a verb demanding action of support and encouragement of a living messenger, rather than utterance of praise for a dead messenger (compare 33:56 to 33:43; 9:103; and 2:157). Despite these verses clarifying the meaning of the word; despite the fact that the Quran does not instruct us to say something, but to do something; despite the fact that the third person pronoun in the phrase indicates that it was an innovation after Muhammed's departure; despite these and many other facts, Sunni and Shiite clerics try hard to find an excuse to continue this form of Muhammed worship. Contradicting the intention and practice of the masses, some clerics even claim this phrase to be a prayer for Muhammed rather than a phrase for his praise. Muhammed, especially the Muhammed of their imaginations, should be the last person who would need the constant prayers of millions. According to them, Muhammed already received the highest rank in paradise, and again according to them he did not commit any sins. Therefore, the addressee of their prayers is wrong. They should pray for themselves, and for each other, not for Muhammed. It is akin to homeless people donating their dimes, several times a day, to the richest person in the world. It is just as absurd.

A great majority of Sunni and Shiite *mushriks* declare their peaceful surrender to God alone while standing in their prayers, but immediately nullify that declaration twice while sitting down. They first tell God, "You alone we worship; you alone we ask for help" and then forget what they just promised God by greeting "the prophet" in the SECOND person, "Peace be on YOU o prophet!" (*as salamu alayKA ayyuhan nabiyyu*), as if he was another omnipotent and omniscient god. They give lip service to monotheism while standing, and they revert to confessing their idolatry when sitting. Those who betray the meaning of "the ruler of the day of judgment" which they utter numerous times in their prayers (1:4; 82:17-19), those who contradict the purpose of the prayer (20:15), are obviously those who are oblivious to their own prayers (107:4-7; 8:35). In defiance of the Quran, many sects and mystic orders competed with each other to put Muhammed in a position that Muhammed unequivocally rejected (39:30 and 16:20, 21). Also see, 2:123,254; 3:80; 5:109; 6:51; 6:70,82,94; 7:53; 9:80; 10:3,18; 13:14-16; 19:87; 21:28; 33:64-68; 34:23,41; 39:3,44; 43:86; 53:19-23; 74:48; 83:11.

After the departure of Jesus, the Pharisee Paul turned Jesus into a divine sacrifice and a middleman between his flock and God: "For there is one God, and one mediator between God and men, the man Christ Jesus; Who gave himself a ransom for all, to be testified in due time" (1 Timothy 2:5-6). The Catholic Church went even further and added Mary and numerous saints to the ranks of the holy power brokers.

The followers of Sunni and Shiite sects follow a similar doctrine. They replace one idol with another and consider Muhammed to be "the mediator" between God and men; they assert that without accepting Muhammed, none can attain salvation! Following the tradition of Paul, and then the tradition of Catholic Church, they too produced their own saints, thereby creating a Pyramid scheme of mediators. Interestingly, some later idols have surpassed the main idol in popularity. For instance, in Iran, people worship the second-generation idol, Ali, more than they worship Muhammed. They invoke Ali's name on almost every occasion. In Syria, a third-generation idol, Hussain, is more popular than both Ali and Muhammed. The long list of idols varies from country to country, from town to town, from order to order, and even includes the names of living local idols as well.

002:051 This event reflects social and psychological weaknesses that may cloud human reasoning. Despite witnessing profound miracles, the Children of Israel turned back to the religion of their oppressors (20:83-98).

002:054 Ego, the self-exaggerating or self-worshiping self, should be avoided while the realist or appreciative self should be nourished. It is astonishing to see that many translations of the Quran render the phrase *uqtulu anfusakum* as "fight/kill each other." How could they not notice thirty verses down, that is, the 84th and 85th verses of this very chapter? The Arabic word *nafs* is a multi-meaning word and its intended meaning can be inferred by considering its proximate context consistent with the entire text of the scripture. The multiple meanings ascribed to the word *nafs* (person) suggest that our personhood is a complex program with multiple layers and one part of it, the ego, needs to be controlled with reason and submission to God alone.

However, there is another way. We should be open to read the text of the scripture without being restricted to the traditionally codified readings. We should be able to read the oldest texts that do not contain dots or vowels, in all possible readings, with the condition that they fit the context well and do not create internal or external contradictions within the *ayat* (signs) of the scripture or the *ayat* of nature. There is a divine blessing and purpose in such flexibility. For instance, we might read the following verses differently. If the alternative readings change the meaning dramatically, they are exclusive. However, sometimes both alternative readings can co-exist at the same time. One of the following, however, is a linguistic marvel; with its four alternative combinations, it excludes and includes at the same time, depending on the reference of the key word (3:7)! The following is a sample list:

- 2:243 *Kharaju* or *Khuriju* (inclusive)
- 2:259 Nunshizuha or Nunshiruha (inclusive)
- 3:146 *Qatala or Qutila* (both inclusive and exclusive)
- 3:7 Putting full stop after the word God and/or not stopping after the word God (both exclusive and inclusive!)
- 5:43; 5:6 *Arjulakum* or *Arjulikum* (exclusive)
- 7:57 *Bushra or Nushra* (inclusive)
- 11:46 Amalun or Amila (inclusive)
- 21:112 *Qala* or *Qul* (exclusive)
- 30:1 *Yaglibun* or *Yughlabun* (exclusive)
- 42:52 *Nashau* or *Yashau* (inclusive)
- 43:19 Ibad or Ind (inclusive)
- 54:3 *Kullu* or *Kulla* (inclusive)
- 74:24 *Yuthir* or *Yuthar* (inclusive)
- *Kitab* or *Kutub* (inclusive or exclusive) in numerous verses

Let's now discuss the alternative reading we are suggesting for 2:54.

The expression *faqtulu anfusakum* is traditionally mistranslated as "fight/kill yourselves" or "fight/kill each other" and it contradicts a proximate verse (2:84); thus, we may choose to translate the word *nafs* as "ego." If we prefer consistency in using "person/self" for translation of *nafs,* then we may follow the following alternative reading: *Faqbilu anfusakum*, that is, "turn to yourselves," or "accept yourselves," or "face yourselves." To discover other examples of different yet consistent and meaningful readings, we are hoping to systematically study the entire Quran in the future.

See 4:66; 7:40. For multi-meaning verses, see 3:7.

002:055 Intriguingly, the verse about people asking for physical evidence for God's existence contains the 19th occurrence of the word *Allah* (God) from the beginning of the Quran. As we know, the number 19 as the prophesied code of the Quran's miraculous mathematical structure, presents us with verifiable and falsifiable physical evidence regarding its divine authorship. See 82:19.

Lightning was not a punishment but was a lesson: if they could not withstand the energy of a lightning bolt, surely, they could not withstand God's physical presence. The following verse supports our inference.

002:059 Today's Bible contains fragments from the books of prophets. Ezra is one of the Jewish Rabbis who changed the words of the Torah. The Pharisee-son-of-a-Pharisee Saul (Paul) and his followers radically distorted the message of Jesus. While Rabbis added Talmudic teachings, equivalent to *Hadith* and *Sunna*, into the Old Testament, Paul added his letters and polytheistic doctrines borrowed from pagans, to lure gentiles. Using God's scientific signs in nature and the revelation of the Quran or using the methodology of Modern Textual Criticism we may be able to detect many of the distortions and additions in the Old and New Testament. For instance, the following verses do not appear to be from a Benevolent and Wise God: Genesis 3:6-16; Exodus 21:7-8; 21-22; 22:18-19; 26-27; Leviticus 12:2-4; 13:6-11; 17:5; 24:13-16; 25:44-46; 34:5-10; Numbers 15:32-36; 31:1-18; Leviticus 20:1-27; 21:14; 21:16-23; 24:13-23; 25:11-12; Deuteronomy 11:25; 12:1-3; 13:5-10; 17:2-7; 20:16-17; 22:23-24; 22:28-30; 25:11-12; Joshua 6:21; 9:6-27; Judges 1:4-12; 3:22,29; Judges 14:18; 15:15-16; 16:1-20; 1 Samuel 15:3; 1 Samuel 18:27; 2 Samuel 3:14; 2 Kings 2:23-24; 2 Chronicles 15:13; 36:17; Psalms 58:10-11; 78:52; 119:176; 137:9;149:6-9; Isaiah 13:13-16; Isaiah 20:1-3; 44:28: 56:11; Jeremiah 6:3; 10:10; 12:1-3; 13:13-15; 16:4; 23:4; 48:10; 49:19, 50:6; 51:10-24; Ezekiel 9:5-6; 23:25; 34:2-10; Zephaniah 3:8; Zechariah 10:2; Matthew 5:17-19, 29-30; Matthew 10:34;19:12; 21:19; John 15:6; 10:14; 1 Corinthians 9:7; 11:6-9; 14:34; Ephesians 6:5; Colossians 3:22; I Timothy 2:11-15; 1 Timothy 6:2; Titus 2:9; Hebrews 13:20; 1 Peter 2:13-14; 1 Peter 2:18.

See 2:79; 5:13-15, 41-44; 9:30-31. Also, see 3:45, 51-52-52, 55; 4:11,157,171; 5:72-79; 6:83-90; 19:36.

002:060 Moses bringing water out of a rock is mentioned in the Old Testament, Numbers 20:7-13.

002:061 After attaining their freedom in Sinai, the Children of Israel started missing the variety of food they were fed by their Egyptian masters. This verse emphasizes the importance of liberty and our lack of appreciation of it when we have it. The book of Exodus in the Old Testament is dedicated to the historic emigration of the Children of Israel that led them to freedom. Compare this verse to the Old Testament, Numbers 11:4-5. For the quail, see Numbers 11:31-33.

A verse in the New Testament is profound in relating freedom to intellect and reality: "And you shall know the truth, and the truth shall make you free" (John 8:32). Also, See Isaiah 61:1; Isaiah 42:6-7.

002:062 Regardless of religion, ritual, language, nationality, and books followed, any individual who fulfills these three requirements attains eternal salvation. Based on these criteria, Socrates who risked his life for promoting dialectic reasoning and rejecting the polytheistic religion of his countrymen was a monotheist peacemaker. Hypatia, too was a monotheist. She was killed by Church for rejecting Trinity. Similarly, Maimonides who considered God as the Prime Mover, Rabi Judah ben Samuel who witnessed one of the greatest divine signs and stood against religious distortion, Leibniz who regarded God as the creator and coordinator of monads, Galileo who studied and appreciated God's signs in the heavens and stood against religious charlatans, Darwin who traveled the world and studied God's biological creation with diligence and open mind, Newton who studied God's laws in the universe and showed the wisdom and courage to reject Trinity and deity of Jesus by arguing that this Christian doctrine to be contradictory to the first Commandment in the scripture, and many of those who fit the description of the Quranic verse 2:62, yes all according to the Quran might be considered muslims.

Those who judge a person's eternal salvation by whether they carry an Arabic name or not, are not following the Quranic criteria; instead they are following diabolic rules made by Arab nationalists.

The conditions listed in this verse are explained in detail within the context of the Quran. For instance, a person is not considered to be affirming God if he or she associates partners or ordained mediators with Him, or follows the teachings and restrictions falsely attributed to Him. Furthermore, such a belief is not accepted if it is mere lip service or to conform to a particular group; it must be based on reason, evidence, and intuition. affirming the hereafter implies affirming the hour, the day of resurrection and the day of judgment in which none except God will be the sole authority. Leading a righteous life is also defined in the Quran. For instance, righteousness requires a desire and action of sharing a portion of one's possessions--be it knowledge, talents, services, or wealth--with others, while appreciating his or her own. It also means maintaining honesty, integrity, and justice with good intention, while fighting against aggression and oppression.

The Quran refers to the followers of the New Testament with the word *Nasara* (Nazarenes), rather than *Masihiyyun* (Christians). The root of the word has several implications. First, it might have originated from the Semitic word *NaSaRa* (to support) and originated from the answer given by the disciples of Jesus when he asked for their support for his cause (61:14). Or, it could have originated from the birthplace of Jesus, Nazareth. Perhaps, it has a linguistic and historical link to both origins. Also, see 5:82.

Knowing the motive of the Gospel authors to establish a stronger messianic link between Jesus and King David, by binding him through genealogy and birthplace, some scholars of theology justifiably question whether Bethlehem was the actual birthplace of Jesus, as is commonly accepted. Though Matthew affirms the fact that Jesus was called Nazarene (Matthew 2:23), both Matthew and Luke mention Bethlehem as his birthplace. However, Mark, which was written earlier, mentions Nazareth as the birthplace of Jesus instead of Bethlehem: "And it came to pass in those days, that Jesus came from Nazareth of Galilee, and was baptized of John in Jordan" (Mark 1:9). For the Biblical verses referring to Nazareth, see (Matthew 2:23; 4:13; 21:11; 26:71; Mark 1:9,24; 10:47; 14:67; 16:6; Luke 1:26; 2:4,39,51; 4:16,34; 18:37; 24:19; John 1:45-46; 18:5,7; 19:19; Acts 2:22; 3:6; 4:10; 6:14; 10:38; 22:8; 26:9).

According to Christian scholars, such as Easton, "The name Christian was given by the Greeks or Romans, probably in reproach, to the followers of Jesus. It was first used at Antioch (Acts 11:26). The names by which the disciples were known among themselves were 'brethren,' 'the faithful,' 'elect,' 'saints,' 'affirmers.' But as distinguishing them from the multitude without, the name 'Christian' came into use, and was universally accepted. This name occurs but three times in the New Testament (Acts 11:26; 26:28; 1Peter 4:16)." It seems that some followers of Jesus adopted the name attributed to them by their enemies, and some continued using one of their earlier names, Nazarenes.

As for the word *Sabiene*, it is mistranslated as a proper name by the majority of commentators. In fact, it derives from the Arabic word *SaBaA*, meaning to be an apostate, or 'the follower of other religions.' *Hadith* books use this word as an accusation of Meccan *mushriks* directed against Muhammed when he started denouncing the religion of his people; they described his conversion to the system of Islam with the verb *SaBaA*.

002:065 Turning to monkeys and swine is most likely a metaphor indicating their spiritual and intellectual regression since verse 5:60 adds another phrase, "Servants of the aggressor", which does not depict a physiological transformation. Also see 7:166. Jesus likens his own people figuratively to swine and dogs (Matthew 7:6; 2 Peter 2:22). Swine was regarded as the filthiest and the most abhorred of all animals (Leviticus 11:7; Isaiah 65:4; 66:3, 17; Luke 15:15-16). See 5:60; 7:166).

002:067-73 Through ordering the Children of Israel to sacrifice a cow, God wanted to remove the influence of the upper-class culture on the Jews who were once the slaves of cow-worshipping Egyptians. Another lesson taught by God through this instruction was to teach them not to complicate God's simple commandments through trivial theological speculations and questions. The resurrection of the cow was to pull their attention to the fact that resurrection is easy for God, just as easy as creation in the first place. The Quran does not repeat the Biblical assertion that the sacrifice was for removing the guilt of murdering an innocent person.

By reminding us of this event, which is narrated in the Old Testament, Numbers 19, we are expected not to repeat the errors committed by the Jews. Ironically, Muslims repeated them to the letter. Muslims, soon after the prophet's death, started doubting the sufficiency and completeness of the Quran and asked many irrelevant and ridiculous questions. To answer those trivial questions, many people fabricated stories and instructions and attributed them to Muhammed and his companions. Two centuries after Muhammed's departure, some ignorant zealots such as *Bukhari*, *Muslim*, *Tirmizi*, *Abu Dawud*, *Nasai*, and *Ibn Hanbal* started collecting and classifying the stories in numerous *hadith* books. These volumes of contradictory "holy fabrications" created a new occupation termed *emamet*, *ijtihad* or *fiqh*. Therefore, many *emams*, *mujtahids*, or *fuqaha* created numerous sects and orders by interpreting, reconciling, refuting and codifying these volumes of contradictory sources. These books fabricated numerous rules that included answers to frivolous questions such as: 1) in what order must one cut his fingernails, 2) with which feet must one enter the bathroom, or 3) with which hand must one eat? The blind followers of *hadith* and *sunna* are hypnotized into believing that they would not even be able to properly clean themselves in the bathroom unless they followed those teachings. Sectarian books contain chapters of extensive, yet primitive and occasionally unhealthy religious instructions on personal hygiene. Thus, God's system was transformed into the religion or sect of this or that scholar, and the medieval Arab culture and borrowings from Christians and Jews were sanctified. Through this story, God pulls our attention to the chronic cause of distortion and corruption in religion. Unfortunately, Muslims repeated the same blunder. See 5:6,101; 23:52-56; 42:21.

002:073 By transferring the DNA of a heifer to old tissues thereby reversing its age, biochemists perhaps have scientifically realized this miraculous event in our times. See 4:82.

002:078 For the meaning of the word *ummy*, see 3:20; 7:157. For the real nature of magic, see 7:116-117.

002:079 Rabbinic Judaism is based not only on the Old Testament, but also on the Talmud. Like the followers of *Hadith*, *Sunna* and sectarian *Sharia*, Orthodox Jews consider the Talmud (*Mishna* hearsay and commentaries called *Gemara*) and *Talmudic Halakha* (*sharia, or rules of life*) to be a supplement or explanation of the Old Testament. The first five books of the Old Testament are attributed to Moses but there are many clues that indicate they were written long after Moses by anonymous authors. For instance, there are many verses in the first five books referring to Moses in the third person, and the last chapter of Deuteronomy is dedicated to Moses' death, which contains the following statements, "And Moses the servant of the Lord died there in Moab… Since then, no prophet has risen in Israel like Moses…" Obviously, a dead person could not have written his own obituary.

This is just one of the many evidences showing that not only the divine nature of the Old Testament, but even the identity of its alleged human authors, is in question. What was the Mysterious Q document? Are the first five books of the Old Testament a mishmash of Yahwist, Elohist, Deuteronomistic, and Priestly Writings? Why are 2 Kings 19 and Isaiah 37 identical word for word? Who wrote the Bible? Why are there so many contradictions in the Bible? These are but a few of the many questions Biblical scholars have wrestled with for centuries. Similar questions are valid for the Gospels. Though they may contain many words from the message delivered by Jesus, they were written about a century after his death and were handpicked by the Trinitarian faction of the Church in the fourth century. Thus, they are not Gospels according to God, or Jesus, but rather Gospels according to this man or that man, people who never met Jesus and who were influenced more by the teaching of St. Paul, a Pharisee who also never met Jesus and was rejected by the real disciples of Jesus. For instance, see Matthew 9:9. See Old Testament, Jeremiah 8:8. Also, see the Quran 2:59; 5:13-15, 41-44; 7:162; 9:30-31.

002:088 According to the Bible, an "uncircumcised heart" is one that is closed and impervious to God's attempts to affect it (See, Acts 7:51-53). For the expression "uncircumcised lips" see, Exodus 6:12,30. As it appears, it is also used by the opponents to mean "our hearts is closed (bag of knowledge)".

002:093 To describe the mindless obsession of the ex-slave Children of Israel with cow worship, the Quran uses the metaphor "they consumed the calf inside their hearts." The more literal translation of this phrase would be: "the love of calf was sipped into their hearts." The similar Biblical metaphor using the key word "sip/drink," however, has been distorted through revisions, commentaries and translation errors, thereby transforming it into a bizarre story falsely attributed to Moses: "And he took the calf which they had made, and burnt it in the fire, and ground it to powder, and strawed it upon the water, and made the children of Israel drink of it" (Exodus 32:20).

See 11:40-44; 16:103.

002:102 Magic, according to the description of the Quran, is not a paranormal phenomenon. Verse 7:116-117 defines magic with two characteristics Magic is a tool employed by people with devious intentions to manipulate individuals and societies via trickery and suggestion. Verses 7:116-117 refer to both its physical and psychological components. Hypnosis and films are two examples of magic. Hypnotists use less illusion more suggestions. Hypnosis can be used to cure phobias or addictions, or to reduce pain during surgeries on patients who have problem with anesthesia. Films are perfect examples of magic for both the tools they employ and the effects they create. Still frames on film are turned into moving images via film projector devices, creating the illusion. The filmmakers, in addition, use special effect, script, plot, actors to create the emotional effect. Thus, films, through illusion and suggestion can create fantastic influence on people, both positive or negative.

002:104 The word *raina* (shepherd us) implies to be led like sheep. Muslims do not and should not follow anyone blindly, including prophets, without using their intelligence, reasoning and senses. However, the Bible uses the shepherd/sheep analogy to depict the relationship between people and their leaders. "And I will set up one shepherd over them, and he shall feed them, even my servant David; he shall feed them, and he shall be their shepherd" Ezekiel 34:23. Jesus Christ is likened to a good shepherd (John 10:14; Heb 13:20). Kings and leaders are compared to shepherds (Isaiah 44:28; Jeremiah 6:3; 49:19). Ministers of the gospel too are likened to shepherds (Isaiah 56:11; Jeremiah 50:6; Jeremiah 23:4;

Ezekiel 34:2, 10). Also, see Zechariah 10:2; Psalms 78:52; Psalms 119:176.

This metaphor would be abused to its full capacity by St. Paul, the dubious figure who distorted the monotheistic message of Jesus after his departure. St. Paul fabricated many stories and practices, including the justification of receiving money for preaching. When the true followers of the *Injeel*, that is, the Good News, criticized him, he defended his "milking" the congregation, thereby twisting the original purpose of this Biblical metaphor. See 1 Corinthians 9:7.

002:106 By declaring the word of God to be vague and ambiguous, early scholars opened the gate for unlimited abuse and distortion. Furthermore, by distorting the meaning of 2:106, they claimed that many verses of the Quran had been abrogated (amended) by other verses or *hadiths*. By this "abrogation theory," they amended verses which they did not understand, or which did not suit their interests, or which contradicted their *hadiths*. Repeating the same error committed by the Children of Israel (2:85), Muslims fulfilled the prophetic description of their action in 15:91-93. Some of them abrogated 5 Quranic verses, some 20 verses and some 50.

They support this claim by distorting the meaning of this verse. The Quran has a peculiar language. The word *Aya* in its singular form occurs 84 times in the Quran and in all cases, means miracle, evidence, or lesson. However, its plural form, *Ayat,* is used both for miracle, evidence, lesson, AND/OR for the language of the revelation that entails or leads to those miracles, evidences, and lessons. The fact that a verse of the Quran does not demonstrate the miraculous characteristics of the Quran supports this peculiar usage or vice versa. There are short verses that are consist of only one or two words, and they were most likely frequently used in daily conversation, letters, and poetries. For instance, the verse "Where are you going?" cannot be called *AYAT* (signs) since it is one verse. This is very appropriate, since that expression was and is used by Arabic speaking people daily, even before the revelation of the Quran.

However, in its semantic and numerical context, that short question is one of the *Ayat* of the Quran. See 55:3; 69:1; 74; 4; 75:8; 80:28; 81:26. Furthermore, we are informed that the minimum unit that demonstrates the Quran's miraculous nature is a chapter (10:38) and the shortest chapter consists of three verses (chapters 103, 108, and 110). The first verse of the Quran, commonly known as Basmalah, though containing independent features, may not be considered a divine evidence/miracle on its own. However, it gains a miraculous nature with its numerical network with other letters, words, verses, and chapters of the Quran. By not using the singular form *Aya* for the verses of the Quran, God also made it possible to distinguish the miracles shown in the language and prophecies of the scripture from the miracles shown in nature. See 4:82 and 16:101 for further evidence that the Quranic verses do not abrogate each other.

Since grammatically we can refer only to three verses with the plural word, *ayat,* and since we are not provided with a word to refer to a single or pair of verses, this unique use might have another implication: are we required to quote the Quranic verses in segments of at least three verses? Will this method eliminate the abuse of Quranic statements by taking them out of their context? I think this question needs to be studied and tested. If quoting verses of the Quran in minimums of three units reasonably eliminates the abuse, then we should adhere to such a rule.

For another example of different meanings assigned to singular and plural forms of the same word, see 4:3. For a detailed discussion of this verse, see the Sample Comparisons section in the Introduction. Also see 13:38-39

002:111-113 It is astonishing that despite verses 2:62 and 5:59, the so-called Muslims repeat the same error.

002:118 Miracles or divine signs strengthen the faith of affirmers and do not guide the unappreciative. God's test is not designed to impose a particular faith. See 74:31.

002:124-126 In the first verse, Abraham makes a wrong addition, and in the second one, he makes a wrong subtraction. This dialogue teaches two facts.

002:132 See 7:143 and 39:12.

002:135 Islam or Peace/Submission is not a proper name. All messengers of God and affirmers since Adam have described their system as Peace/Submission and themselves a Peacemakers or Peaceful Submitters in their languages (2:131; 3:95; 6:161; 7:126; 10:72; 27:31, 32; 28:53). The only system acceptable by God is submission, submission to God alone. Abraham practiced daily contact prayers, purification through charity, fasting and annual gathering (22:78). Muhammed, as a follower of Abraham continued the same practice (16:123).

002:136 See 2:285. The verse is a refutation of the innovation of adding words of praise after Muhammed's name. The verse tells us to mention the names of prophets without praising them. Those who worship Muhammed might seek an excuse in the fact that Muhammed's name is not mentioned here. However, the verse preemptively refutes such an argument by instructing us not to discriminate among God's prophets. Those who praise Muhammed's name with fabricated lengthy phrases are not followers of Muhammed, since Muhammed was a monotheist who invited people to praise God and submit to God alone, not himself. See 2:48; 33:56.

002:142-145 *Qibla* is a point to which we are supposed to turn while observing our daily prayers. *Kaba* played an important role in the Arabian Peninsula, and besides its historic importance, it provided economic and political benefits to the Arabs. *Kaba* is not a *holy* temple or shrine, but an annual gathering place for monotheists to commemorate God, to remember the struggles of monotheists throughout history, to get to know each other, to exchange information, to promote charity, to remember their commonalities regardless of their differences in color, culture and language, to discuss their political and economic issues amicably, and to improve their trade.

The Quran informs us that those who associate other authorities with God continued the tradition of Abraham, only in form. Before the verses ordering muslims to turn to *Kaba* as *qibla* were revealed, muslims, like *mushriks* were turning towards *Kaba*. Muslims suffering from the repression, oppression and torture of Meccan polytheists finally decided to emigrate to Yathrib and there they established a city-state. However, the Meccan theocratic oligarchy did not leave them alone. They organized several major war campaigns. The aggression of the Meccan oligarchy and the improving economic, politic, and social relations between muslims and the people of the book, that is Christians and Jews, led muslims to turn to another uniting point. They picked Jerusalem. Nevertheless, God wanted muslims to turn back to their original *qibla*, *al-Masjid al-Haram* (the restricted place of prostration). We understand this from verses 2:142-145. These verses asking muslims to turn back to their former *qibla* in Mecca created a difficult test for some muslims living in Yathrib.

Those who preferred the advantages of personal, social and economic relations with the Christian and Jewish community in Yathrib and those who could not get over their emotional grievances against the Meccan community could not accept this change and reverted back from islam (2:142-144).

In brief, the verses ordering the change of *qibla* did not abrogate another verse, since there is not a single verse ordering muslims to turn to another *qibla*. If there is an abrogation, the decision of Muhammed and his companions to turn away from *Kaba* was abrogated.

Here are some relevant verses from the Bible: "In those days, and in that time, says the Lord, the children of Israel shall come, they and the children of Judah together, going and weeping: they shall go, and seek the Lord their God. They shall ask the way to Zion with their faces thitherward, saying, Come, and let us join ourselves to the Lord in a perpetual covenant that shall not be forgotten" (Jeremiah 50:4-5). "My soul longs, yea, even faints for the courts of the Lord: my heart and my flesh cries out for the living God. Yea, the sparrow hath found an house, and the swallow a nest for herself, where she may lay her young, even thine altars, O LORD of hosts, my King, and my God. Blessed are they that dwell in thy house: they will be still praising thee. Selah. Blessed is the man whose strength is in thee; in whose heart are the ways of them. Who passing through the valley of Baca make it a well; the rain also fills the pools. They go from strength to strength, every one of them in Zion appears before God. O Lord God of hosts, hear my prayer: give ear, O God of Jacob. Selah. Behold, O God our shield, and look upon the face of thine anointed. For a day in thy courts is better than a thousand. I had rather be a doorkeeper in the house of my God, than to dwell in the tents of wickedness. For the Lord God is a Sun and shield: the LORD will give grace and glory: no good thing will he withhold from them that walk uprightly. O Lord of hosts, blessed is the man that trusts in thee" (Psalms 84:2-12)

002:149 Accepting Muhammed's grave in Yathrib (modern Medina) as the second *al-Masjid al-Haram* (restricted place of prostration) and calling both *masjid* with the fabricated name al-*Masjid al-Haramayn* (the two restricted place of worship) is an innovation, since according to the Quran there is only one restricted place of prostration, and it is in Mecca.

002:152 The word *KaFaRa* is used both as an antonym of *SHaKaRa* (appreciate) and *amana* (affirm). Since it is more frequently contrasted with *amana*, the word *KaFaRa* (not appreciate/become ungrateful) evolves in the Quran and becomes a distinguishing description and descriptive noun for the antagonist and bigoted rejecters of the message. We could not translate the word *kafara* consistently with one English word without adding "don't" or "un-." Our choice of "unappreciative" or "did not appreciate," did not usually fit well in the translation of some verses. Thus, we rendered the meaning of *KaFaRa* interchangeably as "lack of appreciation," or "rejection." See 86:3.

002:154 When monotheist affirmers die, they are disconnected from this world; they continue their existence in another dimension. See 3:169.

002:155 We are tested on this planet whether we devote ourselves to God ALONE or not (29:2).

002:164 Nature is also God's book, written in the language of mathematics and with the letters of elements/atoms. The plural word *Ayat* is used in the Quran for God's law both in nature and in the scripture. There is harmony between God's laws, not contradiction. To witness this harmony, one must study and comprehend both books without adding superstitions and hearsay. Those who study and respect nature are rewarded by God with technology and progress. Those who study and respect the scripture are rewarded by the same God with wisdom and eternal bliss. See 2:106.

002:165 Jesus, Mary, Muhammed, and all friends of God will reject those who idolized them in various ways. See 16:86; 46:5-6; 25:30. Though some versions of the Bible offer a different translation for the beginning of the following verse, the New American Bible by Catholic Press (1970) renders an internally consistent translation: "None of those who cry out, Lord, Lord, will enter into the kingdom of God but only the one who does the will of my Father which is in heaven." (Matthew 7:21-23) The New Jerusalem Bible (1985) translates the key phrase as "It is not anyone who ..." See 2:48. Also, see 2:59; 3:45, 51-52, 55; 4:11,157,171; 5:72-79; 7:162; 19:36.

002:169 See 7:28.

002:172-173 Throughout the Quran, only four dietary articles are prohibited (6:145; 16:115). Other dietary prohibitions are the products of those who have divided themselves into sects and factions by following *hadith* and *sunna.* Prohibiting something that God has not prohibited in the name of God is another way of associating partners with God (6:121,147).

002:174-176 The religious scholars who are trying to hide the miraculous mathematical structure of the Quran and its message demonstrate the attitude criticized by these verses.

002:178-179 The Quran does not encourage the death penalty. Various rules are put in place to save lives, including the lives of murderers. The family of the victim might be satisfied with monetary compensation, which is encouraged by the Quran. Furthermore, the death penalty is not applicable to every murder case. For instance, if a woman kills a man or vice versa, the murderer cannot be sentenced to death; instead, the convict will be punished by society with a lighter punishment. If the murderer is not deemed a danger to society, monetary compensation or mandatory work might be more useful, productive, and rehabilitative than a prison sentence. The death penalty, like all criminal penalties, is not imposed and carried out by individuals, but by society and in accordance with accepted procedural rules. The so-called "honor killing" is a practice of ignorant and dishonorable people and has nothing to do with monotheism or peacemaking.

002:180 The sectarian teachings claim that this verse was abrogated by *hadith* stating, "there is no more leaving inheritance through will for relatives." Prophet Muhammed will complain about those who have traded the Quran with fabricated *hadith* (25:30).

002:183-187 Fasting, like timely prayers, the annual conference in Mecca and purification of blessings via sharing, originates from Abraham (22:73, 78). We are informed of any modification or correction in the ritual. For instance, verse 2:187 gives permission for sexual intercourse during the nights of Ramadan. Furthermore, the sectarian penalty of 60-day fasting for breaking the fast before sunset contradicts the Quran. The Quran makes no such distinction and considers breaking the fast before sunset to be equivalent to not fasting at all; it suggests a one-day-for-one-day make up for the days missed.

002:187 In the language of the Quran, the word *layl* (translated as night) is used to mark the period from sunset to sunrise.

002:188 This verse, together with 42:38 and 59:7 and a few other verses, reminds us one of the most important principles for a just, peaceful, and prosperous society. Monarchies, dictatorships, oligarchies, or regimes where wealth and political power accumulates in the hands of a few, compromise the foundations of justice. A society with no justice cannot enjoy happiness. The biggest challenge for the Western world is the protection of democratic procedures from the corrupting power of money and separation of powers. When a government becomes the government of the wealthy, by the wealthy and for the wealthy, the interest of the nation and indirectly the interest of all of humanity is risked. Today, more than any time in history, there is interdependence among the nations of the world. Regardless of its location, reckless treatment of planet earth poisons everyone's atmosphere. Reckless production of weapons, regardless of its location, risks the security of all nations. The gap between rich and poor nations, and the aggressive militaristic adventures of powerful nations challenge the security of the entire world. This gap is augmented through imperialistic policies that use illicit methods, such as misinformation, jingoistic propaganda, puppet regimes, and overt and covert military operations. Because of technology, the Internet, global economy, and an increase in population, our world is quickly shrinking, and the consequences of neo-colonialism will be different. The oligarchs who have been engaging in state terrorism finally reaped competition, an asymmetrical and unpredictable one. The injustice and corruption enjoyed by the world powers is generating and will continue to generate hatred, frustration, and resentment around the world. Those who have nothing to lose will make sure the rest lose their sleep. The world must take heed.

For a similar reminder, see the Old Testament, Deuteronomy 16:18-20.

002:189 "Do not enter homes from behind" is an idiom. Criticizing the opponent with ambiguous allegations and questions is not an efficient way of communicating; it causes animosity and misunderstanding.

002:190 War is permitted only in self-defense. See 9:5; 5:32; 8:19; 60:7-9.

002:193 God's system is based on freedom of faith and expression. God's system recommends an egalitarian republic, and a federally secular system that allows multiple jurisdictions for different religious or non-religious groups. See 58:12 and 60:8-9.

002:196 *Umra* is visiting the location of the great debate conference to prepare for the incoming debaters and audience during the months other than the four restricted months.

002:197 *Haj,* that is Debate Conference should be performed in the beginning of the restricted months: *Zil Hijja, Muharram, Safar,* and *Rabi Awwal* (2:189). Thus, *Haj* can be performed four times a year. Limiting *Haj* to once a year on a specific day, as traditionalists do, rather than four months, has created negative consequences such as inadequate service, hotels, dirt, and chaos. See 9:37. The story about the black stone is a myth. Showing such a reverence to a stone or a tomb and asking for help from the dead is idolatry. Unfortunately, the International Debate Conference (Hajj) has been distorted into a satanic ritual about 4 stones: kissing a black stone, circumbulating around a cubic building made of stones, and stoning a vertical stone pretending to be Satan. (1:5; 2:24; 10:106; 6:56; 7:194-197; 18:52; 22:73; 26:69-74; 28:88; 35:14,40; 39:38; 40:66; 46:5; 72:18; 2:149-150; 5:3; 16:120; 22:78; 66:6).

002:210 If God and controllers became visible, everyone would be forced to affirm, and the test designed by God would lose its meaning. God wanted to create creatures with free will, and He is testing this program on this planet. Those who make good decisions will join God in eternity and those who make bad decisions will join Satan in hell and there they will vanish eternally.

002:211 The mathematical miracle of the Quran is a great blessing and brings with it more responsibility. Also, see 5:115.

002:213 See 57:22-23.

002:219 The root of the Arabic word we translated as "intoxicants" is *KhaMaRa* and it means "to cover." If its first letter is read with Ha (9) rather than Kh (600), it then means Red, referring to red wine (For the usage of *Hamar* in the Bible, see Deuteronomy 32:14; Isaiah 27:2 Ezra 6:9; 7:22; Daniel 5:1-2,4). We prefer its common pronunciation. Some translations, while accepting the same pronunciation, have erroneously restricted the meaning of the word by translating it as "wine" or "liquor." Consumption of all intoxicants, be they alcoholic beverages, drugs, crack, cocaine, heroin, etc. are covered by this prohibition. The harm inflicted to individual and society by alcoholic beverages, drugs and gambling is enormous. See 5:90-91.

The Quran does not prohibit alcoholic beverages or drugs legally, and thus does not suggest any punishment for mere usage of alcohol. *Hadith* books ordaining severe penalties for the consumption of alcohol contradict the Quranic jurisprudence, since people's personal choices, how bad they may be, cannot be penalized by society. Society can only punish acts that are direct or proximate causes of harm to another person or persons. Besides, putting limitations on individual rights wastes society's resources, increases corruption, hypocrisy, and underground criminal activities, and other crimes. Legally banning and criminalizing the consumption and production of alcohol, a liquid drug, at the turn of 20th century proved to be a bad idea. Similarly, criminalizing the consumption of other drugs is also a bad idea. Society's resources should be used for the prevention and rehabilitation of these kinds of socially and psychologically caused addictions.

However, the Quran prohibits intoxicants to individuals for various reasons, including: moral (the designer and creator of your body and mind asks you not to intentionally harm the body lent to you for a lifetime), intellectual (the greatest gift you have is your brain and its power to make good judgment, so do not choose to be stupid or more stupid than you already are!) and pragmatic (you and your society will suffer grave loss of health, wealth, happiness, and many lives; do not contribute to the production and acceleration of such a destructive boomerang).

The common justification for using alcohol, "I drink in moderation," is not convincing, since almost all those who suffer from alcohol abuse or its consequences such as drunk driving, had started with moderation, sip by sip, and gradually increased the dosage because of peers or uncontrollable events in their lives. Today's moderate drinker may be tomorrow's addict. Why take the chance, especially when it is not a necessity? Stupidity in moderation is not something to be justified, let alone glorified. Rational people should not subject themselves and the society to the grave risks inflicted by alcohol. Besides, moral people should not support an industry that hurts a big segment of the society. The anecdotal "research" results that are occasionally advertised by the media claiming that the usage of wine or alcohol in moderation to be healthy for the heart is suspect, since beer and wine companies are major supporters of the media through commercials.

Many correlations are ignored by the so-called researchers. For instance, studying the impact of wine consumption on health, without taking into account the income, health insurance, diet, lifestyle, genetic make up, and environment of wine-drinkers, will not produce a reliable cause-and-effect relationship. Interestingly, some of the health benefits attributed to alcohol have been listed for grape juice, though in higher doses. Even if alcohol or any other drug had some economic, social and even health benefits, the Quran reminds us of the proven fact that their harms outweigh their benefits.

The Bible contains contradictory messages regarding the use of wine or alcoholic beverages. The Hebrew equivalent of the Quranic word *Sakar* (16:67; 4:43) is *Shekar* (intoxicant) and it is criticized by the Old Testament (Leviticus 10:9; Judges 13:4, Isaiah 28:7; Isaiah 5:11; 24:9; 29:9; 56:12; Joel 3:3; Amos 6:6; Pr 20:1; 31:6; Micah 2:11). Wine impairs the health, judgment and memory (1 Samuel 25:37; Hosea 4:11; Peter 31:4-5), inflames the passions (Isaiah 5:11), and leads to sorrow, contention and remorse (Peter 23:29-32). Wine also cheers God and man (Judges 9:13; Zechariah 9:17; Psalms 104:15; Esther 1:10; Ecclesiastes 10:19) and strengthens the body (2Samuel 16:2; Song 2:5).

Though the consumption of wine is occasionally approved and even encouraged through the daily sacrifice (Exodus 29:40-41), with the offering of the first-fruits (Leviticus 23:13), and with various other sacrifices (Numbers 15:4-10), the Bible also contains some prohibitions. For instance, it prohibits wine for the Rechabites (Jeremiah 35:1-19), the Nazirites (Judges 13:7; Luke 1:15; 7:33), and the priests when engaged in their services (Leviticus 10:1, 9-11). It also lists wine as offerings of idolatry (Deuteronomy 32:37-38).

The Hebrew Bible uses about a dozen words to refer to different kinds of alcoholic beverages (*Ashisha, Asis, Hemer, Enabh, Mesekh, Tirosh, Sobhe, Shekar, Yekebh, Shemarim, Mesek*). However, some Biblical scholars limit the application of this prohibition by translating the *Shekar* as "strong drink." The majority of Christian sects have no religious inhibition regarding the consumption of alcohol. Easton, a Jesus-worshipping tri-theist, writes: "Our Lord miraculously supplied wine at the marriage feast in Cana of Galilee (John 2:1-11)."

The New Testament also contains warnings for people against the excess consumption of intoxicants (Luke 21:34; Romans 13:13; Ephesians 5:18; 1 Timothy 3:8; Titus 1:7). But, in the same volume, we also see wine moving from lip to lip as a sacred drink. According to the Gospels, Jesus turns water to wine (John 4:46) and wine has a prominent place in Passover and the Last Supper. While Mark 15:23 serves Jesus wine in his last moments, Matthew 27:34 serves him vinegar. It is obvious that the Gospel authors could not differentiate between vinegar and wine.

The contradictory position regarding wine or alcoholic beverages reaches absurdity when Jesus allegedly declared wine to be memorials of his body and blood. The Book of Revelation, a theo-fictional nightmare, mixes wine with the numerically defined beast and hell fire. "And the third angel followed them, saying with a loud voice, If any man worship the beast and his image, and receive his mark in his forehead, or in his hand, The same shall drink of the wine of the wrath of God, which is poured out without mixture into the cup of his indignation; and he shall be tormented with fire and brimstone in the presence of the holy angels, and in the presence of the Lamb: And the smoke of their torment ascendeth up for ever and ever: and they have no rest day nor night, who worship the beast and his image, and whosoever receiveth the mark of his name" (Revelations 14:9-11).

It seems that the words grape juice, vinegar, literal wine, and metaphorical wine, may all be mixed up in the Bible due to some intentional tampering during its oral or written transmission, and the errors committed during translations of other translations.

002:222 Prohibition of sexual intercourse during menstruation is for the protection of women from pain and potential infection. Other than that, menstruating women should continue their contact prayers, charity work, fasting, and studying the Quran. Commentaries based on *hadith* and *sunna*, not only mistranslate the word *tahara* but also fabricate a list of prohibitions for women claiming that they are not worthy of communicating with God or performing some good acts. In verse 3:55, the word *tahara* means "to be rescued" or "free from."

Despite the Quranic rule, the followers of *hadith* and *sunna* adopted Jewish laws that consider women unclean and treat them like dirt for fourteen straight days of every month. According to the fabricated rules of the Old Testament, a menstruating woman is considered unclean for seven days, and during that period wherever she sits will be considered unclean; whoever touches her or sits where she sits must wash and bathe. After she finishes the menstruation, she has to wait for seven more days to be considered clean for ceremonial purposes (Leviticus 15:19-33).

002:223 Though this verse does not bring a limitation for sexual positions, in our opinion, it implicitly rules out anal sexual intercourse, since it likens women during the sexual intercourse to farms receptive of seeds.

002:226-230 Sectarian scholars who ignored the Quran and upheld volumes of books of *hadith* and *sunna*, issued laws (*sharia*) allowing the marriage contract to be voided with several words coming from the husband's mouth. This ease and one-sided divorce created miserable marriages and destroyed many families. Many men, who "divorced" their wives by uttering the magical word *talaq* unintentionally or in the heat of anger, desperately looked for a solution (*fatwa*), and found mullahs and religious judges selling fatwas to save their marriage! The class that created the problem in the first place became the benefactor of the solution. See 9:34-35.

Verse 2:228 establishes equal rights to both genders. By associating and even preferring numerous collections of lies and innovations to the Quran, the followers of *hadith* and *sunna* denied Muslim women the right to divorce and turned them into slaves of male despotism. Verse 4:19 clearly recognizes the right of women to divorce.

Divorce is a legal event lasting several months; it is not just an oral declaration of the male spouse. A wife cannot be divorced by announcing, "I divorce you three times." Furthermore, the divorce must be declared by a court, with the participation of both parties and witnesses. In order for three divorces to happen, three marriages must happen. Taking the right of getting married after the first divorce through fabricated *sharia* law is a great injustice. Sunni scholars tried to mend the consequences of the swift divorce by another fabrication by legalizing one-night adultery, *hulla.* After the *hulla* the husband could start the fourth marriage! For the man who somehow changed his mind and did not divorce the one-nighter, some Hanafi scholars made up another law. They claimed that if a man divorced his wife under threat or duress they would be considered divorced. In sum, a patch to solve a self-inflicted problem created another problem. The sectarian divorce law has evolved like a tax code or a bad computer program, each loophole demanding a patch and each patch creating more loopholes or problems. See 33:49.

The living expenses of a divorced women, as a general rule, are expected to be paid by the ex-husband (2:241). Besides, the divorced women cannot be evicted from their houses by their husbands (65:1). Also, see 30:21. However, if after the divorce a woman learns that she is pregnant, she is advised to reconsider reconciliation with her former husband (2:228).

The Old Testament recognizes the right of women to divorce (Proverbs 2:17; Mark 10:12). Also, see Leviticus 21:14; Deuteronomy 24:1-4; Numbers 30:9; Deuteronomy 24:2-4. Though divorce is prohibited in the New Testament (Matthew 5:32; 19:9; Mark 10:2-12; Luke 16:18), because of its unrealistic idealism, many Christians do not adhere to the prohibition of divorce. According to the New Testament, millions of Christians are committing adultery, since divorced men or women are remarrying again.

002:234 The waiting period of married women is the experience of three menstruations. See 2:228.

002:236 Dowry is an economic security and compensation to be paid to the woman. It is the consideration of the marriage contract, since women take more risk than men do by entering such a contract. Divorced women with children usually bear more burdens. The Quran asks us to provide for our ex-wives for a while to allow them to get on their feet. Thus, women will not endure the torture of having to live with a difficult husband for the fear of becoming homeless. Dowry has nothing to do with the families of women; it is the right of women.

The Bible also has a dowry requirement, but it considers it a gift to the family of the bride: Genesis 30:20; 34:12; Exodus 22:17; 1 Samuel 18:25.

002:238 Three timely prayers are mentioned by name: 2:238; 11:114; 17:78 and 24:58. During the revelation of the Quran, like practices such as Debate Conference, fasting, and sharing for betterment, contact prayers too were known. See 8:35; 21:73; 22:78; 107:4. Some people understand the phrase *sala al wusta* to mean "prayer is good" rather than the "middle prayer," and they have therefore inferred two prayers a day, rather than three prayers. However, diligent research on this issue and the waning of the sectarian influence is helping to clarify the issue. See, the Appendix, *Sala Prayer According to the Quran.*

The interruption of the subject on divorce with verses about the contact prayers does not violate textual coherence, since one of the purposes of the prayer is to ask for God's help and guidance during difficult times, and divorce is one of those times. See 2:45; 29:45.

002:239 For instance, one may pray in a sitting position while traveling in buses or airplanes.

002:240 The estate trust recommended by this verse is another layer of protection provided for a divorced woman. This is in addition to the dowry, periodical payment, and inheritance. If there is no will, see 4:12.

002:244 The Quran invites people to peace (2:208). Thus, the universal greeting *salam* or *salamun alaikum* continuously invites people to peace (4:94). War is permitted only in self-defense. For the principles of war, see 60:7-9 and 9:5-29.

002:246 The same event is told in more detail in the Old Testament, 1 Samuel 8:1-18; 9:1-10.

002:247 It seems that knowledge and health are two important characteristics needed for leadership. The expectation of wealth for appointed/elected leaders is criticized. The Old Testament, 1 Samuel 10:27, critically reports the materialistic value system of the society. 1 Samuel 9:7 also reflects the society's expectation of a false value system through the perception of Saul.

002:249 For the Biblical account of this test, see the Old Testament, Judges 7:1-8. For a similar promise, see the Old Testament, Deuteronomy 20:1-4.

002:251 The Old Testament gives contradictory information on this story. According to 1 Samuel 17:4, 50-51, it is David who kills the giant enemy named Goliath, while according to 2 Samuel 21:19, it is Elhanan who does the killing. Biblical scholars who believe in the Bible verbatim, wish for us to believe that there were two different giants called Goliath, simultaneously. Interestingly, both stories are similar, and David appears in both. Some Biblical scholars, however, accept the existence of a textual error.

002:253 See 2:285.

002:254 One of Satan's clever tricks is to attribute the power of intercession to powerless creatures like Jesus, Mary and Muhammed. The Quran repeatedly informs us that there is no intercession beyond affirmation of facts or testimony to the truth. The only intercession of Muhammed referred in the Quran involves his complaint about his people for abandoning the Quran and idolizing him. See 25:30; 72:21; 79:38. Also, 2:48; 2:123; 6:51; 6:70; 19:87; 20:109; 21:28; 39:44. Old Testament, Isaiah 53:12; Jeremiah 7:16; 27:18.

002:255 The verse rejects the distortion seeped into the Old Testament depicting God as resting or sleeping (Genesis 2:2; Exodus 31:17; Psalms 78:65; 44:23). See 7:180.

002:256 Though the Quran denounces imposition of religion, and promotes freedom of religion and expression of thought, the followers of *hadith* and *sunna* created *sharia* laws primarily for the justification of oppressive and dictatorial regimes of the Umayyad and Abbasid caliphs. The Quran rejects the imposition of a religion over another and promotes the Islamic system that can be summarized, in modern terms, as federal secularism (5:43-48). Thus, the city-state of *Yathrib* led by Muhammed, who did not accept any other authority besides the Quran (6:114), established a successful example of a federal secular system by dividing the territory into independent legal jurisdictions to accommodate the diverse religious, social, and political preferences of its communities. The relations between those who accepted the Quran as their jurisdiction and those who followed other laws were determined according to a constitution drafted and signed by all parties.

Some critics take verse 9:3 out of its context and present it as a contradiction with the principle expressed in 2:256 and other verses. Chapter 9 starts with an ultimatum against Meccan *mushriks* who not only tortured, killed, and evicted muslims from their homes but also mobilized several major war campaigns against them while the muslims established a peaceful multi-national and multi-religious community. The beginning of the Chapter refers to their violation of the peace treaty and gives them an ultimatum of four months to stop aggression. Thus, the verses quoted from Chapter 9 have nothing to do with freedom of religion; it is a warning against aggressive, murderous religious fanatics.

As for 9:29, the word *jizya* in that verse has been mistranslated as a taxation of non-muslims, while the word means "compensation" or more accurately, "war reparation," which was levied against the aggressing party who initiated the war. See 9:29; 6:68. Also see 2:193; 4:140; 10:99; 18:29; 74:55; 80:12; 88:21, 22. The death penalty rule in Sunni and Shiite sects contradicts many verses, especially the verse 4:137.

The Bible contains some verses emphasizing the importance of freedom, justice, and human rights. "When all the prisoners in a country are crushed underfoot, when human rights are overridden in defiance of the Most High, when someone is cheated of justice, does not the Lord see it?" (Lamentations 3:34-36). "He has sent me to proclaim liberty to captives, sight to the blind, to let the oppressed go free" (Luke 4:18). On the other hand, the Old Testament contains some verses that show no tolerance for other religions or religious beliefs. For instance, if a man was deemed to blaspheme against God, he was expected to be stoned to death. Even the violation of the Sabbath was considered a crime punishable by death. See, Numbers 15:32-36.

002:258 Abraham's method in this argument is interesting. He moves to another argument without being bogged down with the deceptive language used by the King in his answer to his first argument. See 13:15.

002:259 See 18:19-25.

002:260 Some people refuse to witness the mathematical miracle of the Quran by saying "I already believe in the Quran; I do not need miracles." Do they have stronger faith than Abraham does? Besides, if they really believed in the Quran they would not ignore or belittle its prophecies and mathematical structure. Furthermore, they would not trade it with volumes of man-made teachings. See 60:4. See Bible: Exodus 4:1-5; 8:19; 10:2; Numbers 14:22; 1 Chron 16:12,24; Job 5:9; Psalms 78:10-16; 105:5; 106:7; 118:23; 139:14; Isaiah 29:14; Jeremiah 32:21; John 4:48; 9:3; 12:37; 15:24; Mark 6:52; 16:20; Luke 16:31.

002:263-264 See Old Testament, Catholic version, Sirach 18:14-15.

002:272 See 57:22-23.

002:275-281 It is common wisdom that "high" interest rates are not healthy for a good economy. The interest that the Quran prohibits is not the interest collected from money lent for businesses, but rather the money lent for consumption of necessities. When considered with its context, this prohibition is about usury. The Quran does not treat this subject in the context of business or trade, but in the context of the charity to the needy. Those who exploit the basic needs of individuals may attempt to justify their usury as free market trade. The interest charge on credit cards used by needy consumers should be considered usury, since the money borrowed is not invested for profitable business enterprise but spent on basic needs. Modern banks charge obscene amounts of fees and interest (usury) to their customers who need the money the most. They have developed extensive tricks to steal from their needy victims. Unfortunately, banks have the power to legalize their robberies by buying the elected legislators. Through constant and cunning propaganda, they have even succeeded in legitimizing their exploitation and theft, even to their victims. See 3:130; 4:161.

The Bible prohibits usury (Leviticus 25:36-37) and considers it a great crime (Psalms 15:5; Pr 28:8; Jeremiah 15:10). However, it uses a double standard: "Do not charge your brother interest, whether on money or food or anything else that may earn interest. You may charge a foreigner interest, but not a brother Israelite, so that the Lord your God may bless you in everything you put your hand to in the land you are entering to possess" (Deuteronomy 23:20). Also, see Exodus 22:25; Deuteronomy 23:19-20; Nehemiah 5:7,10; 5:9-13; Isaiah 24:2; Ezekiel 18:8,13,17; 22:12.

002:282 This testimony is limited to business transactions. From this verse, we cannot deduce that women are inferior to men regarding intellect, memory, or trustworthiness. Furthermore, such an interpretation, which relies on *hadith* and *sunna*, contradicts other verses of the Quran (24:6-9; 3:195).

Verse 49:13 unequivocally rejects sexism and racism, and reminds us that neither man nor female, neither this race nor that race is superior to the other. The only measure of superiority is righteousness; being a humble, moral, and socially conscientious person who strives to help others.

The one man plus two women recommendation might be due to the statistical reflection of the generally recommended workplace for both genders. Though the Quran affirms and even encourages women to have her independent savings and income, women are biologically endowed with certain qualities such as pregnancy and breastfeeding that makes them the most suitable person to raise children. This in turn may create a statistical disadvantage regarding the number of women familiar with the terminology of business contracts.

Another reason for this one-man and two-woman arrangement could be the protection of women from being subjected to high pressure by the party breaching the contract. The presence and support of other women might reduce the pressure and possibility of perjury. Modern authorization and confirmation devices have reduced the need for live human witnesses.

The extreme feminist agenda of ignoring such a quality and its benefits to children has harmed both the family structure and the emotional development of millions of children. Despite their prosperous life, many children deprived of close attention of their mothers are suffering from various emotional and mental problems, such as stress, attention deficiency, hyperactivity, and unhappiness.

Instead of demanding respect and monetary compensation for women's work at home, early feminists bought the male chauvinist idea of belittling women's traditional work and tried to put women in competition with men. The feminist assumption that women are exactly equal to men and that they should compete in every area of life and demonstrate statistical equality of accomplishments or failures in every aspect of their lives has pushed women into arenas that men are biologically advantageous in. Expecting equal performance is injustice both to men and women.

From movies to children's stories, from the business world to politics we can witness discrimination on par with the expectations of traditional sex roles. Using the female body as a commercial object, sexual harassment, rape, women battery, and sexist language are endemic in our modern society. The feminist movement is an intelligent, but sometimes highly emotional, protest of this unjust historical treatment.

Men and women, in general, are different by nature, and have different needs and roles. However, some sex roles and inequalities are created by society and exploited by men. In order to let nature and justice prevail over superficiality and injustice, it is imperative to have the following: 1) Equal respect and appreciation of roles regardless of their gender, 2) Equal chance for both males and females to choose their roles freely and responsibly, And 3) Laws to promote and guarantee these two goals.

002:285 Like Christians, the followers of Sunni or Shiite sects have also put their messenger in direct competition with other messengers. Hundreds of *hadiths* have been fabricated to support the claim that Muhammed had a higher position above other messengers of God. For instance, the claim that God created the universe for the sake of Muhammed (*lawlaka lawlaka lama khalaqtu al-aflaka*), was attributed to God under the label *Qudsi hadith* (Holy *hadith*). Many fabricated *hadiths* praise Muhammed for being a superman, thereby contradicting the Quran (18:110). Most *hadith* books, including *Bukhari*, contain numerous records of hearsay that insult Muhammed by depicting him as a sexual maniac with exaggerated stories, such as "praising" his sexual power to be equal to 30 men. The fabricators of those stories and their powerful benefactors tried to justify their own sexual fantasies and practices by attributing them to Muhammed (6:112). Again, in clear contradiction to the Quran (29:50-51), *hadith* narrators fabricated "miracles" for Muhammed, such as splitting the Moon and causing half of it to fall into Ali's backyard, ascending to the seventh heaven and negotiating the numbers of prayers with God with Moses serving as his advisor in the sixth heaven, miraculously crippling a child for passing in front of him while he was praying, etc. Sunni and Shiite *mushriks* went ahead and added Muhammed's name next to God's in the most repeated declaration of faith, the *shahada*, again in clear contradiction of many verses of the Quran (See 3:18; 63:1).

002:286 The word *mawla* (sovereign/patron/lord) occurs 18 times in the Quran, and 13 of them are used for God (2:286; 3:150; 6:62; 8:40; 9:51; 10:30; 22:78; 47:11; 66:2,4); the other 5 are used negatively to criticize the depiction of human idols by this divine title (16:76; 22:13; 44:41; 57:15). Though the Quran clearly warns us not to call anyone besides God *mawlana* (our sovereign/our patron/our lord), ignorant people among Sunni, Shiite and Christian *mushriks* use it as a title for their prophets, saints, or clergymen. In Pakistan and India, Sunni and Shiite religious leaders themselves use this *exclusively divine* title *Mawlana* as a religious title before their names. A simple Internet search of the word *mawlana* will produce thousands of names of false idols, mostly followed by ostentatious Arabic names. See 6:62; 8:40; 9:31; 9:51; 10:30; 22:13,78; 34:41; 42:21; 47:11; 66:2,4. The word *waly* (ally), on the other hand, is used for God and for humans as well. God is *Waly* of monotheists and monotheists are the *waly* of each other.

بسم الله الرحمن الرحيم

3:0 In the name of God, the Gracious, the Compassionate.

3:1 A1L30M40*

3:2 **God**; there is no god, except He; the Living, the Sustainer.

3:3 He sent down to you the book with truth, confirming what is present; and He sent down the Torah and the Injeel...

3:4 Before; He sent down the Criterion as guidance for the people. For those who become unappreciative of **God**'s signs, they will have a severe retribution, and **God** is Noble, exacting in Revenge.

3:5 For **God** nothing is hidden in the earth or in the heaven.

3:6 He is the One who shapes you in the wombs as He pleases. There is no god but He, the Noble, the Wise.

Multi-meaning Verses: Requiring Knowledge and Good Intention

3:7 He is the One who sent down to you the book, from which there are definite signs; they are the essence of the book; and others, which are multi-meaning. As for those who have disease in their hearts, eager to cause confusion and eager to derive their interpretation, they will follow those with multi-meaning. But none knows their meaning except **God** and those who are well founded in knowledge; they say, "We affirm it, all is from our Lord." None will remember except the people of intellect.*

3:8 "Our Lord, do not make our hearts deviate after You have guided us, and grant us from You a mercy; You are the Grantor."

3:9 "Our Lord, You are the Gatherer of the people for a day in which there is no doubt; **God** does not break His promise."

Winners versus Losers

3:10 As for those who did not appreciate, neither their money nor their children will avail them anything from **God**. They are the fuel for the fire.

3:11 Like the behavior of the people of Pharaoh and those before them. They rejected Our signs, so **God** took them to task for their sins. **God** is severe in retribution.*

3:12 Say to those who are ingrates, "You will be defeated and gathered towards hell. What a miserable abode!"

3:13 There was a sign for you in the two groups that met. One was fighting in the cause of **God**, and the other was ingrate. They thus saw them as twice their number with their eyes. **God** supports with His victory whomever He wills. In this is a lesson for those with vision.

3:14 It has been alluring for people to love the desire of women, buildings, ornaments made from gold and silver, trained horses, the livestock, and lands. These are the enjoyment of the world, and with **God** is the best place of return.*

3:15 Say, "Shall I inform you of what is greater than all this? For those who are aware, at their Lord there will be gardens with rivers flowing beneath, abiding there eternally, and purified mates, and an acceptance from **God**. **God** is Seer of the servants."

3:16 The ones who say, "Our Lord, we affirm, so forgive us our sins, and spare us the retribution of the fire."

3:17 The patient, the truthful, the devout, the givers, and the seekers of forgiveness in the late periods of the night.

The Divine System: Peacemaking and Submission to God

3:18 **God** bears witness that there is no god but He; as do the controllers, and those with knowledge; He is standing with justice. There is no god but He, the Noble, the Wise.*

3:19 The system with **God** is peacemaking and peaceful surrendering (Islam). Those who received the book did not dispute except after the knowledge came to them out of jealousy between them. Whoever does not appreciate **God**'s signs, then **God** is swift in computation.*

3:20 If they debate with you, then say, "I have peacefully surrendered myself to **God**, as well as those who follow me." In addition, say to those who were given the book and the Gentiles: "Have you peacefully surrendered?" If they have peacefully surrendered then they are guided; but if they turn away, then you are only to deliver the message. **God** is watcher over the servants.*

3:21 Those who do not appreciate **God**'s signs and fight/kill the prophets unjustly, and fight/kill those who order justice from amongst the people; give them the good news of a painful retribution.

3:22 These are the ones whose works will be lost in this world, and in the Hereafter, they will have no supporters.

3:23 Did you not see those who were given a portion of the book being invited to **God**'s book to judge between them, then a group of them turn away while they are averse?

3:24 That is because they said, "The fire will not touch us except a number of days," and they were arrogant by what they invented in their religious system.*

3:25 How shall it be when We gather them on the day in which there is no doubt; every person shall receive what it has earned, and they will not be wronged.

The Absolute Sovereign

3:26 Say, "Our God, Ruler of sovereignty; you grant sovereignty to whom You please, revoke sovereignty from whom You please, dignify whom You please, and humiliate whom You please; in Your hand is goodness. You are capable of all things."*

3:27 "You blend the night into the day, and blend the day into the night; and you bring the living out of the dead, and bring the dead out of the living; and You provide for whom You please with no computation."

Pick Your Allies Carefully

3:28 Let not those who affirm take the ingrates as allies instead of those who affirm. Whoever does so will have nothing with **God**, for you are to be cautious of them for a particular reason. Moreover, **God** warns you of Himself, and to **God** is the destiny.

3:29 Say, "Whether you hide what is in your chests or reveal it, **God** knows." He knows what is in the heavens and what is in the earth; and **God** is capable of all things.

3:30 The day every person will find what good it had done present, and what bad it had done; it wishes that between them was a great distance. **God** warns you of Himself, and **God** is Kind towards the servants.

3:31 Say, "If you love **God** then follow me so **God** will love you and forgive your sins." **God** is Forgiver, Compassionate.

3:32 Say, "Obey **God** and the messenger." But if they turn away, then **God** does not like the ingrates.*

Family of Imran and Maryam's (Mary) Birth

3:33 **God** has chosen Adam, Noah, the family of Abraham, and the family of Imran over the worlds.

3:34 A progeny each from the other, and **God** is Hearer, Knower.

3:35 When the wife of Imran said, "My Lord, I have vowed to You what is in my womb, dedicated, so accept it from me. You are the Hearer, the Knower."

3:36 So when she delivered, she said, "My Lord, I have delivered a female," and **God** is fully aware of what she delivered, "The male is not like the female, and I have named her Mary, and I seek refuge for her and her progeny with You from the outcast devil."*

3:37 So her Lord accepted her with a good acceptance, and made her grow like a flower, and charged Zechariah with her. Every time Zechariah entered upon her in the temple enclosure, he found provisions with her. He said, "O Mary, from where did you get this?" She said, "It is from **God**, **God** provides for whom He wishes beyond reckoning."

The Birth of Yahya (John)

3:38 It was then that Zechariah called on his Lord, he said, "My Lord, grant me from You a good progeny; You are hearer of the prayers."

3:39 The controllers called him while he was standing and praying in the temple enclosure: "**God** gives you good tidings of John, authenticating a word from **God**, respectable, protected, and a prophet from the reformers."

3:40 He said, "My Lord, how can I have a son when old age has reached me and my wife is sterile?" He said, "Thus it is. **God** does what He pleases."

3:41 He said, "My Lord, make for me a sign" He said, "Your sign is not to speak to the people for three days except by symbol, and remember your Lord greatly, and glorify at dusk and dawn."*

The Birth of Esau (Jesus)

3:42 The controllers said, "O Mary, **God** has selected you and cleansed you, and He has chosen you over the women of the worlds."

3:43 "O Mary, be devoted to your Lord and prostrate and kneel with those who kneel."

3:44 This is from the news of the unseen that We reveal to you. You were not with them when they drew straws as to which one of them will be charged with Mary; you were not with them when they disputed.

3:45 The controllers said, "O Mary, **God** gives good news of a word from Him. His name is the Messiah, Jesus the son of Mary. Honorable in this world and in the Hereafter, and from among those who are made close."*

3:46 "He speaks to the people from the cradle, and as an adult; and he is among the reformers."

3:47 She said, "My Lord, how can I have a son when no human has touched me?" He said, "It is thus that **God** creates what He wills, when He decrees a command, He merely says to it 'Be,' and it becomes."

3:48 He teaches him the book and the Wisdom, the Torah and the Injeel.

3:49 As a messenger to the Children of Israel: "I have come to you with a sign from your Lord; that I create for you from clay the form of a bird, then I blow into it and it becomes a bird by **God**'s leave, and I heal the blind and the lepers, and give life to the dead by **God**'s leave, and I can tell you what you have eaten, and what you have stored in your homes. In that is a sign for you if you have affirmed."

3:50 "Authenticating what is present with me of the Torah, and to make lawful some of what was forbidden to you; and I have come to you with a sign from your Lord, so be aware of **God** and obey me."

3:51 "**God** is my Lord and your Lord, so serve Him, this is a Straight Path."*

The Muslim Disciples

3:52 So when Jesus felt their rejection, he said, "Who are my supporters towards **God**?" The disciples said, "We are **God**'s supporters, we affirm **God** and we bear witness that we have peacefully submitted."*

3:53 "Our Lord, we affirmed what You have sent down; we followed the messenger, so record us with those who bear witness."

The Death of Jesus

3:54 They schemed and **God** schemed, but **God** is the best schemer.

3:55 **God** said, "O Jesus, I will terminate you, and raise you to Me, and cleanse you of those who have rejected, and make those who have followed you above those who rejected, until the day of Resurrection; then to Me is your return so I will judge between all of you in what you were disputing."*

3:56 "As for those who are unappreciative, I will punish them with a severe punishment in this world and in the Hereafter; they will have no supporters."

3:57 "As for those who affirm and promote reforms, We will pay them their reward; **God** does not like the wicked."

3:58 This We recite to you is from the revelation and the wise reminder.

3:59 The example of Jesus with **God** is like that of Adam; He created him from dust, then He said to him, "Be," and he became.*

3:60 The truth is from your Lord; so do not be of those who are doubtful.

3:61 Whoever debates with you in this after the knowledge has come to you, then say, "Let us call our children and your children, our women and your women, ourselves and yourselves, then, let us call out, and we shall make **God**'s curse upon the liars."

3:62 This is the narration of truth, there is no god but **God**; and **God** is the Noble, the Wise.

3:63 If they turn away, then **God** is aware of the corruptors.

Invitation to Unite Around a Common Principle

3:64 Say, "O people of the book, let us come to a common statement between us and you; that we do not serve except **God**, and do not set up anything at all with Him, and that none of us takes each other as lords beside **God**." If they turn away, then say, "Bear witness that we have peacefully surrendered."*

Abraham

3:65 "O people of the book, why do you debate us about Abraham when the Torah and the Injeel were not sent down except after him? Do you not reason?"

3:66 Here you have debated in what you knew; so why then do you debate in what you do not know? **God** knows while you may not know.

3:67 Abraham was neither a Jew nor a Nazarene, but he was a monotheist who peacefully surrendered; he was not of those who set up partners.

3:68 The closest people to Abraham are those who followed him, this prophet, and those who affirmed; and **God** is the ally of those who affirm.

3:69 A group from the people of the book wished that they could misguide you, but they only misguide themselves and they do not notice.

3:70 O people of the book, why do you reject **God**'s signs while you are bearing witness?

3:71 O people of the book, why do you dress the truth with falsehood and conceal the truth while you know?*

3:72 A group among the people of the book said, "affirm what was revealed to those who affirmed during the beginning of the day and reject it by the end of it; perhaps they will return."

3:73 "Do not affirm except for those who follow your system."-- say, "The guidance is **God**'s guidance."-- "That anyone should be given similar to what you have been given, or that they debate with you at your Lord." Say, "The bounty is in **God**'s hand, He gives it to whom He chooses, and **God** is Encompassing, Knowledgeable."

3:74 He specifies His mercy for whomever He chooses, and **God** is with Great Bounty.

Honesty

3:75 Among the people of the book are those whom if you entrust him with a large amount he gives it back to you, and there are those whom if you entrust with one gold coin he will not return it to you unless you are standing over him. That is because they said, "We have no obligation towards the Gentiles." They say about **God** lies while knowing.*

3:76 Indeed, anyone who fulfils his pledge and is conscientious, then **God** loves the conscientious.

3:77 Those who trade **God**'s pledge and their oaths for a small price will have no portion in the Hereafter. **God** will not speak to them nor look at them on the day of Resurrection, nor purify them, and they will have a painful retribution.

Those Who Try to Add to the Book

3:78 From amongst them is a group that twists their tongues with the book so that you may count it from the book, while it is not from the book, and they say it is from **God** while it is not from **God**, and they knowingly say lies about **God**.

3:79 It is not for a human that **God** would give him the book, the authority, and the prophethood, then he would say to the people: "Be servants to me rather than to **God**!", rather: "Be devotees to what you have been taught of the book, and to what you studied."*

3:80 Nor does he order you that you take the controllers and the prophets as lords. Would he order you to reject after you have peacefully surrendered?*

The Messenger after All Prophets

3:81 **God** took a covenant from the prophets: "For what I have given you of the book and wisdom, then a messenger will come to you authenticating what is with you. You will affirm him and support him." He said, "Do you testify, and agree to this burden?" they said, "We testify." He said, "Then bear witness, and I am with you bearing witness."*

3:82 Whoever turns away after that, they are the wicked ones.

3:83 Is it other than **God**'s system that they desire, when those in the heavens and the earth have peacefully surrendered to Him voluntarily or by force? To Him they will be returned.

Do not Discriminate Among Prophets

3:84 Say, "We affirm **God** and what was sent down to us and what was sent down to Abraham, Ishmael, Isaac, Jacob, the Patriarchs, and what was given to Moses, Jesus and the prophets from their Lord. We do not discriminate between them, and to Him we peacefully surrender."

Only One System: Peace and Submission to God

3:85 Whoever follows other than peacemaking and peaceful surrendering as a system, it will not be accepted from him, and in the Hereafter he is of the losers.*

3:86 How would **God** guide a people who have rejected after affirming and they witnessed that the messenger was true, and clear evidence had come to them? **God** does not guide the wicked people.

3:87 For these, the consequence will be that **God**'s curse will be upon them as well as that of the controllers and the people all together!

3:88 Eternally they will abide in it, the retribution will not be lightened for them, nor will they be reprieved;

3:89 Except those who repent after this and reform, then **God** is Forgiver, Compassionate.

Repentance and Criterion for Charity

3:90 Those who denied after their affirmation and then increased in rejection, their repentance will not be accepted; they are the strayers.

3:91 Those who have rejected and died while they were ingrates, if the earth full of gold were to be ransomed with it, it would not be accepted from any of them. For these there will be a painful retribution and they will have no supporters.

3:92 You will not reach piety until you spend from what you love. Whatever you spend, **God** is aware of it.

Prohibitions Falsely Attributed to God

3:93 All the food used to be lawful to the Children of Israel except what Israel forbade for himself before the Torah was sent down. Say, "Bring the Torah and recite it if you are truthful."

3:94 Whoever invents lies about **God** after this, these are the wicked.

3:95 Say, "**God** bears truth. Follow the creed of Abraham, monotheism, and he was not of those who set up partners."

The First Public House (Basis)

3:96 The first house/basis established for the public is the one in Bakka, blessed, and guidance for the worlds.*

3:97 In it are clear signs: the place of Abraham. Whoever enters it will be secure. Debate of the basis is a duty from **God** for the people who can afford a means to it. Whoever rejects, and then **God** has no need of the worlds.

3:98 Say, "O people of the book, why do you reject **God**'s signs, while **God** is witness over what you do?"

3:99 Say, "O people of the book, why do you repel from the path of **God** those who affirm? You wish to twist it while you were witness. **God** is not oblivious to what you do."

3:100 O you who affirm, if you obey a group of those who received the book, they will turn you after your affirmation into ingrates!

Reverting from the Message

3:101 How can you reject when **God**'s signs are being recited to you and His messenger is among you? Whoever holds firmly to **God**, has been guided to the straight path.

3:102 O you who affirm, be aware of **God** as He deserves reverence, and do not die except as ones who have peacefully surrendered.

3:103 Hold firmly to the rope/covenant of **God**, all of you, and do not be disunited. Remember **God**'s blessing upon you when you were enemies and He united your hearts, then you became, with His blessing, brothers; and you were on the verge of a pit of fire, and He saved you from it. Thus, **God** clarifies for you His signs, so that you may be guided.

3:104 Let there be a nation from amongst you that calls towards goodness, and orders recognized norms, and deters from evil. These are the successful ones.

Do not get divided into Sects and Factions

3:105 Do not be like those who separated and differed after the proof had come to them. For them is a painful retribution.*

3:106 The day on which faces will be brightened and faces will be darkened; as for those whose faces will be darkened: "Did you reject after affirming? Taste the retribution for what you rejected."

3:107 As for those whose faces are brightened, they are in **God**'s mercy, in it they abide eternally.

3:108 These are **God**'s signs; We recite them to you with truth. **God** does not want wickedness for the worlds.

3:109 To **God** is all that is in the heavens and in the earth; and to **God** the events will be returned.

3:110 You became the best nation that emerged for the people; you order recognized norms, deter from evil, and trust/ affirm **God**. If the people of the book affirmed, it would have been better for them; from amongst them are affirmers, but most remain wicked.*

3:111 They will not harm you except in being an annoyance, and if they fight you they will turn and flee; then they will not be supported.

3:112 They are stricken with humiliation wherever they are found, except through a covenant from **God** and a covenant from the people. They earned **God**'s wrath and were stricken with humiliation; that is because they were being unappreciative of **God**'s signs and fighting/killing the prophets without just cause. This is the consequence of their disobedience and transgression.

Righteous Jews and Nazarenes

3:113 They are not all the same, from the people of the book are an upright nation; they recite **God**'s signs during parts of the night, and they prostrate.

3:114 They affirm **God** and the Last day, promote recognized norms and deter from evil, and they hasten in goodness; these are of the reformed ones.

3:115 What they do of good will not be turned back, and **God** is aware of the conscientious.

3:116 As for those who were unappreciative, neither their money nor their children will avail them anything from **God**. These are the people of the fire, where they abide eternally.

3:117 The example of what they spend in this world is like a wind in which there is a frost; it afflicts the field of the people who wronged themselves, and thus destroys it. **God** did not wrong them, but it is they who wronged themselves.

Beware of the Hypocrites

3:118 O you who affirm, do not take protection from any besides yourselves; they will only disrupt you greatly. They wish that you suffer. The hatred is spreading from their mouths; but what their chests hide is greater! We have made clear for you the signs if you reason.

3:119 Here you love them while they do not love you, and you affirm the whole book. When they meet you they say, "We affirm," and when they are alone they bite their fingers out of frustration at you. Say, "Die in your frustration, **God** is aware of what is in the chests."*

3:120 If any good befalls you it disturbs them, and if anything, bad befalls you, they rejoice. If you are patient and aware, their planning will not harm you at all. **God** is Encompassing all they do.

The Battle of Badr

3:121 Recall when you departed from your family to prepare for those who affirmed, their stations for battle, and **God** is Hearer, Knowledgeable.

3:122 When the two parties from among you were concerned about their failure, though **God** was their ally. In **God** those who affirm should trust.

3:123 **God** had granted you victory at Badr while you had been the lesser, so be aware of **God** that you may be thankful.

3:124 When you said to those who affirm: "Is it not enough for you that your Lord would supply you with three thousand of the controllers sent down?"

3:125 Indeed, if you are patient and are aware and they come and attack you, He will supply you with five thousand of the well-trained controllers.

3:126 **God** did not give this except as good news to you, and so that your hearts may be assured with it. Victory is only from **God**, the Noble, the Wise.

3:127 Thus, He would sever a group of those who reject, or disgrace them; then they will turn back frustrated.

3:128 You will have no say in the matter, for He may pardon them, or punish them for their wickedness.

3:129 To **God** is what is in the heavens and the earth, He forgives whom He pleases, and He punishes whom He pleases; and **God** is Forgiving, Compassionate.

Do Not Exploit the Needy Through Usury

3:130 O you who affirm, do not consume usury multiplying over, and be aware of **God** that you may succeed.*

3:131 Be aware of the fire that has been prepared for the ingrates.

3:132 Obey **God** and the messenger so that you may obtain mercy.

The Good-doers

3:133 Race towards forgiveness from your Lord and a paradise whose width encompasses the width of the heavens and the earth; it has been prepared for the conscientious.

3:134 The ones who spend in prosperity and adversity, repress anger, and pardon the people; **God** loves the good doers.

3:135 If they commit any evil, or wrong themselves, they remember **God** and seek forgiveness for their sins. Who can forgive the sins except **God**? They do not persist in what they have done while they know.

3:136 To these the reward will be forgiveness from their Lord and gardens with rivers flowing underneath, eternally abiding in it. Excellent is the reward of the workers.

3:137 Many nations have come before you. So, roam the earth and see what the consequence of the liars was.

3:138 This is a clarification for the people and a guidance and advice for the conscientious.

You Will be Tested

3:139 Do not be weak, and do not grieve, for you are superior, if you have affirmed.

3:140 If you are wounded, then know that the other group is also wounded. Such are the days, We alternate them between the people, so that **God** will know those who affirm, and so He may make witnesses from among you; and **God** does not like the wicked.

3:141 **God** will refine those who affirm and He will destroy the ingrates.

3:142 Or did you calculate that you would enter paradise without **God** knowing those who would strive amongst you and knowing those who are patient?

3:143 You used to long for death before you came upon it, and now you see it right in front of you!

Muhammed Was No Different

3:144 Muhammed is but a messenger, like many messengers that have passed before him. If he dies or gets killed, will you turn back on your heels? Whoever turns back on his heels, he will not harm **God** in the least. **God** will reward the thankful.

3:145 It is not permitted for a person to die except by **God**'s leave in an appointed record. Whoever wants the rewards of this world We give him of it, and whoever wants the reward of the Hereafter, We give him of it. We will reward the thankful.

They Did Not Waver, nor Did They Lose Hope

3:146 Many a prophet had numerous devotees fighting with him. They did not waver by what afflicted them in the cause of **God**, nor did they become weak, nor did they become discouraged; and **God** loves the steadfast.

3:147 They said nothing but: "Our Lord, forgive us our sins and our shortcomings in our responsibility, and make firm our foothold, and grant us victory over the ingrates."

3:148 So **God** gave them the reward of this world and the best reward of the Hereafter; and **God** loves the good doers.

3:149 O you who affirm, if you obey those who have rejected, then they will turn you back on your heels and you will turn back as losers.

3:150 It is **God** who is your Patron, and He is the best supporter.*

3:151 We will cast fear in the hearts of those who rejected, because of what they have set up besides **God** while He never sent down any authority to do so, and their destiny is the fire. Miserable is the abode of the wicked.

The Battle of Uhud

3:152 **God** has fulfilled His promise to you, that you would overwhelm them by His leave; but then you failed and disputed in the matter and disobeyed after He showed you what you had sought. Some of you want this world, and some of you want the Hereafter. Then He let you retreat from them that He may test you; and He has pardoned you. **God** is with great favor over the affirmers.

3:153 For you were climbing the hill and would not even glance towards anyone, and the messenger was calling you from behind. Therefore, He gave you worry to replace your worry, so that you would not have sadness by what has passed you, or for what afflicted you, and **God** is Ever-aware of what you do.

3:154 Then after the worry, He sent down to you a peaceful slumber, overtaking a group of you; while another group was worried about themselves; they were thinking about **God** other than the truth, the thoughts of the days of ignorance. They say, "Why are we involved in this affair?" Say, "The entire affair is up to **God**." They hide in their persons what they do not show to you; they say, "If we had a say in this affair then none of us would have been killed here." Say, "If you were inside your own homes, then the ones who have been marked for death would have gone forth to their resting place." **God** will test what is in your chests and bring out what is in your hearts. **God** is Knowledgeable as to what is in the chests.

3:155 The day the two armies met, the devil caused some of you to turn away, because of what they had gained. **God** has pardoned them, for **God** is Forgiver, Compassionate.

3:156 O you who affirm do not be like those who rejected and said to their brothers when they were marching in the land or on the offensive: "If they were here with us they would not have died nor been killed." **God** will make this a source of grief in their hearts, and **God** grants life and death, and **God** is watching over what you do.

3:157 If you are killed in the cause of **God** or die, then forgiveness from **God** and mercy is far greater than all they can put together.

3:158 If you die or are killed, then to **God** you will be gathered.

The Good Example of the Messenger: Caring, Tolerant, Consulting People, and Trusting in God

3:159 It was a mercy from **God** that you were soft towards them; had you been harsh and mean hearted, they would have dispersed from you; so pardon them and ask forgiveness for them, and consult them in the matter; but when you are convinced, then put your trust in **God**; **God** loves those who trust Him.*

3:160 If **God** grants you victory then none can defeat you, and if He abandons you then who can grant you victory after Him? In **God,** those who affirm should put their trust.*

None is Above the Law and Justice

3:161 It was not for any prophet that he should deceit, and whoever deceits, he will be brought with what he has gained on the day of Resurrection; then every person will be given what it has earned without being wronged.

3:162 Is one who follows the pleasure of **God**, as one who draws the wrath of **God** and whose abode is hell? What a miserable destiny!

3:163 They are in different ranks at **God**, and **God** is seer of what they do.

3:164 **God** has bestowed favor upon those who affirm by sending them a messenger from amongst themselves reciting His signs, improving them and teaching them the book and the wisdom, though they were before in manifest confusion.

3:165 It was when you suffered setback. Even though you afflicted them with twice as much setback, you said, "Where is this coming from?" Say, "It is from yourselves." **God** is capable of all things.

3:166 What you suffered on the day the two armies met was by **God**'s leave, and to let those who affirm know.

3:167 To let those who are hypocrites know that they were told: "Come fight in the cause of **God** or defend," they said, "If we knew how to fight, we would have followed you." That day they were closer to rejection than they were to affirmation. They say with their mouths what is not in their hearts, and **God** knows well what they conceal.

3:168 Those who remained and said to their brothers, "If they obeyed us they would not have been killed." Say, "Then avert death away from yourselves if you are truthful!"

Death: Transition to Another Universe

3:169 Do not count that those who are killed in the sake of **God** are dead. No, they are alive with their Lord receiving provisions.*

3:170 Happy with what **God** has granted them from His favor, and they rejoice for those who have yet to follow them. There is no fear over them nor do they grieve.*

3:171 They rejoice with **God**'s blessing and bounty; **God** will not waste the reward of those who affirm.

3:172 For those who have answered **God** and the messenger after they were afflicted with wounds, and for those of them who did good and were aware is a great reward.

Monotheists are Fearless

3:173 The ones who the people said to them: "The people have gathered against you, so be fearful of them," but it only increased their affirmation and they said, "**God** is enough for us, and He is the best defender."

3:174 So they came back with a blessing from **God** and a bounty, no harm would touch them. They had followed the pleasure of **God**, and **God** is the One with great bounty.

3:175 It is only the devil trying to create fear for his allies, so do not fear them, but fear Me if you have affirmed.

3:176 Do not be saddened by those who rush into rejection. They will not harm **God** in the least. **God** does not wish to make for them any share in the Hereafter, and they will have a great retribution.

3:177 Those who have exchanged affirmation for unappreciation will not harm **God** in the least, and they will have a painful retribution.

3:178 Those who do not appreciate should not count that We are providing for them out of the goodness of themselves. We are only providing for them so that they may increase in transgression, and they will have a humiliating retribution.

3:179 **God** was not to leave those who affirm as they were without distinguishing the rotten from the good. **God** was not to let you know the future, but **God** chooses from His messengers whom He wishes; so affirm **God** and His messengers. If you affirm and be aware, then you will have a great reward.

3:180 Those who are stingy with what **God** has given them of His bounty should not count that it is good for them; no, it is evil for them. They will be surrounded by what they were stingy with on the day of Resurrection. To **God** will be the inheritance of the heavens and the earth; and **God** is Ever-aware of what you do.

3:181 **God** has heard the words of those who said, "**God** is poor and we are rich!" We will record what they said as well as their killing of the prophets without just cause. We will say, "Taste the retribution of the fire!"

3:182 This is for what their hands have brought forth, and **God** does not wrong the servants.

3:183 Those who said, "**God** has pledged to us that we should not affirm a messenger unless he brings us an offering which the fire will devour." Say, "Messengers have come to you before me with proof and with what you have said, so why did you fight/kill them if you were truthful?"*

3:184 If they reject you, then messengers before you were also rejected. They came with proof and the Psalms, and the book of enlightenment.

3:185 Every person will taste death, and you will be recompensed your dues on the day of Resurrection. Whoever escaped the fire and entered into paradise, he has indeed won. This worldly life is nothing more than the enjoyment of vanity.

3:186 We will test you with your wealth and with yourselves, and you will hear from those who have been given the book before you and from those who set up partners much annoyance. If you strive and be aware, then these are affairs of great resolve.*

Do Not Hide the Truth

3:187 **God** took the covenant of those who were given the book: "You shall proclaim to the people and not conceal it." However, they threw it behind their backs and exchange it for a petty gain. Miserable indeed is what they have purchased.

3:188 Do not count that those who are happy with what they have been given, and they love to be praised for what they did not do; do not think they are saved from the punishment. For them is a painful retribution.

3:189 To **God** is the sovereignty of heavens and earth, and **God** is capable of all things.

Study Cosmology and Astronomy

3:190 In the creation of heavens and earth, and the difference between night and day, are signs for those with intelligence.*

3:191 Those who remember **God** while standing, and sitting, and on their sides, and they ponder over the creation of the heavens and the earth: "Our Lord you did not create this without purpose, be You glorified, and spare us the retribution of the fire!"*

3:192 "Our Lord, whoever You admit to the fire has been disgraced. The wicked will have no supporters."

"We Heard the Caller"

3:193 "Our Lord, we have heard a caller inviting to the affirmation: 'affirm your Lord,' so we have affirmed. Our Lord forgive us our evil deeds and remit our sins and take us with the obedient ones."

3:194 "Our Lord, also grant us what You have promised through Your messengers, and do not embarrass us on the day of Resurrection. You do not break the promise."

3:195 Their Lord responded them: "I do not waste the work of any worker from among you, be you male or female, you are all equals. For those who emigrated and were driven-out from their homes and were harmed in My cause, and they fought and were killed, I will remit for them their sins and admit them to paradises with rivers flowing beneath; a reward from **God**; and **God** has the best reward."

3:196 Do not be deceived by the success of those ingrates throughout the land.

3:197 A brief enjoyment, then their abode is hell. What a miserable abode!

3:198 As for those who are aware of their Lord, they will have paradises with rivers flowing beneath; eternally they reside in it as a dwelling from **God**. What is with **God** is better for the righteous.

The Righteous Among The People Of The Book, Such As, Jews And Christians

3:199 Among the people of the book are those who affirm **God**, what was sent down to you and what was sent down to them. They revere **God** and they do not exchange **God**'s signs for a petty gain. These will have their reward with their Lord. **God** is quick in computation.

Support Each Other And Strive Individually And Communally

3:200 O you who affirm, be patient, call for patience, bond together, and be aware of **God** that you may succeed.*

ENDNOTES

003:001 A1L30M40. These letters/numbers play an important role in the mathematical system of the Quran based on code 19. See 74:1-56; 1:1; 2:1; 13:38; 27:82; 38:1-8; 40:28-38; 46:10; 72:28.

003:007 The Arabic word we have translated as "multiple meanings" is *mutashabihat*. The word comes from *shabaha* (to became similar), and its singular form is *mutashabih*, which means "similar," "multi-meaning," or "allegorical" (see 2:118; 2:70; 4:157; 6:99; 6:141; 2:25). The verse about the *mutashabih* (allegorical or multi-meaning) verses itself is *mutashabih*. This is one of the most commonly mistranslated verses, and it has crucial implications for understanding the Quran. See 2:106; 16:44; 17:46; 23:14; 41:44; 56:79 for examples of multi-meaning statements. Also, See 39:23.

The word can be confusing for a novice. Verse 39:23, for instance, uses *mutashabihat* for the entire Quran, referring to its overall similarity -- in other words, its consistency. In a narrower sense, however, *mutashabihat* refers to all verses which can be understood in more than one way. The various meanings or implications require some special qualities from the person listening to or reading the Quran: an attentive mind, a positive attitude, contextual perspective, the patience necessary for research, and so forth.

It is one of the intriguing features of the Quran that the verse about *mutashabih* verses of the Quran is itself mutashabih -- that is, it has multiple meanings. The word in question, for instance, can mean "similar", as we have seen; it can mean, "possessing multiple meanings"; it can also mean "allegorical" (where one single, clearly identifiable element represents another single, clearly identifiable element).

As you may have noticed, interpretation of the last part of 3:7 depends on how one punctuates the verse. (There is no punctuation in the original Arabic text.)

If one stops after the word "God", then one will assume, as centuries of Sunni and Shiite scholars have, that even those who possess deep levels of knowledge will never be able to understand the *mutashabih* verses. However, if the sentence does not stop there, the meaning will change to the opposite: Those who possess knowledge *will* be able to understand the meaning of allegorical or multi-meaning verses. For a detailed discussion on this verse, see the Sample Comparisons section in the Introduction.

003:011 The word *aya* in its singular form occurs 84 times in the Quran and, in all of the occurrences, means miracle, evidence, or lesson. However, its plural form *ayat* is used both for miracle/evidence/lesson AND for the language of revelation that entails or leads to those miracles/evidences/lessons. See 2:106.

003:014 The topic of the previous verse is the fighting armies, which were almost all male. "The people" in verse 3:14 refers to adult males who are expected to draft for military service. Verse 3:14 gives a list of weaknesses in the minds of the drafted male population to join the military for defending their community. A short-term preoccupation with those blessings, ironically, could be the cause of long-term destruction and deprivation from all.

003:018 The act of testifying to the oneness of God (*shahada*) is the essential requirement of being a Muslim. The expressions *la ilaha illa Allah* (there is no god, but the god) and *la ilaha illa Hu* (there is no god, but He) occur 30 times in the Quran and never in conjunction with another name. Assuming deficiency in the *shahada* taught by God is a sign of not valuing God as He should be valued. Requiring the addition of another name to God's name implies that God forgot to include Muhammed's name, thirty times (19:64; 6:115). Trying to teach God is the zenith of ignorance and audacity (49:16). God's oneness stands alone. With or without Muhammad, it is a logical and ontological fact. Considering God alone insufficient is the symptom of idolatry (39:45).

The only *shahada* (testimony) about the messengership of Muhammed is mentioned in verse 63:1, and those who feel the need for such a testimony are described as hypocrites. There are different reasons for why a person might be considered a hypocrite while he or she is uttering the expression *ashadu anna Muhammedan rasulullah* (I testify that Muhammed is a messenger of God). Today, the most common hypocrisy is that those who utter this phrase in fact have considered Muhammed to be much more than a messenger, since they reject his message and messengership by not following the verses of the Quran and associating volumes of fabricated narration and sectarian jurisprudence with it. Their testimony regarding Muhammed's messengership is lip service, since they consider him a god by giving him the power of intercession, the power of collaborating with God in decreeing rules for eternal salvation, the power of amending and abrogating God's law, the power of fabricating prohibitions in the name of God, and the authority of explaining the "ambiguous" words in God's book. Despite the words of Muhammed's Lord, they do not consider Muhammed a human being like them (18:110; 41:6). They do not believe what they say when they say, "Muhammed is God's servant" since they consider the title "messenger/mailman" (*rasul*) of God to be an insult. They can utter God's name without phrases of praise, but they cannot utter Muhammed's name without words of praise. Ironically, they establish their custom of praising Muhammed more than God through distorted and abused meanings of certain verses. Besides, they violate the clear Quranic instruction for not discriminating among His messengers; they put Muhammed in competition with other messengers and consider him high above other prophets and messengers. In today's Sunni mosques, unlike the *masjid* of Muhammed's and his monotheist companions' (72:18; 20:14), you will find the names of many idols smirking beside the name of God. In addition to Muhammed's name, one may find the names of Abu Bakr, Omar, Osman, Ali, Hasan, and Husayn. The Shiites have their own set of idols and they too adorn their mosques with their names.

The list of ways Sunnis and Shiites idolize Muhammed can fill an entire book. Nevertheless, they think that they are monotheists, as today's Trinitarian Christians do. Religious leaders dupe their followers by restricting the meaning of 'idols' to pictures and statues; in fact, the idols during Muhammed's time were abstract names and those *mushriks* considered themselves to be monotheists (53:19-28; 6:22-24; 6:148; 16:35). One might ask, "what about adorning the walls of *masjids* with any of the four verses where Muhammed's name is mentioned?" Well, what about hanging the verses about paradise or hell? What about picking any of the 136 verses where Moses' name is mentioned? What about Jesus? Or, how about adorning the walls of the *masjids* with verses about the hypocrites and idolatry? Of course, there is no problem in hanging the Quran upon the walls of *masjids*, but if a particular verse is picked, then the intention or context becomes important. If any verse from the Quran should be picked, I suggest 39:44/45 or 39:11-12, or any other verse that reminds us to be righteous and helpful to other people. Also, See 2:285; 3:64; 39:45; 53:23; 72:18.

Numerous archeological evidences explain how decades after the revelation of the Quran people added Muhammed's name to the original shahada. For instance, see, the gold coin from Umayyad 90 AH, or 80 years after the prophet's departure. Though eighty years after Muhammed, those who worked hard to transform islam to Muhammedanism were not yet able to change the *shahada* (testimony), by adding side notes etc., they gradually replaced the correct *Shahada* with today's most common one, the one that mentions Muhammed's name after Allah.

In the middle of one side of the coin, the expression, "There is no God; He Alone;

He has no partner" is prominent. However, the falsifiers by now were able to insert Muhammed's name in the margin by a patchy quotation from the end of verse 48:28 and the beginning of 48:29, skipping the expression "And God is sufficient as a witness." The expression "*Muhammedun Rasulullah...*" (God's messenger Muhammed...) is a fragment, not a statement, taken out of the context of verse 48:29. They have performed an interesting surgery (deletion) to be able to fit that fragment in the circle. The other side of the coin contains some words from chapter 112, emphasizing God's oneness and rejecting partners. The Umayyad and Abbasid coins later moved Muhammed's name to the center, next to God. This crucial distortion gradually took place in a period of time spanning several generations.

During the years when Muhammad was alive, God was the center, not Muhammad. Thus, there is scant historical reference to Muhammad by his contemporaries. Some opponents who assume the Sunni/Shia obsession with Muhammad to be original, use the lack of widespread reference to

Muhammad to be evidence for a fictional Muhammad. Their trust in Sunni/Shia distortions and stories justify us calling them SunniAteists or ShiaAtheists, SunniChristians, etc.

See, for instance, the Old Testament, Deuteronomy 5:1-11; 6:4-6; 1 Samuel 12:20-21; Psalms 115:4-8.

003:019 Islam is not a proper name; it is a description of the mindset and actions of those who submit themselves to God alone. Islam is the description of the message delivered by all messengers and it reached another level with Abraham (4:125). Islam: is universal (3:83), is the only valid system (3:85), accepts and utilizes diversity (49:13), promotes peace among nations (2:62; 2:135-136), promises justice to everyone (5:8), has the epistemology of requiring objective evidence besides personal experience (3:86; 2:111; 21:24; 74:30), has as its final book a scripture that is numerically coded and protected (15:9; 2:23; 83:9-20), promotes peace and reality (8:19; 60:8,9), rejects holy intermediaries and the clergy class (2:48, 9:31,34), encourages distribution of wealth against monopoly (59:7), does not tolerate an unproductive economy (5:90; 3:130), requires consultation and elections in public affairs (42:38), gives utmost value to the individual (5:32), values women and promotes their rights (3:195; 4:124; 16:97), promotes freedom of expression and tolerance (2:256; 4:140), asks us to be in harmony with nature and respect the ecosystem (30:41), and requires a scientific method utilizing rational and empirical evidences for acquiring knowledge (17:36)... In short, islam is a way of life in accordance with the natural laws and respects the social imperatives and principles dictated to us; the rational, self-interested utility-maximizers. It is also dedicating one's heart, mind, and life to the Lord of the worlds.

003:020 To claim that Muhammed was an illiterate person, the meaning of the word *UMmY* has been distorted by ignorant people as "illiterate." Muhammed was a literate and progressive leader who delivered a book that promotes reading, comprehension, learning, and studying. For a detailed discussion on this verse, see the Sample Comparisons section in the Introduction. See 7:157-158; 2:78; 25:5.

003:024 Those who upheld fabricated *hadith* as partners with the Quran have repeated the same Jewish claim. See 2:80-82. Also, see the Appendix titled: *Merciful God and Eternal Hell*?

003:026 See 57:22-23. Compare this verse to the Old Testament, Judges 1:19, where God is claimed to have been unable to drive out the Canaanites in the valleys of Palestine because they had chariots of iron. See 7:180.

003:032 Obeying the messenger means obeying the message of the messenger, which is submitting oneself to God alone. 6:112-116; 9:1.

003:036 The Arabic root of the word we translated as "the rejected" is *RaJaMa* and it means excommunication or expulsion. We are not sure when the secondary meaning of "stoning" was attached to it, but the Quranic usage of the word is different from its usage in the books of *hadith* and sectarian jurisprudence. Decades after the revelation of the Quran when rational monotheists reverted to the days of ignorance, they started using this word literally to "stone" adulterers to death, or literally to "stone" Satan during *Hajj*. The Quran describes Satan (Satan) with the word *RaJyM* (3:36; 15:17; 15:34; 16:98; 38:77; 81:25). Nevertheless, Satan is not stoned but rejected from heaven. Muslims, with the collection of *hadith* and creation of sects, about two centuries after Muhammed, fully revived the pagan Arab traditions, borrowed superstitions, and backward practices from the distorted teachings of Christianity and Judaism. For instance, while they turned Muhammed's grave into a shrine of idolatry and promoted its visitation among the practice of annual Debate Conference (*Hajj*), they also made a mockery of the phrase "rejection of Satan" by innovating the practice of stoning the Satan, as if Satan was a material being. Every year, hundreds of people are injured or killed during the mayhem caused by this absurd ritual. Some participants throw away their shoes or anything they have in their possession to inflict injury to Satan, unfortunately causing injury to their comrades. If The Satan has fun watching humans do stupid things in the name of God, this ritual is most likely to be one of the most fun. With this reversion, they imported the practice of stoning to death and named it *RaJM*, in direct contradiction to the teachings of the Quran (See 24:1-10). For other usages of the word, *RaJaMa*, see 11:91; 18:20, 18:22; 19:46; 26:116; 36:18; 44:40; 67:5.

003:041 The verse reminds us not to give in to peer pressure and cultural norms when God and truth is the issue. God provides signs to those who sincerely seek them (41:53; 22:52). I, Edip Yuksel, the author of these annotations, after accepting Quran alone on July 1st, 1986, experienced an important theological and intellectual conflict and for days asked for a sign from God. Exactly 114 days after my acceptance of the Quran, I experienced a profound divine sign on November 23rd, 1986, which saved me from a tormenting intellectual problem regarding the implication of code 19 regarding the last verses of Chapter 9. The extraordinary personal experience was followed by a chain of prophetic signs that were partially shared and witnessed by others, and finally was fulfilled on July 1st 1990. See Chapter 74 and the Appendix *On it is Nineteen*.

003:045-47 Since *Esau* (Jesus) *Maseh* (Messiah; Christ) was born through God's word "Be" and not by a father; he is called God's Word. Jesus is mentioned 25 times while his mother Mary's name is mentioned 34 times in the Quran. The Quran uses the name in various combinations of title: Jesus; Jesus, son of Mary; Messiah; Messiah, Jesus, Son of Mary; and Messiah son of Mary. To emphasize his human nature, the Quran frequently refers to Jesus as "Jesus, Son of Mary" in 16 out of 25 occurrences of which 3 are "the Messiah, Jesus, and Son of Mary." The Quran refers to Jesus as "the Messiah, Son of Mary" 5 times, and "Messiah" 3 times.

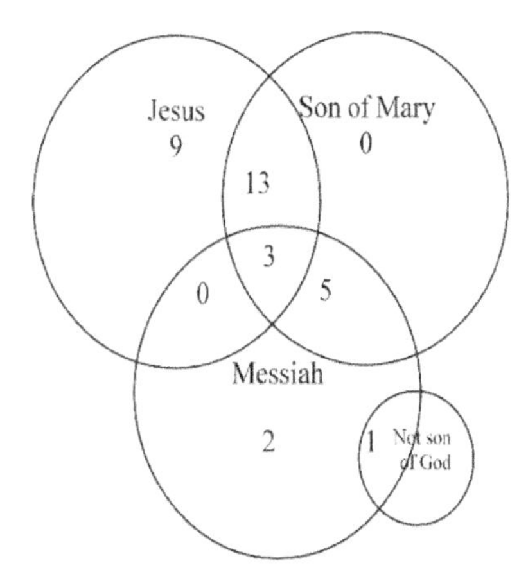

The Quran uses "Son of Mary" to refer to (J)esu(s) in 21 out of 32 total references. The Gospel authors, however, influenced by the misogynistic teachings of Rabbis and the Pharisee-son-of-Pharisee St. Paul, never call their idols with the name of the woman who gave birth to him. Actually, there might be an exception, Mark 6:3, which proves our point. There the "son of Mary" title is used by people who did not expect from wonders from him. Obviously, the authors of the Gospels and their guru St. Paul considered the title "son of Mary" to be demeaning. They transformed the original nickname "son of Mary" or "the son of woman" into "the son of man," one of the oft-used Biblical attributes for Jesus. Gospel authors, however, did not stop there. They managed to promote "the son of man" to "the son of God," and from there one step higher, infinitely higher, to "the only son of God," proving the ingenuity of Satan to carve idols in many forms and shapes. If St. Paul had a miracle, his miracle would be his contribution to the evolution of "son of Mary" to "the only begotten son of God." In the case of Christianity, Satan came up with a new formula, "three gods in one," as well as many other saint-gods, and false teachings bonus, and plenty of guilt and drama as incentive. Since then, the package containing the formula has been a great success in the religious market, where people act according to their primitive feelings rather than thoughts!

Since the Pauline god is a character of fiction whose major deed is his suffering on crucifiction, it is an embodiment of contradiction. It not only continuously oscillates between humanity and divinity, but occasionally becomes a devil (Revelations 22:16 & Isaiah 14:12), accursed of God (Deuteronomy 21:23 & Galatians 3:13 & Matthew 27:38 & Mark 15:27 & Luke 23:33 & John 19:38), infinitely patient and curser of a fig tree (1 Timothy 1:16 v Matthew 21:19), liar (John 7:8-10), a very old sheep (John 1:29 & Exodus 12:5), follower of the laws and abolisher of the laws (Matthew 5:17-19 & Luke 16:17 v Hebrews 7:18-19 & Romans 10:4), teller of everything and not teller of everything (John 15:15 v John 16:12); the peacemaker and the warmonger (John 14:27 & Acts 10:36 & Colossians 1:19-20 v Matthew 10:34), a mighty authority and a powerless man (Colossians 2:10 v John 5:30), part of a Trinity and part of a Quadrinity (Matthew 28:19 v Hebrews 7:1-3), the last and not the last (Revelation 22:13 v John 16:12-13), invisible and visible (John 1:18 v John 18:20), and even one who ends up in hell (1 Peter 3:18-20 & Matthew 25:41).

See the Quran: 25:3; 3:18. Also, see 2:59; 3:51, 55; 4:11,157,171; 5:72-79; 7:162; 19:36.

Jesus is not the original name of the prophet who brought the New Testament two millennia ago. Its original version in Hebrew is Esau, a common Jewish name at the time. In classical times it was spelled Yashua, and then it passed through a series of mutations ending up with what we have now, Jesus, a Western Latin hybrid. If it were not awkward, we would prefer writing the name as (J)esu(s). The same name can be found as Hosea, Hoshea, Jehoshuah, Jeshua, Jeshuah, Osea and Oshea (See Strong's *Exhaustive Concordance of the Bible*). These names are mentioned hundreds of times in the Old Testament. They are all different spellings of the same original expression, meaning, "Yahweh is salvation."

However, Trinitarian Christians use and abuse Matthew 1:1 to create a human-god out of transliteration inconsistencies: "She will give birth to a son, and you are to give him the name Jesus, for he will save his people from their sins." Besides, there is more than one Christ (Anointed) in the Bible. Isaiah 45:1; 1 Samuel 24:6, 10; 1Samuel 26:9-23; Numbers 3:3 are but a few examples. The New Testament refers to the Son of Mary as "Jesus of Nazareth" (John 18:7), and "Jesus the son of Joseph" (John 6:42).

Encyclopedia Americana, in its 1959 print, under the entry of Jesus provides the following information: "Although Matthew 1:21 interprets the name (originally Joshua that is Yahweh is salvation) and finds it especially appropriate for Jesus of Nazareth, it was a common one at the time. Josephus, the Jewish historian, refers to 19 different persons by that name." It is evident from St. Paul's letters, that the use of the name continued among Christians for a while (Colossians 4:11). When St. Paul's polytheistic doctrine became the official creed, Christians mostly abandoned giving this name to their children. With the reputation of Jesus and the rise of Christianity, Jews too gave up using that name, since they considered him a heretic, and the name of a new idol.

Easton provides the following information about the word Jesus: "This is the Greek form of the Hebrew name Joshua, which was originally Hoshea (Numbers 13:8, 16), but changed by Moses into Jehoshua (Numbers 13:16; 1 Corinthians 7:27), or

Joshua. After the Exile it assumed the form Jeshua, whence the Greek form Jesus."

003:051 In contradiction with the Trinitarian doctrine whose seeds were planted by St. Paul and was harvested by the Church at the Nicene conference of 325 AC, the New Testament contains many verses emphasizing the oneness of God and Jesus being at His service.

"Jesus saith unto her, Touch me not; for I am not yet ascended to my Father: but go to my brethren, and say unto them, I ascend unto my Father, and your Father; and to my God, and your God." (John 20:17)

The translation of Matthew 7:21 in The New American Bible (1970) of Catholic Biblical Association of America, reveals an important distortion by other translations:

"None of those who cry out, 'Lord, Lord,' will enter the kingdom of God; but only the one who does the will of my Father in heaven." (Matthew 7:21).

The following verses also support the information given by the Quran regarding the true message of Jesus, not polytheism but un-compromised Monotheism.

"And Jesus answered him, The first of all the commandments is, Hear, O Israel; The Lord our God is one Lord: And thou shalt love the Lord thy God with all thy heart, and with all thy soul, and with all thy mind, and with all thy strength: this is the first commandment. And the second is like, namely this, Thou shalt love thy neighbor as thyself. There is none other commandment greater than these. And the scribe said unto him, Well, Master, thou hast said the truth: for there is one God; and there is none other but he: And to love him with all the heart, and with all the understanding, and with all the soul, and with all the strength, and to love his neighbor as himself, is more than all whole burnt offerings and sacrifices." (Mark 12:29-33).

See 4:171. Also, see 2:59; 5:13-15; 4:11,157,171,176; 5:73-79; 7:162; 19:36.

003:052 Compare it to Matthew 26:56; Mark 14:50. Why would the authors of the Gospels, who were all influenced by the teachings of Paul, depict the disciples of Jesus as cowards? See 5:78 for the answer. Also, see 2:59,165; 3:45, 51, 55; 4:11; 4: 157; 4:171; 5:72-79; 7:162; 19:36.

003:055 During the fertile era of *hadith* fabrication, Christian converts (like their contemporary Jews, Arab nationalists, proponents of various Sunni and Shiite sects, and propagandists of Umayyad or Abbasid caliphs), used *hadith* as a tool to import their religious, cultural or political ideas. The later muslims not only imported them, but by attributing them to the idolized name of prophet Muhammed, baptized them and provided a powerful protective shield against time, reason, facts, and utility.

The Christian converts were successful in introducing their own fabrications regarding the "second coming of Jesus" into islam via *hadith*. Though the verse expressly uses the verb *mutaWaFeKa* (I will terminate you), many commentaries of the Quran ignore the word. The word *WaFaYa* means Complete (11:85; 12:59; 24:39), Fulfill (2:40; 2:177; 3:76; 5:1; 13:20; 22:29; 47:10; 53:37; 76:7), Recompense or Pay (3:57; 4:173; 8:60; 11:15; 11:111; 16:111; 35:30; 46:19). In 25 occurrences, however, it is used in the form *TaWaFa*, and in all of them, it means "terminate" or "put to death" or "take the conscious away." (2:234; 2:240; 3:193; 4:15; 4:97; 6:60-61; 7:37; 7:126; 8:50; 10:46; 10:104; 12:101; 13:40; 16:28; 16:32; 16:70; 22:5; 32:11; 39:42; 40.67, 40.77; 47:27).

Christ's consciousness was uploaded to heaven without possibility of return. His enemies did not crucify him but his unconscious but biologically living body. St. Paul created a pagan religion based on the absurd concept of "God sacrificing his innocent son, or His third personality, to be able to forgive our sins!" See 4:157. Also, see 2:59; 3:45, 51-52, 55; 4:11,171; 5:72-79; 7:162; 19:36.

003:059 The similarity between Jesus and Adam is supported by numbers too. Both names occur in the Quran an equal number of times, each exactly 25.

003:064 Anyone who accepts the principles mentioned in this verse is a Muslim, that is, submitter to God alone. Clergymen and religious leaders have transformed the same divine message into separate and competing religions and sects. A paradigm shift similar to the "Copernican Revolution" is needed in the realm of religion. Instead of Buddha, Jesus, Muhammed centered religions, we should unify in a God-centered system. See 4:171; 9:31. The Quran mentions major distortion in Christianity, such as claiming the exclusive right to enter Paradise (2:111-113, 120, 135), engaging in deceptive evangelical propaganda tactics (3:72), committing thievery and embezzlement (3:75), claiming to be the chosen ones (5:18), not following the laws of their own book (5:47), taking their scholars and priests to be lords besides God (9:31), and hoarding wealth instead of spending in the cause of God, such as for helping the poor.

003:071 Similar accusations and arguments were made by Jews against Jews. See Jeremiah 8:8, Deut 31:29.

003:075 Religious scholars and clergymen have transformed the meaning of the Arabic word *UMmY* from 'gentile' into 'illiterate.' The main reason for this distortion is their attempt to give more credibility to the so-called "literary miracle" of the Quran. As it can be seen in this verse, Jews and Christians were classifying the population of the region into two groups: people with divine books and people without divine books. It is unreasonable to think that in their religious struggle, Jews and Christians divided people into literate and illiterate groups. See 7:157-158; 2:78; 3:20; 25:5.

003:079 The Quran considers slavery to be polytheism. See 4:3,25,92; 5:89; 8:67; 24:32-33; 58:3; 90:13; 2:286; 12:39-42; 79:24

003:080 Those who could not call Muhammed "God's Son" came up with alternative ways of idolizing him and making him a partner with God. They added his name next to God in the statement of testimony (*shahada*); they commemorated his name in *sala* prayers; accepted him as a savior through stories of intercession; considered him infallible and free of sins; considered the fabrications attributed to him as the second source of their religion; claimed that the universe was created because of him; and gave him many exclusive divine attributes. According to the Quran, *mushriks* are ignorant and delusional; even though they have sunk deep into *shirk*, they think they are monotheists (6:23).

003:081 The word *RaSaLa*, with all its derivatives, occurs 513 times in the Quran and the word *RaSuL* (messenger) in this verse is its 19th occurrence. In other words, this verse is the 19th verse where the derivatives of the root *RaSaLa* (send message, messenger) occurs. See 33:7 and 33:40; see 72:24-28; 21:2-3; 4:164; 7:35; 40:28-44; 54:1-2; 74:1-56.

Also, See Malaki 3:1-21, Luka 17:22-36 and Matta 24:27.

003:085 Some critics claim there is a contradiction between 3:85, or 5:72 and 2:62, or 5:69. The critic assumes that peacefully surrendering to God is only possible by uttering a magic Arabic word. Islam is not a proper name. It did not start with Muhammed, nor did it end with Muhammed. Any person, regardless of the name of their religion, who dedicates himself to God alone, affirms the day of judgment and lives a righteous life, is considered Muslim (those who peacefully surrender themselves to God alone and promote peace). Many among the Christians and Jews fit this description.

003:096 It is interesting to see that the name of the city mentioned in 48:24 as Mecca is spelled differently here, as *Bakka*. Considering the role of the alphabet letters initializing this chapter, one understands the relationship of preferring a different spelling of the same name, a spelling without the letter M in a chapter that starts with A.L.M., where the number of M participates in the numerical structure based on code 19. Does the "Valley of Baca" mentioned in Psalms 84:6 refer to the same city?

003:105 See 6:159.

003:110 The verse does not limit the addressee to the era of prophet Muhammed; all monotheists, all submitters to God alone fit the description.

003:119 The word *KiTaB* is singular, but it is used for all divine books. Furthermore, since the original Quran did not contain vowels, this word might be read in its plural form *KuTuB*. All meanings are acceptable as long as they do not create contradictions within the general and specific contexts of the Quran and facts in the natural world.

003:130 See 2:275-281.

003:150 See 2:286.

003:159 The Quran provides us with good examples of Muhammed's attitude, behavior and actions. Here we learn that he was not a harsh man. In *hadith* books, however, we see numerous bad actions attributed to Muhammed. For instance, in a *hadith* narrated by many so-called "authentic" books, including *Bukhari*, Muhammed advised a group from the *Urayna* and *Uqayla* tribes to drink camel urine as medicine, and then upon learning that they killed his shepherd, he cut off their arms and legs, gouged their eyes with hot nails and left them dying under the Sun, denying his companions from giving them water. *Bukhari* and many other *hadith* collections contain many insults to Muhammed. The portrait of Muhammed in *hadith* books is a person with multiple personalities, oscillating between extremes: a merciful person and a violent torturer; a modest person and a sexual maniac; a genius and an imbecile. No wonder the Quran refers to those collections with prophetic language in 6:112-114 as devils' revelation. See 33:21.

003:160 A similar rule is also mentioned in the Old Testament. For instance, see Deuteronomy chapter 28; 2 Chronicles 24:23-25.

003:169 The death of righteous people is a pleasant journey to paradise: see 2:154; 8:24; 16:32; 22:58; 44:56; 36:26-27.

003:170-71 Prophets, those who affirm their mission, those who witness the truth, and righteous people are together in the hereafter (4:69). See Psalms 37:29, Matthew 22:32, Luke 10:25 and John 17:3.

003:183 See 5:27.

003:186 Those who prove their conviction and dedication to God alone by successfully passing through a series of tests attain happiness both in this world and in the hereafter. See 10:62; 24:55 and 29:2, 3.

003:190 The idiosyncratic language of the Quran is interesting. Why does the Quran continuously use the expression "the alteration of night and day" rather than "the setting and rising of the Sun"? See 27:88.

003:191 True monotheists remember God frequently and rejoice when God's name is mentioned alone, without the addition of any other name. Their existence and everything surrounding them reminds them of their Creator, the Lord of the universe. See 13:28; 23:84-89; 33:42; 39:45.

003:200 The teachings of some mystic orders distorted the meaning and implication of the word *Rabitu* (be connected) and fabricated a ritual that required the follower to meditate frequently by thinking of his guru and even visualizing his face in his imagination while commemorating God's name. The Quran condemns religious exploitation. See 6:21-24; 7:37; 10:17-18 and 9:31. The Quran emphases the importance of individual freedom and at the same time it reminds us of the importance of being active members of a community. Social isolation has many side effects, including rise in crime, corruption in politics, troubled children, and decline in charities. In modern societies where family structure and social fabric is harmed, loneliness contributes to unhappiness despite financial prosperity.

بســـــم الله الرحمن الرحيـــــم

4:0 In the name of God, the Gracious, the Compassionate.

4:1 O people, be aware of your Lord who has created you from one person and He created from it its mate and sent forth from both many men and women; and be aware of **God** whom you ask about, and the relatives. **God** is watcher over you.*

Polygamy For Protection Of Orphans

4:2 Give the orphans their money; do not replace the good with the bad, and do not consume their money to your money, for truly it is a great sin!

4:3 If you fear that you cannot be just to fatherless orphans, then marry those whom you see fit from the women, two, and three, and four. But if you fear you will not be fair then only one, or whom you already have contract with. So that you do not commit injustice and suffer hardship.*

4:4 Give the women their property willingly, and if they remit any of it to you of their own will, then you may take it with good feelings.

4:5 Do not give the immature ones your money, which **God** has entrusted to you. Spend on them from it and clothe them, and speak to them nicely.

4:6 Test the orphans until they reach the age of marriage, then if you determine in them sound judgment, then give them their wealth, and do not deliberately consume it wastefully or quickly before they grow up. Whoever is rich, then let him not claim anything, and if he is poor then let him consume only properly. If you give to them their wealth, then make a witness for them, and **God** is enough for Reckoning.

Inheritance Rights

4:7 For the men is a portion from what the parents and the relatives left behind, and for the women is a portion from what the parents and relatives left behind, be it little or much; an ordained portion.*

4:8 If the distribution is attended by the relatives, the orphans, and the needy, then you shall give them part of it and say to them a kind saying.

4:9 What if it was them who had left behind a weak progeny? Would they not be concerned for them? Let them revere **God** and let them say what is appropriate.

4:10 Those who consume the money of the orphans illicitly, in fact are consuming fire in their bellies, and they will end up in a blaze.

4:11 **God** directs you regarding the inheritance of your children, "To the male shall be as that given to two females. If they are only females and more than two, then they will have two thirds of what is inherited. If there is only one female, then she will have one-half, and to his parents each one of them one sixth of what is inherited if he has children. If he has no children and his parents are the heirs, then to his mother is one third; if he has siblings then to his mother is one sixth. All after a will is carried through or debt is paid off. Your parents and your children, you do not know which is closer to you in benefit, a directive from **God**, **God** is Knowledgeable, Wise."*

4:12 For you is half of what your wives leave behind if they have no children; but if they have a child then to you is one quarter of what they leave behind. All after a will is carried through or debt is paid off. To them is one quarter of what you leave behind if you have no child; but if you have a child then to them is one eighth of what you leave behind. All after a will is carried through or debt is paid off. If a man or a woman has no one, but has a brother or sister, then to each one of them is one sixth, but if they are more than this then they are to share in one third. All after a will is carried through or debt is paid off, which does not cause harm. A directive from **God**, and **God** is Knowledgeable, Compassionate.*

4:13 These are **God**'s limits, and whoever obeys **God** and His messenger, He will admit him to gardens with rivers flowing beneath, eternally abiding therein. This is the greatest victory.

4:14 Whoever disobeys **God** and His messenger, and transgresses His limits, He will admit him to a fire in which he abides eternally, and he will have a humiliating retribution.

Quarantine for Prostitutes

4:15 The women who commit lewdness, you shall bring four witnesses over them from amongst you; if they bear witness, then you shall restrict them in the homes until death takes them, or **God** makes for them a way out.*

4:16 The two who commit it from amongst you, then you shall reprimand them. If they repent and amend, then leave them alone. **God** is Redeemer, Compassionate.*

4:17 **God**'s acceptance of repentance is only for those who commit sin out of ignorance and then repent soon after; these will be forgiven by **God**, **God** is Knowledgeable, Wise.

4:18 There will be no repentance for those who commit sin until death comes upon one of them, then he says, "I repent now!" Nor there will be repentance for those who die while they are ingrates. To those We have prepared a painful retribution.*

Treat Women Nicely and Appreciate them

4:19 O you who affirm, it is not lawful for you to inherit the women by force, nor that you become harsh with them to take away some of what you have given them, unless they commit a clear lewdness. Live with them in kindness. If you dislike them, then perhaps you may dislike something and **God** makes in it much good.*

4:20 If you wish to replace one mate instead of another, and you have given one of them a large amount, then do not take anything from it. Would you take it by falsehood while it is clearly a sin?

4:21 How can you take it when you have become intimate with each other, and the women have taken from you a strong covenant?

Incest Prohibited

4:22 Do not marry what your fathers had married from the women, except what has already been done. It is lewdness, abhorrence, and a bad path.

4:23 Forbidden for you are your mothers, your daughters, your sisters, your fathers' sisters, your mothers' sisters, the daughters of your brother, and the daughters of your sister, your foster mothers who suckled you, your sisters from suckling, the mothers of your wives, and your step-daughters residing in your homes from your wives which you have already consummated the marriage with; if you have not consummated the marriage then there is no sin upon you; and the wives of your sons that are from your seed, and that you join between two sisters except what has already been done. **God** is Forgiving, Compassionate.

4:24 The chaste women, except those whom you have contractual rights, are in accordance with **God**'s decree over you. Permitted for you is what is beyond this if you are seeking with your money to be chaste and not for illicit sex. As for those whom you have already had joy with them, then you shall give them their wage as an obligation. There is no sin upon you for what you agree to after the obligation. **God** is Knowledgeable, Wise.*

4:25 Whoever of you cannot afford to marry the affirming chaste women, then from those young women whom you have contractual rights.

God is more aware of your faith; you are equals. You shall marry them with the permission of their family and give them their wage in kindness, to be protected, not for illicit sex or for taking lovers. If they become protected, then any of them who come with lewdness shall have half the punishment of what is for the chaste women of the retribution. This is for those who fear hardship from among you. But if you are patient, it is better for you, and **God** is Forgiver, Compassionate.*

4:26 **God** wants to make clear for you and guide you to the ways of those before you, and pardon you, and **God** is Knowledgeable, Wise.

4:27 **God** wants to pardon you, but those who follow their lusts want that you would be diverted into a great diversion.

4:28 **God** wants to make easy for you; and humankind was created weak.*

Avoid Major Sins

4:29 O you who affirm, do not consume your money between you unjustly, except through a trade, which is mutually agreed by you. Do not fight/kill yourselves; **God** is Compassionate towards you.

4:30 Whoever does so out of animosity and transgression, We will cast him into a fire; and this for **God** is very easy.

4:31 If you avoid the major sins that you are forbidden against, then We will cancel your existing sins and admit you to a generous entrance.

Each Gender Has Own Qualities

4:32 Do not envy what **God** has favored some of you over others. For the men is a portion of what they gained, and for the women is a portion of what they gained. Ask **God** from His favor, **God** is knowledgeable over all things.

4:33 For each We have made inheritors for what was left behind by the parents and the relatives. Those who are dependent on you, you shall give them their portion. **God** is witness over all things.

In Case of Disloyalty

4:34 The men are to support the women by what **God** has gifted them over one another and for what they spend of their money. The reformed women are devotees and protectors of privacy what **God** has protected. As for those women from whom you fear disloyalty, then you shall advise them, abandon them in the bedchamber, and leave them; if they obey you, then do not seek a way over them; **God** is High, Great.*

4:35 If you fear a split between them, then send a judge from his family and a judge from hers. If they want to reconcile, then **God** will bring them together. **God** is Knowledgeable, Ever-aware.*

4:36 Serve **God** and do not set up anything with Him, and be kind to the parents, and the relatives, and the needy, and the neighbor who is of kin, and the neighbor next door, and close friend, and the traveler, and those with whom you have contractual rights. **God** does not like the arrogant, the boastful.*

Philanthropy and God's Justice

4:37 Those who are stingy and order the people to stinginess, and they conceal what **God** has given them from His bounty. We have prepared for these ingrates a painful retribution.

4:38 Those who spend their money to show-off to the people, and they do not affirm **God** or the Last day. Whoever has the devil as his companion, then what a miserable companion!

4:39 What would bother them if they affirmed **God** and the Last day and spent from **God**'s provisions? **God** is aware of them.

4:40 Indeed, **God** does not wrong an atom's weight; and if it is good He will double it. He grants from Himself a great reward.

4:41 How is it then when We bring forth from every nation a witness, and bring you as a witness over these?*

4:42 On that day those who rejected and disobeyed the messenger will wish that the earth would swallow them; but they cannot hide anything said from **God**.

Ablution for Contact prayer (sala)

4:43 O you who affirm, do not come near the contact prayer while you are drunk, until you know what you are saying. Nor if you have had intercourse, unless traveling, until you bathe. If you are ill, or traveling, or one of you come from the bathroom, or you had sexual contact with the women, and could not find water: then you shall seek clean soil and wipe your faces and hands. **God** is Pardoning, Forgiving.*

Taking the Words Out of Context

4:44 Did you not see those who have been given a portion of the book? They purchased straying, and they want you to stray from the path.

4:45 **God** is fully aware of your enemies; and **God** is enough as an ally, and **God** is enough as a supporter.

4:46 From amongst the Jews there are those who take the words out of context, and they say, "We hear and disobey, and listen but let not any listen, and shepherd us," in a twisting of their tongues and as a mockery of the (peacemaking) system! Had they said, "We hear and obey, and listen, and watch over us," it would have been better for them and more upright; but **God** has cursed them for their rejection, they do not affirm except very little.*

4:47 O you who have received the book, affirm what We have sent down authenticating what is with you, before We cast down faces and turn them on their backs or curse them as the people of the Sabbath were cursed. **God**'s will is always done.

The Unforgivable Sin

4:48 **God** does not forgive that any partner be set up with Him, and He forgives what is beside that for whom He wills. Whoever sets up partners with **God** has indeed invented a great sin.*

4:49 Did you not see those who exalt themselves? No, it is **God** who exalts whom He wills, and they will not be wronged in the least.

4:50 See how they invent upon **God** the lies; is that not enough as a clear sin?

4:51 Did you not see those who were given a portion of the book, they trust superstition and aggression, and they say of the ingrates, "Those are better guided than these who affirmed the path."

4:52 These are the ones whom **God** has cursed, and whomever **God** curses, you will not find for him a victor.

4:53 Or would they have a portion of the sovereignty? If so, then they would not give the people the groove of a datestone.

4:54 Or do they envy the people for what **God** has given them of His bounty? We have given the descendants of Abraham the book and the wisdom; We have given them a great power.

4:55 Some of them affirmed it, and some of them turned from it. Hell suffices as a flame.

The Allegory of hell and Heaven

4:56 Those who have rejected Our signs, We will admit them to a fire. Every time their skins are scorched, We replace them with other skin that they may taste the retribution. **God** is Noble, Wise.*

4:57 Those who affirm and promote reforms, We will admit them to gardens with rivers flowing beneath, eternally they abide therein. In it they will have pure spouses, and We will admit them to a vast shade.

Elect/Appoint Qualified People and Observe Justice

4:58 **God** orders you to delegate the responsibilities to those who are qualified. If you judge between the people, then you shall judge with justice. It is always the best that **God** prescribes for you. **God** is Hearer, Seer.*

4:59 O you who affirm, obey **God** and obey the messenger and those entrusted amongst you. But if you dispute in any matter, then you shall refer it to **God** and His messenger, if you affirm **God** and the Last day. That is better and a more suitable solution.*

4:60 Did you not see those who claimed they affirmed what was sent down to you and what was sent before you? They wanted to seek judgment through aggression, while they were ordered to reject it. It is the devil who wants to lead them astray.

4:61 If they are told, "Come to what **God** has sent down and to the messenger," you see the hypocrites turning away from you strongly.

4:62 Why then, when a tragedy befalls them for what their hands have brought forth do they come to you swearing by **God** that they only wanted to do good and reconcile?

4:63 These are a people whom **God** knows what is in their hearts, so do not mind them; advise them, and speak to their persons with a clear saying.

Obey the Messenger; Accept the Rule of Law

4:64 We do not send a messenger except to be obeyed by **God**'s leave. Had they come to you when they had wronged themselves and sought **God**'s forgiveness, and the messenger sought forgiveness for them, they would have then found **God** to be Pardoning, Compassionate.*

4:65 No, by your Lord, they will not affirm until they make you judge in what they dispute with each other, then they will not find in their persons any animosity for what you have decided, and they will yield completely.*

4:66 Had We decreed for them, "Face yourselves," or "Leave your land," they would not have done so except for a few of them. If they had done what they were advised with, it would have been better for them and helped to strengthen them.*

4:67 Then We would have given them from Us a great reward.

4:68 We would have guided them to a Straight Path.

Together in Eternity

4:69 Whoever obeys **God** and the messenger will be among those whom **God** has blessed from the prophets and the truthful and the martyrs and the reformed. What an excellent companionship!*

4:70 That is the bounty from **God**; and **God** suffices as a Knower.

4:71 O you who affirm, take your precaution by going out in clusters, or going out altogether.

4:72 From among you will be those that would lag behind, so that if disaster afflicts you he would say, "**God** has blessed me that I did not witness the event with them!"

4:73 If favor from **God** benefits you, he will speak as if there had been affection between you and him, "Oh, I wish I had been with them so I would win a great prize."

4:74 Let those who seek to trade this world to gain the Hereafter fight in the cause of **God**. Whoever fights in the cause of **God** and is killed or attains victory, then We will grant him a great reward.*

Monotheists Side with the Oppressed

4:75 Why do you not fight in the cause of **God**, when the oppressed amongst the men and women and children say, "Our Lord, bring us out of this town whose people are wicked, and grant us from Yourself an Ally, and grant us from Yourself a Supporter!"

4:76 Those who affirm, fight in the cause of **God**, while those who reject, fight in the cause of aggression; so fight the supporters of the devil, for the planning of the devil is weak.

4:77 Did you not see those who were told, "Restrain yourselves, and observe the *contact prayer*, and contribute towards betterment." But when fighting was decreed for them, a group of them feared the people as much as they feared **God** or even more so. They said, "Our Lord, why did You decree fighting for us? If only You would delay for us till another time." Say, "The enjoyment of this world is little, and the Hereafter is far better for those who are aware; you will not be wronged in the least."

4:78 Wherever you may be, death will find you, even if you are in fortified towers. If any good befalls them, they say, "This is from **God**," and if any bad befalls them, they say, "This is from you!" Say, "All is from **God**;" what is wrong with these people, they barely understand anything said!

4:79 Any good that befalls you is from **God**, and any bad that befalls you is from yourself. We have sent you as a messenger to the people and **God** is enough as a witness.*

4:80 Whoever obeys the messenger has obeyed **God**; and whoever turns away, We have not sent you as a guardian over them.

4:81 They say, "Obedience," but when they emerge from you a group of them prepares for other that what you have said, and **God** records what they planned. So turn away from them and put your trust in **God**. **God** is enough for your trust.

Neither Internal Nor External Contradictions in the Quran

4:82 Do they not reflect on the Quran? If it were from any other than **God** they would have found many contradictions in it.*

When You are Attacked, Do not Panic and Do Not Spread Rumor

4:83 If any matter regarding security, or fear, comes to them they make it publicly known, but if they had referred it to the messenger and to those entrusted from them then it would have been known by those who studied it from them. Had it not been for **God**'s grace upon you and His mercy, you would have followed the devil, except for a few.

4:84 So fight in the cause of **God**. You are not responsible except for yourself. Enjoin those who affirm, "Perhaps **God** will put a stop to the might of those who are ingrates." **God** is far Mightier and far more Punishing.

4:85 Whoever intercedes with a good intercession, he will have a reward of it; and whoever intercedes with an evil intercession, he will receive a share of it. **God** has control over all things.

Promote Friendship and Peace

4:86 If you are greeted with a greeting, then return an even better greeting or return the same. **God** is Reckoning over all things.

4:87 **God**, there is no god but He. He will gather you for the day of Resurrection in which there is no doubt. Who is more truthful in saying than **God**?

4:88 What is the matter with you that you are divided into two groups over the hypocrites, while **God** has allowed them to regress for what they have earned? Do you want to guide those whom **God** misguides? Whoever **God** causes to be misguided, you will never find for him a way.

4:89 They hope that you would reject as they rejected, then you would be the same. Do not take any of them as allies until they emigrate in the cause of **God**. If they turn, then take them and fight/kill them where you find them; and do not take from them any ally or supporter;*

4:90 Except for those who join a people between whom you have a covenant, or those who come to you with reluctance in their chests to fight you or to fight their own people. Had **God** willed He would have given them strength and they would have fought you. But if they retire from you, and did not fight you, and they offer you peace; then **God** does not make for you a way against them.

4:91 You will find others who want to be safe amongst you and safe amongst their own people. Every time they are returned to the test, they fall back in. If they do not withdraw from you, offer you peace, and restrain their hands, then you shall take them and kill them where you encounter them. For these We have given you a clear authority.

Do Not Kill! Monetary Compensation in Cases of Manslaughter

4:92 Those who affirm cannot kill another who has also affirmed except by accident. Whoever kills one who affirmed by accident, then he shall set free an affirming slave, and give compensation to the family; except if they remit it. If he was from a people who are enemies to you, and he had affirmed, then you shall set free an affirming slave. If he was from a people between whom you had a covenant, then compensation to his family, and set free an affirming slave. Whoever does not find, then the fasting of two months sequentially as repentance from **God**; **God** is Knowledgeable, Wise.*

4:93 Whoever kills one who affirmed intentionally, then his reward shall be hell, eternally abiding therein; **God** will be angry with him, curse him, and for him is prepared a great retribution.

Do Not Reject the Offer of Peace

4:94 O you who affirm, if you mobilize in the cause of **God** then make a clear declaration, and do not say to those who greet you with peace, "You are not a affirmer!" You are seeking the vanity of this world; but with **God** are many riches. That is how you were before, but **God** favored you, so make a clear declaration. **God** is Ever-aware of what you do.

4:95 Except the disabled, not equal are those who stayed behind from those who affirm with those who strived in the cause of **God** with their money and lives. **God** has preferred those who strive with their money and lives over those who stayed behind; and to both **God** has promised goodness; and **God** has preferred the strivers to those who stay by a great reward.*

4:96 Grades from Him, forgiveness, and a mercy. **God** is Forgiving, Compassionate.

Either Fight Against Oppression or Seek Asylum Elsewhere

4:97 Those whom the controllers take, while they had wronged themselves; were asked, "What situation were you in?" They responded, "We were oppressed on earth." They asked, "Was **God**'s earth not wide enough that you could emigrate in it?" To these their abode will be hell; what a miserable destiny;

4:98 Except for those men, women, and children who were oppressed and could not devise a plan nor could be guided to a way.

4:99 For these, perhaps **God** will pardon them. **God** is Pardoner, Forgiving.

4:100 Whoever emigrates in the cause of **God** will find in the land many spoils and a bounty. Whoever leaves his home emigrating to **God** and His messenger, then is overcome by death; his reward has fallen to **God**, and **God** is Forgiving, Compassionate.

Keep Contact With God Even During War

4:101 If you are mobilized in the land, then there is no harm that you shorten the contact prayer, if you fear that the ingrates will try you. The ingrates are to you a clear enemy.*

4:102 If you are with them and maintain the contact prayer for them, then let a group from amongst them stand with you and let them bring their weapons; when they have prostrated then let them stand guard from behind; let a group who has not yet contacted come and contact with you, and let them be wary and let them bring their weapons with them. The ingrates hope that you would neglect your weapons and goods so they can come upon you in one blow. There is no sin upon you if you are impeded by rainfall, or if you are ill, that you keep from placing down your weapons. Be wary. **God** has prepared for the ingrates a humiliating retribution.

4:103 So when you are done with the contact prayer, then remember **God** while standing, or sitting, or on your sides. When you are relieved, you shall maintain the contact prayer; the contact prayer for those who affirm is a scheduled event.*

4:104 Do not falter in the pursuit of the remaining group. If you are feeling pain, then they are also feeling pain as you are; and you seek from **God** what they do not seek. **God** is Knowledgeable, Wise.

4:105 We have revealed to you the book with truth that you may judge between the people according to what **God** has shown you, and do not be an advocate for the treacherous.

4:106 Seek forgiveness from **God**; **God** is Forgiver, Compassionate.

The Betrayers

4:107 Do not argue on behalf of those who betray themselves. **God** does not like those who are betrayers, sinners.

4:108 They may conceal this from the people, but they do not conceal it from **God**, and He was with them when they schemed and said the things He does not approve. **God** is Encompassing over what they do.

4:109 Here you are arguing on their behalf in this world, but who will argue on their behalf with **God** on the day of Resurrection? Or who will be their sponsor?

4:110 Whoever does any evil, or wrongs himself, then seeks **God**'s forgiveness; he will find **God** Forgiving, Compassionate.

4:111 Whoever earns any sin, it is he who has brought it on himself. **God** is Knowledgeable, Wise.

4:112 Whoever does a mistake or wrongdoing, then blames it on an innocent person; he has incurred falsehood and a clear sin.

Do not Divert from the Law and Wisdom

4:113 Had it not been for **God**'s favor upon you and His mercy, when a group of them were insistent on misguiding you; they would not have misguided except themselves, nor would they harm you in anything. **God** has sent down to you the book and the wisdom, and He has taught you what you did not know. **God**'s grace upon you is great.

4:114 There is no good in most of their confidential talk, except whoever orders a charity or kindness or reconciliation between the people. Whoever does this seeking **God**'s favor, We will give him a great reward.

4:115 Whoever is hostile to the messenger after the guidance has been made clear to him, and he follows other than the path of those who affirm; We will grant him what he has sought and deliver him to hell; what a miserable destination.

The Unforgivable Sin

4:116 **God** does not forgive to have partners set up with Him, but He forgives other than that for whom He pleases. Whoever sets up partners with **God** has indeed strayed a far straying.*

4:117 They are only calling on females beside **God**. Indeed, they are only calling on a persistent devil.

4:118 **God** has cursed him; and he had said, "I will take from Your servants a sizeable portion."*

4:119 "I will misguide them and make them desire, and I will command them, so that they will mark the ears of the livestock, and I will command them so they will make change to **God**'s creation." Whoever takes the devil as a supporter other than **God**, then he has indeed lost a great loss.*

4:120 He promises them and makes them desire, but what the devil promises them is only vanity.

4:121 For these, their abode shall be hell; they will find no escape from it.

4:122 As for those who affirm and promote reforms, We will admit them into gardens with rivers flowing beneath, eternally they will abide therein. **God**'s promise is truth; and who is more truthful in saying than **God**?

4:123 It will be neither by what you desire, nor by what the people of the book desire. Whoever works evil, he will be paid by it; and he will not find for himself besides **God** any supporter or victor.

4:124 Whoever works good whether male or female, and is an affirmer, then these will be admitted to paradise, and they will not be wronged as much as the tiny groove of datestone.

Abraham: God's Friend

4:125 Who is better in the system than one who peacefully surrenders himself to **God**, and is a good doer, and he followed the creed of Abraham in monotheism? **God** took Abraham as a close friend.*

4:126 To **God** is what is in the heavens and in the earth; **God** is Encompassing over all things.

Support Orphans and their Mothers

4:127 They ask you for divine instruction concerning women. Say, "**God** instructs you regarding them, as has been recited for you in the book about the rights of orphans whose mothers you want to marry without giving them their legal rights. You shall observe the rights of powerless children, and your duty to treat orphans with equity. Whatever good you do, **God** has full knowledge of it.*

4:128 If a woman fears from her husband disloyalty, or estrangement, then there is no sin for them to reconcile between themselves; and reconciliation is good. The persons are brought by need. If you are kind and aware, then **God** is Expert over what you do.

4:129 You will not be able to be fair regarding the women even if you make every effort; so do not sway too greatly and leave her as one hanging in a void. If you reconcile and be aware, then **God** is Forgiving, Compassionate.*

4:130 If they separate, then **God** will provide for each of them from His bounty. **God** is Vast, Wise.

Trust in God

4:131 To **God** is what is in the heavens and the earth; and We have recommended to those who were given the book before you, and you, to be aware of **God**. If you reject, then to **God** is all that is in the heavens and in earth; **God** is Rich, Praiseworthy.

4:132 To **God** is all that is in the heavens and all that is in the earth; and **God** is enough as a Caretaker.

4:133 O you people, if He wills, He could make all of you cease to exist; then He would bring others in your place. **God** is most able to do this.

4:134 Whoever seeks the reward of this world: with **God** is the reward of this world and the Hereafter. **God** is Hearing, Watchful.

Stand for Justice and Truth

4:135 O you who affirm, stand with justice as witnesses to **God**, even if it is against yourselves, or the parents or the relatives, rich or poor, **God** is more worthy of them, so do not follow your desires from being just. If you twist or turn away, then **God** is Ever-aware of what you do.

4:136 O you who affirm; affirm **God** and His messenger, and the book which was sent down to His messenger, and the books that were sent before. Whoever rejects **God**, and His controllers, and His book, and His messengers, and the Last day; then he has strayed a far straying.

4:137 Those who affirm, then reject, then affirm, then reject, then they increase in rejection; **God** was neither to forgive them nor to guide them to the path.

4:138 Give news to the hypocrites that they will have a painful retribution.

4:139 They are the ones who ally themselves with the ingrates besides of those who affirm. "Do they seek glory with them?" All glory belongs to **God**.

When God's Signs/Revelation Were Insulted or Mocked in Your Presence

4:140 It has been sent down to you in the book, that when you hear **God**'s signs being rejected and ridiculed in, then do not sit with them until they move on to a different subject; if not, then you are like them. **God** will gather the hypocrites and the ingrates in hell all together.*

4:141 They linger and observe you. If you have a victory from **God** they say, "Were we not with you?" If the ingrates have success, they say, "Did we not side with you and deter those who affirm from you?" **God** will judge between you on the day of Resurrection, and **God** will not grant the ingrates any success over those who affirm.

The Hypocrites

4:142 The hypocrites seek to deceive **God**, while He is deceiving them; and if they attend to the contact prayer, they do so lazily, only to show the people; they do not remember **God** except very little.

4:143 They are swaying in-between, belonging neither to this group nor to that group. Whomever **God** will misguide, you will not find for him a way.

4:144 O you who affirm, do not take the ingrates as allies instead of those who affirm. Do you want **God** to have a reason against you?

4:145 The hypocrites will be in the lowest level of the fire; and you will not find for them a supporter;

4:146 As for those who repent, amend, hold fast to **God**, and devote their system to **God**; these will be with those who affirm. **God** will grant those who affirm a great reward.

4:147 What would **God** want with your punishment if you only appreciated and affirmed? **God** is Appreciative, Knowledgeable.

Do Not Publicize Personal Negative Events Unless They Cause Injustice

4:148 **God** does not like that any negative sayings be publicized, except if one is wronged. **God** is Hearer, Knowledgeable.

4:149 If you reveal what is good or hide it, or forgive what is bad, then **God** is Pardoner, Omnipotent.

Messengers Do Not Follow Different Teachings than the Teaching of God

4:150 Those who rejected **God** and His messengers, and they want to make a distinction between **God** and His messengers, and they say, "We affirm some and reject some!", and they desire to take a path in-between.*

4:151 These are the true ingrates; and We have prepared for the ingrates a humiliating retribution.

4:152 Those who affirm **God** and His messengers and do not make a distinction between any of them, We will give them their rewards. **God** is Forgiving, Compassionate.

Covenant With the Children of Israel

4:153 The people of the book ask you to bring down to them a book from the heavens. They had asked Moses for even more than that, for they said, "Let us see **God** plainly!", so the lightning bolt took them for their wickedness. Then they took the calf after the proof had come to them, and We pardoned them for this; We gave Moses a clear authority.*

4:154 We raised the mount to be above because of the covenant they took, and We said to them, "Enter the passage by prostrating." We also said to them, "Do not transgress the Sabbath;" and We took from them a solemn covenant.

Crucifixion

4:155 So, for the breaking of their covenant, and their rejection of **God**'s signs, and their killing of the prophets without justice, and their saying, "Our hearts are uncircumcised" Indeed, **God** has stamped upon their hearts because of their rejection; they do not affirm, except for a few.*

4:156 For their rejection and them saying about Mary a great falsehood,

4:157 For their saying, "We have killed the Messiah Jesus the son of Mary, the messenger of **God**!" They did not kill him, nor did they crucify him, but it appeared to them as if they had. Those who dispute this are in doubt of him, they have no knowledge except to follow conjecture; they did not kill him for a certainty.*

4:158 Instead, **God** raised him to Himself; and **God** is Noble, Wise.

4:159 Among the people of the book are few who would have affirmed him before his death, and on the day of Resurrection he will be witness against them.*

4:160 Because of the wickedness from those who are Jews, We made forbidden to them the good things that were lawful to them, and for their deterring many from the path of **God**.

4:161 For practicing usury/interest when they were told not to, and for consuming people's money unjustly. We have prepared for the ingrates amongst them a painful retribution;*

4:162 But those of them who are firm in knowledge, as well as those who affirm, they affirm what was sent down to you and what was sent down before you; and those who observe the contact prayer, and those who contribute towards betterment, and those who affirm **God** and the Last day; to these We will give them their reward greatly.

Messengers of God

4:163 We have revealed you as We have revealed Noah and the prophets after him. We revealed Abraham, and Ishmael, and Isaac, and Jacob, and the Patriarchs, and Jesus, and Job, and Jonah, and Aaron, and Solomon; and We have given David the Psalms.

4:164 Messengers of whom We have narrated to you from before, and messengers We have not narrated to you; and **God** spoke to Moses directly.

4:165 Messengers who were bearers of good news and warners, so that there will be no excuse for the people with **God** after the messengers. **God** is Noble, Wise.

4:166 But **God** bears witness for what He has sent down to you with His knowledge, and the controllers bear witness; and **God** is enough as a witness.*

4:167 Those who have rejected and turned away from the path of **God**, they have strayed a far straying.

4:168 Those who have rejected and did wrong, **God** was not to forgive them, nor guide them to a path;

4:169 Except to the path of hell, in it they will abide eternally. For **God** this is very easy.

4:170 O people, a messenger has come to you with truth from your Lord, so affirm; that is better for you. If you depreciate, then to **God** is what is in the heavens and the earth. **God** is Knowledgeable, Wise.

Trinity: a Polytheistic Fiction

4:171 O people of the book, do not overstep in your system, nor say about **God** except the truth. Jesus the son of Mary was no more than **God**'s messenger and the fulfillment of His word to Mary, and an inspiration from Him. So affirm **God** and His messengers, and do not say, "Trinity." Cease, for it is better for you. **God** is only One god, be He glorified that He should have a son! To Him is all that is in the heavens and what is in the earth. **God** is enough as a caretaker.*

4:172 The Messiah would not be too proud to be a servant to **God**, nor would the controllers who are close to Him. Whoever is too proud from His service, and is arrogant, then He will summon them all before Him.

4:173 As for those who affirm and promote reforms, He will give them their rewards and increase for them from His bounty. As for those who are too proud and arrogant, He will punish them a painful retribution, and they will not find besides **God** any supporter or victor.

4:174 O people, a proof has come to you from your Lord, and We have sent down to you a guiding light.

4:175 As for those who affirm **God** and hold fast to Him, He will admit them into a mercy from Him and a bounty, and He will guide them to Himself, a Straight Path.

Relating Back to the Beginning

4:176 They seek a ruling from you, say, "**God** gives you the ruling for those who have no descendants. If a person passes away and has no children but has a sister, then she shall receive half of what he leaves behind. He will inherit her fully if she has no child. However, if he has two sisters, then they will receive two thirds of what he left behind; and if he has siblings, male and female, then the male shall receive twice what the female receives." **God** makes clear to you that you do not stray; **God** is aware of all things.

ENDNOTES

004:001 Male and female, with little difference, share the same genetic program. The creation of the female from Adam's ribs found in Genesis 2:21-22 does not fit the Quranic description of creation. There is most likely a misunderstanding or deliberate distortion of an original word. On the Biblical account of the creation of Eve, the commentator Matthew Henry makes a beautiful comment, which might give us insight into the original meaning of the text: "This companion was taken from his side to signify that she was to be dear unto him as his own flesh. Not from his head, lest she should rule over him; nor from his feet, lest he should tyrannize over her; but from his side, to denote that species of equality which is to subsist in the marriage state." Ironically, many misogynistic Christian scholars interpreted the Biblical account of Eve's creation to belittle and condemn women. For instance, the 19th century Scottish Presbyterian scholar Easton provides the following information on the verse: "Through the subtle temptation of the serpent she violated the commandment of God by taking of the forbidden fruit, which she gave also unto her husband (1Timothy 2:13-15; 2Corinthians 11:3). When she gave birth to her first son, she said, "I have gotten a man from the Lord" (R.V., "I have gotten a man with the help of the Lord," Genesis 4:1). Thus she welcomed Cain, as some think, as if he had been the Promised One the 'Seed of the woman.'" See 2:36; 49:13.

004:003 Polygamy is allowed only to provide psychological, social, and economic support for widows with orphans (See 4:127). Muhammed's practice of polygamy must have been in accordance with the condition to serve an important social service. Sure, physical attraction of widows might be one of the factors for marriage and there is nothing wrong with that. Those who could afford practicing polygamy, mentally and financially, should try hard to treat them equally, though 4:129 expresses the practical impossibly of attaining that ideal. Additionally, the consent of the first wife is necessary; otherwise, she can always seek divorce. Polygamy is not an ideal form of marriage and an unusual practice allowed for difficult times, such as a dramatic reduction in the male population during wartime. The age gap between marrying men and women creates a surplus of women who will never be able to find a monogamous partner. By a strict prohibition on polygamy, millions of young women are deprived from having a legitimate relationship with men. The only hope for millions of young girls is to get married with already divorced men, perhaps with kids, or to have a relationship born out of promiscuous sexual practices. The Western world does not prohibit polygamy since many males have sexual relationships with more than one woman at the same time. The only thing that modern societies do is deprive those women from the protection of law; they are there to be used, disposed of and recycled by men!

The hypocrisy in the modern attitude becomes clear when homosexuality is defended on the pretext of "consenting adults," but the same standards are not afforded to the polygamists. The conditional permission for polygamy is for the psychological and financial protection of children and their widow mothers, in cases of war or natural disaster.

Polygamy, according to the Old Testament, started with the seventh generation after Cain and continued as a common practice in the patriarchal age together with having concubines (Genesis 4:19; 6:2; 16:1-4; 22:21-24; 28:8-9; 29:23-30, etc.). However, the Old Testament also disapproves of polygamy (Deuteronomy 17:17).

The Old Testament contains numerous exaggerated stories. One is about the number of Solomon's wives. The roundness of the numbers of wives and concubines and their total, the three numbers being perfectly round, indicate an intentional exaggeration. "And he had seven hundred wives, princesses, and three hundred concubines: and his wives turned away his heart" (1 Kings 11:3). It is highly plausible that the word "hundred" was inserted in the text by later scribes to damage the reputation of Solomon for some political agenda. The following verses depict Solomon as an evil person and idolater. The Quran neither accuses Solomon of indulging in a hedonistic sexual life nor of associating partners with God. Ironically, modern Christians are now bashing Solomon for not sticking to monogamy. To be politically correct, modern Christians do not hesitate to condemn the common practice of polygamy among Jews and their prophets.

Contrary to the Quran, which exhorts muslims to help widows, the misogynistic Rabbinical teachings inserted into the Old Testament put them in the category of harlots, and finds them unworthy of marriage by the privileged class: priests (Leviticus 21:14).

The expression *Ma malakat aymanukum* has been translated by most translations as "whom your right hands possess" or "captives" or "concubines." (See the note 23:6). We translated this and similar expressions found in 4:3,24,25,36; 16:71; 23:6; 24:31,33,58; 30:28; 33:50,52,55; and 70:30, as "those with whom you have contractual rights." An example of this category is the wives of the enemy combatants who were persecuted because they affirmed the message of Peacemaking and sought asylum in the Peacemaker community (60:10). Since they did not get through a normal divorce process, an exceptional contract allows them to marry monotheists as free women. Marrying them could create some social, economic, and personal complications for the husband. They have nothing to do with *IBaD* (slaves), as sectarian translations and commentaries state. As we will learn, the Quran categorically rejects slavery and considers it to be the greatest sin (See 3:79; 4:25,92; 5:89; 8:67; 24:32-33; 58:3; 90:13; 2:286; 12:39-42; 79:24). The practice of slavery was justified and resurrected to a certain extent via the influence of Jewish and Christian scholars, as well as fabricated *hadith* and *sharia* laws, decades after Muhammed's departure. Another category of "those with whom you have contractual rights" are those whom you have committed relationship, yet you do not have formal marriage contract. In western legal system this is recognized as "common law marriage."

It is ironic that Jews who suffered the most from slavery and were saved by God through the leadership of Moses (Exodus 1:13-14), later justified enslaving other people, including selling one's own daughter, and inserted that practice into their holy books (Exodus 21:7-8; 21:21-22; 26-27; Leviticus 25:44-46; Joshua 9:6-27).

Though Jesus never condoned slavery, St. Paul, the founder of modern Christianity, once asked the masters to treat their slaves nicely (Colossians 3:22), asked the slaves to be "submissive to your masters with all fear" (1 Peter 2:18; Ephesians 6:5; 1 Timothy 6:2; Colossians 3:22; Titus 2:9) justifying the Marxist maxim, "Religion is the opium of masses." The use of religion by the privileged class to enslave or exploit people is vividly depicted by the South African archbishop Desmond Tutu: "When the missionaries came to Africa, they had the Bible, and we had the land. They said, 'let us close our eyes and pray'. When we opened them, we had the Bible, and they had the land."

The word *YaMYN* means "right hand" or metaphorically "right," "power" or "control." However, its plural form *aYMaN* is consistently mentioned in the Quran to mean not "right hands" but to mean "oaths" or "promises," implying the mutual nature of the relationship (See 4:33 5:89; 9:12; 16:91-94; 2:224-225; 30:28; 66:2; 5:53; 6:109). This unique Quranic usage is akin to the semantic difference between the singular and plural forms of the word *Ayat* (signs) (see 2:106).

The expression in question, thus could be translated as "those whom your oaths/contracts have rights over" or "those whom you hold rights through your contracts," or by reading *aYMaN* (oaths/contracts) as an object rather than a subject, "those who hold/possess your contracts."

The marriage declaration is a mutual partnership between two sexes and is formed by participation of family members. A married woman cannot marry another man without getting divorced from her husband. However, if a woman escapes and joins peacemakers while her husband stayed behind participating in a war against peacemakers, she may marry a Muslim man without actually getting divorced from her combatant husband; she will legally be considered a divorcee (60:10). Since this contract is different from the normal marriage contract, this special relationship is described in different words. The same is valid for a man whose wife allies with the hostile enemy. See 24:31 and 33:55. Those who work for another person according to employment contracts are also referred to with the same expression. See 16:71; 30:28. Also, see 4:25,36; 23:6; 24:58; 33:50; 33:52; 70:30.

The Quran does not demand those who lived together based on a mutual promise (*AYMaN*) during the days of ignorance, without a marriage contract, to get divorced. Similarly, it does not want those who married two sisters before accepting peacemaking (islam) to be a way of life (4:23). This tolerance does not encourage living together without marriage. It only does not want to incur further damage to the family structure and does not want to create hurdles for those who wish to live according to the principles of islam, that is peacemaking.

004:007 Compare this verse to the Old Testament, Numbers 27:8-11.

004:011 In this and following verses and from 2:180 we learn that priority is given to the will in the distribution of inheritance. According to these verses, first the debt is paid and the distribution according to the will is fulfilled. This Quranic rule allows the testator and the testatrix to adjust their will depending on specific conditions and needs of the inheritors or other personal issues. For instance, one may leave more inheritance to a daughter who might be more in need than the others might. The testator or the testatrix might leave more to someone who is sick or handicapped. The Quran, by giving precedence to the will over the default distribution provides flexibility and thus accommodates special circumstances. Unfortunately, unable to comprehend the wisdom behind this divine arrangement, the followers of *hadith* and *sunna* have abrogated these verses via *hadith* fabrications and sectarian rules and thereby have deprived so-called Muslims of God's mercy.

004:012. After paying the debt and distributing the shares according to the will, the inheritance will be distributed if the diseased had a mother, father, or wife, and the rest will be distributed to men and women according to the instructed ratio. It is important to remember that the ratio of shares instructed by the Quran to each other is as important as their ratio to the whole. In brief, the fractions are compared both to the whole and to each other. A deceased person might leave behind hundreds of combinations of relatives, father or no father, mother or no mother, son or no son, one son or more sons, daughter or no daughter, one daughter or more daughters, brother or no brother, one brother or more brothers, sister or no sister, one sister or more sisters, and numerous combinations among them. If the Quran suggested the fractions as merely their ratio compared to the whole inheritance, since for each combination of inheritors the ratios too would need to be changed, we would need hundreds of verses detailing the ratio for each combination. However, when we comprehend that the ratio in suggested distributions also reflect the amount of inheritance in proportion to each other, the verses about inheritance can easily be understood and implemented.

For instance, let us assume that after deducting the debt and other shares, a man left a 50-thousand-dollar inheritance to his wife and father without specially allocating the shares. Their shares should be $1/4^{th}$ and $1/6^{th}$, respectively. Thus, the formula would be (1/4)+(1/6)=50,000. If we find the common denominator, then we see the ratio of inheritance as much clearer. When we take the common denominator, the numerator start to make sense in comparison to the share of each inheritor and to the inheritance: (3/12)+(2/12)=50,000. That means, the wife receives $30,000 and the father receives $20,000 of the inheritance. In sum, to learn the default shares of inheritors the fractions indicating the shares are added up and equated to the inheritance. Let's assume that we are asked to distribute 19 gold pieces among two people; one will receive $1/5^{th}$ and the other $1/3^{rd}$.

$(k1 + k2 + ... + kn) * x = t$
$pn = kn * x$
$k1 = 1/5$
$k2 = 1/3$
$(1/3 + 1/5) * x = 19 \rightarrow x = 19 * 15 / 8$
$p1 = (1/5) * 19 * 15 / 8 \rightarrow p1 = 7 + 1/8$
$p2 = (1/3) * 19 * 15 / 8 \rightarrow p2 = 11 + 7/8$

The Old Testament does deprive daughters of inheritance from their parents and favors the first born against other children. Under the patriarchs, the property of a diseased father was divided among the sons of his legal wives and their concubines would not get a share (Genesis 21:10; 24:36; 25:5). The Mosaic law made specific regulations regarding the distribution of real property, giving the eldest son a larger portion than the rest. Deuteronomy 21:17; Numbers 27:8; 36:6; 27:9-11.

004:015 The relative pronoun in this verse is *allati*, which refers to a group of women. It indicates an organized prostitution, which might pose a grave health problem for the society. In a society that accepted the jurisdiction of the Quran, if a woman is proven four times to be involved in promiscuous sexual activity, she is considered a hazard to the society and should be quarantined. Perhaps there is also an implication for our time, where genetic identification helps us to find the real father of a child. Our genes are made of various combinations of four acid molecules: Adenine, Guanine, Cytosine, and Thiamine. These four molecules, which are made of four different elements, function as witnesses for the identity of individuals to which they belong. The Quran does not necessarily require the witness to be an eyewitness. For instance, we bear witness to the oneness of God via our intelligence. The witness mentioned in 12:26-27 is not considered a witness because he was present during the alleged incident. That witness did not say, "I saw with my eyes." The human witness mentioned in that story is proposing the testimony of circumstantial evidence, and logical conclusions based on both inductive and deductive reasoning. In other words, the Quran accepts the testimony of reliable circumstantial evidence just as it accepts the testimony of a reliable eyewitness.

004:016 The relative pronoun is not marked; the male dual form *allazani* (both) includes the female partner. In this case, it appears that the illegal or extramarital sexual affair is monogamous. For the punishment of the adulterers, see 24:1.

004:018 See 3:90; 10:90; 24:31; 25:71; 42:25.

004:019 Verse 4:19 clearly recognizes the right of women to divorce. According to the Old Testament, a rapist should be forced to marry the girl he violated. This rule punishes the victim and forces her to share the rest of her life with the violent and shameless man who violated her

(Deuteronomy 22:28-30). How can this and many other unjust laws be imposed by a Just God? The Bible also forces the widow to marry another brother (Deuteronomy 25:5).

004:024 For a discussion regarding the expression *Ma malakat aymanukum*, see the note for verse 4:3.

004:024 After listing 13 classes of relatives, our Lord expressly informs us that marrying others is permitted. However, the followers of *hadith* and *sunna*, challenge this divine enumeration and rule. They prohibit marrying, at the same time, an aunt from one's mother side and an aunt from one's father side, and thereby imply that God forgot to add it to the list, and additionally attribute this blasphemy to Muhammed (6:19,38,114; 12:111; 19:64; 25:30). Note that the words sons, daughters, sisters, brothers include grandchildren, aunts and uncles. The Quran contains numerous examples of this genetic generalization, such as, the expression "Children of Adam." Also note, our translation of the word Muhsanat, which differs from hadith-based translations.

004:025 Slavery as an evil practice of Pharaoh and other polytheists is abolished by the Quran (3:79; 4:3,25,92; 5:89; 8:67; 12:39; 24:32-33; 58:3; 90:13; 2:286; 12:39-42; 79:24). No wonder, this verse emphasizes the equality between men and women, between the free and those who were unjustly enslaved by polytheists. See 60:10.

Anyone who owns slaves, declares to be a god besides God, since there is no Rabb (Lord) besides God. See 12:39.

God issues a lighter penalty for formerly slave women who commit adultery because of their past. Their tragic history and experience are considered a mitigating factor. This rule, at the same time, rejects the Sunni and Shiite punishment for adultery (death) since there cannot be half of a capital punishment. See 24:2.

004:028 God created humans in an excellent design (32:7; 64:3; 82:7; 95:4), yet we humans may lower ourselves to the lowest level by following the teachings of devil (95:5). It is interesting to see the passive voice when degeneration of human nature is mentioned. The Quran does not say "we made humans weak," but it says, "humans were made weak." Thus, whenever our weakness, impatience, or pessimism is mentioned the passive voice is used (4:28; 21:37; 70:19). Through this peculiar language, the Best Designer and the Ultimate Creator who created us in best design with freedom of choice, reminds us of the negative impact of wrong choices on our excellent design. See 57:22-23.

004:034 As I discussed extensively, in Turkçe Kuran Çevirilerindeki Hatalar (*Errors in Turkish Translation of the Quran*, Istanbul, 1992-1998) and in English article, *Beating Women or Beating Around the Bush* (Unorthodox Articles, Internet, 1998), four key words or phrases have been mistranslated by traditional translators. To justify the misogynistic and patriarchal practices, deliberately or unknowingly, a majority of translators render the phrase *kawamuna ala al-nisa* as "in charge of women" rather than "providers for women" or "observant of women." Interestingly, the same translators translate the same verb mentioned in 4:135; 5:8; 4:127; 2:229; 20:14; 55:9 as "observe/maintain." When the same verb is used to depict a relationship between man and woman, it somehow magically transforms into a prescription of hierarchy and authority.

The second key word that is commonly mistranslated is *iDRiBuhunna*. In almost all translations, you will see it translated as "scourge," or "beat" or "beat (lightly)". The verb *DaRaBa* is a multi-meaning verb akin to English 'strike' or 'get.' The Quran uses the same verb with various meanings, such as, to travel, to get out (3:156; 4:101; 38:44; 73:20; 2:273), to strike (2:60,73; 7:160; 8:12; 20:77; 24:31; 26:63; 37:93; 47:4), to beat (8:50), to beat or regret (47:27), to set up (43:58; 57:13), to give (examples) (14:24,45; 16:75,76,112; 18:32,45; 24:35; 30:28,58; 36:78; 39:27,29; 43:17; 59:21; 66:10,11), to take away, to ignore (43:5), to condemn (2:61), to seal, to draw over (18:11), to cover (24:31), and to explain (13:17). It is again interesting that the scholars pick the meaning BEAT, among the many other alternatives, when the relationship between man and woman is involved, a relationship that is defined by the Quran with mutual love and care (30:21).

The third word that has been traditionally mistranslated is the word *NuSHuZ* as "rebellion" or "disobedience" or "opposition" to men. If we study 4:34 carefully we will find a clue that leads us to translate that word as embracing a range of related ideas, from "feeling of separation" to "flirting with others" – indeed, any word or words that reflects the range of disloyalty in marriage. The clue is the phrase before *nushuz*, which reads: ". . . they honor them according to God's commandments, even when alone in their privacy." This phrase emphasizes the importance of loyalty in marital life and helps us to make better sense of what follows. Interestingly, the same word, *nushuz*, is used later in the same chapter, in 4:128 – but it is used to describe the misbehavior of husbands, not wives, as it was in 4:34. In our view, the traditional translation of *nushuz*, that is, "opposition," will not fit in both contexts. However, the understanding of *nushuz* as separation of heart or disloyalty, in a variety of forms, *is* clearly appropriate for both 4:34 and 4:128. A careful study of verses where the word NaShaZa is used in different contexts clarifies the meaning. See: 2:259 (bones are separated from each other), 58:11 (separate or break up to open space for newcomers).

The fourth word is the word *QaNiTat*, which means "devoted to God," and in some verses it describes both man and woman (2:116,238; 3:17,43; 16:120; 30:26; 33:31,35; 39:9; 66:5,12). Though this word is mostly translated correctly as "obedient," when read in the context of the above-mentioned distortion it conveys a false message as if to imply that women must be "obedient" to their husbands as they were inferior, while the word refers to obedience to God's law. The word is mentioned as a general description of Muslim women (66:12), and more interestingly the description of Mary who, according to the Quran, did not even have a husband! (66:12).

The traditional distortion of this verse was first questioned by Edip Yuksel in his book, "*Kuran Çevirilerindeki Hatalar*" (Errors in Turkish Translations) (1992, Istanbul). For a detailed discussion on verse 4:34, see the Sample Comparisons Section in the Introduction.

After the revelation of the Quran, Muslim scholars turned back to the days of ignorance, and they were supported by some Jewish and Christian scholars who apparently converted to Islam yet did not experience a paradigm change. These semi-converts and those Arabs who longed for the old culture of ignorance combined their forces together to take back the rights of women recognized and promoted by the Quran. The rights of women in the West have been recognized through the separation of church and state; however, the culture is still basically a male dominant one and thus western women are objectified and exploited tremendously in the business world. The western culture is deeply influenced by the teaching of Christianity originating from the misogynistic authors of Old Testament and St. Paul (not Jesus) who subordinates women to men. For instance, see Ephesians 5:22-33; Colossians 3:18-19; 1 Peter 3:1-7.

"Let your women keep silent in the churches, for they are not permitted to speak; but they are to be submissive, as the law also says" (I Corinthians 14: 34). "For a woman is not covered, let her also be shorn. But if it is shameful for a woman to be shorn or shaved, let her be covered. For a man indeed ought not to cover his head, since he is the image and glory of God; but woman is the glory of man. For man is not from woman, but woman from man. Nor was man created for the woman, but woman for the man" (I Corinthians 11:6-9). "Let a women learn in silence with all submission. And do not permit a woman to teach or to have authority over a man, but to be in silence. For Adam was formed first, then Eve. And Adam was not deceived, but the woman being deceived, fell into transgression. Nevertheless, she will be saved in childbearing if they continue in faith, love and holiness, with self-control" (I Timothy 2:11-15).

St. Paul's misogynistic teaching is a reflection and extension of a historical trend. The Old Testament contains many man-made misogynist teachings. For instance, a woman is considered unclean for one week if she gives birth to a son, but unclean for two weeks if she gives birth to a daughter (Leviticus 12:1-5).

The Quran prohibits a sexual relationship with a menstruating woman, not because she is dirty, but because menstruation is painful (2:222). The purpose is to protect women's health from being burdened by the sexual desires of their husbands. However, the male authors of the Old Testament, exaggerated and generalized this divine prohibition so much so that they turned menstruation into a reason for a woman's humiliation, isolation, and punishment (Leviticus 15:19-33).

Christianity puts all the blame on the shoulders of women for the troubles in this world. Yet, according to the Quran, we are created from one person (*nafs*), not one man (4:1). Furthermore, it was not Eve, but it was both Adam and his spouse who were deceived in the Paradise (2:30-39; 7:19-27).

The Old Testament contains hyperbolic exaggerations and bizarre practices. For some examples, See the footnote of 4:119.

004:035 Divorce does not occur by some magic words uttered by the husband. Furthermore, woman is mentioned as one of the parties whose wish is equally important in case of divorce. See 2:226-230. Verse 4:19 clearly recognizes the right of women to divorce.

004:036 Keeping the family ties, cooperation, and humbleness are among the traits recommended by God (2:83; 3:92; 17:22-37).

004:041 For Muhammed's testimony, see 25:30.

004:043 The followers of *hadith* and *sunna* reject this statement by claiming the presence of a contradiction between this verse and other verses prohibiting the intoxicants. Their rejection of God's verse and their allegation of contradiction in God's book are due to their ignorance of the fact that there could be individuals among muslims who consume alcoholic beverages just as there could be those who steal, commit adultery, or commit false accusation. affirmation of the imperfect nature of a Muslim community does not mean that God justifies sins and crimes. Our Lord informs a Muslim who is doing wrong to himself or herself by consuming alcohol not to commit additional disrespect to God by addressing God in a drunken state of mind. Additionally, this verse indirectly gives permission to the community to not to allow their intoxicated brothers and sisters into the places of prostration. Another lesson derived from this commandment is that we should know what we are saying or reciting during our prayers. Reciting Arabic words without knowing their meanings is like praying while intoxicated. See 2:106; 4:82.

Bleeding, passing gas, or shaking hands with women and other reasons mentioned in *hadith* and sectarian books, do not void the ablution. Despite this clear verse, Sunni and Shiite sects contain many contradictory rules about ablution. The relationship between washing hands or faces with water and mental readiness to speak to God is not scientifically obvious, but, according to a recent study conducted at the University of Toronto, washing hands may clean of a guilty conscience. See 5:6.

The language of the phrase regarding *tayammum* (ablution without water) appears to be difficult to understand; but expressing it in symbolic logic makes it much easier to understand:

"... If you are ill (p), or traveling (q), or one of you come from the bathroom (r), or you had sexual contact with the women (s), and could not find water (t): then you shall seek clean soil and wipe your faces and hands (u)"

$\{(pVq) \text{ V } [(rVs) \bullet t]\} \supset u$

Thus, in case of illness or travelling, even if there is water, one still does not need to wash.

004:046 See 2:104.

004:048 Before death, one may reform from associating partners to God (4:18; 40:66). A majority of the so-called Muslims, like a majority of Christians, have fallen into the trap of *Shirk* (associating other partners with God) by following the sectarian teachings concocted by their clergymen and scholars (9:31; 12:40; 42:21; 6:145-150).

There is no contradiction between 4:48 and 4:153. The Quran contains numerous verses regarding polytheists or *mushriks* accepting the message of islam (42:25). Most of the supporters and companions of messengers and prophets were associating partners with God before they repented and accepted the message. For instance, the Quran informs us that even Muhammed was an idolater before he received revelation, and obviously, after his affirmation of the truth, he repented from his ignorance and God forgave him (40:66; 42:52; 93:7; 48:2).

004:056-57 We learn that the descriptions of Paradise and Hell are metaphors. See 2:24-26; 13:35; 47:15.

004:058 It is our responsibility to inform ourselves about politics and current affairs so that we can elect trustworthy and qualified people to the public offices.

004:059 Obeying a messenger while he is a leader or obeying elected or appointed officers does not mean taking their words or commands as "God's commandment" or absolute, infallible, unchangeable, and universal law.

One of the most frequently cited Quranic instructions by the followers of *hadith* and *sunna* is "obey God and His Messenger" (4:59). Let us briefly discuss this example to see better the nature of abuse, and how thousands of *hadith* rabbits are produced from empty hats. Obeying Bukhari, a narrator of hearsay, is not obeying the messenger. Obeying the messenger is obeying the complete, perfect and fully detailed Quran. Verse 25:73 describes the attitude of affirmers towards God's revelations. But the followers of *hadith* and *sunna* are very good in ignoring them. They do not see 6:19, 7:3, and 50:45, which say that the only teaching delivered by God's messenger was the Quran. They do not think that Muhammed practiced the Quran, and the Quran alone (5:48, 49). They do not hear Muhammed's only complaint about his people (25:30). They do not understand that Muhammed disowns those who do not understand that the Quran is enough and fully detailed (6:114). The first verse of Chapter 9 states that an ultimatum is issued from God and His messenger. Muslims affirm that verses about the ultimatum are entirely from God. God did not consult Muhammed about the ultimatum. Muhammed's only mission was to deliver God's message (16:35; 24:54). Thus, the reason that God included the messenger in 9:1 is not because he was another authority in issuing it, but because he participated as the deliverer of the ultimatum. Similarly, because people receive God's message through messengers, we are ordered to obey the messengers. We also know that the Quran is a permanent messenger (65:11), and the Quran is a reminder and deliverer of good news (41:4; 11:2).

004:064 The divine message delivered by the messenger should be affirmed and obeyed. Additionally, when the messenger is alive as an elected leader, his instructions and judgment should be obeyed, as long as his instructions are in conformity with known facts (60:12). The live messenger, as a fallible human, must consult people around him (3:159). The messenger of God cannot issue rules in the name of God, and like everyone else he is also responsible for following God's judgment (5:48-50). We may pray for the forgiveness of our friends and subordinates if their mistake involves us. Also, see 12:98.

004:065 See 5:48.

004:066 The phrase traditionally being translated as "kill yourselves" can be read as "*faqbilu anfusakum*", that is "face yourselves" or "accept yourselves." Here, it is an alternative to leaving the land. It means, either look at the mirror and change yourselves or leave the land. See 2:54,84.

004:069 It is not necessarily that prophets or messengers have the highest rank above those who accept their message and support them. Thus, monotheists who live a righteous life will spend eternity together with prophets and messengers.

004:074 Attacking atheists, agnostics or the members of other religions and sects by chanting God's name is not "fighting in the cause of God." The fight that is permitted by the Quran is the one done for the purpose of defending self and community against aggressive parties. See 9:3-29.

004:079 Everything is under God's control (4:78; 8:17) and bad things happen to us because of our wrong choices (42:30; 64:11). For instance, one may harm himself by putting his hand in the fire, and the harm is invited by his action. However, since the fire operates and burns according to God's law, it is from God. It is noteworthy that the word *indi* (side) in verse 4:78, which indicates intervening causes, does not exist in verse 4:79. In the former the indirectness of the relationship and, in the latter, the directness of the relationship is implied. Also, see 57:22-23.

004:082 The followers of *hadith* and *sunna* deny the truth-value of this verse by claiming numerous contradictions (*nasikh-mansukh*) in the Quran (see 2:106). It is a proof of divine authorship that, though the Quran was written in a time when superstitions and mythologies were popular, it does not contain even a trace of them. Comparing the verses of the Quran to the commentaries written centuries after it, such as Qurtubi, Ibn Kasir, Tabari, Nasafi, or to the *hadith* books, will highlight the uniqueness of the Quran. The Quran will stand out as a book that does not contain any falsehood.

The Quranic description of the earth, the solar system, the cosmos, and the origin of the universe is centuries ahead of the time of its first revelation. For instance, the Quran, delivered in the seventh century C.E., states or implies that: Time is relative (32:5; 70:4; 22:47). God created the universe from nothing (2:117). The earth and heavenly bodies were once a single point, and they were separated from each other (21:30). The universe is continuously expanding (51:47). The universe was created in six days (stages) and the conditions that made life possible on earth took place in the last four stages (50:38; 41:10). The stage before the creation of the earth is described as a gas nebula (41:11). Planet earth is floating in an orbit (27:88; 21:33; 36:40). The earth is round (10:24; 39:5; 55:33) and resembles an egg (10:24; 39:5; 79:30). The universe is also round (55:33). Oceans have subsurface wave patterns (24:40). Earth's atmosphere acts like a protective shield (21:32). Wind also pollinates plants (15:22). The bee has multiple stomachs (16:69). The workers in honeybee communities are females (16:68-69). After years of disappearance, Periodical Cicadas emerge all together with a cacophony of songs, testifying a similitude of resurrection for those who appreciate God's signs in the nature (54:7). The creation of living creatures follows an evolutionary system (7:69; 15:28-29; 24:45; 32:7-9; 71:14-17). The earliest biological creatures were incubated inside flexible layers of clay (15:26). The stages of human development in the womb are detailed (23:14). Our biological life span is coded in our genes (35:11). Photosynthesis is a recreation of energy stored through chlorophyll (36:77-81). Everything is created in pairs (13:3; 51:49; 36:36). The atomic number, atomic weight and isotopes of Iron are specified (57:25). Atoms of elements found on earth contain a maximum of seven energy layers (65:12). There will be new and better transportation vehicles beyond what we know (16:8). The edges of land, that is shores, will be reduced in size because of human's reckless behavior (13:41). The sound and vision of water and the action of eating dates (which contain oxytocin) reduce labor pains (19:24-25). There is life (not necessarily intelligent) beyond earth (42:29). The Quran correctly refers to Egypt's ruler who made Joseph his chief adviser as king (*malik*), not as Pharaoh (12:43,50,54,72). Many of the miracles mentioned in the Quran, for instance, represent the ultimate goals of science and technology. The Quran relates that matter (but not humans) can be transported at the speed of light (27:30-40); that smell can be transported to remote places (12:94); that extensive communication with animals is possible (27:16-17); that sleep, in certain conditions, can slow down metabolism and increase life spans (18:25); and that the vision of blind people can be restored (3:49). The number of months in a year is stated as 12 and the word Month (*shahr*) is used exactly twelve times. The number of days in a year is not stated, but the word Day (*yawm*) is used exactly 365 times. The frequency of the word year (*sana*) in its singular form occurs 7 times and plural form 12 times and together 19 times; each number relating to an astronomic event. A prophetic mathematical structure based on the number 19 implied in chapter 74 of the Quran was discovered in 1974 by the aid of computers shows that the Quran is embedded with an interlocking extraordinary mathematical system, which was also discovered in the original parts of the Old Testament in the 11th century. And there's more – much more. See 68:1; 79:30; 74:1-37. Also see Appendix titled: *There is No Contradiction in the Quran.*

004:089-91 Note the exception in 4:90. The apostates cannot be harmed unless they participate in a war against peacemakers (2:256). For the basic principles of war see 8:19; 60:8,9, and 9:29.

004:092 See 4:25.

004:095 War is permitted only for self-defense. See 60:7-9.

004:101 See 2:239.

004:103 To remain under God's domain and receive his blessings and support, the Quran teaches us to commemorate God constantly (2:152,200; 3:191, 33:41,42). However, a great majority of the human population does not use their innate critical thinking and reasoning skills, and thus they believe in superstitions and false teachings that lead them to associate partners with God (12:106; 23:84-89; 29:61-63; 31:25; 39:38; 43:87). Also, see 3:191.

004:116 Many of those who associate partners with God in his powers, such as His creative, legislative, guiding, or judiciary powers, do not accept that they are indeed polytheists who associate partners with God (6:22-23). See 4:48; 4:153.

004:118 Majority of people, in fact do not appreciate or affirm God (12:103). A majority of those who affirm God set up various idols, such as, prophets, messengers, religious or political leaders, sectarian teachings as partners with Him (12:106).

004:119 Modifying God's creation for religious purposes is considered evil. Though genetically modified foods may have unintended negative consequences, circumcision is a prime example of violating this divine commandment. Foreskin is not an abnormality in God's creation; it is the norm. Hence, attempting to change such a creation through surgery to attain salvation is superstition (13:8; 25:2; 32:7; 40:64; 64:3; 82:6-9).

Sunni sources report many contradictory stories regarding circumcision. For instance, Ahmed B. Hanbal in his *Musnad* reports that Usman bin el-As refused to participate in a circumcision ceremony, since he considered circumcision an innovation. The Sunni historian Taberi reports that Caliph Abd al-Aziz rejected the suggestion of his advisors that the people of *Khurasan* should be circumcised; they were converted to "Islam" to avoid paying extra tax! *Bukhari* gives contradictory numbers for the year Abraham was allegedly circumcised, 80 versus 120. *Bukhari* who reports hearsay regarding the circumcision of converts and women, also reports that when Greeks and Abyssinians embraced islam they were not examined at all by Muhammed.

Hadith books, including *Bukhari*, contain numerous *hadiths* promoting circumcision including female circumcision, which is a torturous mutilation. However, *hadith* fabricators somehow forgot to fabricate *hadiths* about the circumcision of prominent figures during the time of Muhammed. More interestingly, since the practice of circumcision was adopted centuries later, they missed the opportunity to attribute this practice to Muhammed himself. Sunni scholars, therefore, came up with another so-called miracle: Muhammed was born circumcised. This would answer those who wondered about the absence of such an "important" record in the books of *hadith* and *sunna.*

The Quran never mentions Abraham practicing circumcision. If indeed Abraham did such a surgery on himself, perhaps he wanted to eliminate infection, and the blind followers who later idolized him turned his personal deed into a religious ritual. Looking at the history of the Jewish people and their trials and tribulations, it is more likely that this is an invention of Rabbis to mark the endangered race and protect it from extinction. Introducing innovations in religious communities may need some "holy stories" to attribute the innovation to historical idols.

The Quran never mentions the adventures of the Biblical character Samson who had a bizarre hobby of collecting the foreskins of the thousands of people he killed by the jaw of an ass (Old Testament Judges 15:16).

The Old Testament contains hyperbolic exaggerations and bizarre practices. For instance, ignoring the discrepancy in the number of mutilated penises read the following verses from Bible:

"So David rose and he and his men went and struck down among the Philistines two hundred men, and David came bringing their foreskins and giving them in full number to the king, to form a marriage alliance with the king. In turn Saul gave him Michal, his daughter, as a wife." (1 Samuel 18:27).

"Then David sent messengers to Ish-Bosheth son of Saul, demanding, 'Give me my wife Michal, whom I engaged to myself for a hundred foreskins of the Philistines" (2 Samuel 3:14).

Using a bundle of foreskins of mutilated genitals of the dead bodies of enemy as the symbolic show of manhood, and literally using them in exchange for a woman is appalling and insulting to women.

Samson's obsession and adventure with Philistine girls is similarly strange (Judges 14). When Samson is betrayed by his wife, Timnah, or his heifer (Judges 14:18!), he loses the bet during the seven days of the feast. This time thirty men from Ashkelon have to lose their lives. Later, Samson torches Philistine grain fields with torches tied to the tails of foxes; kills a thousand Philistines with a "donkey's jawbone," and prays to God not to let him die in the hands of the "uncircumcised" (Judges 15:15-16). This Biblical hero, in his bloody pursuit of another wife, spends a night with a prostitute (Judges 16:1) and later another wife, Delilah, by shaving his hair, the source of his extraordinary power, betrays him (Jud 16:18-20). Samson dies after killing more Philistines. The story can be outlined in a few words: Marriage, Feast, Foreskins, Slaughtering, Torching, Betraying, Heifer, Prostitute, Superstition, Killing, and Killing more!

Muslims, long after the revelation of the Quran and departure of Muhammed, acquired from Jews the bizarre obsession with hair and foreskins! If someone converts to the Sunni or Shiite religions, one of the first troubles he finds himself in is to undergo a small surgery on the foreskin of his penis and a holy recommendation to grow a long beard. Additionally, he will exchange his original name with an Arabic one! This marks just the beginning of becoming a Jewish Arab who follows a concocted culture from the medieval ages.

004:125 All the messengers of God delivered the same message: Free yourself from idols; appreciate God's blessings, trust in God and His justice; live with the knowledge of one day being judged in the Day of Judgment; do not follow your parents and crowds blindly; engage in philosophical inquiry; be charitable; do good deeds; be optimistic; respect diversity; be humble, stand up for the oppressed; stand against racism, misogynistic ideas and practices, etc. All messengers and their supporters were muslims, that is peacemakers and peaceful surrenderers to God. The word *islam* is not a proper name, but an Arabic descriptive word meaning: to submit and peacefully surrender to God, to be in peace and make peace. Both these meanings apply to rational monotheists, while only the second apply to non-monotheistic peacemakers. See 3:19.

004:127 This verse has been commonly mistranslated to justify marrying young orphan girls rather than marrying their widow mothers. The mistranslation is so obvious, that it is curious how those who had knowledge of Arabic did not notice it. Unfortunately, this mistranslation helped the justification of marrying girls at a very young age.

Though the Quran permits polygamy to men (4:3), it strictly discourages its actual practice by requiring certain significant preconditions: men may marry more than one wife only if the later ones are widows with children, and they should treat each wife equally and fairly (See 4:19-20; 127-129.) Unfortunately, verse 4:127 has been traditionally misinterpreted and mistranslated in such a way as to suggest that God permits marriage with juvenile orphans. This is clearly not the case.

The Arabic expression *yatam al-nisa illati* in 4:127 has been routinely mistranslated as "women orphans, whom..." The expression is also sometimes translated as "orphans of women whom..." This later translation, though accurate, makes the crucial reference of the objective pronoun "whom" ambiguous: Does the phrase after "whom" describe orphans or women?

As it happens, the Arabic plural pronoun in this verse is the female form, *allaty* (not the male form *allazyna*), and it can *only* refer to the women just referenced, not to the orphans. This is because the Arabic word *yatama* (orphans) is grammatically male in gender!

All the English translations of the Quran that we have seen have mistranslated this passage. This is remarkable because the correct translation requires only an elementary knowledge of Arabic grammar. This error is thus much more than a simple grammatical slip; it is, we would argue, willful misrepresentation. The traditional interpretation of this passage offers an apparent justification for marriage with children, which flatly contradicts the Quran.

Like so many passages in the Quran, 4:127's meaning was severely distorted in order to gain the favor of rich, dominant males. Over the centuries, male scholars with active libidos used fabricated *hadith* to pervert the meaning of this and other Quranic verses relating to marriage and sexuality (See 66:5). For a comparative discussion on verse 4:127, see the Sample Comparisons section in the Introduction.

004:129 For the conditions suggested for polygamy, see 4:2-3.

004:140 We are not permitted to kill or punish people for their insults and mockery of God's revelation and signs. Any aggressive behavior against those people is against God's law that recognizes freedom of choice, opinion and their expression (2:256; 4:90; 4:137, 140; 10:99; 18:29; 25:63; 88:21,22. Also see 28:54). In verse 4:140, the Quran recommends us to protest passively those who indulge in mockery of our faith by leaving their presence. Furthermore, it recommends us not to cut our relationship with them; we should turn back in peace and continue our dialogue when they come to their senses and are able to engage in a rational discourse.

Those who react with violence against those who insult the tenets and principles of peacemaking (islam), are not following the very system they claim to defend. The Quran does not recognize the "fighting words" exception recognized by Western jurisprudence. Specific false accusation against a person, however, is not included in expression of opinion, since it is defamation, and it can harm a person in many ways. Insult to someone's values or heroes, however, only harms the person who indulges in such an ignorant and arrogant action.

The only unforgivable sin, according to the Quran is the sin of associating other partners/idols with God. God allows this biggest sin to be committed in this world. He fulfills his promise to test humans by giving them free choice. He condemns those who deprive others from exercising that freedom of choice. Who then, in the name of the same God, can force others from any expression of their belief or disbelief?

In contrast to what the warmongering Crusaders, Sunnis and Shiites wish to portray, Muhammed was not a man of violence but a man of reason and peace. Numerous verses of the Quran and a critical study of history will reveal that the portrait of Muhammed depicted in Sunni or Shiite hearsay books is fictional; a fiction created by the propagandists of rulers of the Umayyad and Abbasid dynasties to justify their atrocities and aggression! He and his supporters were threatened and tortured in Mecca for their criticism of their corrupt and unjust theocratic system. They were forced to leave everything behind and emigrate to Yathrib (today's Medina). There they established a peaceful city-state, a federal secular democracy, among its multi-religious diverse citizens. Nevertheless, the Meccan oligarchy did not leave them alone to enjoy peace and freedom; they organized several major war campaigns against the coalition of Peacemakers consisting of Monotheists, Christians, Jews and Pagans united under the leadership of Muhammed. In all the wars, including Uhud, Badr, and Handaq (Trenches), the monotheist reformers fought for self-defense. They even dug trenches around the city to defend themselves from the aggressive religious coalition led by Mecca's theocratic oligarchy. Muhammed's message, which promoted reason, freedom, peace, justice, unity of children of Adam, appreciation of diversity, rights of women and slaves, and social consciousness, soon received acceptance by the masses in the land. Yet, after ten years in exile, when Muhammed and his supporters finally returned to Mecca as victors, he declared amnesty for all those oppressors and warmongers who inflicted on them great suffering, who maimed and murdered many of their comrades, all because the rational monotheists had questioned the teachings and culture they inherited from their parents. However, guided by the teachings of the Quran, Muhammed chose forgiveness and peace; he did not punish any of his bloody enemies. After all, he was one of the many messengers of peace and submission to God alone.

004:150 Those who claim that God is represented by the Quran, and the messenger is represented by *hadith* hearsay collection are warned. See 9:1.

004:153 See 4:48.

004:155 For the usage of this Biblical expression see the endnote for verse 2:88.

004:157-158 We understand that Jesus was not conscious when they crucified his body. Jesus' person was already terminated, and he was with his Lord.

St. Paul created a pagan religion based on the absurd concept of "God sacrificing his innocent son, or His third personality, to be able to forgive our sins!" Christianity is a religion that is based on a "forsaken God." In *19 Questions for Scholars,* I examined the Christian version of crucifixion and argued that it is no more than a "crucifiction"! See, "A Forsaken God?" in Appendices section.

004:159 The majority of the people of the book, unfortunately, do not affirm Jesus who prophetically warned them against the teaching of Pharisees. They have accepted being the sheep of St. Paul, who never met Jesus, yet had intense arguments with Peter (Cephas) and Barnabas (Acts 15:36-41; Galatians 2:11-14; 4:10-14), and who was also not trusted by disciples who escaped from him (Acts 9:26). The disciples knew that he was a wolf in the midst of sheep. They knew that he was a Pharisee (Acts 23:6), and they also knew well the repeated warning of the Messiah (Matthew 16:11-12; 23: 13-33; Luke 12:1-2). It was the self-appointed disciple Saul (The Acts 26:16-19) who after changing his name to Paul (Acts 8:3; 22:3-10), changed the name of the supporters of Jesus too (Acts 24:5; Acts 11:26). It was Paul who provided grounds for the polytheistic Doctrine of the Trinity, created the story of Jesus' sacrifice for our redemption, raised Jesus to his "Father's right side" for judgment, nailed the written code to the cross (Colossians 2:14), thereby dismissing Jesus (Matthew 19:16-19). He perverted the main message of Jesus, which was to worship God alone. He cunningly cursed him (Galatians 3:13 and Deuteronomy 21:23), and he became "all things to all men in order to win them" (1. Corinthians 9:20-22). Indeed, it is not Jesus, the son of Mary, but it is this dubious character that most Christians are now following.

004:161 See 2:275.

004:166 Learning the testimony about the Quran itself might at first sound like an empty claim. Yet, the extraordinary knowledge contained in the Quran and the extraordinary design in its physical structure provides evidence from within (See 4:82; 74:30).

004:171 Millions of Christians believe in the "Holy Trinity" on faith. Through this formula, they transformed Jesus, the son of Mary, into the "Son of God", and even God himself. However, history, logic, arithmetic, the Old Testament, and the New Testament prove the contrary: Jesus was not Lord; he was a creation of God like Adam was.

The doctrine of the Trinity is found in many pagan religions. Brahma, Shiva, and Vishnu are the Trinitarian godhead in Indian religions. In Egypt there was the triad of Osiris, Isis and Horus; in Babylon, Ishtar, Sin, Shamash; in Arabia, Al-Laat, Al-Uzza, and Manat. The Encyclopedia Britannica (1975) gives a critical piece of information:

"Trinity, the doctrine of God taught by Christians that asserts that God is one in essence but three in 'person,' Father, Son, and Holy Spirit. Neither the word Trinity, nor the explicit doctrine as such, appears in the New Testament, nor did Jesus and his followers intend to contradict the scheme in the Old Testament: 'Hear, O Israel: The Lord our God is one Lord' (Deuteronomy 6:4)"

This information on the Trinity contradicts the faith of most Christians. They believe that Matthew 28:19 and John1:1 and some other verses clearly provide a basis for the doctrine of the Trinity. However, the *New Catholic Encyclopedia* (1967 edition, Vol: 14, p. 306) affirms that the Trinity doctrine does not exist in the Old Testament, and that it was formulated three centuries after Jesus. The Athanasian Creed formulated a polytheistic doctrine with the following words: "We worship one God in Trinity, and Trinity in Unity; neither confounding the Persons, nor dividing the Substance (Prayer Book, 1662). The Father is God, the Son is God, and the Holy Spirit is God, and yet there are not three Gods but one God." It is unanimously accepted that the doctrine of the Trinity is the product of the Nicene Conference (325 AC). The renowned Christian commentator Easton explains Trinity in one paragraph:

"Trinity: a word not found in Scripture; but used to express the doctrine of the unity of God as subsisting in three distinct Persons. This word is derived from the Gr. trias, first used by Theophilus (A.D. 168-183), or from the Lat. trinitas, first used by Tertullian (A.D. 220), to express this doctrine. The propositions involved in the doctrine are these: 1. That God is one, and that there is but one God (De 6:4; 1Ki 8:60; Isa 44:6; Mr 12:29,32; Joh 10:30). 2. That the Father is a distinct divine Person (hypostasis, subsistentia, persona, suppositum intellectuale), distinct from the Son and the Holy Spirit. 3. That Jesus Christ was truly God, and yet was a Person distinct from the Father and the Holy Spirit. 4. That the Holy Spirit is also a distinct divine Person."

Questions such as, "How could the Father, the Son, and the Holy Spirit be totally different and yet participate in the one undivided nature of God?" have given Christian scholars a hard time for centuries. To explain the nature of the Trinity, they have written volumes of books full of interpretations and speculations ending up with a divine paradox, or a divine mystery, which amount to no more than holy gobbledygook. So, it is not worthwhile to question the meaning of the Trinity further as the answer, ultimately, will be that it is a divine mystery which cannot be understood. Instead, we will question the compatibility of the doctrine with the Bible.

The Trinity is not taught in any of the thirty-nine books of the Old Testament. None of the Biblical prophets, such as Noah, Abraham, Moses, or David mention the Trinity. To the contrary, they emphasized God's oneness (Deuteronomy 4:39; 6:4; 32:39. Exodus 20: 2-3. 1 Samuel 2:2. 1 Kings 8:60, Isaiah 42:8; 45:5-6).

The concept of the Trinity was fabricated over several centuries through gradual distortion and exaggeration of the powers of the hero. Initially, there were many Christian communities rejecting the idea of the deity of Jesus, such as the Ebonites, but ultimately, the followers of St. Paul were victorious against the true supporters of Jesus, and they used force, occasionally in a very cruel way, to impose their authority.

The Bible contains many verses rejecting the doctrine of the Trinity. For instance, according to the Bible, we all can become the children of God, that is, the followers of God, thereby contradicting the idea that Jesus was the only one (Matthew 5:9; 6:14, Luke 20:36; John 8:47, 1 John 5:18,19). According to the Bible, Jesus rebukes someone calling him "Good," by asking him rhetorically, "Why do you call me good?" and then answering the question, "Only God is truly good!" (Mark 10:18-19). Furthermore, to those who asked about the time of the end of the world, Jesus rejected the concept of the Trinity, which equates God with Jesus: "No one knows about that day or hour, not even the controllers in heaven, nor I myself, but only the Father." (Mark 13:32) If Jesus were Lord, as St. Paul's followers assert, how could he not know the future? Many verses in the Gospel reject the concept of the deity of Jesus and promote strict monotheism. "Jesus replied, 'The one that says, 'Hear, O Israel! The Lord our God is the one and only God. And you must love him with all your heart and soul and mind and strength.'" (Mark 12:29). Also, see 4:10; 6:24, Mark 10:18, Luke 18:19.

Another verse quoting from the Old Testament depicts Jesus as a Servant of God (Matthew 12:17-18). Isn't there a difference between God and the Servant of God? The polytheist Christians will find it very difficult to answer this simple question and they will seek refuge in their dark cave called "mystery." They will do anything to continue their belief in the fabricated doctrine of the "Pharisee-son-of-a-Pharisee" as their prime holy teaching (Matthew 16:11-12; 23:13-33; Luke 12:1-2; Acts 23:6). Jesus, according to the Gospels, was a "messenger of God" (Matthew 21:11,46; Luke 7:16; 24:19; John 4:19; 6:14), yet to Jesus-worshipping polytheists there is not much difference between the "messenger of God" and "God." If you bring them a dictionary and demand some rational justification, you will hit the wall of "mystery." In St. Paul's Wonderland, Biblical words either do not have much meaning, or they mutate and transform in many incredible ways. Whether Jesus saw God or not, might be another important questing in refuting the Trinity, but verse John 1:18 provides two contradictory answers in two different versions.

The Trinity is not a logical or rational theory; its absurdity can be expressed in mathematical terms as 1=1+1+1. Although it is logically and mathematically false, it survives in the imaginations of people who have faith in the wishful thinking of idolized clergymen. One may find hundreds of Biblical verses rejecting the deity of Jesus and still be unable to convince a Christian inflicted with this virus. The doctrine is based on revering and consecrating a clear logical contradiction: A being who was a creature, a human being, and at the same time a non-human being, a creator, the creator! It ignores the fact that nothing can be both a man and a God, according to the very definition of the word "man" and "God." God is not created, but man is created; God does not need food, but man does; God is eternal, but man is mortal; the eternal cannot be mortal. So on and so forth. The First Cause, or the uncreated cannot be, at the same time, created.

Therefore, the doctrine of the Trinity is a virus that attacks and destroys the immunity system of the brain first. A person who received such a virus "on faith," by blindly following the teaching of a particular church or the proximate crowd, will not be healed by the rules of logic, mathematics, history, archeological findings, or scientific evidence. Nothing can sober them up from their intoxication. The virus is made even more invincible by the first planter of the seed of this virus, Paul, who boasted to his "flock" that he becomes everything and anything, all things to all men just to win them over (1 Corinthians 9:20-22):

"But even if we, or an angel from heaven, preach any other gospel to you than what we have preached to you, let him be accursed. As we have said before, so now I say again, if anyone preaches any other gospel to you than what you have received, let him be accursed." (Galatians 1:8-9).

No wonder, the idol-carver Pharisee-the-son-of-a-Pharisee has plenty of contempt for those who reason. He wrote the most impressive eulogy for foolishness and packaged it as part of the "gospel" together with an invincible bug against reason, evidence, and even God's controllers and messengers. Tertullian, the one who gave birth to the doctrine of the Trinity, wrote one of the fanciest defenses for dogmatism, bigotry, and narrow-mindedness, that later gave fruit to indulgences, inquisition, witch-hunts, crusades, and many other tragedies. Tertullian tried to banish reason through lousy reasoning:

"These are human and demonic doctrines, engendered for itching ears by the ingenuity of that worldly wisdom which the Lord called foolishness, choosing the foolish things of the world to put philosophy to shame. For worldly wisdom culminates in philosophy with its rash interpretation of God's nature and purpose. It is philosophy that supplies the heresies with their equipment... After Jesus Christ we have no need of speculation, after the Gospel no need of research. When we come to believe, we have no desire to believe anything else; for we begin by believing that there is nothing else which we have to believe."

Ironically, a handful of verses are abused to justify the Trinity, and ALL of them are questionable. For instance, many Christian scholars affirm that the crucial word, "only begotten" in John 1:14,18 and John 3:16,18 does not exist in the original manuscripts. The phrase "son of God," through various versions, and translation of translations mutated into "the only begotten Son of God." The difference in the number of words is small, but in theology, the difference is as big as the difference between monotheism and polytheism.

Many Western theologians and historians have come to the conclusion that today's Christianity is the product of St. Paul, not (J)esu(s). Here are a few of those books investigating the origin of today's Christianity: The Mythmaker: Paul and Invention of Christianity, Hyam Maccoby, Barnes and Noble, 1986. When Jesus Became God: The Epic Fight over Christ's Divinity in the Last Days of Rome, Richard Rubenstein, Harvest/HBJ Book, 2000. Who wrote the Bible?, Richard E. Friedman, Harper San Francisco, 1997. The Doctrine of the Trinity: Christianity's Self-Inflicted Wound, Sir Anthony Buzzard and Charles Hunting, University Press of America, 1998. Misquoting Jesus: The Story Behind Who Changed the Bible and Why, Bart D. Ehrman, HarperSanFrancisco, 2005. The Orthodox Corruption of Scripture: The Effect of Early Christological Controversies on the Text of the New Testament, Bart D. Ehrman, Oxford University Press, 1996. Lost Scriptures: Books That Did Not Make It into the New Testament, Bart D. Ehrman, Oxford University Press, 2003. Truth in Translation: Accuracy and Bias in English Translations of the New Testament, Jason Beduhn and Jason David Beduhn, University Press of America, 2003. We also recommend the following books: Is the Bible God's Word?, Ahmed Deedat, African Islamic Mission Publication, 1990. Jesus: Myths & Message, Lisa Spray, Universal Unity, 1992. Losing Faith in Faith: From Preacher to Atheist, Dan Barker, Freedom From Religion Foundation, 1994. 200+ Ways the Quran Corrects the Bible, Mohamed Ghounem, MNMC, 2004. ZEALOT: The Life and Times of Jesus of Nazareth, Reza Aslan, Random House, 2013. 19 Questions for

Christians, Edip Yuksel, BrainbowPress, 2019.

Also, see 2:59; 3:45,51-52,55; 4:11,157; 5:72-79; 7:162; 19:36.

5:0 In the name of God, the Gracious, the Compassionate.

5:1 O you who affirm, honor your contracts. Made lawful for you are all the animals of the livestock, except what is being recited to you, and what you are not allowed to hunt of the game while you are under restriction. **God** decrees as He pleases.

5:2 O you who affirm, do not violate **God**'s decrees, nor the restricted month, nor the donations, nor the offerings, nor maintainers of the Regulated Sanctuary who are seeking a bounty from their Lord and a blessing. When it is permitted for you, then you may hunt. Let not the hatred of another people, because they had barred you from the Restricted Temple, tempt you to aggress. Bond together in piety and righteousness, and do not bind together in sin and animosity. Be aware of **God**, for **God**'s retribution is severe.*

Only Four Dietary Prohibitions

5:3 Forbidden to you is what is already dead, and the blood, the meat of pig, and what was dedicated to other than **God**. The strangled, killed by a blow, fallen from height, gored, eaten by wild animals except what you managed to rescue, and those slaughtered on altars... Also what you divide through gambling. This is all vile. Today the ingrates have given up from your system, so do not revere them, but revere Me. Today I have perfected your system for you, completed My blessings upon you, and I have accepted peaceful surrender as the system for you. So, whoever is forced by severe hunger and not seeking sin, then **God** is Forgiving, Compassionate.*

5:4 They ask you what was made lawful to them, say, "All the good things have been made lawful for you, and what the trained dogs and birds catch; you teach them from what **God** teaches you." So eat from what they have captured for you and mention **God**'s name upon it, and be aware of **God**. **God** is quick in reckoning.*

5:5 Today, the good things have been made lawful to you, the food of those who have been given the book is lawful for you, and your food is lawful for them; and the affirming chaste women, and the chaste women from the people of the book before you, when you have given them their wages, you are chaste, not fornicating, and not seeking to take lovers. Whoever rejects affirmation, then his work has fallen, and in the Hereafter, he is of the losers.

Ablution for Sala

5:6 O you who affirm, when you rise to attend the contact prayer, then wash your faces and your hands up to the elbows; and wipe your heads and your feet to the ankles; and if you have had intercourse, then you shall bathe. If you are ill, or traveling, or had defecation, or you have had sexual contact with the women, and you could not find water, then you shall do ablution from clean soil; you shall wipe your faces and your hands. **God** does not want to make any hardship over you, but He wants to cleanse you and to complete His blessings upon you that you may be appreciative.*

5:7 Recall **God**'s blessings over you and His covenant that He has bound you with, for which you have said, "We hear and obey," and be aware of **God**; for **God** knows what is inside the chests.

Justice

5:8 O you who affirm, stand for **God** as witnesses for justice, and let not the hatred towards a people make you avoid being just. Be just, for it is closer to awareness, and be aware of **God**. **God** is Aware over what you do.

5:9 **God** has promised those who affirm and do right that they will have forgiveness and a great reward.

5:10 Those who reject and deny Our signs, they are the dwellers of hell.

5:11 O you who affirm, recall **God**'s blessings upon you when a group desired to extend their hands against you, and He restrained their hands from you. So, revere **God**; and in **God** those who affirm should put their trust.

5:12 **God** took the covenant of the Children of Israel and sent from them twelve representatives, and **God** said, "I am with you if you maintain the contact prayer, contribute towards betterment, affirm My messengers, support them, and give **God** a loan of righteousness; then I will cancel your sins and admit you into gardens with rivers flowing beneath." Whoever rejects after this from you, then he has strayed from the path.

5:13 But because of them breaking their covenant, We have cursed them, made their hearts become hardened. They take the words out of context; and they forgot much of what they were reminded of. You will still discover betrayal in them except for a few; so pardon them and overlook; **God** loves the good doers.

5:14 From those who have said, "We are Nazarenes," We have taken their covenant and they have forgotten much of what they were reminded of; so We planted between them animosity and hatred until the day of Resurrection; and **God** will inform them of what they had done.

See the Light and Accept the Truth

5:15 People of the book, Our messenger has come to you to proclaim for you much of what you were hiding from the book, and to ignore over much. A light has come to you from **God** and a clarifying book.

5:16 **God** guides with it whoever seeks His approval, to the ways of peace;

and He brings them out of the darkness and into the light by His leave; and He guides them to a Straight Path.

5:17 Ingrates indeed are those who have said, "**God** is the Messiah the son of Mary." Say, "Who has any power against **God** if He had wanted to destroy the Messiah son of Mary, and his mother, and all who are on earth!" To **God** is the sovereignty of heavens and earth and all that is in-between; He creates what He pleases. **God** is capable of all things.

5:18 The Jews and the Nazarenes said, "We are **God**'s children and His loved ones." Say, "Then why does He punish you for your sins?" No, you are merely humans from what He has created. He forgives whom He pleases, and He punishes whom He pleases. To **God** is the sovereignty of heavens, earth, and all that is in-between, and to Him is the destiny.

5:19 O people of the book, Our messenger has come to clarify for you after a period of absence of messengers; so that you cannot say, "No bearer of good news or warner has come to us;" for a bearer of good news and a warner has come to you; and **God** is capable of all things.

The Holy Land

5:20 Moses said to his people, "My people, remember **God**'s favor upon you that he made amongst you prophets, and made you sovereigns, and he gave you what He had not given any from the worlds."

5:21 "O my people, enter the holy land that **God** Has decreed for you, and do not turn your backs, or you will become losers."

5:22 They said, "O Moses, in it are a mighty people, and we will not enter it until they leave. So when they leave, then we will enter it."

5:23 Among those who were afraid, two men whom **God** had blessed, said, "Enter upon them through the gate; when you enter it then you will be the victors. You shall put your trust in **God** if you had affirmed."

5:24 They said, "O Moses, we will never enter it as long as they are in it, so go you and your Lord and fight, we will stay right here!"

5:25 He said, "My Lord, I do not control except myself and my brother, so separate between us and between the wicked people."

5:26 He said, "Then it has become forbidden for them for forty years, and they will be lost in the land." Do not be sorrowful over the wicked people.*

Each Human is as Important as All of Humanity

5:27 Recite for them the news of Adam's two sons in truth. They had both made an offering, and it was accepted from one of them, and not accepted from the other. He said, "I will kill you!"; he said, "**God** only accepts from the righteous."*

5:28 "If you stretch your hand to kill me, I will not stretch my hand to kill you, for I fear **God**, the Lord of the worlds."

5:29 "I want you to have both my sin as well as your own sin, and you will then be among the dwellers of the fire. Such is the reward of the wicked."

5:30 So he found it in himself to kill his brother, and he killed him. He thus became one of the losers.

5:31 So **God** sent forth a raven to scratch the land and show him how to bury his brother's body. He said, "Woes on me, I failed to be like this raven and bury my brother's body!" So he became of those who regretted.*

5:32 It is because of this that We have decreed for the Children of Israel, "Anyone who kills a person who has not committed murder, or who has not committed mischief in the land; then it is as if he has killed all the people! Whoever spares a life, then it is as if he has given life to all the people." Our messengers had come to them with clarification, but many of them are, after this, still transgressing on the earth.

Live by the Sword, Die by the Sword; Karma

5:33 The recompense of those who fight **God** and His messenger and seek to corrupt the land, is that they are killed or crucified or that their hands and feet are cut off on alternate sides or that they are banished from the land. That is a disgrace for them in this world. In the Hereafter, they will have a great retribution;*

5:34 Except for those who repent before you overpower them, then know that **God** is Forgiving, Compassionate.

5:35 O you who affirm, be aware of **God** and seek a way to Him; strive in His cause so that you may succeed.

5:36 Those who have rejected, if they had all that is on earth and the same again with it to ransom against the retribution of the day of Resurrection, it will not be accepted from them; and they will have a painful retribution.

5:37 They want to get out of the fire, but they can never leave it; and they will have a lasting retribution.

Penalty for thieves

5:38 The male thief, and the female thief, you shall mark, cut their hands/means as a punishment for their crime, and to serve as a deterrent from **God**. **God** is Noble, Wise.*

5:39 Whoever repents after his wrongdoing and makes reparations, then **God** will accept his repentance. **God** is Forgiving, Compassionate.

5:40 Did you not know that **God** possesses the sovereignty of heavens and earth? He punishes whom He wills, and He forgives whom He wills; and **God** is capable of all things.

Corrupt People Distort the Message

5:41 O messenger, do not be saddened by those who compete in rejection from among those who said, "We affirm" with their mouths while their hearts did not affirm. From among the Jews, there are those who listen to lies; they listen to people who never came to you; they distort the words from their context, and they say, "If you are given this, then take it; but if you are given anything different, then beware!" Whomever **God** wants to test; you can do nothing for him against **God**. These are the ones whose hearts **God** did not want to cleanse; in this world, they will have humiliation, and in the Hereafter, they will have a great retribution.

5:42 They listen to lies and consume money illicitly. If they come to you, then you may judge between them or turn away from them. If you turn away from them then they cannot harm you in the least; and if you judge then you should judge between them with justice. **God** loves those who are just.*

5:43 How can they make you their judge when they have the Torah, containing **God**'s judgment; then they turn away after that. They do not affirm.

5:44 We have sent down the Torah, in it is guidance and a light; the prophets who have peacefully surrendered judged with it for those who are Jews, as well as the Rabbis, and the Priests, for what they were entrusted of **God**'s book, and they were witness over it. So do not revere the people but revere Me; and do not exchange My signs for a petty gain. Whoever does not judge with what **God** has sent down, then these are the ingrates.*

5:45 We have decreed for them in it that a life for a life, and an eye for an eye, and a nose for a nose, and an ear for an ear, and a tooth for a tooth, and wounds to be similar; and whoever remits anything of it, then it will cancel sins for him. Whoever does not judge by what **God** has sent down, then these are the wicked.*

The Injeel (Good News) and the Quran

5:46 We followed their teaching with Jesus the son of Mary, authenticating what was present with him of the Torah. We gave him the Injeel, in it is guidance and light, and to authenticate what is present with him of the Torah, and a guidance and lesson for the righteous.

5:47 Let the people of the Injeel judge with what **God** has sent down in it. Whoever does not judge by what **God** has sent down, then these are the vile ones.

5:48 We have sent down to you the book with truth, authenticating what is present of the book and superseding it. So judge between them by what **God** has sent down, and do not follow their desires from what has come to you of the truth. For each of you We have made laws, a structure. Had **God** willed, He would have made you all one nation, but He tests you with what He has given you, so advance the good deeds. To **God** you will return all of you, and He will inform you regarding that in which you dispute.*

5:49 You should rule among them by what **God** has sent down, and do not follow their wishes, and beware lest they divert you away from some of what **God** has sent down to you. If they turn away, then know that **God** wants to inflict them with some of their sins. Indeed, many among the people are corrupt.

5:50 Is it the judgment of the days of ignorance that they seek? Who is better than **God** as a judge for a people that comprehend?

Warmongers among the People of the Book

5:51 O you who affirm, do not take the Jews and the Nazarenes as allies, for they are allies to one another; and whoever takes them as such from amongst you is one of them. **God** does not guide the wicked people.*

5:52 You will see those who have a disease in their hearts hurrying to them, saying, "We fear that a circle of fate will befall us!" Perhaps **God** will bring a victory or a decree from Himself, then they will become regretful over what they had kept hidden within themselves.

5:53 Those who affirm said, "Were these the ones who swore oaths by **God** that they were with you?" Their works have collapsed, and they have become losers.

5:54 O you who affirm, whoever from among you turns away from His system, then **God** will bring a people whom He loves and they love Him; humble towards those who affirm, dignified towards the ingrates; they strive in the cause of **God** and do not fear the blame of those who blame. This is **God**'s grace, He bestows it upon whom He wills; **God** is Encompassing, Knowledgeable.

5:55 Your ally is **God**, His messenger and those who affirm, who maintain the contact prayer, and contribute towards betterment, and they are kneeling.

5:56 Whoever allies with **God**, His messenger and those who affirm; then the party of **God** is the ones who will be victorious.

5:57 O you who affirm, do not take as allies those who have taken your system as fun and games from among those who have been given the book before you and the ingrates. Be aware of **God** if you are those who affirm.

5:58 If you call to the contact prayer, they take it as fun and games. That is because they are a people who do not reason.

5:59 Say, "O people of the book, do you hate us simply because we affirm **God** and what was sent down to us and what was sent down before?" Alas, most of them are wicked.

5:60 Say, "Shall I inform you of worse than this as a punishment from **God**? Those whom **God** cursed and displeased with them, and He made from them apes, pigs and servants of the aggressor. Those have a worst place and are more astray from the right path."*

Sinful, Hateful, Greedy, and yet Self-Righteous

5:61 If they come to you, they say, "We affirm," while they had started with rejection and exited with the same. **God** is aware of what they were hiding.

5:62 You see many of them hasten to sin, animosity, and consuming money illicitly. Miserable indeed is what they were doing.

5:63 Shouldn't the Rabbis and Priests deter them from speaking evil and consuming money illicitly? Miserable indeed is what they had done.

5:64 The Jews said, "The hand of **God** is tied-down!" Their hands will be tied-down, and they will be cursed for what they have said. No, His hands are wide open, spending as He wills. For many of them, what has been sent down to you will increase them in aggression and rejection; and We have cast between them animosity and hatred until the day of Resurrection. Every time they ignite the fire of war, **God** puts it out; and they seek to make corruption in the land; and **God** does not like the corrupters.

5:65 If the people of the book only would have affirmed and been aware We would have cancelled for them their sins and admitted them to gardens of paradise.

5:66 If they had upheld the Torah, the Injeel, and what was sent down to them from their Lord, they would have been rewarded from above them and below their feet. From among them is a pious nation, but many of them commit evil.

5:67 O messenger, deliver what was sent down to you from your Lord; if you do not, then you have not delivered His message. **God** will protect you from the people. **God** does not guide the rejecting people.

5:68 Say, "O people of the book, you are not upon anything until you uphold the Torah and the Injeel and what was sent down to you from your Lord." For many of them, what was sent down to you from your Lord will only increase them in transgression and rejection. So do not feel sorry for the ingrates.

Requirements for Attaining Eternal Salvation

5:69 "Those who affirm, those who are Jewish, those who are the followers of other religions, and the Nazarenes… Whoever affirms **God** and the Last day and does good works, then they will have nothing to fear, nor will they grieve."*

5:70 We have taken the covenant of the Children of Israel and We sent to them Our messengers. Every time a messenger came to them with what their minds did not desire, a group of them they would reject, and a group of them they would kill.

5:71 They did not consider that it might be a test, so they turned blind and deaf. Then **God** accepted the repentance from them. But again, many of them turned blind and deaf. **God** is watcher over what they do.

Many Modern Christians are Not the Followers of Jesus, the Messiah

5:72 Ingrates indeed are those who have said, "**God** is the Messiah son of Mary!" Although the Messiah had said, "O Children of Israel, serve **God**, my Lord and your Lord. Whoever sets up partners with **God**, then **God** will forbid paradise for him, and his destiny will be the fire; and the wicked will have no supporters."*

5:73 Ingrates indeed are those who have said, "**God** is the third of trinity!" There is no god but One god. If they do not cease from what they are saying, then those who reject from among them will be afflicted with a painful retribution.

5:74 Would they not repent to **God** and seek His forgiveness? **God** is Forgiving, Compassionate.

5:75 The Messiah son of Mary, was no more than a messenger; like the messengers before him. His mother was trustworthy, and they used to eat the food. See how We clarify the signs for them, then see how they deviate.

5:76 Say, "Do you serve other than **God** what cannot harm you or benefit you?" **God** is the Hearer, the Knower.

5:77 Say, "O followers of the book, do not be extreme in your system other than the truth, and do not follow the desires of a people who have been misguided before, and they misguide many; and they strayed from the right path."

5:78 Cursed are those who have rejected from among the Children of Israel by the tongue of David and Jesus son of Mary. That is for what they have disobeyed, and for what they transgressed.*

5:79 They would not stop each other from doing sin. Wickedness is what they used to do.

5:80 You see many of them allying the ingrates. Wicked is what they have provided for themselves, for **God**'s wrath is upon them, and in the retribution they will abide.

5:81 If they had only affirmed **God** and the prophet, and what was sent down to him, then they would not have taken them as allies; but many of them are corrupt.

5:82 You will find the people with greatest animosity towards those who affirm are the Jews and those who set up partners; and you will find the closest in affection to those who affirm are those who said, "We are Nazarenes;" that is because amongst them are Priests and Monks, and they are not arrogant.*

5:83 If they hear what was sent down to the messenger you see their eyes flooding with tears, for what they have known as the truth, they say, "Our Lord, we affirm, so record us with the witnesses."

5:84 "Why shouldn't we affirm **God** and what has come to us of the truth? We yearn that our Lord admits us with the righteous people."

5:85 So **God** rewarded them for what they have said with paradises that have rivers flowing beneath them, in them they shall abide; such is the reward of the good doers.

5:86 Those who rejected and denied Our signs, they are the dwellers of hell.

Do not Transgress by Fabricating Prohibitions in the Name of God

5:87 O you who affirm, do not forbid the good things that **God** has made lawful to you, and do not transgress; **God** does not like the transgressors.*

5:88 Eat from what **God** has provided for you, good and lawful; and be aware of **God** in whom you affirm.

5:89 **God** will not hold you for your unintentional oaths, but He will hold you for what oaths you have made binding with consideration. Its cancellation shall be the feeding of ten poor people from the average of what you feed your family, or that you clothe them, or that you free a slave; whoever cannot find shall fast for three days; as an atonement when you swear. You shall fulfill your oaths. It is such that **God** clarifies for you His signs that you may be thankful.*

Stay Away from Alcohol, Drugs, Gambling

5:90 O you who affirm, intoxicants, and gambling, and altars, and arrows of chance are tools of affliction used by the devil. You shall avoid him so that you may be successful.*

5:91 The devil only wants to cause strife between you through intoxicants and gambling, and to repel you away from remembering **God** and from the contact prayer. Will you then desist?

5:92 Obey **God** and obey the messenger; and beware. If you turn away, then know that the duty of the messenger is clear conveyance.

5:93 There is no sin upon those who affirm and promote reforms for what they eat when they are aware, affirm, and promote reforms; then they are aware and affirm, then they are aware and do good; **God** loves the good doers.

Respect the Environmental Restrictions

5:94 O you who affirm, **God** will test you with some of the game for hunting, coming within reach of your hands and your spears, so that **God** will know who reveres Him when unseen. Whoever transgresses after this, then he will have a painful retribution.

5:95 O you who affirm, do not kill any game while you are restricted; and whosoever of you kills it deliberately, then the recompense is to value what was killed against the livestock, which shall be judged by two equitable persons from you, and to make such as a donation to reach the Quadrangle. Or, its expiation shall be in using it to feed the needy ones, or by an equivalent fast; to taste the consequence of his deed; **God** forgives what has past. Whosoever reverts, then **God** will avenge it. **God** is Noble, Avenger.

5:96 Lawful for you is the catch of the sea, to eat it as enjoyment for you and for those who travel; and forbidden for you is the catch of the land as long as you are under restriction; and be aware of **God** to whom you will be summoned.

5:97 **God** has made the Quadrangle, the Restricted Sanctuary, a stance for the people; and the restricted month, the donations, the offerings so that you know that **God** knows what is in heavens and what is in earth, and that **God** is aware of all things.*

5:98 Know that **God** is powerful in retribution, and that **God** is Forgiving, Compassionate.

5:99 The only duty of the messenger is to deliver the message. **God** knows what you reveal and what you conceal.

Trivial Questions Create an Abundance of False Teachings

5:100 Say, "The bad and the good are not equal, even if the abundance of the bad pleases you." So be aware of **God**, O you who have intelligence, that you may succeed.

5:101 O you who affirm, do not ask about things, which if revealed to you, would harm you. If you ask about them after the Quran has been sent down, then they will become clear to you. **God** pardons for them, and **God** is Forgiving, Compassionate.*

5:102 A community before you had asked the same, then they became ingrates in it.*

Superstitions, Responsibility and Will

5:103 **God** did not prohibit livestock for begetting five or seven, for oaths, for begetting twins, for fathering ten; but the people who rejected invented lies to **God**, and most of them do not reason.

5:104 If they are told, "Come to what **God** has sent down, and to the messenger;" they say, "We are content with what we found our fathers doing." What if their fathers did not know anything nor were they guided?

5:105 O you who affirm, you are responsible for yourselves. None of the misguided can harm you when you are guided. To **God** is your return, all of you, and then He will inform you of what you have done.

5:106 O you who affirm, witnessing a will when death is approaching to one of you shall be done by two who are just amongst you. Or, if you have ventured out in the land, then two others may suffice if death is approaching you; you will hold them after the contact prayer, and let them swear by **God** if you have doubt, "We will not sell it for any price, even if he was a near relative, and we will not conceal the testimony of **God**, or else we are then of the sinners."

5:107 In case it is noticed that they were biased, then two others belonging to the discriminated party shall take their place. They should swear by **God**, "Our testimony is more truthful than their testimony, and if we transgress, then we are of the wicked."

5:108 This is best so that they would bring the testimony on its face value, for fear that their oaths would be discarded after being made. Be aware of **God** and listen; **God** does not guide the wicked people.

Dead Messengers and Saints Cannot Hear You

5:109 The day **God** will gather the messengers and He will say, "What was the response you received?"; they said, "We have no knowledge, you are the Knower of all the unseen."*

Jesus and His Disciples Were Muslims

5:110 **God** said, "O Jesus son of Mary, recall My blessings upon you and your mother that I supported you with the Holy Spirit; you spoke to the people in the cradle and in old age; and I taught you the book and the wisdom, and the Torah, and the Injeel; and you would create from clay the shape of a bird, then blow into it and it becomes a bird by My leave; and you heal the blind and the leper by My leave; and you brought out the dead by My leave. I have restrained the Children of Israel from you, that you came to them with proofs; but those who rejected amongst them said, "This is an obvious magic!""

5:111 I revealed the disciples that you shall affirm Me and My messenger; they Said, 'We affirm, and bear witness that we have peacefully surrendered.'

5:112 When the disciples said, "O Jesus son of Mary, can your Lord send down for us a feast from heaven?" he said, "Be aware of **God** if you are those who affirm."

5:113 They said, "We want to eat from it and to reassure our hearts, and so we know that you are truthful, and so that we can become a witness over it."

Witnessing Miracles Raises Responsibility

5:114 Jesus son of Mary said, "Our Lord and god, send down to us a feast from heaven, that it becomes a joy to us from the first to the last, and a sign from You, and provide for us, you are the best provider."*

5:115 **God** said, "I will send it down to you, but whoever rejects after this from amongst you, I will punish him a punishment like no one else from the worlds will be punished!"

5:116 **God** will say, "O Jesus son of Mary, did you tell the people to take you and your mother as gods besides **God**?" He said, "Be you glorified, I cannot say what I have no right to. If I had said it then You know it, You know what is in my mind while I do not know what is in Your mind. You are the Knower of the unseen."

5:117 "I only said to them what You commanded me to say, that you shall serve **God** my Lord and your Lord; and I was witness over them as long as I was with them, but when You took me, You were watcher over them. You are witness over all things."

5:118 "If You punish them, then they are Your servants, and if You forgive them, then You are the Noble, Wise."

5:119 **God** will say, "This is a day when the truth will benefit the truthful." They will have gardens with rivers flowing beneath; eternally they will abide therein. **God** has become pleased with them, and they are pleased with Him. Such is the greatest victory.

5:120 To **God** is the sovereignty of heavens, earth, and all that is in them; and He is capable of all things.

ENDNOTES

005:002 The prohibition on hunting during the Debate Conference not only provides a more peaceful social and psychological environment but also protects the ecosystem at a time when so many people visit Mecca from around the world. Without such a hunting prohibition during the time of such an international conference, the area would lose all its wild animals (2:196).

005:003 This verse does not add another item to the list in 6:145. It repeats the list and explains the two items, carcasses and those which are slaughtered for other than God. The Quran is explained by its author and verses of the Quran explain and detail each other (11:1; 6:119; 75:19).

005:004 We should remember God's name or attributes before eating. This remembrance could be verbal or mental and in any language we understand. There is no condition that this remembrance, commonly practiced as uttering *Bismillah,* should be done when the animals are slaughtered. In the following verse, we are told that the food of the People of the Book is permitted for us. We cannot expect them to remember God's name as we remember and the way we remember.

005:006 The Quran, which is complete, and sufficiently detailed, describes ablution in four steps. We should not add more steps to it. The Arabic word, which is translated "your feet," can be read in two different ways, *arjulakum* or *arjulikum*, changing its reference verb and thus changing the meaning from "wash your feet" to "wipe your feet." One may prefer one of these readings or follow both depending on occasion. See 4:43.

The Bible contains various references to ablution (Deuteronomy 23:10-15; Exodus 30:17-21; Leviticus 8:6; Psalms 26:6; Hebrews 10:22). The Bible also refers to Jesus being critical of Pharisees for being too scrupulous regarding the formalities of ablution, exactly like today's Sunnis and Shiites are (Matthew 23:25).

005:026 For the same event, see the Old Testament, Numbers 14:33-34; Deuteronomy 8:2-3.

005:027 Killing a human intentionally without justice, is equivalent to killing all of humanity (5:32), since if one loses his respect for the life of one human, he will lose his respect for the basic principle of the unity of humanity. For the Biblical version of this story, see Genesis 4:1-25; Joshua 15:57; Hebrews 11:4; 1 John 3:12; Jude 1:11.

005:031 Burying the body is not the only way of eliminating the health and psychological harms of decaying corpses. In some countries, burning the corpse is more economical than burying. Some scientists studying the intelligence level of birds consider the crow to be the bird with the highest I.Q (The I.Q. level of those scientists has not yet been investigated, though). For the Quranic relationship between sin and body, see 7:20.

005:033-34 The Quran contains hundreds of direct instructions to various groups. However, the followers of hadith and sunna miss this well-established Quranic language and ignore the obvious linguistic pattern. In this verse, the repeated use of the passive voice is not a coincidence; it is to indicate that the acts are not instructions, but statements of fact. We are reminded this point four times. In other words, those who roam the earth to promote and commit atrocities and bloodshed are going to get what they promote. Those who live by the sword die by the sword. (See: 6:65; 16:93; 2:251) No wonder, that the consequences listed in this verse is mentioned in the Quran twice, and in both cases as the punishment practiced by Pharoah, the enemy of freedom (7:123-124; 26:49). However, only by becoming peacemakers, people may avoid these painful and tragic consequences.

The Arabic word *fasad* means destruction, mischief, discord, warmongering, corruption. It is frequently contrasted with *islah* and its derivatives, which mean reform or promoting peace (7:56,85). *Fasad* is not mere faith or opinion; it refers to the acts of corruption, aggressive and destructive actions (See 2:30,205,251; 5:64; 10:91; 18:94; 21:22; 22:40; 28:4; 33:71; 89:12; 2:256. Contrast with 4:140 and 60:7-9). The Bible has a similar statement: "those who kill by the sword must die by the sword." See Matthew

26:52; Revelation 13:10. Also see the Quran 9:3-29.

005:038 The Arabic word we translated as "cut," occurs 36 times with all its derivatives. Nineteen of these are read in triliteral form *QaTaA*, and with the exceptions of 5:38 and 13:4, in all occurrences mean the non-physical or metaphorical action of "cutting off relationship" or "ending" (2:27; 3:127; 6:45; 7:72; 8:7; 9:121; 10:27; 11:81; 15:65; 15:66; 13:25; 22:15; 27:32; 29:29; 56:33; 59:5; 69:46). In 13:4, the noun form is used to describe "pieces," and its usage in 69:46 is understood by some scholars to be physical too. The derivatives that are read in the *QaTTaa* form occur 17 times. This form, which expresses intensity or frequency of the action, is used both to mean physical cutting off (5:33; 7:124; 20:71; 26:49; 13:31) and metaphorical cutting off (2:166; 6:94; 7:160; 7:167; 9:110; 47:15; 47:22; 21:93; 22:19; 23:53) as well as physically cutting or marking something (12:31; 12:50).

Verse 12:31 mentions women gathered for a party cutting their hands with knives, for they were awed by the physical attraction of Joseph. Obviously, those women did not cut off their hands.

Thus, the verse recommending punishment for theft or burglary, in the context of the Quran and its terminology (and not the terminology or interpretation attributed to Mu-hammed or his followers) provides us with a single verb … but one that God has permitted to incorporate a range of possible penalties. For instance:

1. Cutting or marking the person's hands as a means of public humiliation and identification;
2. Physically cutting off the person's hands; or
3. Cutting off a person's means and resources to steal and burglarize (presumably through rehabilitation or imprisonment).

The act of imposing any of these penalties, or any of their combinations, would of course depend on the facts of each case, the culpability and mental capacity of the accused, and the ability of the society as a whole to act in accordance with God's other instructions in the Quran. Note, for instance, that a Muslim society cannot punish a hungry person for stealing food, since letting a member of the society go hungry is a much bigger crime than the act of stealing food. Such a society demonstrates the characteristics of a society of unappreciative people! (See 107:1-7; 89:17-20; and 90:6-20). Considering theft solely as an individual crime and advocating the severest possible interpretation of the language in rendering punishment, is neither fair nor consistent with the scripture promoting mercy (7:52). Cutting off organic hands is permanent, and in case of error it is an irreversible injustice. Thus, a peacemaker society will not cut off organic hands, but will cut off causal hands of thieves by eliminating causes related to poverty, mental health, education, culture, politics, and economic inequalities.

Interestingly, Chapter 12, which uses the same expression for little cut inflicted on one's hand while peeling fruits (12:31); yes, this very chapter provides us an example of punishment for committing theft: detention and prison! It is also instructive, since this punishment (arrest and temporary prison) is practiced by a Prophet, namely, Joseph. See 12:70-76.

It is not right to see theft merely as an individual crime. Family structures, social norms, economic systems, social institutions, and values are also responsible for this crime. A society might reduce the crime of theft and burglary by increasing the strength of the family structure and ties, by establishing social and charity institutions, prohibiting usury, changing their wasteful lifestyle, and distributing wealth in a way that is more egalitarian. Punishment alone is not the right way to fight crimes; preventive measures need to be taken.

According to the Bible, thieves were punished by punitive restitution (Exodus 22:1-8; 2 Samuel 12:6). If the thief could not pay the fine, he was to be sold to a Hebrew master till he could pay (Exodus

22:1-4). A night-thief might be smitten till he died (Exodus 22:2).

005:042-43 This verse alone is an answer to those who consider Muhammed to be God's partner in legislation (*hukm*); the messenger was not a co-legislator (*shari*) of Islam. A group of Jews who sought Muhammed to judge among them was criticized for not referring to the Bible. The criticism, as it seems, had two reasons: Muhammed was not a legislator in the divine sense. Moreover, they should not jurisdiction-shop; they should stick with the laws they had agreed to apply in their community. It seems that no Muslim during the time of Muhammed demonstrated the ignorance of seeking Muhammed's judgment on matters expressed in the Quran. No wonder, 6:114; 98:5 and many other verses emphasize that the authority in Islamic system is God alone. However, when issues do not involve universal and absolute divine rules, peacemakers should refer to their elected officials and their appointees. Since Muhammed was the elected leader of the city-state of Yathrib, which had multiple jurisdictions accommodating a racially and religiously diverse population, he was in position to make judgments in daily affairs and the application of its secular Constitution. See 4:60. In brief, God alone has the power to legislate the universal laws of peacemaking, and the leaders elected by peacemakers should follow those laws. They may issue laws, as long as they do not contradict universal laws issued by God, after consultation, and subject to change whenever the society wishes so.

For Biblical verses on "laws," see Genesis 26:5; Exodus 16:28; 18:16,20; Leviticus 26:46; Ezra 7:25; Nehemiah 9:13-14; Esther 1:19; Psalms 105:45; Isaiah 24:5; Ezekiel 43:11; 44:5,24; Daniel 7:25; 9:10; Hebrews 8:10; 10:16.

005:044 The Torah is known as the name of the book given to Moses. However, this common knowledge might be wrong. The Quran nowhere mentions that the Torah was given to Moses. The word Torah occurs 18 times in the Quran, and it refers to the collection of books given to Jewish prophets excluding the *Injeel* delivered by Jesus. Though the name of the book(s) given to Moses is not mentioned, it is described with various adjectives such as *Furqan* (the distinguisher), *Emam* (the leader), *Rahma* (mercy), *Noor* (light) and *Huda* (guidance).

There is ample scholarly work indicating that current books of the Old Testament were extensively tampered with and written by Ezra and other rabbis. See 2:59; 2:79; 9:30-31. Also, see the Old Testament, Jeremiah 8:8.

005:045 Compare it to Exodus 21:24-36. Though the Old Testament version contains many more instructions, it does not mention the Quranic version reminding the Children of Israel that remission or giving up the punishment is a better choice. The Quranic version of the Old Testament rule is the Golden-plated Brazen rule, or "Generous tit-for-tat" which is the most effective moral rule in the real world. See 9:2-29. Also see 8:61; 4:90; 41:34.

005:048-50 Prophet Muhammed was not judging among people according to divine instructions besides the Quran. The Quran was the only divine authority. However, decades after Muhammed, those who turned away from the Quran turned Muhammed himself into a competing, supplementing, or contradicting authority through *hadith* collections and thereby regressed to the days of ignorance. The system belongs to God alone. Muhammed's sole mission was to deliver God's system to humanity. See 6:19, 112-116.

005:051-57 Verse 5:57 provides the characteristics of the people of the book, whom we are instructed not to ally ourselves with. Muslims should not ally themselves with those who insult their values or promote hatred against them. Obviously, this instruction is in the context of religious conflict, not in the context of social, personal, or financial relationship. There are righteous people among the people of the book (3:113; 3:199: 5:69). Eating their food or marriage with them is permitted (5:5). Verses 5:57 and 60:7-9 are among the verses that provide us with basic principles in dealing with others.

005:060 See 2:65 and 7:166.

005:069 See 2:62.

005:072-76 Despite the Pauline Church's polytheistic teaching that transformed ONE God into a fictional Triune god with contradictory multiple personalities, the Gospels are filled with verses rejecting the deity of Jesus. See John 7:16,28-29; 5:30; 14:28; 12:44-50; 20:127; Mark 10:17-18; 12:29; Matthew 7:21. One of the chapters in 19 Questions for Christians is dedicated to the discussion of the few Biblical verses abused to promote polytheism. See 4:171. Also, see 2:59; 3:45,51-52,55; 4:11,157; 7:162; 19:36.

005:078 The following Biblical verses where David and Jesus condemn the unappreciative people among the Children of Israel fit the Quranic reference: Psalms 109; Matthew 16:11-12; 23:1-39; Luke 12:1-2.

"Woe to you, teachers of the law and Pharisees, you hypocrites! You shut the kingdom of heaven in men's faces. You yourselves do not enter, nor will you let those enter who are trying to. [15]"Woe to you, teachers of the law and Pharisees, you hypocrites! You travel over land and sea to win a single convert, and when he becomes one, you make him twice as much a son of hell as you are. [16]"Woe to you, blind guides! You say, 'If anyone swears by the temple, it means nothing; but if anyone swears by the gold of the temple, he is bound by his oath.' [17]You blind fools! Which is greater: the gold, or the temple that makes the gold sacred? [18]You also say, 'If anyone swears by the altar, it means nothing; but if anyone swears by the gift on it, he is bound by his oath.' [19]You blind men! Which is greater: the gift, or the altar that makes the gift sacred? [20]Therefore, he who swears by the altar swears by it and by everything on it. [21]And he who swears by the temple swears by it and by the one who dwells in it. [22]And he who swears by heaven swears by God's throne and by the one who sits on it. [23]"Woe to you, teachers of the law and Pharisees, you hypocrites! You give a tenth of your spices—mint, dill and cumin. But you have neglected the more important matters of the law—justice, mercy and faithfulness. You should have practiced the latter, without neglecting the former. [24]You blind guides! You strain out a gnat but swallow a camel. [25]"Woe to you, teachers of the law and Pharisees, you hypocrites! You clean the outside of the cup and dish, but inside they are full of greed and self-indulgence. [26]Blind Pharisee! First clean the inside of the cup and dish, and then the outside also will be clean. [27]"Woe to you, teachers of the law and Pharisees, you hypocrites! You are like whitewashed tombs, which look beautiful on the outside but on the inside are full of dead men's bones and everything unclean. [28]In the same way, on the outside you appear to people as righteous but on the inside you are full of hypocrisy and wickedness. [29]"Woe to you, teachers of the law and Pharisees, you hypocrites! You build tombs for the prophets and decorate the graves of the righteous. [30]And you say, 'If we had lived in the days of our forefathers, we would not have taken part with them in shedding the blood of the prophets.' [31]So you testify against yourselves that you are the descendants of those who murdered the prophets. [32]Fill up, then, the measure of the sin of your forefathers! [33]"You snakes! You brood of vipers! How will you escape being condemned to hell? (Matthew 23:13-33).

To learn the identity of a "Pharisee" and a "hypocrite" who turned the above Biblical condemnation into a prophecy, see the New Testament, Acts 23:6; 1. Corinthians 9:20-22; Acts 15:36-41; Galatians 2:11-14; 4:10-14; Galatians 3:13 & Deuteronomy 21:23. See the Quran 2:59; 3:45,51-52,55; 4:11,157,171; 5:72-79; 7:162; 19:36.

005:082 The word *Masihiy* (Christian) is not used in the Quran. The Quran refers to the supporters of Jesus with the Arabic word, *Nasara*, literally meaning "supporters (2:62,111,113,120,135,140; 5:14,18, 51,69,82; 9:30; 22:17). When Jesus asked them who would be his supporters in the cause of God, his disciples answered, "We are supporters of God" (3:52; 61:14). It is interesting that Jesus was called Jesus the Nazarene in the Bible, from the name of the town of *Nasara*. The name of the town *Nasara* (supporters) perhaps comes from what Jesus' disciples called themselves. Ironically, the name of the son of Mary was not Jesus either. Christianity, as the product of the Pharisee-son-of-a-Pharisee is a polytheistic religion with a fabricated name and falsely attributed to a fabricated idol (See 3:45 and 2:62).

When we read this verse within context, starting from verse 5:77, it becomes clear that the Jews referred to in verse 5:82 are not all Jews, but those who are considered unappreciative of God's blessings and message.

005:087 The Quran condemns those who decree laws in the name of God as the worst people. Throughout history, Satan used clergymen of all sorts as his favorite tool to fight against God's system by introducing a myriad of man-made rules and prohibitions in the name of God. They turned God's system into a theo-fiction consisting of superstitions, collections of silly rituals, frivolous rules, absurd doctrines, numerous prohibitions, opium of masses, intellectual slavery and wishful thinking. See 6:21,145; 7:17,30,31,37; 42:21.

005:089 Slavery is prohibited by the Quran. See 4:3,25;92; 3:79; 5:89; 8:67; 24:32-33; 58:3; 90:13; 2:286; 12:39-42; 79:24.

005:090 *Hamr* is derived from the verb that means, "to cover," and is used for all intoxicants and drugs that cover the mind. Translating it as "wine" limits its intended reference. While accepting some benefits of intoxicants, the Quran prohibits its intake (2:219). This prohibition, as I understand, is categorical, and the Quran does not recommend the slippery and deceptive measure known as "drinking in moderation." The Quran does not legally prohibit alcohol though. It is up to society to regulate its consumption. The societies that lack strict moral or religious norms against alcohol pay a hefty price in terms of loss and problems with economy, social life, relationships, health, and personal happiness. See 2:219.

The word *ansab* denotes altars and shrines. For instance, worshipping *Kaba*, which was originally a temple built to worship God alone violates the principles of monotheism. *Kaba* is not a holy building, but merely a historic point for the annual conference of Monotheists. Visiting a place to commemorate God does not mean idolizing such a place. The black stone next to *Kaba*, which was perhaps originally used as a marker to indicate the start of circumvolution, is another big idol with millions of worshipers. Muhammed's grave, which did not exist during the revelation of the Quran, is too one of the most popular idols in the world. All these shrines and stones are within the reference set of the word *ansab*. Contrary to unequivocal verses of the Quran, millions of Sunnis and Shiites worship rocks, stones, shrines, and the rotten bones of their idols. See 2:48; 25:3.

005:097 The Quran mentions only one restricted *masjid*; yet the followers of *hadith* and *sunna* centuries after the revelation of the Quran increased the number to two by calling Muhammed's tomb the second "restricted masjid."

005:101 The Quran explains and details all that we need for attaining salvation. Those who were not satisfied with the details provided by God, asked many trivial questions like some Jews asked before (2:67-71), and since they could not find answers in the Quran to their questions, they were attracted to *hadith*. Because of the great demand for more and more details on trivial issues, *hadith* narration and collection became a booming business.

To reject numerous verses informing us that the Quran is easy to understand, fully detailed, complete, and sufficient for guidance; and to add volumes of fabricated religious rules made up by Ibn Filan and Abu Falan, Sunni and Shiite mushriks ask us: "If the Quran is complete, fully detailed and clear as God says, can you tell us how we can pray by using Quran alone?"

They are so confident as if their hadith books contain video clips of their idol, Muhammad while praying. Furthermore, their prayer is not based on description of hadiths either, since hadith books contain numerous contradictory descriptions of prayer, and the number of *rak'as* (units) which they are so obsessed with.

Ironically, the information regarding the details of prayer is not their only problem. They have bigger problem, much bigger. Without their fabricated *hadiths, sunna* and *sharia,* they cannot eat, they cannot sleep, and they cannot even go to bathroom!

We are told many things ranging from, which order you should clip your fingernails in, how long you should grow your beard and moustache, on which side you should sleep, to which foot you should use to first step into the bathroom, and how many rocks you should use to clean your buttock. You will find long chapters in hadith books dedicated to details of potty training. (9:31; 42:21).

005:102 Jewish Rabbis produced piles of Mishnah (words) and Gemara (practice) to explain, detail, and supplement God's book, which contained sufficient explanation and details (5:13-15; 6:154; 7:145; 37:117). Thus, they distorted God's system. Muslims too repeated the exact same blunder and produced piles of *Hadith* (words) and *Sunna* (practice), to explain, detail, and supplement God's book, which contained sufficient explanation and details (6:38; 114; 12:111; 16:89). As a result, they divided themselves into sects and factions (23:52-56; 25:30). Many people with sound minds have chosen to stay away from this mishmash collection of nonsense. Agnostics are much closer to God than the majority of people who follow manmade teachings in the name of God.

005:109 What about those who believe that Muhammed would save them on the Day of Judgment? They will be rejected by their idol on that day, and they will be surprised to hear his intercession in the form of testimony to the truth (25:30). Also, see 2:48.

005:114-115 The responsibility of those who witness the mathematical system of the Quran is similarly great since the mathematical system is one of the greatest prophecies and divine signs (74:35). It is also of note, that (J)esu(s) is frequently referred to as "son of Mary" in the Quran.

بسم الله الرحمن الرحيم

6:0 In the name of God, the Gracious, the Compassionate.

6:1 Praise be to **God** who created the heavens and the earth, and Who made darkness and light. Yet those who have been unappreciative continue to equate others with their Lord.

6:2 He is the One who created you from clay, then decreed a fixed time period for you; a time defined by Him. Yet you are still skeptical.

6:3 He is **God** in the heavens and the earth. He knows your secrets and what you reveal; and He knows what you earn.

They do not Witness the Signs

6:4 Whenever a sign came to them from their Lord, they turned away from it.

6:5 They have denied the truth when it came to them. The news will ultimately come to them of what they were mocking.

6:6 Did they not see how many generations We have destroyed before them? We granted them dominance over the land more than what We granted you, We sent the clouds to them abundantly, and We made rivers flow beneath them; then We destroyed them for their sins, and established after them a new generation.

6:7 If We had sent down to you a book already written on paper, and they touched it with their own hands, then those who have rejected would say, "This is but clear magic!"

6:8 They said, "If only a controller were sent down to him?" But if We had sent down a controller, then the matter would be over, and they would not be reprieved.

6:9 If We had chosen an controller, We would have made him appear as a man, and We would have confused them in what they already are confused.

6:10 Messengers before you were ridiculed; but those who mocked them suffered the consequences of what they ridiculed.

6:11 Say, "Roam the earth, then see what the punishment of the ingrates was!"

God's Compassion

6:12 Say, "To whom is all that is in heavens and earth?" Say, "To **God**." He has decreed mercy on Himself, that He will gather you to the day of Resurrection in which there is no doubt. Those who have lost themselves, they do not affirm.*

6:13 For Him is what resides in the night and in the day; and He is the Hearer, the Knower.

6:14 Say, "Shall I take other than **God** as an ally; the creator of heavens and earth; while He feeds and is not fed!" Say, "I have been commanded to be the first who peacefully surrenders. Do not be of those polytheists."

6:15 Say, "I fear if I disobey my Lord, the retribution of a great day!"

6:16 Whoever He spares from it, then He had treated him with mercy; and that is the obvious success.

6:17 If **God** were to touch you with harm, then none can remove it except Him; and if He were to touch you with good, then He is capable of all things.

6:18 He is the Supreme over His servants; and He is the Wise, the Ever-aware.*

Muhammed was Given Only the Quran

6:19 Say, "Which is the greatest testimony?" Say, "**God** is witness between me and you, and He has revealed to me this Quran that I may warn you with it and whomever it reaches. Do you bear witness that along with **God** are other gods?" Say, "I do not bear witness!" Say, "He is only One god, and I am innocent of your setting up partners!"*

6:20 Those to whom We have given the book know it as they know their children. Those who lost themselves, they do not affirm.

6:21 Who is more wicked than one who invents lies about **God** or denies His signs! The wicked never succeeds.

Polytheists Deny Their Polytheism

6:22 The day We gather them all, then We say to those who set up partners: "Where are your partners whom you used to claim?"*

6:23 Then, their only blunder was to say, "By **God**, our Lord, we did not set up partners!"

6:24 See how they lied to themselves; and what they invented deserted them.

Hypocrites Call the Divine Signs Mythology

6:25 Among them are those who listen to you; and We have made covers over their hearts to prevent them from understanding it, and deafness in their ears. If they see every sign they will not affirm; even when they come to you, they argue. Those who reject say, "This is nothing but the tales from the past!"*

6:26 They are deterring others from it, and keeping away themselves; but they will only destroy themselves, yet they do not notice.

6:27 If you could see when they are standing over the fire, they say, "If only we were sent back, we would not deny the signs of our Lord, and we would be those who affirm!"

6:28 No, it is merely what they have been concealing that has now become evident to them; and if they were sent back they would return to what they were deterred from doing, they are but liars!

6:29 They had also said, "There is only this worldly life, and we will not be resurrected!"

6:30 If you could see them when they are standing at their Lord? He said, "Is this not the truth?" They said, "Yes, by our Lord." He said, "Then taste the retribution for what you have rejected."

6:31 Losers are those who have denied their meeting with **God**; until the Moment comes to them suddenly, then they say, "We deeply regret what we have wasted in it," and they will carry their burdens on their backs; miserable indeed is their burden.

Don't be Distracted by the Trivial Issues

6:32 This worldly life is nothing more than games and a diversion, and the abode of the Hereafter is far better for those who are aware. Do you not understand?

6:33 We know that you are saddened by what they say; they are not rejecting you, but it is **God**'s signs which the wicked disregard.

6:34 Messengers before you were denied, but they were patient for what they were denied, and they were harmed until Our victory came to them; nothing changes the words of **God**. News of the messengers has come to you.

Warning to the Messenger: Do not Be Ignorant

6:35 If their aversion has become too much for you, then perhaps you could make a tunnel in the earth, or a ladder to the heavens, and bring them a sign. Had **God** willed, He would have summoned them to the guidance; so do not be of the ignorant ones.

6:36 Only those who listen will respond. As for the dead, **God** will resurrect them, then to Him they will return.

Everything is Recorded

6:37 They said, "If only a sign was sent to him from his Lord!" Say, "**God** is able to send a sign, but most of them would not know."

6:38 There is not a creature in the earth, or a bird that flies with its wings, but are nations like you. We did not leave anything out of the book; then to their Lord they will be summoned.*

6:39 Those who deny Our signs, they are deaf and dumb, in darkness. Whoever **God** wishes He misguides, and whoever He wishes He makes him on a straight path.

6:40 Say, "We will see when **God**'s retribution comes to you, or the Moment comes to you, if you will still call on any other than **God**. If you are truthful."

6:41 "No, it is Him alone you will call on; then He will remove what you called Him for, if He wills; you will forget what you set up."

6:42 We have sent others to nations before you. We then took them with adversity and hardship, perhaps they would implore.

6:43 If only they had implored when Our punishment came to them; but their hearts hardened, and the devil adorned for them what they used to do.

The System

6:44 So when they forgot what they had been reminded of, We opened for them the gates of all opportunities; and when they became haughty with what they were given, We took them suddenly! They were confounded.

6:45 It was thus that the remainder of the wicked people was wiped out; and praise be to **God**, the Lord of the worlds.

6:46 Say, "Do you see that if **God** were to take away your hearing and your eyesight, and He seals your hearts; which god besides **God** can bring it to you?" Look how We cite the signs, but then they turn away.

6:47 Say, "Do you see if **God**'s retribution comes to you suddenly, or gradually, will anyone be destroyed except the wicked people?"

Messengers are not Supermen

6:48 We do not send the messengers except as bearers of good news and warners; whoever affirms, and reforms, then there is no fear for them, nor will they grieve.

6:49 Those who deny Our signs, the retribution will touch them for their wickedness.

6:50 Say, "I do not say to you that I possess **God**'s treasures, nor do I know the future, nor do I say to you that I am an controller. I merely follow what is revealed to me." Say, "Are the blind and the seer the same? Do you not think?"

6:51 Warn with it those who realize that they will be summoned to their Lord; they do not have besides Him any protector nor intercessor; so that they may be righteous.*

6:52 Do not turn away those who call on their Lord at dawn and dusk seeking His presence; you are not responsible for their account, nor are they responsible for your account; if you turn them away, then you will be of the wicked.

6:53 Thus, We test them with one another, so that they may say, "Are these the ones whom **God** has blessed from amongst us?" Is **God** not aware of the thankful?

6:54 If those who affirm Our signs come to you, then say, "Peace be upon you, our Lord has decreed mercy upon Himself, that any of you who commits sin out of ignorance and then repents afterwards and reforms, then He is Forgiving, Compassionate."

6:55 It is such that We explain the signs, and make clear the way of the criminals.

6:56 Say, "I am warned to stop serving those you call upon other than **God.**" Say, "I will not follow your desires; otherwise, I would go astray and I would not be of those guided."*

6:57 Say, "I am on a proof from my Lord and you have denied it. I do not have what you hasten towards; the judgment is with **God** only; He carves out the truth, and He is the best analyzer."*

6:58 Say, "If I had what you are hastening for, then the matter between us would have been resolved. **God** is fully aware of the wicked."

God the Omniscient, the Omnipotent

6:59 With Him are the keys to the unseen, none knows them except He. He knows what is in the land and in the sea; and not a leaf falls except with His knowledge; nor a seed in the darkness of the earth; nor anything moist or anything dry; all is in a clear book.*

6:60 He is the One who takes you at night, and He knows what you have done during the day, then He sends you back to it until a fixed span; then to Him is your return and He will inform you of what you used to do.*

6:61 He is the Supreme over His servants, and He sends over you guardians. So that when the time of death comes to one of you, Our messengers take him, and they do not neglect any.*

6:62 Then they are returned to **God**, their true Patron; to Him is the judgment and He is the swiftest in reckoning.

6:63 Say, "Who rescues you from the darkness of the land and the sea?," you call on Him openly and secretly: "If You save us from this, we will be of the thankful!"

6:64 Say, "**God** will save you from it and from all distresses, yet you still set up partners!"

6:65 Say, "He is able to send retribution from above you or from below your feet, or He will make you belong to opposing factions, then He will let you taste the might of each other." See how We cite the signs, perhaps they may comprehend.

Some of the News of the Quran left for Future Generations

6:66 Your people denied it, while it is the truth. Say, "I am not a guardian over you!

6:67 For every news there is a time, and you will come to know."

If they Make Ridicule God's Signs

6:68 If you encounter those who make fun of Our signs, then turn away from them until they move on to a different topic; and if the devil lets you forget, then do not sit after remembering with the wicked people.*

6:69 Those who are aware are not held accountable for them, but a reminder may help them to be aware.

6:70 Leave alone those who have taken their system as a game and diversion, and this worldly life has tempted them. Remind with it that a person will suffer for what s/he has earned, s/he will not have besides **God** any supporter nor intercessor; and even if s/he offers every ransom, none will be accepted from it. These are the ones who have suffered with what they have earned; for them will be a boiling drink, and a painful retribution for what they had rejected.

6:71 Say, "Shall we call upon other than **God** what cannot benefit us or harm us, and we turn back on our heels after **God** has guided us?" This is like the one whom the devils have managed to mislead on earth, he is confused, having friends who call him to the guidance: "Come to us!" Say, "The guidance is **God**'s guidance, and we have been ordered to peacefully surrender to the Lord of the worlds."

6:72 Maintain the contact prayer and be aware; and to Him you will be summoned.

6:73 He is the One who created the heavens and the earth with truth, and the day He says: "Be," then it is! His saying is truth; and to Him is the sovereignty, the day the trumpet is blown. Knower of the unseen and the seen; He is the Wise, the Ever-aware.

Abrahams Debates (Hajj) with People
The Abrahamic Method of Reasoning

6:74 Abraham said to his father, Azar: "Will you take idols as gods? I see you and your people clearly misguided."*

6:75 Thus We showed Abraham the kingdom of heavens and earth, so that he will be of those who have certainty.

6:76 When the night covered him, he saw a planet, and he said, "This is my Lord." But when it disappeared, he said, "I do not like those that disappear."

6:77 So when he saw the Moon rising, he said, "This is my Lord." But when it disappeared, he said, "If my Lord will not guide me, then I will be amongst the wicked people!"

6:78 So when he saw the Sun rising, he said, "This is my Lord, this is bigger." But when it disappeared, he said, "My people, I am innocent of what you have set up."

6:79 "I shall turn my face to the One who created the heavens and the earth, as a monotheist, and I am not of those who set up partners."*

6:80 His people argued with him. He said, "Do you argue with me regarding **God**, when He has guided me? I do not fear what you have set up except if my Lord wishes it so; my Lord encompasses all things with knowledge; will you not remember?"

6:81 "How can I possibly fear what you have set up; yet you do not fear that you have set up partners with **God**, for which He has not sent down to you any authority! So, which of our two groups is more worthy of security if you know?"

Pure Monotheism

6:82 Those who affirm and do not dress their affirmation with inequity; they will have security, and they are guided.*

6:83 Such was Our argument that We gave Abraham over his people; We raise the degree of whom We please. Your Lord is Wise, Knowledgeable.*

Prophets

6:84 We granted him Isaac and Jacob, both of whom We guided; and Noah We guided from before; and from his progeny is David, Solomon, Job, Joseph, Moses, and Aaron. It is such that We reward the good doers.

6:85 And Zachariah and John, and Jesus, and Elias; all were of the reformed.

6:86 Also Ishmael, Elisha, Jonah, and Lot; each We have distinguished over the worlds.

6:87 From their fathers and their progeny and their brothers We have also chosen; and We guided them to a straight path.

6:88 Such is the guidance of **God**, He guides with it whom He pleases of His servants. If they set up partners, then all that they had worked would fall away from them.*

6:89 Those to whom We have given the book, the law, and the prophethood, if they reject it, then We will en+1trust it to a people who will not reject it.

6:90 These are the ones guided by **God**; so let their guidance be an example. Say, "I do not ask you for a wage; this is but a reminder for the worlds."

The Most Wicked People

6:91 They did not value **God** as He deserves to be valued, for they said, "**God** has never sent down anything to any human being." Say, "Who then has sent down the book which Moses had come with, a light and guidance for the people? You treat it just as scrolls of paper; you show some of it and conceal much, though you were taught what neither you nor your fathers knew." Say, "**God** has." Then leave them playing in their folly.

6:92 This too is a book which We have sent down, blessed, authenticating what is before it, that you may warn the capital city and those around it. Those who affirm the Hereafter, will affirm this, and they are dedicated to their contact prayer.

6:93 Who is more wicked than one who invents lies about **God**, or says: "It has been revealed to me," when no such thing revealed to him; or who says: "I will bring down the same as what **God** has sent down." If you could only see the wicked at the moments of death when the controllers have their arms opened: "Bring yourselves out, today you will be given the disgraceful punishment for what you used to say about **God** without truth, and you used to be arrogant towards His signs."

Don't Expect to be Saved by an Intercessor

6:94 "You have come to Us individually, just as We had created you the first time; and you have left behind you all that We have provided for you. We do not see your intercessors with you that you used to claim were with you in partnership; all is severed between you, and what you have alleged has abandoned you."*

The Creator, Noble, Subtle, Ever-aware

6:95 **God** is the splitter of the seeds and the grains; He brings the living out from the dead and He is the bringer of the dead out from the living. Such is **God**; so why are you deluded?

6:96 Initiator of morning. He made the night for rest, and the Sun and the Moon as a calculation device. Such is the measure of the Noble, the Knowledgeable.

6:97 He is the One who made for you the stars, to guide you by them in the darkness of the land and the sea. We have explained the signs to a people who know.

6:98 He is the One who initiated you from one person; secured and reposited. We have explained the signs to a people who comprehend.

6:99 He is the One who sent down water from the sky, and We brought out with it plants of every kind. We brought out from it the green, from which We bring out multiple seeds; and what is from the palm trees, from its sheaths hanging low and near; and gardens of grapes, olives and pomegranates, similar and not similar. Look at its fruit when it blossoms and its ripeness. In this are signs for a people that affirm.

No appreciation of God

6:100 Yet, they made partners with **God** from among the Jinn, while He had created them. They invented for Him sons and daughters without having any knowledge! Be He glorified and far above what they describe.

6:101 Originator of the heavens and the earth, how can He have a son when He did not take a wife? He created all things, and He is knowledgeable over all things.

6:102 Such is **God**, your Lord, there is no god but He; creator of all things, so serve Him. He is caretaker over all things.

6:103 The eyesight cannot reach Him, yet He can reach all eyesight; and He is the Subtle, the Ever-aware.

6:104 "Visible proofs have come to you from your Lord; so whoever can see, does so for himself, and whoever is blinded, does the same. I am not a guardian over you."

6:105 It is thus that We cite the signs and that they may say, "You have studied," and We will make it clear for a people who know.

6:106 Follow what is revealed to you from your Lord; there is no god but He, and turn away from those who have set up partners.

6:107 If **God** had willed, they would not have set up partners. We did not place you over them as a guardian nor are you their lawyer.

Do not Provoke the Ignorant People

6:108 Do not insult those whom they invoke other than **God**, lest they insult **God** without knowledge. We have similarly adorned for every nation their works; then to their Lord is their return and He will inform them of what they had done.

Bigotry and Prejudice

6:109 They swore by **God** using their strongest oaths; that if a sign came to them they would affirm it. Say, "The signs are from **God**; and how do you know that once it comes, that they would not reject?"

6:110 We divert their hearts and eyesight, as they did not affirm it the first time; and We leave them wandering in their transgression.*

6:111 If We had sent down to them the controllers, and the dead spoke to them, and We had gathered before them everything, they still would not affirm except if **God** wills. But most of them are ignorant.

Hadith: Satanic Inspiration

6:112 We have permitted the enemies of every prophet, human and Jinn devils, to reveal in each other fancy words in order to deceive. Had your Lord willed, they would not have done it. You shall disregard them and their fabrications.*

6:113 That is so the hearts of those who do not affirm the Hereafter will listen to it, and they will accept it, and they will take of it what they will.

6:114 "Shall I seek other than **God** as a judge when He has sent down to you this book sufficiently detailed?" Those to whom We have given the book know it is sent down from your Lord with truth; so do not be of those who have doubt.*

6:115 The word of your Lord has been completed with truth and justice; there is no changing His words. He is the Hearer, the Knower.

6:116 If you obey the majority of those on earth, they will lead you away from **God**'s path; that is because they follow conjecture, and that is because they only guess.*

6:117 Your Lord is fully aware of who strays from His path, and He is fully aware of the guided ones.

Dietary Prohibitions

6:118 So eat from that on which **God**'s name has been mentioned; if you indeed affirm His signs.*

6:119 Why should you not eat that on which **God**'s name has been mentioned, when He has sufficiently detailed to you what has been made forbidden; except what you are forced to? Many misguide by their desires without knowledge; your Lord is fully aware of the transgressors.*

6:120 Leave what is openly a sin as well as what is discreet; those who earn sin will be punished for what they have taken.

6:121 Do not eat from what **God**'s name has not been mentioned, for it is vile. The devils reveal to their supporters to argue with you; and if you obey them, then you have set up partners.*

6:122 Is one who was dead and to whom We gave life, and We made for him a light to walk with amongst the people, as one whose example is in darkness and he will not exit from it? It is such that the work of the ingrates has been adorned for them.

6:123 As such, We have permitted in every town the influential from its criminals, to scheme in it. They only scheme their own selves without perceiving.

Blind Followers do not Appreciate Originality

6:124 If a sign comes to them they say, "We will not affirm until we are given the same as what **God**'s messengers were given!" **God** is fully aware of where He makes His message; those criminals will have debasement with **God** and a painful retribution for what they had schemed.*

6:125 Whomever **God** wishes to guide, He will open his chest towards peacefully surrender; and whomever He wishes to misguide, He will make his chest tight and constricted, as one who is climbing towards the sky. It is such that **God** afflicts those who do not affirm.

6:126 This is your Lord's straight path. We have sufficiently detailed the signs to a people who take heed.

6:127 They will have the abode of peace with their Lord; He is their supporter because of what they used to do.

6:128 The day He summons them all: "O tribes of Jinn, you have managed to take many humans." Their supporters from the humans said, "Our Lord, we have indeed enjoyed one another, and we have reached our destiny to which You delayed us." He said, "The fire is your dwelling, in it you shall abide eternally, except as your Lord wishes." Your Lord is Wise, Knowledgeable.

6:129 It is such that We make the wicked

as supporters to one another for what they had earned.

God Does not Commit Injustice

6:130 "O you tribes of Jinn and humans, did not messengers come to you from amongst you and relate to you My signs, and warn you of the meeting of this day?" They said, "Yes, we bear witness upon ourselves;" and the worldly life deceived them, and they bore witness on themselves that they were ingrates.

6:131 That is because your Lord was not to destroy any town because of its wickedness while its people were unaware.

6:132 But there are degrees for what they had done; and your Lord is never unaware of what they do.

6:133 Your Lord is Rich, full of Compassion. If He wished, He could remove you and bring after you whom He pleases, just like He established you from the seed of another people.

6:134 What you have been promised will come, you cannot escape it.

6:135 Say, "My people, do your best, and so will I. You will then know those who deserve the ultimate abode. The wicked will not succeed."

Prohibitions in the Name of God

6:136 Of crops and the livestock He provided, they set aside a portion for **God**, they said, "This is for **God**," according to their claim, "This is for our partners." However, what was for their partners did not reach **God**, while what was for **God** reached their partners! Miserable is how they judged.*

6:137 Thus, for those who set up partners their partners adorned the killing of born children, to turn them and to confuse their system for them. Had **God** willed they would not have done this, so ignore them and what they invent.*

6:138 They said by asserting that: "These are livestock and crops that are restricted, and none shall eat from them except as we please." Even, livestock whose backs are forbidden, and livestock over which they do not mention **God**'s name as invention against Him. He will recompense them for what they invented.

Further Discrimination against Females

6:139 They said, "What is in the bellies of these livestock is purely for our males and forbidden for our spouses, and if it comes out dead, then they will be partners in it." **God** will recompense them for what they describe. He is Wise, Knowledgeable.

6:140 Losers are those who have killed their born children foolishly, without knowledge, and they forbade what **God** had provided for them by attributing lies to **God**. They have strayed and they were not guided.

Purify your Income by Sharing it Immediately With the Needy

6:141 He is the One Who initiated gardens; both trellised and untrellised; palm trees, and plants, all with different tastes; and olives and pomegranates, similar and not similar. Eat from its fruit when it blossoms and give its due on the day of harvest; and do not waste. He does not like the wasteful.*

6:142 From the livestock are those for burden, and also for clothing. Eat from what **God** has provided you and do not follow the footsteps of the devil; he is to you a clear enemy.

6:143 Eight pairs: from the sheep two, and from the goats two; say, "Is it the two males that He forbade or the two females, or what the womb of the two females bore? Inform me if you are truthful!"*

6:144 From the camels two, and from the cattle two; say, "Is it the two males that He forbade or the two females, or what the womb of the two females bore? Or were you witnesses when **God** ordered you this?" Who is more wicked than one who invents lies about **God** to misguide the people without knowledge? **God** does not guide the wicked people.*

Enumerated List of Dietary Prohibitions; Adding Extra Prohibitions in the Name of God is Polytheism

6:145 Say, "I do not find in what is revealed to me forbidden except that it be already dead, or blood poured forth, or the meat of pig, for it is foul; or what is corruptly dedicated to other than **God**." Whoever is forced without seeking disobedience or transgression, then your Lord is Forgiving, Compassionate.*

6:146 For those who are Jewish, We forbade all that have claws; and from the cattle and the sheep We forbade their fat except what is attached to the back, or entrails, or mixed with bone. That is a punishment for their rebellion, and We are truthful.

6:147 If they deny you, then say, "Your Lord has vast Compassion, but His Might will not be turned away from the criminal people."

6:148 Those who set up partners will say, "If **God** wished, we would not have set up partners, nor would have our fathers, nor would we have forbidden anything." Those before them lied in the same way, until they tasted Our might. Say, "Do you have any knowledge to bring out to us? You only follow conjecture, you only guess."*

6:149 Say, "**God** has the conclusive argument. If He wished, He would have guided you all."*

6:150 Say, "Bring forth your witnesses who bear witness that **God** has forbidden this." If they bear witness, then do not bear witness with them, nor follow the desires of those who deny Our signs, and those who do not affirm the Hereafter; and they make equals with their Lord!*

6:151 Say, "Come let me recite to you what your Lord has forbidden for you: that you should not set up anything with Him; and be kind to your parents; and do not kill your born children for fear of poverty, We provide for you and for them; and do not come near lewdness, what is plain of it or subtle; and do not kill the person which **God** has forbidden, except in justice. That is what He enjoined you that you may comprehend."

6:152 "Do not approach the wealth of the orphan, except for what is best, until he reached his maturity; and give honestly full measure and weight equitably. We do not burden a person beyond its capacity. When you speak you must observe justice, even if against a relative. You shall also observe the pledges you made to **God**. He has enjoined you this so that you may take heed."

6:153 This is My path, a Straight One. "So follow it, and do not follow the other paths lest they divert you from His path. That is what He has enjoined you so that you may be aware."

God's Signs Only Help the Open-Minded

6:154 Then We gave Moses the book, complete with the best and sufficiently detailing all things, and a guide and mercy that they may affirm the meeting of their Lord.

6:155 This too is a blessed book that We have sent down. So follow it and be aware, that you may receive mercy.

6:156 Lest you say, "The book was only sent down to two groups before us, and we were unaware of their study!"

6:157 Or you say, "If the book was sent to us we would have been more guided than they!" Proof has come to you from your Lord, guidance and a mercy. Who is more wicked than one who denies **God**'s signs and turns away from them? We will punish those who turn away from Our signs with the pain of retribution for what they turned away.

6:158 Do they wait until the controllers will come to them, or your Lord comes, or some signs from your Lord? The day some signs come from your Lord, it will do no good for any person to affirm if s/he did not affirm before, or s/he gained good through his/her affirmation. Say, "Wait, for we too are waiting."

Sunni and Shiite Sects Have Nothing to do with the Prophet

6:159 Those who have divided their system and become sects, you have nothing to do with them. Their matter will be with **God**, then, He will inform them of what they had done.*

6:160 Whoever comes with a good deed, he will receive ten times its worth, and whoever comes with a sin, he will only be recompensed its like; they will not be wronged.

6:161 Say, "My Lord has guided me to a Straight Path. An upright system, the creed of Abraham the monotheist; he was not from those who set up partners."

6:162 Say, "My contact prayer, my devotion and offerings, my life, and my death, are all to **God** the Lord of the worlds."

6:163 "He has no partner, and this is what I was commanded; I am the first of those who have peacefully surrendered."

6:164 Say, "Shall I seek other than **God** as a Lord when He is the Lord of everything?" Every person earns what is for it, and none will bear the burden of another. Then to your Lord is your return and He will inform you regarding your disputes.*

6:165 He is the One who made you successors on earth, and He raised some of you over others in grades, to test you with what He had given you. Your Lord is quick to punishment, and He is Forgiving, Compassionate.

ENDNOTES

006:012 The most repeated attributes of God are those that emphasize His mercy and care towards His creation.

006:018 See 53:42-62.

006:019 Accepting other authorities besides the Quran to define God's system is considered polytheism (6:112-115; 7:2-3; 9:31; 17:46; 16:89; 45:6; 42:21).

006:022-24 Those who associate man-made religious teachings and the opinions of their religious scholars with God's word and judgment are *mushriks* (polytheists), and yet they deny their own *shirk* (polytheism). The relationship between the verses is obvious (Also, see 3:80; 6:148; 16:35; 53:19-23).

006:025 The followers of *hadith* and *sunna* are implied by this verse, because of their depiction of the Quran's code 19 as a "myth." A book written by four top Sunni scholars in Turkey, including Edip Yuksel's father, picked the "The Myth of 19" as the title of their book. Their choice of words to reject God's signs, demonstrates their intellectual and spiritual genetic link with their unappreciative ancestors; this is no coincidence (47:29-30; 74:30-31).

006:038 This verse is primarily about the master record that records everything we do. However, with its secondary meaning it implies the Quran too, since it contains all we need to attain eternal salvation (16:89).

006:051 See 2:48.

006:056 Also, see 42:52; 93:7.

006:057 The fanatics who are slaves of their environment, traditions, irrational interests, and unguided emotions, cannot and would not want to see/appreciate God's signs.

006:059 Said Nursi, an influential Sunni scholar and political figure, in his incomplete commentary of the Quran, *Isharat ul-'Icaz* draws our attention to the implication of verses narrating the miracles of previous messengers. He argued that the Quran, expressly or implicitly, symbolically, or indicatively contains all the information in the world. Though this fancy assertion is not plausible, his argument regarding the purpose of the miracles mentioned in the Quran makes sense: to support their missions in their lifetime and to show us the ultimate goals of scientific and technological progress, and to inspire us to repeat them by acquiring the knowledge of divine laws embedded in nature. See 6:38; 16:89.

006:060 See 39:42.

006:061 God uses protectors'/preservers' powers to download the program/data information containing conscious personalities from human brains and upload them to the master record. These protectors/preservers are mentioned in plural, and thus the verse rejects the traditional belief about *Azrail*, the so-called sole controller of death.

006:068 God has given ultimate freedom of thought, belief and expression (2:256; 4:137, 140; 6:110; 10:99; 18:29; 88:20-21). Those who threaten the lives of those who criticize or insult their faith, ignore God's advice in this and other verses. If God asks us to ignore those who insult and mock God's revelations and signs by protesting passively, how can we get enraged and mad in the name of the same God? We should enlighten ignorant people without falling to their level. The technique known as the Socratic Method is an efficient and positive method for discussing issues involving religion and faith. The following verses, 6:74-83, will give us an example from Abraham's method for pulling the attention of the faithful people to engage in a rational discussion.

"The Lord said to Moses: 'Take the blasphemer outside the camp. All those who heard him are to lay their hands on his head, and the entire assembly is to stone him. Say to the Israelites: If anyone curses his God, he will be held responsible; anyone who blasphemes the name of the Lord must be put to death. The entire assembly must stone him. Whether an alien or native-born, when he blasphemes the Name, he must be put to death.... Then Moses spoke to the Israelites, and they took the blasphemer outside the camp and stoned him. The Israelites did as the Lord commanded Moses" (Leviticus 24:13-16).

"If your very own brother, or your son or daughter, or the wife you love, or your closest friend secretly entices you, saying, 'Let us go and worship other gods' ... Show him no pity. Do not spare him or shield him. You must certainly put him to death. Your hand must be the first in putting him to death, and then the hands of all the people. Stone him to death, because he tried to turn you away from the Lord your God, who brought you out of Egypt, out the land of slavery. Then all Israel will hear and be afraid, and no one among you will do such an evil thing again" (Deuteronomy 13:6-11).

006:074-83 Abraham was a rational monotheist. Before his messengership, Abraham, as a young philosopher, reached the idea of the "greatest" by a series of hypothetical questions. His method of proving the existence of the creator of all things was both empirical and rational. He invited people to observe the heavenly bodies and then deduce the existence of an absolute creator from their contingent characteristics. God supports this empirical and rational methodology.

Abraham not only supported his monotheistic faith through rational arguments, but also refuted the claims of his opponents via rational arguments by breaking the little statutes of his pagan people and sparing the biggest one. When the polytheists inquired about the "disbeliever" who committed such a blasphemous act to their idols, Abraham stood up and pointed at the biggest statue, thereby forcing his people to reflect and examine their religious dogmas (21:51-67).

The Quran provides a rational argument for why God cannot have partners or equals. The argument in verse 21:21-22 is a logical argument called Modus Tollens or Denying the Consequent. Thus, it is no wonder that the Quran invites us not to be gullible. We should not follow anything without sufficient knowledge, including belief in God (17:36).

Abraham's methodology of emphasizing reason, valiantly examining the religious dogmas of his people, engaging his opponents with intellectual dialogues, and occasionally modeling the thinking process for his audience, was later made famous by

Socrates, a messenger of God to Greece. Socrates, like Abraham criticized the superstitions and polytheistic religious dogmas of his people through dialogues where he subjected them to a series of questions and inferences. Socrates, similarly, incurred the wrath of the ruling class, whose authority depended on the ignorance and cowardice of their subjects. The Abrahamic method has been used by many messengers of God. Thus, we are asked to follow Abraham's example (60:4).

006:079 The word *fatara* which is translated as "create," or "initiate" contains the meaning of "splitting/cracking" (82:1). We learn from embryology and cosmology, that the creation of new bodies usually involves a splitting and diversification process. The universe was created approximately 13.7 billion years ago with a big separation (21:30) and it will collapse back as it started (21:104). See 4:82.

006:082 *Shirk*, that is associating equal or lower-level partners to God, is injustice (31:13). See 2:48,286; 3:18.

006:083-90 In these verses, 18 names of prophets are mentioned, and they are praised by God. Though the Quran refers to some of their errors and weaknesses, and underscores their human nature, they are depicted as role models. Nevertheless, through using their intellect and determination, we are informed that they worked hard to avoid temptations, sins, and injustices. The Old Testament, which was extensively distorted by Ezra and numerous Rabbis, contains many wicked and despicable acts, some of which should have been X-rated, to these role models. The New Testament's hero, however, is treated to the contrary; he is transformed into a human-god, into an oxymoron, or more accurately into a logical contradiction. Evangelicals try to defend those shameful stories as a merit, as a sign of a realistic account of the Old Testament. Knowing that numerous disgraceful and wicked stories were falsely attributed to Muhammed to justify the actions of corrupt and oppressive rulers who came after him, we are justified in speculating that the real motive behind the X-rated and V-rated Biblical stories was most likely to reduce the guilt from the fabricators of the stories who themselves either committed or contemplated similar acts of sexual abuse and violence, or to reduce the guilt of their criminal friends and rulers. Indeed, this is a cynical view, and the Bible contains more than enough reasons to be cynical. For instance, according to the Old Testament, Lot had incest with his daughters while he was drunk (Genesis 19:30-38); Judah too committed incest (Genesis 38:15-30); David tempted and committed adultery with someone else's wife and had her husband killed (2 Samuel 11:1-27); David's son rapes his sister Tamar, causing interfamily conflict. (2 Samuel 13:1-34). See the Quran 2:59.

006:088 The Arabic text supports two meanings: (1) God guides whomever He wishes. (2) God guides whoever wishes so. It seems that both meanings are equally intended. See 57:22-23; 6;110; 7:30; 14:4; 42:13...

006:094 See 2:48.

006:110 Individuals makes their initial decision with the freedom of choice given to them (18:29) and God leads them according to their choice. Those who have decided deep in their minds to reject God's signs/miracles with prejudice, will not see, witness, affirm, or appreciate those signs/miracles (7:146). These people label the divine signs and miracles as "myth" or "illusion/magic" (6:7,25). To witness a 3-D stereogram made with computer generated random dots, one needs to have an open mind, open both his eyes, and follow certain rules. Recent studies by Mahzarin Banaji Anthony Greenwald and other scientists using Implicit Association Test, or IAT show that our unconscious attitudes create strong biases. Bruce Bower, in April 22, 2006, issue of Science News described these hidden biases with the following statements: "It lurks in the mind's dark basement, secretly shaping our opinions, attitudes, and stereotypes. This devious manipulator does its best to twist our behavior to its nefarious end. Its stock in trade: stirring up racial prejudice and a host of other pernicious preconceptions about members of various groups. Upstairs, our conscious mind ignores this pushy cellar dweller and assumes that we're decent folk whose actions usually reflect good intentions." Inductive reasoning, which creates stereotyping or generalization is an important tool for our survival. Occasionally we need to react fast in events when we sense risk to our lives or well-being. However, the quick retrieval of stored conclusions of inductive reasoning, has also the chance to be wrong. Thus, when we have more than few seconds to reflect, to reach an accurate or beneficial judgment it is always important to examine our biases. To overcome against this stealthy prejudice, the Quran recommends us to engage in self-examination, critical thinking, and deliberate and constant struggle to be open-minded (53:33; 54:43; 17:36; 39:18).

006:112-113 None who affirms the reality of life after death would be attracted to the fancy words presented as divine inspiration or revelation. Those who idolized (J)esu(s), a human messenger, by alleging him to be literally the "son of God," are in fact, his ardent enemies. Comparing the hearsay stories in *Hadith* books with the precepts in the Quran, we can see why and how their narrators and collectors were indeed the enemies of Muhammed, another human messenger and servant of God.

006:114-116 Knowing that the enemies of Jesus fabricated the doctrine of the Trinity and associated him as a partner with One God by creating multiple personalities for God, similarly those who gave lip service to rational monotheism or peacemaking (*islam*) also fabricated *hadith* books and associated Muhammed as partner with God in His judgment. The Quran informs us about many ways of setting up partners with God, or polytheism. A careful reader will notice that the verses quoted above reject all the major excuses used by the proponents of *hadith* and *sunna*. These verses describe the adherents of fancy *hadiths* with the following qualities:

Despite their lip service, they do not appreciate God's signs/miracles (We have witnessed a modern example of this lack of appreciation regarding the prophetic fulfillment of the number 19 mentioned in chapter 74).

- Most of them are ignorant, blind followers.
- They tell each other fancy *hadiths* presenting them as divine inspiration or revelation.
- The *hadiths* that they are narrating to each other are fabrications.
- Despite their lip service, they do not appreciate the hereafter.
- They seek Muhammed and other idols as partners with God's judgment regarding *islam*, the system of peacemaking.
- They do not accept that the Quran is sufficiently detailed.
- They do not accept that that God's word is complete.
- They put too much confidence in the numbers of those who follow their sect; they follow the crowd.
- They follow conjecture.

See 17:39; 33:38; 49:1

006:118-119 One of the common characteristics of those who associate religious leaders as partners with God is to prohibit things that God has not prohibited. Thus, the description of *mushriks* (polytheists) starting from 6:112, continues with a specific example until 6:155. The followers of the same *mushriks* might deceive themselves by assuming that those verses have no application in our times; that they were about some people who lived and perished 15 centuries ago! They ignore the warning of the Quran (36:70).

006:121 We should commemorate God's name while slaughtering animals and eating their product. We are not asked to monitor or even investigate whether the person who slaughtered it commemorated God's name or not. However, if we know that the animal was slaughtered according to the rules of a religion falsely attributed to God, then we should not consume it since it would legitimize that blasphemy. This verse is a commonly misunderstood one, and it creates problems for muslims who live in foreign countries. If we evaluate this verse under the light of 6:145 and similar verses, its meaning gains clarity. This verse does not prohibit eating the animal food slaughtered by non-muslims or atheists. This verse asks muslims to protest the *mushriks* who did not eat the meat of the animals over which they mentioned God's name when they slaughtered; they let it go to waste for the sake of God. In order to make the animal clean for consumption, when they slaughter them, they deliberately and specifically did not mention God's name. This tradition was based on the fabricated rules of their religion, and it was thus, for other than God. Verse 6:119 refers to this tradition. In brief, if we read 6:119-121,145 and 22:37 carefully, we will learn that the prohibition in this verse is about the animals sacrificed by *mushriks* (polytheists) who deliberately did not mention God's name, because of false religious teachings and traditions.

The Quran requires offering of domesticated animals only during the *Haj* (Debate Conference) as a penalty for violation of the conference rules (2:196; 5:2,95-97; 48:25), and it has practical purposes: to regulate the conference in peaceful manner and feed participants of the annual international conference where hunting is prohibited to protect the wildlife and environment.

The Old Testament requires offerings for many different reasons and contains complicated instructions, which were partially adopted by Sunni and Shiite *mushriks* through *hadith*, *sunna*, and *sharia* (Genesis 8:20; 12:7; 13:4,18; 15:9-11; 22:1-18; Ex 12:3-27; Leviticus 23:5-8; Nu 9:2-14). Torrey's index lists the following verses under the title, "Offerings": To be made to God alone (Ex 22:20; Jg 13:16); Its origin (Genesis 4:3-4). DIFFERENT KINDS OF; Burnt (Leviticus 1:3-17; Ps 66:15); Sin (Leviticus 4:3-35; 6:25; 10:17); Trespass (Leviticus 5:16-19; 6:6; 7:1); Peace (Leviticus 3:1-17; 7:11); Heave (Ex 29:27; 7:14; Nu 15:19); Wave (Ex 29:26; Leviticus 7:30); Meat (Leviticus 2:1-16; Nu 15:4); Drink (Genesis 35:14; Ex 29:40; Nu 15:5); Thank (Leviticus 7:12; 22:29; Ps 50:14); Free-will (Leviticus 23:38; De 16:10; 23:23); Incense (Ex 30:8; Mal 1:11; Lu 1:9); First-fruits (Ex 22:29; De 18:4); Tithe (Leviticus 27:30; Nu 18:21; De 14:22); Gifts (Ex 35:22; Nu 7:2-88); Jealousy (Nu 5:15); Personal, for redemption (Ex 30:13,15); Declared to be most holy (Nu 18:9), REQUIRED TO BE Perfect (Leviticus 22:21); The best of their kind (Mal 1:14); Offered willingly (Leviticus 22:19); Offered in righteousness (Mal 3:3); Offered in love and charity (Mt 5:23-24); Brought in a clean vessel (Isa 66:20); Brought to the place appointed of God (De 12:6; Ps 27:6; Heb 9:9); Laid before the altar (Mt 5:23-24); Presented by the priest (Heb 5:1); Brought without delay (Ex 22:29-30) Unacceptable, without gratitude (Ps 50:8,14 (Could not make the offeror perfect (Heb 9:9); THINGS FORBIDDEN AS; The price of fornication (De 23:18); The price of a dog (De 23:18); Whatever was blemished (Leviticus 22:20); Whatever was imperfect (Leviticus 22:24); Whatever was unclean (Leviticus 27:11,27); Laid up in the temple (2Ch 31:12; Ne 10:37); Hezekiah prepared chambers for (2Ch 31:11); THE JEWS OFTEN Slow in presenting (Ne 13:10-12); Defrauded God of (Mal 3:8) Gave the worst they had as (Mal 1:8,13); Rejected in, because of sin (Isa 1:13; Mal 1:10); Abhorred, on account of the sins of the priests (1 Samuel 2:17); Presented to idols (Eze 20:28).

006:124 Those who followed their ancestors and religious leaders blindly opposed the messengers of God throughout history, by using the past and the number of the crowd as a justification for their lack of appreciation. Ignorant people rejected and opposed Muhammed who claimed to not bring any evidence, sign, or miracle except the Quran (29:50-51); they expected evidences/miracles which were only similar to those given to the previous messengers (17:90-93). What they cared about was not the truth, but conformity with the past and preservation of the status quo. They would reject any evidence without evaluating it, with prejudice and in a reflexive manner to keep conformity with their proximate group. The same unappreciative mind now is rejecting the mathematical signs and miracles of the Quran. Verses 38:1-15, and

26:1-6 criticizes these conservative conformists.

006:136 The Quran was not revealed to warn only the Meccan polytheists; it is a warning and good news for all humanity. Thus, we notice that the characteristics of medieval *mushriks* are found in modern *mushriks*. The verse refers to the following common claim of the followers of manmade religious teachings: "According to God this is *haram* (prohibited), according to Prophet Muhammed (peace and blessings be upon him), or according to this or that *emam/alim* (may God be pleased from them), these and those are *haram*."

006:137 The word "*awlad*" is a derivative of the verb "*WaLaDa*" and means "born children." The verse condemns the tradition of infanticide based on sex discrimination. For abortion, see 46:15; 31:14; 22:5; 23:14; 16:58-59 and 17:31

006:141 Betterment by sharing (*zaka*) is so important that God specifically mentions it in connection with His mercy (7:156). Sunni and Shiite *mushriks* reduced the universal implication of *zaka*, which means, purification by sharing a portion of any blessing one may have, such as talents, professional skills, knowledge, presence. Not only did they restrict the meaning of *zaka* with alms *(sadaqa)* or feeding the poor (*infaq*), but they also limited it to an annual practice. The amount needed for purification is left up to us to decide within guidelines. See 2:219; 17:29.

006:143-144 The triviality and complexity of cleric-made prohibitions is reminded by directing more detailed questions to the proponents of those prohibitions. The verse uses an excellent rhetorical device to debunk the claims of those who follow man-made fabrications based on their common excuse, "God's word is not detailed." God informs us that His word is sufficiently detailed, and there is no limit for details if one wishes to see more of them by asking further questions. Bring me any *hadith* or sectarian practice you claim to be detailed, I will find deficiency in their explanation by just asking a few more questions. For instance, you claim that you follow *hadith* books because they taught you how to hold your hands during prayers; well, what about how to breathe during prayers? From your nose or mouth, or both? What about the color of clothes you should wear during prayers? Red, green, yellow, blue, white, black or a particular combination? Thus, your very excuse (seeking more and more details) for following hearsay and manmade religions, will come back and haunt you. The moment you wish to rebuttal and come up with answers to the demands for further details, you have to start fabricating new rules; thereby proving the origins of your previous details: man-made fabrications. However, if you reject answering those questions by saying, "Your question will make prayers more difficult to fulfill. *Hadith, sunna* and sectarian jurisprudence provide us with sufficient details that we need. We are not held responsible for following further details." That would be the best answer. Now look at the mirror: when you demand more details from people who follow the Quran alone, they too would respond to you with the same answer by just replacing the word *hadith, sunna and sectarian jurisprudence* with the Quran! "The Quran provides us with the sufficient details that we need. We are not held responsible for following further details. Your question will make prayers more difficult to fulfill."

006:145-146 The Quran enumerates the list of the dietary prohibitions and states clearly that only the enumerated ones are prohibited by God. It is up to your experience, taste, cultural cuisine, or doctor to pick what to eat or not to eat. You may choose to not eat frog legs, but you cannot attribute your personal taste to God and criticize others in the name of God. In verses 145 and 146, God indirectly tells us that He is able to distinguish meat from fat, as opposed to clergymen's claim, and if He wishes to prohibit fat alone, He could express that. Those who do not appreciate God's wisdom and ability to use the language to communicate His will, despite these clear verses, follow numerous contradictory prohibitions falsely attributed to Muhammed, who followed the Quran alone. See 10:59-60; 16:112-116. Similar prohibitions are found in the Old Testament, Leviticus 17:11-15; Deuteronomy 12:24; 14:8; 14:21. Deuteronomy 14:7 mentions the additional prohibition.

006:148 Fabricating prohibitions and attributing them to God through his messenger is *shirk*, that is associating partners with God. Modern *mushriks*, like their predecessors, blame God for their *shirk*. The portrait of *mushriks* drawn by the Quran is universal. See 3:18; 6:22-24; 16:35; 10:59-60.

006:149 God's evidence is in his mathematically coded and detailed book; not in *hadith* books that contain hearsay stories and contradictory prohibitions.

006:150 This verse provides an excellent question for cross-examination and impeachment. When discussing this issue with clerics who claim many more prohibitions such as mussel, crab, or lobster, start reading from five verses before and stop at the question in this verse by directing the question personally to them while looking in their eyes. Wait for their response. If they list the names of *hadith* books and sect leaders, then convict them with the rest of the verse. You will be surprised to see that many of them will incriminate themselves. Also, see 9:31; 42:21.

006:159 The followers of hearsay, and the *sharia* concocted by clerics, divided themselves into many hostile sects, factions, and orders. See 3:105; 23:52-56.

006:164 None can save another from their responsibility. Those who mislead others will get a share in the crimes of those who were misled. None can assume someone else's responsibility; however, if someone misleads another one, he will also be held responsible for that. See 16:25.

7:0 In the name of God, the Gracious, the Compassionate.

7:1 A1L30M40S90*

7:2 A book that has been sent down to you, so let there not be any burden in your chest from it, that you may warn with it; and a reminder to those who affirm.

Do Not Follow Any Other Source Besides God

7:3 Follow what was sent down to you all from your Lord, and do not follow besides Him any supporters. Little do you remember!

7:4 How many a town have We destroyed; for Our punishment came to them while sleeping, or while resting.

7:5 Then their only saying when Our punishment came to them was: "We were wicked!"

7:6 We will ask those who received the message, and We will ask the messengers.

7:7 We will narrate to them with knowledge; We were not absent.

7:8 The scales on that day will be the truth. Whoever has heavy scales, these are the successful ones.

7:9 Whoever has light scales, these are the ones who lost themselves for they wrongfully treated Our signs.*

The Test

7:10 We granted you dominion on earth, and made for you in it a habitat; little do you give thanks!

7:11 We created you, We shaped you, and then We said to the controllers: "Submit to Adam;" so they submitted except for Satan, he was not of those who submitted.

7:12 He said, "What has prevented you from submitting when I have ordered you?" He said, "I am better than him, You created me from fire and created him from clay!"*

7:13 He said, "Descend from it, it is not for you to be arrogant here; depart, for you are disgraced."

7:14 He said, "Grant me respite until the day they are resurrected?"

7:15 He said, "You are granted respite."*

7:16 He said, "For You having me sent astray, I will stalk for them on Your Straight Path."*

7:17 "Then I will come to them from between their hands, and from behind them, and from their right, and from their left; You will not find most of them to be appreciative."

7:18 He said, "Get out from this, you are despised and banished. As for those among them who follow you, I will fill hell with you all!"

7:19 "Adam, reside with your mate in the paradise, and eat from wherever you please; but do not approach this tree or you will be of the wicked."*

7:20 But the devil whispered to them, to reveal their bodies/transgressions which was hidden from them; and he said, "Your Lord did not forbid you from this tree except that you would become controllers, or you would be immortals."*

7:21 He swore to them: "I am giving good advice."

7:22 So he misled them with deception; and when they tasted the tree, their bodies became apparent to them, and they rushed to cover themselves with the leaves of the paradise; and their Lord called to them: "Did I not forbid you from that tree, and tell you that the devil is your clear enemy?"*

7:23 They said, "Our Lord, we have wronged ourselves and if You do not forgive us and have mercy on us, then we will be of the losers!"

7:24 He said, "Descend as enemies to one another; on earth you will have residence and provisions until the appointed time."

7:25 He said, "In it you will live and in it you will die, and from it you will be brought forth."

7:26 O Children of Adam, We have sent down for you garments to cover your bodies, as well as ornaments; and the garment of awareness is the best. That is from **God**'s signs, perhaps they will remember.*

7:27 O Children of Adam, do not let the devil afflict you as he evicted your parents from the paradise; removing from them their garments to show them their bodies. He and his tribe see you from where you do not see them. We have made the devils as allies for those who do not affirm.*

Examine Claims about God with Knowledge

7:28 When they commit evil acts, they say, "We found our fathers doing such, and **God** ordered us to it." Say, "**God** does not order evil! Do you say about **God** what you do not know?"*

7:29 Say, "My Lord orders justice, and that you be devoted at every temple, and that you call on Him, while dedicating to Him in the system; as He initiated you, so you will return."*

7:30 A group He has guided, and a group has deserved misguidance; that is because they have taken the devils as allies besides **God**; and they think they are guided!

7:31 O Children of Adam, dress nicely at every temple, and eat and drink and do not indulge or waste; He does not like the indulgers and the prodigals.

Do Not Prohibit in the Name of God

7:32 Say, "Who has forbidden the nice things that **God** has brought forth for His servants and the good provisions?" Say, "They are in this worldly life for those who affirm, and they will be exclusive for them on the day of Resurrection" We thus explain the signs for those who know.*

7:33 Say, "My Lord has forbidden all lewd action, what is obvious from them and what is subtle, and sin, aggression without cause, your setting up partners with **God** that were never authorized by Him, and saying about **God** what you do not know."

7:34 For every nation will be an allotted time; when their time is reached, they will not delay by one moment nor advance it.

When Messengers Come to You

7:35 O Children of Adam, when messengers come to you from amongst yourselves and narrate My signs to you; then for those who are aware, and reform, there will be no fear over them nor will they grieve.

7:36 As for those who reject Our signs, and become arrogant towards them; then these are the dwellers of the fire, in it they will abide"

The Worst People

7:37 Who is more wicked than one who invents lies about **God**, or denies His signs? These will receive their recompense from the book; so that when Our messengers come to take them, they will say, "Where are those whom you used to call on besides **God**?" they said, "They have abandoned us!" and they bore witness upon themselves that they were ingrates.

7:38 He said, "Enter with the multitude of nations before you from humans and Jinn to the fire!" Every time a nation entered, it cursed its sister nation, until they are all gathered inside it; then the last of them says to the first: "Our Lord, these are the ones who have misguided us, so give them double the retribution of the fire!" He replied: "Each will receive double, but you do not know."*

7:39 The first of them said to the last: "You have no preference over us, so taste the retribution for what you earned!"

Do Not Be Unappreciative of God's Signs

7:40 Those who have denied Our signs, and reacted to them with arrogance, the gates of the sky will not open for them, nor will they enter paradise until the camel passes through the eye of a needle. It is such that We recompense the criminals.*

7:41 They will have hell as an abode, and from above them will be barriers. It is thus We recompense the wicked.

7:42 As for those who affirm and do good; We do not burden a person beyond its capacity; those are the dwellers of paradise; in it they will abide.

7:43 We removed what was in their chests of jealousy and hidden enmity; rivers will flow beneath them; and they will say, "Praise be to **God** who has guided us to this, and we would not have been guided unless **God** guided us. The messengers of our Lord had come with truth." It will be announced to them: "This is paradise; you have inherited it for what you have done."

People of Paradise and Hell

7:44 The companions of paradise called the dwellers of the fire: "We have found what our Lord promised us to be true; did you find what your Lord promised to be true?" They said, "Indeed!" Then a caller announced between them: "**God**'s curse is on the wicked."

7:45 "The ones who hinder from the path of **God**, and seek to distort it, are the rejecters of the Hereafter."

Identification Station

7:46 A barrier separates them. At the identification station there are people who recognize others by their features. They called out to the dwellers of paradise: "Peace be upon you!" They have not yet entered it, but they are hoping.*

7:47 When their eyes are turned towards the dwellers of the fire, they say, "Our Lord, do not make us with the wicked people!"

7:48 Those attending the identification station called on men they recognized by their features, they said, "What good did your large number do for you, or what you were arrogant for?"*

7:49 "Weren't these the ones whom you swore **God** would not grant them of His mercy? "Enter paradise, there is no fear for you nor will you grieve."

7:50 The dwellers of the fire called on the dwellers of paradise: "Give us some water, or what **God** has provided for you?" They said, "**God** has forbidden it for the ingrates."

7:51 The ones who took their system as diversion and games; and were preoccupied with this worldly life. Today we ignore them as they ignored their meeting on this day, and they denied Our signs.

The Quran is Detailed in Knowledge

7:52 We have brought them a book which We have detailed with knowledge; a guide and a mercy to those who affirm.*

7:53 Are they waiting for its fulfillment? On the day which its fulfillment happens, those who previously forgot it will say, "Our Lord's messengers have come with truth! Are there any intercessors to intercede for us? Or could we be returned so that we may act differently than what we did?" They have lost themselves and what they have invented has abandoned them.

7:54 Your Lord is **God** who created the heavens and earth in six days, and then He established the authority. The night covers the day, which seeks it continually; and the Sun and the Moon and the stars are all subjected to His law; to Him is the creation and the law. Glory be to **God**, the Lord of the worlds.*

7:55 Call on your Lord openly and in secret. He does not like the aggressors.

Do Not Corrupt the Earth

7:56 Do not corrupt the earth after it has been reformed; and call on Him out of fear and want. **God**'s mercy is near to the righteous.

7:57 He is the One who sends the winds to be a goodnews/distributer between His hands of mercy; so, when it carries a heavy cloud, We drive it to a dead town, and We send down the water with it and We bring forth fruits of all kind. Thus, We will bring out the dead, perhaps you may remember.*

7:58 As for the good town, its plants are produced by its Lord's permission, while the inferior land does not produce except very little. It is such that We cite the signs for a people who are thankful.

Noah

7:59 We have sent Noah to his people, so he said, "My people, serve **God**, you have no god besides Him. I fear for you the retribution of a great day!"

7:60 The leaders from his people said, "We see that you are clearly misguided."

7:61 He said, "My people, I am not misguided, but I am a messenger from the Lord of the worlds."

7:62 "I deliver to you my Lord's messages, I advise you, and I know from **God** what you do not know."

7:63 "Are you surprised that a reminder has come to you from your Lord through a man from amongst you to warn you so that you become aware, and that you may receive mercy?"

7:64 They denied him, so We saved him and those with him in the ship, and We drowned those who denied Our signs; they were a blind people.

Hud

7:65 To Aad We sent their brother Hud, he said, "My people, serve **God**, you have no god besides Him. Will you not be aware?"

7:66 The leaders who rejected from among his people said, "We see you in foolishness, and we think you are one of the liars"

7:67 He said, "My people, there is no foolishness in me, but I am a messenger from the Lord of the worlds."

7:68 "To deliver to you my Lord's messages, and to you I am a trustworthy advisor."

7:69 "Are you surprised that a reminder has come to you from your Lord through a man from amongst you to warn you? Recall that he made you successors after the people of

Noah, and He improved you in creation. So, recall **God**'s blessings that you may succeed."*

7:70 They said, "Have you come to us to serve **God** alone and abandon what our fathers had served? Bring us what you promise if you are of the truthful ones!"

7:71 He said, "An affliction and wrath shall befall you from your Lord. Do you argue with me over names which you and your fathers have created with no authority being sent down by **God**? Wait then, and I will wait with you."*

7:72 We saved him and those with him by a mercy from Us, and We destroyed the remnant of those who rejected Our signs and did not affirm.

Saleh

7:73 To Thamud, We sent their brother Saleh, he said, "My people, serve **God**, you have no god besides Him; proof has come to you from your Lord. This is **God**'s camel, in her you have a sign, so leave her to eat in **God**'s land, and do not harm her, else a painful retribution will take you."

7:74 "Recall that He made you successors after Aad, and He established you in the land so that you make palaces on its plains, and you carve homes in the mountains. So, recall **God**'s grace, and do not roam the earth as corrupters."

7:75 The leaders who were arrogant from among his people said to those who had affirmed among the powerless: "How do you know that Saleh was sent from his Lord?" They said, "We affirm the message he has been sent with."

7:76 Those who were arrogant said, "We reject what you affirm!"

7:77 They thus killed the camel and defied their Lord's command, and they said, "O Saleh, bring us what you promised us if you are of the messengers!"

7:78 The earthquake took them, so they fell motionless in their homes.

7:79 Thus he turned away from them, and said, "My people, I had delivered to you the message of my Lord and advised you; but you do not like the advisers."

Lot

7:80 And Lot, when he said to his people: "Do you commit lewdness to an extend that none in the worlds had surpassed you?"

7:81 "You come to men with desire instead of women; you are a transgressing people."

7:82 The only response of his people was, "Drive them out of your town; they are a people who wish to be pure!"

7:83 We saved him and his family, except for his wife; she was of those who lagged.

7:84 We showered them with a shower, so see what the punishment for the criminals was.

Shuayb

7:85 To Midian, their brother Shuayb, he said, "My people, serve **God**, you have no god besides Him. Proof has come to you from your Lord, so give full weight and measure, and do not hold back from the people what belongs to them, and do not make mischief on the earth after it has been reformed. That is better for you if you affirm."

7:86 "Do not stand on every road to threaten and repel from the path of **God** those who affirmed Him, by attempting to distort it. Recall that you were few and He multiplied you; and see what the retribution of the troublemakers is."

7:87 "If a group of you affirms what I have been sent with, and another group rejects, then wait until **God** judges between us. He is the best of judges."

7:88 The leaders who became arrogant from among his people said, "We will evict you Shuayb, along with those who affirmed, or else you will return to our creed!" He said, "Even though we despise?"

7:89 "We would then be inventing lies about **God** if we return to your creed after **God** had saved us from it; and we would not return to it except if **God** our Lord wishes. Our Lord encompasses everything with knowledge. In **God** we put our trust. Our Lord, separate between us and between our people with truth, you are the best separator."

7:90 The leaders who rejected from among his people said, "If you follow Shuayb, then you are losers."

7:91 The earthquake took them, thus they fell motionless in their homes.

7:92 Those who denied Shuayb, it is as if they never prospered therein. Those who denied Shuayb, they were the losers.

7:93 Thus he turned away from them, and said, "My people, I had delivered to you the messages of my Lord and advised you. How can I feel sorry over a rejecting people?"

A Majority Fail the Test That Determines Their Eternity

7:94 Whenever We sent a prophet to any town, We would afflict its people with hardship and adversity so that they may implore.

7:95 Then We replaced the bad with the good until they had plenty and they said, "Our fathers were the ones afflicted by hardship, and adversity." We then take them suddenly, while they do not perceive.

7:96 If the people of the towns had only affirmed and been aware, then We would have opened for them blessings from the sky and the land; but they denied, so We took them for what they used to earn.

7:97 Are the people of the towns sure that Our punishment will not come to them at night while they are sleeping?

7:98 Or are the people of the towns sure that Our punishment will not come to them during the late morning while they are playing?

7:99 Have they become sure about **God**'s scheming? None feels secure about **God**'s scheming except the losers.

7:100 Is it not clear for those who inherited the land after them, that if We wished, We could have punished them immediately for their sins and We could have stamped on their hearts, so they would not hear.

7:101 These are the towns whose stories We relate to you; their messengers had come to them with proofs, but they would not affirm what they had denied before. It is such that **God** stamps on the hearts of the ingrates.

7:102 We did not find most of them up to their pledge, rather, We found most of them corrupt.

Moses

7:103 Then after them We sent Moses with Our signs to Pharaoh and his entourage. But they treated them with prejudice, so see what the end of the corrupters is.

7:104 Moses said, "O Pharaoh, I am a messenger from the Lord of the worlds."

7:105 "It is not proper for me to say about **God** except the truth; I have come to you with proof from your Lord. So send with me the Children of Israel."

7:106 He said, "If you have come with a sign then bring it, if you are of the truthful?"

7:107 He threw down his staff and there it was an obvious serpent.

7:108 He drew out his hand, and it became pure white for the spectators.

7:109 The commanders among the people of Pharaoh said, "This is a knowledgeable magician!"

7:110 "He wishes to drive you out of your land; what do you recommend?"

7:111 They said, "Delay him and his brother, and send gatherers to the cities."

7:112 "They will come to you with every knowledgeable magician."

7:113 The magicians came to Pharaoh and said, "There should be a reward for us if we are the victors."

7:114 He said, "Yes, and you will be made close to me."

7:115 They said, "O Moses, either you cast, or shall we cast?"

Magic Has No Objective Reality: It Is Merely the Product Of Illusion and Suggestion

7:116 He said, "You cast." So when they cast, they deceived the people's eyes, intimidated them, and they produced a great magic.*

7:117 We revealed Moses: "Throw down your staff;" and soon it was swallowing what they had invented.

7:118 Thus the truth was confirmed, and what they were doing was falsified.

7:119 They were thus defeated there, and they turned in disgrace.

7:120 The magicians fell down prostrating.

7:121 They said, "We affirm the Lord of the worlds."

7:122 "The Lord of Moses and Aaron."

7:123 Pharaoh said, "Did you affirm him before I have given you permission? This is surely some scheme which you have schemed in the city to drive its people out; but soon you shall know."

7:124 "I will cut off your hands and feet from alternate sides; then I will crucify you all."

7:125 They said, "To our Lord shall we return."

7:126 "You are only seeking revenge on us because we affirmed our Lord's signs when they came to us. Our Lord, grant us patience and let us die having peacefully surrendered."

7:127 The leaders among the people of Pharaoh said, "Will you let Moses and his people corrupt the land, and abandon you and your gods?" He said, "We will kill their children and shame their women; we will be supreme over them."

7:128 Moses said to his people: "Seek help with **God**, and be patient; for the land is **God**'s. He will inherit it to whom He pleases of His servants; and the victory belongs to the righteous."

7:129 They said, "We were being harmed before you came to us and since you have come to us." He said, "Perhaps your Lord will destroy your enemy, and make you successors in the land, so He sees how you behave."

Disasters

7:130 We then afflicted the people of Pharaoh with years of drought, a shortage in crops, so that they might take heed.

7:131 When any good comes to them, they say, "This is ours," and when any bad afflicts them, they blame it on Moses and those with him. Their blame is with **God**, but most of them do not know.

7:132 They said, "No matter what you bring us of a sign to bewitch us with, we will never affirm you."

7:133 So We sent them the flood, the locust, the lice, the frogs, and the blood; all detailed signs; but they turned arrogant, they were a criminal people.*

7:134 When the plague befell them, they said, "O Moses, call on your Lord for what pledge He gave you. If you remove this plague from us, then we will affirm you and we will send with you the Children of Israel."

7:135 So when We removed from them the plague until a future time; they broke their pledge.

Children of Israel Tested After Receiving Divine Support

7:136 For their denial and disregard of Our signs, We thus inflicted them with punishment by drowning them in the sea,

7:137 We let the people who were oppressed inherit the east and the west of the land which We have blessed, and good word of your Lord was completed towards the Children of Israel. For they were patient, and We destroyed what Pharaoh and his people were doing, and what they contrived.

7:138 We let the Children of Israel cross the sea, then they passed by a people who were devoted to idols they had. They said, "O Moses, make for us a god like the gods they have." He said, "You are an ignorant people!"

7:139 "These people are ruined for what they are in, and worthless is what they do."

7:140 He said, "Shall I seek other than **God** as a god for you when He has preferred you over the worlds?"

7:141 When We saved you from the people of Pharaoh, who were afflicting you with the worst punishment, killing your children and shaming your women; and in that was a great trial from your Lord.

Moses Receives The Commandments

7:142 We appointed for Moses thirty nights and completed them with ten, so the appointed time of his Lord was completed at forty nights. Moses said to his brother Aaron: "Be my successor with my people and reform, and do not follow the path of the corrupters."*

7:143 So when Moses came to Our appointed time, and his Lord spoke to him, he said, "My Lord, let me look upon you." He said, "You will not see Me, but look upon the mountain, if it stays in its place then you will see Me." So, when his Lord manifested Himself to the mountain, He caused it to crumble; thus Moses fell unconscious. When he recovered, he said, "Glory be to You, I repent to You and I am the first of those who affirm."*

7:144 He said, "O Moses, I have chosen you over people with My message and My words; so, take what I have given you and be of the thankful."

7:145 We wrote for him on the tablets from everything a lesson and detailing all things. You shall uphold them firmly and order your people to take the best from it. I will show you the abode of the corrupt.*

Those Who Deny Signs and Miracles

7:146 I will divert from My signs those who are arrogant on earth unjustly, and if they see every sign they do not affirm it, and if they see the path of guidance they do not take it as a path; and if they see the path of straying, they take it as a path. That is because they have denied Our signs and were heedless of them.*

7:147 Those who deny Our signs and the meeting of the Hereafter, their work has collapsed. Will they not be rewarded except for what they used to do?

Children of Israel are Tested with the Symbol of Their Former Masters

7:148 The people of Moses, in his absence, made from their ornaments the statue of a calf which had a sound. Did they not see that it could not speak to them, nor guide them to any path. They took it and turned wicked.*

7:149 When they were smitten with remorse and they saw that they had gone astray, they said, "If our Lord will not have mercy on us and forgive us, then we will be of the losers!"

7:150 When Moses returned to his people, angry and grieved, he said, "Miserable is what you have done after me; do you wish to hasten the action of your Lord?" He cast down the tablets, and took his brother by his head dragging him towards him. He said, "Son of my mother, the people overpowered me and nearly killed me, so do not make the enemies rejoice over me, and do not count me with the wicked people."*

7:151 He said, "My Lord, forgive me and my brother, and admit us in your mercy; you are the most Compassionate of the compassionate."

7:152 Those who took up the calf will be dealt with a wrath from their Lord and a humiliation in this worldly life. We thus punish the fabricators.

7:153 As for those who commit sin but then repent afterwards and affirm;

your Lord afterward is Forgiving, Compassionate.

7:154 When the anger subsided from Moses, he took the tablets; and in its inscription was a guidance and mercy for those who revere their Lord.

7:155 Moses selected from his people seventy men for Our appointed time; so, when the quake seized them, he said, "My Lord, if You wished You could have destroyed them before this, and me as well. Will you destroy us for what the foolish amongst us have done? It is all Your test, You misguide with it whom You please and You guide with it whom You please. You are our supporter, so forgive us and have mercy on us; You are the best forgiver."*

Importance of Charity and Social Work

7:156 "Decree for us good in this world, and in the Hereafter; we have been guided towards You." He said, "I will afflict with My punishment whom I chose, but My mercy encompasses all things. I will thus decree it for those who are aware and contribute towards betterment, and those who affirm Our signs."*

Muhammed: The Gentile Prophet

7:157 "Those who follow the gentile prophet whom they find written for them in the Torah and the Injeel; he orders them to goodness, deters them from evil, he makes lawful for them the good things, he forbids for them the evil, and he removes their burden and the shackles imposed upon them. So those who affirmed him, honored him, supported him, and followed the light that was sent down with him; these are the successful."*

7:158 Say, "O people, I am **God**'s messenger to you all. The One who has the sovereignty of heavens and earth, there is no god but He; He gives life and causes death." So, affirm **God** and His gentile prophet, who affirms **God** and His words; and follow him that you may be guided.*

7:159 Among the people of Moses are a nation who guide with truth and with it they become just.

7:160 We separated them into twelve tribes as nations; and We revealed Moses when his people wanted to drink: "Strike the rock with your staff," thus twelve springs burst forth. Every group knew where from to drink. We shaded them with clouds and We sent down to them manna and quail: "Eat from the good things that We have provided for you." They did not wrong Us, but it was themselves that they wronged.

Children of Israel Trade Humility and Peace with Violence and Arrogance

7:161 When they were told: "Reside in this town and eat from it as you please, talk amicably and enter the passage by prostrating, We will then forgive for you your wrong doings. We will increase for the good doers."

7:162 Those who were wicked amongst them altered what was said to them with something different; so We sent to them a pestilence from the sky because they were wicked.

7:163 Ask them about the town which was by the sea, after they had transgressed the Sabbath; their fish would come to them openly on the day of their Sabbath, and when they were not in Sabbath, they would not come to them! It is such that We afflicted them for what they corrupted.

7:164 A nation from amongst them said, "Why do you preach to a people whom **God** will destroy or punish a painful retribution?" They said, "To fulfill our duty to your Lord, and perhaps they may become aware."

7:165 So when they forgot what they were reminded of, We saved those who desisted from evil, and We took those who transgressed with a grievous retribution for what they were corrupting.

7:166 When they persisted in what they

had been forbidden from, We said to them: "Be despicable apes!"*

7:167 Furthermore, your Lord declared that He will rise against them people who would inflict severe persecution on them, until the day of Resurrection. Your Lord is quick to punish, and He is Forgiving, Compassionate.

7:168 We divided them through the land as nations. From them are the reformed, and from them are other than that. We tested them with good things and bad, perhaps they will return.

7:169 A generation came after them who inherited the book, but they indulged in the petty materials of this world by saying, "It will be forgiven for us." They continued favoring ephemeral goods, whenever they were given a chance. Did they not make a covenant to uphold the book that they would only say the truth about **God**? They studied what was in it; but the abode of the Hereafter is better for those who are aware. Do you not reason?

7:170 As for those who adhere to the book, and they maintain the contact prayer; We will not waste the reward of the reformers.

7:171 When We raised the mountain above them as if it were a cloud, and they thought it would fall on them: "Take what We have given you with strength and remember what is in it that you may be aware."

We Are Born with the Genetic Program to Infer the Creator of Universe

7:172 When your Lord took from the children of Adam from their backs, their progeny; and He made them witness over themselves: "Am I not your Lord?" They said, "Yes, we bear witness." Thus, you cannot say on the day of Resurrection that you were unaware of this.*

7:173 Nor can you say, "It was our fathers who set up partners before and we were simply a progeny who came after them. Would You destroy us for what the innovators did?"

7:174 We thus explain the signs, perhaps they will return.*

7:175 Relate to them the news of the person whom We gave him Our signs, but he withdrew from them, and thus the devil followed him, and He became of those who went astray.

7:176 Had We willed, We could have elevated him by it, but he stuck to the earth and he followed his wishful thinking. His example is like the dog; if you scold him, he pants, and if you leave him he pants; such is the example of the people who deny Our signs. Relate the stories, perhaps they will think.

7:177 Miserable is the example of the people who denied Our signs, and it was themselves that they had wronged.

7:178 Whoever **God** guides, then he is the guided one; and whoever He misguides, then these are the losers.

Satan Hypnotizes His Followers

7:179 We have directed to hell many Jinns and humans. They have hearts, yet they do not comprehend; they have eyes yet they do not see; they have ears yet they do not hear. They are like cattle; no, they are even more astray. These are the heedless ones.

Beautiful Names Belong to God

7:180 To **God** belong the beautiful names, so call Him by them; and disregard those who blaspheme in His names. They will be punished for what they used to do.*

7:181 From among those We created is a community who guides with truth, and with it they establish justice.

7:182 As for those who deny Our signs, We will gradually lead them from where they do not know.

7:183 I will respite them, for My scheming is formidable.

7:184 Do they not reflect that their companion is not crazy; but he is a clear warner.

7:185 Do they not look at the dominion of heavens and earth, and all that **God**

has created, and perhaps their time is drawing near? Which *hadith* after this one will they affirm?*

7:186 Whoever **God** misguides, then there is none to guide him; and He leaves them blundering in their transgression.

Only God Can Reveal the Time of the End; Messengers Are Humans Like You

7:187 They ask you regarding the Moment: "When will be its time?" Say, "Its knowledge is with my Lord, none can reveal its time except Him. It is heavy through the heavens and the earth; it will not come to you except suddenly." They ask you, as if you are too curious about it! Say, "Its knowledge is with **God**, but most people do not know."*

7:188 Say, "I do not possess for myself any benefit or harm, except what **God** wills. If I could know the future, then I would have increased my good fortune, and no harm would have come to me. I am but a warner and a bearer of good news to a people who affirm."*

7:189 He is the One who created you from one person, and He made from it its mate to attain tranquility. When he covered her, she became pregnant with a light load, and she continued with it. When it became heavy, they called on **God**, their Lord: "If You give us a healthy child, then we will be among the thankful."

7:190 But when He granted them a healthy child, they made others partners with Him regarding what He granted them. **God** be exalted above what they set up as partners.

7:191 Do they set up those who do not create anything, while they are created?

7:192 They cannot help them, nor can they help themselves?

7:193 If you invite them to the guidance they will not follow you. It is the same whether you invite them or simply remain silent.

Prophets, Messengers, Saints Whom You Implore Are Powerless Humans Like You

7:194 Those whom you call on besides **God** are servants like you; so let them answer for you if you are truthful.

7:195 Do they have feet to walk with? Or do they have hands to strike with? Or do they have eyes to see with? Or do they have ears to hear with? Say, "Call on your partners, then scheme against me with no respite."

7:196 "My supporter is **God** who sent down the book; and He takes care of the good-doers."

7:197 As for those whom you call on beside Him, they cannot help you, nor can they help themselves.

7:198 If you invite them to the guidance, they do not listen; you see them looking at you, while they do not see.

Be Tolerant And Promote Recognition Of Diversity

7:199 You shall resort to pardoning, and advocate mutually agreed facts, and turn away from the ignorant ones.

7:200 If a provocation from the devil incites you, then seek refuge with **God**. He is the Hearer, the Knower.

7:201 Those who are aware, when a visit from the devil touches them, they remember; soon they are seers.

7:202 But their brethren plunge them into error; they do not cease.

7:203 Since you do not bring them a sign, they say, "If only you had brought one." Say, "I only follow what is revealed to me from my Lord. These are enlightenments from your Lord, a guide and a mercy to a people who affirm."

Constantly Remember Your Lord

7:204 If the Quran is being recited, then listen to it and pay attention, so that you may receive mercy.

7:205 Remember your Lord in yourself out of humility and fear, and without being loud during the morning and the evening. Do not be of the heedless ones.

7:206 Those who are at your Lord, they are never too proud to serve Him, and they glorify Him, and to Him they prostrate.

ENDNOTES

007:001 A1L30M40S90. These letters/numbers play an important role in the mathematical system of the Quran based on code 19. See 74:1-56; 1:1; 2:1; 13:38; 27:82; 38:1-8; 40:28-38; 46:10; 72:28.

007:009 Our personality weakens by not following the instruction of its creator.

007:012 It is interesting that God does not kick out the Satan from His presence as soon as he disobeys the divine order; by directing him to defend his action, God wants to inform us about the source of his disobedience. Satan demonstrated an unjustified pride in something that was not his own work. Racism finds its roots in such a diabolic false pride. See 15:28.

007:015 The so-called "problem of evil" has created a great challenge for theologians and philosophers who accept a Benevolent and Omnipotent God. The Christian medieval philosopher St. Augustine, in Enchiridion, has an interesting argument regarding the existence of Satan: "He used the very will of the creature which was working in opposition to the Creator's will as an instrument for carrying out his will..."

007:016 Satan (Satan) is a convicted liar; his followers too (7:20; 6:22-23).

007:019-27 The Quran describes the tree as the tree of eternity. Satan tempted Adam and his mate with the false promise of eternity. The Old Testament, in Genesis 2:1-25 and 3:1-24, narrates the same event with some differences. According to Genesis, the tree is about the knowledge of evil and good, and the one who was first tempted is Eve. The Biblical accusation of Eve for the failure of Adam in this major event, would later be exploited fully by misogynistic clergymen. St. Paul justifies his male chauvinist teaching based on women's serpentine-like role in the original sin. The other major difference in the account of this event is the depiction of the tree. The Bible describes the tree as the tree of knowledge. This depiction might be one of the causes of dogmatic and anti-scientific attitude developed by the Church. Sunni and Shiite *mushriks* imported and adopted this Judeo-Christian distortion regarding creation and the role of man and woman in our failure, through *hadith* narrations. These misogynistic ideas later were sneaked into the commentaries of the Quran that relied on *hadiths*. See 2:36; 20:115-123.

007:020 The word *SeWAt* (body/sins) is generally misinterpreted as "private parts" while the 12 derivatives from the root *SWA* is mentioned more than 160 times, almost all meaning evil, sin, misdeed. (See 40:58; 30:10; 39:35; 41:27; 41:46; 53:31; 9:102; 17:38; 35:43; 17:7; 9:98 ... and as body (5:31; 7:20- 27; 20:121). The Quran calls this world *DuNYa* (lowly) universe, and we are sent down here for our disobedience to God. To be tested for 40+ years on this planet earth, our *nafs* (mind) is paired with a lowly body. God, the Compassionate has given us *Rouh* (logos) to help us in our choices.

007:022 Our creation on earth with material bodies starts with our failure of a divine test. We believed in Satan's promise, and we ended up with a temporal body rather than an eternal one. We are given a second chance in this temporal body, which is designed to fail in a short time. The temptation of attaining eternity, shame, body, and covering the body is interpreted to imply that the temptation was sexual intercourse between man and woman. In this second test, however, sexual attraction and intercourse per se is not prohibited; to the contrary, it is considered a divine blessing (30:21; 33:52). Nevertheless, the purpose of the test is the same: we should peacefully surrender ourselves to God alone, lead a righteous life to redeem ourselves, and expect that there will be a Day after death in which God alone will judge our intentions, attitude, and actions, which will determine our eternal salvation or damnation. See 7:26 and 15:29.

007:026 Some translators and commentators render the word *sawa* as "sin" or "ugly parts" or "genitals." Though the word is etymologically related to "sin" or "shame," we prefer translating it as "body," which appeared to us after our sinful and shameful failure of the first test. For the other usages of the same word, see 5:31; 7:20,22,27; 20:121.

007:027 There is no contradiction between 7:27 and 7:30. While the Quran states that every event happens in accordance with God's design and permission (8:17; 57:22-25), the Quran also informs us regarding our freedom to choose our path (6:110; 13:11; 18:29 42:13,48; 46:15).

007:028 In Euthyphro, Plato narrates a conversation between Socrates, God 's messenger to Greek, and a pious priest on the meaning of virtue and the baffling question, "is it right because God says so or does God says so because it is right?" We know from the Quran that Truth or Justice are among God's attributes and thus their existence is the essence of God's self. In other words, the answer to the Socratic question should be "both are true: it is right because God says so; and God says so because it is right." See 37:105-107.

007:029 See 2:139; 39:2,11,14; 40:14,65; 98:5.

007:032 This verse implicitly criticizes the clerics who prohibited gold and silk for men.

007:038 The last word could also be read *la ya'lamun* (they do not know). See 6:31; 16:25. For examples of different readings, see 2:54.

007:040 If the word *jamal* is read as *jummal* (rope), then it changes the meaning of the phrase to "until the rope passes through the eye of a needle." The Bible uses the same metaphor. "And again, I say unto you, It is easier for a camel to go through the eye of a needle, than for a rich man to enter into the kingdom of God." (Matthew 19:24; Mark 10:25; Luke 18:25). Ironically, despite these Biblical verses, many modern evangelical Christians, who worship the fictional Jesus, have become avid proponents of wild capitalism and big corporations.

007:046-48 The word *araf* is a plural noun derived from the verb *ARaFa* (recognize, discern, know, identify). The usage of both forms in this very verse clarifies the meaning of *araf* as identification station, where people are separated according to their failure and success in their test during their lives on the planet earth. Traditional translations render it as "purgatory, limbo" or an elevated third place between paradise and hell. We translated the word *rijal* as "people" rather than "man," since the verse is not in the context of relations between male and female. Unlike *zakar* (male), *rijal* (men, walkers on feet; humans) does not necessarily exclude women (See, 9:108; 16:43; 17:64; 21:7; 38:62; 64:6; 72:6). At the identification station, people are recognized and separated based on certain marks recognized by detectors commissioned by God. See 2:26; 13:35; 47:15. Also, see 12:109; 33:4; 39:29; 72:5.

The idea of limbo has been much more important among Catholics. However, in 2006, after a week-long deliberation, the Theological Commission recommended to Pope Benedict XVI that he abolish *limbus infantium* which has been in existence for the last 700 years. Below is an excerpt from the Internationalist Humanist News authored by Babu Gogineni about this "radical restructuring of the heavens":

"Since early days, Christians have wrestled with the thorny question of the fate of unbaptized prophets, as well as the fate of children who die before they are baptized. Catholics believe that all human beings, with the exception of Mary and Jesus, are born in original sin and that only the ritual of baptism will cleanse their sins and will redeem them. Without baptism, there can be no union of the believer with Christ in His death, burial and resurrection, and there can be no holy communion with God, either in this life or in the next. Such a person cannot go to heaven even if he or she has never committed sin.

"To help solve the problem, St. Gregory of Nazianzus (329-390 AD) proposed that the unbaptized should neither be punished, nor could they access the full glory of God. However, the hardliner St Augustine of Hippo (354-430 AD) rejected this idea, insisting that baptism was necessary for salvation, and that even babies would be consigned to hell if they were not baptized. Though St. Augustine made the generous concession that their torment would be the mildest of all of hell's residents, this torture of the innocent was unacceptable to St Thomas Aquinas (1226-1274 AD), the first major theologian to speculate about the existence of a place called limbo where these souls would be lodged forever. Limbo is now a part of Canon Law. …

"This safe passage to heaven that the Catholic Church now assures children who are dying young is a significant step in the right direction. One should now hope and pray that in the Catholic heaven the children will also receive adequate protection from sexual abuse by Catholic priests. Since the Church did not deem sexual abuse by its clergy a matter worthy of punishment in this world, they will all no doubt now be going to heaven too. The Church needs to take immediate steps to ensure that the millions of little children who are now being admitted to heaven are adequately protected in line with the Holy See's international obligations as an early ratifier of the 1990 Convention on the Rights of the Child."

On the other hand, a Lutheran theologian Martin E Marty, in an article published at Christian Post, criticized the timing of this papal amendment and its ramification:

"The post-limbo announcement awakens all kinds of responses, many of them easily accessible on the internet or in the press. Taunters who have heard that Catholicism does not change now taunt, "Here's a change." Catholic pastors who have always found the reference to limbo, a place of non-descriptness and non-happening, to be more chilling than comforting to parents of unbaptized children can be relieved of the charge to pass on word about it. Catholics who lean toward a most expansive Catholic view of salvation and tend toward universalism cheer, for this proclamation that unbaptized infants can go straight to heaven might open the door for Catholic witness that some non-infants could have the same experience. "Pro-choice" Catholics are coming on record as seeing that this can fortify their cause: If fetuses are babies, and they no longer go to limbo but can go to heaven, then abortion may not be as dire a fate as it is often pictured to be. Abort and send them prematurely to heaven. Catholic traditionalists – you'll find plenty of them – rage at Pope Benedict and others involved in this announcement, seeing them as traitors to the Catholic cause: If this can change, can't other things? Relativism, which the pope abhors, will take over.

"Pope Benedict made clear in his announcement that limbo was never an infallible teaching and was not even a formal doctrine of the church. We wonder whether generations of parents who suffered endlessly as they imagined their infants endlessly denied the vision of God or much of any other kind of vision knew of that nuance. Those of us who are not Catholic, and who care about Catholic teaching and Catholic parents, but cannot appreciate all the niceties of gradation of authority among "infallible" and "not quite infallible" and "traditional" and "easy to change" teachings, will look for clarification.

"So will Catholics of many stripes, including some who had not thought about limbo for a long time, but in response to press coverage now find themselves in an intellectual limbo."

007:052 See 11:1.

007:054 This verse, like many other verses informs us that God is employing evolution and stages in His creation (41:9-10; 7:69; 15:28-29; 24:45; 29:18-20; 32:7-9; 71:14-17). Furthermore, creation in six days provides a yardstick regarding the times allocated for the creation of the earth and the universe. According to the Quran, the creation of earth took four out of six days. If universe started 13.7 billion years ago, ignoring the possibility of early inflation, then 4/6th of this, would be approximately 9 billion years, which is about 4 billion years more than currently estimated age.

According to the Bible, God created the universe for a special purpose (Psalms 135:6), and creation shows the power of God (Isaiah 40:26,28), His glory and handiwork (Psalms 19:1), and the wisdom of God (Psalms 104:24; 136:5). In six days. Exodus 20:11; 31:17). First day, making light and dividing it from darkness (Genesis 1:3-5; 2Co 4:6). Second day, making the firmament or atmosphere, and separating the waters (Genesis 1:6-8). Third day, separating the land from the water, and making it fruitful (Genesis 1:9-13). Fourth day, placing the Sun, Moon, and stars to give light (Genesis 1:14-19). Fifth day, making birds, insects, and fish (Genesis 1:20-23). Sixth day, making beasts of the earth, and man (Genesis 1:24,28). God rested on the seventh day (Genesis 2:2-3). Those who distorted the revelation in the Bible misunderstood the meaning of the phrase "six days" as "six 24-hour days." According to the Bible, the earth is only 5700-plus years old.

The Quran rejects the assertion of God "resting" on the seventh day (50:38).

07:057 Two different readings, Nushra (dispersed) or Bushra (Goodnews) are inclusive.

007:069 The word *Basta* is erroneously misspelled in modern manuscripts. Some manuscripts indicate a correction through a tiny letter S (Sin) on top of the erroneous letter Š (Sad). This spelling error came to our attention when we noticed that the frequency of the letter/number Š90 in the three Š90-initialed chapters were 153, rather than the expected 152 (19x8). Our study of the oldest available manuscripts supported the implied correction of the mathematical system. The mathematical system of the Quran is a verifiable and falsifiable scientific theory that is supported by textual facts, historical events, and fulfilled predictions and prophecies. See 74:1-56; 1:1; 2:1; 13:38; 27:82; 38:1-8; 40:28-38; 46:10; 72:28.

Another error correction based on code 19 occurred after Edip Yuksel's debate with a hadith professor on February 18, 2013, in Bursa. The opponent accused us of adding extra letter (Alif) to the word *WAHiD*. The spelling of the word *WAHiD* (One) was later supported by more than two dozen oldest Quranic pages and fragments held in various museums, collections, and libraries. In the oldest available manuscripts, the word is spelled with *Alif*, making its numerical value 19. You may find a video at Edip Yuksel's YouTube channel showing the archeological evidence supporting the code.

This verse also informs us of the evolutionary stage or genetic improvement in our creation after Noah. Chapter 71, which focuses on the story of Noah, informs us of our creation on this planet in terms of evolution rather than an instantaneous one (71:14-17). See 7:69; 15:28-29; 24:45; 32:7-9).

007:071 Those who follow the teachings of Bukhari, Muslim, Ibn Maja, Ibn Hanbal, Tirmizi, Abu Dawud, Imam Azam, Imam Shafi, Imam Yusuf, Imam Rabbani, Abdulqadir Gaylani, Naqshibendi, Gazzali, Said Nursi, and many more, besides the Quran are addressees of this verse.

007:116-117 Magic, according to the description of the Quran, is not a paranormal phenomenon. Magic is a tool abused by people with devious intentions to manipulate individuals and societies via trickery and suggestion. These verses refer to both its physical and psychological components. Hypnosis and films are two examples of magic. See: 2:102.

007:133 We are receiving warning signs in our times too. AIDS and many other sexually transmitted diseases that became epidemics as a consequence of our transgression and indulgence in sexual promiscuity; deformed frogs providing signs of the hazards of toxic chemicals contaminating our lakes, groundwater, and environment; the greenhouse effect and related climatic disasters caused by our wasteful lifestyle and irresponsible consumption of coal, petroleum and forests; putting the entire planet at risk of destruction by stockpiling nuclear weapons in a frenzy; these and many other problems are divine warnings for our irresponsible and unappreciative ways of life.

The Quran mentions numerous animals, each being witnesses to God's creative and intelligent design and signs, which started from a single cell: mosquito (2:26); calf (2:51,54,92,93; 4:153; 7:148,152; 11:69; 20:88; 51:26); monkey (2:65; 5:60; 7:166); cow, heifer (2:67-71; 6:144,146; 12:43,46); pig (2:173; 5:3; 6:145; 16:115); donkey (2:259; 16:8; 31:19; 62:5); horse (3:14; 8:60; 16:8; 17:64; 59:6); falcon (5:4); dog, (5:4; 7:176; 18:18-22); crow (5:31); bird (6:38; 27:20); goat, sheep (6:143); camel (6:144; 88:17); adult he camel (7:40); she camel (7:73,77; 11:64; 17:59; 26:155; 54:27; 91:13); snake (7:107; 26:32), locust (7:133; 54:7); lice (7:133); frog (7:133); fish, whale (7:163; 18:61,63; 37:142; 68:48); wolf (12:17); bee (16:68); sheep/goat/ewe (20:18); fly (22:73); ant (27:18); hoopoe (27:20); spider (29:41); termite (34:14); ewe/sheep (38:23-24); lion (74:51); zebra (74:50); butterfly (101:4); elephant (105:1).

007:142 One of the reasons that the number "forty" is expressed curiously as "thirty plus ten" is the role that numbers play in the numerical structure of the Quran.

007:143 People in paradise will see God (39:69; 75:23, 89:22). Some critics claim that the last statement of this verse contradicts 39:12 and 2:132. If we check a search engine with the following words: *Olympic first place 100-meters*, we will find many names of athletes who got first place. If we used the logic of the critics, we would think there is a great confusion and contradictory claims regarding the first place winner of the 100-meters race. What is wrong with that logic? Obviously, we need to consider the two important dimensions: time and space! Abraham was the first muslim (submitter and promoter of peace) in his time and location. Similarly, Moses and Muhammed were pioneer muslims too, of their times and location.

007:145 See the Old Testament, Deuteronomy 5:1-22 for the Ten Commandments. Also see Deuteronomy 6:4-5.

007:146 Religious clerics who oppose the numerical miracle of the Quran are one example. See 74:1-56; 6:110.

007:148 The verse does not say the calf made a sound; but a calf that had a sound, indicating its mechanical function. This sound was most likely created by wind passing through its holes. Indeed, archeologists discovered in Egyptian temples calf statues that created sound by the help of wind. See 20:83-99.

007:150 The tablets containing God's commandments were not sacred or important but the commandments themselves were. Thus, as a reaction to the Children of Israel's disobedience to the first commandment, Moses reacted in anger. His throwing away the physical medium carrying the commandments and interrogating his brother for letting his people regress into idolatry has a message. There are those who ignore many commandments of the Quran, including its first commandment, yet demonstrate extreme respect to the paper and ink comprising the manuscript of the Quran. This behavior is materialistic and totemic.

007:155 Compare it with the Old Testament, Numbers 11:16-30.

007:156 The importance of *zaka* (purification and betterment) through giving from blessings, is emphasized. *Infaq* is the financial *zaka* given away privately. The amount of financial charity (2:219; 17:26,29), its time (6:141), the list of recommended recipients (2:215), how it will be given (2:274; 13:22), and why it should be given (30:39) are all clarified.

007:157 The prophetic statements in Deuteronomy 18:15-19, and John 14:16-17 and 16:13 might be understood to be about Muhammed.

007:158 By distorting the meaning of *ummi,* clerics turned Muhammed into an illiterate and incompetent person who could not learn reading and writing 28 Arabic letters while dictating Quran for 23 years! For the meaning of *ummi*, see 3:20 and 2:78. For a detailed discussion of this verse, see the Sample Comparisons section in the Introduction.

007:166 See 2:65 and 5:60.

007:172-173 The past tense denotes the creation of human prototype. We translated the verse by taking its whole meaning. The dialog here is not necessarily verbal; it is through divinely established natural laws. We, the humans, are all born with innate ability to infer God's existence and power. In other words, by the very possession of DNA we are born with a divine contract not to corrupt our person, discover our creator and devote ourselves to Him alone. Compare this to verse 41:11.

007:174 This life is our last chance to return to God.

007:180 The Quran uses more than a hundred attributes for God, and attributes indicate continuity. Not every verb used for God can be considered an attribute. For instance, not every person who writes can be called a "writer." Furthermore, God's attributes are not necessarily Arabic. God sent messengers in many different languages to each nation and informed them about His attributes in their languages. Thus, the Quran teaches us that to God belong all beautiful attributes. However, *hadith* books list 99 attributes of which some cannot be considered "beautiful." The list, which is very popular among Sunni and Shiite *mushriks,* and which many people memorize, includes "bad" names such as *al-Dar* (the one who harms). Quran tells us otherwise (42:30).

Those who have confused Arab nationalism with islam might criticize our use of the English word God in the English text, rather than the word Allah. We would like to bring the attention of those who are not intoxicated with *hadith* and *sunna* that promotes Arab culture to the following points: the word Allah is not a proper name; it is an Arabic word contraction of the article *Al* (the) and *Elah* (god). Also, see 2:165; 3:26; 6:12; 17:110; 20:52; 42:11; 58:7.

The Quran informs us that God has been sending messengers to every nation in their own language (14:4). In each language, names or attributes represented by different sounds and symbols are used for the creator. For instance, the Old Testament uses *Yehovah* or *Elohim.* The New Testament quotes from Jesus addressing to God as *Eloi* (my lord), which is very close to the Arabic word *Elahi* (my lord) (Mark 15:34).

Through distortion and mistranslations, some Biblical verses depict God as less than a perfect being. For instance, Judges 1:19 (powerless); Genesis 6:6-7 (fallible); Psalms 13:1; Lamentations 5:20 (forgetful); Genesis 3:8-10 (can't see); 1 Samuel 15:2-3 (cruel). For more on divine attributes in the Quran and the Bible, see 59:22-24.

The Quran contains more than a hundred attributes of God, and they are designed letter by letter in accordance with the mathematical structure based on code 19. The studies of Prof. Adib Majul, which was continued by Edip Yuksel on the attributes of God, demonstrate an interlocking system. For instance, among the attributes of God, the frequency of only four of them are multiples of 19. They are *Shahyd* (Witness) 19, *Allah* (God) 2698, *Rahman* (Gracious) 57, *Rahym* (Compassionate) 114 times. When we analyze the attributes of God in according to their numerical values, we learn that only four of them are multiples of 19 and each correspond to the frequency of the other four: Wahid (One) 19, *Zulfadl-il Azym* (The Possessor of Great Bounty) 2698, *Majyd* (Glorious) 57, and *Jami* (Editor) 114. The details of this extraordinary and intricate mathematical design are discussed in, Nineteen: God's Signature in Nature and Scripture," Edip Yuksel, Brainbowpress, 2009. See: 74:30.

007:185 Whenever the word *hadith* is used for a word other than the Quran, it is used in a negative context. God knew that those who would revert back to their polytheistic religion would call their false teachings *hadith* (12:111; 31:6; 33:38; 35:43; 45:6; 52;34; 77:50).

007:187 God revealed that knowledge in 1980. See 20:15; 15:87; 72:27. From this and similar verses, we may infer that monotheists will not experience the horror of the end of the world. (See 6:31,44; 12:107; 21:40; 22:55; 26:202; 29:53; 43:66; 47:18).

007:188 None knows the future, including Muhammed (6:50; 7:188; 10:20; 27:65; 81:24). Only God knows the future and this knowledge can be attained only through divine revelations/signs (3:44; 11:49; 12:102; 30:2; 72:27).

بسم الله الرحمن الرحيم

8:0 In the name of God, the Gracious, the Compassionate.

8:1 They ask you regarding the spoils of war, say, "The spoils of war are for **God** and the messenger." So be aware of **God**, and peacefully resolve issues among you; and obey **God** and His messenger if you are those who affirm.*

8:2 The affirmers are those whom, when **God** is mentioned, their hearts tremble; when His signs are recited to them, it strengthens their affirmation; and they put their trust in their Lord.

8:3 They maintain the contact prayer, and from Our provisions to them they spend.

8:4 These are the true affirmers; they will have ranks at their Lord, forgiveness, and a generous provision.

8:5 As your Lord made you go out from your home with the truth, but a party from among those who affirm opposed this.

8:6 They argue with you about the truth when it has been made clear; as if they were being herded towards death while they are watching!

8:7 **God** promises you that one of the two parties will be defeated by you; yet you wish that the one least armed be the one. But **God** wishes that the truth be manifest with His words, and that He eliminates the remnant of the ingrates.

8:8 So that truth will be manifest, and the falsehood will be falsified; even if the criminals oppose it.

The Invisible Soldiers of God

8:9 You implored your Lord and He answered you: "I will provide you with one thousand f as defenders."

8:10 **God** did not do this except to give you good news, and that your hearts may be assured by it. Victory is only from **God**; **God** is Noble, Wise.

8:11 A peaceful sleep from Him overcame you, and He sent down to you water from the sky to cleanse you with it. He caused the affliction of the devil to leave you so that He may fortify your hearts and set firm your feet.

Defending People Against an Aggressors

8:12 Your Lord revealed to the controllers: "I am with you so keep firm those who affirm. I will cast fear into the hearts of those who have rejected; so strike above the necks, and strike every finger."*

8:13 That is because they have aggressed against **God** and His messenger. Whoever aggresses against **God** and His messenger, then **God** is severe in retribution.

8:14 This is for you to taste; and for the ingrates will be retribution of fire.

8:15 O you who affirm; when you encounter those who have rejected on the battlefield, do not flee from them.

8:16 Whoever on that day flees from them; unless it is part of the battle strategy or if he is retreating to his group; then he has drawn **God**'s wrath upon him, and his abode will be hell. What a miserable destiny.

8:17 It was not you who killed them, but it was **God** who killed them. It was not you who threw when you did, but it was **God** who threw. So that those who affirm would be tested well by Him. **God** is Hearer, Knowledgeable.

8:18 That, and **God** weakens the plots of the ingrates.

8:19 If you sought conquest, then conquest has come to you. But if you cease, then it is better for you. If you return, then We will also return, and your group will avail you nothing even if it is many. **God** is with those who affirm.

8:20 O you who affirm, obey **God** and His messenger, and do not turn away from him while you have heard.

8:21 Do not be like those who have said, "We hear," but they do not hear.

8:22 The worst creatures with **God** are the deaf and dumb who do not reason.

8:23 If **God** had found any good in them, then He would have made them listen. But if He made them listen, they would still turn away, averse.

Messenger is title of mission; not person

8:24 O you who affirm, answer the call of **God** and His messenger when He calls you to what will grant you life. Know that **God** comes between a person and his heart, and that to Him you will be gathered.*

8:25 Be aware of a test that will not only afflict those of you who were wicked; and know that **God** is severe in retribution.

8:26 Recall when you were but a few who were overpowered in the land, you were fearful that men might capture you. But He sheltered you, and He supported you with His victory, and He provided you with good provisions, so that you may be thankful.

8:27 O you who affirm, do not betray **God** and the messenger, nor betray your responsibilities, while you know.

8:28 You should know that your wealth and your children are a test, and that **God** has the greatest reward.

8:29 O you who affirm, if you are aware of **God**, He will endow you with criterion, and He will cancel your sins and forgive you. **God** is possessor of great favor.

8:30 The ingrates plot against you to confine you, to kill you, or to expel you. They plot, and **God** plots, and **God** is the best of plotters.

8:31 When Our signs are recited to them, they say, "We have listened, and if we wish, we could have said the same thing. This is nothing but tales of the ancients!"

8:32 They said, "Our God, if this is the truth from You, then send down upon us a rain of stones from the sky or bring on us a painful retribution."

8:33 But **God** was not to punish them while you are with them, nor will **God** punish them while they continue to seek forgiveness.

8:34 Why should **God** not punish them when they are turning others away from the Restricted Temple, and they were never its protectors! Its protectors are the righteous. But most of them do not know.

Meccan Polytheists Were Formally Following Abraham's Practices

8:35 Their contact prayer at the sanctuary was nothing but deception/noise and aversion. Taste the retribution for what you have rejected.*

8:36 Those who have rejected, they spend their money to turn others away from the path of **God**. They will spend it, then it will become a source of regret for them, then they are defeated. Those who rejected will be summoned to hell.

8:37 This is so that **God** will distinguish the bad from the good, and so that the bad will be gathered into one heap; then He will cast it all into hell. These are the losers.

8:38 Say to the ingrates: "If they cease, then what has passed before will be forgiven to them, and if they return to it, then the example of the previous generations has already been given."*

Fight Oppressive Powers for Freedom

8:39 You should fight them all until there is no more oppression, and so that the entire system is **God**'s. But if they cease, then **God** is seer of what they do.

8:40 If they turn away, then know that **God** is your Patron. What an excellent Patron, and what an excellent Victor.*

8:41 You should know: "Of anything you gain, that one-fifth shall go to **God** and the messenger: to the relatives, the orphans, the poor, and the wayfarer." You will do this if you affirm **God** and in what We revealed to Our servant on the day of distinction, the day the two armies clashed. **God** is able to do all things.*

Trust in God

8:42 When you were on the near side, and they were on the far side, then the supply line became directly beneath you. Had you planned for this meeting, you would have disagreed on its timing, but **God** was to enforce a command that was already done. So that He would destroy those to be destroyed with proof, and to let those who will live be alive with proof. **God** is Hearer, Knowledgeable.

8:43 **God** shows them to you as being few in your dream, and had He shown them to be many, then you would have failed and you would have disputed in the matter; it was **God** who saved you. He is the Knower of what is inside the chests.*

8:44 He showed them to you when you met as being few to your eyes, and He made you appear as being fewer to their eyes. That was so **God**'s decree would come to be; and to **God** all matters are returned.

8:45 O you who affirm, when you meet an armed group, stand firm and mention **God** frequently, that you may succeed.

8:46 You should obey **God** and His messenger, and do not dispute; else you will fail and your strength will depart. You should also patiently persevere. **God** is with the ones who patiently persevere.

8:47 Do not be like those who came out from their homes to boast and to be seen by men, and they repel others away from the path of **God**. **God** encompasses of what they do.

8:48 The devil adorned their work for them, and he said, "None from the people can defeat you today, and I am by your side." But when the two forces came together, he turned on his heels to flee and he said, "I am innocent from you! I see what you do not see. I fear **God**, and **God** is severe in punishment."*

8:49 The hypocrites and those who have a disease in their hearts said, "These people have been deceived by their system." But whoever puts his trust in **God**, then **God** is Noble, Wise.

8:50 If you could only see as the controllers take those who have rejected, they strike their faces and their backs: "Taste the punishment of the blazing fire!"

8:51 "This is for what your hands have put forth, and **God** does not wrong the servants."

8:52 Like the behavior of the people of Pharaoh, and those before them; they rejected **God**'s signs, so **God** took them for their sins. **God** is Strong, severe in punishment.

To Change the Negative Conditions, You Need First To Change Yourself

8:53 That is because **God** was not to change anything He bestowed to a people, unless they change what is in themselves. **God** is Hearer, Knowledgeable.

8:54 Like the behavior of Pharaoh's people and those before them. They denied the signs of their Lord, so We destroyed them by their sins, and We drowned the people of Pharaoh; all of them were wicked.

8:55 The worst creatures to **God** are those who reject, for they do not affirm.

War and Peace

8:56 The ones whom you made a pledge with them, then they break their pledge every time, and they do not care.

8:57 So, when you encounter them in battle, set them as an example to those who would come after them; perhaps they may remember.

8:58 If you are being betrayed by a people, then you shall likewise move against them. **God** does not like the betrayers.

8:59 Let not those who have rejected think that they have escaped; they will never avail themselves.

8:60 Prepare for them all that you can of might, and from horse powers/virtual networks, that you may instill fear with it towards **God**'s enemy and your enemy, and others beside them whom you do not know but **God** knows them. Whatever you spend in the cause of **God** will be returned to you, and you will not be wronged.

8:61 If they seek peace, then you also seek it, and put your trust in **God**. He is the Hearer, the Knowledgeable.

8:62 If they wish to deceive you, then **God** is sufficient for you. He is the One who supported you with His victory and with those who affirm.

8:63 He made unity between their hearts. Had you spent all that is on earth, you would not have united between their hearts, but **God** united between them. He is Noble, Wise.

8:64 O prophet, **God** is sufficient for you and those who affirmed among your followers.*

8:65 O prophet, urge those who affirm to battle. If there are twenty of you who are patient, they will defeat two hundred. If there are one hundred of you, they will defeat one thousand from amongst those who reject; that is because they are a people who do not reason.

8:66 Now, **God** has lightened for you, for He knows that there is weakness in you. If there are one hundred of you who are patient, they will defeat two hundred. If there are a thousand of you, they will defeat two thousand by **God**'s leave. **God** is with the patient.

Slavery Prohibited

8:67 It was not for any prophet to take prisoners unless it was in battle. You desire the materials of this world, while **God** wants the Hereafter for you. **God** is Noble, Wise.*

8:68 Had it not been previously ordained from **God**, then a severe punishment would have afflicted you for what you took.

8:69 So consume what you have gained, lawful and good, and be aware of

God. **God** is Forgiving, Compassionate.

Treat Prisoners of War with Dignity

8:70 O prophet, say to those prisoners whom you hold: "If **God** finds in your hearts any good, He will grant you better than what He took from you, and He will forgive you. **God** is Forgiving, Compassionate."

8:71 If they want to betray you, they have already betrayed **God** before, and He overpowered them. **God** is Knowledgeable, Wise.

International Treaties Trump Other Allegiances

8:72 The affirmers and emigrated and strived with their money and lives in the cause of **God**, and those who sheltered and supported; these are the allies of one another. Those who affirmed but did not emigrate, you do not owe them any allegiance until they emigrate. But if they seek your help in the system, then you must support them, except if it is against a people with whom there is a treaty between you. **God** is watcher over what you do.*

8:73 As for those who reject, they are allies to one another. If you do not do this, then there will be oppression on earth and great corruption.

8:74 The affirmers and emigrated and strived in the cause of **God**, and those who sheltered and supported, these are truly those who affirm. They will have forgiveness and a generous provision.

8:75 Those who affirmed afterwards and emigrated and strived with you, then they are from you. The relatives by birth are also supportive of one another in **God**'s book. **God** is aware of all things.

ENDNOTES

008:001 See 8:41.

008:012-16 Wars are subject to the general principles spelled out in 60:8-9. Also, see 9:29.

008:024 After the expression "God and messenger" the verb "Da'akum" (he calls you) is used. If Messenger was considered as second source besides God, as Sunni and Shiite *mushriks* claim, then the dual form of the verb "Da'ukum" (they call you) would have been used. Also see, 4:150-152 and 9:1

008:035 The Quran informs us that the practices of Islam were initially revealed to Abraham. Abraham and his followers were observing the *sala* prayer (21:73). Meccan polytheists never accepted that they were polytheists (6:23,148; 16:35). Since they believed in the intercession of some holy people, and since they falsely attributed numerous prohibitions and rules to God, the Quran considered them polytheists. Those so-called Muslims who reverted to the belief and practices of the era of ignorance after the revelation of the Quran, tried to create some artificial differences between them and Meccan polytheists. They distorted the meaning of the verse into "their prayer is only clapping hands and whistles" so that Meccan polytheists will be seen to be bizarre and different. The word *muka* means whistle, noise, or hypocrisy. The celebrated Arabic dictionary <u>*Lisan al-Arab,*</u> under the definition of this word, lists underground escape holes dug by moles. This resembles the word *nafaqa* (hypocrisy), which relates a semantic relation between physical holes and abstract hypocrisy. As for the word *tasdiya*, either means aversion or it is a form of *tasdidah*, which simply means repelling. In fact, the following verse uses its verb form, supporting our understanding.

See 53:19-23. Also, see chapter 107 and verse 9:54.

008:038 The expression of *sunnat al-awalyn* (the path/laws of previous generations) reminds God's punishment to the past communities who transgressed. The word *sunna* means "law," and there is only one valid *sunna*, God's *sunna* (33:38,62; 35:43; 40:85; 48:23). Muhammed could not have his own *sunna* (law) besides God's *sunna* in the Quran.

008:040 See 2:286.

008:041 Distorting the meaning of this verse, Shiite clergymen have created an entitlement for themselves. The followers of Shiite sects donate one fifth of their annual income (*khumus*) to their religious leeches.

008:043 See 25:47; 2:255; 12:6; 30:23; 39:42; 78:9; 37:102.

008:048 Satan's fear (*khafy*) is without reverence (*khashya*).

008:064 Traditional translations dilute the sufficiency of God by including those who affirm as the second subject besides God, rather than another object besides Muhammed. They translate the verse as "God and the affirmers who follow you suffice for you."

008:067 This verse prohibits slavery. Prisoners of war are released after the war ends (47:4). Also, see 4:3,25; 90:1-20.

008:072 This verse unequivocally states that the rule of law is above any other affiliation. Islam emphasizes the importance of the rule of law, justice and peace (16:91,92).

****

****

9:1 This is an ultimatum from **God** and His messenger to those who set up partners with whom you had entered a treaty.*

9:2 Therefore, roam the land for four months and know that you will not escape **God**, and that **God** will humiliate the ingrates.

9:3 A declaration from **God** and His messenger to the people, on this, the peak day of the Debate Conference: "That **God** and His messenger are free from obligation to those who set up partners." If you repent, then it is better for you, but if you turn away, then know that you will not escape **God**. Promise those who have rejected of a painful retribution;*

9:4 Except for those with whom you had a treaty from among those who have set up partners if they did not reduce anything from you, nor did they plan to attack you; you shall fulfill their terms until they expire. **God** loves the righteous.

9:5 So when the restricted months have passed, then fight those who have set up partners wherever you find them, take them, surround them, and stand against them at every point. If they repent, maintain the contact prayer, and contribute towards betterment, then you shall leave them alone. **God** is Forgiving, Compassionate.*

Do Not Let Those Who Violate the Peace Treaty Succeed

9:6 If any of those who have set up partners seeks your protection, then you may protect him so that he may hear the words of **God**, then let him reach his sanctuary. This is because they are a people who do not know.

9:7 How can those who have set up partners have a pledge with **God** and with His messenger? Except for those with whom you made a pledge near the Regulated Temple, as long as they are upright with you, then you are upright with them. **God** loves the righteous.

9:8 How is it that if they ever defeated you they would neither respect neither any rights of kinship nor any pledge. They seek to please you with their words, but their hearts deny, and most of them are wicked.

9:9 They exchanged **God**'s signs for petty gain, so they turn others from His path. Evil indeed is what they used to do.

9:10 They neither respect the ties of kinship nor a pledge for any those who affirm. These are the transgressors.

9:11 If they repent, and they observe the support activities, and they contribute towards betterment, then they are your brothers in the system. We explain the signs for a people who know.

9:12 If they break their oaths after their pledge, and they taunt and attack your system; then you may kill the chiefs of rejection. Their oaths are nothing to them, perhaps they will then cease.

9:13 Would you not fight a people who broke their oaths and intended to expel the messenger, especially while they were the ones who attacked you first? Do you fear them? It is **God** who is more worthy to be feared if you are those who affirm.

9:14 Fight them; perhaps **God** will punish them by your hands, humiliate them, grant you victory over them and heal the chests of an affirming people,

9:15 To remove the anger from their hearts; **God** pardons whom he pleases. **God** is Knowledgeable, Wise.

9:16 Or did you think that you would be left alone? **God** will come to know those of you who strived and did not take other than **God** and His messenger and those who affirm as helpers. **God** is Ever-aware in what you do.

9:17 It was not for those who have set up partners to maintain **God**'s temples while they bear witness over their own rejection. For these, their works have fallen, and in the fire they will abide.

9:18 Rather, the temples of **God** are maintained by the one who affirms **God** and the Last day, holds the contact prayer, contributes towards betterment, and he does not fear except **God**. It is these that will be of the guided ones.

9:19 Have you made serving drink to the Debaters and the maintenance of the Restricted Temple the same as one who affirms **God** and the Last day, who strives in the cause of **God**? They are not the same with **God**. **God** does not guide the wicked people.

9:20 Those who affirmed, emigrated, strived in the cause of **God** with their wealth and their lives are in a greater degree with **God**. These are the winners.

9:21 Their Lord gives them good news of a Mercy from Him, acceptance, and gardens that are for them in which there is permanent bliss.

9:22 They will abide in it eternally. **God** has a great reward.

9:23 O you who affirm, do not take your fathers nor brothers as allies if they prefer rejection to affirmation. Whoever of you takes them as such, then these are the wicked.

Dedicate Yourself to Establish Peace and Liberty

9:24 Say, "If your fathers, your sons, your brothers, your spouses, your clan, and money which you have gathered, a trade in which you fear a decline, and homes which you enjoy; if these are dearer to you than **God** and His messenger and striving in His cause, then wait until **God** brings His decision. **God** does not guide the wicked people."

9:25 **God** has granted you victory in many battlefields. On the day of Hunayn, when you were pleased with your great numbers but it did not help you at all, and the land became tight around you for what it held, then you turned to flee.

9:26 Then **God** sent down tranquility upon His messenger and those who affirm, and He sent down soldiers which you did not see. He thus punished those who rejected. Such is the recompense of the ingrates.

9:27 Then **God** will accept the repentance of whom He pleases after that. **God** is Forgiving, Compassionate.

9:28 O you who affirm, those who have set up partners are impure, so let them not approach the Restricted Temple after this calendar year of theirs. If you fear poverty, then **God** will enrich you from His blessings if He wills. **God** is Knowledgeable, Wise.

9:29 Fight those who do not affirm **God** nor the Last day among the people who received the book; they do not forbid what **God** and His messenger have forbidden, and they do not uphold the system of truth; until they pay the reparation, in humility.*

Do Not Accept Religious Leaders As Authorities Besides God

9:30 The Jews said, "Ezra is **God**'s son," and the Nazarenes said, "The Messiah is **God**'s son." Such is their utterances with their mouths; they imitate the sayings of those who rejected before them. **God**'s curse be on them. How deviated are their minds!*

9:31 They took their scholars and priests to be lords besides **God**, and the Messiah son of Mary, while they were only commanded to serve One god, there is no god but He, be He glorified for what they set up.

9:32 They want to extinguish **God**'s light with their mouths, but **God** refuses such and lets His light continue, even if the ingrates hate it.

9:33 He is the One who sent His messenger with guidance and the system of truth, to make it manifest above all other systems, even if those who set up partners hate it.*

Beware of Religious Scholars and Leaders Who Exploit In God's Name

9:34 O you who affirm, many of the scholars and priests consume people's money in falsehood, and they turn away from the path of **God**. Those who hoard gold and silver, and do not spend it in the cause of **God**, give them news of a painful retribution.

9:35 On the day when they will be seared in the fires of hell, their foreheads, sides and backs will be branded with it: "This is what you have hoarded for yourselves, so taste what you have hoarded!"

9:36 The count of the months with **God** is twelve months in **God**'s book the day He created the heavens and the earth; four of them are restricted. This is the correct system; so do not wrong yourselves in them. Fight those who set up partners collectively as they fight you collectively. Know that **God** is with the righteous.*

9:37 Know that accelerating the intercalary is an addition in rejection; that those who have rejected may misguide with it. They make it lawful one calendar year, and they forbid it one calendar year, so as to circumvent the count that **God** has made restricted; thus they make lawful what **God** made forbidden! Their evil works have been adorned for them, and **God** does not guide the rejecting people.*

Do Not Accept Aggression and Oppression

9:38 O you who affirm, what is wrong with you when you are told: "March forth in the cause of **God**," you become heavy on earth. Have you become content with this worldly life over the Hereafter? The enjoyment of this worldly life compared to the Hereafter is nothing.

9:39 If you do not march forth, then He will punish you with a painful retribution, He will replace you with another people, and you do not bother Him in the least. **God** is capable of all things.

9:40 If you do not help him, then **God** has helped him. When those who rejected expelled him, he was one of only two, and when both were in the cave, he said to his friend: "Do not grieve, for **God** is with us." So **God** sent down tranquility over him and He supported him with soldiers that you did not see, and He made the word of those who rejected be the lowest, and **God**'s word be the highest. **God** is Noble, Wise.

9:41 March forth in light gear or heavy gear, and strive with your money and lives in the cause of **God**. That is best if you knew.

9:42 If it were a near gain, or an easy journey, they would have followed you; but the distance was too much for them. They will swear by **God**: "If we could have, we would have come with you." They destroy themselves, and **God** knows they are liars.

9:43 **God** pardons you; why did you grant them leave before it became clear to you who are truthful, and who are lying?

9:44 Those who affirm **God** and the Last day will not ask leave. They strive with their money and their lives. **God** is aware of the righteous.

9:45 Those who ask leave are the ones who do not affirm **God** and the Last day, and their hearts are in doubt. In their doubts they are wavering.*

9:46 If they had gone with you then they would have taken all preparation for it, but **God** disliked their being sent forth, so He hindered them, and they were told: "Stay with those who have stayed."

9:47 Had they come out with you they would have added nothing but disorder, and they would have hurried about seeking a test among you. There are some amongst you who listen to them. **God** is aware of the wicked.

9:48 They wanted to test from before, and they turned matters upside down for you until the truth came and **God**'s command was revealed, while they hated it.

9:49 Some of them say, "Grant me leave, and do not test me." But it is in the test that they have indeed fallen, and hell is surrounding the ingrates.

9:50 When any good befalls you, it upsets them, and if any bad befalls you, they say, "We have taken our precautions beforehand," and they turn away rejoicing.

9:51 Say, "Nothing will befall us except what **God** has decreed for us; He is our Patron." In **God** those who affirm shall put their trust.*

9:52 Say, "While we wait for you to be afflicted by **God** with retribution from Him, or at our hands you can only expect for us one of two good things. So wait, we are with you waiting."

9:53 Say, "Spend willingly or unwillingly, it will not be accepted from you. You are a wicked people."

9:54 What prevented the acceptance of their spending was that they rejected **God** and His messenger, and they do not attend to the contact prayer except lazily, and they do not spend except unwillingly.*

Worldly Gains

9:55 So do not be impressed by their wealth or children; **God** only wishes to punish them with it in the worldly life, and so that their lives will end while they are ingrates.

9:56 They swear by **God** that they are with you, while they are not with you; but they are timid and divisive people.

9:57 If they could find a refuge, a cave, or any place to enter, then they would have run to it, rushing.

9:58 Some of them are those who criticize you regarding the charities. If they are given from it, they are content; but if they are not given from it, they become enraged!

9:59 If only they were content with what **God** and His messenger had given them, and had said, "**God** suffices us; **God** will give us from His bounty, and so His messenger; it is to **God** that we desire."

Distribution of Charities

9:60 The charities are to go to the poor, the needy, those who work on their collection, those whose hearts are to be reconciled, free the slaves, those in debt, in the cause of **God**, and to the wayfarer. A duty from **God**; **God** is Knowledgeable, Wise.

9:61 Among them are those who hurt the prophet, and they say, "He only listens!" Say, "What he listens to is best for you. He affirms **God**, he trusts those who affirm, and he is a mercy to those who affirm among you." Those who hurt **God**'s messenger will have a painful retribution.

9:62 They swear to you by **God** in order to please you; while **God** and His messenger is more worthy to be pleased if they were those who affirm.

Hypocrites and Opponents

9:63 Did they not know that whoever is hostile towards **God** and His messenger, he will have the fire of hell to abide in. Such is the greatest humiliation.

9:64 The hypocrites fear that a chapter will be sent down against them exposing what is in their hearts. Say, "Mock, for **God** will bring out what you fear."

9:65 If you ask them they say, "We were only jesting and playing." Say, "Is it **God** and His signs and His messenger you were mocking?"

9:66 Do not apologize, for you have rejected after your affirmation. Even if We pardon one group from you, We will punish another group, because they were criminals.

9:67 The hypocrite men and the hypocrite women, they are the same. They order evil, deter from good, and they are tightfisted. They ignored **God**, so He ignored them; the hypocrites are the wicked.

9:68 **God** has promised the hypocrite men and women and the ingrates a fire in hell, in it they will abide; it suffices them. **God** has cursed them and they will have a lasting retribution.

9:69 Like those before you; they were more powerful than you, and had more wealth and offspring. They enjoyed their lot, and you enjoyed your lot as those before you enjoyed their lives; and you indulged as they indulged. These are those whose works crumbled in this world and the hereafter, and they were the losers.

9:70 Had not the news of those before them come to them, the people of Noah and Aad and Thamud. The people of Abraham, and the dwellers of Midian, and those overthrown. Their messengers came to them with proofs; it was not **God** who wronged them, but it was themselves that they wronged.

9:71 The affirming men and women, they are allies to one another. They order good and deter from evil, and they maintain the contact prayer, and they contribute towards betterment, and they obey **God** and His messenger. **God** will have mercy on them; **God** is Noble, Wise.

9:72 **God** has promised the affirming men and women gardens with rivers flowing beneath it, in which they will abide, and pleasing homes in the everlasting gardens. The acceptance from **God** is the most important; such is the greatest success.

9:73 O prophet, strive against the ingrates and the hypocrites and be firm against them. Their dwelling is hell, what a miserable destiny!

9:74 They swear by **God** that they did not say it, while they had said the word of rejection, and they rejected after they had peacefully surrendered, and they were concerned with what they could not possess; and they could not find any fault except that **God** and His messenger had enriched them from His bounty. If they repent it is better for them, and if they turn away, then **God** will punish them severely in this world and the next. They will not have on earth any ally or supporter.

9:75 Some of them pledged to **God**: "If He gives us from His bounty, then we will affirm and we will be amongst the reformed."

9:76 Yet when He gave them of His bounty, they became stingy with it, and they turned away in aversion.

9:77 Thus, they ended up with hypocrisy in hearts until the day they meet Him; that is for breaking what they promised to **God**, and for what they were lying.

9:78 Did they not know that **God** knows their secrets, their conspiracies, and that **God** is the knower of all the unseen?

9:79 **God** mocks those who criticize the generous affirmers for giving too much; and disdain those who do not have anything to give but their effort, **God** disdains them, and they will have a painful retribution.

Muhammed Did Not Have Power of Intercession

9:80 Whether you seek forgiveness for them, or do not seek forgiveness for them. If you seek forgiveness for them seventy times, **God** will not forgive them. That is because they have rejected **God** and His messenger; and **God** does not guide the wicked people.*

9:81 Those who were left behind rejoiced their staying behind **God**'s messenger, and they disliked to strive with their wealth and lives in the cause of **God**. They said, "Do not march out in the heat." Say, "The fire of hell is far hotter." If they could only understand!

9:82 Let them laugh a little, and cry a lot, as a recompense for what they had earned.

9:83 If **God** returns you to a group of them, and they ask your permission to come with you, then say, "You will not come with me ever; nor will you fight any enemy with me. You had accepted staying behind the first time, so stay with those who remain behind."

9:84 Do not show support for anyone of them who dies, nor stand at his grave. They have rejected **God** and His messenger and died while they were wicked.

Do Not Be Impressed by Their Wealth

9:85 Do not also be impressed by their wealth and their children; **God** only wishes to punish them with it in this world, and their selves will vanish while they are ingrates.

9:86 If a chapter is sent down: "That you shall affirm **God** and strive with His messenger," those with wealth and influence ask your permission and they say, "Let us be with those who remain behind."

9:87 They were content to stay with those who remained behind, and it was stamped on their hearts, for they do not reason.

Strive For Good Causes

9:88 But the messenger and those who affirm with him have strived with their money and their lives. For them will be the good things, and they are the successful ones.

9:89 **God** has prepared for them gardens with rivers flowing beneath, in them they shall abide. Such is the great gain.

9:90 Those who made excuses from among the Arabs came to get permission. Thus, those who denied **God** and His messenger stayed behind. Those who rejected from them will be inflicted with a painful retribution.

Handicapped People Are Protected

9:91 There is no blame to be placed upon the weak, the sick, or those who do not find anything to spend on, as long as they are sincere to **God** and His messenger. There is no argument against the good doers; and **God** is Forgiving, Compassionate.

9:92 Nor upon those who come to ride out with you, while you said, "I do not have any mounts for you;" they turned away while their eyes were flooded with tears out of sadness, for they could not find anything to contribute.

9:93 Indeed, the argument is against those who sought your permission to stay while they had the means. They accepted to be with those who remained behind; **God** stamped their hearts, for they do not know.

Ignore the Hypocrites

9:94 They will apologize to you when you return to them. Say, "Do not apologize; we will no longer trust you, for **God** has told us of your news." **God** will see your work, as will His messenger, then you will be returned to the knower of the unseen and the seen, He will inform you of what you did.

9:95 They will swear by **God** to you when you return to them, so that you may overlook them. So overlook them, for they are tainted, and their destiny is hell as a recompense for what they earned.

9:96 They swear to you so that you would accept them. Even if you accept them, **God** does not accept the wicked people.

Good and Bad Arabs

9:97 The Arabs are the worst in rejection and hypocrisy, and more likely not to know the limits of what **God** has sent down upon His messenger. **God** is Knower, Wise.

9:98 Some from Arabs are who look upon what they spend as a fine, and wait for turns of disasters to befall you. They will have a turn of disaster, and **God** is Hearer, Knowledgeable.

9:99 From the Arabs there is he who affirms **God** and the Last day, and considers his spending as a means towards **God**, and the support of the messenger. It indeed makes them closer. **God** will admit them in His mercy; **God** is Forgiving, Compassionate.*

9:100 The pioneers of the emigrants and the supporters, and those who followed them in kindness; **God** has accepted them, and they have accepted Him; and He prepared for them gardens with rivers flowing beneath in which they will abide eternally. Such is the great success.

9:101 Among the Arabs around you are hypocrites, as well as from the city people; they are audacious in hypocrisy. You do not know them, but We know them. We will punish them twice, then they will be returned to a great punishment.

9:102 Others who have affirmed their sins, they have mixed good work with bad. Perhaps **God** will pardon them. **God** is Forgiving, Compassionate.

9:103 Take from their money a charity to purify them and improve them with it, and support them. Your support is tranquility for them; and **God** is Hearer, Knowledgeable.

9:104 Did they not know that it is **God** who accepts repentance from His servants, and He takes the charities, and that **God** is the Pardoner, Compassionate.

9:105 Say, "Work, and **God** will see your work and so His messenger and those who affirm. You will be sent back to the knower of the unseen and the seen, and He will inform you of what you did."

9:106 Others are waiting for **God**'s decision; either He will punish them, or He will pardon them. **God** is Knowledgeable, Wise.

Mosques That Promote Division and Falsehood

9:107 There are those who establish a temple to do harm and cause rejection, to cause division among those who affirm, and as an outpost for those who fought **God** and His messenger before. They will swear that they only wanted to do good, but **God** bears witness that they are liars.

9:108 You shall never stand there. A temple that is founded on righteousness from the first day is more worthy of your standing; in it are men who love to be cleansed. **God** loves the cleansed.

9:109 Is one who lays his foundation on obtaining awareness from **God** and His acceptance better, or one who lays his foundation on the edge of a cliff which is about to crumble, so that it crumbled with him into the fires of hell? **God** does not guide the wicked people.

9:110 Such a structure that they erect will never cease to cause doubt in their hearts, until their hearts are severed. **God** is Knowledgeable, Wise.

The Monotheists

9:111 **God** has bought from the affirmers their lives and their money in exchange for paradise. They fight in the cause of **God**; they kill and are killed. A promise that is true upon Him in the Torah, the Injeel and the Quran. Whoever fulfills his pledge to **God**, then have good news of the deal which you concluded with. Such is the supreme success.

9:112 They are the repenters, the servers, the praisers, the activists, the bowing, the prostrating, the advocators of good, the forbidders of evil, and the keepers of **God**'s ordinance. Give good news to those who affirm.

9:113 It is not for the prophet and those who affirm that they should seek forgiveness for those who have set up partners, even if they are relatives, after it has been made clear to them that they are the dwellers of hell.

9:114 Abraham seeking forgiveness for his father was only because of a promise he made to him. But when it became clear that he was **God**'s enemy, he disowned him. Abraham was kind, compassionate.

9:115 **God** will never mislead a people after He guided them, until He makes clear to them what they should be aware. **God** is knower of all things.

9:116 To **God** is the sovereignty of heavens and earth, He causes life and death. You have none besides **God** as a supporter or victor.

God is Supporting the Monotheists

9:117 **God** has pardoned the prophet and the emigrants and the supporters that followed him in the darkest moment, even though the hearts of some of them nearly deviated, but then He pardoned them. He is towards them Kind, Compassionate.

9:118 Also upon the three who were left behind until the land, as vast as it is, became strained to them, and their very selves became strained; they thought there is no shelter from **God** except to Him. Then He pardoned them that they might repent. **God** is the Redeemer, the Compassionate.

9:119 O you who affirm, be aware of **God** and be with the truthful.

9:120 It is not advisable for the city dwellers and those around them of the Arabs that they should lag behind after **God**'s messenger, nor should they yearn for themselves above him. That is because any thirst, fatigue, or hunger that afflicts them in the cause of **God**, or any step that they take which will annoy the ingrates, or any gain they have over any enemy; it will be recorded as a good deed for them. **God** does not waste the reward of the good doers.

9:121 Anything small or large they spend, or any valley they cross, it will be recorded for them. **God** will recompense them with the best of what they did.

Importance of Education; Restriction on Draft

9:122 It is not advisable for those who affirm to march out in their entirety. For every battalion that marches out, let a group remain to study the system, and warn their people when they return to them, perhaps they will be cautious.

9:123 O you who affirm, fight the ingrates who gird you about, and let them find strength in you; know that **God** is with the righteous.

Conspiracy of Hypocrites Increases Only Impurity

9:124 When a chapter is revealed, some of them say, "Whose affirmation has this increased?" For those who affirm, it increased their affirmation, and they rejoice.*

9:125 As for those who have a disease in their hearts, it only increased foulness to their foulness, and they died as ingrates.

9:126 Do they not see they are tested every calendar year once or twice? Yet, they do not repent, nor do they take heed.

9:127 When a chapter is revealed, they looked at one another: "Does anyone see you?", then they turn away. **God** turns away their hearts, for they are a people who do not comprehend.*

ENDNOTES

009:* The reason why this chapter does not contain Basmalah was questioned for centuries, and the conclusive answer came with the discovery of the code 19, prophesized in chapter 74. See 74:1-56; 1:1; 2:1; 13:38; 27:82; 38:1-8; 40:28-38; 46:10; 72:28.

009:001 Those who abuse the Quranic statement "obey God and His messenger," to mean obeying the Quran and numerous books of hearsay, are refuted by this verse. By claiming that God is represented by the Quran, but the messenger is represented by hearsay collections produced centuries after Muhammed, they separate God and messenger from each other (4:150-152). Verse 9:1 attributes the ultimatum issued in this chapter to "God and His messenger." Since God did not author the Quran in consultation with Muhammed, the only reason why the *messenger* was mentioned after God, is that the messenger will deliver the ultimatum as it is, like a trustworthy mailman. Thus, we should use similar inferences when we see verses telling us to "obey God and His messenger." Those verses do not mean God and His messenger represent different authorities or sources, but simply mean, "Obey God by obeying the message brought to you through His messenger." In the Quran we do not see a single verse instructing us to "obey Muhammed" or "obey Moses;" the verses always refer to obeying the messenger, since messengers' sole mission is to deliver God's message without distorting, adding, or subtracting from it. The messenger, on the Day of Judgment (25:30), will reject those who distort, add to, and subtract from God's message by hearsay.

It is also noteworthy that the verb used for God and His messenger in 8:24 is in its singular form. Remind verse 6:114 to the polytheists who wish to raise Muhammed from the position of messengership to the position of legislator next to God. Also, see 49:1.

009:003-029 The verse 9:5 does not encourage muslims to attack those who associate partners to God, but to attack those who have violated the peace treaty and killed and terrorized people because of their belief and way of life.

According to verses 9:5 and 9:11, the aggressive party has two ways to stop the war: reinstate the treaty for peace (*silm*), which is limited in scope; or accept the system of peace and submission to God (islam), which is comprehensive in scope; it includes observation of *sala* and purification through sharing one's blessings. These two verses refer to the second alternative. When, accepting islam (system of peace and submission) as the second equally acceptable alternative and when the first alternative involves only making a temporary peace, then none can argue for coercion in promoting the Din.

The Quran does not promote war but encourages us to stand against aggressors on the side of peace and justice. War is permitted only in self-defense (See 2:190,192,193,256; 4:91; 5:32; 8:19; 60:7-9). We are encouraged to work hard to establish peace (47:35; 8:56-61; 2:208). The Quranic precept promoting peace and justice is so fundamental that a peace treaty with the enemy is preferred to religious ties (8:72).

Please note that the context of the verse is about the War of *Hunain*, which was provoked by the enemy. The verse 9:29 is mistranslated by almost every translator.

Furthermore, note that we suggest "reparation," which is the legal word for compensation for damages done by the aggressing party during the war, instead of the Arabic word *jizya*. The meaning of *jizya* has been distorted as a perpetual tax on non-Muslims, which was invented long after Muhammed to further the imperialistic agenda of Sultans or Kings. The origin of the word that we translated as Compensation is *JaZaYa*, which simply means compensation, not tax. Because of their aggression and initiation of a war against muslims and their allies, after the war, the allied community should require their enemies to compensate for the damage they inflicted on the peaceful community. Various derivatives of this word are used in the Quran frequently, and they are translated as "compensation" for a particular deed.

Unfortunately, the distortion in the meaning of the verse above and the practice of collecting a special tax from Christians and Jews, contradicts the basic principle of the Quran that there should not be compulsion in religion and that there should be freedom of belief and expression (2:256; 4:90; 4:137, 140; 10:99; 18:29; 88:21,22). Since taxation based on religion creates financial duress on people to convert to the privileged religion, it violates this important Quranic principle. Dividing a population that is united under a social contract (constitution) into privileged groups based on their religion contradicts many principles of the Quran, including justice, peace, and brotherhood/sisterhood of all humanity. See 2:256. For a comparative discussion of this verse, see the Sample Comparisons section in the Introduction.

Moral rules involving retaliation can be classified under several titles:

- **The Golden Rule**: Do unto others as you would have them do unto you.
- **The Silver Rule**: Do not do unto others what you would not have them do unto you.
- **The Golden-plated Brazen Rule:** Do unto others as they do unto you; and occasionally forgive them.
- **The Brazen Rule**: Do unto others as they do unto you.
- **The Iron Rule**: Do unto others as you like before they do it unto you.

Empirical studies on groups have shown that the golden-plated brazen rule is the most efficient in reducing negative behaviors in a community abiding by the rule, since the rule has both deterrence and guiding components. Game theorists call this rule "Generous Tit-For-Tat" (GTFT), a rule that leads one party to cooperate as long as the other party does too. If the other party cheats or hurts the party adopting GTFT, then the GTFT-adopting party stops cooperation and retaliates, while demonstrating willingness to forgive the wrongdoing and start a new stage of cooperation. This positive tilt successfully leads the other party to seek cooperation of the GTF party and ultimately adopt the same rule of engagement. The Golden rule, on the other hand, does not correspond to the reality of human nature; it rewards those who wish to take advantage of the other party's niceness. Therefore, though the golden rule is the most popular rule on the lips of people, it is the least used rule in world affairs. It might have some merits in small groups with intimate relations, but we do not have evidence for that.

The Quran is a book of reality, and its instructions involving social issues consider the side effects of freedom. Thus, the Quran recommends us to employ the golden-plated brazen rule. "If the enemy inclines toward peace, do you also incline toward peace" (8:61; 4:90; 41:34). Other verses encouraging forgiveness and patience in the practice of retaliation (2:178; 16:126, etc.), make the Quranic rule a "Golden-plated Brazen Rule," the most efficient rule in promoting goodness and discouraging crimes.

Sunni and Shiite *mushriks* inherited many vicious laws and instructions of violence through *hadith* books, most of which scholars trace their roots to the influence of Jewish Rabbis and Christian priests who supposedly converted to Islam. Here are some samples of terrifying and bloody instructions found in The Old Testament. We recommend the reader to study them in their context:

"The Lord said to Moses: 'Take the blasphemer outside the camp. All those who heard him are to lay their hands on his head, and the entire assembly is to stone him. Say to the Israelites: If anyone curses his God, he will be held responsible; anyone who blasphemes the name of the Lord must be put to death. The entire assembly must stone him. Whether an alien or native-born, when he blasphemes the Name, he must be put to death.... Then Moses spoke to the Israelites, and they took the blasphemer outside the camp and stoned him. The Israelites did as the Lord commanded Moses" (Leviticus 24:13-16).

"Now kill all the boys. And kill every woman who has slept with a man. But save for yourselves every girl who has never slept with a man" (Numbers 31:18).

"And they utterly destroyed all that was in the city, both man and woman, young and old, and ox, and sheep, and ass, with the edge of the sword" (Joshua 6:21).

"Now go and smite Amalek, and utterly destroy all that they have, and spare them not; but slay both man and woman, infant and suckling, ox and sheep, camel and ass" (1 Samuel 15:3).

"And as David returned from the slaughter of the Philistine, Abner took him, and brought him before Saul with the head of the Philistine in his hand" (1 Samuel 17:57).

"Thus the Jews smote all their enemies with the stroke of the sword, and slaughter, and destruction, and did what they would unto those that hated them" (Esther 9:5).

"Why do the wicked prosper and the treacherous all live at ease?... But you know me, Lord, you see me; you test my devotion to you. Drag them away like sheep to the shambles; set them apart for the day of slaughter" (Jeremiah 12:1-3).

"A curse on all who are slack in doing the Lord's work! A curse on all who withhold their swords from bloodshed!" (Jeremiah 48:10).

Chapter 20 of Leviticus contains a list of very severe punishments for various sins. For instance, cursing one's own father or mother would prompt the death penalty. A man marrying a woman together with her daughter must be burned in the fire. Homosexual men must be put to death. Those who commit bestiality must be put to death together with the animals. And many more deaths and burning penalties.

009:024 Most of the world's organized religious institutions collect money from their people. Most of that money is wasted through promotion of lies and ridiculous beliefs. In 2007, the approximately nineteen thousand Catholic parishes raised about 6 billion dollars. Every year, millions of dollars, a great percentage of that money, is stolen by thief priests. See 57:27; 3:71; 4:161; 9:24; 2:42; 2:188.

009:030-31 The system of islam belongs to God alone (98:5). Accepting the *fatwas* of this or that cleric as islam is setting up partners with God. See 42:21. Also, see 2:59; 2:79; 5:41-44.

009:033 Islam, which has been subjected to a terrible deformation and transformation process for centuries, has long ceased being islam (peacemaking), but rather became a man-made doctrine attributed to scholars, such as Shafii, Hanbali, Maliki, Hanafi, Jafari, Vahhabi, Naqshi, Qadiri, Alawi, Muhammedi, etc. By God's will, the fulfillment of the great prophecy (74:1-6) and the unfolding global events are signs that God has started a new era in which Islam will again provide humanity with a chance to reform their minds, their actions, and their world. See Chapter 110.

009:036 Verse 9:36 relates the number of months to the day of the creation of the earth. Perhaps in the early days of the creation, the rotation of the Moon and Earth were synchronized and there were exactly 12 lunar months in a solar year. The increase in the length of the year or the decelerating earth is causing problems for modern technology which is very sensitive to small changes in the measurement of time, such as computers, electric grids, GPS systems, and missiles. In mid-20th century, scientists learned that the rotation of Earth was not sufficiently uniform as a standard of time. Besides solar gravity, there are many factors influencing the rotation of the earth. For instance, though in minor way, the Moon's gravity through tides causes deceleration, and melting of ice in the North Pole causes little acceleration. To compensate this net deceleration, almost every year we are adding one second to our atomic clocks, which defines and measures a second as *9,192,631,770* electromagnetic radiation or oscillation of Cesium 133. Since 1972 we added 23 leap seconds until 2006. However, the rate of the deceleration does not show a pattern. In recent years, the earth's rotation has been on schedule. See. 9:122

009:037 Today's Sunni and Shiite *mushriks* consider *Rajab*, *Zul-Qada*, *Zul-Hijja*, and *Muharram* to be the Restricted Months (the 7th, 11th, 12th, and 1st months of lunar calendar). However, when we study 2:197,217; 9:2,5,36 and the names of known months, we will discover that real Restricted Months must be four consecutive months and they are *Zul-Hijja, Muharram, Safar, Rabi ul-Awal* (12th, 1st, 2nd, and 3rd months). The very name of the 12th month, *Zul-Hijja* (Contains *Haj*), gives us a clue that it is the first month of *haj* Debate Conference. When we start from the 12th month, the last of the Restricted Months becomes the 3rd month, *Rabi ul-Awal* (First Fourth), and the very name of this month also is revealing; it suggests that it is the fourth month of the Restricted Months. Then, why the qualification "First"? Well, the name of the following month provides an explanation: *Rabi ul-Akhir* (Second Fourth), which is the fourth month from the beginning of the year. In other words, we have two months called "Fourth," the first one referring to the fourth restricted moth, with the second fourth referring to the fourth month of the lunar year. The word *rabi* (fourth) is used for seasons because each season is one fourth of a year. It is interesting that many of the crimes and distortions mentioned in the Quran have been committed by those who have abandoned the Quran for the sake of *hadith* and *sunna*; they repeated the same blunder of their polytheist ancestors.

009:045 Hypocrites and polytheists claim that they believe in God (23:84-90). Indeed, with their faith in the intercession of prophets and saints, they profess faith in the hereafter (10:18). Nevertheless, they have doubts in their compromised faith, which is the result of joining the bandwagon or wishful thinking (9:45; 11:110; 14:9; 34:21; 41:45; 44:9). We are told that a conviction/ affirmation based on knowledge is not contaminated with doubts (49:15). Ironically, many polytheists are in denial of their polytheism (6:23).

009:051 See 2:286.

009:054 This verse is another piece of evidence that *sala* prayers were known since Abraham and were practiced by Meccan polytheists (8:35; 21:73). This is another answer for those who reject Quran's assertion that it is complete, perfect, and sufficiently detailed, and then ask, "Where can we find the details of *sala* prayers in the Quran?" (6:19, 38,114). One can find the details of *sala* prayer for themselves by just studying the sixty plus verses that mention *sala* prayer.

009:080 How can Muhammed intercede on behalf of complete strangers while he could not even intercede on behalf of his uncles and his closest relatives? Abraham could not intercede on behalf of his own father (60:4), Noah could not save his own son (11:46). Also, see 2:48.

009:099-103 This verse exposes one of the most popular distortions made in the meaning of *salla ala* (to support, encourage), which has been transformed into glorifying Muhammed's name day and night. See 33:43,56 and 2:136.

009:124 The language of this and following verses that repeatedly use the word "increase" and the key words, "chapter" and "affirmation" is prophetic since it depicts a conspiracy of hypocrites after the revelation of the Quran who attempted to add two verses to the end of this chapter. For the example of a chapter that has increased the faith of those who affirm and tested the hypocrites, see Chapter 74.

009:127 The numerical structure of the Quran leads us to accept this chapter having 127 verses, rather than 129, and it rejects the additional statements found in modern manuscripts, which are written here in italics. This major correction is imposed on us by the Quran's own testimony, by its authenticating system. The correction made by the numerical structure of the Quran is also supported by numerous historical events and reports. See, Chapter 74 and the Appendix *On it Nineteen.*

بسم الله الرحمن الرحيم

10:0 In the name of God, the Gracious, the Compassionate.

10:1 A1L30R200. These are the signs of the book of wisdom.*

10:2 Is it a surprise for the people that We would reveal a man from amongst them: "Warn people and give good news to those who affirm, that they will have a footing of truth with their Lord." The ingrates said, "Evidently, this is a magician!"

10:3 Your Lord is **God** who created the heavens and the earth in six days, then He settled over the throne; He handles all affairs. There is no intercessor except after His leave. Such is **God** your Lord, so serve Him. Would you not take heed?

10:4 To Him is your return, all of you, for **God**'s promise is true. He initiates the creation then He repeats it to recompense with justice those who affirmed and did good work. As for those who rejected, they will have a boiling drink, and a painful retribution for what they had rejected.

10:5 He is the One who made the Sun an illuminator, and the Moon a light, and He measured its phases so that you would know the number of the years and the calculation. **God** has not created this except for truth. He details the signs for a people who know.

10:6 In the alternation of night and day and what **God** has created in heavens and earth are signs for a people who are aware.

10:7 As for those who do not wish to meet Us, while pleased and content with the worldly life, they are unaware of Our signs.

10:8 To these will be the destiny of the fire for what they earned.

10:9 Those who affirm and promote reforms, their Lord will guide them by their affirmation. Rivers will flow beneath them in the gardens of paradise.

10:10 Their prayer in it is: "Glory be to our God!" and their greeting in it is: "Peace," and the end of their prayer is: "Praise be to **God**, the Lord of the worlds!"

10:11 If **God** were to hasten for people the evil as they desire the hastening of good for them, then they would have been ruined. We thus leave those who do not wish to meet Us wandering in their transgression.

10:12 If any adversity inflicts people, then he calls upon Us on his side or sitting or standing. But when We remove his adversity from him, he goes on as if he never implored Us because of an adversity afflicted him! Thus, the works of the transgressors are adorned for them.

10:13 We have destroyed the generations before you when they transgressed, and their messengers came to them with clear proofs, but they were not to affirm. It is such that We will recompense the criminal people.

10:14 Then We made you successors on earth after them to see how you would perform.

Denying God's Signs/Miracles; Inventing Lies

10:15 When Our clear signs are recited to them, those who do not wish to meet Us said, "Bring a Quran other than this, or change it!" Say, "It is not for me to change it from my own accord, I merely follow what is revealed to me. I fear if I disobey my Lord the retribution of a great day!"

10:16 Say, "Had **God** wished, I would not have recited it to you, nor would you have known about it. I have been residing amongst you a lifetime before this; do you not reason?"

10:17 Who is more wicked than one who invents lies about **God** or denies His signs? The criminals will never succeed.

Intercession: A Polytheistic Fiction

10:18 They serve besides **God** what does not harm them or benefit them, and they say, "These are our intercessors with **God**." Say, "Are you informing **God** of what He does not know in the heavens or in the earth?" Be He glorified and high from what they set up.

10:19 The people were but one nation, and then they differed. Had it not been for a previous command from your Lord, the matter would have been immediately judged between them for what they differed.

The Sign of the Quran Would Be Revealed After Muhammed

10:20 They say, "If only a sign was sent down to him from His Lord." Say, "The future is with **God**, so wait, and I will wait with you."*

10:21 When We let the people taste a mercy after some harm had afflicted them, they take to scheming against Our signs! Say, "**God** is faster in scheming;" Our messengers record what you scheme.

10:22 He is the One who carries you on land and on the sea. When you are on the ships and We drive them with a good wind which they rejoice with, a strong gust comes to it and the waves come to them from all sides, and they suppose that they are overwhelmed, they implore **God** devoting the system to Him: "If You save us from this, we will be of the thankful."*

10:23 But when He saves them, they then traverse through the land with injustice. O people, your rebellion is only against yourselves. What you seek out is only the consumption of this worldly life, then to Us is your return and We will inform you of all that you had done.

When Humans Are Intoxicated With Arrogance For Their Technological Achievements

10:24 The example of the worldly life is like water which has come down from the sky; it mixed with the plants of the earth from what people and the livestock eat. Then the earth takes its attractions and becomes glamorous, and its inhabitants think that they have mastered it. Then Our judgment comes by night or by day, so We make it a wasteland as if it never prospered yesterday! It is such that We clarify the signs to a people who think.*

10:25 **God** calls to the abode of peace, and He guides those whom He wishes to a Straight Path.

10:26 For those who promote reforms will be good and more, and their faces will not be darkened or humiliated. These are the people of paradise; in it they will abide.

10:27 As for those who earn evil, the recompense of evil will be evil like it, and they will be humiliated. They do not have any besides **God** as a protector. It is as if their faces have been covered by a piece of darkness from the night. These are the people of the fire; in it they will abide.

Idolized Leaders Reject Their Servants

10:28 On the day We gather them all, We will say to those who have set up partners: "Take your place, you and your partners," then We separate between them, and their partners will say, "It was not us that you served."*

10:29 "**God** is sufficient as a witness between us and you, we were unaware of your serving of us."

10:30 Then and there every person will know what it has done, and they are returned to **God** their true patron, and what they had invented has left them.*

The Affirmers of False Religious Teachings

10:31 Say, "Who provides for you from the sky and the land? Who possesses the hearing and the eyesight? Who brings forth the living from the dead and brings forth the dead from the living? Who controls all affairs?" They will say, "**God**." Say, "Why then you do not show awareness!"

10:32 Such is **God**, your true Lord. So what is after the truth except straying! How then you are turned away!

10:33 It was thus that your Lord's command came against those who were wicked, for they do not affirm.

10:34 Say, "Are there any from those whom you have set up as partners who can initiate the creation and then return it?" Say, "**God** initiates the creation and then returns it." How are you deluded!

Follow the Message/the Messenger

10:35 Say, "Are there any of those whom you set up as partners who can guide to the truth?" Say, "**God** guides to the truth. Is He who guides to the truth more worthy of being followed, or the one who does not guide except after he is guided? What is wrong with you, how do you judge?"*

10:36 Most of them only follow conjecture. While conjecture does not avail against the truth in anything. **God** is knower of what they do.

Bigots Who Criticize the Quran Without Comprehending Its Knowledge

10:37 This Quran could not have been produced without **God**, but it is to authenticate what is already present, and to provide detail to the book in which there is no doubt, from the Lord of the worlds.

10:38 Or do they say he invented it? Say, "Then bring a chapter like it, and call upon whoever you can besides **God** if you are truthful!"*

10:39 No, they have lied about the things they did not have comprehensive knowledge of, and before its explanation came to them. Similarly, those before them denied, so see what the retribution of the wicked was!

10:40 Some of them affirm it, and some of them do not affirm it. Your Lord best knows the corrupters.

10:41 If they deny you, then say, "My works are for me, and your works are for you. You are innocent from what I do, and I am innocent from what you do."

10:42 Among them are some who listen to you; but can you make the deaf hear, if they do not reason?

10:43 Among them are some who look at you; but can you guide the blind, even though they will not see?

10:44 **God** does not wrong the people in the least, but it is the people who wrong themselves.

10:45 The day We gather them, it will be as if they have slept for only a moment of a day. They will get to know one another. Losers are those who denied meeting **God**, and they were not guided.

10:46 Whether We show you some of what We promise them, or We let you pass away, then to Us is their return and **God** is witness over what they do.

For Every Nation

10:47 For every nation is a messenger; so when their messenger comes, the matter is decreed between them with justice, and they are not wronged.*

10:48 They say, "When is this promise, if you are truthful?"

The Messenger Has No Special Power

10:49 Say, "I do not possess for myself any harm or benefit except what **God** wills. For every nation is a time. When their time comes, they cannot delay it one moment nor advance it."

10:50 Say, "Do you see if His retribution will come to you by night or by day, then which portion would the criminals hasten in?"

10:51 "When it occurs, would you then affirm it? While now you are hastening it on!"

10:52 Then it will be said to the wicked: "Taste the everlasting retribution. You are only recompensed for what you have earned!"

10:53 They seek news from you: "Is it true?" Say, "Yes, by my Lord it is true, and you cannot escape from it."

10:54 If every person that wronged had possessed all that is on earth, it would have attempted to ransom it. They declared their regret when they saw the retribution, and it was judged between them with fairness. They are not wronged.

God's Grace and Mercy

10:55 To **God** is what is in the heavens and the earth. **God**'s promise is true, but most of them do not know.

10:56 He causes life, and He causes death, and to Him you will return.

10:57 O people, advice has come to you from your Lord and a remedy for what is in the chests, and a guidance and mercy for those who affirm.*

10:58 Say, "By **God**'s grace and His mercy." For that let them rejoice, that is better than all that they gather.

10:59 Say, "Have you seen what **God** has sent down to you from provisions, then you made some of it forbidden and some lawful?" Say, "Did **God** authorize you, or did you invent lies against **God**?"*

10:60 What will those who invent lies against **God** think on the day of Resurrection? **God** is with great bounty to people, but most of them are not thankful.

God's Allies

10:61 You do not engage in any business, nor do you recite any from the Quran, nor do you do any work; without us being witnesses over you when you undertake it. Nothing is hidden from your Lord, not even an atom's weight on earth or in the heavens, nor smaller than that nor larger, but is in a clear book.

10:62 Indeed, for **God**'s allies, there is no fear over them nor will they grieve;*

10:63 Those who affirmed and were aware.

10:64 For them are glad tidings in the worldly life and in the Hereafter. There is no changing the words of **God**. Such is the supreme success.

10:65 Do not be saddened by their statements, for all glory is to **God**. He is the Hearer, the Knowledgeable.

10:66 Certainly, to **God** belongs all who are in the heavens and those who are on earth. As for those who call on partners besides **God**, they only follow conjecture, and they only guess.

10:67 He is the One who made the night for your rest, and the day to see. In that are signs for a people who listen.

10:68 They said, "**God** has taken a son." Be He glorified! He is the Rich. To Him is what is in the heavens and what is in the earth. Do you have proof for this? Or do you say about **God** what you do not know?

10:69 Say, "Those who invent lies about **God**, they will not be successful."

Noah

10:70 A short pleasure in this world Then to Us is their return and We will make them taste the severe retribution for what they were rejecting.

10:71 Recite for them the news of Noah as he said to his people: "My people, if my position has become too troublesome for you, and my reminding you of **God**'s signs, then in **God** I place my trust. So, gather your action and your partners together, then make certain your action does not cause you regret, then come to judge me, and do not hold back."

10:72 "But if you turn away, then I have not asked you for any reward, for my reward is with **God**. I have been commanded to be of those who have peacefully surrendered."

10:73 They denied him, so We saved him and those with him in the Ship, and We made them to succeed each other, and We drowned those who denied Our signs. So see how the punishment of those who were warned was!

Moses and Aaron Against A Racist Tyrant

10:74 Then, We sent messengers after him to their own people, so they came to them with proofs. But they did not want to affirm what they had already denied beforehand. It is such that We stamp on the hearts of the transgressors.

10:75 Then, We sent Moses and Aaron after them with Our signs to Pharaoh and his entourage, but they turned arrogant, they were a criminal people.

10:76 So when the truth came to them from Us, they said, "This is clearly magic!"

10:77 Moses said, "Would you say this about the truth when it came to you? Is this magic? The magicians will not be successful."

10:78 They said, "Have you come to us to turn us away from what we found our fathers upon, and so that you two would have greatness in the land? We will not affirm you."

10:79 Pharaoh said, "Bring me every knowledgeable magician."

10:80 When the magicians came, Moses said to them: "Cast what you will cast."

10:81 So when they cast, Moses said, "What you have brought is magic, **God** will falsify it. **God** does not set right the work of the corrupters."*

10:82 "So that **God** will verify the truth with His words, even if the criminals dislike it."

10:83 Because of their fear from Pharaoh and his entourage that he would persecute them, none affirmed Moses from his people except some of their descendants. Pharaoh was high in the land, and he was of the tyrants.*

10:84 Moses said, "O my people, if you affirm **God**, then put your trust in Him if you have peacefully surrendered."

10:85 They said, "In **God** we put our trust. Our Lord, do not make us a test for the wicked people."

10:86 "Save us by Your mercy from the rejecting people."

10:87 We revealed to Moses and his brother: "Let your people leave their homes in Egypt, and let these homes be your focal point and maintain the contact prayer. Give good news to those who affirm."

10:88 Moses said, "Our Lord, you have given Pharaoh and his entourage adornments and wealth in this worldly life so that they will misguide from Your path. Our Lord, wipe-out their wealth and bring grief to their hearts so that they will not affirm until they see the painful retribution."*

10:89 He said, "The prayer of both of you has been accepted, so keep straight and do not follow the path of those who do not know."

10:90 We helped the Children of Israel cross the sea, and Pharaoh and his soldiers followed them out of hatred and animosity. But when drowning overtook him, he said, "I affirm that there is no god except the One in whom the Children of Israel affirm, and I am of those who have peacefully surrendered."

10:91 Right now? But before you disobeyed and were of the corrupters!

10:92 This day, We will preserve your body, so that you become a sign for those after you. But many people are oblivious to Our signs!*

10:93 We helped the Children of Israel reach a place of sanctity, and We provided them from the good things, and they did not differ until the knowledge came to them. Your Lord will judge between them on the day of Resurrection for what they differed in.

When In Doubt, Question and Study

10:94 If you are in doubt regarding what We have sent down to you, then ask those who have been studying the book from before you. The truth has come to you from your Lord, so do not be of those who doubt.

10:95 Do not be of those who denied **God**'s signs, for you will be of the losers.

10:96 Those who deserved your Lord's word against them do not affirm,

10:97 Even if every sign were to come to them; until they see the painful retribution.

Appreciation of God's Signs and Blessings Brings Prosperity

10:98 Any town, surely benefitted from its affirmation, such as The People of Jonah. When they affirmed, We removed from them the retribution of disgrace in this worldly life, and We let them enjoy until a time.

10:99 Had your Lord willed, all the people on earth in their entirety would have affirmed. Would you force the people to make them affirm?*

10:100 It is not for a person to affirm except by **God**'s leave. He casts the affliction upon those who do not reason.

10:101 Say, "Look at all that is in the heavens and the earth." But what good are the signs and warnings to a people who do not affirm?

10:102 Are they waiting for the days like those who passed away before them? Say, "Wait, for I am with you waiting."

Put Your Full Trust In God

10:103 Then We will save Our messengers and those who

affirm. It is thus binding upon Us that We save those who affirm.

10:104 Say, "O people, if you are in doubt of my system, then I do not serve those that you serve besides **God**. But I serve **God** who terminates you, and I have been commanded to be of those who affirm."

10:105 Set your direction to the system of monotheism, and do not be of those who set up partners.

10:106 Do not call upon other than **God** what does not benefit you or harm you; if you do, then you are of the wicked.

10:107 If **God** afflicts you with any harm, then none can remove it except Him; and if He wanted good for you, then none can turn away His grace. He bestows it to whom He wishes of His servants. He is the Forgiver, the Compassionate.

10:108 Say, "O people, the truth has come to you from your Lord, so whosoever is guided is guided for himself, and whosoever is misguided is misguided against himself. I am not a caretaker over you."

10:109 Therefore, follow what is being revealed to you and be patient until **God** judges. He is the best of judges.

ENDNOTES

010:001 A1L30R200. These letters/numbers play an important role in the mathematical system of the Quran. See 74:1-56; 1:1; 2:1; 13:38; 27:82; 38:1-8; 40:28-38; 46:10; 72:28.

010:020 Keeping "one of the greatest" signs of the Quran (74:30-37) for fourteen centuries as a secret and unveiling it in the computer era is God's plan.

010:022 "At first sight it may appear hopelessly garbled, but the three consecutive pronominal shifts are all perfectly logical. The shift from the second person plural to the third person plural objectifies the addressees and enables them to see themselves as God sees them, and to recognize how ridiculous and hypocritical their behavior is. The shift back to the second person plural marks God's turning to admonish them. Finally the speaker's shift from the third person singular to the first person plural expresses His majesty and power, which is appropriate in view of the allusion to the resurrection and judgment." Neal Robinson, Discovering The Qur'ān: A Contemporary Approach To A Veiled Text (1996, SCM Press Ltd. p. 252). For the literary aspect and purpose of shift in person, see the footnote of verse 39:53.

010:024 Surely, God knows the exact time of the Hour. But, since the world is round, when that moment comes, half of the world will be experiencing night and the other half daylight. See 4:82.

010:028-29 The affirmers of intercession do not really worship those they hope will intercede, since their idols are neither aware of their polytheistic belief nor would they approve of it. Since polytheists follow the teachings of Satan, knowingly or unknowingly, they serve Satan (16:60; 40:74; 36:60; 4:48).

010:030 Those who affirm the truth only accept the Truth as their *mawla* (master/patron/lord). See 2:286.

010:035-36 These verses also convict those who associate Muhammed as a partner with God in *hukm* (legislation/judgment). They confuse following the message of the messenger with following Muhammed. Verse 36, points at the source of their partnership: conjecture or hearsay. See 6:112; 12:111; 9:31.

010:038-48 Study the relationship between these and 27:77-85.

010:047 See 16:36; 40:78.

010:057 Some clerics abused the meaning of "cure" in this verse and sold the verses of the Quran as an amulet, or as a drug for their physical problems.

010:059-60 Attributing to God man-made prohibitions is considered polytheism, idolatry. According to the Quran, the religious leaders and clerics who commit this crime are the worst enemies of God. For instance, recently, Iranian *mullahs* prohibited fish with no scales. Some Turkish clerics prohibit lobster, mussel, crab, shrimp and many other excellent sources of nutrition; they attribute their prohibitions to God. See 6:145-155.

010:062-64 The Quran guarantees perfect and eternal happiness for those monotheists who pass certain tests.

010:081 For the definition of magic, see 7:116-117.

010:083 The word *musrifin* means "the transgressors," or "those who exceed the limits." Considering the context here, it is best translated as a single word, "tyrant" or "despot."

010:088 Those who are subjected to injustice are excused if they ask for divine punishment or use bad words. See 4:148.

010:092 God preserved the bodies of the Pharaohs by giving them the special knowledge of mummification.

010:099-101 God does not interfere with our initial choice. Those who decide to reject the truth, later in their lives are condemned to dogmatism and bigotry (13:11; 18:29; 42:13; 46:15; 57:22).

بسم الله الرحمن الرحيم

11:0 In the name of God, the Gracious, the Compassionate.

11:1 A1L30R200. A book whose signs have been ascertained, then detailed, from One who is Wise, Ever-aware.*

The Quran as a Warner and Good News

11:2 "That you shall serve none other than **God**. I am to you from Him a warner and a bearer of good news."

11:3 "Seek forgiveness from your Lord, then repent to Him; He will make you enjoy an enjoyment until a predetermined period. He bestows privilege for those who are privileged. If you turn away, then I do fear for you the retribution of a great day!"

11:4 "To **God** is your return, and He is capable of all things."

11:5 Alas, they folded their chests to hide from Him. Even when they hide themselves under their outer garments, He knows what they keep secret and what they declare. He is the knower of all that is in the chests.*

11:6 There is not a creature on earth except that its provision is due from **God**. He knows their habitat and their depository. All is in a clear record.*

11:7 He is the One who created the heavens and earth in six days, and His dominion was upon the water, and to test who from amongst you works the best. When you say, "You will be resurrected after death." those who rejected will say, "This is but clear magic!"*

11:8 If We delay for them the retribution to a near period, they will say, "What has kept it?" Alas, on the day it comes to them, nothing will turn it away from them, and what they used to mock will catch up with them.

11:9 If We give the human being a taste of mercy from Us, then We withdraw it from him; he becomes despairing, rejecting.

11:10 If We give him the taste of a blessing after hardship had afflicted him, he will say, "Evil has gone from me!" he becomes happy, boastful;

11:11 Except for those who are steadfast and promote reforms; these will have a pardon, and a great reward.

11:12 So perhaps you wish to ignore some of what has been revealed to you, and you are depressed by it, because they say, "If only a treasure was sent down with him, or a controller had come with him!" You are but a warner, and **God** is caretaker over all things.

The Quran is a Unique Book with Extraordinary Features

11:13 Or do they say, "He invented it!" Say, "Bring ten invented chapters like it, and call on whom you can besides **God** if you are truthful."*

11:14 If they do not respond to you, then know that it was sent down with **God**'s knowledge, and that there is no god but He. Will you then peacefully surrender?

11:15 Whoever wants the worldly life and its adornments, then We will grant them their works in it, and they will not be short changed in it.

11:16 These will have nothing but fire in the Hereafter, and what they have done will be in vain, and evil is what they have worked.

Similar Proof in the Book Given to Moses

11:17 As for those who got proof from their Lord and are followed by a testimony from Him; and before it was the book of Moses as a guide and a mercy; they will affirm it. Whoever rejects it from amongst the parties, then the fire is his meeting place. So do not be in any doubt about it. It is the truth from your Lord, but most people do not affirm.*

The Most Wicked: Rabbis, Priests, Mullahs, Monks Who Attribute Lies to God

11:18 Who is more wicked than one who invents lies about **God**? They will be brought before their Lord, and the witnesses will say, "These are the ones who lied about their Lord." Alas, **God**'s curse will be upon the wicked.

11:19 Those who repel others from the path of **God** and seek to twist it; and regarding the Hereafter they are in denial.

11:20 These are the ones who will not escape on earth, nor do they have besides **God** any allies. The retribution will be doubled for them. They were not able to hear, nor could they see.

11:21 They are the ones who lost their selves, and what they had invented has abandoned them.

11:22 There is no doubt, that in the Hereafter they are the greatest losers.

11:23 Those who affirm and promote reforms, and are humble towards their Lord; they are the dwellers of paradise, in it they will remain.

11:24 The example of the two groups is like the blind and deaf, the seer and hearer. Are they equal when compared? Do you not take heed?

Noah and His Aristocrat Opponents

11:25 We had sent Noah to his people: "I am to you a clear warner!"

11:26 "Do not serve except **God**. I fear for you the retribution of a painful day."

11:27 The leaders who rejected from amongst his people said, "We do not see you except as a human like us, and we see that only the lowest amongst our people with shallow opinion have followed you. We do not see anything that makes you better than us; in fact, we think you are liars."

11:28 He said, "My people, do you not see that I have proof from my Lord, and He gave me mercy from Himself, but you are blinded to it? Shall we compel you to it while you are averse to it?"

11:29 "My people, I do not ask you for money, my reward is with **God**. Nor will I turn away those who affirm, for they will meet their Lord. But I see that you are a people who are ignorant."

11:30 "My people, who will give me victory against **God** if I turn them away? Will you not reflect?"

11:31 "Nor do I say to you that I have the treasures of **God**, nor do I know the future, nor do I say that I am a controller, nor do I say to those whom your eyes look down upon that **God** will not grant them any good. **God** is more aware of what is in them; in such case I would be among the wicked."

11:32 They said, "O Noah, you have argued with us, and continued arguing with us, so bring us what you promise us if you are of the truthful ones."

11:33 He said, "It is **God** who will bring it to you if He wishes; you will not have any escape."

11:34 "My advice will not benefit you if I wanted to advise you and **God** wanted that you should stray. He is your Lord, and to Him you will return."

11:35 Or do they say, "He invented it?" Say, "If I invented it, then I am responsible for my crime, and I am innocent from your crimes"

11:36 It was revealed to Noah: "No more of your people will affirm except those who have already affirmed. So do not be saddened by what they have done."

11:37 "Construct the Ship under Our eyes and Our inspiration, and do not speak to Me regarding those who are wicked. They will be drowned."

11:38 As he was constructing the Ship, every time any cluster from his people passed by, they mocked him. He said, "If you mock us, then we also mock you as you mock."

11:39 "You will know to whom the retribution will come to disgrace him, and upon him will be a lasting punishment."

11:40 So, when Our command came and the volcano erupted. We said, "Carry in it two from every pair, and your family; except those against whom the word has been issued; and whoever affirmed." But those who affirmed with him were few.*

11:41 He said, "Climb inside, in the name of **God** shall be its sailing and its anchorage. My Lord is Forgiving, Compassionate."

11:42 While it was running with them in waves like mountains, Noah called to his son, who was in an isolated place: "My son, ride with us, and do not be with the ingrates!"

11:43 He said, "I will take refuge to the mountain which will save me from the water." He said, "There is no savior from **God**'s decree except to whom He grants mercy." The wave came between them, so he was one of those who drowned.

11:44 It was said, "O land, swallow your water, and O sky, cease." The water was diminished, and the matter concluded. It came to rest on the Judea, and it was said, "Away with the wicked people."

No Nepotism, No Intercession

11:45 Noah called on his Lord, and he said, "My Lord, my son is from my family, and your promise is the truth, and you are the Wisest of all Judges."

11:46 He said, "O Noah, he is not from your family, he committed sin, so do not ask what you have no knowledge of. I advise you not to be of the ignorant."*

11:47 He said, "My Lord, I seek refuge with You that I would ask You what I did not have knowledge of. If You do not forgive me and have mercy on me, I will be of the losers!"

11:48 It was said, "O Noah, descend with peace from Us, and with blessings upon you and upon nations to come from those with you. As for other nations whom We will first grant them pleasure, then they will receive a painful retribution from Us."

11:49 This is from the news of the unseen that We reveal to you. Neither did you nor your people know this, so be patient. The ending is always in favor of the righteous.

Hood and the People of Aad

11:50 To Aad was sent their brother Hud. He said, "My people, serve **God**, you have no god besides Him; you are simply conjecturing."

11:51 "My people, I do not ask you for any wage, my wage is from the One who initiated me. Will you not reason?"

11:52 "My people, seek forgiveness from your Lord, then repent to Him; He will send the sky to you abundantly, and He will add might to your might. So do not turn away as criminals."

11:53 They said, "O Hud, you have not come to us with any proof, nor will we leave our gods based on what you say. We will not affirm you."

11:54 "All we can say is that perhaps some of our gods have possessed you with evil." He said, "I make **God** my witness, and all of you witness that I disown what you have set up as partners--

11:55 "-- besides Him, so scheme against me all of you, then do not give me respite."

11:56 "I have put my trust in **God**, my Lord and your Lord. There is not a creature except He will seize it by its frontal lobe. My Lord is on a Straight Path."

11:57 "If you turn away, then I have delivered what I was sent to you with. My Lord will bring after you a people who are not like you, and you will not harm Him in the least. My Lord is Guardian over all things."

11:58 When Our command came, We saved Hood and those who affirmed with him by a mercy from Us, and We saved them from a harsh retribution.

11:59 Such was the case of Aad. They disregarded the signs of their Lord, and they disobeyed His messengers, and they followed the lead of everyone powerful and stubborn.

11:60 They were followed by a curse in this world and on the day of judgment, for Aad rejected their Lord. There is no more Aad, the people of Hud.

Saleh to Thamud

11:61 To Thamud was sent their brother Saleh. He said, "My people, serve **God**, you have no god besides Him. He established you in the land and gave you control over it, so seek His forgiveness, then repent to Him. My Lord is Near, Responsive."

11:62 They said, "O Saleh, you were well liked amongst us before this. Do you deter us from serving what our fathers served? We are in serious doubt as to what you are inviting us."

11:63 He said, "My people, what do you think if I was on clear evidence from my Lord, and He gave me from Him a mercy. Who would then support me against **God** if I disobey Him? You would only increase me in loss!"

11:64 "My people, this is **God**'s camel, in her you have a sign. So, leave her to eat from **God**'s land freely, and do not harm her, or else a close retribution will take you."

11:65 But they slaughtered her. So he said, "You will only have three days of enjoyment in your homes. This is a promise that will not be denied."

11:66 So, when Our command came, We saved Saleh and those who affirmed with him by a mercy from Us against the disgrace of that day. Your Lord is the Powerful, the Noble.

11:67 Those who wronged were taken by the scream, thus they lay motionless in their homes.

11:68 It is as if they never lived there. For Thamud rejected their Lord. There is no more Thamud.

Abraham and Lut

11:69 Our messengers came to Abraham with good news, they said, "Peace" He said, "Peace," and it was not long before he came back with a roasted calf.

11:70 But when he saw that their hands did not go towards it, he mistrusted them, and felt a fear of them. They said, "Have no fear, we have been sent to the people of Lot."

11:71 His wife was standing, so she laughed when We gave her good news of Isaac, and after Isaac, Jacob.

11:72 She said, "O my! how can I give birth while I am an old woman, and here is my husband an old man? This is indeed a strange thing!"

11:73 They said, "Do you wonder at the decree of **God**? **God**'s Compassion and Blessings are upon you O people of the sanctuary. He is Praiseworthy, Glorious."

11:74 So when the shock left Abraham, and the good news was delivered to him, he began to argue with Us for the people of Lot.

11:75 Abraham was compassionate, kind.

11:76 O Abraham, turn away from this. Your Lord's command has come, and a retribution that will not be turned back is coming for them.

Homosexual Aggression is Condemned

11:77 When Our messengers came to Lot, he was grieved on their account and he felt discomfort for them and said, "This is a distressful day."

11:78 His people came rushing towards him, and were accustomed to committing sin. He said, "My people, these are my daughters, they are purer for you, so be aware of **God** and do not disgrace me regarding my guests. Is there no reasonable man among you?"

11:79 They said, "You know we have no interest in your daughters, and you are aware of what we want!"*

11:80 He said, "If only I had strength against you, or I could find for myself some powerful support."

11:81 They said, "O Lot, we are your Lord's messengers; they will not be able to harm you. So travel with your family during the cover of the night and let not any of you look back except for your wife; she will be afflicted with what they will be afflicted. Their appointed time will be the morning. Is the morning not near?"

11:82 So when Our command came, We turned it upside down, and We rained on it with hardened fiery projectiles.

11:83 Marked from your Lord, and they are never far from the wicked.

Shuayb to Midian

11:84 To Midian was their brother Shuayb, he said, "My people, serve **God**, you have no god besides Him, and do not give short in the measure and weight. I see you in prosperity, and I fear for you the retribution of a day that is surrounding."

11:85 "My people, give full measure and weight with justice, do not hold back from the people what is theirs, and do not roam the land corrupting."

11:86 "What will remain for you with **God** is far better if you are those who affirm. I am not a guardian over you."

11:87 They said, "O Shuayb, does your contact prayer order you that we leave what our fathers served, or that we do not do with our money as we please? You are the compassionate, the sane!"*

11:88 He said, "O my people, do you see that if I am on clear evidence from my Lord, and He has provided me with good provision from Him, then I would not want to contradict by doing what I forbid you from. I only want to fix what I can, and my guidance is only with **God**. To Him I place my trust, and to Him I repent."

11:89 "My people, let not your hatred towards me incriminate you that you suffer the fate of what afflicted the people of Noah, or the people of Hud, or the people of Saleh; and the people of Lot were not far off from you."

11:90 "Seek forgiveness from your Lord then repent to Him. My Lord is Compassionate, Loving."

11:91 They said, "O Shuayb, we do not understand most of what you say, and we see you as weak amongst us. If it were not for who your family is, we would have stoned/rejected you, and you would not be proud against us."

11:92 He said, "My people, is my family more important to you than **God**, while you have cast Him away behind your backs? My Lord is Encompassing over what you do."

11:93 "My people, continue to act as you do, and I will act. You will then come to know to whom the humiliating retribution will come and who is the liar. Watch then, and I will watch with you."

11:94 When Our command came, We saved Shuayb and those who affirmed with him by a mercy from Us; and the scream took those who had wronged, so they lay motionless in their homes.

11:95 It is as if they never lived there. Away with Midian as it was away with Thamud.

Moses to Egypt

11:96 We sent Moses with Our signs and a clear authority.

11:97 To Pharaoh and his entourage; but they followed the command of Pharaoh, and Pharaoh's command was not wise.

11:98 He will be at the head of his people on the day of Resurrection, and he will lead them to the fire. What a miserable place they are lead in!

11:99 They were followed by a curse in this, and on the day of Resurrection. What a miserable path to follow!

11:100 That is from the news of the towns which We relate to you; some are still standing, and some have been wiped-out.

11:101 We did not wrong them, but they wronged themselves. Their gods that they called on besides **God** did not rescue them at all when your Lord's command came, and they only added to their destruction.

11:102 Such is the taking of your Lord when He takes the towns while they are wicked. His taking is painful, severe.

11:103 In this is a sign for he who fears the retribution of the Hereafter. That is a day to which all the people will be gathered, and that is a day which shall be witnessed.

11:104 We do not delay it except to a term already prepared for.

11:105 On the day it comes, no person will speak to another except by His leave. Some of them will be distraught, some will be happy.

11:106 As for those who are distraught, they will be in the fire; in it for them is a sighing and a wailing.

11:107 They will abide in it as long as the heavens and earth exist, except for what your Lord wishes. Your Lord does as He pleases.

11:108 As for those who are fortunate, they will be in paradise; in it they will abide as long as the heavens and earth exist, except for what your Lord wishes, a giving without end.*

11:109 So do not be in doubt as to what these men serve. They serve nothing but what their fathers before them served. We will give them their recompense in full.

11:110 We have given Moses the book, yet they disputed in it; and had it not been for a word which was already given by your Lord, their case would have been judged immediately. They are in grave doubt concerning it.

11:111 To each your Lord will recompense their works. He is Ever-Aware of what they do.

11:112 So stand straight as you were commanded together with those who repented with you and do not transgress. He is watcher over what you do.

11:113 You shall not lean towards those who have wronged, else you will incur the fire; and you will not have besides **God** any allies, and you will not be victorious.

Times of the Contact Prayers

11:114 You shall maintain the contact prayer at both ends of the day, that is, section of the night. The good deeds take away the bad. This is a reminder for those who remember.*

11:115 Be patient, for **God** does not waste the reward of the good doers.

11:116 If only there was from the previous generations a people with wisdom who deterred from the corruption on earth, except the few that We saved of them. Those who were wicked followed the enjoyment they were in, and they were criminal.

11:117 Your Lord would not destroy the towns wrongfully, while its people were good doers.

11:118 Had your Lord wished, He could have made all the people one nation, but they still would continue to disagree;

11:119 Except whom your Lord has mercy upon; and for that He has created them. The word of your Lord came true: "I will fill hell with the Jinn and the humans together!"*

11:120 All the news of the messengers that we relate to you is to strengthen your heart. In this has come to you the truth and a lesson and a reminder for those who affirm.

11:121 Say to those who do not affirm: "Continue to do what you will, we will also do."

11:122 "Wait, for we are also waiting."

11:123 To **God** is the unseen of heavens and earth, and to Him all matters return. So serve Him and put your trust in Him. Your Lord is not unaware of what you all do.

ENDNOTES

011:001 A1L30R200. These letters/numbers play an important role in the mathematical system of the Quran based on code 19. See 74:1-56; 1:1; 2:1; 13:38; 27:82; 38:1-8; 40:28-38; 46:10; 72:28.

Verses of the Quran are treasures of details, and they contain even further details when evaluated within their semantic network. For instance, when we study 46:15; 31:14; 22:5; 23:14; 16:58-59 and 17:31 carefully together with God's signs in the nature, they provide specific guidelines regarding abortion. When we study the verse recommending the punishment for theft (5:38) together with other verses using the same key word, we learn a flexible legal system. When we study the multi-meaning key words in 4:34, we come across a different picture, much different than what the all-male scholars have taught us. The details related to the mathematical system prophesied in chapter 74, themselves can fill many volumes of books.

011:006 Starvation and famine are the consequences of ignorance, wars, laziness, monopoly, myopic selfishness, dictatorship, wasting natural resources, injustice, and apathy. Communities that follow God's system have little chance of experiencing famine and starvation. See 17:31.

011:007 The expression "six days" provide comparison. For instance, we learn that though the creation of galaxies took two days, the creation of earth to be habitable for life took four days (41:10-12). In other words, the creation of earth started 13.7 billion x 4/6 years ago. As for time, the Quran informs us that it is relative (32:5 and 70:4). The earth was initially covered with water; lands emerged later.

011:005 Such as robes, indicating religious rank or holiness.

011:013 The mathematical structure of the Quran is too intricate to be imitated.

011:017 *Bayyina* means "proof" and it is repeated in the Quran 19 times. The book given to Moses too was designed according to a mathematical structure based on the number 19 (46:10), which was discovered by a French Rabbi, Judah, in the eleventh century.

011:040-44 Noah's ark was a simple watercraft made of logs connected with ropes (54:12). The flood was limited to the Dead Sea region and Noah's people (7:59; 9:70; 11:25,36,89; 22:42; 25:35-39; 38:12; 40:5,31; 50:12; 51:46; 53:52; 54:9; 71:1-11). The animals taken to the watercraft were a few domesticated animals in Noah's farm. The holy storytellers mutated the story, and with time, it was exaggerated into a global event. Perhaps public interest encouraged the storytellers to exaggerate a bit more, until it became a mythology about a worldwide flood.

The mythology found its way to the Old Testament; three chapters of Genesis, chapters 6-8, are allocated to the story of a universal flood (See Genesis 7:21-23). The letters of the New Testament also refer to a worldwide flood (1 Peter 3:18-20; 2 Peter 2:5; 2 Peter 3:6). Also, see 2:93; 16:103; 26:101-105; 71:1-28.

011:046 Intercession is a false hope, and it contradicts the monotheistic theology advocated by the Quran. Neither Abraham could help his father, Noah his son, nor Muhammed his relatives (2:48,123,254; 6:70,94; 7:53; 9:80; 10:3; 39:44; 43:86; 74:48; 82:17-19).

011:079 Note that these are not people who are committing sin in their private lives; they are aggressive homosexuals who are proud of their behavior and show the audacity to harass men around them.

011:087 *Sala* prayer observed properly keeps away from immorality and crimes (29:45).

011:108 After the end of the world, the earth together with the universe will be recreated anew (14:48).

011:114 In the Quranic Arabic, the word *layl* denotes the period from sunset to sunrise (2:187). The times of the evening and dawn prayers extend from the proximate edges of night to both ends of the day. The Quran provides detailed information about *sala* prayer. This verse, according to the popular reading, refers to three *sala* prayers. Some students of the Quran understand five times rather than three times for contact prayers. There are some who have inferred two prayers a day. However, as the detected and undetected influence of sectarian teachings and practices wanes with time, the disagreement on this issue may lead to a better understanding. We should respect the differences in our understanding, as long as they are based on the Quran under the light of reason. See the Appendix *Sala Prayer According to the Quran.*

011:119 God has given us a second chance to redeem ourselves in this world. Those who enter choose it by their own will.

بسم الله الرحمن الرحيم

12:0 In the name of God, the Gracious, the Compassionate.

12:1 A1L30R200. These are the signs of the clarifying book.*

12:2 We have sent it down an Arabic Quran/recitation, perhaps you will reason.

12:3 We relate to you the best stories through revelation of this Quran; before it you were of those who were unaware.

An Adventure Starting with a Dream

12:4 When Joseph said to his father: "My father, I have seen eleven planets and the Sun and the Moon, I saw them prostrating before me."*

12:5 He said, "O my son, do not relate your dream to your brothers, or they will scheme against you. The devil is to human being a clear enemy."

12:6 As such, your Lord has chosen you, and He teaches you the interpretation of dreams, and He completes His blessings upon you and upon the descendants of Jacob, as He completed it for your fathers before that, Abraham and Isaac. Your Lord is Omniscient, Wise.*

12:7 It is thus that in Joseph and his brothers are signs for those who seek.

12:8 For they said, "Joseph and his brother are more loved by our father than us, while we are a gang. Our father is clearly misguided."

The Evil Whisperer

12:9 "Kill Joseph or cast him in the land, then your father's favor will be all yours, and after that you will be a reformed people."

12:10 One amongst them said, "Do not kill Joseph, but if you are going to do anything, then cast him into the bottom of the well, so that anyone traveling by will pick him up."

12:11 They said, "Our father, why do you not trust us with Joseph, we are to him well wishers."

12:12 "Send him with us tomorrow to enjoy and play, and we will take care of him."

12:13 He said, "It saddens me that you should take him, and I fear that the wolf would eat him if you would be absent from him."

12:14 They said, "If the wolf eats him, while we are a gang, then we are the losers."

12:15 So, when they went with him they had agreed to place him at the bottom of the well. We revealed him: "You will inform them of this act of theirs while they will not expect it."*

12:16 They came to their father at dusk crying.

12:17 They said, "Our father, we went to race and left Joseph by our things, and the wolf ate him! But you would not trust us even if we are truthful."

12:18 They came with his shirt stained in false blood. He said, "You have invented this tale yourselves. Patience is good, and **God**'s help is sought against what you describe."

Joseph is Taken to Egypt

12:19 A traveling caravan came and they sent their water-drawer. When he let down his bucket he said, "Good news, there is a boy!" So they hid him as merchandise. **God** knows what they do.

12:20 They sold him for a low price, a few silver coins, and they regarded him as insignificant.

12:21 The one from Egypt who bought him said to his wife: "Host him honorably. Perhaps he will benefit us or we may adopt him as a son." It was thus that We established Joseph in the land and to teach him the interpretation of dreams. **God** has full power over matters, but most of the people do not know.*

Tested with Sexual Temptation

12:22 When he reached his maturity, We gave him position and knowledge. It is thus that We reward the good doers.

12:23 The woman, in whose house he was staying, attempted to seduce him. She closed the doors and said, "I have prepared myself for you." He said, "I seek refuge with **God**, He is my Lord, He made good my residence; the wicked do not succeed."*

12:24 She desired him and he desired her, had it not been that he saw His Lord's manifest evidence; thus We turned evil and lewdness away from him, as one of Our loyal servants.

12:25 As they rushed towards the door, she tore his shirt from behind; and they found her noble husband at the door. She said, "What is the punishment of he who wanted to molest your family? Is it not that he be jailed or punished painfully?"

Circumstantial Evidence Exonerates Joseph

12:26 He said, "She is the one who tried to seduce me," A witness from her family gave testimony: "If his shirt was torn from the front, then she is truthful, and he is the liar."*

12:27 "If his shirt is torn from behind, then she is lying, and he is truthful."

12:28 So when he saw that his shirt was torn from behind, he said, "This is from your female scheming, your female scheming is indeed great!"

12:29 "Joseph, turn away from this. You woman, seek forgiveness for your sin; you were of the wrongdoers."

12:30 Some women in the city said, "The wife of the Governor is trying to seduce her young man; she is taken by love. We see her clearly misguided."

Physical Attraction

12:31 So when she heard of their scheming, she sent for them and prepared a banquet for them, and she gave each one of them a knife. She said, "Come out to them," so when they saw him, they exalted him and cut their hands, and they said, "**God** be praised, this is not a human, but a blessed controller!"*

12:32 She said, "This is the one whom you blamed me for, and I have seduced him, but he refused. If he does not do as I order him, he will be imprisoned, and he will be one of those disgraced."

12:33 He said, "My Lord, prison is better to me than what they are inviting me to do. If You do not turn their scheming away from me, I will fall for them and be of the ignorant."

12:34 So his Lord responded to him, and He turned away their scheming from him. He is the Hearer, the Knowledgeable.

12:35 But it appealed to them, even after they had seen the signs, to imprison him until a time.

In the Prison

12:36 With him in the prison entered two young men. One of them said, "I dreamt that I was pressing wine," and the other said, "I dreamt that I was carrying bread on top of my head, and that the birds were eating from it." "Tell us what this means, for we see that you are of the good doers."

12:37 He said, "There is not any provision of food that will come to you except that I will tell you of its interpretation before it comes. That is from what my Lord has taught me. I have just left the creed of a people who do not affirm **God**, and they are rejecting the Hereafter."

12:38 "I followed the creed of my fathers: Abraham, Isaac, and Jacob. It was not for us to set up partners with **God** at all. That is **God**'s blessings over us and over the people, but most of the people are not thankful."

There is no lord Besides God

12:39 "O my fellow inmates, are various lords better, or **God**, the One, the Omniscient?"*

12:40 "What you serve besides Him are nothing but names which you have fabricated you and your fathers. **God** did not send down any authority for such. The judgment is for none but **God**. He ordered that none be served but He. That is the true system, but most of the people do not know."

12:41 "My fellow inmates, one of you will be serving wine for his lord, while the other will be crucified so that the birds will eat from his head. The matter which you have sought is now concluded."

12:42 He said to the one whom he thought would be saved of them: "Mention me to your lord." But the devil made him forget to mention to his lord, so He remained in prison for a few years.*

King's Dream

12:43 The King said, "I continue to dream of seven fat cows which are being eaten by seven thin ones, and seven green pods and others which are dry. O you chiefs, tell me what my vision means if you are able to interpret the visions."*

12:44 They said, "It is nothing but medley dreams; and we are not knowledgeable in the interpretation of dreams."

12:45 The one of them who had been saved and remembered after all this time said, "I will tell you of its interpretation, so send me forth."

12:46 "Joseph, O truth one, explain to us the matter regarding seven fat cows being eaten by seven thin ones, and seven green pods and others which are dry? Then perhaps I may go back to the people so they will know."

12:47 He said, "You will plant regularly for seven years, and whatever you harvest you must leave it in its pods, except for the little that you will eat."

12:48 "Then after that will come seven ones severe in drought, which will consume all that you planted except for what you have stored."

12:49 "Then after that, will come a calendar year in which the people will have abundant rain and which they will be able to produce once again."

12:50 The King said, "Bring him to me." When the messenger came to him, he said, "Go back to your lord and ask him what the matter was regarding the women that cut their hands? My Lord is well aware of their scheming."

Governor's Wife Tells the Truth

12:51 He said, "What is your plea that you tried to seduce Joseph?" They said, "**God** forbid that we would do any harm to him." The wife of the Governor said, "Now the truth must be known, I did seek to seduce Joseph and he is of the truthful ones…"

12:52 "That is so he knows that I will not betray him while he is not present, and that **God** does not guide the scheming of the betrayers."

12:53 "I do not make myself free of blame, for the person is inclined to sin, except what my Lord has mercy on. My Lord is Forgiving, Compassionate."

Joseph is Appointed to a Top Position

12:54 The King said, "Bring him to me so that I may employ him." So, when he spoke to him, he said, "Today you are with us in high rank and trusted."

12:55 He said, "Make me keeper over the granaries of the land, for I know how to keep records and I am knowledgeable."

12:56 Thus, We gave Joseph authority in the land, to travel in it as he pleases. We bestow Our mercy upon whom We please, and We do not waste the reward of the good doers.

12:57 The reward of the Hereafter is better for those who affirm and were aware.

Joseph's Brothers Travel to Egypt for Grain

12:58 Joseph's brothers came and entered upon him, and he recognized them, but they did not recognize him.

12:59 So, when he furnished them with their provisions, he said, "Bring me a step-brother of yours who is from your father. Do you not see that I give full measure of grain and that I am the best of hosts?"*

12:60 "But if you do not bring him to me, then there shall be no measure of grain for you with me, and do not come near me."

12:61 They said, "We will try to get him away from his father, and we shall do it."

12:62 He said to his servants: "Return their goods back into their bags, so that they will recognize it when they return to their family and they will come back again."

12:63 So when they returned to their father, they said, "Our father, we have been banned from getting anymore measure of grain, so send our brother with us so we may be given a measure of grain, and we will be his guardians."

12:64 He said, "Shall I trust him with you as I trusted you with Joseph before that? **God** is the best guardian, and He is the Most Compassionate of those who show compassion."

12:65 So when they opened their bags, they found their goods had been returned to them, and they said, "Our father, what more can we desire, this is our goods returned to us, so we can get more for our family, and be guardians over our brother, and increase a measure of grain to load a camel. That is truly an easy measure!"

12:66 He said, "I will not send him with you until you give me a pledge before **God** that you will bring him back unless you are completely overtaken." So, when they gave him their pledge, he said, "**God** is a guard over what we say."

Risk Reduction

12:67 He said, "My sons, do not enter from one gate, but enter from separate gates; and I cannot avail you anything against **God**, for the judgment is to **God**. In Him I put my trust, and in Him those who put their trust should trust."

12:68 When they entered from where their father had commanded them, it would not have availed them in the least against **God**, but it was out of a concern in Jacob's person. Since We have taught him, he had certain knowledge; but most people do not know.*

12:69 When they entered upon Joseph, he called his brother to himself and said, "I am your brother, so do not be saddened by what they have done."

Joseph Keeps His Brother from the Same Mother

12:70 When he furnished them with their provisions, he placed the measuring bowl in his brother's bag. Then a caller cried out: "O you in the caravan, you are thieves!"

12:71 They said, turning towards them: "What is it you are missing?"

12:72 He said, "We are missing the measuring bowl of the King, and whoever finds it shall receive a camel-load; I guarantee this."

12:73 They said, "By **God**, you know we did not come to cause corruption in the land, and we are no thieves!"

12:74 He said, "What shall be the punishment, if you are not truthful?"

12:75 They said, "The punishment is that the person who has it in his bag shall himself be held as penalty. It is such that we punish the wicked."

12:76 So he began with their bags before his brother's bag. Then he brought it out of his brother's bag. It was such that We planned for Joseph, for he would not have been able to take his brother under the King's system, except that **God** wished it so. We raise the degrees of whom We please, and over every one of knowledge is the All Knowledgeable.*

12:77 They said, "If he has stolen, there was a brother of his before who also had stolen." Joseph kept this all inside himself, and did not reveal anything to them. He said, "You are in a worse position, and **God** best knows what you describe."

12:78 They said, "O Governor, he has an elderly father, so take one of us in his place. Indeed we see you as one of the good doers."

12:79 He said, "**God** forbid that we would take anyone except he whom we found our belongings with. Indeed, we would then be wrong doers."

12:80 So when they gave up from him, they held a conference in private. The eldest of them said, "Did you not know that your father has taken a covenant from you before **God**, and the past, you also failed in your duty with Joseph? I will not leave this land until my father permits me to do so or that **God** will judge for me. He is the best of judges."

12:81 "Return to your father, and tell him: "Our father, your son has stolen, and we did not witness except what we learned, and we could not know the unseen!"

12:82 "Ask the people of the town which we were in, and the caravan which we have returned with. We are being truthful."

12:83 He replied: "No, you conspired to commit this work. So, patience is good; perhaps **God** will bring them all to me. He is the Knowledgeable, the Wise."*

Psychosomatic Blindness

12:84 He turned away from them and said, "Oh, my grief over Joseph." His eyes turned white from sadness, and he became blind.

12:85 They said, "By **God**, will you never cease to remember Joseph until you become ill, or you are dead!"

12:86 He said, "I merely complain my grief and sorrow to **God**, and I know from **God** what you do not know."

12:87 "My sons, go and inquire about Joseph and his brother, and do not give up from **God**'s Spirit. The only people who would give up from **God**'s Spirit are the ingrates of affirmation."*

The Family Secret is Unveiled

12:88 So when they entered upon him, they said, "O Governor, we have been afflicted with harm, us and our family, and we have come with poor goods to trade, so give us a measure of grain, and be charitable towards us, for **God** does reward the charitable."

12:89 He said, "Do you know what you have done with Joseph and his brother, during your ignorance?"

12:90 They said, "Are you indeed Joseph?" He said, "I am Joseph, and this is my brother. **God** has been gracious to us. For anyone who is righteous and patient, then **God** will not waste the reward of the good doers."

12:91 They said, "By **God, God** has indeed preferred you over us and we were wrongdoers."

12:92 He said, "There is no blame on you this day, may **God** forgive you, and He is the most Compassionate of those who show compassion."*

12:93 "Take this shirt of mine and cast it over my father's face, and he will become with sight; and bring to me all your family."

12:94 When the caravan departed, their father said, "I do indeed feel the scent of Joseph, except that you may think me senile."

12:95 People said, "By **God**, you are back to your old misguidance."

12:96 Then, when the bearer of good news came, he cast it over his face and his sight was restored. He said, "Did I not tell you that I know from **God** what you do not know?"

12:97 They said, "Our father, ask forgiveness for our sins, indeed we have been wrong."

12:98 He said, "I will ask forgiveness for you from my Lord, He is the Forgiving, the Compassionate."*

Dream is Fulfilled; Family Reunion in Egypt

12:99 Then, when they entered upon Joseph, he took his parents to him and he said, "Enter Egypt, **God** willing, in security."

12:100 He raised his parents on the throne, and all fell in prostration before Him. He said, "My father, this is the interpretation of my old dream. My Lord has made it true, and He has been good to me that he took me out of prison and brought you out of the wilderness after the devil had placed a rift between me and my brothers. My Lord is kind to whom He wills. He is the Knowledgeable, the Wise."

12:101 "My Lord, you have given me sovereignty and taught me the interpretations of events and utterances. Initiator of the heavens and earth, you are my protector in this world and the Hereafter. Take me as one who has peacefully surrendered, and join me with the good doers."

12:102 That is from the news of the unseen that We reveal to you. You were not amongst them when they arranged their plan and were scheming.

Majority of People are Unappreciative

12:103 Most people will not affirm, even if you wish eagerly.

12:104 Though you do not ask them for a wage for it; it is but a reminder to the worlds.

12:105 How many a sign in the heavens and the earth do they pass by, while they are turning away from it.

12:106 Most of them will not affirm **God** without setting up partners.

12:107 Are they secure against the coming of a cover of retribution from **God**, or that the Moment would come to them suddenly while they do not perceive?

12:108 Say, "This is my way, I invite to **God** in full disclosure, myself and whoever follows me. Glory be to **God**. I am not of those who set up partners."

12:109 We have not sent before you except men, to whom We gave inspiration, from the people of the towns. Would they not roam the earth and see what the punishment of those before them was? The abode of the Hereafter is far better for those who are aware. Do you not reason?*

12:110 Then, when the messengers gave up, and they thought that they have been denied, Our victory came to them. We then save whom We wish, and Our punishment cannot be swayed from the wicked people.

The Quran is not a Fabricated Hadith

12:111 In their stories is a lesson for the people of intelligence. It is not a *hadith* that was invented, but an authentication of what is already present, a detailing of all things, and a guidance and mercy to a people who affirm.*

ENDNOTES

012:001 A1L30R200. These letters/numbers play an important role in the mathematical system of the Quran based on code 19. See 74:1-56; 1:1; 2:1; 13:38; 27:82; 38:1-8; 40:28-38; 46:10; 72:28.

012:002 For an alternative understanding of this verse see 43:3.

012:004 The dream perhaps has another prophetic meaning besides the one related to Joseph's family. For Joseph and his family, the dream was fulfilled about 2600 years ago with a good ending (12:100). As for the other meaning of the dream, it is still waiting to be fulfilled. Does this allegory give us a clue regarding the number of planets in the solar system? The number of planets in our solar system may ultimately be determined by the conventional definition of astronomers and remain controversial for long time to come. Interestingly, the differences between planets and the controversy around them too resemble the differences between brothers and half-brothers and their acceptance and dynamics in the family. The Old Testament narrates the same dream; however, because of transnational errors, the word "planet" in the dream is translated as "star" (Genesis 37:9). The story of Joseph starting with his dream when he was seventeen years-old, takes several chapters in Genesis, from chapter 37 to 50, its last chapter.

012:006 From Joseph's dream we learn that he will have a bright future. That is, when his jealous brothers were plotting against him, his future was already determined by God. How can God's knowledge of a future event leave any room for free choice? See 57:22-23. For sleep, see 25:47.

Do we receive symbolic information regarding future events through our dreams? We may infer such a function from these and other episodes reported in this chapter, or we may understand it to be a series of extraordinary events which were created by God specifically to help the Children of Israel in that period. It is obvious that interpretation of dreams was a divinely inspired "knowledge" specifically given to Joseph. However, being the subject of prophetic dreams, as it seems, is not limited to being a messenger of God.

The Book of Genesis of the Old Testament allocates several chapters, from 37 to 50, to the story of Joseph, starting with his dream and the following events. The Quranic story has many parallel lines with the Biblical story, though with some variations. The Bible contains many references to dreams, especially as a way of divine communication. For instance, Jacob (Genesis 28:12; 31:10), Laban (Genesis 31:24), Joseph (Genesis 37:9-11), Gideon (Judge 7:1-25), and Solomon (1King 3:5). Abimelech (Genesis 20:3-7), Pharaoh's chief butler and baker (Genesis 40:5), Pharaoh (Genesis 41:1-8), the Midianites (Judge 7:13), Nebuchadnezzar (Daniel 2:1; 4:10,18), the wise men from the east (Matthew 2:12), and Pilate's wife (Matthew 27:19).

012:015 Consensus (*ijma*) has been accepted by those who deserted the Quran to be one of the main authorities besides and even above the Quran. As the usage of *hadith* and *sunna* in the Quran is prophetic (6:112-116; 7:185; 8:38; 12:111; 33:38; 45:6), the usage of this word too is prophetic. The word *ijma* and *ijtima* are mentioned six times in the Quran and, in all cases, are used to depict the conspiracy of unappreciative opponents of God's message (10:71; 12:102; 20:64; 17:88; 22:73). See 33:38.

012:021 The Old Testament has a contradictory account regarding whom Joseph was sold to. Genesis 37:28-36 sells Joseph to Midianites, while Genesis 39:1 sells him to Ishmailites.

012:023 Joseph is using a statement with multiple meanings. The woman most likely understood it as a reference to her husband. In fact, Joseph, as a monotheist, could not have meant that; with the phrase "my Lord" he meant his Creator. Joseph cannot be referring to the person who purchased him as slave (12: 39,42,50; 6:164; 9:31).

012:026-28 This is an example of circumstantial evidence. Circumstantial evidence based on genetic testing is now more reliable than witness testimony. See the endnote for verse 4:15 for discussion on circumstantial evidence.

012:031 There is an alternative understanding, which creates a parallel scenario, but I am not sure yet about it. The word *sikkyn* (knife) could also mean tranquilizer from root *SKN*. Their anxiety or excitement is remedied through herbs.

According to this understanding, women did not literally cut their hands, but they laid off their hands, that is, they stopped from blaming her. See 33:52. Also, see 5:38.

012:039 See 4:25; 79:24.

012:042 The ending of the verse could also be understood as, "When Satan made him (Joseph) to forget remember his Lord (God), he remained in prison a few more years." Both meanings might be intended at the same time: When asking his friend to remind the King about him, Joseph forgot to remember his Lord (God), and his friend, in turn, forgot to mention him to the king whom he falsely considered "his lord." Verses 12:39-40 and 12:50 clearly indicate that Joseph was a monotheist and he would never call anyone other than God as his lord.

The Quran reminds us that only God can save us from difficulties. Those who affirm the truth, trust God (1:5; 6:17; 8:17; 10:107; 26:77-80).

012:043 The Quran refers to the Egyptian leader contemporary of Moses as Pharaoh. However, it refers to Egypt's ruler who appointed Joseph to be his chief adviser as king (*malik*), not as Pharaoh (12:43,50,54,72). The Old Testament erroneously refers to the Egyptian king during the time of Joseph as Pharaoh (See Genesis 41:14,25,46). Joseph lived approximately in 1600 BC, and according to archeological evidence, Egyptian rulers were not yet called Pharaoh then. The title Pharaoh started with the 18th Dynasty, about a hundred years after Joseph. For the Old Testament account of this story, see Genesis chapter 37 to the end of chapter 50.

012:059 Joseph was hiding his identity from his brothers. Wouldn't his knowledge about their missing brother give him away? Perhaps, Joseph prepared them for this by letting them talk about their family before; or it was known by his brothers that he had official records about their family.

012:068 Perhaps, Jacob thought that a crowded group would pull the attention of officers or spies on the Egyptian border. Putting the eggs in different baskets is one of the ways of risk management.

012:076 This verse raises the dilemma of choosing between legal and moral alternatives. If laws are based on reason and universal principles, they usually do not create such dilemmas. However, even the best and most just laws, occasionally deny justice in some special cases. Then, a person might find himself in a conundrum: the legal and moral virtue of applying the law consistently, or the moral imperative of suspending the application of the law to bring about justice. Here, we are informed that as an officer of the Egyptian government, Joseph is using the laws of Israel rather than the Egyptian law. This was perhaps illegal, but was it also immoral? Joseph could defend his action by claiming that his illegal action did not harm anyone, and he had a good intention. Indeed, this intentionally good "trick" ends the tragedy and pain of a big family.

012:083 Reading the father's answer immediately after reading the recommended message of the big brother is an intriguing and economic style of narrating a story. Through this technique, the repetition of what the message of the big brother in verses 81-82 was, and an introductory statement such as "When they returned to Palestine, they reported their situation according to what the recommendation of their older brother was," are avoided. The conversation among the brothers while they were in Egypt, with the 83rd verse, suddenly jumps over three dimensions (place, time, and context) and transforms into a conversation between them and their father in Palestine. See 20:47-49; 26:16-18.

012:087 Muslims are resolute optimists. Pessimism, as a self-fulfilling negative mindset, closes the door to God's blessings. See 15:56; 30:36; 39:53; 41:49.

012:092 What a beautiful example of clemency!

012:098 This does not justify the idea of intercession of a messenger. It is normal for those to ask someone to pray for their forgiveness if their act or crime negatively affects that person.

012:109 The word *rijal*, which we translated as "man" has multiple meanings; it is also used to mean, "walker/pedestrian." See 2:239; 7:46-48; 22:27; 39:29; 72:5.

012:111 In this verse, God the Wise, rejects both the "*hadith*" and the basic excuse for accepting it as a source of Islam. No excuse is accepted from the followers of *hadith* in this world, nor on the Day of Judgment. The followers of fabricated *hadiths* claim that the Quran is not sufficiently detailed! They thus reject God's repeated assertion that the Quran is "complete, perfect, and sufficiently detailed" (6:19,38,114), and thereby justify the creation of 60 volumes of *hadith*, and a library full of contradictory teachings that are supposed to complete the Quran. By reflecting on 12:111 above, one can see God's answer to those fabricators and their followers. God informs us that we do not need fabricated *hadith*; that the Quran as a sufficiently detailed guide, is all we need. The Quran is the only "*ahsan al-hadith*" (best statement) to be followed (39:23). See 6:112-116; 11:1; 31:6; 33:38; 45:6; 52:34.

13:0 In the name of God, the Gracious, the Compassionate.

13:1 A1L30M40R200. These are the signs of the book. What has been sent down to you from your Lord is the truth; but most of the people do not affirm.*

13:2 **God**, is the one who raised the skies without any pillars that you can see, then He established the authority, and He commissioned the Sun and the moon; each moving for a fixed term. He manages all affairs, and He details the signs so that you will be aware of the meeting with your Lord.

13:3 He is the One who stretched out the land; placed on it stabilizers and rivers. From every fruit He made a pair of two. The night covers the day. In that are signs for a people who will think.

13:4 On earth are neighboring pieces of land with gardens of grapes, plants and palm trees, some of which may be twins sharing the same root, or single, even though they are being watered with the same water source. We choose some of them over others in what they consume. In this there are signs for a people who reason.

13:5 If you wonder, then what is more wondrous is their saying: "Can it be that when we are dust, we will be created anew!" These are those who rejected their Lord, and they will have chains around their necks, and they are the dwellers of the fire, in it they will abide.

13:6 They ask you to hasten with the doom rather than the good, yet the examples of those before them have already been given. Your Lord has forgiveness for people despite their transgression, and your Lord is severe in retribution.

13:7 Those who reject say, "If only a sign was sent down to him from his

Lord." You are but a warner, and to every nation is a guide.

13:8 **God** knows what every female carries, and how short her pregnancy or how long. Everything with Him is measured.*

13:9 The knower of the unseen and the seen, the Great, the most High.

13:10 It is the same whether any of you conceals the word or openly declares it; whether he is hiding in the night or going openly in the day.

Will to Change the Condition of a Society

13:11 As if he has followers in his presence and behind him guarding him from **God**'s command! **God** does not change the condition of people until they change what is within themselves. If **God** wanted to harm a people, then there is no turning Him back, nor will they have any protector against Him.*

13:12 He is the One who shows you the lightning, giving you fear and hope. He establishes the heavy clouds.

13:13 The thunder glorifies with His praise, the controllers are in awe of Him, and He sends the thunderbolts, thus striking with them whomever He wills. Yet they are still arguing regarding **God**, while He is severe in punishment.

Imploring Saints, Shrines, Prophets, etc.

13:14 To Him is the call of truth. Those who are called on besides Him, they will not respond to them in anything. It is like one who places his hands openly in the water to drink, but it never reaches his mouth. The call of the ingrates is nothing but in misguidance.

13:15 To **God** prostrate all who are in the heavens and the earth, willingly and unwillingly, as do their shadows in the morning and the evening.*

13:16 Say, "Who is the Lord of the heavens and earth," Say, "**God**." Say, "Have you taken others besides Him as your allies who do not possess for themselves any benefit or harm?" Say, "Is the blind and the seer the same? Or does the darkness and the light equate?" Or have they made partners with **God** who have created like His creation, so the creations all seemed the same to them? Say, "**God** has created all things, and He is the One, the Supreme."

13:17 He sent down water from the sky, so valleys flowed according to their capacity, and the flood produces foam. From what they burn to smelt jewelry or goods similar foam is produced. It is in such a manner that **God** strikes the falsehood with the truth. As for the foam, it passes away; as for what benefits the people, it remains in the earth. It is such that **God** cites the examples.

13:18 For those who responded to their Lord is goodness. As for those who did not respond to Him, if they had all that is in the earth twice over, they would offer it to be saved. To these will be a terrible reckoning, and their abode is hell; what a miserable abode.

13:19 Is one who knows that the truth has been sent down to you from your Lord like one who is blind? Only those with understanding will remember.

13:20 Those who fulfill **God**'s covenant, and they do not break the Covenant.

13:21 Those who deliver what **God** has ordered be delivered, and they revere their Lord, and they fear the terrible reckoning.

13:22 Those who are patient seeking their Lord's direction; and they maintain the contact prayer, and they spend from what We bestowed upon them secretly and openly, and they counter sin with good; these will have an excellent abode.

13:23 The gardens of Eden, they will enter it with those who are good doers from their fathers and their mates and their progeny. The controllers will enter upon them from every gate:

13:24 "Peace be upon you for what you

have been patient for. Excellent indeed is the final abode."

13:25 As for those who break the pledge of **God** after making its covenant, and they sever what **God** ordered that it be joined, and they corrupt in the earth; to those is a curse and they will have a miserable abode.

13:26 **God** extends the provisions for whom He wishes or restricts them. They celebrated the worldly life, but the worldly life compared to the Hereafter was nothing but a brief enjoyment.

13:27 Those who have rejected say, "If only a sign were sent down to him from his Lord!" Say, "**God** misguides whom He wishes and guides to Him whoever is repenting."

13:28 The ones who affirmed and their hearts are satisfied by the remembrance of **God**; for in **God**'s remembrance the hearts are satisfied.

13:29 Those who affirmed, and promoted reforms, there will be happiness for them and a good abode.

Dogmatism and Bigotry of Ingrates

13:30 As such, We have sent you to a nation as other nations that have come before, so that you may recite for them what has been revealed to you; while they are still rejecting the Gracious. Say, "He is my Lord, there is no god but He; in Him I place my trust and to Him is my repentance."

13:31 If there were a Quran/recitation with which mountains were moved, the earth was sliced, or the dead were made to speak... Nay, but **God** controls all issues. Did not those who affirm know that if **God** wished He would have guided all the people? As for those who reject, until **God**'s promise comes true, a disaster will continue to strike them or alight at near their homes because of what they do. **God** does not breach the appointment.

13:32 Messengers before you were ridiculed, but I gave time to those who rejected, then I seized them. Then how was My punishment?

13:33 Who is the One standing over every person for what it has earned? Yet they make partners with **God**. Say, "Name them? Or are you informing Him of what He does not know on earth? Or is it just a show of words?" But to the ingrates, their scheming is made to appear clever, and they are turned away from the path. Whoever **God** misguides will have no guide.

13:34 They will have a punishment in the worldly life, and the punishment of the Hereafter is more difficult. They will have no protector against **God**.

Heaven is Described Allegorically

13:35 The allegory of paradise that the righteous have been promised is that rivers flow beneath it, and its provisions are continuous as is its shade. Such is the abode of those who were righteous, while the abode of the ingrates is the fire.

13:36 Those to whom We have previously given the book rejoice at what has been sent down to you, but there are some of the groups that reject parts of it. Say, "I am only ordered to serve **God** and not to associate any partner with Him. To Him I pray and to Him is the return."

The Miraculous Signs for Future Generations

13:37 Thus, We have sent it down a law in Arabic. If you follow their desires after what has come to you of the knowledge, then you will not have any ally or protector against **God**.

13:38 We have sent messengers before you and We have made for them mates and offspring. It was not for a messenger to come with any sign except by **God**'s leave, but for every era there is a decree.*

13:39 **God** erases what He wishes and affirms, and with Him is the source of the book.

13:40 If We show you some of what We promise them or if We let you pass away, for you is only to deliver, while for Us is the reckoning.*

Consequences of Arrogance and Indulgence

13:41 Do they not see that We come to the land and reduce from its edges? **God** gives judgment and there is none to override His judgment. He is quick reckoning.*

13:42 Those before them have schemed, but to **God** is all scheming. He knows what every person earns and the ingrates will come to know to whom is the ultimate abode.

13:43 Those who reject say, "You are not a messenger." Say, "**God** is sufficient as a witness between me and you, and those who have the knowledge of the book."

ENDNOTES

013:001 A1L30M40R200. These letters/numbers play an important role in the mathematical system of the Quran based on the number nineteen. See 74:1-56; 1:1; 2:1; 13:38; 27:82; 38:1-8; 40:28-38; 46:10; 72:28.

013:008 See 4:119. The verse does not assert that only God knows these. See 42:49; 31:34.

013:011 This verse is traditionally rendered as "For each there are succeeding controllers, before and behind him, guarding him by God's command." However, the context of the verse led us to prefer a different understanding.

013:015 The unappreciative opponents of the monotheistic message also prostrate for God; in other words, willingly or unwillingly they follow God's laws in nature. Unappreciative people are obliged to follow the laws of nature. Those who do not cognitively wish to accept God's sovereignty might speculate about their ability to control their biological bodies, but the Quran challenges them with a simple example that they can never escape: your shadow obeys God. The shade created by sunlight follows a precise path depending on geographic coordinates, hour of the day, and the day of the year. In other words, the motion of your shadow follows the laws of reflection of light that comes from the Sun and changes in accordance to earth's motion. As long as you live on this moving planet and under this illuminating and radiating Sun, you cannot change the motion of your shade. See 2:258, 16:48-49.

013:038 The Quran informs us about many miracles/evidences (*aya/beyyinat*) given to previous messengers, as extraordinary evidences for their extraordinary claims. However, many dogmatic skeptics or dogmatic followers of a particular teaching always managed to find excuses to blind themselves to those evidences. They considered them to be magic, tricks, illusions, or mythology. Some modern Muslim scholars, perhaps because of their lack of appreciation of God's power, tried very hard to deny the existence of miracles; they tried to transform them into natural events, or symbolic statements. Those who have witnessed the perpetual mathematical miracle of the Quran, will have no problem in accepting the examples of miracles given in the Quran. Miracles remove the doubts of those who sincerely seek for truth and strengthen those who already affirmed the message. The Quran repeatedly informs us that God does not change His *sunna* (law); and sending messengers with supporting miracles/signs has been God's law. Below is a list of miracles/signs mentioned in the Quran:

- Some of the Children of Israel are resurrected on Earth after being killed (2:56).
- Moses causes 12 springs to gush forth (2:60).
- A man dies and is resurrected after 100 years (2:259).
- Abraham kills four birds and brings them back to life (2:260).
- Zechariah's wife gives birth while infertile (3:40).
- Mary gives birth while being a virgin (3:47).
- Jesus heals the blind, lepers (3:49).
- Jesus raises the dead (3:49).
- Controllers descend to fight alongside the prophet (3:125).

- God speaks to Moses (4:164).
- Jesus brings down a feast from heaven (5:112).
- Moses turns his staff into a serpent (7:107).
- Moses' hand becomes pure white (7:108).
- God reveals Himself physically to a mountain (7:143).
- Abraham's wife gives birth while beyond menstruation (11:72).
- The sleepers of the cave survive 309 years in sleep (18:25).
- Jesus speaks from birth (19:29).
- Abraham survives being burned alive by the fire turning cold (21:69).
- Solomon commands the winds (21:81).
- Moses parts the sea and saves the Children of Israel (26:63).
- Solomon speaks to the birds (27:16).
- Solomon moves a palace across time and space (27:40).

013:040-43 These verses contain one of the many implied prophecies about the mathematical miracle of the Quran, which was designed by God to be unveiled in 1974. The Arabic word we translated with the word "reckoning" is *hesab*, which also means "to calculate" or "to count." The Arabic of the Quran contains many such rich meanings, which are usually lost in translation. In the endnotes, we have pulled the attention to a few of them. Furthermore, the reduction of the land from its edges might be a prophetic reference to the feared consequence of global warming. Melting of the icebergs can increase the height of ocean and claim the shores of many regions, thereby reducing the land from its edges.

بسم الله الرحمن الرحيم

14:0 In the name of God, the Gracious, the Compassionate.

14:1 A1L30R200. A book which We have sent down to you so that you may take the people out of the darkness and into the light, by their Lord's leave, to the path of the Noble, the Praiseworthy.*

14:2 **God**, to whom belongs all that is in the heavens, and all that is in the earth. Woe to the ingrates from a painful retribution.

14:3 The ones who have preferred the worldly life over the Hereafter, and they repel away from the path of **God**, and they seek its distortion. Those are the ones who are in misguidance.

The Language of the Messenger

14:4 We did not send any messenger except in the language of his people, so he may proclaim to them. But **God** misguides whom He wills, and He guides whom He wills. He is the Noble, the Wise.*

Moses

14:5 We sent Moses with Our signs that you should bring your people out from the darkness and into the light and remind them of **God**'s days. In this are signs for any who are patient and thankful.

14:6 Moses said to his people: "Remember **God**'s blessings upon you that He saved you from the people of Pharaoh. They used to inflict the worst punishment upon you, and they used to murder your children, and shame your women. In that was a great trial from your Lord."

14:7 When your Lord proclaimed: "If you give thanks, then I will increase for you, but if you turn unappreciative, then My retribution is severe."

14:8 Moses said, "If you reject, you and all who are on earth together, yet **God** is Rich, Praiseworthy."

14:9 Did not news come to you of those before you: the people of Noah, and Aad, and Thamud. Those after them, whom none know but **God**? Their messengers came to them with proof, but they placed their hands to their mouths and said, "We are rejecting what you have been sent with, and we are in grave doubt as to what you are inviting us to."

Blind Inheritors of Faith

14:10 Their messengers said, "Is there doubt regarding **God**, the initiator of the heavens and the earth? He invites you so that He may forgive some of your sins and grant you until a predetermined time." They said, "You are but humans like us, you wish to turn us away from what our fathers used to serve. So come to us with clear authority."

14:11 Their messengers said, "We are but humans like you, but **God** will bestow His grace upon whom He pleases from His servants. It is not up to us to bring you an authorization except by **God**'s leave. In **God** those who affirm should place their trust"

14:12 "Why should we not place our trust in **God**, when He has guided us to our paths. We will be resolute against the harm you inflict upon us. In **God** those who trust should put their trust."

14:13 Those who rejected said to their messengers: "We will drive you out of our land, or you will return to our creed." It was then that their Lord revealed to them: "We will destroy the wicked."

14:14 "We will let you reside in the land after them. That is for those who fear My majesty and fear My threat."

14:15 They sought victory, and every arrogant tyrant then failed.

14:16 From behind him is hell, and he will be served repulsive water.

14:17 He tries to drink it, but cannot swallow it, and death comes to him from everywhere, but he will not die; and after this is a powerful retribution.

14:18 The example of those who reject their Lord is that their works are like ashes, on which the wind blows strongly on a stormy day; they cannot get anything of what they earned. Such is the farthest straying.

14:19 Did you not see that **God** created the heavens and the earth with truth? If He wished, He would do away with you and bring a new creation.

14:20 That for **God** is not difficult to do.

14:21 They all appeared before **God**. The weak ones said to those who were arrogant: "We were following you, so will you avail us anything from **God**'s retribution?" They said, "If **God** had guided us, then we would have guided you. It makes no difference if we rage or endure, for we have no refuge."

14:22 The devil said when the matter was complete: "**God** had promised you the promise of truth, and I promised you and broke my promise. I had no power over you except that I invited you and you responded to me. So do not blame me, but blame yourselves; I cannot help you nor can you help me. I reject that you have set me as a partner before this; the wicked will have a painful retribution."

14:23 Admission was given to those who affirmed and promoted reforms to gardens with rivers flowing beneath them. In them they will abide by the leave of their Lord. Their greeting therein is "Peace."

The Fruits of Truth

14:24 Have you not seen how **God** cites the example of a good word is like a good tree, whose root is firm and its branches in the sky?

14:25 It bears its fruit every so often by its Lord's leave. **God** cites the examples for the people, perhaps they will remember.

14:26 The example of a bad word is like a tree which has been uprooted from the surface of the earth, it has nowhere to settle.

14:27 **God** makes firm those who affirm with firm sayings in the worldly life, and in the Hereafter. **God** sends the wicked astray, and **God** does what He wishes.

14:28 Did you not see those who replaced **God**'s blessings with rejection, and they caused their people to dwell in the abode of destruction?

14:29 Hell is where they will burn, what a miserable place to settle.

14:30 They made equals to **God** in order that they may divert from His path. Say, "Enjoy, for your destiny is to the fire."

14:31 Say to My servants who affirm, that they should maintain the contact prayer, and spend from provisions We granted them, secretly and publicly, before a day comes when there is no trade therein, nor will there be any friends.

14:32 **God** is the One who created the heavens and the earth. He sent down water from the sky and brought out fruits as provisions for you. He directed for you the ships to run in the sea by His command, and He directed for you the rivers.

14:33 He directed for you the Sun and the Moon, both in continuity; and He directed for you the night and the day.

14:34 He gave you all that you have asked Him. If you were to count **God**'s blessings, you will never enumerate them. The human is indeed transgressing, unappreciative.

Abraham

14:35 When Abraham said, "My Lord, make this town a sanctuary, and keep me and my sons away from serving idols."

14:36 "My Lord, they have misguided many from amongst the people. So, whoever follows me, then he is of me, and whoever disobeys me, then You are Forgiving, Compassionate."

14:37 "My Lord, I have settled a part of my progeny in an uncultivated valley near your Restricted Sanctuary. My Lord, so that they may maintain the contact prayer. So let the hearts of the people incline towards them. Provide them with fruits that they may appreciate."

14:38 "Our Lord, you know what we hide and what we declare. Nothing is hidden from **God** in the earth nor in the heavens."

14:39 "Praise be to **God** who has granted me, despite my old age, Ishmael and Isaac. My Lord is the hearer of prayers."

14:40 "My Lord, let me maintain the contact prayer, and also from my progeny. Our Lord, accept my prayer."

14:41 "Our Lord, forgive me and my parents, and those who affirm on the day the reckoning is called."

14:42 Do not think that **God** is unaware of what the wicked do. He is merely delaying them to a day when all eyes are watching.

14:43 They will approach with their heads bowed, their eyes will not blink, and their hearts will be void.

Warning

14:44 Warn the people of the day when the retribution will come to them, and those who were wicked will say, "Our Lord, delay this for us until a short time, and we will respond to Your call and follow the messengers!" Did you not swear before this that you would last forever?

14:45 You even resided in the homes of those who had wronged themselves, and it was made clear to you what We did to them. We had cited the examples to you.

14:46 Indeed they schemed their scheming, and their scheming was at **God**; though their scheming was enough to eliminate mountains.

14:47 So do not think that **God** will fail to keep His promise to His messengers. **God** is Noble, able to seek revenge.

New Earth and New Heavens

14:48 The day the earth is replaced with another earth, as are the heavens, and they will appear before **God**, the One, the Irresistible.*

14:49 You will see the criminals on that day held by restraints.

14:50 Their clothes will be of tar, and the fire will overwhelm their faces.

14:51 Thus **God** will recompense every person for what it earned. **God** is quick in reckoning.

14:52 This is a proclamation for the people and so that they are warned with it, and so that they know that there is but One God, and so that those with intelligence will remember.

ENDNOTES

014:001 A1L30R200. These letters/numbers play an important role in the mathematical system of the Quran based on code 19. See 74:1-56; 1:1; 2:1; 13:38; 27:82; 38:1-8; 40:28-38; 46:10; 72:28.

014:004 This verse rejects the Arab nationalists who glorify Arabic and claim its superiority over other languages. God can communicate His message to any nation in their language. After all, God communicates with all his creatures through the language they understand.

The frequently used phrases *yudillu man yasha* can be understood in two ways: "He misleads whomever He wills," OR "He misleads whoever wills to be misled." The same is also true for *yahdi man yasha*: "He guides whomever He wills," OR "He guides whomever wills to be guided." See 57:22-23; 6:88.

014:048 See the Old Testament, Isaiah 65:17 & 66:22.

15:0 In the name of God, the Gracious, the Compassionate.

15:1 A1L30R200 these are the signs of the book, and a clear Quran.*

15:2 Those who have rejected will often wish they were of those who peacefully surrendered!

15:3 Leave them to eat and enjoy; let them be preoccupied with hope. They will come to know.

15:4 We have not destroyed any town except that it had an appointed time.

15:5 No nation can quicken its fate, nor can they delay.

15:6 They said, "O you upon whom the Reminder has been sent down, you are crazy."

15:7 "Why not bring us the controllers if you are of the truthful ones?"

15:8 We do not send down the controllers except with truth, and then they would have no more respite.

Divine Protection

15:9 We, indeed We, it is We who have sent down the Reminder, and indeed it is We who will preserve it.*

15:10 We have sent before you to the factions of old.

15:11 Any messenger that came to them, they would mock him.

15:12 We thus let this attitude sneak into the hearts of the criminals.

15:13 They do not affirm it, while the examples of the early generations have been given to them.

15:14 If We opened for them a gate in the sky and they were to continue ascending to it,

15:15 They would have said, "Our sight has been distorted. No, we are a people who are being bewitched!"

15:16 We have placed galaxies in the heavens, and We have made them pleasant to the onlookers.

15:17 We have guarded them from every outcast devil.

15:18 Except he who manages to eavesdrop, he will be pursued clearly by a flaming meteor.

15:19 As for the land, We have stretched it, placed stabilizers in it, and We have planted in it everything in balance.

15:20 We made for you in it a habitat, as well as those whom you are not required to provide for.

15:21 There is not a thing, except that We have vaults of it, yet We only send it down in a measured amount.

Wind: Pollinator of Clouds and Plants

15:22 We have sent the winds to pollinate, and We sent down water from the sky for you to drink. It is not you who store it up.

15:23 It is indeed We who give life and death, and We are the inheritors.

15:24 Certainly We know those among you who are progressive, and We know those of you who are regressive.

15:25 It is your Lord that will gather them. He is Wise, Knowledgeable.

Creation of Humans

15:26 We have created the human being from hardened clay of aged mud.*

15:27 The Jinn, We created them before that from the flames of fire.

15:28 Your Lord said to the controllers: "I am creating a human from hardened clay of aged mud."

15:29 "So when I perfect him, and blow of My Spirit in him, you shall fall prostrate to him."*

15:30 Thus, all of the controllers fell prostrate to him,

15:31 Except for Satan, he refused to be with those who prostrated.

15:32 He said, "O Satan, what is the matter that you are not with the prostrators?"

15:33 He said, "I am not to prostrate to a human You have created from a hardened clay of aged mud."

15:34 He said, "Exit from here, you are cast out."

15:35 "A curse shall be upon you until the day of judgment."

15:36 He said, "My Lord, respite me until the day they are resurrected."

15:37 He said, "You are given respite,"

15:38 "Until the day of the appointed time."

Satan's Limited Power

15:39 He said, "My Lord, you sent me astray, I will make the earth appear beautiful for them, and I will mislead them all,"

15:40 "Except Your devoted servants."

15:41 He said, "This is a straight path to Me."

15:42 "For My servants, you shall have no authority over them, except those who are misled and follow you."

15:43 "Hell, shall be the appointed place for them all."

15:44 "It has seven gates, for every gate will be an assigned segment from them."

15:45 The righteous will be in paradises and springs.

15:46 "Enter it in peace and security."

15:47 We removed all negative feelings in their chests; they are brothers in quarters facing one another.

15:48 No fatigue shall touch them, nor will they be made to leave from it.

15:49 Inform My servants that I am the Forgiver, the Compassionate.

15:50 That My punishment is a painful retribution.

Controllers Give Good and Bad News

15:51 Also inform them of Abraham's guests.

15:52 That they entered upon him, they said, "Peace." He said, "We are apprehensive of you."

15:53 They said, "Do not worry, we bring you good news of a knowledgeable son."

15:54 He said, "What good news can you bring me when old age has come upon me? Is that your good news?"

15:55 They said, "We have brought you good news with truth, so do not be of those in denial."

15:56 He said, "Who would deny the mercy of his Lord except the misguided ones!"*

15:57 He said, "What then is your business here, O messengers?"

15:58 They said, "We have been sent to a people who are criminals;"

15:59 "Except for the family of Lot, we will save them all;"

15:60 "Except for his wife, we have measured that she will be with those destroyed."

Lot and the Aggressive Homosexuals

15:61 So when the messengers came to the family of Lot.

15:62 He said, "You are a people unknown to me."

15:63 They said, "Alas, we have come to you with what they are in doubt."

15:64 "We have come to you with the truth, and we are forthcoming."

15:65 "So let your family leave during the last moments of the night, and you follow just behind them, and do not let any of you look back, and go to where you are commanded."

15:66 We conveyed him the decree, that the remnants of these people will be cut off in the morning.

15:67 The people of the city came seeking good news.

15:68 He said, "These are my guests, so do not embarrass me!"

15:69 "Be aware of **God**, and do not disgrace me!"

15:70 They said, "Did we not forbid you from strangers?"

15:71 He said, "Here are my daughters if you are going to act."*

15:72 By your life, they are in their drunkenness, blundering.

15:73 So the scream took them at sunrise.

15:74 Thus We turned it upside down and rained upon them with fiery projectiles.

15:75 In this are signs for those who distinguish.

15:76 It is on an established path.

15:77 In that is a sign for those who affirm.

15:78 The dwellers of the forest were wicked.

15:79 So, We sought revenge from them. They were both on an open plain.

15:80 The dwellers of the cavern rejected the messengers.

15:81 We gave them Our signs, but they turned away from them.

15:82 They used to sculpt from the mountains dwellings that were secure.

15:83 But the blast took them in the morning.

15:84 What they earned did not benefit them.

The End of the World and Seven Pairs

15:85 We did not create the heavens and the earth and what is in between except with the truth. The moment is coming, so overlook their faults gracefully.*

15:86 Your Lord is the Creator, the Knower.

15:87 We have given you seven of the pairs and the great Quran.*

15:88 So do not linger with your eyes on what We have bestowed upon some couples from them, do not grieve for them, and lower your wing for those who affirm.

15:89 Say, "I am a manifest warner."

The Sectarians

15:90 As We have sent down on the sectarians.*

15:91 The ones who have taken the Quran apart.

15:92 By your Lord, We will question them all.

15:93 Regarding what they used to do.

15:94 So proclaim what you have been commanded and turn away from those who set up partners.

15:95 We will relieve you from the mockers.

15:96 Those who set up with **God** another god; they will come to know.

15:97 We know that your chest is strained by what they say.

15:98 So glorify with the praise of your Lord; be of those who prostrate.

15:99 Serve your Lord until certainty comes to you.

ENDNOTES

015:001 A1L30R200. These letters/numbers play an important role in the mathematical system of the Quran based on code 19. See 74:1-56; 1:1; 2:1; 13:38; 27:82; 38:1-8; 40:28-38; 46:10; 72:28.

Quran's divine nature, authenticity and the promise regarding the protection of its message (*zikr*) have been supported by the mathematical structure of the Quran discovered in 1974. See 74:1-56; 1:1; 2:1;

13:38; 27:82; 38:1-8; 40:28-38; 46:10; 72:28.

015:009 The Quran is protected through an interlocking error-sensitive mathematical code. The emphasis in this short verse is unique; with its four references to God and his controllers, it makes it clear that it is not the humans who would protect it, but the One who revealed the message. The word *zikr* (reminder) establishes a link between the preservation of the Quran and its mathematical system, which is described with the grammatically feminine of the same word, *zikra* (74:31). See 9:127; 41:41-42.

015:026 Creation "from" clay has two meanings. (1) clay as the substance of origin, and (2) clay as a place of origin. Both meanings could be true at the same time, though we learn from other verses that clay is not the only substance used in our creation; water too is a vital ingredient. Our Creator started the biological evolution of microscopic organisms within layers of clay. Recent scientific research led some scientists to consider clay as the origin of life, since clay is a porous network of atoms arranged geodesically within octahedral and tetrahedral forms. This design creates sliding and flexible layers that catalyze chemical reactions. Humans are the most advanced fruits of organic life started millions of years ago from layers of clay. See 29:18-20; 41:9-10; 7:69; 24:45; 32:7-9; 71:14-17.

015:029 The word *rouh*, does not mean soul or spirit, as commonly thought. In the terminology of the Quran *rouh* means revelation, inspiration, information, program or commands. The word *nafs* is used to mean person or consciousness (39:42). If we try to describe the human genetic makeup and the program etched into the brain in computer language, it is like a piece of hardware that facilitates communication, processing, analysis and synthesis of information through an intelligent system program containing God's knowledge/commands/code/logic (*rouh*) in its core. Person (*nafs*) is analogous to an open system program that can be improved through new application programs, addition of new features, and memory (experience) data. This system program (person) can be put in hibernation (sleep), or terminated by the turning off of the switch, or by the destruction of its essential hardware (death) (39:42; 21:35; 29:57; 2:28; 10:56; 32:11; 45:15).

Contradictory and irrational actions that are the products of reflexive choices, gullibility, superstitious ideas, intoxicants, religious dogmas, peer pressure, ignorance, etc., might make the system program vulnerable to the attacks of evil viruses and Trojan horses (6:116; 10:36,42; 17:36; 53:28; 2:170-171; 3:154; 8:22; 10:100; 12:40; 13:19; 28:63; 35:28; 39:9,18; 41:46; 45:15; 79:40; 9:31). If due to prejudice and acquired bad habits, they do not use vigilant virus protection programs called critical thinking or logic, numerous harmful and fatal viruses might infect and corrupt system programs (persons) beyond reform (2:231; 4:110; 7:160; 17:7; 31:13; 65:1; 91:7, and 6:110). Good intentions, thoughts, deeds, habits, good networks, and numerous tests advance the system program (person) to new levels (2:272; 23:60). Then the improved or regressed system program is separated from the hardware (death) and uploaded (recorded in the master record/*ummul-kitab*) for evaluation (18:49; 6:38; 54:53). Those that kept the system program free from infection and corruption, and improved it through various tests, are beamed to a higher dimension for eternal enjoyment (89:27-30). The system programs (persons) that had progressed (through affirmation and good deeds) and passed all major tests (through a humble and appreciative attitude during blessings and tragedies) in its previous hardware (2:155; 3:186; 29:1-3); these few successful system programs will be matched with an indestructible hardware (body) to live in eternal bliss (81:7; 3:30,161,185; 40:17; 45:22; 50:21; 74:38; 89:27-28; 2:123,233,281). Also, see 17:85; 39:42.

015:056 See 12:87; 30:36; 39:53; 41:49.

015:071 Did prophet Lot mean his own daughters or the daughters in his town? Regardless, he was reminding them to follow God's design, rather than choosing a lifestyle that contradicts it. The distorted Bible contains many false accusations against prophets. For instance, it depicts Lot as a drunkard who committed incest with his daughters (Genesis 19:30-38).

015:085-88 The Quran contains some interesting hints regarding the end of the world. See 20:15; 72:27.

015:087 The meaning of *saban minal mathani* (seven from pairs) has been interpreted by the majority of scholars as "oft-repeated seven," implying the most repeated chapter of the Quran, the seven-versed first chapter. However, if the first chapter of the Quran is part of the Quran, it surely is the part of the "great Quran." If the arrangement of the words were different, "we gave you the great Quran and seven from pairs," then, such an interpretation could be more plausible since a part of the whole can be repeated after the whole for the purpose of emphasis.

If the "great Quran" is the entire Quran, which is the most reasonable understanding, then the "seven pairs" or "fourteen" (of something) might have implications beyond the Quran. There is a hint in the verse: it might refer to the life span of Muhammed's people and its relation to the end of the world.

015:090-93 These verses criticize the followers of sects, those who claim the existence of abrogation in the Quran, and those who ignore the entire context of the Quran when trying to understand a particular Quran statement or term. See 2:106.

بسم الله الرحمن الرحيم

16:0 In the name of God, the Gracious, the Compassionate.

16:1 **God**'s command has come; so do not hasten it. Glorified is He and High above the partners they have set up.*

16:2 He sends down the controllers with the Spirit by His command upon whom He wishes of His servants: "That you shall warn that there is no god but I, so be aware of Me."

16:3 He created the heavens and the earth with the truth. He is High above the partners they have set up.

16:4 He created the human being from a seed, but then he becomes a clear opponent.

16:5 The livestock He created for you, in them is warmth and benefits, and from them you eat.

16:6 For you in them is attraction, when you relax and when you travel.

16:7 They carry your loads to a place you would not have been able to reach except with great strain. Your Lord is Kind, Compassionate.

16:8 The horses, the mules, and the donkeys, that you may ride them and as an adornment. He creates what you do not know.*

16:9 To **God** shall be the path, but some are misled. If He wished, He could have guided you all.

16:10 He is the One who sent down water from the sky for you, from it you drink, and from it emerge the trees that you benefit from.

16:11 He brings forth with it vegetation, olives, palm trees, grapes and from all the fruits. In that are signs for a people who think.

16:12 He directed in your service the night, the day, the Sun, the moon. Also the stars are directed by His command. In that are signs for a people who reason.

16:13 What He has placed for you on earth in various colors. In that are signs for a people who remember.

16:14 He is the One who directed the sea, that you may eat from it a tender meat, and that you may extract from it pearls that you wear. You see the ships flowing through it, so that you may seek from His bounty, and that you may be thankful.

16:15 He has cast into the earth stabilizers so that it does not sway with you, and rivers, and paths, perhaps you will be guided.

16:16 As well as landmarks, and by the stars people navigate.

16:17 Is the One who creates the same as one who does not create? Will you not remember?

16:18 If you count the blessings of **God** you will not be able to fathom them. **God** is Forgiving, Compassionate.

16:19 **God** knows what you hide and what you declare.

The Dead-servants

16:20 As for those they call on besides **God**, they do not create anything, but are themselves created!

16:21 They are dead, not alive, and they will not know when they are resurrected.

16:22 Your god is One god. Those who do not affirm the Hereafter, their hearts are denying, and they are arrogant.

16:23 Certainly, **God** knows what they hide and what they declare. He does not like the arrogant.

16:24 If they are told: "What has your Lord sent down?" They say, "Fairytales of the past."*

16:25 They will carry their burdens in full on the day of Resurrection, and also from the burdens of those whom they misguided without knowledge. Evil indeed is what they bear.*

16:26 Those before them have schemed, but **God** came to their buildings from the foundation, thus the roof fell on top of them, and the retribution came to them from where they did not know.

16:27 Then, on the day of Resurrection He will humiliate them, and say, "Where are My partners whom you used to dispute regarding them?" Those who have received the knowledge said, "The humiliation today and misery is upon the ingrates."

16:28 Those whom the controllers take while they had wronged themselves: "Amnesty, we did not do any evil!" "Alas, **God** is aware of what you had done."

16:29 "So enter the gates of hell, in it you shall reside; such is the abode of the arrogant."

16:30 It was said to those who were righteous: "What has your Lord sent down?" They said, "All goodness." For those who have done good in this world there is good, and the Hereafter is even better. Excellent indeed is the home of the righteous.

16:31 The gardens of Eden, which they will enter, with rivers flowing beneath, in it they will have what they wish. It is such that **God** rewards the righteous.

16:32 Those whom the controllers take, while they had been good, they will say, "Peace be upon you, enter paradise because of what you have done."

16:33 Are they waiting for the controllers to come for them, or a command from your Lord? It was exactly the same as what those before them did. **God** did not wrong them, but it was their own selves that they wronged.

16:34 Thus, the evil of their work afflicted them, and they were surrounded by what they used to make fun of!

They Ascribe Their Polytheism to God

16:35 Those who set up partners said, "If **God** had wished it, we would not have served anything besides Him; neither us nor our fathers; nor would we have forbidden anything without Him." Those before them did the exact same thing; so are the messengers required to do anything but deliver with proof?*

16:36 We have sent a messenger to every nation: "You shall serve **God** and avoid transgression." Some of them were guided by **God**, and some of them deserved to be misguided. So travel in the land, and see how the punishment was of those who denied.

16:37 If you are concerned for their guidance, **God** does not guide those who chose falsehood. They will have no supporters.

16:38 They swore by **God**, in their strongest oaths, that **God** will not resurrect whoever dies. No, it is a promise of truth upon Him, but most people do not know.

16:39 So that He will make clear for them that in which they have disputed, and so that those who have rejected will know that they were liars.

16:40 When we want something, our word is only to say to it: "Be," and it is.*

Do not Accept Oppression

16:41 Those who have emigrated for **God**, after they were oppressed, We will grant them good in the world, and the reward of the Hereafter will be greater, if they knew.

16:42 Those who are steadfast and put their trust in their Lord.

16:43 We did not send any except men/walkers before you whom We revealed to, so ask the people who received the Reminder if you do not know.

16:44 With proof and the scriptures. We sent down to you the Reminder to proclaim to the people what was sent to them, and perhaps they would think.*

16:45 Have those who schemed evil guaranteed that **God** will not make the earth swallow them, or the retribution come to them from where they do not expect?

16:46 Or that He will take them in their actions as they cannot escape?

16:47 Or that He will take them while they are in fear? Your Lord is Kind, Compassionate.

16:48 Did they not see that what **God** has created, its shadow inclines to the right and the left prostrating to **God**, willingly?

16:49 To **God** prostrate all those in the heavens and all those on the earth, from the creatures as well as the controllers, and they are not arrogant.*

16:50 They fear their Lord from above them, and they do what they are commanded.

16:51 **God** said, "Do not take-up two gods, there is only One god, so it is Me that you shall revere."

The Only Source of the System

16:52 To Him is what is in the heavens and the earth, and the system shall always be to Him. Is it other than **God** that you shall be aware of?*

16:53 Any blessings that are with you are from **God**. Then, when harm afflicts you, to Him you cry out.

16:54 Then, when He removes the harm from you, a group of you set up partners with their Lord!

16:55 So they reject what We have given them. Enjoy, for you will come to know.

16:56 They allocate a portion from what We provide to them to what they did not have knowledge of. By **God**, you will be asked about the lies you have invented!

Misogynistic Culture Condemned

16:57 They assign their daughters to **God**; be He glorified; and to them is what they desire.

16:58 When one of them is given news of a female, his face becomes darkened, and he is in grief!*

16:59 He hides from his people because of the bad news he has received. Shall he keep her with dishonor, or bury her in the sand?

Miserable indeed is how they judge!

16:60 For those who do not affirm the Hereafter is the worst example, and for **God** is the highest example, and He is the Noble, the Wise.

16:61 If **God** were to immediately call the people to account for their transgression, then He would not leave a single creature standing. But He delays them to a determined time; so when that time comes to any of them, they cannot delay it by one moment or advance it.

16:62 They attribute to **God** what they hate, and their tongues assert lies that they will get the best. No doubt they will have the fire, for they have rebelled.

16:63 By **God**, We have sent others to nations before you, but the devil made them proud with their work. So, he is their ally today, and they will have a painful retribution.

16:64 We did not send down the book to you except that you may proclaim to them that in which they disputed, and as a guidance and mercy to a people who affirm.*

16:65 **God** has sent down water from the sky, so He revives the land with it after its death. In that is a sign for a people who listen.

16:66 For you there is a lesson in the livestock; We give you to drink from what is in its stomachs between the digested food and the blood, pure milk which is relieving for the drinkers.

16:67 From the fruits of the palm trees and the grapes you make wine as well as a good provision. In that is a sign for a people who reason.*

The Bee

16:68 Your Lord revealed to the bee: "You shall take homes of the mountains and of the trees and of what they erect."

16:69 Then you shall eat from every fruit, so seek the path your Lord has made easy. From her stomachs will emerge a liquid that has different colors, wherein there is a healing for the people. In that is a sign for a people who will think.*

16:70 **God** created you, then He will terminate you. Some of you will continue to the most miserable age so that he will not know anything after knowledge. **God** is Omniscient, Omnipotent.

Share your blessings with the needy

16:71 **God** has endowed some of you over others in provision. Yet, those who have been endowed are unwilling to share their provision with those whom they have contractual rights so that they may become equal in it. Are they denying the favor of **God**?*

16:72 **God** has made for you mates from your own kind, and He has made from your mates children and grandchildren, and He has provided you from the good provisions. Do they affirm falsehood, while in **God**'s favor they are denying?

16:73 They serve besides **God** what does not and cannot possess anything of the provisions from the sky or the land.

16:74 So do not give parables to **God**. **God** knows while you do not know.

Richness and Freedom, Poverty and Dependence

16:75 **God** cites the example of a slave who is owned and cannot control anything, against one whom We have provided a good provision which he spends of it secretly and openly. Are they the same? **God** be praised, but most of them do not know.

16:76 **God** puts forth the example of two men, one of them is mute and he cannot control anything, and he is a burden to his master. Wherever he points him, he does not come with any good. Is he the same as one who orders good, and he is on a straight path?*

16:77 To **God** is the mystery of the heavens and the earth, and the matter of the Moment is like the blink of the eye or nearer. **God** is capable of all things.

16:78 **God** brought you out of your mothers' wombs while you knew nothing. He made for you the hearing and the eyesight and the heart, perhaps you would be thankful.

16:79 Did they not look to the birds held in the atmosphere of the sky? No one holds them up except **God**. In that are signs for a people who affirm.

16:80 **God** has made your homes a habitat, and He made for you from the hides of the livestock shelter which you find light for travel and when you stay, and from its wool, fur, and hair you make furnishings and goods, for a while.

16:81 **God** has made for you shade from what He created. He made from the mountains a refuge for you. He made for you garments which protect you from the heat, and garments which protect you from attack. It is such that He completes His blessing upon you, that you may peacefully surrender.

16:82 So if they turn away, then you are only required to deliver clearly.

16:83 They recognize **God**'s blessing, then they deny it. Most of them are ingrates.

16:84 The day We send from every nation a witness, then those who have rejected will not be given leave, nor will they be allowed to make amends.

16:85 Then those who were wicked will see the retribution; it will not be lightened for them, nor will they be given respite.

Idolized Messengers and Saints Will Disown Those Who Consider them Partners with God

16:86 When those who set up partners saw the partners they made, they said, "Our Lord, these are our partners that we used to call upon besides You." But they returned in answer to them: "You are liars!"*

16:87 They peacefully surrendered to **God** on that day, and what they had invented abandoned them.

16:88 Those who rejected, and repelled others from the path of **God**, We have increased the retribution for them over the retribution for what they had corrupted.

16:89 The day We send to every nation a witness against them from themselves, and We have brought you as a witness against these. We have sent down to you the book as a clarification for all things, a guide, mercy and good tidings for those who have peacefully surrendered.*

16:90 **God** orders justice and goodness, and that you shall help your relatives, and He forbids from evil, vice, and transgression. He warns you that you may remember.

Keep Your Words

16:91 You shall fulfill your pledge to **God** when you pledge so, and do not break your oath after making it, for you have made **God** a guarantor over you. **God** is aware of what you do.

16:92 Do not be like the one who unraveled her knitting after it had become strong, by breaking your oaths as a means of deception between you, because a nation is more numerous than another nation. **God** only puts you to the test by it. He will show you on the day of Resurrection what you were disputing in.

16:93 Had **God** wished, He would have made you one nation, but He misguides whom He wishes, and He guides whom He wishes. You will be asked about what you used to do.*

16:94 Do not use your oaths as a means of deception between you, that a foot will falter after it had been firm, and you will taste the evil of turning away from the path of **God**, and you will have a great retribution.

16:95 Do not trade your pledge with **God** for petty gains. What is with **God** is far better for you if you know.

16:96 What you have will run out, while what **God** has will remain. We will compensate those who are patient with the best of what they used to do.

16:97 Whoever does good work, be he male or female, and is an affirmer, We will give him a good life and We will reward them their dues with the best of what they used to do.*

16:98 So, when you study the Quran, you shall seek refuge with **God** from Satan the outcast.*

16:99 He has no authority over those who affirm, and who put their trust in their Lord.

16:100 His authority is over those who follow him, and those who set him up as a partner.

Divine Signs for Contemporary People

16:101 If We exchange a sign in place of another sign; and **God** is more aware of what He is revealing; they say, "You are making this up!" Alas, most of them do not know.

16:102 Say, "The Holy Spirit has sent it down from your Lord with truth, so that those who affirm will be strengthened, and as a guidance and good news for those who have peacefully surrendered."

16:103 We are aware that they say, "A human is teaching him." The language of the one they refer to is foreign, while this is a clear Arabic language.*

16:104 Those who do not affirm **God**'s signs, **God** will not guide them, and they will have a painful retribution.

16:105 Making up lies is only done by those who do not affirm **God**'s signs, and these are the liars.

God Knows Your Intentions

16:106 Whoever rejects **God** after having affirmed; except for one who is forced while his heart is still content with affirmation; and has opened his chest to rejection, then they will have a wrath from **God** and they will have a great retribution.

16:107 That is because they preferred the worldly life over the Hereafter, and **God** does not guide the rejecting people.

16:108 Those are the ones whom **God** has stamped on their hearts, their hearing, and their sight; such are the heedless.

16:109 Without doubt, in the Hereafter they are the losers.

16:110 Your Lord is to those who emigrated after they were persecuted, then they strived and were patient, your Lord after that is Forgiving, Compassionate.

16:111 The day every person will come to argue for herself, and every person will be paid in full for what she did, and they will not be wronged.*

Prohibiting Blessings Invites Disasters

16:112 **God** cites the example of a town which was peaceful and content, its provisions were coming to it bountifully from all places, but then it rebelled against **God**'s blessings and **God** made it taste hunger and fear for what they used to do.

16:113 A messenger came to them from themselves, but they denied him. So, the punishment took them while they were wicked.

16:114 So eat from what **God** has provided for you, what is good and lawful, and thank the blessings of **God**, if it is indeed He whom you serve.*

A Very Limited Prohibition

16:115 He only made forbidden for you what is already dead, blood, the meat of pig, and what was sacrificed to any other than **God**. But whoever is forced to, without disobedience or transgression, then **God** is Forgiving, Compassionate.

16:116 You shall not invent lies about **God** by attributing lies with your tongues, saying: "This is lawful and that is forbidden." Those who invent lies about **God** will not succeed.

16:117 A small enjoyment, and they will have a painful retribution.

16:118 For those who are Jews, We forbade what We narrated to you before. We did not wrong them, but it was themselves they used to wrong.

16:119 Yet, your Lord, to those who repent and reform after committing evil out of ignorance, your Lord after that is Forgiving, Compassionate.

16:120 Abraham was a righteous leader, devoted to **God**, a monotheist, and he was not of those who set up partners.

16:121 Because he was thankful for His blessings, He chose him and guided him to a Straight Path.

16:122 We gave him good in this world, and in the Hereafter, he is of the reformed ones.

Follow Abraham

16:123 Then We revealed to you: "You shall follow the creed of Abraham, monotheism, and he was not of those who set up partners."*

16:124 The Sabbath was only decreed for those who disputed in it, and your Lord will judge between them for what they disputed in.

Engage in Philosophical Discourse

16:125 Invite to the path of your Lord with wisdom and good advice and argue with them in the best possible manner. Your Lord is fully aware of who is misguided from His path, and He is fully aware of the guided ones.

16:126 If you penalize, then punish with an equivalent punishment inflicted on you. If you show patience then it is better for the patient ones.*

16:127 Be patient, and your patience is only possible by **God**. Do not grieve for them, and do not be depressed by what they scheme.

16:128 **God** is with those who are aware and are good doers.

ENDNOTES

016:001 See 57:22.

016:008 God creates new means of transportation we do not yet know about. According to the teaching of the Quran, God is the ultimate creator of everything, including our actions (37:96). We are only given a limited freedom of choice. This verse refers to all transportation vehicles that were not available during the time of revelation, such as, trains, automobiles, airplanes, helicopters, rockets, and spaceships. Since we are also the Quran's audience, it refers to many more to come in the future.

016:024 The language of the verse is interesting. It refers to the unappreciative people who reject the message of the Quran by labeling it a myth, and it also refers to the modern polytheists who follow *hadith* mythologies and consider them to be "revelation."

016:025 See 6:164.

016:035 See 6:23,148; 53:19-23.

016:040 Indeed the entire universe came into existence with an "abrupt" big bang. See 21:30.

016:044 The Arabic word *BaYaNa* means (1) explain an ambiguous message; or (2) declare a hidden message. This multi-meaning verse is one of the most abused verses by the followers of *hadith* hearsay. They choose the first meaning. Their rendering, however, contradicts many verses of the Quran: the Quran is explained not by Muhammed, but by its author, God Himself (75:19). Thus, it is described as a "clarified/explained book" (5:15; 12:1; 26:195; 44:6), and we are reminded over and over that it is "easy to understand" (54:17,22,32,40). Thus, the second alternative is the intended meaning, since the Quran was received by Muhammed in a private session called revelation, and his job was to deliver and declare the message he received. God orders Muhammed to proclaim the revelation which is revealed to him personally. Indeed, this is the whole mission of the messengers (16:35). Thus, the word "*litubayyena*" of 16:44 is similar to the one in 3:187. Verse 3:187 tells us that the people who received the revelation should "proclaim the scripture to the people, and never conceal it." See 2:159,160; 3:187 and 16:64. For a comparative discussion on verse 16:44, see the Sample Comparisons section in the Introduction.

The Quran is simple to understand (54:17). Whoever opens his/her mind and heart as a monotheist and takes the time to study it, will understand it. This understanding will be enough for salvation. Beyond this, to understand the multi-meaning verses you do not need to be a messenger of God. If you have a good mind and have studied the Quran through reason, then you will be able to understand the true meanings of multi-meaning verses. The verse 3:7, which is about the multi-meaning verses, points this fact in a multi-meaning way (this is an interesting subject warranting another article): ". . . No one knows their true meaning except God and those who possess knowledge. . ."

016:049 The human body, whether belonging to an appreciative or unappreciative person, has peacefully surrendered to its Creator, and thus is a muslim (submitter/peaceful surrenderer to God).

It is true that *SaJDa* means obedience, submission to God's law (16:49). From this verse we should not necessarily infer that everything has a mind like humans. Even if we subscribe to such a fancy assumption, mental events too are physical events; they are the consequence of interactions of electrons. In fact, the prostration of all the creatures we perceive is *physical*. Any event that can be sensed by our five senses is a physical event. Atoms peacefully surrender to God's laws when they interact with each other. When water freezes it peacefully surrenders to God's law and expands; when it is heated to a certain degree it evaporates. When Hydrogen, Carbon, Nitrogen, and Oxygen atoms come together in a particular order they create bases called Adenine, Cytosine, Guanine, and Thymine; and when these bases join each other in a particular order, pre-ordained by God, they create DNAs that produce new qualities and events, including life. Every cell and organ in our bodies submits to God's law. These are all physical demonstration of *SaJDa* to God. No wonder, the entire universe is stated to have submitted to God (3:83; 13:15). Interestingly, in these two verses, the same idea is depicted; yet in 3:83 the verb *aSLaMa* (submitted) is used while in 13:15 the verb *yaSJuDu* (prostrating) is used. This is strong evidence that both verbs belong to the same semantic field. As we know, submission to God is not just a mental activity but also a physical action, such as working hard, delivering God's message, feeding the poor, etc.

The word *SaJDa* in some verses means ONLY mental (in quantum, electronic or neural level) prostration while in other verses it means ONLY bodily (atomic and molecular level) prostration. The prostration mentioned in 27:20-24 must be bodily, and the one in verse 77:48 most likely refers to mental prostration.

As for verse 22:18, it describes both mental and bodily submission to God. One might argue, "Has anyone ever seen any of these creations bow and prostrate on the ground (animals excluded)? Since all things in heavens and earth do not prostrate to God by casting themselves face down on the ground in humility, then we should not prostrate that way either." This conclusion is based on a false assumption that ALL creatures prostrate the same way. Every creature in the heavens and the earth glorifies God (24:41; 57:1; 59:1,24; 61:1; 62:1; 64:1). But we may not understand their glorification (17:44). We are also instructed to glorify our Lord, day and night, not like birds or planets, but like humans, in our language and according to our natural design (3:41; 5:98; 7:206;

19:11; 20:130; 32:15; 40:55; 50:39-40; 52:48-49; 56:74,96; 69:52; 76:26; 87:1; 110:3).

Hence: "Do you not see that everything in the heavens and the earth glorify God, including the birds in columns. Each knows his/her/its prayer (*sala*) and glorification. God is fully aware of everything you do." (24:41).

The following verse is also important: "Are they seeking other than God's system, when everything has submitted (*aSLaMa*) to Him in the heavens and the earth, willingly or unwillingly, and to Him they will be returned?" (3:83). Now those who reject that *sala* prayer includes bodily bowing and prostration should ask a similar question about how to be a muslim just as they ask about how to perform prostration: "Has anyone ever seen any of these creations give charity or deliver God's message or study the Quran?" Perhaps, according to those who object physical prostration besides mental one, we should not give charity, should not deliver the message, should not study the Quran, and should not use computers!

One might refer to 2:62 and ask: "No, not all the listed groups bow and prostrate on the ground. Will they lose? No, it is the people who do not submit to God and follow His commands that will deserve His punishment." This is a hasty conclusion. The groups mentioned in the verse (affirmers, Jews, Nazarenes, and those from other religions) attain salvation as long as they believe in God, do righteous work and believe in the hereafter. Perhaps not all bow and prostrate, and perhaps not all believe and follow the Quran, either. First, we should remember that God holds each responsible depending on their circumstances. As those who witnessed Moses' miracles will be held responsible for their reaction to them, similarly those who are instructed to bow and prostrate will be held responsible for their reaction. Each nation received a messenger, and each received their instruction (10:47; 22:34; 40:28).

Referring to verse 18:50, one might conclude: "There is no need to get on the ground to worship your Lord because He has not commanded us to do this. He wants us to bow and prostrate our wills to that of His will." This is a reasonable conclusion. However, knowing that the word prostration was also used to mean humbly getting on the ground, we cannot equate the prostration in this verse with all others. The verse refers to an event which took place before the creation of life on earth. Besides, the instruction is to Satan, a creature made of energy, not matter like us. The nature of the prostration asked of Satan, or the controllers might be a little different than the one asked of us; or, more accurately, the form, way or style of our prostration might not necessarily be expected to be the same as that of the controllers.

David, who was observing *sala* prayer, is described by the Quran as: "... He then implored his Lord for forgiveness, bowed down (*KhaRra RuKka'An*), and repented." (38:24). David's repentance is clearly described as bowing down, both mentally and bodily.

We should remind those who are trying hard to eliminate the formal component of *sala* prayer of the following verse: "... When the revelations of the Gracious are recited to them, they fall prostrate (*KhaRRu SuJjaDan*), weeping" (19:58). See 41:37; 13:15; 3:83.

016:052 See 39:11. Compare it to the New Testament: "And I say unto you, my friends, Be not afraid of them that kill the body, and after that have no more that they can do. But I will forewarn you whom ye shall fear. Fear Him, which after He hath killed, hath the power to cast into Hell. Yea, I say unto you, fear Him!" (Luke 12:4-5).

016:058 See 46:15.

016:064 The multi-meaning word *litubayyina* is translated falsely to justify the hearsay collections called *Hadith*. See 2:159-160; 3:187 and 16:44.

016:067 It is up to us to benefit from God's blessings or get harmed by abusing them. We can use nuclear energy to destroy cities or to generate electricity; we can burn our hands in fire or cook our food with it.

The Quran brings our attention to the importance of our choices, by comparing alcoholic beverages to fruit juice. While fruit juice is a healthy nutrition, by fermenting them we transform it into a very harmful substance.

016:069 Besides containing numerous vitamins and nutrients, honey is also considered a medicine for some allergies. Pure natural honey is known to be the only food that will never spoil; thus, it is called "miracle food." In low humidity, no organism can live in it and it never decomposes. If it crystallizes, warming it up and mixing it will restore its original texture. Bees are superb architects and engineers; they design and manufacture their honeycomb in hexagonal shapes, which provide optimal efficiency in using wax. Bees are brilliant pharmacists; they produce medicine called honey and poison made of acid and alkaline in their labs. Bees are excellent navigators and talented choreographers; they communicate the distance, the amount, and the direction of flowers, by dancing. Bees are admirable citizens; they work together with thousands of other bees in overpopulated hives in peace and harmony. Bees have great work ethics; they work hard and do not cheat each other. Bees are both pilots and suicide stinger jets; they strike bravely those who occupy their territories. This little insect, with so many qualities bestowed by its Creator, provides us with many signs and lessons.

The Bible contains numerous references to honey and bees. It is found in rocks, (Deuteronomy 32:13; Psalms 81:16); woods (1 Samuel 14:25-26); and strangely, in carcasses of dead animals (Judges 14:8). Its sweetness and nutritional value is praised (Proverbs 16:24; 24:13; 25:16,27); was eaten together with other nutrients, including locusts (Matthew 3:4; Mark 1:6); was used as a religious offering (2Chronthians 31:5); and honey was likened to wisdom and nice words (Proverbs 16:24; Proverbs 24:13-14); to the word of God (Psalms 19:10; 119:103); and to the lips of a strange woman (Proverbs 5:3). Also, see Leviticus 2:11; Genesis 43:11; 1Kings 14:3; Ezekiel 27:17

016:071 What about translating it as, "Yet, those who are privileged do not share their provisions with others. While in fact, they are all equal in terms of provision"? God gives us in different proportions, and He asks us to share it with others. See 30:28.

016:076 See 2:286.

016:086 See 10:28-29.

016:089 For the messenger's complaint about those who do not find God's book sufficient for eternal salvation, see 25:30.

016:093 The middle statement could be understood as "He misleads those who wish so" or "He misleads those He wishes." In fact, both fit the paradox emerging from having freedom and God's knowledge and control. God guides those who seek the truth with honesty through reason and misleads those who do not demonstrate honesty and pure intention, by depriving them from witnessing His great signs in the scripture and in nature (6:110; 7:146; 21:51; 37:84). See 57:22.

016:097 Unless pronouns are restricted by expression or context, the pronoun *Hu* and *Hum* refer to both male and female.

016:098 Why was this instruction not placed in the beginning of the Quran or every chapter? Is it because we are not asked to utter some words but to acquire a proper mental state before studying the Quran?

016.101 See 2:106.

016:103 Holy books besides the Quran have been tampered by clergymen. If we compare the Old and New Testaments, the originality of the Quran will become evident. The Quran might report similar historical events, yet it never reports contradictions or silly stories like the ones inserted into the Bible. For instance, compare Quran's 2:93 with the Bible's Exodus 32:20; or compare Quran's 11:40-44 to the Bible's Genesis, Chapters 6-8 and 1Peter 3:18-20; 2Peter 2:5; 2Pe 3:6. Dr. Maurice Bucaille's book, "*The Bible, The Quran and Science*" contains numerous comparisons between the Quran and the Bible. (See 95:2).

We might translate the end of the verse as, "The language of the one they refer to is unintelligible, while this is perfect and clear" See 43:3.

016:111 In Arabic the pronoun "he" or "she" does not necessarily mean male or female, though the use of "she" is more restricted (for instance, see 16:97,111). Like French, all Arabic words are grammatically feminine or masculine.

016:114 Those who serve God alone do not fabricate religious prohibitions. But those who set up prophets, messengers, popes, rabis, saints, mullahs, sheiks, imams, clerics, and muftis to God, accept and follow many prohibitions falsely attributed to God and His messenger. Thus, they deprive themselves and others from many blessings of God. See 6:145-150 and 42:21.

016:123 See 21:73; 22:78

016:126 See 9:3-29.

بسم الله الرحمن الرحيم

17:0 In the name of God, the Gracious, the Compassionate.

17:1 Glory be to the One who took His servant by night from the Restricted Temple to the most distant temple which We had blessed around, so that We may show him of Our signs. Indeed, He is the Hearer, the Seer.*

17:2 We gave Moses the book and We made it guidance for the Children of Israel: "Do not pick any protector besides Me."

17:3 They were the progeny of those whom We carried with Noah; he was a thankful servant.

17:4 We decreed to the Children of Israel in the book, that you will make corruption twice on earth, and that you will reach the zenith of arrogance.

17:5 So, when the promise of the first one comes to pass, We would send against you servants of Ours who are very powerful, thus they managed to breach your very homes, and this was a promise which has come to pass.

17:6 Then We gave back to you your independence from them, and We supplied you with wealth and children, and We made you more influential.

17:7 If you do good, then it will be good for you, and if you do bad, then so be it. But when the promise of the second time comes, they will make your faces filled with sorrow and they will enter the Temple as they did the first time, and they will strike down all that was raised up.

17:8 Perhaps your Lord will have mercy on you, and if you revert then so will We. We made hell a prison for the ingrates.

17:9 This Quran guides to what is more upright, and it gives glad tidings to those who affirm, those who promote reforms that they will have a great reward.

17:10 For those who do not affirm the Hereafter, We have prepared for them a painful retribution.

17:11 The human asks for evil as he asks for 'good'; the human is prone to be hasty.

17:12 We made the night and the day as two signs, so We erased the sign of night and We made the sign of day manifest, that you may seek bounty from your Lord, and that you may know the number of the years and the count. Everything We have detailed meticulously.

17:13 We have attached every human's deed to his own neck, and We bring forth for him a book on the day of Resurrection which he will find on display.*

17:14 Read your book! Today, you suffice as a reckoner against yourself.

17:15 Whoever is guided is guided for himself, and whoever is misguided is for his own loss. No person shall carry the load of another, and We were not to punish until We send a messenger.*

17:16 If We wish to destroy a town, We allow its privileged hedonists to rule it, then they commit evil in it, then it deserves the punishment, then We destroy it completely.

17:17 How many a generation have We destroyed after Noah? It is enough for your Lord to have knowledge and sight over the sins of His servants.

17:18 Whoever seeks the life of haste, We will hasten for Him what he wishes, then We will make hell for him a place which he will reach disgraced and rejected.

17:19 Whoever seeks the Hereafter and strives for it as it deserves, and is an affirming person, then their effort is appreciated.

17:20 For both groups We will bestow from the bounty of your Lord. The bounty of your Lord is never restricted.

17:21 See how We have preferred some of them over the others; and in the Hereafter are greater levels, and greater preference.

17:22 Do not set up with **God** another god; or you will find yourself disgraced, abandoned.

17:23 Your Lord decreed that you shall not serve except Him; and do good to your parents. When one of them or both reaches old age, do not say to them a word of disrespect nor raise your voice at them, but say to them a kind saying.

17:24 Lower for them the wing of humility through mercy, and say, "My Lord, have mercy upon them as they have raised me when I was small."

17:25 Your Lord is fully aware of what is in yourselves. If you are good, then He is to those who repent a Forgiver.

17:26 Give the relative his due, and the poor, and the wayfarer; and do not waste excessively.

17:27 Those who waste excessively are brothers to the devils, and the devil was an unappreciative of his Lord.

17:28 If you turn away from them to seek a mercy from your Lord which you desire, then say to them a gentle saying.

17:29 Do not make your hand stingy by holding it to your neck, nor shall you make it fully open, so you become in despair and regret.

17:30 Your Lord lays out openly the provision for whom He wishes, and He is able to do so. He is Ever-aware and Watcher to His servants.

17:31 Do not kill your born children out of fear of poverty; We shall provide for you and them. The killing of them was a big mistake.*

17:32 Do not go near adultery, for it is a sin and an evil path.

17:33 Do not kill, for **God** has made this forbidden, except in the course of justice. Whoever is killed unjustly, then We have given his heir authority. Since he received help let him not transgress in the taking of a life.

17:34 Do not go near the orphan's money, except for what is best, until he reaches maturity. Fulfill your oath, for the oath brings responsibility.

17:35 Give full measure when you deal, and weigh with a balance that is straight. That is good and better in the end.

17:36 Do not uphold what you have no knowledge of. For the hearing, eyesight, and mind, all these are held responsible for that.*

17:37 Do not walk in the land arrogantly, for you will not penetrate the earth, nor will you reach the mountains in height.

17:38 All of this is bad and disliked by your Lord.

17:39 That is from what your Lord has revealed to you of the wisdom. Do not make with **God** another god, or you will be cast into hell, blameworthy and rejected.*

17:40 Has your Lord preferred for you the males, while He takes the females as controllers? You are indeed saying a great thing!

17:41 We have cited in this Quran so they may remember, but it only increases their aversion!

17:42 Say, "If there had been gods with Him as they say, then they would have tried to gain a way to the Possessor of Authority."

17:43 Be He glorified and high above from what they say.

17:44 Seven heavens and the earth glorify him and who is in them. There is not a thing, but glorifies his praise, but you do not understand their glorification. He is Compassionate, Forgiving.

17:45 When you study the Quran, We place between you and those who do not affirm the Hereafter an invisible barrier.

Quran Alone

17:46 We place shields over their hearts, that they should not understand it, and deafness in their ears. If you mention your Lord in the Quran alone, they run away turning their backs in aversion.*

17:47 We are fully aware of what they are listening to when they are listening to you, and when they conspire secretly the wicked say, "You are but following a bewitched man!"

17:48 See how they cite the examples for you. They have gone astray, and cannot come to the path.

17:49 They said, "When we are bones and fragments, will we then be resurrected to a new creation?"

17:50 Say, "Even if you be stones or iron."

17:51 "Or a creation that is held dear in your chests." They will say, "Who will return us?" Say, "The One who initiated you the first time." They will shake their heads to you and say, "When is this?" Say, "Perhaps it is near."

17:52 The day He calls you and you respond by His grace, and you think that you only stayed a little while.

17:53 Say to My servants to speak what is best. The devil plants animosity between them. The devil was to the people a clear enemy.

17:54 Your Lord is fully aware of you, if He wishes He will have mercy on you, or if He wishes He will punish you. We have not sent you as a guardian over them.

17:55 Your Lord is fully aware of who is in the heavens and the earth. We have preferred some prophets over others, and We gave David the Psalms.*

17:56 Say, "Call on those you have claimed besides Him. For they have no power to remove harm from you or shift it."

17:57 The ones they call on, they are themselves seeking a path to their

Lord which is nearer, and they desire His mercy, and they fear His retribution. The retribution of your Lord is to be feared!

17:58 There is not a town before the day of Resurrection that We will not destroy it, or punish it a severe punishment. This has been written in the record.

17:59 The rejection of previous people did not stop Us from sending the signs. We sent to Thamud the camel with foresight, but they did her wrong. We do not send the signs except to alert.*

17:60 We had said to you: "Your Lord has encompassed the people." We did not make the vision that We showed you except as a test for the people, and the tree that was cursed in the Quran. We alert them, but it only increases their transgression.*

17:61 We said to the controllers: "Fall prostrate to Adam." So they fell prostrate except for Satan, he said, "Shall I prostrate to one you have created from clay!"

17:62 He said, "Shall I show You this one whom You have preferred over me, that if You respite me until the day of Resurrection, I will destroy his progeny, except for a few."

17:63 He said, "Go, and whoever follows you from them. Hell shall be the reward to you all, a reward well deserved."

17:64 "Mobilize whoever you can with your voice, and mobilize all your forces and men against them, and you may share their money and children, and promise them." But the devil promises nothing but deceit.

17:65 "As for My servants, you will have no power over them." Your Lord suffices as a Caretaker.*

17:66 Your Lord is the One Who drives the ships for you in the sea so that you may seek of His bounty. He is Compassionate towards you.*

17:67 When harm should afflict you at sea, then all those whom you called on besides Him suddenly vanish from you except for Him. So when He saves you to dry land, you turn away. The people are ever ungrateful.

17:68 Are you confident that He will not cause this side of the land to swallow you up, or that He would not send a violent storm against you? Then you will find no caretaker for yourselves.

17:69 Or are you confident that He would not send you back again in it, then He would send against you a violent wind and cause you to drown for your rejection? Then you will not find a pursuer against Us.

17:70 We have honored the Children of Adam and carried them in the land and the sea, and We have provided for them of the good things, and We have preferred them over many of those We created.

17:71 The day We call every people by their record. Then, whoever are given their book by their right, they will read their book, and they will not be wronged in the least.

17:72 Whoever is blind in this, then he will be blind in the Hereafter and more astray from the path.

Even Muhammed Was Tempted to Follow Other than the Quran

17:73 They nearly diverted you from what We revealed to you so that you would fabricate something different against Us, and then they would have taken you as a friend!

17:74 If We had not made you stand firm, you were about to lean towards them a little bit.

17:75 Then, We would have made you taste double the retribution in this life and double the retribution in death. Then you would not find for yourself any supporter against Us.

17:76 They nearly enticed you to drive you out of the land. But in that case, they would have shortly been destroyed after you were gone.*

17:77 This has been the way for Our messengers We sent before you. You will not find any change in Our way.

17:78 You shall maintain the contact prayer at the setting of the Sun, until the darkness of the night; and the Quran at dawn. The Quran at dawn has been witnessed.*

17:79 From the night you shall reflect upon it additionally for yourself, perhaps your Lord would grant you to a high rank.

17:80 Say, "My Lord, admit me an entrance of truth and let me exit an exit of truth, and grant me from Yourself a victorious authority."

17:81 Say, "The truth has come and falsehood has perished. Falsehood is always bound to perish!"

The Same Book, Yet Two Different Effects

17:82 We send down from the Quran what is a healing and mercy to those who affirm. It only increases the wicked in their loss.*

17:83 When We bless the human being, he turns away and turns his side. But when adversity touches him, he is ever in despair!

17:84 Say, "Let each work according to his own. Your Lord is fully aware of who is best guided to the path."

17:85 They ask you concerning the Spirit. Say, "The Spirit is from the command of my Lord, and the knowledge you were given was but very little."*

17:86 If We wished, We would take away what We have revealed to you. Then you would not find for yourself with it against Us a caretaker.

17:87 Except for a mercy from your Lord. His favor upon you has been great.

Fanatic Skeptics Do not Appreciate the Intellectual Challenge

17:88 Say, "If all the humans and the Jinn were to gather to bring a Quran like this, they could not come with it, even if they were helping one another."*

17:89 In this Quran We have cited every example for the people; but most of the people refuse to be anything but ingrates!

17:90 They said, "We will not affirm you until you cause a spring of water to burst from the land"

17:91 "Or that you have a garden of palm trees and grapes, and you cause gushing rivers to burst through it."

17:92 "Or that you make the sky fall upon us in pieces as you claimed, or that you bring **God** and the controllers before us."

17:93 "Or that you have a luxurious mansion, or that you can ascend into the heavens. We will not affirm your ascension unless you bring for us a book that we can study." Say, "Glory be to my Lord. Am I anything other than a human messenger!"

17:94 Nothing stopped the people from affirming when the guidance came to them, except they said, "Has **God** sent a human messenger?"

17:95 Say, "If the earth had controllers walking about at ease, We would have sent down to them from heaven an controller as a messenger."

17:96 Say, "**God** suffices as a witness between me and you. He is Ever-aware and Watcher over His servants."

17:97 Whomever **God** guides is the truly guided one. Whomever He misguides then you will not find for them any allies except for Him. We gather them on the day of Resurrection on their faces, blind, mute, and deaf; their abode will be hell. Every time it dies down, We increase for them the fire.

17:98 Such is their recompense that they rejected Our signs, and they said, "If we are bones and fragments, will we be sent into a new creation?"

17:99 Did they not see that **God** who has created the heavens and the earth is able to create their like? He has made an appointed time for them in which there is no doubt. But the wicked refuse anything except rejection.

17:100 Say, "If you were the ones possessing the vaults of my Lord's mercy, you would have held back for fear of spending. The human being became stingy!"

Nine Miracles to Moses

17:101 We had given Moses nine clear signs. So ask the Children of Israel, when he came to them Pharaoh said, "I think that you, Moses, are bewitched!"

17:102 He said, "You know that no one has sent these down except for the Lord of the heavens and the earth as visible proofs. I think that you Pharaoh are doomed!"

17:103 So he wanted to entice them out of the land. But We drowned him and all those with him.

Relating Back to the Beginning of the Chapter

17:104 We said after him to the Children of Israel: "Dwell in the land, then, when the time of the second promise comes, We will bring you all together as a mixed crowd."*

17:105 It is with truth that We have sent it down, and with truth it came down. We have not sent you except as a bearer of good news and a warner.

17:106 A Quran/Recitation that We have separated, so that you may relate it to the people over time; and We have brought it down gradually.

17:107 Say, "affirm it or do not affirm. Those who have been given the knowledge before it, when it is recited to them, they fall to their chins prostrating.

17:108 They say, "Praise be to our Lord. Truly, the promise of our Lord was fulfilled."

17:109 They fall upon their chins crying, and it increases them in humility.

17:110 Say, "Call on **God** or call on the Gracious. Whichever it is you call on, to Him are the best names." Do not be loud/public in your contact prayer, nor quiet/private; but seek a path in between.*

17:111 Say, "Praise is to **God,** who has not taken a son, nor does He have a partner in sovereignty, nor does He have an ally out of weakness." Glorify Him greatly.

ENDNOTES

017:001 *Hadith* books contain many hearsay stories on this issue. See the note for verse 20:41.

017:013 All our intentions, words, and actions are recorded in a record. We will be impeached by this record on the Day of Judgment. See 57:22.

017:015 This verse confirms the Biblical verse Ezekiel 18:20. However, the Bible also contains verses claiming collective punishment: Exodus 20:5; Lamentations 5:7; Psalms 109:9-10.

017:031 The Arabic word we translated as "children" is *aWLaD* (plural of *WaLaD*) and it literally means "those who are born." The story of pagans killing their daughters by burying them alive years after they were born is a mythology. They were simply practicing infanticide, that is, they would murder their daughters as soon as they learned their sexes. For abortion, see 46:15.

017:036 This verse instructs us to use both our reason and senses to examine all the information we receive. It warns us against blindly following a religious teaching or political ideology. It warns us not to be hypnotized by the charisma of leaders nor by the social conventions. A society comprised of iconoclastic individuals that value rational and empirical inquiry will never become the victim of religious fanaticism or tragedies inflicted by charismatic politicians. A religion or sect that glorifies ignorance and gullibility can be very dangerous for its followers and others. As the Physicist Steven Weinberg once put profoundly, "With or without religion, you would have good people doing good things, and evil people doing evil things. But for good people to do evil things, that takes religion." See 6:74-83.

The following advice is attributed to Buddha. We are not quoting this because we consider his name to be an authority, nor because we are sure that it is uttered or written by a man called Buddha; but because it articulates a profound fact: "Do not put faith in traditions, even though they have been accepted for long generations and in many countries. Do not believe a thing because many repeat it. Do not accept a thing on the authority of one or another of the sages of old, nor on the ground of statements as found in the books. Never believe anything because probability is in its favor. Do not believe in that which you yourselves have imagined, thinking that a god has inspired it. Believe nothing merely on the authority of the teachers or the priests. After examination, believe that which you have tested for yourself and found reasonable, which is in conformity with your well-being and that of others." Interestingly, this is the summary of many verses in the Quran! If the above wise statements were indeed made by Buddha, then he would be the first one who would reject the allegiance of the so-called Buddhists.

The Biblical, "Know the truth, and the truth will set you free!" (John 3:24), is a powerful statement against idolatry and ignorance. St. Paul and his followers, such as Tertullian and many other church leaders turned the wisdom preached by Jesus into bigotry and dogmatism that considered philosophy and philosophers the enemy. Most faithful of religions nod positively at Voltaire's depiction of their understanding: "The truths of religion are never so well understood as by those who have lost the power of reasoning."

If I really want to name myself with suffixes such as an –ist, –ite, or –an, then I should call myself Truthist, Truthite, or Truthian! Or, maybe just, Godist, Godite, Godian!

The Quran repeatedly advises us to use our intellect, to reason, to be open-minded, to be the seekers of truth, to be philosophers, to be critical thinkers, and not to be the followers of our wishful thinking or a particular crowd. For instance, see 2:170, 171, 242, 269; 3:118, 190; 6:74-83; 7:169; 8:22; 10:42, 100; 11:51; 12:2, 111; 13:4, 19; 16:67; 21:10, 67; 23:80; 24:61; 29:63; 30:28; 38:29; 39:9, 18, 21; 40:54; 59:14. See also 6:110.

017:039 The followers of *hadith* and *sunna* claim that Muhammed was given both the Quran and Wisdom, the latter being delivered in *hadith* books. For those who know the nature of *hadith* collections, this is clearly not a wise statement. The verse expressly states that the Wisdom is not separate from the Quran; it is in the Quran. We cannot divorce the Quran from wisdom and make a mockery of "wisdom" by using it to sanctify contradictory hearsay accounts and medieval culture. Besides, the fabricators of *hadith* were not "wise" enough to name their fabrications "Wisdom," since they picked a wrong word, *hadith*, which is negatively treated by the Quran when it is used for any other utterance besides the Quran (6:112-116; 33:38). Furthermore, one of the attributes of the Quran is *hakym*, which means "wise." See 9:31; 17:46; 18:26; 36:1; 42:21.

017:046 Those who do not affirm the hereafter with certainty will not understand the Quran, and they will claim that the Quran is difficult or even impossible to understand on its own. See 54:17,22,32,and 40.

Traditional translations and commentaries somehow separate the word "Quran" from the adjective (*wahdahu* = alone, only) that follows it. They translate it as "Lord alone in the Quran." Though there are many verses emphasizing God's oneness (see

39:45), this verse could be understood as another one emphasizing that message. However, this could be only a secondary meaning of this particular verse, since the adjective *wahdehu* is used not after the word *Rab* (Lord) but after the word Quran. In Arabic if one wants to say, 'Quran alone,' the only way of saying it is "*Quranun wahdahu.*" The mathematical structure of the Quran too confirms our translation. For instance, the word *wahdahu* is used for God in 7:70; 39:45; 40:12,84 and 60:4. If we add these numbers, we get 361, or 19x19. However, if we add 17:46, where the word *wahdahu* is used for the Quran, the total is not in harmony with the great mathematical system.

017:055 The Creator and Master of prophets and messengers has given each special qualities and skills. It is not up to us to put them in a competition that ultimately produces fancy lies and idols. See 2:285.

017:059 Traditional translations render the meaning of "Ma" as a relative pronoun rather than a negative particle, "What stopped us from sending...." The traditional translation implies that God initially sent signs/miracles to convert people and had an intention to send more signs, but He is now disappointed to learn that His signs did not work as expected. Such an implication contradicts the attributes of God detailed in the Quran. Besides, we are repeatedly reminded that the purpose of sending signs/miracles has never been to make people affirm the message, but to support the open-minded people in their rational inquiries. The traditional translation also contradicts other verses of the Quran that remind us that messengers were supported by signs/miracles. Muhammed was not given a sign like previous messengers, but was rather given a unique sign, the Quran, a perpetual sign for every generation until the end of the world. See 29:51. Also, see 7:73-77; 26:155; 54:27; 91:13.

017:060 This might be a reference to Muhammed's journey to receive revelation (17:1 and 53:1-18). The diabolically poisonous tree, *zaqqum*, grows inside hell (37:62-66) and we are informed that it will confuse and expose the hypocrites as unappreciative people since they cannot understand that the hellfire is a metaphor. With their shallow literal understanding, they solidify their dogmatic fanaticism by scoffing at the Quran, "How can a tree grow in the middle of fire?" God uses examples with multiple purposes, thereby guiding those who are appreciative and misleading those who have already made their minds and chosen the wrong path (2:26; 3:7; 17:82; 74:31). No wonder, the Quran is a mercy and reminder for those who wish to take heed (6:157; 7:203; 12:111; 17:82; 29:51; 31:3; 45:20).

017:065 The most efficient weapon against Satan is *la ilaha illa allah*, that there is no god but the one God. See 15:40.

017:066 Water, with its unique chemical and physical properties, is crucial for life. For instance, the maximum density of water being at +4 Celsius degrees, in other words, its expansion in solid form, prevents oceans from freezing from the bottom up. Thus, oceans, with their diverse organisms, function as giant livers cleaning the air in the atmosphere.

017:076 Indeed, soon after being forced to emigrate to Yathrib and turning it into Medina (Civilization), Muhammed and his followers would return to their hometown Mecca, in peace, without bloodshed.

017:078 The time for evening (*esha*) prayer starts from the declining of the Sun in the horizon and lasts until the darkness of the night. The Quran contains sufficient details about the *sala* prayer. See 24:58; 11:1124; 38:32.

017:082 The meaning of this verse has been distorted by clerics so that they could write, package, and sell verses of the Quran as amulets, drugs, aspirin, or band-aids. The placebo effect was enough to keep the market, and Muslim masses started considering the ink and paper where the Quran written as totems. The Quran is not a drug, but a prescription for universal salvation and eternal happiness, including this world. See 10:57.

017:085 Influenced by the Christian scholars who were influenced by Greek mythology, the word *rouh* is commonly mistranslated as "soul." In the Quranic terminology, *rouh* means divine information and commandments. The thing that disconnects from our body or quits functioning during sleep or death is not *rouh* but *nafs*, that is, personality and consciousness (39:42). The Quran refers to the controller that delivers the divine revelation as *Jibreel* or *Rouh-ul Quds*, meaning "Holy Revelation" or "Holy Spirit" (6:102). God endowed the human species with a program called consciousness and personality via a special revelation (15:29; 38:72; 33:9). We can reform or restore our innate program if it is infected by viruses of ignorance, indulgence in sins, or self-inflicted abuses, by re-installing the program called *rouh*. The *rouh* in revelation revives those who have lost the function of the *rouh* in their genetic makeup (6:122; 8:24). See 15:29.

Charlatan psychics or so-called mediums cannot bring the *nafs* (person) that is uploaded to God's master record, back to the world. Communication with the person of the dead is a third-rate fraud and it haunts the gullible who take the pretender serious.

Is it possible to record or transfer the unique neural connections and memories recorded in a brain throughout its life into another organic or inorganic medium, such as a computer disk? In other words, can we one day isolate the information recorded in the brain in terms of a holographic network of connections from its biological material and then transfer, preserve and duplicate it as data? Considering the speed of technological advances, this possibility might even become a probability. The atheists who scoff at God's promise of resurrection by asking, "How can rotten bones be revived?" might learn this lesson in their labs: that with the biological death everything is not lost, and humans and everything in the universe are no more than "bundles of information," and information is not dependent on a particular organic medium; its transferable and theoretically eternal. However, looking at history, it is a safe bet that instead of regretting their bigotry, and affirming their Lord who declared this fact millennia ago, they will continue their opposition, doubts, and arrogant attitudes. Of course, this self-deception will end one day (10:24). See 15:29.

017:088-95 See 29:50-51; 74:1-56.

017:104 Compared to their small population, the Jewish influence is immense in its global platform, financially, politically and culturally. Disproportional to their population, Jews have exhibited astonishing examples in both good and bad, in both success and blunder, and they have enjoyed a vivid presence in world politics for millennia. This explains why the Quran mentions them so frequently both in negative (3:71; 4:46; 5:78-82; 5:60-65; 7:166) and positive terms (2:47,62; 3:33; 5:20).

After being subjected to genocide, atrocious, and tortures by European fascist forces, Jews were scattered around the world as immigrants. Yet, they did not disappear from the global scene or take centuries to recover, as many other nations would do. A prophecy of the Old Testament, Deuteronomy 30:4-20 was fulfilled. Not surprisingly, with the help of the major powers of the time, they were able to establish their own independent state in 1948, soon after their almost utter annihilation; a state not in Germany, but in their historical land, which has once again become the focal point of a global conflict, stirring the world by showcasing human aggression, greed, hatred, cruelty, racism, and terror. As it seems, some are determined to use any means possible to get the promised lands of their ancestors (Genesis 12:1-3; Deuteronomy 1:6-8; 30:4-5; Joshua 1:1-5), while disregarding the warnings and conditions related to the promise (Joshua 1:6-7; Deuteronomy 16:18-20; 30:15-20; 31:16-17).

017:110 The word *ism* comes from the root *SMY* and means "attribute, quality." Regardless of the language, all beautiful attributes may be used for God (the god). Divine attributes do not necessarily define or describe God in an objective sense but define and describe our relationship with our Creator and the mutual expectations. The word *allah* (the god) is a contraction of *al* (the) and *elah* (god). The Quran uses the word *elah* (god) in many verses for *allah* (the god) in the proper context. For instance, in the last chapter of the Quran, God is referred to as "lord of the people." The famous Quranic declaration *la ilaha illa allah*, therefore, can be translated into English as "there is no god but the god". See 7:180, 26:198; 41:44.

The Quran is correcting the tone of the recitation and manner of the *sala* prayer performed by Meccan polytheists. Despite this verse reminding us of a moderate tone of voice and moderate display of prayer, the followers of *hadith* and *sunna*, like their ancestors, either pray loudly (nighttime) or quietly (daytime). Some even use prayer as political demonstration. This verse simply reminds us not to be ashamed of our prostration before God, nor use it as a means of showing off.

بسم الله الرحمن الرحيم

18:0 In the name of God, the Gracious, the Compassionate.

18:1 Praise be to **God** who sent down the book to His servant, and He did not allow any flaw in it.

18:2 Straightforward; giving warning of the severe punishment from Him; and it gives glad tidings to those who affirm and promote reforms that they will have a fair reward.

18:3 In which they will abide eternally.

18:4 To warn those who said, "**God** has taken a son."

18:5 They have no knowledge of this, nor do their fathers. Horrendous indeed is the word coming out of their mouths. They are but telling a lie!

18:6 Perhaps you will torment yourself in grief over them, because they will not affirm this saying at all.

The End of the World

18:7 We have made what is on earth an adornment for it, so that We will test them as to who is better in deeds.

18:8 We will then make what is on it a barren wasteland.*

Prophetic Numerical Signs Hidden in the Cave

18:9 Did you reckon that the dwellers of the cave and the numerals were of Our wondrous signs?*

18:10 When the youths hid in the cave, and they said, "Our Lord, bring us a mercy from Yourself, and prepare for us a right direction in our affair!"

18:11 So We sealed upon their ears in the cave for a number of years.

18:12 Then We roused them to know which of the two groups would be best at calculating the duration of their stay.

18:13 We narrate to you their news with truth. They were youths who affirmed their Lord, and We increased them in guidance.

18:14 We made firm their hearts when they stood and said, "Our Lord, the Lord of heavens and earth, we will not call besides Him any god. Had We said so then it would be far astray."

18:15 "Here are our people, they have taken gods besides Him, while they do not come with any clear authority. Who then is more wicked than one who invents lies about **God**?"

18:16 So when you withdraw from them and what they serve besides **God**, seek refuge in the cave, and your Lord will spread His mercy upon you and prepare for your problem a solution.*

18:17 You see the Sun when it rises, visiting their cave from the right, and when it sets, it touches them from the left, while they are in its spacious hallow. That is from **God**'s signs. Whomever **God** guides is the guided one, and whomever He misguides, you will not find for him any guiding friend.

18:18 You would reckon they are awake while they are asleep. We turn them on the right-side and on the left-side, and their dog has his legs outstretched by the entrance. If you looked upon them, you would have run away from them and you would have been filled with fear of them!

18:19 Thus We roused them so they would ask themselves. A speaker from amongst them said, "How long have you stayed?" They said, "We stayed a day or part of a day." He said, "Your Lord is surely aware how long you stayed, so send one of you with these money of yours to the city, and let him see which is the tastiest food, and let him come with a provision of it. Let him be courteous and let no one take notice of you."

18:20 "If they discover you, they will stone you or return you to their creed. Then you will never be successful."

18:21 Thus We let them be discovered so that they would know that **God**'s promise is true and that there is no doubt regarding the Moment. They argued amongst themselves regarding them, so they said, "Erect a monument for them!" Their Lord is fully aware of them. Those who managed to win the argument said, "We will construct a temple over them."*

18:22 They will say, "Three, the fourth is their dog." They say, "Five, the sixth is their dog," guessing at what they do not know. They say, "Seven, and the eighth is their dog." Say, "My Lord is fully aware of their number, none know them except for a few." So, do not debate in them except with proof, and do not seek information regarding them from anyone.*

18:23 Do not say of anything: "I will do this tomorrow;"

18:24 "Except if **God** wills." Remember your Lord if you forget and say, "Perhaps my Lord will guide me closer than this to the truth."

18:25 They remained in their cave for three hundred years, and increased nine.*

18:26 Say, "**God** is fully aware how long they remained, to Him is the unseen of heavens and earth, He sees and hears. They do not have besides Him any ally, nor does He share in His judgment with anyone."

18:27 Recite what has been revealed to you from your Lord's book, there is no changing His words, and you will not find besides Him any refuge.

Be in Touch with God-conscious People

18:28 Have patience upon yourself regarding those who call on their Lord at dawn and dusk seeking His presence. Let not your eyes overlook them that you seek the attraction of this worldly life. Do not obey the one whom We have made his heart heedless of Our remembrance and he followed his desire, and his case was lost.

Freedom of Opinion and Expression

18:29 Say, "The truth is from your Lord, whoever desires may affirm, and whoever desires may reject." We have prepared for the wicked a fire whose walls will be surrounding them. If they cry out, they are given a liquid like boiling oil which burns their faces. What a miserable place!*

18:30 Those who affirm and promote reforms, We do not waste the reward of those who have done well.

18:31 They will have the gardens of Eden with rivers flowing beneath them, and they will be adorned with bracelets of gold and they will wear green garments of fine silk. They will sit in it on raised thrones. Beautiful is the reward, and beautiful is the dwelling place.

Idolizing One's Wealth

18:32 Give them the example of two men. We made for one of them two gardens of grapes, We surrounded them with palm trees, and We made between them a green field.*

18:33 Both gardens brought forth their fruit, and none failed in the least. We caused a river to pass through them.

18:34 He had abundant fruit, so he said to his friend while conversing with him: "I am better off than you financially, and of great influence."

18:35 He went back into his garden while he had wronged himself. He said, "I do not think that this will ever perish.

18:36 I do not also think that the Moment is coming. If I am indeed returned to my Lord, then I will surely find even better things for me."

18:37 His friend said to him while conversing with him: "Have you rejected the One who created you from dirt, then from a seed, and then He completed you to a man?"

18:38 "But He is **God**, my Lord, and I do not place any partners at all with my Lord."

18:39 "When you entered your garden, you should have said, "This is what **God** has given, there is no power except by **God**." You may see me as having less wealth and fewer children than you."

18:40 "Yet, perhaps my Lord will give me better than your garden, and send upon it a reckoning from the sky, so it becomes completely barren."

18:41 "Or that its water becomes deep under-ground, so you will not be able to seek it."

18:42 So his fruits were ruined, and he began turning his hands for what was destroyed upon its foundations though he has spent on it. He said, "I wish I did not make any partner with my Lord!"

18:43 He had no group which could help against **God**, and he would not have had victory.

18:44 Such is the true authority of **God**. He is best to reward, and best to punish.*

18:45 Give them the parable of this worldly life, like water which We have sent down from the sky, so that the plants of the earth mix with it and it turns dry twigs scattered by the wind. **God** is capable of all things.

18:46 Wealth and buildings are the attraction of this life. But the good deeds that remain behind are better at your Lord for a reward, and better for hope.

18:47 The day We move the mountains, and you see the earth barren, and We gather them, not leaving out anyone of them.

18:48 They are displayed before your Lord as a column: "You have come to Us as We had created you the first time. No, you claimed We would not make an appointment for you!"

18:49 The book was displayed, so you see the criminals fearful of what is in it. They say, "Woe to us!, what is wrong with this book that it does not leave out anything small or large except that it has counted it." They found what they had done present. Your Lord does not wrong anyone.

Do not Follow Your Ardent Enemy

18:50 When We said to the controllers: "Peacefully surrender to Adam." They all surrendered peacefully except for Satan, he was of the Jinn, so he disobeyed the order of his Lord. "Will you take him and his progeny as allies besides Me, while they are your enemy?" Miserable for the wicked is the substitute!*

18:51 I did not make them witness the creation of heavens and earth, nor the creation of themselves. Nor do I take the misleaders as helpers.

18:52 The day when He says: "Call on your partners that you had claimed." So they called them, but they did not respond to them. We made between them a place of destruction.

18:53 The criminals saw the fire; realizing that they would be placed in it. They did not find any way to avert it

The Quran Contains Every Example for Our Salvation

18:54 We have cited in this Quran every example for the people. But the human being is always most argumentative.

18:55 When the guidance came to them nothing prevented the people from affirming, and seeking forgiveness from their Lord, except they sought to receive the ways of the previous people, or receive the retribution face to face.

18:56 We do not send the messengers except as bearers of good news and warners. But those who reject will argue using falsehood to overshadow the truth with it. They took My signs and what they have been warned by as a joke!

18:57 Who is more wicked than one who is reminded of his Lord's signs but he turned away from them, and he forgot what his hands had done. We have made veils upon their hearts from understanding them, and deafness in their ears. If you invite them to the guidance, they will never be guided.

18:58 Your Lord is forgiving, with mercy. If He were to judge the people for what they had already earned, He would hasten for them the retribution. No, they have an appointment, beyond which they will find no escape.

18:59 Such are the towns which We had destroyed when they transgressed. We made for their destruction an appointed time.

A Controller Teaches Lessons to Moses

18:60 Moses said to his youth: "I will not stop until I reach the junction of the two seas, or I spend a lifetime trying."

18:61 But when they did reach the junction between, they forgot their fish, and it was able to make its way back to the sea in a stream.

18:62 When they had passed further on, he said to his youth: "Bring us our lunch; we have found much fatigue in this journey of ours."

18:63 He said, "Do you remember when we rested upon the rock? I forgot the fish, and it was the devil that made me forget to remember it. It made its way back to the sea amazingly!"

18:64 He said, "That is what we have been seeking!" So they went back retracing their steps.

18:65 So they came upon a servant of Ours whom We had given him compassion from Us and We taught him knowledge from Us.

18:66 Moses said to him: "Can I follow you so that you will teach me from the guidance you have been taught?"

18:67 He said, "You will not be able to have patience with me."

18:68 "How can you be patient about what you have not been given any news?"

18:69 He said, "You will find me, **God** willing, to be patient. I will not disobey any command of yours."

18:70 He said, "If you follow me, then do not ask about anything until I relate it to you."

18:71 So they ventured forth until they rode in a boat and he made a hole in it. He said, "Did you make a hole in it to drown its people? You have done something dreadful!"

18:72 He said, "Did I not tell you that you will not be able to have patience with me?"

18:73 He said, "Forgive me for what I forgot, and do not be hard upon my request with you."

18:74 So they ventured forth until they came upon a youth, and he killed him. He said, "Have you killed an innocent person without justice? You have truly come with something awful!"

18:75 He said, "Did I not tell you that you will not be able to have patience with me?"

18:76 He said, "If I ask you about anything after this, then do not keep me in your company. You will then have a reason over me."

18:77 So they ventured forth until they came to the people of a town. They requested food from its people but they refused to host them. Then they found a wall which was close to collapsing, so he built it. He said, "If you wished, you could have asked a wage for it!"

18:78 He said, "For this, we will now part ways. I will inform you of the meanings of those things that you could not have patience over."

18:79 "As for the boat, it belonged to some poor people who were working the sea, so I wanted to damage it as there was a king coming who takes every boat by force."

18:80 "As for the youth, his parents were those who affirm, so we feared that he would oppress them by his transgression and denial."*

18:81 "So we wanted their Lord to replace for them with one who is better than him in purity and closer to mercy."

18:82 "As for the wall, it belonged to two orphaned boys in the city, and underneath it was a treasure for them, and their father was a good man, so your Lord wanted that they would reach their maturity and take out their treasure as a mercy from your Lord. None of what I did was of my own accord. That is the meaning of what you could not have patience for."

The Leader of Two Generations

18:83 They ask you about the one who is from Two Eras, say, "I will recite to you a memory from him."

18:84 We had facilitated for him in the land, and We had given him the means of everything.

18:85 So he followed the means.

18:86 Until he reached the setting of the Sun, and he found it setting at a black water, and he found near it a people. We said, "O Two Eras, either you are to punish, or you are to do them good."*

18:87 He said, "As for he who has transgressed, we will punish him then he will be returned to his Lord and He will punish him an awful punishment."

18:88 "As for he who affirms and does good, then he will have the reward of goodness, and we will speak to him simply of our plan."

18:89 Then he followed the means.

18:90 So when he reached the rising of the Sun, he found it rising on a people whom We did not provide them with coverings against it.*

18:91 So it was, and We knew ahead of time about what he intended.

18:92 Then he followed the means.

18:93 Until he reached the area between the two barriers, he found no one beside it except a people who could barely understand anything said.

The Aggression of Gog and Magog

18:94 They said, "O Two Eras, Gog and Magog are destroyers of the land, and so shall we make a tribute for you that you will make between us and them a barrier?"*

18:95 He said, "What my Lord has given me is far better. So help me with strength and I will make between you and them a barrier."

18:96 "Bring me iron ore." Until he leveled between the two walls, he said, "Blow," until he made it a furnace, he said, "Bring me tar so I can pour it over."

18:97 So they could not come over it, and they could not make a hole in it.

18:98 He said, "This is a mercy from my Lord. But when the promise of my Lord comes, He will make it rubble. The promise of my Lord is truth."

18:99 We left them till that day to surge like waves on one another. The horn was blown so We gathered them together.

18:100 We displayed hell openly on that day to those who do not appreciate.

18:101 Those whose eyes were closed from My remembrance, and they were unable to hear.

18:102 Did those who reject think that they can take My servants as allies besides Me? We have prepared hell for the ingrates as a dwelling place.

18:103 Say, "Shall we inform you of the greatest losers?

18:104 Those whose efforts in the worldly life were wasted while they thought they were doing good!"

18:105 These are the ones who rejected the signs of their Lord and His meeting. So their works were in vain, and We will not give them any value on the day of Resurrection.

18:106 That is their recompense, hell; for what they rejected and for taking My signs and My messengers for mockery!

18:107 Those who affirm and promote reforms, they will have gardens of paradise as a dwelling place.

18:108 Abiding therein. They will not want to be moved from it.

The Quran Has All We Need

18:109 Say, "If the sea were an inkwell for the words of my Lord, then the sea would run out before the words of my Lord run out;" even if We were to bring another like it for its aid.

18:110 Say, "I am but a human being like you, it is revealed to me that your god is One god. So, whoever looks forward to meeting his Lord, then let him promote reforms and not set up any partner in the service of his Lord."

ENDNOTES

018:008-12 The young monotheists, wrongly known as the "seven sleepers," are closely connected to the end of the world, as it is expressed in 18:9-21.

018:009-27 These 19 verses contain numerous key words that strongly imply that the story of the cave has multi-meaning like the verses in chapter 74, and the second meaning is about a prophetic event. I have already observed examples of a hidden mathematical design, and God the Wise will unveil it when the right time comes. When the secrets contained in the cave is revealed by divine permission, it will prove once again the truth in the assertion at 18:27.

018:016-20 If these youngsters escaped from the persecution of the Church that made the polytheistic doctrine of the Trinity the official doctrine of Christianity through the Nicene conference in 325 AC, then they woke up just after Muhammed and his supporters immigrated to Yathrib.

018:021 The numbers related to this event shed light on the question regarding the end of the world. See 18:25.

018:022 The popular number, seven, is not the correct one. Also note that dog is treated as a member of the group, which contradicts numerous hadith reports that depict dog as animals to be shunned and even killed.

018:025 Through a unique expression, 300+9, God provides us with the period of their stay in the cave in both the solar and lunar calendars. Three hundred solar years corresponds to three hundred and nine lunar years. See 15:87 and 72:27.

018:029 Islam recognizes the freedom to reject the message of the Quran. This freedom is guaranteed. See 2:193,256; 10:99; 80:12; 88:21,22.

018:032-42 The Quran gives examples of many false gods, idolized by people, such as children (7:90), religious leaders and scholars (9:31), money and wealth (18:42), dead saints and prophets (16:20,21; 35:14; 46:5,6), and wishful thinking (25:43, 45:23). Satan, in order to infect his subject with polytheism, first introduced a deficient and false definition of polytheism (*shirk*) thereby eliminating any potential self-criticism mechanism. Therefore, polytheists, generally reject the fact that they are polytheists (6:23; 6:148; 16:35). See 19:81.

018:044 In opposition to this verse, Shiite teachings require people to accept the authority of mullahs under the term "*walayat-I faqih,*" that is "the authority of religious scholars."

018:050-51 Perhaps the events that tested God's creation in paradise (38:69), resulted in their classification as controllers, *jinns* and humans. See 2:34.

018:080-81 The lesson we are taught is that we should not be hasty in making judgment of certain events based on limited information. We cannot make a right judgment of characters of an action movie just by looking at a clip of a few seconds.

018:086 This verse is among the handful verses cited by careless critics of the Quran as an example of scientific inaccuracies in the Quran. However, they ignore the obvious fact that the Quran describes the setting of the Sun through the eyes of *Zul-Qarnain*. In other words, it is not used within the context of an objective scientific description but merely a perception of a person. A little search of the Internet will bring millions of occurrences of expressions such as "sun set" or "sun will set" or "sun sets". Of course, none accuses the author of those expressions of believing that the Sun is rotating around the earth. Besides, the Arabic word for "sank" is "*gharaqa*" not "*gharaba*" (set) as is used in the verse. See 4:82.

018:090 Is it a description of a desert?

018:094-99 Because of the link with the end of the world, the verses can be understood as prophetic narratives of future events, rather than historical information. The Quran occasionally uses the past tense as a style of telling future events, to emphasize their inevitability. Can we understand this passage as an international civilization led by a just and peaceful leader that lasted for two generations, and just before the end of the world the two hostile nations will regain their power and attack that civilization?

بسم الله الرحمن الرحيم

19:0 In the name of God, the Gracious, the Compassionate.

19:1 K20H5Y10**A**70**S**90*

Zachariah

19:2 A reminder of your Lord's mercy to His servant Zachariah.

19:3 When he called out to his Lord secretly.

19:4 He said, "My Lord, my bones have gone frail, and my hair has turned white, and I have never been mischievous in calling to You my Lord."

19:5 "I fear the kinfolk after I am gone, and my wife is infertile, so grant me from Yourself an ally."

19:6 "To inherit from me and inherit from the descendants of Jacob. Make him, my Lord, well pleasing."

Yahya (John)

19:7 "O Zachariah, We give you glad tidings of a son whose name/attribute is Yahya (John). We have not used that name/attribute for anyone before."*

19:8 He said, "My Lord, how can I have a son when my wife is infertile, and I have reached a very old age?"

19:9 He said, "It is such that your Lord has said. It is an easy thing for Me. I had created you before when you were not even anything."

19:10 He said, "My Lord, make for me a sign." He said, "Your sign is that you will not speak to the people for three nights consecutively."

19:11 So he went out amongst his people from the temple enclosure, and he indicated to them that they should glorify Him at dawn and dusk.

19:12 "O John, take the book with confidence." We gave him authority while in his youth.

19:13 He had kindness from Us and purity, and he was ever righteous.

19:14 Dutiful to his parents, and never was he a tyrant or disobedient.

19:15 Peace be upon him the day he was born, and the day he dies, and the day he is resurrected alive.*

Maryam (Mary)

19:16 Relate in the book Mary, when she withdrew herself from her family to a place which was to the east.

19:17 She took to a barrier which separated her from them, so We sent Our Spirit to her, and he took on the shape of a human in all similarity.

19:18 She said, "I seek refuge with the Gracious from you if you are a person of affirmation."

19:19 He said, "I am the messenger of your Lord, that He may grant you the gift of a pure son."

19:20 She said, "How can I have a son when no human has touched me, nor do I desire such?"

19:21 He said, "It is such that your Lord has said, it is easy for Me. We shall make him a sign for the people and a mercy from Us. It is a matter already ordained."

The Birth of Esau (Jesus)

19:22 So she was pregnant with him, and she went to deliver in a far place.

19:23 Then the birth pains came to her, by the trunk of a palm tree. She said, "I wish I had died before this, and became totally forgotten!"

19:24 But then he called to her from beneath her: "Do not be sad, your Lord has made below you a stream."*

19:25 "Shake the trunk of this palm tree, it will cause ripe dates to fall upon you."

19:26 "So eat and drink and be happy. If you see any human being, then say, 'I have declared a fast for the Gracious, and I will not talk today to any of the people.'"

19:27 Then she came to her people carrying him. They said, "O Mary, you have come with something totally unexpected!"

19:28 "O sister of Aaron, your father was not a bad man, and your mother was never unchaste!"

19:29 So she pointed to him. They said, "How can we talk to someone who is a child in a cradle?"

19:30 He said, "I am **God**'s servant, He has given me the book and made me a prophet."

19:31 "He made me blessed wherever I was, and He charged me with the contact prayer and towards betterment as long as I am alive."

19:32 "To be dutiful to my mother, and He did not make me a rebellious tyrant."

19:33 "Peace be upon me the day I was born, and the day I die, and the day I am resurrected alive."

19:34 Such was Jesus, the son of Mary, and this is the truth of the matter in which they doubt.

19:35 **God** was never to take a son, be He glorified. If He decrees a matter, then He simply says to it: "Be," and it is.

19:36 **God** is my Lord and your Lord, so serve Him. This is a straight path.*

19:37 So the parties disputed between them. Therefore, woe to those who have rejected from the sight of a terrible day.

19:38 Listen to what they say and watch on the day they come to Us. But the wicked today are in clear misguidance.

19:39 Warn them of the day of remorse. When the matter is decided while they are oblivious, and they do not affirm.

19:40 It is We who will inherit the earth and all that is on it. To Us they will return.

Ibrahim (Abraham), Ishaq (Isaac), Yacob (Jacob)

19:41 Recall in the book Abraham. He was a sincere prophet.

19:42 When he said to his father: "O father, why do you serve what does not hear or see, nor help you in anything?"

19:43 "My father, knowledge has come to me which did not come to you. So follow me that I will guide you to a good path."

19:44 "My father, do not serve the devil. For the devil was ever disobedient to the Gracious."

19:45 "My father, I fear that a retribution will inflict you from the Gracious and that you will become an ally to the devil."

19:46 He said, "Have you abandoned my gods O Abraham? If you do not stop this, I will disown you. Leave me alone."*

19:47 He said, "Peace be upon you, I will ask forgiveness for you from my Lord. He has been most kind to me."

19:48 "I will abandon you and what you call on besides **God**. I will call on my Lord. I hope that I will not be mischievous in calling upon my Lord."

19:49 So when he abandoned them and what they serve besides **God**, We granted him Isaac and Jacob, and We made each a prophet.

19:50 We granted them from Our mercy, and We made for them a tongue of truth to be heard.

Musa (Moses), Ismail (Ishmael), Idris (Enoch), Nuh (Noah)

19:51 Recall in the book Moses; he was loyal, and he was a messenger prophet.

19:52 We called him from the right side of the mount, and We brought him close to talk with.

19:53 We granted him from Our mercy his brother Aaron as a prophet.

19:54 Recall in the book Ishmael; he was truthful to his promise, and he was a messenger prophet.

19:55 He used to instruct his family with the contact prayer and towards betterment, and his Lord was pleased with him.

19:56 Recall in the book Enoch; he was truthful, a prophet.

19:57 We raised him to a high place.

19:58 Those are the ones whom **God** has blessed from amongst the prophets from the progeny of Adam, and those We carried with Noah, and from the progeny of Abraham and Israel, and from whom We have guided and chosen. When the signs of the Gracious are recited to them, they fall down prostrating, and in tears.

19:59 Then generations came after them who lost the contact prayer, and followed desires. They will find their consequences.

19:60 Except for whoever repents and affirms and does good work; they will be admitted to paradise, and they will not be wronged in the least.

19:61 The gardens of Eden, which the Gracious had promised His servants in the unseen. His promise must come to pass.

19:62 They will not hear in it any nonsense, only peace. They will have their provision in it morning and evening.

19:63 Such is paradise that We inherit to any of Our servants who are righteous.

19:64 "We are not sent except by the command of your Lord. To Him is our present and our future, and all that is in-between. Your Lord was never to forget."*

19:65 "The Lord of the heavens and the earth and what is in-between them. So serve Him and be patient in His service. Do you know anything that is like Him?"

Resurrection and Judgment

19:66 The human says: "Can it be that when I am dead, I will be brought out alive?"

19:67 Does the human being not remember that We created him before that and he was nothing at all?

19:68 By your Lord, We will gather them and the devils, then We will place them around hell on their knees.

19:69 Then We will drag out from every group which of them was the worst opposition to the Gracious.

19:70 Then, it is We who are best aware of those who deserve to be burnt therein.

19:71 Every single one of you must pass by it. This for your Lord is a certainty that will come to pass.

19:72 Then We will save those who were righteous, and We leave the wicked in it on their knees.

19:73 When Our clear signs were recited to them, those who rejected said to those who affirmed: "Which of the two groups of ours is more in authority and more prosperous?"

19:74 How many generations have We destroyed before them? They had more wealth and more influence.

19:75 Say, "Whoever is in misguidance, then the Gracious will lead them on." Until they see what they have been promised, either the retribution or the Moment. Then they will know who is in a worst place and weaker in power.

19:76 **God** increases the guidance of those who are guided. The lasting good deeds is better with your Lord as a reward and a far better return.

19:77 Did you see the one who rejected Our signs and said, "I will be given wealth and children."

19:78 Did he look into the future? Or has he taken a pledge with the Gracious?

19:79 No, We will record what he says, and We will increase for him the retribution significantly.

19:80 Then We inherit from him all that he said, and he shall come to Us all alone.

Serving Heroes

19:81 They have taken gods besides **God** to bestow them glory.*

19:82 On the contrary, they will reject their service of them, and they will be standing against them.*

19:83 Did you not see that We send the

devils upon the ingrates to drive them into evil?*

19:84 So do not be impatient; for We are preparing for them by counting.

19:85 The day We gather the righteous to the Gracious as a delegation.

19:86 We drive the criminals to hell as a herd.

19:87 None will possess intercession, except for he who has taken a pledge with the Gracious.*

19:88 They said the Gracious has taken a son!

19:89 You have come with a gross blasphemy.

19:90 The heavens are about to shatter from it, and the earth crack open, and the mountains fall and crumble.

19:91 That they claimed that the Gracious had a son!

19:92 What need does the Gracious have to take a son?

19:93 When all there is in the heavens and the earth will but come to the Gracious as servants.

19:94 He has encompassed them, and counted them one by one.

19:95 All of them will come to Him on the day of Resurrection, all alone.

Appreciative People are Blessed with Love

19:96 As for those who affirm and promote reforms, the Gracious will bestow them with love.

19:97 Thus We have made this easy in your tongue so that you may give good news with it to the righteous and that you may warn with it the quarrelsome people.*

19:98 How many a generation have We destroyed before them? Do you perceive any of them or hear from them a sound?

ENDNOTES

019:001 K20H5Y10A70S90. These combinations of five letters/numbers plays an important role in the mathematical system of the Quran based on code 19. See 74:1-56; 1:1; 2:1; 13:38; 27:82; 38:1-8; 40:28-38; 46:10; 72:28.

019:007 The word *samiyya* has been mistranslated as "name" while it means "likeness" or "similar." The verse does not state the uniqueness of Yahya (John) as a name, but the uniqueness of Yahya's person. See 19:65.

Luke 1:61 makes an assertion regarding the uniqueness of the name, but Matt 11:11 refers to the uniqueness of John (Yahya) in terms of personal qualities, which is what the Quran refers to.

019:015 The repetition of the word day (*yawm*) three times, besides emphasis, is also linked to the numerical structure of the Quran. The word day (*yawm*), in its singular form, occurs exactly 365 times in the Quran. See 9:36.

019:024-25 The recommendation to Mary has medical benefits. The sound of water and mature dates have a regulating effect on human muscles and thus reduce the birth pain. Indeed, some modern birth clinics provide birth-in-water services. The oxytocin found in dates regulates contraction and stimulates lactation, and thus is given to pregnant women as a drug to ease their labor. See 4:82.

This verse gives a clue regarding the season in which Jesus was born. In the Middle East, dates get ripe in the end of September, and in the beginning of October.

Jesus was born in Nazareth, not Bethlehem (*beit*/house+*lahm*/meat) as the majority of Christians believe today. The information in the Gospels regarding the origin of his parents is contradictory; Luke points to Nazareth, while Matthew points to Bethlehem (Matthew 1:18-25; 2:1-12; Luke 2:4-7). Jesus was called the Nazarene in the New Testament (Mark 1:24,10:47,14:67,16:6), yet Christian scholars are not sure why he was called Nazarene. Perhaps, this is because the Old Testament does not mention Nazareth, but mentions Bethlehem (Genesis 35:16,19; 48:7; Ruth 4:11; 1 Samuel 17:12; 2 Samuel 23:13-17). Since we see the Gospel authors trying hard to find as many as prophecies from the Old Testament regarding the coming Messiah, the discrepancy and "mystery" between the name and the birthplace might be explained. The Old Testament praises Bethlehem for being the birthplace of the awaited ruler (Micah 5:2).

019:036 The New Testament contains many verses from the mouth of Jesus rejecting his deity or multiple personality. For instance, in the following verse he could not be clearer in asserting his human nature, since he is equating his position with his audience in relation to God:

"Jesus said unto her, Touch me not; for I am not yet ascended to my Father: but go to my brethren, and say unto them, I ascend unto my Father, and your Father; and to my God, and your God" John 20:17.

Though the four Gospels were selected among hundreds of gospels by the Nicene conference to be the most in harmony with the polytheistic teaching of St. Paul, they still contain many more verses rejecting the idea of the Trinity or deity of Jesus than the few ones considered to be supportive of such blasphemous ideas.

"I cannot do anything of myself. I judge as I hear, and my judgment is honest because I am not seeking my own will but the will of Him who sent me" (John 5:30).

"Jesus said: 'My doctrine is not my own; it comes from Him who sent me'" (John 7:16).

". . . If you loved me, you would be glad that I am going to the Father, for the Father is greater than I" (John 14:28).

"As Jesus started on his way, a man ran up to him and fell on his knees before him. 'Good teacher,' he asked, 'what must I do to inherit eternal life?' Jesus answered, 'Why do you call me good? No one is good--except God alone'" (Mark 10:17-18; Matthew 19:16-17; Luke 18:18-19).

"Whoever welcomes me welcomes, not me, but Him who sent me" (Matthew 10:40; Mark 9:37; Luke 9:48 & John 13:20).

". . . I have not come of myself. I was sent by One who has the right to send, and Him you do not know. I know Him because it is from Him I come; he sent me" (John 7:28-29).

"'The most important one,' answered Jesus, 'is this: Hear, O Israel, the Lord our God, the Lord is one' " (Mark 12:29).

"None of those who call me 'Lord' will enter the kingdom of God, but only the one who does the will of my Father in heaven" (Matthew 7:21).

" . . . Go to my brothers and tell them, 'I am ascending to my Father and your Father, to my God and your God' " (John 20:17).

Also, see 2:59; 3:45,51-52-52,55; 4:11,157,171; 5:72-79; 7:162; 19:36.

019:046 Polytheist (*mushrik*) leaders mentioned in the Quran are belligerent and aggressive characters. Abraham's father excommunicated Abraham and his people tried to kill him; Pharaoh and his people wished to kill Moses, Jews Jesus, *Abu Jahl* and his followers Muhammed. The Quran informs us that many messengers were killed by *mushriks*. Mushrik leaders who know well that their doctrine cannot survive the scrutiny of thinking and questioning minds, have always waged war and terror against messengers of God who invited to reason, questioned traditions and superstitions, and promoted peaceful coexistence. It is no coincidence that similar authoritarian, aggressive, warmongering characteristics are also common among modern polytheists, regardless of their religion.

019:064 See 20:52 and 7:51

019:081 Serving statues is only one of the many ways of committing idolatry or polytheism (7:138). Enjoying mentioning other names besides God's name (39:45), following the religious teachings of clergymen and scholars (9:31; 42:21; 6:148), expecting the intercession of messengers, prophets, and "saints" (2:48; 10:18), and believing that messengers and prophets were infallible (18:110), also fall into the category of idolatry. Despite these clear warnings, since *mushriks* are in a trance under the hypnosis of Satan, they still wish to deceive themselves as monotheists, since they follow a distorted popular message attributed to a popular monotheist (6:23). See 18:32-42.

019:082 This verse has multiple meanings. See 6:23 and 10:28.

019:083 See 14:22.

019:087 The Quran categorically rejects the idea of intercessor as an angel or human "savior" commonly believed by the majority of religious people (2:48; 10:18). In the Day of Judgment, the judgment completely belongs to God. The intercession that the Quran approves is no more than testifying the truth (78:38). Messengers and righteous people cannot save anyone who deserves punishment (9:80; 74:48). Muhammed's only intercession, ironically, will be a negative testimony about those who deserted the Quran by hoping for his intercession (25:30). See 2:48.

019:097 God uses our language, which is limited to our senses and experiences, to communicate His message to us, so that we understand it. Translations made by monotheists serve this purpose. See 41:44.

بسم الله الرحمن الرحيم

20:0 In the name of God, the Gracious, the Compassionate.
20:1 T9H5*
20:2 We did not send down to you the Quran so you may suffer.
20:3 It is but a reminder for the one who takes heed.
20:4 Sent down from the One who created the earth and the heavens above.
20:5 The Gracious, on the throne He settled.
20:6 To Him is what is in the heavens and what is in the earth, and what is in between, and what is underneath the soil.
20:7 If you declare openly what you say, He knows the secret and what is hidden.
20:8 **God**, there is no god but He, to Him are the beautiful names.*

Moses is Commissioned

20:9 Did the news of Moses come to you?
20:10 When he saw a fire, so he said to his family: "Stay here, I have seen a fire, perhaps I can bring from it something, or find at the fire some guidance."
20:11 So when he came to it he was called: "O Moses."
20:12 "I am your Lord, so take off your slippers, you are in the holy valley Tawa."
20:13 "I have chosen you, so listen to what is being revealed."
20:14 "I am **God**, there is no god but Me, so serve Me and maintain the contact prayer for My remembrance."*

The Time of the End is Hinted

20:15 "The moment is coming, I am almost keeping it hidden, so that every person will be recompensed with what it strived."*
20:16 "So do not be deterred from it by he who does not affirm it and followed his desire and perished."

20:17 "What is in your right hand O Moses?"*
20:18 He said, "It is my staff, I lean on it, and I guide my sheep with it, and I have other uses in it."
20:19 He said, "Cast it down O Moses."
20:20 So he cast it down, and it became a moving serpent!*
20:21 He said, "Take it and do not be fearful, We will turn it back to its previous form."
20:22 "Place your hand under your arm, it will come out white without blemish, as another sign."
20:23 "This is to show you Our great signs."
20:24 "Go to Pharaoh, for he has transgressed."

Moses Prays for Eloquence and Addresses Pharaoh

20:25 He said, "My Lord, relieve for me my chest."
20:26 "Make my mission easy."
20:27 "Remove the knot in my tongue."
20:28 "So they can understand what I say."
20:29 "Allow for me an advisor from my family."
20:30 "Aaron, my brother."
20:31 "So that I may strengthen my resolve through him."
20:32 "Share with him my mission."
20:33 "So that we may glorify You plenty."
20:34 "Remember You plenty."
20:35 "You Have been watcher over us."
20:36 He said, "You have been given what you asked O Moses."
20:37 "We Have graced you another time."
20:38 "When We revealed to your mother what was needed to be revealed."
20:39 "That she should cast him in the basket, and cast the basket in the sea, so the sea will place him on the shore, where an enemy of Mine and his will take him. I placed upon you a love from Me and that you shall be raised under My eye."
20:40 "That your sister should follow, and say, "Shall I guide you to a person who will nurse him?" Thus We returned you to your mother, so that she may be pleased and not be sad. You killed a person, but We saved you from harm and We tested you greatly. So it was that you stayed with the people of Midian for many years, then you came here by fate O Moses."
20:41 "I Have crafted you for Myself."*
20:42 "Go, you and your brother with Our signs, and do not linger from My remembrance."
20:43 "Go, both of you, to Pharaoh, for he has transgressed."
20:44 "So say to him soft words, perhaps he will remember or take heed."
20:45 They said, "Our Lord, we fear that he would let loose upon us, or transgress."
20:46 He said, "Do not fear, I am with you, I hear and I see."
20:47 So come to him and say, "We are messengers from your Lord, so send with us the Children of Israel, and do not punish them. We have come to you with a sign from your Lord, and peace be upon those who follow the guidance."*
20:48 "It has been revealed to us that the retribution will be upon he who denies and turns away."
20:49 He said, "So who is the lord of you both O Moses?"
20:50 He said, "Our Lord is the One who gave everything its creation, then guided."
20:51 He said, "What then has happened to the previous generations?*
20:52 He said, "Its knowledge is with my Lord, in a record. My Lord does not err or forget."
20:53 The One who made for you the earth habitable, and He made ways for you in it, and He brought down water from the sky, so We brought out with it pairs of vegetation of all types.
20:54 Eat and raise your livestock, in

that are signs for those of thought.

20:55 From it We created you and in it We return you, and from it We bring you out another time.

Ingrates Confuse Divine Signs with Magic

20:56 We showed him Our signs, all of them, but he denied and refused.

20:57 He said, "Have you come to us to take us out of our land with your magic O Moses?"

20:58 "We will bring you a magic like it, so let us make an appointment between us and you which neither of us will break, a place where we both agree."

20:59 He said, "Your appointment is the day of festival, and when the people start crowding during the late morning."

20:60 So Pharaoh went away, and he gathered his scheming then he came.

20:61 Moses said to them: "Woe to you, do not invent lies about **God**; else the retribution will take you, and miserable is the one who invents."

20:62 So they disputed in their matter between themselves, and they kept secret their council.

20:63 They said, "These are but two magicians who want to take you out of your land with their magic, and they want to do away with your ideal way.*

20:64 "So agree upon your scheme, then come as one front. Whoever wins today will succeed."

20:65 They said, "O Moses, either you cast down or we will be the first to cast down."

20:66 He said, "No, you cast down." So their ropes and staffs appeared from their magic as if they were moving.*

20:67 Moses held some fear in himself.

20:68 We said, "Do not fear, you will prevail."

20:69 "Cast down what is in your right hand; it will consume what they have made. They have only made the work of a magician, and the magician will not succeed no matter what he does."

20:70 So the magicians went down in prostration. They said, "We affirm the Lord of Aaron and Moses."

20:71 He said, "Have you affirmed him before taking my permission? He is surely your great one who has taught you magic. So, I will cut off your hands and feet from alternate sides, and I will crucify you on the trunks of the palm trees, and you will come to know which of us is greater in retribution and more lasting!"

The Determination and Courage of Muslims

20:72 They said, "We will not prefer you over the proofs that have come to us, and over the One who initiated us. So, issue whatever judgment you have, for you only issue judgment in this worldly life."

20:73 "We have affirmed our Lord that He may forgive us our sins, and what you had forced us into doing of magic. **God** is better and everlasting."

20:74 He who comes to his Lord as a criminal, he will have hell, where he will neither die in it nor live.

20:75 He who comes to Him as an affirmer doing good works, for those will be the highest ranks.

20:76 The gardens of Eden with rivers flowing beneath them, abiding therein. Such is the reward of he who is developed.

20:77 We revealed to Moses: "Take My servants out, and make for them through the sea a path. You shall not fear being overtaken, nor be afraid."

20:78 So Pharaoh followed them with his soldiers, but the sea came over them and covered them.

20:79 Thus, Pharaoh misled his people; he did not guide.

20:80 O Children of Israel, We have saved you from your enemy, and We summoned you at the right side of the mount, and We sent down to you manna and quail.

20:81 Eat from the good things that We have provided for you and do not transgress in this, else My wrath will be upon you. Whomever has incurred My wrath is lost.

20:82 I am forgiving for whomever repents and does good work, then is guided.

Children of Israel Follow the Polytheistic Teaching of a Religious Leader

20:83 "What has caused you to rush ahead of your people O Moses?"*

20:84 He said, "They are following my teachings, and I came quickly to you my Lord so you would be pleased."

20:85 He said, "We have tested your people after you left, and the Samarian misguided them."

20:86 So Moses returned to his people angry and disappointed. He said, "My people, did not your Lord promise you a good promise? Has the waiting been too long, or did you want that the wrath of your Lord to be upon you? Thus you broke the promise with me."

20:87 They said, "We did not break the promise by our own will, but we were loaded down with the jewelry of the people, so we cast them down, and it was such that the Samarian suggested."

20:88 He then produced for them a statue of a calf that emitted a sound. So they said, "This is your god and the god of Moses, but he had forgotten!"

20:89 Did they not see that it did not respond to them? Nor did it possess for them any harm or benefit?

20:90 Aaron had told them before: "My people, you are being tested with it. Your Lord is the Gracious, so follow me and obey my command!"

20:91 They said, "We will remain devoted to it until Moses comes back to us."

20:92 He said, "O Aaron, what prevented you when you saw them being astray?

20:93 "Do you not follow me? Have you disobeyed my command?"

20:94 He said, "O son of my mother, do not grab me by my beard nor by my head. I feared that you would say that I divided Children of Israel, and that I did not follow your orders."

20:95 He said, "So what do you have to say O Samarian?"

20:96 He said, "I noticed what they did not notice, so I took a portion from the teaching of the messenger, and I cast it away. This is what my person inspired me to do."*

20:97 He said, "Then be gone, for you will have it in this life to say, "I am not to be touched." You will have an appointed time which you will not forsake. Look to your god that you remained devoted to, we will burn him, then we will destroy him in the sea completely."

Only One God

20:98 Your god is **God**; whom there is no god but He. His knowledge encompasses all things.

20:99 It is such that We relate to you the news of what has passed. We have given you from Us a remembrance.

At God's Presence

20:100 Whoever turns away from it, then he will carry a load on the day of Resurrection.

20:101 They will remain therein, and miserable on the day of Resurrection is what they carry.

20:102 The day the horn is blown, and We gather the criminals on that day bleary-eyed.

20:103 They whisper amongst themselves: "You have only been away for a period of ten."

20:104 We are fully aware of what they say, for the best in knowledge amongst them will say, "No, you have only been away for a day."

20:105 They ask you about the mountains, say, "My Lord will annihilate them completely."

20:106 "Then He will leave it as a smooth plain.

20:107 "You will not see in it any crookedness or curves."

20:108 On that day, they will follow the caller, there is no crookedness to him. All voices will be humbled for the Gracious, you will not be able to hear except whispers.

20:109 On that day, no intercession will be of help, except for he whom the Gracious allows and accepts what he has to say.*

20:110 He knows their present and their future, and they do not have any of His knowledge.

20:111 The faces shall be humbled for the Eternal, the Living. Whoever carried wickedness with him has failed.

20:112 Whoever does any good works, while he is an affirmer, then he should not fear injustice nor being given less than his due.

20:113 It was such that We sent it down as an Arabic compilation, and We cited in it many warnings, perhaps they will become aware or it will cause for them a remembrance.*

Do not Rush in Understanding the Quran

20:114 Then High above all is **God**, the King, the Truth. Do not be hasty with the Quran before its inspiration is completed to you, and say, "My Lord, increase my knowledge."*

Adam Deceived by Satan

20:115 We had made a pledge to Adam from before, but he forgot, and We did not find in him the will power.*

20:116 We said to the controllers: "Submit to Adam." They all submitted except for Satan, he refused.

20:117 So We said, "O Adam, this is an enemy to you and your mate. So do not let him take you out from the paradise, else you will have hardship."

20:118 "There you will not go hungry nor need clothes."

20:119 "There you will not go thirsty nor suffer from heat."

20:120 But the devil whispered to him, he said, "O Adam, shall I lead you to the tree of immortality and a kingdom which will not waste away?"

20:121 So they both ate from it, and their sin became apparent to them, and they began to place leaves on themselves from the paradise. Adam had disobeyed his Lord, and had gone astray.

20:122 Then his Lord recalled him, and He forgave him and gave guidance to him.

20:123 He said, "Descend from this, all of you, for you are enemies to one another. So, when My guidance comes to you; then, whoever follows My guidance, he will not go astray nor suffer."

Ingrates Create a Miserable Life for Themselves

20:124 "Whoever turns away from My remembrance, then he will have a miserable life, and We will raise him blind on the day of Resurrection."

20:125 He said, "My Lord, you have raised me blind while I used to be able to see?"

20:126 He said, "It was the same when Our signs came to you, you ignored them, and similarly today you will be ignored."

20:127 It is such that We recompense he who transgresses, and did not affirm His Lord's signs. The retribution of the Hereafter is more severe and lasting.

20:128 Is it not a guide to them how many generations We had destroyed before them, which they are walking now in their homes? In that are signs for the people of understanding.

20:129 Had it not been for a word already given by your Lord, they would have been held to account immediately.

20:130 So be patient to what they are saying and glorify the grace of your Lord before the rising of the Sun, and before its setting, and from the early part of the night glorify, and at the edges of the day that you may be content.

20:131 Do not strain your eyes on what We have given other people as luxuries of this worldly life. We are testing them with it; and the provision of your Lord is better and more lasting.

20:132 Instruct your family with the contact prayer, and be patient for it. We do not ask you for provision, for We provide for you. The good end is for the righteous.

Proof and Messengers

20:133 They said, "If only he would bring us a sign from his Lord!" Did not proof come to them from what is in the previous book?*

20:134 If We had destroyed them with retribution before this, they would have said, "Our Lord, if only You had sent us a messenger so we could follow Your signs before we are humiliated and shamed!"

20:135 Say, "All are waiting, so wait, and you will come to know who the people upon the balanced path are and who are guided."

ENDNOTES

020:001 T9H5. This combination of two letters/numbers plays an important role in the mathematical system of the Quran based on code 19. See 74:1-56; 1:1; 2:1; 13:38; 27:82; 38:1-8; 40:28-38; 46:10; 72:28.

020:008 See 7:180.

020:014 This short verse not only informs us that *sala* prayer did not start with Muhammed, but also provides important guidelines for correcting the many errors committed in traditional prayers. For instance, in our *sala* prayer we should not mention Muhammed or other names.

020:015 The information regarding end of the world is given by God (15:87).

020:017 Surely, God knew what was in Moses' hand. The following verses show that the question was a rhetorical one. See the Old Testament, Exodus 4:1-9.

020:020 Where is the Moses who we met at 18:60-82, who was so inquisitive? Here, when he hears God commanding him to throw away his valuable possession, his staff, he does not even utter one word demanding an explanation. What could be the reason for this change?

020:041 If similar statements used for Moses were used for Muhammed, they would become the most popular Quranic statements and would adorn the walls of every mosque. The Sunni and Shiite worshipers of Muhammed try hard to put Muhammed above all other prophets and messengers through fabricated *hadith*. According to a very popular *hadith*, God supposedly had said to Muhammed, *Lawlaka lawlaka lama khalaqtul aflaka* (if it were not you, if it were not you, I would not have created the universes). Under the name *Qudsi Hadith* (Holy *Hadith*), clerics and their blind followers attribute many false statements directly to God. If you study the longest story told in Bukhari, there you will find Muhammed ascending and descending numerous times between the sixth and seventh heavens, between Moses and God, trying to get as much of a discount possible, in the number of prayers. This bargain between God and Muhammed, which was supervised by Moses, is reminded to the public at least once a year during the celebration of the "Night of Miraj." The message given to the audience is simple and polytheistic: "If it were not for Muhammed's bargain on your behalf, you would end up praying 50 times a day (one session every 28 minutes, day and night); and you would suffer greatly." Whether this story was originally made up by a Jew to insult the intelligence of Muhammed and exalt Moses, it is adopted by gullible Muhammedans to praise the mercy of their prophet (more merciful than God!), and to justify their belief in his power of intercession and negotiation with

God on their behalf. The reporters and collectors of this story, including Bukhari, were the archenemies of prophet Muhammed and his monotheistic message. See 2:285; 3:159; 6:112; 68:42.

For the struggle, leadership, and message of this brave servant of God, see the books of Exodus and Deuteronomy in Old Testament. Psalms and Prophets too contain numerous references to Moses. For instance, see Jeremiah 1:5. The New Testament quotes Jesus comparing people's reaction to his message to their reaction to the message delivered by Moses (John 5:46). However, Luke, in Acts 7:37, tries to make a connection between Jesus and "the prophet like Moses" prophesied in Deuteronomy 18:15,18-19. A comparative study on the history of Moses, Jesus, and Muhammed will show that Muhammed was much more like Moses than Jesus was. From their birth to their political leadership, Muhammed resembles Moses much more than Jesus does.

020:047-49 Without repeating the divine instructions dictated to Moses, we are presented with Pharaoh's answer. The language and narrative of the Quran is deliberately left with gaps, so that the reader can actively participate in its comprehension through sound inferences and harmless imagination. Thus, not only is "unnecessary" repetition avoided, but the brain of the reader is stimulated. The Quran does not like a passive reader, and a lazy reader does not enjoy the Quran either. For other examples of scene overlaps, see 12:83; 26:16-18.

020:051-52 Modern polytheists and unappreciative people, when they do not have the intellectual courage to accept truth, in order to avoid affirming the truth, ask exactly the same question Pharaoh asked millennia ago. While delivering the monotheistic message of the Quran, while sharing the signs and miracles of the Quran, if you encounter similar excuses, you should use the same answer given by Moses in verse 20:52.

Verse 20:52 also rejects the distorted or mistranslated Biblical verse depicting God to be like a fallible human being regretting his mistakes or forgetting things (Genesis 6:6-7; Psalms 13:1; Lamentations 5:20). See 7:180.

020:063 Disinformation and misinformation is a common propaganda tool of all despots and corrupt leaders.

020:066 For the definition of magic, see 7:116 and 2:102.

020:083-98 Aaron is criticized for being a passive bystander here. However, the Biblical version of the story puts Aaron in the rank of idolaters. See Old Testament, Exodus 32:2.

020:096 Ignoring common sense, the context of the verse, and the semantics of the Quran, traditional translations and commentaries create a bizarre story through the mistranslation of the words *AThaR*, and *NaBaZa.* To accommodate the mistranslation, they also add several non-existing and non-implied words, such as, "Muhammed," "Gabriel," "Dust," and "into calf." For a detailed discussion of this verse, see the Sample Comparisons section in the Introduction.

020:109 This verse rejects the "savior" role attributed to prophets and righteous people. 2:48; 19:87.

020:113 For another meaning of this verse, see 43:3.

020:114 A Quranic statement or phrase that we do not understand upon a first reading might become clear upon the second or later readings. To understand some verses, one might need to read the entire Quran and get its perspective. We might sometimes understand a verse years later with an instantaneous inspiration. If we use the analogy of a radio, the Quran is like a radio station with a powerful and clear broadcast, and we are like a radio device receiving its message. The false doctrines and practices we inherit from the past and the environment might create static interference or background noise and prevent us from hearing the divine message more clearly. Besides, our tuning, the condition of the transistor, the power of our battery, and the direction of our antenna, all the factors that are affected by our attitude, prejudice, interest and knowledge level, might prevent us from hearing well, or understanding well. The radios that are made by "*hadith* and *sunna*" have defective designs, and the radios that automatically tune to the "*hadith* and *sunna*" station while also trying to tune to the Quran station, have too much noise; they all condemn themselves to receive the message with plenty of noise and interruptions (2:20;).

020:115-123 See 2:36-39; 7:19-25; 20:115-123.

020:133 The truth-value of the Quran's message and its historical dimension provide evidence for its divine nature. The occurrence of the singular word *bayyina* (evidence/proof) in the Quran being exactly 19 times and the existence of the code 19 in previous scriptures is another fulfillment of this verse. See 46:10.

بسم الله الرحمن الرحيم

21:0 In the name of God, the Gracious, the Compassionate.

21:1 The judgment of people has come near, while they are turning away unaware.

21:2 When a reminder comes to them from their Lord that is new, they listen to it while playing.*

21:3 Their hearts are preoccupied, and those who are wicked confer privately: "Is he not a human being like you? Would you accept this magic while you know?"

21:4 Say, "My Lord knows what is said in the heavens and in the earth, and He is the Hearer, the Knower."

21:5 They said, "No, these are just bad dreams; no, he made it up; no, he is a poet. So let him bring us a sign like those who were sent before."

21:6 None of the towns which We destroyed before them had affirmed. Will they affirm?

21:7 We did not send before you except people whom We revealed to. So ask the people of the remembrance if you do not know.*

21:8 We did not make for them bodies that do not need to eat, nor were they immortal.

21:9 Then We fulfilled the promise to them, so We saved them and whom We pleased, and We destroyed those who transgressed.

21:10 We have sent down to you a book in which is your remembrance. Do you not reason?

21:11 How many a town have We destroyed because of its wrongdoing, and We established after them a different people?

21:12 So it was that when they

perceived Our power, they were running from it.

21:13 "Do not run; come back to your lavish lifestyle and your homes, so that you will be questioned!"

21:14 They said, "Woe to us, we have been wicked!"

21:15 So that remained as their cry until We took them all, and they became still.

21:16 We did not create the heavens and the earth and what is between them for entertainment.

21:17 If We wanted to be amused, We could have done so from what is already with Us, if that is what We wished to do.

21:18 No, We cast the truth upon the falsehood, so it crushes it, and then it is refuted. Woe to you for what you have described.

God has no Partner in His Dominion

21:19 To Him is whoever is in the heavens and in the earth. Those who are near Him are not too proud to serve Him, nor do they complain.

21:20 They glorify in the night and the day; they do not cease.

21:21 Or have they taken gods from the earth who can resurrect?*

21:22 If there were gods in them except for **God**, then they would have been ruined. Glory be to **God**, the Lord of the dominion from what they describe.

21:23 He is not questioned about what He does, while they will be questioned.

21:24 Or have they taken gods besides Him? Say, "Bring your proof. This is a reminder for those with me and a reminder for those before me." But most of them do not know the truth, so they turn away.

21:25 We did not send any messenger before you except that We revealed to him: "There is no god but Me, so serve Me."

21:26 They said, "The Gracious has taken a son!" Be He glorified; they are all but honored servants.

21:27 They do not speak ahead of Him, and on His command they act.

21:28 He knows their present and their future, and they cannot intercede unless it is for those whom He is pleased with. They stand in awe and reverence of Him.*

21:29 Whoever of them says: "I am a god besides Him," then We will punish that person with hell. It is such that We punish the wicked.

The Big Bang; Motion and Stability

21:30 Did those who reject not see that the heavens and the earth were one mass, and We tore them apart? That We made from the water everything that lives. Will they not affirm?*

21:31 We made on the earth stabilizers so that it would not tumble with you, and We made in it wide paths that they may be guided.

21:32 We made the sky a protective ceiling. Yet they are turning away from Our signs!*

21:33 He is the One who created the night and the day, and the Sun and the Moon, each swimming in an orbit.*

Mortality

21:34 We did not give immortality to any human that came before you. If you are going to die, would they be immortal?

21:35 Every person will taste death. We burden you with evil and good as a test, and it is to Us that you will return.

21:36 When those who reject see you, they take you for mockery: "Is this the one who speaks about your gods!" While in the remembrance of the Gracious they are ingrates!

21:37 The human being is made of haste. I will show you My signs, so do not be in a rush.*

21:38 They say, "When will this promise come to pass if you are truthful?"

21:39 If only those who reject knew, that they will not be able to ward off the fire from their faces, nor from their backs, nor will they be helped.

21:40 No, it will come to them suddenly. They will not be able to turn it away, nor will they be delayed.

21:41 Messengers before you have been mocked, but those who mocked were then surrounded by the object of their mockery!

21:42 Say, "Who can protect you during the night and the day from the Gracious?" No, they are turning away from the remembrance of their Lord.

21:43 Or do they have gods that will protect them from Us? They cannot help themselves, nor can they be protected from Us.

21:44 It was Us who gave luxury to these and their fathers, until they grew old with age. Do they not see that We come to the land and make it shrink from its edges? Will they be able to win?

Merely a Warner

21:45 Say, "I am merely warning you with the inspiration." But the deaf do not hear the call when they are being warned.

21:46 If a breath of your Lord's torment touches them, they will say, "Woe to us, we have been wicked!"

21:47 We will place the scales of justice for the day of Resurrection, so that no person will be wronged in the least. Even if it was the weight of a mustard seed, We will bring it. We are sufficient as a Reckoner.

21:48 We had given Moses and Aaron the Criterion, a shining light, and a reminder for the righteous.

21:49 Those who revere their Lord, even when unseen, and they are wary of the Moment.

21:50 This is a blessed reminder which We have sent down. Will you be deniers of it?

Young Abraham Teaches Through a Prank

21:51 Before that We gave Abraham his understanding, and We were aware of him.*

21:52 As he said to his father and people: "What are these statues to which you are devoted?"

21:53 They said, "We found our fathers serving them."

21:54 He said, "You and your fathers have been clearly misguided."

21:55 They said, "Have you come to us with the truth, or are you simply playing?"

21:56 He said, "No, your lord is the Lord of heavens and earth, the One who initiated them. I bear witness to such."*

21:57 "By **God**, I will scheme against your statues after you have gone away and given your backs."

21:58 So he broke them into pieces except for the biggest of them, so that they may turn to him.*

21:59 They said, "Who has done this to our gods? He is surely one of the wicked."

21:60 They said, "We heard a young man mentioning them. He was called Abraham."

21:61 They said, "Bring him before the eyes of the people so that they may bear witness."

21:62 They said, "Did you do this to our gods O Abraham?"

21:63 He said, "It was the biggest one of them here who did it, so ask them, if they do speak!"

21:64 So they turned and said to themselves: "It is indeed ourselves who have been wicked!"

21:65 Then they reverted to their old ideas: "You know that they do not speak!"

21:66 He said, "Do you serve besides **God** what does not benefit you at all nor harm you?"

21:67 "I am fed-up with you and to what you serve besides **God**! Do you not reason?"

21:68 They said, "If you are to do anything, then burn him, and give victory to your gods."

21:69 We said, "O fire, be cool and safe upon Abraham."

21:70 They wanted to harm him, but We made them the losers.

Lot, Isaac, and Jacob

21:71 We saved him and Lot to the land which We have blessed for the worlds.

21:72 We granted him Isaac and Jacob as a gift, and each of them We made a good doer.

21:73 We made them leaders who guide by Our command, and We revealed to them to promote reforms and maintain the contact prayer and contribute towards betterment, and they were in service to Us.*

21:74 And Lot, We gave him wisdom and knowledge, and We saved him from the town that used to do vile things. They were a people of evil, corrupt.

21:75 We admitted him into Our mercy. He was of the good doers.

Noah

21:76 And Noah when he called out before that, thus We responded to him, and We saved him and his family from the great distress.

21:77 We granted him victory against the people that denied Our signs. They were a people of evil, so We drowned them all.

David and Solomon

21:78 And David and Solomon, when they gave judgment in the case of the crop that was damaged by the sheep of the people, and We were witness to their judgment.

21:79 So We gave Solomon the correct understanding. We have given wisdom and knowledge to each of them. We directed the mountains with David to praise as well as the birds. This is what We did.

21:80 We taught him the making of armor for you to protect you from your enemy. Are you then thankful?

21:81 To Solomon the gusting winds run by his command all the way to the land which We have blessed. We were aware of everything.

21:82 From the devils are those who dive for him, and they perform other tasks, and We were guardian over them.

21:83 And Job when he called his Lord: "I have been afflicted with harm, and you are the Compassionate!"

21:84 So We answered him, and We removed what was afflicting him, and We brought him back his family and others with them as a mercy from Us and a reminder to those who serve.

Ishmail, Enoch, Isaiah, Jonah, and Zachariah

21:85 Ishmael and Enoch and Isaiah, all of them were patient.

21:86 We admitted them in Our mercy, they were of the good doers.

21:87 The One Who Has the N (JoNah) when he went off in anger, and he thought that We would not be able to take him. Then he called out in the darkness: "There is no god but You! Glory to You, I was of the wicked!"*

21:88 So We responded to him, and We saved him from distress. It is such that We save those who affirm.

21:89 And Zachariah when he called out to his Lord: "My Lord, do not leave me without an heir, and You are the best inheritor."

21:90 So We responded to him, and We granted him John, and We cured his wife. They used to hasten to do good, and they would call to Us in joy and in fear. To Us they were reverent.

21:91 The one who protected her chastity, so We blew into her from Our Spirit, and We made her and her son a sign for the worlds.

21:92 This is your nation, one nation, and I am your Lord so serve Me.

21:93 They disputed in the matter amongst themselves. Each of them will be returned to Us.

21:94 So whosoever does good work and he is an affirmer, then his efforts will not be rejected and We will record it for him.

Gog and Magog

21:95 It is forbidden for a town that We destroy that they would return;

21:96 Until Gog and Magog is opened, and from every elevated place they will race forth.

21:97 The promise of truth draws near. Then, when it is seen by the eyes of those who rejected: "Woe to us, we have been oblivious to this. Indeed, we were wicked!"*

21:98 Both you and what you serve besides **God** shall be fuel for hell; you will enter it.

21:99 If these were gods, then they would not have entered it! All will abide therein.

21:100 They will be breathing heavily in it, and they will not be heard therein.

21:101 As for those who deserved good from Us, they will be removed far away from it.

21:102 They shall not hear the slightest sound from it, and they will be in what their person desires abiding therein.

21:103 The great horror will not sadden them, and the controllers will receive them: "This is your day which you have been promised."

The Universe Will Collapse Back into a Singularity

21:104 On the day when We roll up the heavens like a scroll of books is rolled up. As We initiated the first creation, so shall We return it. It is a promise of Ours that We will do this.

21:105 We have written in the Psalms: 'After the reminder, the earth will be inherited by My servants who do good.'*

21:106 In this is a proclamation for a people who serve.

21:107 We have not sent you except as a mercy to the worlds.*

21:108 Say, "It is revealed to me that your god is but One god, so will you peacefully surrender to Him?"

21:109 So if they turn away, then say, "I have given you notice sufficiently, and I do not know if what you are promised is near or far."

21:110 "He knows what is spoken publicly and He knows what you keep secret."

21:111 "For all I know, it may be a test for you and an enjoyment for a while."

21:112 Say, "My Lord, judge with truth. Our Lord, the Gracious, is sought for what you describe."*

ENDNOTES

021:002-3 The Quran is the only miracle given to Muhammed (29:51). Muhammed's *mushrik* companions could not comprehend that a book could be a miracle, and they wanted miracles "similar" to the ones given to the previous prophets (11:12; 17:90-95; 25:7,8; 37:7-8). Modern *mushriks* also demonstrated a similar reaction when God unveiled the prophesied miracle in 1974. When the miracle demanded from them the dedication of the system to God alone, and the rejection of all other "holy" teachings they have associated with the Quran, they objected, "How can there be mathematics in the Quran; the Quran is not a book of mathematics" or, "How can there be such a miracle; no previous messenger came up with such a miracle!" When a monotheist who was selected to fulfill the prophecy and discover the code started inviting his people to give up polytheism and the worship of Muhammed and clerics, he was declared an apostate by Muslim scholars in 1989, and within a year he was assassinated by a group linked to al-Qaida in early 1990 in Tucson, Arizona. See 3:81; 40:28-38; 72:24-28; 74:1-56. For the prophetic use of the word reminder (*ZKR*), see 15:9; 21:2; 21:24,105; 26:5; 29:51; 38:1,8; 41:41; 44:13; 72:17; 74:31,49,54.

021:007 See the note for 7:46 for why we preferred *people* rather than *men*. The Old Testament mentions several women messengers, such us, Ruth.

021:021-22 See 6:74-83.

021:028 The common belief in

intercession is a polytheistic doctrine. See, 2:48; 43:86.

021:030 The creation of the universe from a singularity about 13.7 billion years ago with a Big Bang has been supported by many scientific observations and calculations since the beginning of the early 20th century. The Quran stated this fact unequivocally 14 centuries before. Furthermore, the Quran supports the closed model of an expanding universe (21:104; 51:47). We are also informed that the early stage of the universe was gas (41:11). The vitality of water for carbon-based biological life is evident. See 4:82.

021:032 If the word *sama* refers to space, then we should reflect on the force of gravity and counter forces that keep billions of galaxies containing billions of stars and planets in harmonious motion in space. If the word refers to the atmosphere, then we may understand it as the transparent blanket that protects us from extreme heat and cold, a transparent armor that protects us from meteorites, and a transparent filter that protects us from harmful rays. The iron core of the earth works like a dynamo and creates a powerful magnetic field that holds the atmosphere. See 4:82

021:033 This verse clearly informs us that all (*kul*) celestial bodies rotate in an orbit. Interestingly, the earth's rotation around its axis is also implied by the expression "night and day." See 27:88; 36:40; 39:5; 68:1; 79:30.

021:037 See 4:28.

021:051 The planet earth was created to provide a new chance for us to redeem ourselves. When controllers questioned the wisdom in our creation, God told them that He knows what they did not know (2:30).

021:056 The testimony mentioned here is testimony through reason and evidence. See 12:26-28.

021:058 The breaking of all the statues (with the exception of the biggest one) by the young Abraham was not a tyrannical action to deprive the freedom of idol worshipers, nor was it an expression of his hate. He was not in a position of authority to be a tyrant, and if he was pretending to be one, he would not have spared the biggest statue. He was using this tactic for didactical purposes. Being a member of a polytheistic society that regressed into the darkness of superstitions and dogmatism, perhaps Abraham did not have any other choice but to draw their attention with some action. Instead of showing the courage and wisdom of reflecting on that youngster's rational argument and questioning their inherited religion, the polytheists reacted with the common reaction of polytheists: group-think, threat, excommunication, and violence. See 6:74-83; 29:25.

021:073 During the revelation of the Quran, with some distortions, the rituals and duties revealed to Abraham were known knowledge. The religious people in Mecca had the knowledge of fasting, *sala*, *zaka*, and Debate Conference, since the Quran uses the words in their language. See 2:128; 16:123; 22:78.

021:087 Yunus' (Jonah) name occurs in the Quran four times as Jonah. However, in the chapter that starts with the letter N or number 50, Jonah is referred to by the expression *Sahib ul-Hut*, that is, The Companion of the Fish (68:48). Referring to a messenger who has the letter N in his name, with an expression that does not contain the letter N, in a chapter where the letter N is counted (133 = 19 x 7), is a part of the Quran's mathematical system. Indeed, a unique expression in verse 21:87 supports this implied message.

021:097 In the light of the last verses of chapter 18, can we speculate the following: in the future some Muslim nations will create a union under the leadership of a messenger, forming a federal government, creating a peaceful, free and just society under Islamic principles (18:83-101; 9:33; 41:53; 48:28; 61:9). Before the end of the world, two rival nations will attack the Muslim community, thereby fulfilling the last condition for the end of the world (15:87; 18:94; 21:96,97).

021:104 In the beginning, the entire universe was an infinitely dense small seed.

At that point, there was no atom, the building block of matter, no time, and no space. Everything came into existence from a singularity, or nothingness, about 13.7 billion years ago. Verse 30 of this chapter informed us that the universe was created after a cosmic rapture, which we call the Big Bang, and now 74 verses later, we are informed about its ending, in a big crunch in reverse motion. This verse supports the model known as "closed universe." See 4:82.

021:105 This earth and these heavens will be replaced by new ones and will be given to the service of the righteous people (14:48). The verse may also be translated as, "We wrote in Psalms after the Reminder, that my righteous servants will inherit the land." The derivatives of the word "reminder" (*ZKR*) are used to describe the message of the Quran and previous books, and they also have prophetic implications. For instance, see 15:9; 21:2; 21:24; 26:5; 29:51; 38:1,8; 41:41; 44:13; 72:17; 74:31,49,54.

021:112 The mathematical system of the Quran leads us to read the word *QWL* not as a past tense verb, but as a declarative one. Our choice is supported by the original spelling of the word, which has no letter "A" after the first letter. Furthermore, when read starting from verse 108, our choice fits best to the flow of the language. This is also supported by the numerical semantics (nusemantics) of the Quran: the frequency of the word *QaLu* (they said), as used for God's creation, is equal to the frequency of the divine command *QuL* (say), both being 332 times. In other words, for each of what "they say," we have a refutation. These two words are not necessarily used in successive pairs.

بسم الله الرحمن الرحيم

22:0 In the name of God, the Gracious, the Compassionate.

22:1 O people, be aware of your Lord, for the quaking of the Moment is a terrible thing.

22:2 The moment you see it, every nursing mother will leave her suckling child, and every pregnant one will miscarry, and you will see the people drunk while they are not drunk, but the retribution of **God** is most severe.

22:3 Among the people there are those who argue regarding **God** without knowledge and follow every rebellious devil.

22:4 It was decreed for him that any who follow him, he will mislead him, and guide him to the retribution of fire.

22:5 O people, if you are in doubt as to the resurrection, then We have created you from dirt, then from a seed, then from an embryo, then from a fetus developed and undeveloped so that We make it clear to you. We settle in the wombs what We wish to an appointed time, then We bring you out a child, then you reach your maturity, and of you are those who will pass away, and of you are those who are sent to an old age where he will not be able to learn any new knowledge after what he already has. You see the land still, but when We send down the water to it, it vibrates and grows, and it brings forth of every lovely pair.*

22:6 That is because **God** is the truth, and He gives life to the dead, and He is capable of all things.

22:7 The moment is coming, there is no doubt in it, and **God** will resurrect those who are in the graves.

22:8 From people are those who argue regarding **God** with no knowledge nor guidance nor an enlightening book.

22:9 Bending his side to misguide from the path of **God**. He will have humiliation in the world and We will make him taste on the day of Resurrection the retribution of burning.

22:10 That is for what your hands have brought forth, and **God** does not wrong the servants.

Those Who Conditionally Serve God

22:11 Among the people there is he who serves **God** hesitantly. So if good comes to him, he is content with it; and if an ordeal comes to him, he makes an about-face. He has lost this world and the Hereafter. Such is the clear loss.

22:12 He calls upon besides **God** what will not harm him and what will not benefit him. Such is the far straying.

22:13 He calls on those who harm him more than they benefit him. What a miserable patron, and what a miserable companion.*

22:14 **God** admits those who affirm and promote reforms to gardens with rivers flowing beneath them. **God** does as He wishes.

Put in God Your Full Trust

22:15 Whosoever thinks that **God** will not grant him victory in this world and the Hereafter, then let him extend his request to the heaven via some means, then let him cut off and see whether this will remove the cause of his anger.*

22:16 As such, We have sent down clear signs, and **God** guides whomever He wishes.*

22:17 Those who affirm, the Jews, the Converts, the Nazarenes, the Zoroastrians, and those who have set up partners; **God** will separate between them on the day of Resurrection. For **God** is witness over all things.

22:18 Did you not see that to **God** submit what is in the heavens and what is in the earth, and the Sun and the Moon and the stars and the mountains and the trees and what moves, and many people, and many who have deserved the retribution. Whoever **God** disgraces, then none can honor him. **God** does what He pleases.

22:19 Here are two opponents who have disputed regarding their Lord; as for those who rejected, outer garments made from fire are cut out for them, and boiling water is poured from above their heads.*

22:20 It melts the inside of their stomachs and their skin.

22:21 They will have hooked rods of iron.

22:22 Every time they want to escape the anguish, they are returned to it. Taste the retribution of the burning!

22:23 **God** will admit those who affirm and promote reforms to gardens with rivers flowing beneath them, wherein they will be adorned with bracelets of gold and pearl, and their garments will be of silk.

22:24 They are guided to the good sayings and guided to the path of the Praiseworthy.

22:25 Those who have rejected and repel from the path of **God** and the Regulated Temple that We have made for people, for the dweller or the visitor to it; and whoever inclines to evil action in it, We will let them taste a painful retribution.

Debate Conference: An International Conference for the Commemoration of Monotheistic Basis, Promotion of Charity, Peace, and Unity of Humanity

22:26 We have appointed to Abraham the location of the sanctuary: "Do not set up anyone with Me, and purify My sanctuary for those who will partake, and those who will enforce, and those who kneel and prostrate."

22:27 Call out to people to the Debate Conference, they will come to you walking and on every transport, they will come from every deep enclosure.

22:28 So that they may witness benefits for themselves and recall **God**'s name in the appointed days over what He has provided for them of the animal livestock. So eat from it and feed the needy and the poor.

22:29 Then let them complete their duties and fulfill their vows, and let them partake at the ancient sanctuary.

22:30 Thus, and whosoever honors **God**'s restrictions, then it is better for him with His Lord. The livestock is made lawful for you except what is recited upon you. So avoid the foulness of idols and avoid saying false statements.

22:31 Monotheists for **God**, not setting up anything with Him. Whosoever sets up partners with **God**, then it is as if he has fallen from the sky and the birds snatch him or the wind takes him to a place far off.

22:32 Thus, and whoever honors the decrees of **God**, then it is from the piety of the hearts.

22:33 In them are benefits to an appointed time, then their place is to the ancient sanctuary.

22:34 For every nation We have made a rite that they may mention **God**'s name over what He has provided for them of the animal livestock. So your god is One god, so to Him peacefully surrender and give good tidings to those who obey.

22:35 Those who, when **God** is mentioned, their hearts revere, and they are patient to what befalls them, and they maintain the contact prayer, and from what We provide them they give charity.

22:36 The plump animals for offering, We have made them for you as decrees from **God**, in them is goodness for you. So mention **God**'s name upon them while being orderly, so once their body becomes still, then eat from them and give food to the poor and the needy. It was thus that We have made them in service to you, that you may be thankful.

22:37 Neither their meat nor their blood reaches **God**, but what reaches Him is the righteousness from you. It was thus that He made them in service to you, so that you may glorify **God** for what He has guided you to, and give news to the good doers.*

22:38 **God** defends those who affirm. **God** does not like any betrayer, ingrate.

Defend Yourself against Tyranny and Aggression

22:39 It is permitted for those who have been persecuted to fight. **God** is able to give them victory.

22:40 The ones who were driven out of their homes without justice, except that they said, "Our Lord is **God**!" If it were not for **God** defending people against themselves, then many places of gathering, markets, support centers, and temples where the name of **God** is frequently mentioned, would have been destroyed. **God** will give victory to those who support Him. **God** is Powerful, Noble.

22:41 Those whom if We allow them authority in the land, they maintain the contact prayer, and they contribute towards betterment, and deter from evil. To **God** is the conclusion of all matters.

22:42 If they deny you, then denied also before them did the people of Noah and Aad and Thamud.

22:43 The people of Abraham and the people of Lot.

22:44 The dwellers of Midian; and Moses was denied. So I granted respite to the ingrates then I took them, how then was My punishment?

22:45 So how many a town have We destroyed while it was doing wrong, so that it is laying in ruins with its wells abandoned, and empty palaces.

22:46 Have they not roamed the land and had hearts with which to reason and ears with which to hear? No, it is not the sight, which is blind, but it is the hearts that are in the chests that are blind.

22:47 They seek you to hasten the retribution; and **God** will not break His promise. A day with your Lord is like one thousand of the years you count.*

22:48 Many a town I have given it respite while it was doing wrong, then I seized it! To Me is the destiny.

22:49 Say, "O people, I am but a clear warner to you!"

22:50 As for those who affirm and promote reforms, for them is forgiveness and a great provision.

22:51 Those who seek to obstruct Our signs, those are the dwellers of hell.

Testing People through Fallible Human Messengers

22:52 We did not send before you any messenger or prophet, without having the devil interfere with his wishes. **God** then overrides what the devil has cast, and **God** secures His signs. **God** is Knower, Wise.*

22:53 That He may make what the devil has cast as a test for those who have a disease in their hearts and those whose hearts are hardened. The wicked are far away in opposition.

22:54 To let those who have been given knowledge know that it is the truth from your Lord, and they will affirm it, and their hearts will soften to it. **God** will guide those who affirm to a straight path.

22:55 Those who have rejected will remain in doubt from it until the Moment comes to them suddenly, or the retribution will come to them on a day which will stand still.

When the Time Given to Satan Expires

22:56 The sovereignty on that day is to **God**, He will judge between them. As for the affirmers and promoted reforms, they are in the gardens of Paradise.

22:57 Those who did not appreciate and rejected Our signs, those will have a humiliating retribution.

22:58 Those who emigrated in the cause of **God**, then they were killed or died, **God** will provide them with a good provision, and **God** is the best of providers.

22:59 He will admit them an entrance that they will be pleased with, and **God** is Knowledgeable, Compassionate.

22:60 It is decreed that whoever retaliates with equal measure as was retaliated against him, then he was persecuted for this, **God** will give him victory. **God** is Pardoning, Forgiving.

22:61 That is because **God** merges the night into the day, and He merges the day into the night. **God** is Hearer, Seer.

22:62 That is because **God** is the truth, and what they call on besides Him is falsehood. **God** is the High, the Great.

22:63 Did you not see that **God** sends down water from the sky, and then the land becomes green? **God** is Compassionate, Ever-aware.

22:64 To Him is what is in the heavens and what is in the earth. **God** is the Rich, the Praiseworthy.

22:65 Did you not see that **God** commits to you what is in the land, and the ships which sail in the sea by His leave? He holds the sky so that it would not collapse upon the earth, except by His leave. Indeed, **God** is Kind towards people, Compassionate.

22:66 He is the One who gave you life, then He makes you die, then He gives you life. But the human being is always rejecting!

22:67 For every nation We have made rites which they will fulfill. So do not let the matter fall into dispute. Call upon your Lord, for you are on a guidance which is straight.

22:68 If they argue with you, then say, "**God** is fully aware of what you do."

22:69 "**God** will judge between you on the day of Resurrection in what you dispute therein."

22:70 Did you not know that **God** knows what is in the heavens and the earth? All is in a record. All that for **God** is easy.

22:71 They serve besides **God** what He did not send any authority upon, and what they have no knowledge of. The wicked will not have any helper.

Ingrates are Aggressive

22:72 If Our clear signs are recited to them, you see hatred in the faces of those who have rejected. They are nearly close to attacking those who are reciting to them Our signs! Say, "Shall I inform you of what is worse than this? The fire, which **God** has promised to those who have rejected. What a miserable destiny!"

22:73 O people, an example is being cited so listen to it: those you call upon besides **God** will not create a fly even if they all gathered to do so. If the fly takes anything from them, they will not be able to return it from him. Weak is both the seeker and the sought!

22:74 They have truly underestimated **God**'s power. **God** is Powerful, Noble.

22:75 **God** chooses messengers from amongst the controllers and from amongst people. **God** is Hearer, Seer.

22:76 He knows their present and their future. To **God** all matters are returned.

22:77 O you who affirm, kneel and prostrate and serve your Lord and do good that you may succeed.

22:78 Strive in the cause of **God** properly. He is the One who has chosen you, and He has made no hardship for you in the system, the creed of your father Abraham; He is the One who named you 'those who have peacefully surrendered' from before as well as in this. So let the messenger be witness over you and you be witness over people. So maintain the contact prayer and contribute towards betterment and hold tight to **God**, He is your patron. What an excellent Patron, and what an excellent Supporter.*

ENDNOTES

022:005 According to modern biology, the normal period for pregnancy, starting from fertilization until birth is 38 weeks and 266 days. See 46:15.

022:013 See 2:286.

022:015 The traditional rendering of verse 22:15 is so bad that it turns into an absurdity, into a joke. The amazing thing is that anyone who studies the Quran should easily understand its meaning, since the expressions are used in other verses and contexts. Instead of first looking at the usage of words and expression in other parts of the Quran, the traditional translators look for inspiration form the early commentators who mostly relied heavily on *hadith* hearsay. Regardless of the source, except for a few, such as Muhammed Asad, Muhammed Ali, and Rashad Khalifa, many translations have duplicated the bizarre and absurd traditional rendering. For a detailed discussion of this verse, see the Sample Comparisons section in the Introduction. See 6:41.

022:016 We wished to reflect both equally plausible meanings. See 16:93; 57:22.

022:019-22 It is interesting that the Quran usually mentions hell and paradise after each other. While the appreciative see and enjoy the promise of heaven, the unappreciative people complain about how bad hell is. Ironically, they try very hard to enter the place they are so critical of. Though many people will end up in hell forever, it does not mean that hell is eternal. Similarly, verses informing us about ingrates or idol worshipers staying in Hell eternally, does not necessarily mean that Hell is eternal, unless we are informed that Hell itself is eternal. It simply means that ingrates and idol worshipers would end up in Hell and nothing else. If Hell together with its inhabitants one day ceased to exist, then the ingrates or idol worshipers would have stayed in Hell eternally. Their punishment would be the entire life of Hell.

In fact, the Quran informs us that both the eternal punishment in Hell and reward in Paradise is conditioned on their life span (11:107). Let's reflect on verse 40:11 and 67:2. The first verse refers to two creations and two deaths and the second refers to creation of death and life. These two verses can be understood better if we know that in the Quranic language, death cannot exist without life and vice versa. They exist together, since Death is the permanent halt of the brain's conscious activity (39:42; 16:21) and a temporal stage to be followed by resurrection (29:57; 10:56; 22:6). Death is a process which leads to life. A living creature will die and a dead creature will get a new life (22:66). Vegetation experiences successive lives and deaths through seasons (2:64; 3:27; 6:95; 16:65; 22:5-6; 30:19-50; 35:9). After the first creation there was no death nor life; we just existed. But God decided to create death and life (67:2). Creation, death, life (current), death, life (resurrection). In other words, two deaths and two lives (40:11). The word *HaLaKa*, on the other hand, is occasionally used to describe the death of an individual (40:34), but usually irreversible destruction and annihilation, or total existential extinction of an entity (5:17; 6:6; 6:47; 8:42; 20:128; 21:95; 22:45; 28:59; 36:31; 69:5; 77:16).

The Quran informs us that the Earth and the Heavens will be changed with a different Earth and Heavens (14:48). If the re-creation referred to in this verse is the one before the Day of Judgment, then there is the possibility of another re-creation. There are indications that Paradise will be preserved or created again. For instance, the word *KHuLD* (eternal, everlasting) is not used for Hell, but it is used as an adjective to describe Paradise (25:15). On the other hand, the same adjective is not used to describe Hell, but to describe the punishment *in* Hell (10:52; 41:28).

Verse 8:42 does not only refer to the loss of lives and surviving a particular battle, but also to a higher cosmic event: ingrates will perish forever since they relied on falsehood, while affirmers will last forever since they relied on a clear argument. No wonder life and death are used as metaphors for attaining truth or falsehood (6:122). In fact, witnessing and affirming the truth leads to life, metaphorically and literally (8:24). From the above verse, it is fair to infer that those who reject the divine message will lose eternal life. Will they die in Hell? The answer is No: "Anyone who comes to His Lord guilty will deserve Hell, wherein he never dies, nor stays alive." (20:74).

Then, the alternative is obvious: total annihilation, ontological extinction together with Hell. Those persons/programs with free choice that chose to corrupt themselves with the worst diabolic viruses (such as associating partners with God or killing an innocent program) will be sentenced to an eternal punishment: after resurrection they will experience a period of diagnosis, justice, regret and then with the creation of a new earth and heaven, they will be hurled to non-existence together with Hell. Perhaps, their memories too will be erased from the minds of their relatives who chose eternal life by dedicating their system to God alone and by leading a righteous life with the day of judgment in mind. It is interesting that the Quran refers to this annihilation through ultimate deletion from the ultimate record. God's deliberate deletion of their existence from His mind. (32:14). The Quran repeatedly comforts us by reminding us of God's perfect justice. The following verse provides us with a precise idea of God's justice and mercy: "For those who did good work there will be the best and more. . . As for those who earned evil, they will receive equivalent evil (10:27).

Suffering in an eternal Hell creates a contradiction between this divine justice, since eternity cannot be equal to an evil committed during a limited human lifespan. However, eternal punishment for the lifetime of a non-eternal Hell avoids such a contradiction. After receiving an equivalent punishment, the chief evildoers like those referred to in verse 4:48, will be eliminated from existence. They will end up in Hell and Hell will end up in nothingness. No wonder the first and the most repeated verse of the Quran reminds us repeatedly that God is Gracious, Merciful (1:1).

022:037 See 6:119-121.

022:047 Time is relative. See 32:5; 70:4.

022:052-55 The weakness and deficiencies demonstrated by messengers and prophets provide excuses for those who try to criticize them and reject their message. On the other hand, those deficiencies and weaknesses reject the common Sunni and Shiite belief that the messengers are "innocent" or "sinless" people. Those who affirm and appreciate God's message are capable of distinguishing revelation from personal errors.

022:078 All messengers delivered the message of "serve only one God," and declared that they were muslims, that is peaceful surrenderers to God. Except for the intentional or unintentional rejection of some humans and jinns, everything in the universe are muslims, that is, behave according to the laws of God imposed on nature (41:11). For instance, Noah, Abraham, Moses, Jesus, and his supporters are all described with the same word, muslim: (10:72; 2:128; 10:84; 27:31; 5:111; 72:14). The rituals of Islam first came through Abraham (16:123; 21:73). Also, see 3:19.

بسم الله الرحمن الرحيم

23:0 In the name of God, the Gracious, the Compassionate.
23:1 Those who affirm are indeed successful.
23:2 Those who are humble in their contact prayer.
23:3 They abstain from vain talk.
23:4 They are active towards betterment.
23:5 They guard their private parts.
23:6 Except around their mates, or those whom they have contractual rights, they are free from blame.*
23:7 But whosoever seeks anything beyond this, then these are the transgressors.
23:8 Those who are true to what they have been entrusted and their pledges.
23:9 Those who maintain their support.
23:10 These are the inheritors.
23:11 The ones who shall inherit Paradise, in it they will abide.

Creation on Earth

23:12 We have created people from an extract of clay.
23:13 Then We made him a seed in a safe lodging.
23:14 Then We created the seed into an embryo, then We created the embryo into a fetus, then We created the fetus into bone, then We covered the bone with flesh, then We brought forth a new creation. So glory be to **God**, the best of creators.*
23:15 Then after that, you will die.
23:16 Then you will be resurrected on the day of Resurrection.
23:17 We have created above you seven passageways, and We were never unaware of the creations.
23:18 We sent down from the sky water in due measure, then We let it reside in the land, and We can take it away.
23:19 So We brought forth for you gardens of palm trees and grapes,

for which you will find many fruits and from it you will eat.

23:20 A tree which emerges from the mount of Sinai, it grows oil and is a relish for those who eat.

23:21 In the livestock are lessons for you. We give you to drink from what is in its bellies, and you have many other benefits from them, and of them you eat.

23:22 On them and on the ships you are carried.

Noah

23:23 We have sent Noah to his people, so he said, "O my people, serve **God**, and you have no other god besides Him. Will you not take heed?"

23:24 But the leaders who rejected from among his people said, "What is this but a human like you? He wants to make himself better than you! If it was indeed **God**'s will, He would have sent down controllers. We did not hear such a thing among our fathers of old."*

23:25 "He is no more than a man who has madness in him. So keep watch on him for a while."

23:26 He said, "My Lord, grant me victory for what they denied me."

23:27 So We revealed to him: "Construct the Ship under Our eyes and Our inspiration. Then, when Our command comes and the volcano erupts, then you shall take on the Ship two from every pair, and your family; except for those of them upon whom the word has come. Do not address Me regarding those who have done wrong, for they will be drowned."*

23:28 So when you and those who are with you have embarked on the Ship, then say, "All praise is due to **God** who has saved us from the wicked people."

23:29 Say, "My Lord, cause me to embark upon a blessed place, for You are the best for those who embark."

23:30 In this are signs, and We will always test.

Leaders Reject the Message of Monotheism

23:31 Then We raised after them a different generation.

23:32 So We sent a messenger to them from amongst them: "Serve **God**, you have no other god besides Him. Will you not take heed?"

23:33 The leaders from amongst his people who rejected and denied the meeting of the Hereafter; and We indulged them in this worldly life; said, "What is this but a human like you? He eats from what you eat and he drinks from what you drink."

23:34 "If you obey a human like you, then you will indeed be losers."

23:35 "Does he promise you that if you die and become dust and bones that you will be brought out?"

23:36 "Farfetched is what you are being promised."

23:37 "There is nothing but our worldly life, we die and we live, and we will not be resurrected."

23:38 "He is but a man who has invented a lie against **God**, and we will not affirm him."

23:39 He said, "My Lord, grant me victory for what they denied me."

23:40 He replied: "In a little while they will be in regret."

23:41 So the explosion took them with justice, and We made them as dead plants. So away with the wicked people.

23:42 Then We rose after them different generations.

23:43 No nation can hasten its appointed time, nor can it delay.

23:44 Then We sent Our messengers in succession. Every time there came to a nation their messenger, they denied him. So We made them follow one another, and We made them history. So away with a people who do not affirm!

Moses and Aaron

23:45 Then We sent Moses and his brother Aaron. With Our signs and a clear authority.

23:46 To Pharaoh and his entourage. But they became arrogant, for they were a high and mighty people.

23:47 So they said, "Shall we affirm two humans like us, while their people are servants to us?"

23:48 So they denied them, and they became of those who were destroyed.

23:49 We have given Moses the book, so that they may be guided.

23:50 We made the son of Mary and his mother a sign, and We gave them refuge on high ground, a place for resting and with a flowing stream.

Those who Find God's Word Insufficient are Divided into Hostile Factions

23:51 "O messengers, eat from what is good, and do righteous work. I am aware of what you do."

23:52 "This is your nation, one nation, and I am your Lord so be aware."

23:53 But the affair was disputed between them into segments. Every group happy with what it had.*

23:54 So leave them in their error until a time.

23:55 Do they not think why We are extending them with wealth and children?

23:56 We are quick to give them the good things. But they do not perceive.

23:57 As for those who are in reverence from the awareness of their Lord.

23:58 They affirm the signs of their Lord.

23:59 They do not set up anything with their Lord.

23:60 They give of what they were given, and their hearts are full of reverence that they will return to their Lord.

23:61 These are the ones who race in doing good, and they are the first to it.

23:62 We do not burden a person except with what it can bear. We have a record that speaks with the truth, they will not be wronged.

23:63 No, their hearts are unaware of this! They have deeds besides this which they are doing.

23:64 Until We take their carefree people with the retribution, then they will shout for help.

23:65 Do not shout for help today, for you will not be helped against Us.

23:66 My signs were recited to you, but you used to turn back on your heels.

23:67 You were too proud from it, talking evil about it; and you defiantly disregarded it.

23:68 Did they not ponder the words, or has what come to them not come to their fathers of old?

23:69 Or did they not know their messenger? For they are in denial of him.

23:70 Or do they say that there is a madness in him? No, he has come to them with the truth, but most of them are hateful of the truth.

23:71 If the truth were to follow their desires, then the heavens and the earth and all who are in them would have been corrupted. No, We have come to them with their reminder, but from their reminder they are turning away.

23:72 Or do you ask them for a wage? The wage of your Lord is best, and He is the best of providers.

23:73 You are inviting them to a Straight Path.

23:74 Those who do not affirm the Hereafter, they are deviating away from the path.

23:75 If We were to have mercy on them and remove what distress was upon them, they would still return to persist in their transgression, wandering blindly.

23:76 When We seized them with the punishment, they did not humble themselves to their Lord nor did they invoke Him.

23:77 Until We open for them a door of immense retribution, then they become in sorrow and despair.

God

23:78 He is the One who established for you the hearing, and the sight, and the minds. Little do you give thanks.

23:79 He is the One who multiplied you on the earth, and to Him you will be returned.

23:80 He is the One who gives life and brings death, and to Him is the alteration of the night and the day. Do you not reason?

23:81 Indeed, they have said the same as what the people of old had said.

23:82 They said, "If we are dead and become dust and bones, will we then be resurrected?"

23:83 "We, as well as our fathers, have been promised this from before. This is nothing but the tales of the ancients."

23:84 Say, "To whom the earth is and whosoever is in it if you know?"*

23:85 They will say, "To **God**." Say, "Will you then not remember!"

23:86 Say, "Who is the Lord of the seven heavens and the Lord of the great dominion?"

23:87 They will say, "To **God**." Say, "Will you then not take heed?"

23:88 Say, "In whose hand is the sovereignty of all things, and He protects while there is no protector against Him, if you know?"

23:89 They will say, "To **God**." Say, "Then why are you deceived?"

23:90 In fact, We have come to them with the truth, and they are truly liars.

23:91 **God** has not taken a son, nor is there with Him any god. If it were so, then every god would have taken what He created and they would have tried to overtake each other. **God** be glorified against what they describe!

23:92 The knower of the unseen and the seen. Be He exalted above what they set up.

Repel Evil with Good Work

23:93 Say, "My Lord, if You show me what they are promised."

23:94 "My Lord, then do not leave me amongst the wicked people."

23:95 We are able to show you what We promise them.

23:96 Repel evil with what is better. We are better aware of what they describe.

23:97 Say, "My Lord, I seek refuge with you from the whispers of the devils."

23:98 "I seek refuge with you O Lord that they should come near."

23:99 Until death comes to one of them, he says: "My Lord, send me back."

23:100 "So that I may do good in what I have left behind." No, it is but a word he is speaking. There is a barrier to prevent them from going back until the day they are all resurrected.*

23:101 So when the horn is blown, then there will be no kinship between them on that day, nor will they ask for one another.

23:102 So, those whose weight is heavy on the scales, then those are the successful ones.

23:103 Those whose weight is light on the scales, those are the ones who lost themselves, in hell they will abide.

23:104 The fire will scorch their faces, and in it their grin will be with displaced lips.

23:105 "Were My signs not recited to you, then you denied them?"

23:106 They said, "Our Lord, our wickedness overcame us, and we were a misguided people."

23:107 "Our Lord, bring us out of it, and if we return to this then we are wicked."

23:108 He said, "Be humiliated therein and do not speak to Me."

23:109 There was a group from among My servants who used to say, "Our Lord, we have affirmed, so forgive us, and have mercy upon us, for You are the best of those who show mercy."

23:110 But you mocked them, so much so that they made you forget My remembrance, and you used to laugh at them.

23:111 I have rewarded them today for their patience, they are indeed the winners.

23:112 Say, "How long have you stayed on earth in terms of years?"

23:113 They said, "We remained for a day, or for part of a day. So ask those who kept count."

23:114 He said, "You have remained for very little, if you knew."

23:115 "Did you think that We have created you without purpose, and that you would not return to Us?"

23:116 So, exalted is **God**, the true King, there is no god but He, the Lord of the supreme throne.

23:117 Whosoever calls on another god with **God** when he has no proof for such, then his judgment is with his Lord. The ingrates will never succeed.

23:118 Say, "My Lord, forgive and have mercy, and you are the best of those who are merciful."

ENDNOTES

023:006 For discussion on the Quranic expression *ma malakat aymanukum* see the notes on verses 4:3,25; 90:1-20.

023:014 Embryo is the name of the stage when the fertilized egg hangs on the wall of uterus. The word *alaq*, which we translated as embryo has three meanings in Arabic, (1) blood clot; (2) a hanging object; (3) leech.

It is normal for medieval commentators who were ignorant of embryology, to render the meaning of the word *alaq* as "clot," since it is an organic ingredient of the human body. However, it is no excuse for modern commentators and translators who blindly copy from medieval books, to render this word as "clot," since there is not a stage, from fertilization of the embryo until birth, that could be described as "clot." See 46:15; 95:2.

023:024 The most ardent opponents of God's messengers have always been the blind followers of the traditions of their ancestors. The opposition against those who invite to the "Quran alone" in modern times demonstrates a similar mindset and reaction. Throughout history, the modus operandi of polytheists, have not changed much. They show the same weakness, the same knee-jerk reaction, the same ancestor-worship, the same gullibility, the same belligerent attitude, the same bigotry, the same ignorance and arrogance, and the same apathy and fear towards learning the truth.

Polytheists are the riders of twin horses, named Ignorance and Arrogance. To protect themselves from being torn apart, they spend a lot of effort to keep the two competing twin horses at the same speed and distance.

023:027 Storytellers have exaggerated the size and scope of the flood, turning it into a mythology. Noah's ark was not a colossus ship, but a boat comprising several logs connected with ropes (54:13); the flood was not global but limited to Noah's people around the Dead Sea region, the lowest point on the land; the animals taken to the boat, were not members of all species, but the domesticated animals in Noah's farm. This major event in the region, since then, was incorporated into mythologies of many neighboring nations, including Sumerians.

023:052-56 The Quran strictly condemns sectarian division and tells us that the struggle for power and idol-worship are the sources of sectarian division. See 6:153-159; *30:32*; 98:4; 21:92; 23:52; 2:285.

023:053 54 See 30:32; 3:188. Compare it to the New Testament: "And I say unto you, that many shall come from the east and west, and shall sit down with Abraham and Isaac and Jacob in the kingdom of heaven. But those who believe they own the kingdom of heaven shall be cast out into the outer darkness. There shall be weeping and gnashing of teeth." (Matthew 8:11-12)

023:084-89 affirming God means accepting all of God's attributes as defined by God. Thus, those who do not affirm God's absolute control over His creation (8:17), His being the only source of law (12:40), are in fact not affirming God.

023:100 This verse rejects the idea of reincarnation and the spirits/souls of dead people roaming on earth. Communication with the dead is not allowed.

24:0 In the name of God, the Gracious, the Compassionate.

24:1 A chapter which We have sent down and imposed, and We have sent down in it clear revelations that you may remember.

Adultery

24:2 The adulteress and the adulterer, you shall lash each of them one hundred lashes, and do not let any pity take you over **God**'s system if you affirm **God** and the Last Day. Let a group of the affirmers witness their punishment.*

24:3 The adulterer will only marry an adulteress or one who is an idolatress. The adulteress, she will only be married to an adulterer or an idolater. This has been forbidden for the affirmers.

24:4 Those who accuse the chaste women and they do not bring forth four witnesses, then you shall lash them eighty lashes and do not ever accept their testimony. Those indeed are the evildoers;*

24:5 Except those who repent after this and do right, then **God** is Forgiving, Compassionate.

24:6 Those who accuse their wives, but they have no witnesses except for themselves, then the testimony of one of them should be four testimonies, swearing by **God** that he is being truthful.

24:7 The fifth shall be **God**'s curse upon him if he is one of those lying.

24:8 The punishment will be averted from her if she testifies four times by **God** that he is of the liars.

24:9 The fifth shall be that **God**'s curse is upon her if he is speaking the truth.

24:10 All this is from **God**'s favor upon you and His mercy. **God** is Forgiving, Wise.

24:11 Those who have brought forth the false accusation were a group from within you. Do not think it is bad for you, for it is good for you.

Every person amongst them will have what he deserves of the sin. As for he who had the greatest portion of it, he will have a great retribution.*

24:12 If only when you heard it the affirming men and women should have thought good of themselves and said: "This is an obvious lie!"

24:13 If only they had brought four witnesses to it. If they did not have the witnesses, then these with **God** are the liars.

24:14 Had it not been for **God**'s favor upon you and His mercy upon you in this world and the Hereafter, a great retribution would have touched you for what you have spoken.

24:15 For you have cast it with your tongues, and you say with your mouths what you have no knowledge of, and you think it is a minor issue, while with **God** it was great.*

24:16 When you heard it you should have said: "It was not right for us to speak of this. Glory to You, this is a great lie."

24:17 **God** warns you not to repeat something like this ever, if you are affirmers.

24:18 **God** clarifies for you the revelations; and **God** is Knower, Wise.

Beware of Sexual Promiscuity

24:19 As for those who love to see immorality spread amongst the affirmers, they will have a painful retribution in the world and the Hereafter. **God** knows while you do not know.

24:20 All this is from **God**'s favor upon you and His mercy. **God** is Kind, Compassionate.

24:21 O you who affirm, do not follow the footsteps of the devil. Whosoever follows the footsteps of the devil should know that he advocates immorality and evil. Had it not been for **God**'s favor upon you and His mercy, not one of you would have ever been purified, but **God** purifies whom He wishes, and **God** is Hearer, Knower.

24:22 Let not those amongst you who have been blessed with abundance refuse to give to the relatives, the needy, and those who have immigrated in the cause of **God**. Let them pardon and forgive. Would you not like **God** to forgive you? **God** is Forgiving, Compassionate.

24:23 Those who accuse the chaste, innocent, affirming women, they will be cursed in this world and the Hereafter, and they will have a painful retribution.

24:24 On the day when their tongues, and their hands, and their feet will bear witness against them for what they used to do.

24:25 On that day, **God** will pay them what they are owed in full, and they will know that **God** is the Truth Manifested.

24:26 The wicked women are for the wicked men, and the wicked men are for the wicked women. The good women are for the good men, and the good men are for the good women. These people are innocent from what statements have been made, and for them is forgiveness and a generous provision.

Respect Privacy Rights

24:27 O you who affirm, do not enter any homes except your own unless you are invited and you greet the people in them. This is best for you, perhaps you will remember.

24:28 But, if you do not find anyone in them then do not enter until you are given permission. If you are told to go back, then return for it is purer for you. **God** is aware of what you do.

24:29 There is no sin upon you that you enter homes which are abandoned if in them there are belongings of yours. **God** knows what you reveal and what you conceal.

24:30 Tell the affirming men to lower their gaze and guard their private parts, for that is purer for them.

God is fully aware of what you do.*

24:31 Tell the affirming women to lower their gaze and guard their private parts, and that they should not show off their attraction except what is apparent and let them cast their clothes over their cleavage. Let them not show off their attraction except to their estranged husbands, or their fathers, or fathers of their husbands, or their sons, or the sons of their husbands, or their brothers, or the sons of their brothers, or the sons of their sisters, or their children that come after them, or those who are still their dependents, or the male servants who are without need, or the child who has not yet understood the nakedness of women. Let them not strike with their feet in a manner that reveals what they are keeping hidden of their attraction. Repent to **God**, all of you, affirmers, so that you may succeed.*

Help the Singles Marry

24:32 Marry off those among you that are single/widow, and the reformed ones who are slaves among you and maids. If they are poor, then **God** will grant them from His grace. **God** is Encompassing, Knowledgeable.*

24:33 Let those who are not able to marry continue to be chaste until **God** enriches them of His Bounty. If those whom you have personal contractual rights seek official recording, then sign with them if you find that they are ready and give them from the wealth of **God** which He has bestowed upon you. Do not force your young daughters into prostitution when they have desired independence, in order that you may make a gain in the goods of this worldly life. But if anyone has compelled them, then considering their compulsion, **God** is Forgiving, Compassionate.*

24:34 We have sent down to you clarifying revelations and an example of those who came before you and a lesson for the righteous.

God's Message: a Universal Light

24:35 **God** is the Light of the heavens and the earth. The example of His light is like a concave mirror within a lamp, the lamp is within a glass, the glass is like a radiant planet, which is lit from a blessed olive tree that is neither of the east nor of the west, its oil nearly radiates light even if not touched by fire. Light upon light. **God** guides to His light those whom He pleases. **God** sets forth parables for mankind; **God** is aware of all things.*

24:36 In houses/bases that **God** has allowed to be raised and His name is mentioned in them. He is glorified therein morning and evening.

24:37 People who are not distracted by trade or sale from the remembrance of **God**, holding the contact prayers, and contributing towards betterment. They fear a day when the hearts and sight will be overturned.

24:38 **God** will reward them for the best of what they did, and He will increase for them from His grace. For whom He pleases, **God** provides without price.

24:39 As for those who reject, their works are like a mirage in the desert. A thirsty person thinks it is water, until he reaches it, he finds it is nothing; he finds **God** there and He pays him his due, and **God** is swift in judgment.

24:40 Or like the darkness out in a deep ocean in the midst of violent waves, with waves upon waves and dark clouds. Darkness upon darkness, if he brings out his own hand, he could barely see it. For whomever **God** does not make a light, he will have no light.

God

24:41 Do you not see that everything in the heavens and the earth, even the birds in formation, glorifies **God**? Each knows its contact prayer and its glorification. **God** is fully aware of everything they do.

24:42 To **God** is the sovereignty of the heavens and the earth, and to **God** is the final destiny.

24:43 Do you not see that **God** drives the clouds, then He gathers them, then He piles them upon each other, then you see the soft rain coming out of them? He sends down hail from the sky from the mountains to afflict whomever He wills, and He diverts it from whomever He wills; the brightness of the snow almost blinds the eyes.

24:44 **God** rotates the night and the day. In that is a lesson for those who have insight.

24:45 **God** created every moving creature from water. So, some of them move on their bellies, and some walk on two legs, and some walk on four. **God** creates whatever He wills. **God** is capable of all things.*

24:46 We have sent down to you clarifying revelations, and **God** guides whoever He wills to a straight path.

The Messenger as a Living Leader

24:47 They say: "We affirm **God** and the messenger, and we obey," but a group of them turn away after that. These are not affirmers.

24:48 If they are invited to **God** and His messenger to judge between them, a party of them turn away.

24:49 However, if the judgment is in their favor, they come to it willingly with submission/peace!

24:50 Is there a disease in their hearts, or are they doubtful? Or do they fear that **God** and His messenger would wrong them in the judgment? In fact, they are the wrong doers.

24:51 The utterance of those who affirm when they are invited to **God** and His messenger to judge in their affairs is to say: "We hear and obey." These are the winners.

24:52 Whosoever obeys **God** and His messenger, and reveres **God** and is aware of Him, then these are the winners.

24:53 They swear by **God** with their strongest oaths that if you would only command them, they would mobilize. Say: "Do not swear, for obedience is an obligation. **God** is Ever-aware of what you do."

24:54 Say: "Obey **God** and obey the messenger." But if they turn away, then he is only responsible for his obligation, and you are responsible for your obligations. If you obey him, you will be guided. The messenger is only required to deliver clearly.

24:55 **God** promises those among you who affirm and promote reforms, that He will make them successors on earth, as He made successors of those before them, He will enable for them their system which He has approved for them, and He will substitute security for them in place of fear, that they serve Me and do not associate anything with Me. As for those who reject after that, they are the transgressors.

24:56 Maintain the contact prayer and contribute towards betterment, and obey the messenger, that you may attain mercy.

24:57 Do not think that those who reject will remain in the land. Their abode is Hell. What a miserable destiny!

Privacy; Dawn and Evening Prayers

24:58 O you who affirm, let those whom are under your contract and have not yet attained puberty request your permission regarding three times: Before the dawn contact prayer, and when you take off your attire from the heat of the noon, and after the dusk contact prayer. These are three private times for you. Other than these times, it is not wrong for you or them to intermingle with one another. **God** thus clarifies the revelations for you. **God** is Knowledgeable, Wise.*

24:59 When the children among you reach puberty, then let permission be sought from them like it was sought from those before them. **God** thus clarifies His revelations for you. **God** is Knowledgeable, Wise.

24:60 The women who are past childbearing and who do not seek to get married have no sin upon them if they put-off their garments, provided they do not display their attraction. If they remain as they were, then it is better for them. **God** is Hearer, Knower.

Do not Discriminate against Handicapped People; Share Food with Extended Family

24:61 There is no blame upon the blind, nor is there any blame upon the crippled, nor is there any blame upon the ill, nor is there any blame upon yourselves, if you eat at your homes, or the homes of your fathers, or the homes of your mothers, or the homes of your brothers, or the homes of your sisters, or the homes of your uncles, or the homes of your aunts, or the homes of your mothers' brothers, or the homes of your mothers' sisters, or those to which you possess their keys, or that of your friends. You commit nothing wrong by eating together or as individuals. When you enter any home, you shall greet each other a greeting from **God** that is blessed and good. **God** thus explains the revelations for you that you may reason.*

Etiquettes in Public Meetings

24:62 The affirmers are those who affirm **God** and His messenger, and when they are with him in a meeting, they do not leave him without permission. Those who ask permission are the ones who do affirm **God** and His messenger. If they ask your permission, to tend to some of their affairs, you may grant permission to whomever you wish, and ask **God** to forgive them. **God** is Forgiver, Compassionate.

24:63 Do not let the invitation of the messenger between you be as if you are inviting each other. **God** is fully aware of those among you who slip away under lame excuses. Let those who oppose his command beware, for an ordeal may strike them, or a severe retribution.*

24:64 Certainly, to **God** belongs all that is in the heavens and the earth. Surely, He knows what your condition is; and on the day when they are returned to Him, He will inform them of everything they had done. **God** is aware of all things.

ENDNOTES

024:002 Public shaming, though it has long been abandoned in Western world, is one of the tools we may use against crimes. The Quran gives high importance to the institution of marriage. Social and psychological problems experienced among millions of children because of the breakdown of the marriage institution in otherwise prosperous Western countries highlight the importance of marriage. Additionally, the harms of a promiscuous lifestyle to the health and wellbeing of the society are another reason why the Quran takes the marriage contract so seriously and issues a harsh penalty for its violation.

The verse emphasizes public shame rather than inflicting physical pain. The word *JaLDa* does not mean a stick, but something that will touch the skin, since the word *jalda* is a derivative of a word that is used for skin (*JeLD*).

Shaming is a social and psychological penalty. For such a penalty to be effective, researchers suggest the fulfillment of the following conditions: (1) The convict must be a member of a particular group; (2) The group must approve of the penalty; (3) The shaming penalty should be delivered to the group and group should ignore the convict for a period physically, emotionally, and economically; (4) The convict who is to be shamed must fear the group's decision to cut relations; and (5) The convict must have a means to regain the trust of the group.

First, we should be reminded that there is no stoning-to-death punishment for adultery in Islam. The punishment of stoning was introduced to the Islamic law long after prophet Muhammed through fabricated *hadith*. It is the followers of

Sunni and Shiite sects that adopted the Jewish practice and fabricated the most ridiculous stories, such as the story of a hungry holy goat eating a verse about stoning, etc. The followers of *hadith* and *sunna* accept the Jewish practice and the related silly stories by claiming that "these verses are not clear about adultery; thus, *hadith* and *sunna* explain these verses." Knowing the excuse of those who associate other partners with God, the first two verses rebut the claim of the unappreciative people by asserting the clarity of its language.

The origin of stoning is found in the Old Testament. For instance:

"But if this thing be true, and the tokens of virginity be not found for the damsel: Then they shall bring out the damsel to the door of her father's house, and the men of her city shall stone her with stones that she die: because she hath wrought folly in Israel, to play the whore in her father's house: so shalt thou put evil away from among you." (Deuteronomy 22:20-21)

According to the Old Testament, a rapist should be forced to marry the girl he violated. This rule punishes the victim to share the rest of her life with the violent and shameless man who violated her (Deuteronomy 22:28-30). How can this and many other unjust laws be imposed by a Just God? (See 2:59; 2:79; 5:13-15; 5:41-44; 7:162; 9:31).

See Leviticus 20:2,27; 24:14-16; Numbers 15:35; 35:17; Deuteronomy 13:10; 17:5; 22:21-25; Joshua 7:25; 1 Kings 21:1-29. Also, see the Quran: 3:36.

024:004 See 4:15.

024:011 By not mentioning the person's name in this historical event, the Quran wants to keep its universal application clear. See 111:1

024:015 We are instructed not to disseminate rumors or hearsay statements about a person without proof or strong evidence. This warning is about any person. But, what about the prophet? What about collecting rumors and hearsay statements about a prophet long after he passed away? What about believing those rumors and hearsay statements, and even sanctifying them as a part of the religion for future generations? What about going even further by preferring them to a book believed to be the verbatim word of God? And what about even going further, and declaring those who reject them to be apostates, deserving to be killed? It is these "holy" rumors and hearsay statements that have condemned Muslim societies to ignorance, oppression, and backwardness. Those who consider *hadith* and *sunna*, which would not be admitted in any court of law as evidence, as an important source of their system, have no respect for God and his messengers; since they follow conjecture and contradictory hearsay reports (6:116). See 6:148; 10:36,66; 53:28.

024:030-31 Righteous men and women should not dress provocatively. An affirming woman should wear a dress (33:59) and cover her chest. God uses flexible language, to allow culture, time, climate, age, and social dynamics and other factors play a role in the decision. The underlying reason for this recommendation is the protection of women from potential harassment by males. There is nothing in the Quran that instructs the government or the society to force women, in the name of God, to cover themselves. A punishment neither in this world nor in hereafter is issued. While the male clerics and the followers of their misogynistic teachings are sunk up to their eyebrows in the mud of *shirk*, the only unforgivable sin, somehow, they are scrupulously obsessed with imposing dress codes on women. When they attain power, they bury women alive in black sacks. Does sanctifying their dress code of Christian nuns have anything to do with their psychological and sexual problems? The verses recommending that monotheist women cover themselves aim to protect them from the harassment of unrighteous men. Ironically, women are now harassed more by the self-righteous men.

It is up to women, not men, to determine the length of their dresses, and whether they will cover their breasts or not. It is one thing to remind a monotheist woman nicely to be modest for her protection and to maintain social order. But to use this issue to patronize and subjugate women is not what the Quran expects from us. Even worse is to impose this recommendation on women who are not affirmers of the Quran, since that violates additional Quranic principles.

The male clerics who covered their intelligence by their obsessive interference with women's hair and dress, do not even notice the lesson in verse 7:26. Despite the verse informing us that women used to walk in the street with their faces uncovered in front of Muhammed (33:52), they still have the audacity to preach to women to cover their faces with veils. See 33:55.

For the Biblical account of women's dress, which is mostly ignored by Christians, see Genesis 24:65; Jeremiah 4:30; Ezekiel 16:10; Zephaniah 1:8 ; 1Timothy 2:9; 1 Peter 3:3; 1Corintions 11:15.

024:032 The reference pronoun of the word "*IBaDukum*" is misunderstood, and contrary to the clear verses of the Quran; it is abused to justify slavery. The expression "your slaves" should be understood as "slaves among you" or "slaves from your group" not as "the slaves that you own." For instance, the expression "feed your poor" does not mean "poor that you own." Similarly, "your ingrates" means, "the ingrates among you." (54:43; also see: 72:4). It would be literally awkward to repeat the preposition "from" (min) twice. (For a discussion on slavery, see the endnote for verse 4:3)

024:033 What about if they don't want to be chaste, then should they be forced? Of course not. For instance, "Don't force your children to eat meat, if they chose to be vegetarians" does not mean "Force your children to eat meat if they don't want to be vegetarians." The word *FaTaYa* has been usually mistranslated as slaves. It simply means young girls. Its masculine form is used for Abraham and many other free people. For its meaning through its Quranic usages please see: 12:30; 12:36; 18:10; 18:13; 18:60; 21:60; and 24:33.

024:035 This example highlights some of the main characteristics of the Quran. God is the source of enlightenment (light). The Quran radiates God's message (lamp). Though it is protected with a mathematical code, it has a transparent language (glass), reflecting a small portion of divine knowledge (radiant planet). Its message is universal; not limited to any language, race, or geography (neither of the east nor of the west). Those with open eyes may get its message even without showing much effort (gives off light even if not touched by fire). It has meaning inside meaning, a message inside the message (light upon light). Receiving its message is a blessing and its teacher is God Himself (God guides whomever/whoever wish(es) to His light). The word *nur* (light), in the normative case, occurs 33 times and is never used in plural from in the Quran, while its antonym is always used in its plural form, *zulumat* (darknesses), in 23 occurrences, and never in singular form. This implies singularity of the source of truth, which is God, and diversity of the sources of falsehood, which could be Satan, polytheist clergymen, ignorance, peer pressure, wrong choices, acquired weaknesses and addictions, etc.

According to the Bible, God created the light first (Genesis 1:3). Light is also used as a metaphor for divine enlightenment or instruction (Psalms 119:105; Isaiah 8:20; Matthew 4:16; Matthew 5:16; John 5:35). God is described as "the Father of lights" (James 1:17). The Bible and its appendices use the light metaphor for many creatures too, such as John the Baptist (John 5:35), Jesus (Luke 2:32; John 1:7-9), his disciples, (Matthew 5:14), and angels/controllers (2 Corinthians 11:14).

024:045 It is noteworthy that humans are included in the classification of living creatures according to physiological characteristics. This is in line with other verses that indicate an evolutionary method in creating Adam's biological body. Millions of years ago, the ability of a mammalian to walk on two feet is considered a critical point in the evolution of the human brain and the creation of homo sapiens. Walking on two feet might seem like a simple orthopedic change but its effect on the neuropsychological transformation is huge. See 29:18-20; 41:9-10; 7:69; 15:26; 32:7-9; 71:14-17.

Like the Quran, the Bible mentions water and ground as the main origin or ingredients of life (Genesis 1:20-21; 2:19).

024:058 The Quran provides sufficient details regarding how to perform *sala* prayer. See 11:114; 17:78; 38:32; 16:49. Taking short naps during midday is recommended since it improves people's performance in work substantially.

024:061 The Quran finds no problem in eating together but religious scholars who follow other sources besides the Quran, discourage the intermingling of men and women. The Quran does not discriminate against handicapped people. The only criterion that makes a person above another person in the sight of God is righteousness (49:13). Compare this to the instruction inserted into the Old Testament at Leviticus 21:16-24, which deems handicapped or disfigured people spiritually inferior to healthy ones. Also, see 48:17.

0024:063 This is a commonly mistranslated and abused verse. Muhammed was an elected leader of a multi-religious and multi-racial city-state, and the city was under constant military attack by neighboring polytheists. The verse is asking affirmers to take seriously the invitation of the messenger to a public gathering for a consultation limited to the community of affirmers. However, those who idolize Muhammed take the verse out of its context and interpret it to mean we should mention Muhammed's name with ostentatious titles. See 33:56; 2:48.

بسم الله الرحمن الرحيم

25:0 In the name of God, the Gracious, the Compassionate.

25:1 Blessed is the One who sent down the Statute Book to His servant, so he can be a warner to the world.

25:2 The One who has the sovereignty of the heavens and the earth; He did not take a son, and He does not have any partner in kingship. He created everything and measured it precisely.*

25:3 They took besides Him gods who do not create anything and are themselves created! They do not possess power to harm nor benefit, nor do they possess death or life, nor resurrection.*

The Secret Evidence

25:4 Those who rejected said, "This is but a falsehood that he invented and other people have helped him with it; for they have come with what is wrong and fabricated."

25:5 They said, "Mythologies of the ancient people; he wrote them down while they were being dictated to him morning and evening."*

25:6 Say, "It was sent down by the One who knows the secrets in the heavens and the earth. He is always Forgiving, Compassionate."*

25:7 They said, "What is with this messenger that he eats the food and walks in the marketplaces? If only a controller were sent down to him so that he would jointly be a warner."

25:8 "Or that a treasure is given to him, or that he has a paradise that he eats from." The wicked said, "You are but following a man bewitched!"

25:9 See how they cited the examples for you, for they have strayed, and they cannot find a path.

25:10 Blessed is the One who if He wishes can make for you better than that. Paradises with rivers flowing beneath them, and He will make for you palaces.

25:11 But they have denied the Moment, and We have prepared for those who deny the Moment a flaming fire.
25:12 When it sees them from a far place, they hear its raging and roaring.
25:13 When they are cast into it from a tight crevice, in chains, they call out their remorse.
25:14 Do not call out one sorrow today but call out many sorrows.
25:15 Say, "Is that better or the Paradise of eternity that the righteous have been promised?" It is their reward and destiny.
25:16 In it they will have what they wish eternally. It is upon your Lord an obligated promise.

The Answer of the Idolized Heroes

25:17 On the day when We gather them together with what they served besides **God**; then He will say, "Did you misguide My servants here, or did they stray from the path?"
25:18 They said, "Glory be to You, it was not for us to take besides You any allies; indeed, You gave them and their fathers luxury until they forgot the remembrance, and they were a lost people."
25:19 They have refuted what you claimed, so you will not find any excuse nor any victor. Whosoever has done wrong among you, We will let him taste a great retribution.

Messengers: Humans Like Us

25:20 We have not sent before you any messengers except that they ate the food and walked in the marketplaces. We have made some of you as a trial for others to see if you will have patience. Your Lord was ever seeing.
25:21 Those who do not expect to meet Us said, "If only the controllers were sent down upon him, or that we see our Lord." They have become arrogant in themselves, and they have produced a great blasphemy!
25:22 On the day they see the controllers, that is not good news for the criminals. They will say, "Away to a place that is sealed off."
25:23 We turned to the works that they did, and We made it dust in the wind.
25:24 The dwellers of Paradise on that day are in the best abode and the best destiny.
25:25 The day when the sky will be filled with clouds, and the controllers will be sent down in succession.
25:26 The true kingship on that day will be to the Gracious. It is a day which will be very hard on the ingrates.
25:27 That day the wrongdoer will bite on his hand and say, "I wish I had taken the path with the messenger!"
25:28 "Woe unto me, I wish I did not take so and so as a friend!"
25:29 "He has misguided me from the remembrance after it came to me, and the devil was always a betrayer of people!"

The Intercession (Testimony) of the Messenger on Behalf of His People

25:30 The messenger said, "My Lord, my people have deserted this Quran."*
25:31 It is such that for every prophet we allow enemies from criminals. Your Lord suffices as a Guide and a Victor.*
25:32 Those who rejected said, "If only the Quran was sent down to him in one go!" It was done as such so We could strengthen your heart with it, and We arranged it accordingly.
25:33 For every example they come to you with, We bring you the truth and the best explanation.
25:34 Those who will be gathered to hell on their faces, these are the most evil and the most strayed from the path.
25:35 We gave Moses the book and We made his brother Aaron a minister with him.
25:36 So We said, "Go both of you to the people who have denied Our signs." So, We then destroyed them utterly.
25:37 The people of Noah, when they denied the messengers We drowned them, and We made them a lesson

for people. We have prepared for the wicked a painful retribution.

25:38 And Aad, Thamud, the dwellers of Al-Raas, and many generations in between.

25:39 For each We cited the examples, and each We destroyed utterly.

25:40 They have come upon the town that was showered with a miserable shower. Did they not see it? No, they do not expect any resurrection.

25:41 If they see you, they take you for a mockery: "Is this the one that **God** sent as a messenger?"

25:42 "He nearly diverted us from our gods had we not been patient for them." When they see the retribution, they will learn who is on a path most astray.

25:43 Have you seen the one who has taken his ego as his god? Will you be a caretaker over him?

25:44 Or do you think that most of them hear or understand? They are just like cattle. No, they are worse off.

God's Blessings

25:45 Did you not see to your Lord how He casts the shadow? If He wished, He could have made it still, then We would have made the Sun as a guide to it.

25:46 Then We withdraw it to us a simple withdrawal.

25:47 He is the One who made for you the night as a cover, sleep for resting, and He made the day to move about in.*

25:48 He is the One who sent the winds to spread between the hands of His mercy. We sent down from the sky water which is pure.

25:49 So that We can revive a dead land with it and We supply drink for many what we created; the livestock and people.

25:50 We have distributed it amongst them so that they may remember, but most people refuse to be anything but ingrates.

25:51 If We wish, We could send to every town a warner.*

25:52 So do not obey the ingrates and strive against them with it in a great striving.

25:53 He is the One who merges the two bodies of water. This is fresh and palatable, and this is salty and bitter. He made between them a partition and an inviolable barrier.*

25:54 He is the One who created a human being from the water, so He made for him blood kin and mates. Your Lord is Omnipotent.

25:55 They serve besides **God** what does not benefit them nor harm them. The ingrate is always set against his Lord.

25:56 We have not sent you except as a bearer of good news and a warner.

25:57 Say, "I do not ask you for a wage; but for whoever chooses to make a path to his Lord."

25:58 Put your trust in the Eternal who does not die; glorify His praise. None is aware of the sins of His servants as He.

25:59 The One who created the heavens and the earth and what is between them in six days, then He settled to the throne. The Gracious; so ask Him for He is Ever-aware.*

25:60 If they are told: "Prostrate to the Gracious." They say, "What is the Gracious? Shall we prostrate to what you order us?" It increases their aversion.*

25:61 Blessed is the One Who made galaxies in the universe, and He made in it a beacon and a shining moon.

25:62 He is the One Who made the night and the day in succession, for those who wish to remember or wish to be thankful.

The Servants of God

25:63 The servants of the Gracious who walk on the earth in humility and if the ignorant speak to them, they say, "Peace."

25:64 Those who stay awake for their Lord, in prostration and standing.

25:65 Those who say, "Our Lord, avert from us the retribution of hell. Its retribution is terrible."

25:66 "It is a miserable abode and dwelling."

25:67 Those who when they give they are

not excessive nor stingy, but they are in a measure between that.

25:68 Those who do not call on any other god with **God**, nor do they kill the life that **God** has made forbidden except in justice, nor do they commit adultery; and whosoever does that will receive the punishment.

25:69 The retribution will be doubled for him on the day of Judgment, and he will abide in it eternally in disgrace.

25:70 Except for the one who repents and affirms and does good work, for those **God** will replace their sins with good; **God** is Forgiving, Compassionate.

25:71 Whosoever repents, and does good, then he shall repent towards **God** a true repentance.

25:72 Those who do not bear false witness, and if they pass by vain talk they pass by with dignity.

25:73 Those who when they are reminded of their Lord's signs, they do not fall on them deaf and blind.

25:74 Those who say, "Our Lord, grant us from our mates and our progeny what will be the comfort of our eyes, and make us role models for the righteous."

25:75 These will be rewarded with a dwelling for what they have been patient for, and they will find in it a greeting and peace.

25:76 In it they will abide, what an excellent dwelling and position.

25:77 Say, "My Lord would not care about you except for your imploring. But you have denied, so it will be your destiny."

ENDNOTES

025:002 The planet earth is a self-contained spaceship and each of us astronauts were beamed down here into the bodies with our memories erased. Everything we need for life: food, water, and oxygen is provided for us. We are also provided with hundreds of elements to tinker with, underground fossils to use, and we are rewarded with technological candies each time we learn a letter or sentence from the book of nature. Everything is interconnected with a complex network according to a well calculated plan.

025:003 The majority of humans on earth, carve idols from dead prophets, saints, cult leaders, historical heroes, and founding fathers. The living leaders of polytheistic societies gain more power and authority by exploiting the exaggerated power of dead idols that are monopolized and used as the trademark of a particular ruling class. Turning the tombs or relics of dead idols into shrines or memorials are characteristics of polytheistic societies. Around the names of common idols, certain doctrines and practices are baptized and transformed into religious or political dogmas, thereby turning them into taboos. This might provide short-term security, unity and tranquility in the society, but in the long run, when circumstances change and old realities are replaced with new realties, the society that has turned fallible humans, institutions, and teachings into absolute facts, start having problems with the new realities. The tension between new and old, between religion and science, between reality and fiction, between the pragmatic and symbolic, occasionally may lead to a clash and revolution.

025:005 Contemporaries of Muhammed knew that Muhammed was a literate person. Prophet Muhammed wrote God's message with his own hand. Also, see 7:157-158; 2:78; 3:20.

025:006 The answer contains a secret or a hidden word, an implication of the hidden code of the mathematical structure, which was unveiled in 1974, that is, 1406 lunar years after the revelation of the Quran. See 74:1-56.

025:030 Soon after Muhammed's death, hypocrites and those who reverted to polytheism engaged in the oral transmission of hearsay, thereby creating another source besides the Quran. Centuries later, they were collected, written and codified as the teachings of various sects. Those who follow these hearsays and fabrications pretend to follow Muhammed, while they abandon the only book delivered by him. They will be surprised on the Day of Judgment to see that Muhammed will reject their fabrications and idolatry, by following the rules expressed in 78:38 and 19:87. Muhammed will intercede for them by fulfilling the prophecy in verse 25:30.

025:031 For the identity of the ardent enemies of the messenger, see 6:112-116.

025:047 Sleep is as essential as drinking water and consuming food. Sleep deprivation reduces mental and bodily functions and after a certain point, the brain crashes. After 24 waking hours, the brain's ability to use glucose goes down dramatically. The purpose and meaning of sleep and dreaming is still the subject of scientific debate and research. There are many theories regarding why we sleep. Some researchers believe that we sleep and dream for rest, or for the brain to restore and rejuvenate itself. Others claim that sleep and dreams allow the brain to perform a series of repeated cycles of pruning and strengthening of neural connections that enable the person to learn new information without forgetting old information. Some claim that the brain needs downtime to detoxify itself from free radicals generated by our metabolism. Perhaps, sleep allows the brain to organize the accumulated data, sort them, place them in their proper memory folders, and then associate them with little tags for easy retrieval. The brain is like a busy office where the secretary cannot keep up with processing the numerous documents, faxes, and boxes that are sent and received during daytime. The clutter caused by this high traffic rate reduces performance in the office, and during the night shift, another work starts to sift, clean, and organize the office. Similarly, our brains need such a cleaning and organization period. Dreams perhaps happen when new files are organized and associated, folders are filled, and unused data are discarded. See 2:255; 8:43; 12:6; 30:23; 39:42; 78:9; 37:102.

025:051 Every town (*qarya*) may not receive a warner, but every community/nation (*umma*) receives a warner. See 35:24.

025:053 Could this barrier be evaporation? The water of lakes is maintained by rain. Through evaporation, the salty ocean water is filtered from its minerals. Evaporation creates a physical barrier between salty water and drinkable water.

025:059 For the evolutionary creation of the universe, see 29:18-20; 41:9-10; 7:69; 15:26; 24:45; 32:7-9; 71:14-17.

025:060 It seems that the Arabs of the day had some aversion towards the divine attribute *Rahman*. See 17:110.

بسم الله الرحمن الرحيم

26:0 In the name of God, the Gracious, the Compassionate.
26:1 T9S60M40*
26:2 These are the signs of the manifest/clarifying book.
26:3 Perhaps you are grieving yourself that they do not affirm.
26:4 If We wish, We could send down for them from the heavens a sign, to which they would bend their necks in humility.

Aversion to a New Message

26:5 Not a new reminder comes to them from the Gracious, except that they turn away from it.
26:6 They have denied, thus the news will come to them of what they used to ridicule.
26:7 Did they not look to the earth, how many plants have We raised in it, from each a good pair.
26:8 In that is a sign, but most of them are not those who affirm.
26:9 Your Lord is the Noble, the Compassionate.

Moses and Aaron Risk Their Lives to Free the Children of Israel

26:10 When your Lord called Moses: "Go to the people who are wicked."
26:11 "The people of Pharaoh. Will they not be righteous?"
26:12 He said, "My Lord, I fear that they would deny me."
26:13 "My chest would become tight, and my tongue would not be able to express; so send for Aaron."
26:14 "They have charges of a crime against me, so I fear they will kill me."
26:15 He said, "It will not be. Go both of you with Our signs. I am with you listening."
26:16 "So both of you go to Pharaoh and say, 'We are messengers of the Lord of the worlds…'"*
26:17 "'So send with us the Children of Israel'"
26:18 He said, "Did we not raise you amongst us as a newborn, and you stayed with us for many of your years?"
26:19 "You did the deed what you did, and you are of the ingrates."
26:20 He said, "I did it, and I was of those misguided."
26:21 "So I ran away from you all when I feared you. So my Lord granted me judgment, and made me of the messengers."
26:22 "That was a favor that you held against me, so you could continue to enslave the Children of Israel."
26:23 Pharaoh said, "What is the Lord of the worlds?"*
26:24 He said, "The Lord of the heavens and the earth and all that is between them, if you are aware."
26:25 He said to those around him: "Do you hear that?"
26:26 He said, "Your Lord and the Lord of your forefathers!"
26:27 He said, "This messenger of yours who has been sent to you is mad!"
26:28 He said, "The Lord of the east and the west, and what is between them, if you understand."
26:29 He said, "If you take a god other than me, then I will put you among the prisoners."
26:30 He said, "What if I brought you proof?"
26:31 He said, "Then bring it forth if you are of the truthful ones."
26:32 So he cast his staff, then it manifested into a serpent.
26:33 He drew out his hand, then it became white to the onlookers.
26:34 He said to the commanders around him: "This is a knowledgeable magician."
26:35 "He wants to bring you out of your land with his magic. So, what will you decide?"
26:36 They said, "Delay him and his brother, and send people in the cities to gather."
26:37 "They will come to you with every knowledgeable magician."

26:38 So the magicians were gathered to an appointed day.

26:39 It was said to the people: "Will you also gather?"

26:40 "Perhaps we can follow the magicians if they are the winners."

26:41 So when the magicians came, they said to Pharaoh: "We should be rewarded if we are the winners."

26:42 He said, "Yes, and you shall also be near to me."

26:43 Moses said to them: "Cast down what it is you will cast."

26:44 So they cast down their ropes and staffs and they said, "By the might of Pharaoh, we will be the winners."

26:45 So whenever Moses cast down his staff, it ate up all that they showed!

26:46 So the magicians went down prostrating.

26:47 They said, "We affirm the Lord of the worlds!"

26:48 "The Lord of Moses and Aaron."

26:49 He said, "Have you affirmed him before I permitted you? He is surely your great one who has taught you magic. So you shall come to know. I will cut off your hands and feet from alternate sides, and I will crucify you all."

26:50 They said, "There is no worry, for we are all returning to our Lord."

26:51 "We hope that our Lord will forgive us our sins, as we are the first to affirm."

26:52 We revealed to Moses: "Take away My servants, for you will be followed."

The Attempt to Maintain Slavery Fails

26:53 So Pharaoh sent gatherers to the cities.

26:54 "These are but a small band."

26:55 "They have done what has enraged us."

26:56 "We are all gathered and forewarned."

26:57 So, We evicted them out of gardens and springs.

26:58 Treasures and an honorable dwelling.

26:59 As such, We gave it to the Children of Israel.

26:60 So they were pursued at sunrise.

26:61 But when the two groups saw each other, the companions of Moses said, "We are caught!"

26:62 He said, "No, my Lord is with me and He will guide me."

26:63 So We revealed to Moses: "Strike the sea with your staff." So it split into two, each side like a great mountain.

26:64 We then brought them to the other side.

26:65 We saved Moses and all those with him.

26:66 Then We drowned the others.

26:67 In that is a sign, but most of them are not those who affirm.

26:68 Your Lord is the Noble, the Compassionate.

Abraham Reasons With His People

26:69 Recite to them the news of Abraham.

26:70 When he said to his father and his people: "What do you serve?"

26:71 They said, "We serve statues, thus we remain devoted to them."

26:72 He said, "Do they hear you when you pray to them?"

26:73 "Or do they benefit you or harm you?"

26:74 They said, "No, but we found our fathers doing the same."

26:75 He said, "Do you see what you have been serving."

26:76 "You and your fathers of old."

26:77 "They are enemies to me, except for the Lord of the worlds."

26:78 "The One who created me, He will guide me."

26:79 "He is the One who feeds me and gives me drink."

26:80 "If I am sick, He is the One who cures me."

26:81 "The One who will make me die and then bring me to life."

26:82 "The One whom I hope will forgive my faults on the day of Judgment."

26:83 "My Lord, grant me authority and join me with the good doers."

26:84 "Give me a tongue that is true for those who will follow."
26:85 "Make me of the inheritors of the gardens of bliss."
26:86 "Forgive my father, for he was of those misguided."
26:87 "Do not disgrace me on the day when they are resurrected."
26:88 "The day when no money or children can help."
26:89 "Except for he who comes to **God** with a pure heart."
26:90 Paradise was made near for the righteous.
26:91 Hell was displayed for the wrongdoers.
26:92 It was said to them: "Where is what you used to serve?"
26:93 "Without **God**, can they help you or help themselves?"
26:94 So they were thrown in it on their faces, them, and the wrongdoers.
26:95 As well as all the soldiers of Satan.
26:96 They said, while they were disputing therein:
26:97 "By **God**, we were clearly misguided."
26:98 "For we equated you with the Lord of the Worlds!"
26:99 "None misled us except the criminals."
26:100 "So we have none to intercede for us."
26:101 "Nor a close friend."
26:102 "If only we could have another chance, we would be among those who affirm."
26:103 In that is a sign, but most of them are not those who affirm.
26:104 Your Lord is the Noble, the Compassionate.

Noah and his Arrogant People

26:105 The people of Noah rejected the messengers.
26:106 When their brother Noah said to them: "Will you not be righteous?"
26:107 "I am to you a trustworthy messenger."
26:108 "So be aware of **God** and obey me."
26:109 "I do not ask you for a wage, for my reward is upon the Lord of the worlds."
26:110 "So be aware of **God** and obey me."
26:111 They said, "Shall we affirm you when the lowest type of people has followed you?"
26:112 He said, "What knowledge do I have of what they used to do?"
26:113 "Their judgment is on my Lord, if you could know."
26:114 "I will not drive away those who affirm."
26:115 "I am but a clear warner."
26:116 They said, "If you do not cease, O Noah, you will be among those who are stoned."
26:117 He said, "My Lord, my people have denied me!"
26:118 "So open between me and them; save me and those affirmers who are with me."
26:119 So We saved him and those who were with him in the charged Ship.
26:120 Then after that We drowned the rest.
26:121 In that is a sign, but most of them are not affirmers.
26:122 Your Lord is the Noble, the Compassionate.

Hood to Aad

26:123 Aad denied the messengers.
26:124 For their brother Hood said to them: "Will you not be righteous?"
26:125 "I am to you a clear messenger."
26:126 "So be aware of **God** and obey me."
26:127 "I do not ask you for a wage, for my reward is upon the Lord of the worlds."
26:128 "Do you build on every high place a symbol, for the sake of vanity!"
26:129 "You take for yourselves strongholds, perhaps you will live forever?"
26:130 "If you attack, you strike ruthlessly?"
26:131 "So be aware of **God** and obey me."

26:132 "Be aware of the One who provided you with what you know."
26:133 "He provided you with livestock and shelters."
26:134 "Gardens and springs."
26:135 "I fear for you the retribution of a great day"
26:136 They said, "It is the same whether you preach or do not preach."
26:137 "This is nothing but an invention by the people of old."
26:138 "We are not going to be punished."
26:139 So they denied him, and We destroyed them. In that is a sign, but most do not affirm.
26:140 Your Lord is the Noble, the Compassionate.

Saleh to Thamud

26:141 Thamud denied the messengers.
26:142 For their brother Saleh said to them: "Will you not be righteous?"
26:143 "I am to you a clear messenger."
26:144 "So be aware of **God** and obey me."
26:145 "I do not ask you for a wage, for my reward is upon the Lord of the worlds."
26:146 "Will you be left secure in what you have here?"
26:147 "In gardens and springs."
26:148 "Green crops and palm trees laden with fruit."
26:149 "You carve homes out of the mountains with great skill?"
26:150 "So be aware of **God** and obey me."
26:151 "Do not obey the command of the carefree."
26:152 "The ones who corrupt in the land and are not reformers."
26:153 They said, "You are but one of those bewitched!"
26:154 "You are but a human being like us. So bring a sign if you are of the truthful ones."
26:155 He said, "This is a female camel which will have to drink and you will drink on a different day."*
26:156 "Do not afflict her with harm, else the retribution of a great day will seize you."
26:157 But they slaughtered her, and they became regretful.
26:158 So the retribution took them. In that is a sign, but most of them are not those who affirm.
26:159 Your Lord is the Noble, the Compassionate.

Lot and Transgressing Homosexuals

26:160 The people of Lot denied the messengers.
26:161 For their brother Lot said to them: "Will you not be righteous?"
26:162 "I am to you a trustworthy messenger."
26:163 "So be aware of **God** and obey me."
26:164 "I do not ask you for a wage for my reward is upon the Lord of the worlds."
26:165 "Are you attracted to the males of the worlds?"
26:166 "You leave what your Lord has created for you of mates? You are an aggressive people!"
26:167 They said, "If you do not cease O Lot, you will be among those driven out."
26:168 He said, "I am in severe opposition to your acts!"
26:169 "My Lord, save me and my family from what they do."
26:170 So We saved him and his entire family.
26:171 Except for an old woman who remained.
26:172 Then We destroyed the others.
26:173 We rained upon them a rain. Miserable was the rain to those who had been warned.
26:174 In that is a sign, but most of them are not those who affirm.
26:175 That your Lord is the Noble, the Compassionate.

Shuayb Fights against Corruption and Fraud

26:176 The dwellers of the Woods rejected the messengers.
26:177 For Shuayb said to them: "Will you not be righteous?"
26:178 "I am to you a trustworthy messenger."
26:179 "So be aware of **God** and obey me."

26:180 "I do not ask you for a wage, for my reward is upon the Lord of the worlds."
26:181 "Give full measure and do not be of those who cause losses."
26:182 "Weigh with the balance that is straight."
26:183 "Do not defraud the people of their belongings, and do not venture into the land corrupting."
26:184 "Be aware of the One who created you and the generations of old."
26:185 They said, "You are but one of those bewitched."
26:186 "You are nothing but a human being like us; we think you are one of those who lie."
26:187 "So let pieces from the sky fall upon us if you are of those who are truthful!"
26:188 He said, "My Lord is most aware of what you do."
26:189 But they denied him, so the retribution of the day of shadow took them. It was the retribution of a terrible day.
26:190 In that is a sign, but most of them are not those who affirm.
26:191 Your Lord is the Noble, the Compassionate.

The Quran

26:192 This is a revelation from the Lord of the worlds.
26:193 It was sent down with the trusted Spirit.
26:194 Upon your heart, so that you would be of the warners.
26:195 In a clear Arabic tongue.
26:196 It is in the Psalms of old.
26:197 Was it not a sign for them that the scholars of the Children of Israel knew it?
26:198 Had We sent it down upon some of the foreigners,
26:199 He related it to them; they would not have affirmed it.
26:200 As such, We diverted it from the hearts of the criminals.
26:201 They do not affirm it until they see the painful retribution.
26:202 So it will come to them suddenly, while they do not perceive it.
26:203 Then they would say, "Can we be given more time?"
26:204 Was it not their wish that Our punishment be hastened?
26:205 Do you see that if We gave them luxury for years.
26:206 Then what they were promised came to them.
26:207 The luxury they were given will not avail them.
26:208 We have not destroyed any town except after having warners.
26:209 By way of a reminder, and We were never unjust.
26:210 It is not the devils that have brought this down.
26:211 Nor would they, nor could they.
26:212 They are blocked from overhearing.
26:213 So do not call upon any other god with **God**; else you will be with those punished.
26:214 Warn your closest kin.
26:215 Lower your wing for any who follow you of those who affirm.
26:216 Then if they disobey you, then say, "I am innocent from what you do."
26:217 Put your trust in the Noble, the Compassionate.
26:218 The One who sees you when you stand.
26:219 Your movements amongst those who prostrate.
26:220 He is the Hearer, the Knowledgeable.
26:221 Shall I inform you on whom the devils descend?
26:222 They descend on every sinful liar.
26:223 They claim to listen, but most of them are liars.
26:224 The poets, are followed by the strayers.
26:225 Do you not see that they traverse every valley?
26:226 That what they say, is not what they do?
26:227 Except for those who affirm, and promote reforms, and remember **God** greatly, and were victorious after they were wronged. As for those who did wrong, they will know which fate they will meet.

ENDNOTES

026:001 T9S60M40. This combination of letters/numbers plays an important role in the mathematical system of the Quran based on code 19. See 74:1-56; 1:1; 2:1; 13:38; 27:82; 38:1-8; 40:28-38; 46:10; 72:28.

026:001-6 See 6:124 and 38:1-15.

026:016-18 God's instruction to Moses, without being repeated by Moses, is followed by Pharaoh's answer, as if Moses had already told Pharaoh what was asked of him. This overlapping of scenes is one of the techniques used in the Quran to convey events. For other examples, see 12:83; 20:47-49.

026:023 See 2:60.

026:155 The miracles given to messengers relate to the interest and knowledge of their times. Perhaps, a tribal agrarian community that is knowledgeable about animals and their behavior would not appreciate a mathematical sign.

27:0 In the name of God, the Gracious, the Compassionate.

27:1 T9S60. These are the signs of the Quran and a clear book.*

27:2 A guide and good news to those who affirm.

27:3 Those who maintain the contact prayer, and contribute towards betterment, and regarding the Hereafter they are certain.

27:4 As for those who do not affirm the Hereafter, We have made their work appear pleasing to them, so they walk around blind.

27:5 They will have the worst retribution, and in the Hereafter, they are the biggest losers.

27:6 You are receiving the Quran from One who is Wise, Knowledgeable.

Moses

27:7 Moses said to his family: "I have seen a fire; I will bring you from there some news or I will bring you a burning piece so that you may be warmed."

27:8 So when he came to it he was called: "Blessed is the One at the fire and whoever is around it, and glory be to **God**, the Lord of the worlds."

27:9 "O Moses, it is I, **God**, the Noble, the Wise."

27:10 "Throw down your staff." So, when he saw it vibrated as if it were a Jinni, he ran away and would not turn back. "O Moses, do not fear, for My messengers shall have no fear from Me."

27:11 "Except one who transgresses. But then if he replaces the evil deed with good, then I am Forgiving, Compassionate."

27:12 "Place your hand into your pocket; it will come out white with no blemish, one of nine signs to Pharaoh and his people."

27:13 So when Our signs came to them for all to see, they said, "This is clearly magic!"

27:14 They rejected them; they justified transgression and arrogance for themselves. So see how it ended for the wicked.

David and Solomon

27:15 We bestowed upon David and Solomon knowledge, and they both said, "Praise be to **God** who preferred us over many of His affirming servants."

27:16 Solomon inherited from David, and he said, "O people, we have been taught how to understand the speech of birds, and we have been given from everything. This is indeed an evident grace."

27:17 Solomon's soldiers were gathered, comprising of humans and Jinn and birds, for they were to be spread out.

27:18 Until they came to a valley of ants, a female ant said, "O ants, enter your homes else you will be crushed by Solomon and his soldiers while they do not notice."*

27:19 He then smiled, amused by what she said. He said, "My Lord, help me to be thankful for the blessings You have bestowed upon me and upon my parents, and that I may promote reforms that please You, and admit me by Your Compassion with Your righteous servants."

27:20 He inspected the birds, then said, "Why do I not see the hoopoe, or is he among those absent?"

27:21 "I will punish him severely, or I will slaughter him, unless he gives me a clear excuse."

27:22 But the hoopoe did not stay away long, then he said, "I have seen what you do not know, and I have come to you from Sheba with news which is certain."

27:23 "I found them owned by a woman, and she was given all possession, and she had a great throne."

27:24 "I found her and her people prostrating to the Sun instead of **God**! The devil had made their work appear good to them, so he kept them away from the path, for they are not being guided."*

27:25 "Will they not prostrate to **God** who brings out what is hidden in the heavens and the earth, and He knows what you hide and what you declare?"

27:26 "**God**, there is no god but He, the Lord of the great throne."

27:27 He said, "We will see if you are being truthful or are one of those who lie."

27:28 "Take this letter of mine and deliver it to them, then withdraw from them and observe what they respond with."

27:29 She said, "O commanders, a noble letter has been delivered to me."

Solomon too Received the Numerically Coded Message

27:30 "It is from Solomon, and it reads: In the name of **God**, the Gracious, the Compassionate"*

بسم الله الرحمن الرحيم

27:31 ""Do not be arrogant toward me and come to me peacefully surrendering""*

27:32 She said, "O commanders, advise me in this matter of mine, for I will not take a decision until you give testimony."

27:33 They said, "We are people of strength and mighty in power. But the decision is yours, so see what you will command."

27:34 She said, "When the kings enter a town they destroy it and make its most noble people humiliated. It is such that they do."*

27:35 "I will send to them a gift, then I will see with what the messengers will return."

27:36 So when they came to Solomon he said, "Are you providing me with wealth? What **God** has provided for me is far better than what He has given you. Now you are happy with your gift!"

27:37 "Return to them. For we shall come to them with soldiers the like of which they have never seen, and we will drive them out humiliated, while they are feeble."

Teleportation

27:38 He said, "O commanders, which of you can bring me her throne before they come to me peacefully surrendering?"

27:39 A powerful being from among the Jinn said, "I will bring it to you before you rise from your station. For I am strong and trustworthy."

27:40 One who had knowledge from the book said, "I will bring it to you before you blink." So when he saw it resting before him, he said, "This is from the grace of my Lord, so that He tests me whether I am thankful or whether I reject. As for he who is thankful, he is thankful for himself, and as for he who rejects, then my Lord is Rich, Bountiful."

27:41 He said, "Disguise her throne so we may see if she will be guided or if she will be of those who are not guided."

27:42 So when she came, it was said, "Is your throne like this?" She said, "It appears to be similar." We were given knowledge before her, and we had peacefully surrendered.

27:43 She was prevented by what she served besides **God**. She was of the people who were ingrates.

27:44 It was said to her: "Enter the palace." So when she saw it she thought there was a pool, and she uncovered her legs. He said, "It is a palace paved with crystal." She said, "My Lord, I have wronged myself; and I peacefully surrender with Solomon to **God**, the Lord of the worlds."

Saleh

27:45 We have sent to Thamud their brother Saleh: "You shall serve **God**." But they became two disputing groups.

27:46 He said, "My people, why do you hasten with evil ahead of good? If you would only seek **God**'s forgiveness, perhaps you will receive mercy."

27:47 They said, "You have an ill omen with you and those with you." He said, "Your ill omen is with **God**, but you are a people that are being tested."

27:48 In the city were nine thugs who were causing corruption in the land, and they were not reforming.

27:49 They said, "Swear by **God** to one another that we will attack him and his family at night, and we will then say to his supporters: 'We did not witness who murdered his family, and we are being truthful.'"

27:50 They plotted a scheme and We plotted a scheme, while they did not notice.

27:51 So see what the result of their planning was! We destroyed them and their people together.

27:52 So these are their homes, ruined, for what they transgressed. In that is a sign for a people who know.

27:53 We saved those who affirmed and were righteous.

Lot

27:54 And Lot, when he said to his people: "Are you committing lewdness while you can clearly see?"

27:55 "You are approaching the men out of lust instead of the women! No, you are an ignorant people!"

27:56 But the reply of his people was that they said, "Expel the family of Lot from your town, for they are a people who are being pure!"

27:57 So We saved him and his family, except for his wife, for We found her to be of those who will remain.

27:58 We rained down a rain upon them. Miserable was the rain of those warned.

God

27:59 Say, "Praise be to **God**, and peace be upon His servants whom He has selected." Is **God** better, or what you set up?

27:60 The One who created the heavens and the earth, and He sent down water from the sky for you, so We cause gardens to grow with it that are full of beauty. It is not your ability to cause the growth of its trees. Is there a god with **God**? No. But they are a people who ascribe equals!

27:61 The One who made the earth a habitat, and He made in it rivers and He made for it stabilizers, and He made between the two seas a barrier. Is there a god with **God**? No. But most of them do not know.

27:62 The One who answers the distressed when he calls Him, and He removes the evil, and He makes you successors on earth. Is there a god with **God**? Little do you remember!

27:63 The One who guides you in the darkness of the land and the sea, and He sends the winds to spread between the hands of His mercy. Is there a god with **God**? **God** be exalted above what they set up!

27:64 The One who initiates the creation then He returns it, and He provides for you from the sky and the land. Is there a god with **God**? Say, "Bring your proof if you are being truthful."

27:65 Say, "None in the heavens or the earth know the unseen except **God**. They do not perceive when they will be resurrected."

27:66 No, they have no knowledge of the Hereafter. No, they are in doubt regarding it. No, they are blind to it.*

27:67 Those who rejected said, "When we have become dust, as our fathers, shall we be brought out?"

27:68 "We have been promised this, both us and our fathers before. This is nothing but tales of the ancients!"

27:69 Say, "Roam the earth and see what the end of the criminals was."

27:70 Do not be saddened for them, and do not be distraught for what they plot.

27:71 They say, "When is this promise if you are truthful?"

27:72 Say, "You may already be being followed by some of what you hasten."

27:73 Your Lord has made favors for people, but most of them are not thankful.

27:74 Your Lord knows what is concealed in their chests and what they reveal.

27:75 There is not a thing hidden in the heavens or the earth, but is in a clear record.

27:76 This Quran narrates to the Children of Israel most of what they are in dispute over.

27:77 It is a guidance and mercy for those who affirm.

27:78 Your Lord decides between them by His judgment. He is the Noble, the Knowledgeable.

27:79 So put your trust in **God**, for you are clearly on the truth.

27:80 You cannot make the dead hear, nor can you make the deaf hear the call when they turn their backs and flee.

27:81 Nor can you guide the blind from their misguidance. You can only make those who affirm Our signs hear you, for they have peacefully surrendered.

Computer Speaks

27:82 When the punishment has been deserved by them, We will bring out for them a creature made of earthly material, it will speak to them that the people have been unaware regarding Our signs.*

27:83 The day We gather from every nation a party that denied Our signs, then they will be driven.

27:84 Until they have come, He will say, "Have you denied My signs while you had no explicit knowledge of them? What were you doing?"*

27:85 The punishment was deserved by them for what they transgressed, for they did not speak.

27:86 Did they not see that We made the night for them to rest in, and the day to see in? In that are signs for a people who have faith.

27:87 On the day when the horn is blown, then those in the heavens and the earth will be in terror except for whom **God** wills. All shall come to Him humbled.

Earth's Motion

27:88 You see the mountains, you think they are solid, while they are passing by like the clouds. The making of **God** who perfected everything. He is Ever-aware of what you do.*

27:89 Whoever comes with a good deed will receive better than it, and from the terror of that day they will be safe.

27:90 Whoever comes with the bad deed, their faces will be cast in hell. Are you not being rewarded for what you used to do?

27:91 "I have been ordered to serve the Lord of this town that He has made restricted, and to Him are all things, and I have been ordered to be of those who peacefully surrender."

27:92 "That I recite the Quran." He who is guided is guided for himself, and to he who is misguided, say, "I am but one of the warners."

For the Future

27:93 Say, "Praise be to **God**, He will show you His signs and you will know them. Your Lord is not unaware of what you do."*

ENDNOTES

027:001 **T**9S60. This combination of letters/numbers plays an important role in the mathematical system of the Quran based on code 19. See 74:1-56; 1:1; 2:1; 13:38; 27:82; 38:1-8; 40:28-38; 46:10; 72:28.

027:018 Solomon's demonstration of miracles makes sense within the context of the Quran, but how can we explain the ants' recognition of Solomon and his troops and communicating this information with each other? Were ants also showing miracles?! Do all ants have such a secret talent? Did they recognize Solomon from the smell of his feet? 20:114.

027:024 For prostration, see 16:49.

027:030 The Basmalah in this verse, compensates the missing Basmalah from the beginning of chapter 9, exactly 19 chapters before. This extra Basmalah here, thus, completes the number of Basmalahs being 114 (19x6). See 1:1; 74:1-56.

027:031-37 Did Solomon violate the principle of "no compulsion in religion" by issuing such a threat? Or was he just bluffing? Even if he did not mean what he said, didn't he risk the possibility of a war? Solomon could not have violated the divine principles, such as freedom of choosing one's own religion (2:256), the freedom to reject the truth (2:193; 10:99; 18:29), and the promotion of peace among people (2:208).

The above questions raised in the first edition of the QRT drew the attention of some other students of the Quran. For instance, El Mehdi Haddou, reminded me the word *taMLiKuhum* in verse 27:23 which means, "own them, possess them." He proposed "owning" instead of "leading." Indeed, the derivatives of this word occur 44 times in the Quran, and all are related to ownership and possession. I would like to quote from Mehdi's article on this verse, which is published in an anthology containing articles of thirty authors, Critical Thinkers for Islamic Reform, Brainbow Press, 2009:

"What really made Solomon angry is the state of slavery that affects the people owned by the Queen of Sheba and not their religion. Solomon's reaction against slavery and oppression should be an example for every leader who has the power, the means and the potential to free slaves and fight for the rights of oppressed people even if they are living in another country. What Solomon did in the past is what the United States should do in our time. Their resemblance is reflected in their military power, but the difference resides in the fact that only Solomon has submitted to God's laws. If Solomon was living in the 21st century, who would you think will be at the top of his blacklist? Yes, many from the so-called Muslim countries."

027:034 It is interesting that Queen is criticizing Kings for their corruption and aggression. The Queen is depicted by the Quran as someone who is not a tyrant, but a reasonable leader who consults others in her affairs, which is one of the main principles in the Quran (42:38). The Queen of Sheba tested Solomon by sending him some gifts. When she learned that he was not interested in her wealth or her country's resources, she became curious about Solomon and accepted his invitation. The event implies more communication between the two leaders and the emergence of trust between them. Ultimately, she realizes that Solomon is not one of those corrupt kings, but a servant of God with great morals.

The Bible mentions this meeting, but it contains some differences. For instance, in I Kings 10:1-13 we learn that Solomon accepted her valuable gifts, but the Quran tells us that Solomon showed no interest. The same Biblical passage is repeated at II Chronicles 9:1-12.

027:066 Having doubts about the hereafter is similar to not accepting it. The Quran does not accept "faith" as conjecture; the Quran expects us to affirm the hereafter as a fact, by reflecting on circumstantial evidence abundant around us and within us. According to the Quran, those who do not find the Quran sufficient for guidance, those who do not enjoy mentioning God's name alone, and those who expect the intercession of a savior, do not really affirm the hereafter. See 6:113,150; 27:66; 34:21; 39:45; 74:46-48.

027:082 Note that the creature mentioned here is described as *dabbatan min al-ard* (creature from earth), which means either a creature coming out of the earth, or a creature made of earthly materials. Could it refer to the computer that became the tool for the discovery of the code 19 and declaration of its message?

Unlike the creatures made of water (24:45), this one is made from earthly elements. After the discovery of the code 19, and from the context of this verse, this creature (*DaBBa*) might be understood as a reference to the computer that uses silicon as its essential part. In traditional books, there are bizarre descriptions of this prophesied creature. Contrary to the Quran's positive depiction of the earth-based creature, *hadith* books contain negative descriptions.

Some may object to our interpretation by saying that the word *DaBBa* implies something alive that moves. This objection is reasonable. But we prefer the computer to an animal since we can understand the phrase *dabbatan min al-ard* not as a creature geographically from earth, but rather a creature made of earth. Thus, we can accept some differences in mobility between the "*dabba* from water" and "*dabba* from earth." Since a computer has many moving parts, from its hard disk to the information carried by trillions of electrons, it is not a far-fetched understanding of *DaBBa*. See 72:24-28.

027:084 See the relationship between this and verse 10:38-48.

027:088 Interspersed among the verses about the hereafter, the verses 86 and 88 remind us of examples of God's power that we witness. During the revelation of the Quran, not many people considered the earth a moving object; perhaps none knew about the tectonic movement of mountains. Early commentators of the Quran also could not comprehend this fact. So, they struggled to make sense out of this bizarre claim in their holy book. They claimed that the verse refers to the end of the world. However, they ignored that the verses describing the last Hour depicted chaos and destruction. The end of the world is a calamity of cosmic proportions, which creates an utmost shock and panic among those who will experience it (See chapter 81 and 101 of the Quran). Thus, the unnoticed event described in 27:88 cannot be about the chaos created by the destruction of the universe. Indeed, the end of the verse 27:88 leaves no room for such confusion: the making of God who perfected everything. The end of the world is neither the perfection of art, nor an imperceptible event. The early audience of the Quran, unable even to imagine the motion of the mountains and earth, tried hard to reconcile it with the verses describing the destruction, chaos, and panic of the end times. See 21:33; 36:40; 39:5; 68:1; 79:30.

Since the Bible has been tampered by clergymen (See Quran 2:59), it contains numerous false information about nature and scientific issues. Christendom, up to the 17th century, firmly believed that the earth was a firm center of the universe. The Church persecuted many scientists, including Galileo, for asserting that the earth rotates around the sun. See Psalms 93:1; 96:10; 104:5; 1 Chronicles 16:30.

027:093 See 41:53; 74:30-35.

28:0. In the name of God, the Gracious, the Compassionate.

28:1 T9S60M40.*

28:2 These are the signs of the clear book.

Moses

28:3 We recite to you from the news of Moses and Pharaoh with truth, for a people who affirm.

28:4 Pharaoh became mighty in the land, and he turned its people into factions, he oppressed a group of them by killing their children and shaming their women. He was of those who corrupted.

28:5 We wanted to help those who were oppressed in the land, and to make them role models, and to make them the inheritors.

28:6 To enable them in the land, and to show Pharaoh and Haman and their troops what they had feared.

28:7 We revealed to Moses' mother: "Suckle him, and when you become fearful for him, then cast him off in the sea, and do not fear nor grieve. We will return him to you and We will make him of the messengers."

Raised by the Pharaoh's Family

28:8 Then the family of Pharaoh picked him up, so he would be an enemy to them and a source of sadness. Certainly, Pharaoh and Haman and their troops were doing wrong.

28:9 Pharaoh's wife said, "A pleasure to my eye and yours, so do not kill him, perhaps he will benefit us or we may adopt him as a son!", while they did not perceive.

28:10 Moses' mother's heart became anxious, that she nearly revealed her identity. But We strengthened her heart, so that she would be of those who affirm.

28:11 She said to his sister: "Follow his path." So, she watched him from afar, while they did not notice.

28:12 We forbade him from accepting all the nursing mothers. Then his sister said, "Shall I lead you to a family that can nurse him for you, and take good care of him?"

28:13 Thus, We returned him to his mother, so that she may be pleased and not be saddened, and to let her know that **God**'s promise is the truth. However, most of them do not know.

28:14 When he reached his maturity and strength, We bestowed him with wisdom and knowledge. We thus reward the good doers.

Manslaughter

28:15 He entered the city unexpectedly, without being noticed by the people. He found in it two men who were fighting, one was from his own race, and the other was from his enemy's. So, the one who was from his own race called on him for help against his enemy, whereby Moses punched him, killing him. He said, "This is from the work of the devil; he is an enemy that clearly misleads."*

28:16 He said, "My Lord, I have wronged my person, so forgive me." He then forgave him, for He is the Forgiver, the Compassionate.

28:17 He said, "My Lord, for what blessings you have bestowed upon me, I will never be a supporter of the criminals."

28:18 So he spent the night in the city, afraid and watchful. Then the one who sought his help yesterday, was asking again for his help. Moses said to him: "You are clearly a troublemaker."

28:19 But when he was about to strike their common enemy, he said, "O Moses, do you want to kill me, as you killed that person yesterday? Obviously, you wish to be a tyrant on earth; you dó not wish to be of the righteous."

28:20 A man came running from the far side of the city, saying: "O Moses, the commanders are plotting to kill you, so leave immediately. I am giving you good advice."

28:21 He exited the city, afraid and watchful. He said, "My Lord, save me from the wicked people."

28:22 As he traveled towards Midyan, he said, "Perhaps my Lord will guide me to the right path."

28:23 When he arrived at the watering hole of Midyan, he found a crowd of people watering. He noticed two women waiting on the side. He said, "What is holding you back." They said, "We cannot draw water until the shepherds finish, and our father is an old man."

28:24 So he drew water for them, then he turned to a shaded area, and he said, "My Lord, I am in dire need of any provision You may sent down for me."

28:25 So one of the two women approached him, shyly, and said, "My father invites you to reward you for watering for us." So when he came to him, and told him his story, he said, "Do not fear, for you have been saved from the wicked people."

Moses Marries the Woman Who Chooses Him

28:26 One of the two women said, "O my father, hire him. For the best to be hired is one who is strong and honest."

28:27 He said, "I wish you to marry one of my two daughters, on condition that you work for me through eight periods of Debate Conference; if you complete them to ten, it will be voluntary on your part. I do not wish to make this matter too difficult for you. You will find me, **God** willing, of the righteous."

28:28 He said, "It is then an agreement between me and you. Whichever period I fulfill, you will not be averse to either one. **God** is the guarantor over what we said."

Back to Egypt

28:29 So when Moses fulfilled his obligation, he was traveling with his family, and he saw a fire from the slope of the mount. He said to his family: "Stay here. I have seen a fire. Perhaps I can bring to you news from there, or a portion of the fire to warm you."

28:30 When he reached it, he was called from the tree at the edge of the right side of the valley on blessed ground: "O Moses, it is Me, **God**, the Lord of the worlds."

28:31 "Now, throw down your staff." So, when he saw it vibrate as if it were a Jinni, he turned around to flee and would not return. "O Moses, come forward and do not be afraid. You are of those who are safe."

28:32 "Place your hand into your pocket. It will come out white without any blemish, and fold your hand close to your side against fear. These are two proofs from your Lord, to Pharaoh and his entourage. They have been a wicked people."

28:33 He said, "My Lord, I have killed a person from them, so I fear that they will kill me."

28:34 "My brother Aaron is more eloquent in speech than I. So send him with me to help and to confirm me. For I fear that they will deny me."

28:35 He said, "We will strengthen you with your brother, and We will provide you both with authority. Thus, they will not be able to touch either one of you. With Our signs, the two of you, along with those who follow you, will be the victors."

28:36 So when Moses went to them with Our clear signs, they said, "This is nothing but fabricated magic; nor did we hear of this from our fathers of old!"

28:37 Moses said, "My Lord is fully aware of who has brought the guidance from Him, and who will have the best deal in the Hereafter. The wicked never succeed."

Pharaoh Mocks Moses

28:38 Pharaoh said, "O commanders, I have not known of any god for you other than me. Therefore, O Haman, fire up the bricks and build me a platform, so perhaps I may take a look at the god of Moses! I am certain he is a liar."

28:39 He and his soldiers were arrogant in the land without any right, and they thought that they would not be returned to Us.

28:40 So We took him and his soldiers, and We cast them into the sea. So see what the end of the wicked was!

28:41 We made them role models for inviting to the fire. On the day of Resurrection, they will not be supported.

28:42 We made them followed by a curse in this world, and on the day of Resurrection they will be despised.

28:43 We had given Moses the book after We had destroyed the earlier generations, as an example for people and a guidance and mercy, perhaps they will take heed.

28:44 You were not on the western slope when We decreed the command to Moses. You were not a witness.

28:45 We established many generations, and many ages passed them by. You were not living among the people of Midyan, reciting Our signs to them. But We did send messengers.

28:46 Nor were you on the side of the mount when We called. But it is a mercy from your Lord, so that you may warn a people who received no warner before you, perhaps they may take heed.

28:47 Thus, if any disaster strikes them because of their own deeds they cannot say, "Our Lord, if only You sent a messenger to us, so that we would follow Your signs, and we would be of those who affirm."

Religious People do not Appreciate New Signs

28:48 But when the truth came to them from Us, they said, "If only he was given the same that was given to Moses!" Had they not rejected what was given to Moses before? They had said, "Two magicians assisting one another." They said, "We reject all these things."

28:49 Say, "Then bring forth a book from **God** that is better in guidance so I may follow it, if you are truthful."

28:50 But if they fail to respond to you, then know that they follow only their desires. Who is more astray than one who follows his desire, without guidance from **God**? **God** does not guide the wicked people.

The People of the Book Should Recognize the Message

28:51 We have delivered the message, perhaps they may take heed.

28:52 Those whom We gave the book to before this, they will affirm it.

28:53 If it is recited to them, they say, "We affirm it. This is the truth from our Lord. Indeed, we had already peacefully surrendered before this."

28:54 To these We grant twice the reward for that they have been patient. They counter evil with good, and from Our provisions to them, they give.*

28:55 If they come across vain talk, they disregard it and say, "To us are our deeds, and to you is your deeds. Peace be upon you. We do not seek the ignorant."

Only God Guides

28:56 You cannot guide whom you love. But it is **God** who guides whom He wills; and He is fully aware of those who receive the guidance.*

28:57 They said, "If we follow the guidance with you, we will be deposed from our land." Did We not establish for them a safe territory, to which all kinds of fruits are offered, as a provision from Us? Indeed, most of them do not know.

28:58 How many a town have We destroyed for turning boastful of their lifestyles. So here are their homes, nothing but uninhabited ruins after them, except a few. We were the inheritors.

28:59 Your Lord never annihilates the towns without sending a messenger to their capital, to recite Our signs to them. We never annihilate the towns unless their people are wicked.

28:60 Anything that is given to you is only the materials of this world, and its glitter. What is with **God** is far better, and everlasting. Do you not reason?

28:61 Is one whom We promised a good promise, and it will come to pass, equal to one whom We provided with the goods of this life, then on the day of Resurrection he is of those who will suffer?

28:62 On the day when He calls upon them, saying: "Where are My partners whom you used to claim?"

28:63 Those who have deserved the retribution will say, "Our Lord, these are the ones we misled; we misled them only because we ourselves were misled. We seek to absolve ourselves to You, it was not us that they served."

28:64 It will be said, "Call upon your partners," but they will not respond to them. They will see the retribution. If only they were guided!

28:65 On that day, He will ask them and say, "What was the answer you gave to the messengers?"

28:66 They will be shocked by the news on that day, they will be speechless.

28:67 As for the one who repents, affirms, and does good works; perhaps he will be with the winners.

28:68 Your Lord creates what He wills, and He selects; it is not for them to select. Glory be to **God**; He is far above the partners they set up.

28:69 Your Lord knows what is concealed in their chests and what they declare.

28:70 He is **God**, there is no god but He. To Him belongs all praise in the first and in the last, and judgment belongs with Him, and to Him you will be returned.

28:71 Say, "Have you noted: what if **God** made the night eternal, until the day of Resurrection? Which god, other than **God**, can bring you with light? Do you not listen?"

28:72 Say, "Have you noted: what if **God** made the daylight eternal, until the day of Resurrection? Which god, other than **God**, can provide you with a night for your rest? Do you not see?"

28:73 From His mercy is that He created for you the night and the day, to rest in it, and that you may seek of His provisions. Perhaps you may be appreciative.

28:74 The day He will call them and say, "Where are My partners whom you had claimed?"*

28:75 We will extract from every nation a witness, then We will say, "Bring forth your proof." They will then realize that all truth belongs with **God**, and what they had invented will abandon them.

Qarun: A Hedonist and Arrogant Capitalist

28:76 Qarun was from among Moses' people, but he betrayed them. We gave him such treasures that the keys thereof were almost too heavy for the strongest person. His people said to him: "Do not enjoy, for **God** does not like hedonists."*

28:77 "Seek with the provisions bestowed upon you by **God** the abode of the Hereafter, and do not forget your share in this world, and do good as **God** has done good to you. Do not seek corruption in the land. **God** does not like the corruptors."

28:78 He said, "I have attained all this only because of my own knowledge." Did he not realize that **God** had annihilated before him generations that were much stronger than he, and greater in riches? The transgressors were not asked about their crimes.

28:79 Then he came out among his people draped in his ornaments. Those who preferred this worldly life said, "Oh, if only we were given similar to what Qarun has been given. Indeed, he is very fortunate."

28:80 Those who were blessed with knowledge said, "Woe to you! The reward from **God** is far better for those who affirm and promote reforms. None attains it except the steadfast."

28:81 We then caused the earth to swallow him and his mansion. He had no group that could protect him against **God**; nor would he be victorious.

28:82 Those who wished they were in his place the day before said, "Indeed it is **God** Who provides or restricts for whomever He chooses from among His servants. Had it not been for **God**'s grace towards us, He could have caused the earth to swallow us as well. We now realize that the ingrates never succeed."

28:83 Such will be the abode of the Hereafter; We reserve it for those who do not seek prestige on earth, nor corruption. The end belongs to the righteous.

28:84 Whoever brings forth a good deed, he will receive a better reward. Whoever brings forth a sin then the retribution for their sins will be to the extent of their deeds.

28:85 Surely, the One who decreed the Quran to you will summon you to a predetermined appointment. Say, "My Lord is fully aware of who it is that brings the guidance, and who has gone astray."

28:86 Nor did you expect this book to come your way; but this is a mercy from your Lord. Therefore, you shall not side with the ingrates.

28:87 Nor shall you be diverted from **God**'s signs, after they have come to you. Invite to your Lord, and do not be of those who set up partners.

28:88 Do not call besides **God** any god, there is no god but He. Everything will fade away except His presence. To Him is the judgment, and to Him you will be returned.

ENDNOTES

028:001 T9S60M40. This combination of letters/numbers plays an important role in the mathematical system of the Quran based on code 19. See 74:1-56; 1:1; 2:1; 13:38; 27:82; 38:1-8; 40:28-38; 46:10; 72:28.

028:015 Moses attributes this event to the devil. The reason is that Moses acted with racist emotions by trusting the account of his tribesman without cross examining him or without questioning the other party about the issue. This is a common sin we tend to commit. We make judgments about other people just by listening to one party. If a wise man like Moses (28:14; 26:21) fell into the trap of tribalism and racism, we should investigate matters thoroughly before siding with any party against another. We are expected to uphold justice, even against ourselves and our family members. (See also: 38:24 and 4:135).

028:054 Countering evil with goodness is a common trait of muslims or rational monotheists. See, 4:140.

028:056 Compare this verse to 42:52.

028:074-75 The followers of man-made stories and teachings that are peddled in the name of God's messengers will be rejected by those very messengers who will serve as witnesses against them. Muhammed will reject Sunni and Shiite *mushriks* for turning the monotheistic system into various sects concocted by a unanimous "holy" corporation consisting of Muhammed and troops of scholars and religious leaders.

028:076 In our times, the world has hundreds of obscenely rich people. Many of them accumulated their wealth through corporate laws designed by the wealthy or their proxies to maintain and accumulate more wealth for the rich. While a select few get richer, the majority of the world population is living in poverty. The alarming rate of widening gap between the haves and have-nots is created by corrupt business conduct, government subsidies for the rich, dubious contract deals, monopolistic practices, corporate termites lobbying the so-called elected officials, war profiteering, currency speculations, credit card usury, banking regulations, unjust takeovers by the governments, and ignorance and apathy of the masses.

بسم الله الرحمن الرحيم

29:0 In the name of God, the Gracious, the Compassionate.

29:1 A1L30M40*

29:2 Did the people think that they will be left to say, "We affirm" without being put to the test?

29:3 While We had tested those before them, so that **God** would know those who are truthful and so that He would know the liars.

29:4 Or did those who sinned think that they would be ahead of Us? Miserable indeed is their judgment!

29:5 Whosoever looks forward to meeting **God**, then **God**'s meeting will come. He is the Hearer, the Knowledgeable.

29:6 Whosoever strives then he is only striving for himself; for **God** is in no need of the worlds.

29:7 Those who affirmed and promoted reforms, We will cancel their sins and We will reward them in the best for what they did.

Honor Your Parents

29:8 We instructed the human being to be good to his parents. But if they strive to make you set up partners with Me, then do not obey them. To Me are all your destinies, and I will inform you of what you used to do.

29:9 Those who affirmed and promoted reforms, We will admit them with the righteous.

29:10 Among the people are those who say, "We affirm **God**," but if he is harmed in the sake of **God**, he equates the persecution inflicted by the people with **God**'s punishment! If a victory comes from your Lord, he says: "We were with you!" Is **God** not fully aware of what is inside the chests of the worlds?

29:11 **God** is fully aware of those who affirmed, and He is fully aware of the hypocrites.

29:12 Those who rejected said to those who affirmed: "Follow our path and we will carry your sins." But they cannot carry anything from their sins, they are liars!

29:13 They will carry their own burdens as well as burdens with their burdens, and they will be asked on the day of Resurrection regarding what they fabricated.*

Noah

29:14 We had sent Noah to his people, so he stayed with them one thousand years less fifty calendar years. Then the flood took them while they were wicked.*

29:15 So We saved him and the people on the Ship, and We made it a sign for the worlds.*

Abraham, the Legendary Iconoclast

29:16 And Abraham when he said to his people: "Serve **God** and be aware of Him; that is better for you if you knew."

29:17 "You are serving nothing but idols besides Him, and you are creating fabrications. Those that you serve besides **God** do not possess for you any provisions, so seek with **God** the provision and serve Him and be thankful to Him; to Him you will return."

29:18 "If you reject, then nations before you have also rejected." The messenger is only required to deliver clearly.

29:19 Did they not observe how **God** initiates the creation then He returns it? All that for **God** is easy to do.*

29:20 Say, "Roam the earth and observe how the creation was initiated. Then **God** will establish the final design. **God** is capable of all things."*

29:21 "**God** will punish whom He wills and He will have mercy on whom He wills, and to Him you will return."

29:22 "You will not be able to challenge this fact, on earth or in the heavens, nor do you have besides **God** any ally or supporter."

29:23 Those who rejected **God**'s signs and in meeting Him, they have forsaken My mercy, and they will have a painful retribution.

29:24 But the only response from his people was their saying: "Kill him, or burn him." But **God** saved him from the fire. In this are signs for a people who affirm.

29:25 He said, "To preserve some friendship among you in this worldly life, you have taken idols besides **God**. But then, on the day of Resurrection, you will disown one another, and curse one another. Your destiny is hell, and you will have no supporters."*

29:26 Thus, Lot affirmed with him and said, "I am emigrating to my Lord. He is the Noble, the Wise."

29:27 We granted him Isaac and Jacob, and We assigned prophethood and the book to his progeny. We gave him his reward in this world, and in the Hereafter, he is among the righteous.

Lot

29:28 And Lot, when he said to his people: "You commit a lewdness that no others in the world have done before!"

29:29 "You sexually approach men, you commit highway robbery, and you bring all vice to your society." But the only response from his people was to say, "Bring us **God**'s retribution, if you are being truthful!"

29:30 He said, "My Lord, grant me victory over the wicked people."

29:31 When Our messengers came to Abraham with good news, they then said, "We are to destroy the people of such a town, for its people have been wicked."

29:32 He said, "But Lot is in it!" They said, "We are fully aware of who is in it. We will save him and his family, except his wife; she is of those doomed."

29:33 Thus when Our messengers came to Lot, they were mistreated, and he was embarrassed towards them. They said, "Do not fear, and do not be saddened. We will save you and your family, except for your wife; she is of those doomed."

29:34 "We will send down upon the people of this town an affliction from the sky, as a consequence of their wickedness."

29:35 We left remains of it as a clear sign for a people who understand.

Shuayb

29:36 To Midian was their brother Shuayb. He said, "O my people, serve **God** and seek the Last day, and do not roam the earth corrupting."

29:37 But they denied him, thus the earthquake took them; they were then left dead in their homes.

29:38 And Aad and Thamud. Much was made apparent to you from their dwellings. The devil had adorned their works in their eyes, thus he diverted them from the path, even though they could see.

29:39 And Qarun, and Pharaoh, and Haman; Moses went to them with clear proofs. But they became arrogant in the land, and they were not the first.*

29:40 We took each by his sins. Some of them We sent upon violent winds, some of them were taken by the scream, some of them We caused the earth to swallow, and some of them We drowned. **God** is not the One who wronged them; it is they who wronged themselves.

The Spider

29:41 The example of those who take allies besides **God** is like the spider, how it makes a home; and the less reliable home is the home of the spider, if they only knew.*

29:42 **God** knows that what they are calling on besides Him is nothing. He is the Noble, the Wise.

29:43 Such are the examples We cite for the people, but none reason except the knowledgeable.

29:44 **God** created the heavens and the earth, with truth. In that is a sign for those who affirm.

The Importance of the Contact Prayers and Call for Unity

29:45 Recite what is revealed to you of the book, and maintain the contact prayer; for the contact prayer prohibits evil and vice, and the remembrance of **God** is the greatest. **God** knows everything you do.*

29:46 Do not argue with the people of the book except in the best manner; except for those who are wicked amongst them; and say, "We affirm what was revealed to us and in what was revealed to you, and our god and your god is the same; to Him we peacefully surrender."

29:47 Similarly, We have sent down to you the Book. Thus, those whom We have given the book will affirm it. Also, some of your people will affirm it. The only ones who mock Our signs are the ingrates.

The Sign Given to Muhammed

29:48 You were not reciting any book before this, nor were you writing one down by your hand. In that case, the doubters would have had reason.

29:49 In fact, it is a clear revelation in the chests of those who have been given knowledge. It is only the wicked who doubt Our signs.

29:50 They said, "If only signs would come down to him from his Lord!" Say, "All signs are with **God**, and I am but a clear warner."*

29:51 Is it not enough for them that We have sent down to you the book, being recited to them? In that is a mercy and a reminder for people who affirm.

29:52 Say, "**God** is enough as a witness between me and you. He knows what is in the heavens and the earth. As for those who affirm falsehood and reject **God**, they are the losers."

They Are Already Experiencing Hell

29:53 They hasten you for the retribution! If it were not for a predetermined appointment, the retribution would have come to them. It will come to them suddenly when they do not expect.

29:54 They hasten you for the retribution; while hell surrounds the ingrates.

29:55 The day will come when the retribution overwhelms them, from above them and from beneath their feet; and We will say, "Taste the results of what you used to do!"

Emigrate for Your Freedom

29:56 "O My servants who affirmed, My earth is spacious, so serve only Me."

29:57 Every person will taste death, then to Us you will be returned.

29:58 Those who affirm and promote reforms, We will settle them in mansions in Paradise, with rivers flowing beneath. Eternally they abide therein. Excellent is the reward for the workers.

29:59 They are the ones who were patient, and put their trust in their Lord.

29:60 Many a creature does not carry its provision; **God** provides for it, as well as for you. He is the Hearer, the Knowledgeable.*

29:61 If you ask them: "Who created the heavens and the earth, and put the Sun and the Moon in motion?" They will say, "**God**." Why then did they deviate?

29:62 **God** expands the provision for whomever He chooses from among His servants, and withholds it. **God** is fully aware of all things.

29:63 If you ask them: "Who sends down water from the sky, thus reviving the land after its death?" They will say, "**God**," Say, "Praise be to **God**." But most of them do not understand.

Do not Trade Eternal Life with the Worldly Life

29:64 This worldly life is no more than fun and distraction, while the abode of the Hereafter is the reality, if they only knew.

29:65 When they ride on a ship, they call on **God**, devoting the system to Him. But as soon as He saves them to the shore, they set up partners.

29:66 Let them reject what We have given them, and let them enjoy; for they will come to know.

29:67 Have they not seen that We have established a safe territory, while all around them the people are in constant danger? Would they still affirm falsehood, and reject **God**'s blessings?

29:68 Who is more evil than one who fabricates lies and attributes them to **God**, or denies the truth when it comes to him? Is there not a place in hell for the ingrates?

29:69 As for those who strive in Our cause, We will guide them to Our paths. For **God** is with the pious.

ENDNOTES

029:001 A1L30M40. This combination of letters/numbers plays an important role in the mathematical system of the Quran based on code 19. See 74:1-56; 1:1; 2:1; 13:38; 27:82; 38:1-8; 40:28-38; 46:10; 72:28.

029:013 None will be held responsible for the sin of another person (6:164). A person aiding another person in committing a sin or evil, shares the responsibility together with the person who committed it (4:85).

029:014 Until God revealed the secret code prophesied in chapter 74, we did not have a satisfactory answer to the question, why "one thousand minus fifty instead of the conventional expression, nine hundred and fifty?" Every element of the Quran, its chapters, verses, words, and letters, including its numbers, plays a role in the mathematical structure of the Quran. For instance, the sum of all the different numbers in the Quran without repetition gives us 162146 (19x8534). If the expression in the verse was "nine hundred and fifty," then the sum would be 900 extra and would not participate in the numerical system based on code 19. Furthermore, the age of Noah being a multiple of 19 also supports this relationship.

It seems too difficult to accept such a long lifespan for a human being. Since, the Quran contains its own scientific authentication, we have no choice but to trust this apparently incredible information.

The Bible too gives 950 as Noah's age (Genesis 9:29). Chapter eleven of Genesis mentions nine names who lived more than a hundred years, such as Shem (600), Arphaxad (438), Salah (433), Eber (464),

Peleg (239), Reu (239), Serug (230), Terah (205),

029:015 See 11:40-44.

029:019-20 The verse implies that evolution is a divinely designed and guided method of creation.

A critic citing 96:2; 15:26; 3:59; 19:67; and 16:4 may ask the following question: "What was man created from: blood, clay, dust, or nothing?" The criticism is a classic example of the either/or fallacy, or the product of a frozen mind that does not consider or perceive time and evolution as reality. If the critic uses the same standard, the critic of these verses will find contradictions in almost every book. If he looks into biology books, he will similarly get confused. In one page he will learn that he is made of atoms, in other made of cells, of DNA, sperm, egg, embryo, earthly materials, etc. He would express his disbelief and confusion with a similar question. A careful and educated reading of the Quran will reveal the following facts about creation:

1. There were times when man did not exist. Billions of years after the creation of the universe, humans were created. In other words, we were nothing before we were created (19:67).

2. Humans were created according to a divinely guided evolution (29:19-20; 71:14-17).

3. Creation of man started from clay (15:26).

Our Creator started the biological evolution of microscopic organisms within the layers of clay. Donald E. Ingber, professor at Harvard University, in an article titled "The Architecture of Life" published as the cover story of *Scientific American* stated the following:

"Researchers now think biological evolution began in layers of clay, rather than in the primordial sea. Interestingly, clay is itself a porous network of atoms arranged geodesically within octahedral and tetrahedral forms. But because these octahedra and tetrahedra are not closely packed, they retain the ability to move and slide relative to one another. This flexibility apparently allows clay to catalyze many chemical reactions, including ones that may have produced the first molecular building blocks of organic life."

Humans are the most advanced fruits of organic life started millions of years ago from layers of clay.

4. Human being is made of water (21:30).

The verse above not only emphasizes the importance of water as an essential ingredient for organic life, but also clearly refers to the beginning of the universe, which we now call the Big Bang. The Quran's information regarding cosmology is centuries ahead of its time. For instance, verse 51:47 informs us that the universe is continuously expanding. Furthermore, the Quran informs us that the universe will collapse back to its origin, confirming the closed universe model (21:104).

Verse 24:45 repeats the role of water in creation and follows with a classification of animals according to their ability to move around and mentions "two legs." Walking on two legs is a crucial point in the evolution of humanoids. Walking on two feet may initially appear to be insignificant in the evolutionary process, yet many scientists discovered that walking on two feet had a significant contribution in human evolution by enabling Homo Erectus to use tools and gain consciousness, thereby leading to Homo Sapiens.

5. The human being is made of dust, or earth, which contains the essential elements for life (3:59).

6. The human being is the product of long-term evolution, and when human sperm and egg, which consist of water and earthly elements, meet each other in right conditions, they evolve into embryo, fetus, and finally, after 266 days into a human being (75:37; 16:4; 96:2; 22:5).

We do not translate the Arabic word *alaq* as "clot"; since neither in interspecies evolution nor intraspecies evolution, is there a stage where the human is a "clot". This is a traditional mistranslation of the word, and the error was first noticed by the medical doctor Maurice Bucaille. Any decent Arabic dictionary will give you three definitions for the word *alaq*: (1) clot; (2) hanging thing; (3) leech. Early commentators of the Quran, lacking the knowledge of embryology, justifiably picked the "clot" as the meaning of the word. However, the author of the Quran, by this multi-meaning word was referring to embryo, which hangs to the lining of uterus and nourishes itself like a leech. In modern times, we do not have an excuse for picking the wrong meaning. This is one of the many examples of the Quran's language in verses related to science and mathematics: while its words provide a kind of understanding to former generations, its real meaning shines with knowledge of God's creation and natural laws. See 95:2.

029:020 Scientists who followed God's instruction in this verse gathered many evidences regarding the evolution of human creation. Archeological findings show that, following the laws of an intelligent design, life started from simpler organisms and evolved through genetic mutations and cumulative selection. See 7:69; 15:26-28; 24:45; 71:14-17.

029:025 Peer pressure, especially on religious and cultural issues, creates formidable psychological walls, social, political and economic deterrence that discourages the individual from questioning, reasoning and examining the religious teachings and cultural norms dictated onto him/her. That is why there is a relation between religion and geography. Abraham was one of those rare brave individuals who stood on his own feet, broke the walls of conformity, and rejected the notion that society, or the number of followers, is the criterion for deciding between truth and falsehood. Only philosophical inquiry can free the person from the piles of falsehood and superstition that accumulate in the culture of societies because of ruling classes who wish to keep the status quo. We salute Abraham. We salute all those who question assumptions and traditions of organized religions. Blessed are those who seek truth, who do not rush to conclusion, who examine assumptions and assertions, and do not hesitate to embrace truth when it becomes evident. See 6:74-83; 21:58.

029:039 The three institutions/powers that comprised the fascist Egyptian regime and many other systems in history are referred by three characters: Pharaoh represents politics, Qarun represents economic monopoly, and Haman represents the military. What about magicians? Which institution/power do they represent? How do they contribute to the system that aims to accumulate power and wealth in the hands of the few? See 7:112-116; 28:38; 40:24-36.

For the magicians of the New Testament, see, 1 Peter 2:13-21; Ephesians 6:5-9; Timothy 2:11-14.

The word Haman is not a proper name, but a title of Egyptian origin, a contraction of *Ham-Amon*, meaning high priest of the god, Amon. The Haman mentioned in the Quran has nothing to do with the Haman mentioned in the Old Testament (Esther 3:1; 7:10).

029:041-43 The example is interesting. The spider is not the male one (*ankab*), but a female one (*ankabut*). Thus, it might be referring to spiders like the Black Widow that eats their males after having an intimate relationship. (The grammatical structure of the verse is similar to that of 40:20).

029:045 *Sala* prayer is a periodic mental connection and communication with God. We should commemorate God alone in our prayers (20:14). Mentioning the names of Muhammed, Ali, Wali, or any other creature contradicts the purpose of *sala* prayer as defined by God.

029:050-51 God unveiled the prophecy of the hidden code 19 in chapter 74, in year 1974 to a biochemist that made the first computerized study of the Quran (27:82; 74:1-56). This was exactly 1406 (19x74) lunar years after the revelation of the Quran. The fulfillment of the prophecy opened a new era and started a powerful reformation movement. For the word *ZKR* (reminder), see 15:9; 21:2; 21:24,105; 26:5; 38:1,8; 41:41; 44:13; 72:17; 74:31,49,54.

029:060 When considered in the context of immigration, those who emigrate in the cause of God are advised not to worry about their provisions; God will provide for them.

بسم الله الرحمن الرحيم

30:0 In the name of God, the Gracious, the Compassionate.

30:1 A1L30M40*

30:2 The Romans have won,*

30:3 At the lowest point on Earth. But after their victory, they will be defeated.

30:4 In a few more years. The decision before and after is for **God**, and on that day those who affirm will rejoice.

30:5 With **God**'s victory. He gives victory to whom He wishes; He is the Noble, the Compassionate.

30:6 Such is **God**'s pledge, and **God** does not break His pledge, but most people do not know.

Do not be Arrogant

30:7 They only know the outside appearance of the worldly life; and regarding the Hereafter, they are ignorant.

30:8 Did they not reflect upon themselves? **God** did not create the heavens and the earth and what is between them except by truth and an appointed term. But most of the people are in denial regarding their meeting with their Lord.

30:9 Did they not roam the earth and see how it ended for those before them? They were more powerful than them, and they cultivated the land and they built in it far more than these have built, and their messengers came to them with the proofs. **God** was not to wrong them, but it was they who wronged themselves.

30:10 Then the end of those who did evil was evil, that is because they denied **God**'s signs, and they used to mock them.

30:11 **God** initiates the creation, then He repeats it, then to Him you will return.

30:12 The day when the Moment will be established, the criminals will be in despair.

30:13 They did not have any intercessors from the partners they set up, and they will reject such partners.
30:14 On the day the Moment is established, they will be separated.
30:15 Then as for those who affirmed and promoted reforms, they will be delighted in a luxurious place.
30:16 As for those who rejected and denied Our signs and the meeting of the Hereafter, they shall be brought forth for the retribution.
30:17 So, glory be to **God** when you retire and when you wake.
30:18 To Him is all praise in the heavens and in the earth, and in the evening and when you go out.
30:19 He brings the living out of the dead, and He brings the dead out of the living. He revives the land after its death. Similarly, you will be brought out.

God's Signs in Creation and Diversity

30:20 From His signs is that He created you from dust, then you become human beings spreading out.
30:21 From His signs is that He created for you mates from yourselves that you may reside with them, and He placed between you affection and compassion. In that are signs for a people who reflect.*
30:22 From His signs are the creation of the heavens and the earth, and the difference of your languages and your colors. In that are signs for the world.*
30:23 From His signs is your sleep by night and day, and your seeking of His bounty. In that are signs for a people who listen.
30:24 From His signs is that He shows you the lightning, by way of fear and hope, and He sends down water from the sky, and He revives the land with it after its death. In that are signs for a people who understand.
30:25 From His signs is that the heavens and the earth will rise by His command. Then, when He calls you forth from the earth, you will come out.
30:26 To Him is all that is in the heavens and the earth. All are subservient to Him.
30:27 He is the One who initiates the creation, then He repeats it; this is easy for Him. To Him is the highest example in the heavens and the earth. He is the Noble, the Wise.*

Monotheism: A Natural and Rational System

30:28 An example is cited for you from among yourselves: are there any from among those who are still dependent that are partners to you in what provisions We have given you that you become equal therein? Would you fear them as you fear each other? It is such that We clarify the signs for a people who have sense.
30:29 No, those who were wrong followed their desires without knowledge. So, who can guide one whom **God** misguides? They will have no supporters.
30:30 So establish yourself to the system of monotheism. It is the nature that **God** has made the people on. There is no changing in **God**'s creation. Such is the pure system, but most people do not know.*
30:31 Turn to Him, be aware of Him, maintain the contact prayer, and do not be of those who set up partners.
30:32 Or like those who split up their system, and they became sects; each group happy with what it had.*
30:33 If harm afflicts the people, they call out sincerely to their Lord. But then, when He gives them a taste of His mercy, a group of them set up partners with their Lord!
30:34 So as to reject what We gave them. Enjoy then, for you will come to know.
30:35 Or have We sent down to them a proof, which speaks to them of what they have set up?
30:36 If We grant the people a taste of mercy, they become happy with it, but if evil afflicts them for what they have done, they become in despair!*

30:37 Did they not see that **God** grants the provisions for whom He wills, and He is able? In that are signs for a people who affirm.

30:38 So give the relative his due, and the poor, and the wayfarer. That is best for those who seek **God**'s presence, and they are the successful ones.

30:39 Any financial interest you have placed to grow in the people's money, it will not grow with **God**. But, any contribution that you have placed seeking His presence; those will be multiplied.

30:40 **God** who created you, then He provided for you, then He puts you to death, then He brings you to life. Is there any among the partners you set up that can do any of this? Be He glorified and exalted above what they set up.

Destruction of Earth Because of Polytheism, Ignorance, Nationalism, Oppression, Wars, Greed, and Waste

30:41 Pollution has appeared in the land and the sea by the hands of people for what they earned. He will make them taste some of what they have done, perhaps they will revert.

30:42 Say, "Roam the earth and see how the end was for those before. Most of them had set up partners."

30:43 So establish yourself to the system, which is straight, before a day comes from **God** that none can avert. On that day they shall be separated.

30:44 Whosoever rejects, then he will suffer his rejection; and whosoever does good works, then it is for themselves that they are preparing a good place.

30:45 So that He will reward those who affirm and promote reforms from His grace. He does not love the ingrates.

30:46 Among His signs is that He sends the winds with glad tidings, and to give you a taste of His mercy, and so that the ships may sail by His command, and that you may seek of His bounty; perhaps you will be thankful.

30:47 We have sent before you messengers to their people, so they came to them with clear proofs, then We took revenge on those who were criminals, and it is an obligation for Us to grant victory to those who affirm.

30:48 **God** is the one who sends the winds, so they raise clouds, and He spreads them in the sky as He wishes, then He turns them into joined pieces, then you see the rain drops come forth from their midst. Then when He makes them fall on whomever He wills of His servants, they rejoice!

30:49 Even though just before He sent it down to them, they were in despair!

30:50 So look at the effects of **God**'s compassion, how He revives the land after its death. Such is the One who will revive the dead, and He is capable of all things.

30:51 If We chose to send a wind and they see it turn yellow, then they will continue to be ingrates after it.

To Receive God's Message

30:52 You cannot make the dead listen, nor can you make the deaf hear the call when they have turned their backs.*

30:53 Nor can you enlighten the blind from their straying; but you can only make those who affirm Our signs listen, for they have peacefully surrendered.

30:54 **God** is the One Who created you from weakness, then He made strength after the weakness, then He makes after the strength a weakness and gray hair. He creates what He wills and He is the Knowledgeable, the Able.

30:55 The day the Moment is established, the criminals will swear that they have remained only for a moment! Thus they were deluded.

30:56 Those who were given knowledge and affirmation said, "You have remained according to **God**'s decree until the day of Resurrection; so this is the day of

Resurrection, but you did not know."

30:57 So on that day, the excuses of those who transgressed will not help them, nor will they be allowed to revert.

The Quran Contains all the Examples Necessary for our Salvation

30:58 We have cited for the people in this Quran of every example. If you come to them with a sign, those who rejected will say, "You are bringing falsehood!"

30:59 It is thus that **God** seals the hearts of those who do not know.

30:60 So be patient, for **God**'s promise is true, and do not be dissuaded by those who do not have certainty.

ENDNOTES

030:001 A1L30M40. These letters/numbers play an important role in the mathematical system of the Quran based on code 19. See 74:1-56; 1:1; 2:1; 13:38; 27:82; 38:1-8; 40:28-38; 46:10; 72:28.

030:002-05 You might have noticed that we translated the reference of the verb "*GHaLaBa*" differently than the traditional translations. Instead of reading the verb in 30:2 as "*ghulibat*" (were defeated) we read as "*ghalabat*" which means just the opposite, "defeated." Similarly, we also read its continuous/future tense in the following verse differently. The prophecy of the verse was realized in August of 636, four years after the death of Muhammed, when Peacemakers confronted the army of Byzantine Empire around Yarmouk (or Yarmuk) River, in one of the most significant battles in history.

Historically, the first major battle between Peacemakers and Byzantine army occurred in September of 629, in the village of Mutah, east of the Jordan River. In the Battle of Mutah, Peacemakers suffered a major defeat and retreated to the city called Civilization (Medina).

About seven years after that painful defeat, under the command of Khalid bin Walid, the Peacemaker army beat the Christian imperial army of four or more times their numbers. The six-day war, Battle of Yarmouk, occurred in area near the Sea of Galilee and Dead Sea, which is the lowest land depression on earth, 200-400 meters below sea level. The Battle of Yarmouk is taught in military academies as one of the major historical battles.

The orthodox reading and translation of the verse, reflects an interesting historical conflict and is a prime example of the abuse of a divine book to justify nationalistic wars and imperialistic ambitions. By changing the verse's original reading to the opposite, the Umayyad and Abbasid dynasties were able to depict Persians as the major enemy. Thus, they justified aggression against Persian Empire.

030:021 Marriage must be based on mutual love and care between man and woman. Maintaining and improving love and care requires steady attention, compromise, respect and patience. One spouse should not try to control or suppress the individuality or identity of the other. Marrying through arranged marriages, without the bride and groom seeing and meeting each other, is against God's law in nature, since God must have had a purpose to create man handsome and woman beautiful. Our brain considers the geometric average or the composite of all faces we see as a visual measure for attractiveness. God did not design the package to be ignored at the time it is needed the most. Indeed, physical attraction is mentioned as a main factor in marriage (33:52; 24:30). For divorce, see 2:226-230 and 4:137. For celibacy, see 57:27.

030:022 In the past, and even in the present, some paranoid majorities banned the language of minorities, because of their differences in color, race, or culture. Totalitarian or racist regimes that ban the language and identity of a group will suffer the consequences of fighting against God's signs. The tradition of changing the names of converts to Arabic names was a diabolic invention started centuries later as an expression of Arab nationalism and cultural imperialism. See 33:5; 49:13.

030:023 See 25:47.

030:027 For attributes of God, see 59:22-24.

030:028 See 16:71.

030:030 Accepting God alone as the only Lord and the only Master is a fact that can be attained through philosophical inquiry. However, the weaknesses and ignorance we learn throughout our life cover that innate faculty. See 7:172-173.

030:032 This verse convicts the clerics who introduce this imam's sect or that imam's sect as God's system. History bears witness those professional religious leaders are the most ardent enemies of God and monotheism. Each messenger or prophet found the greatest opposition to their message from the religious leaders of their time. See 23:53.

Prophet Muhammed was neither Sunni nor Shiite. He was a monotheist, a muslim, a peacemaker, a peaceful submitter to God alone. All other prophets and messengers before him were also monotheists and muslims. All monotheists, starting from Adam until the end of the world, belong to one group. The Quran strictly condemns sectarian division and tells us that power struggle and idol-worship as the source of sectarian division (6:153-159; 98:4; 21:92; 23:52; 2:285).

The sectarian division came after the departure of the prophet Muhammed. Mostly, out of tribal and political division. When there was a power struggle between Ali, Muhammed's son-in-law, and Uthman, Muhammed's father-in-law, people were divided. Uthman was a weak leader, and unfortunately history shows that he was favoring his tribe (Umayyad) and hiring them to key government positions. This corruption created resentment in other tribes and resurrected the tribal conflict that was almost eradicated by the light of the Quran.

After Uthman, though Ali was elected as the fourth Caliph (president), soon he was assassinated. Ali ruled only for two years. The division among tribes turned to a bloody strife. There were basically two parties. The party who supported Ali and his two sons, Hassan and Hussain, and those who sided with the Umayyad dynasty. The former, started calling themselves Shiite of Ali (Ali's Party) and the latter, which was joined by some former Ali supporters, called themselves Ahl-i *Sunna* (Followers of Tradition). The history of sectarian conflict is a horrible power struggle. The Umayyads turned the federally secular republican that was established by the prophet Muhammed in Yathrib (today's Medina) into a theocratic monarchy. The kingdom of Umayyad committed many atrocities. The most memorable one is their massacre in Karbala; they killed Muhammed's grandson Hussain with his 72 supporters in a dessert near Baghdad.

The division between the Shiite and Ahl-i *Sunna* grew and transformed into religious and theological matters. Here I will focus on Shiite, since we have numerous times referred to Sunni sects and their distortions. The early Shiites were generally righteous people. They were against nepotism, tribalism, and corruption. However, the following generation went to extreme and lost the original noble goal. They turned to a cult who identified themselves with Ali and his descendants and they started worshiping them by giving them authority in designing and defining God's religion. Their clergymen fabricated doctrines thereby sanctifying the descendants of the Prophet Muhammed; they claimed infallibility for the descendants of Muhammed through Ali and Fatima. Volumes of false stories were fabricated to promote this cult. God was no more the focus. Instead, the so-called *imam-i masumin* (innocent imams), became the focus.

Shiite mosques display the names of those twelve idols and Muhammed besides God's name. Little bricks made of dirt from Karbala is used to put one's forehead on during *sala* prayers. Some pricks contain inscriptions with the names of their idols besides God: Allah, Muhammed, Ali, Fatima, Hassan and Hussein. Sunni mosques, on the other hand, display the names of five (four Caliphs and Muhammed) or seven (four Caliphs plus Hassan and Hussain plus Muhammed) idols besides God's name. The fight between the sects boils down over the non-common idols, as it is the case between Protestants and Catholics. Shiite clerics distorted the meaning of some verses, such as the meaning of 42:23 to establish authority besides God for the descendants of Prophet Muhammed. In Iran now there are thousands of mullahs with black turbans claiming and enjoying this privilege.

After the disappearance (the death) of the 12th Imam in a lost battle with the Sunnite faction, Shiite scholars fabricated stories around his disappearance. He was not dead according to those stories, but he was alive somewhere in a cave. When the right time comes, he would return to lead Shiites to an ultimate victory against Sunnis. This story which was initially fabricated to preserve the spirits of the Shiite party, turned into a mythology with the passing of time. Despite centuries, the members of the Shiite cult are still waiting for the return of their 12th imam, the imam who went into hiding in a cave after he lost the battle! Researchers can easily trace the source of this mythology to the teaching of St. Paul who made up a story about the return of Jesus Christ. Denying the death of leaders is a cult syndrome and is one of the signs of idol worshipers; they cannot survive without their idols.

The followers of the Shiite sect developed tact of survival under the oppression of Sunni rulers. This tactic is known as "*taqiyya*" (hiding one's faith). Shiite clergymen distorted the meaning of 3:28 which permitted affirmers to hide their faith when they were weak and threatened by their enemy. They broadened this permission and turned it to a lifestyle. Furthermore, they turned this permission to an obligation. Throughout history, the followers of Shiite sect developed a hypocritical religious life even there was no compelling reason to do so. Waiting for the return of 12th imam coupled with this religious mask (taqiyya) turned the Shiite masses to obedient constituencies. They survived under Sunni and oppressive Sultans, Kings and Shahs with no claim for religious governance. However, Ayatullah Khomaini, a charismatic leader, modified this traditional belief. He claimed that even in the absence of the 12th Imam, his substitute could still act and fight against the enemy. With several other minor theological tinkering he released the long-locked anger into demonstrations against Reza Shah's corrupt and oppressive regime. The taqiyya erupted like a volcano. Unfortunately, the revolution did not live up to its promises regarding freedom and justice; it devolved into another corrupt and oppressive system. The ousted Shah was replaced by an infallible mullah, the *Velayet-i Faqih*, with much greater authority: in the name of revolution, in the name of imams, in the name of the lost imam, and in the name of God. The monarchy turned into a mollarchy!

As for the Sunnis, they also fabricated many *hadith*s to justify their history filled with corruption and injustice. They even justified the atrocities of Umayyad dynasty by choosing silence over the early conflicts between Ali and Uthman and later Muawiya. They fabricated religious amnesty and immunity for every corrupt and cruel *sahaba* (Muhammed's comrades) or *tabeen* (the comrades of those comrades). Even the drunken murderer of prophet's grandson, Yazid, the governor of Damascus, was given religious immunity against criticism.

You can find many similarities and differences between Sunni and Shiite sects. In fact, there are more than a dozen Shiite sub-sects which they differ greatly in religious matters and there are dozens of Sunnite sub-sects which also differ significantly. Nevertheless, today only four Sunni and a few Shiite sects survive. Their common characteristic is that all hate to accept God's word alone as their divine guide and invoke the names of saints and intercessors in their prayers. They all follow conjecture and fabricated sectarian *hadith*s to justify their dogmas, complicated rituals, and backward practices.

030:036 See 12:87; 15:56; 39:53; 41:49.

030:052 Just as we cannot hear or understand a radio broadcast via a radio with dead batteries, or with its switch off, or not being tuned to the correct wavelength, similarly we cannot hear or understand God's message if we do not use our reasoning faculties, and if we do not open our minds to new ideas.

31:0 In the name of God, the Gracious, the Compassionate.

31:1 A1L30M40*

31:2 These are the signs of the book of wisdom.

31:3 A guide and a mercy for the good doers.

31:4 Those who maintain the contact prayer and contribute towards betterment and regarding the Hereafter they are certain.

31:5 These are on guidance from their Lord, and they are the successful ones.

The Followers of Baseless Teachings

31:6 Among the people, there are those who accept baseless *hadiths* to mislead from the path of **God** without knowledge, and they take it as entertainment. These will have a humiliating retribution.

31:7 When Our signs are recited to him, he turns away arrogantly as if he did not hear them, as if there is deafness in his ears. So announce to him a painful retribution.

31:8 Those who affirm and promote reforms, for them will be gardens of bliss.

31:9 Abiding eternally therein with certainty, as promised by **God**. He is the Noble, the Wise.

31:10 He created the skies without pillars that you can see. He placed in the earth firm stabilizers so that it would not tumble with you, and He spread on it all kinds of creatures. We sent down water from the sky, thus We caused to grow all kinds of good plants.

31:11 This is **God**'s creation, so show me what those besides Him have created? Indeed, the transgressors are far astray.

Luqman Shares Wisdom with His Son

31:12 We had given Luqman the wisdom: "You shall be thankful to **God**, and whoever is thankful is being thankful for his own good. As for whosoever rejects, then **God** is Rich, Praiseworthy."

31:13 And Luqman said to his son, while he was advising him: "O my son, do not set up any partners with **God**; for setting up partners is an immense wrongdoing."

31:14 We enjoined the human being regarding his parents. His mother bore him with hardship upon hardship, and his weaning takes two calendar years. You shall give thanks to Me, and to your parents. To Me is the final destiny.*

31:15 If they strive to make you set up any partners besides Me, then do not obey them. But continue to treat them amicably in this world. You shall follow only the path of those who have sought Me. Ultimately, you all return to Me, then I will inform you of everything you have done.

31:16 "O my son, God will bring out anything, be it as small as a mustard seed, be it deep inside a rock, or be it in the heavens or the earth, **God** will bring it. **God** is Sublime, Ever-aware."

31:17 "O my son, maintain the contact prayer and advocate righteousness and forbid evil, and bear with patience whatever befalls you. These are the most honorable traits."

31:18 "Do not turn your cheek arrogantly from people, nor shall you roam the earth insolently. For **God** does not like the arrogant showoffs."

31:19 "Be humble in how you walk and lower your voice. For the harshest of all voices is the donkey's voice."

31:20 Did you not see that **God** has committed in your service everything in the heavens and the earth, and He has showered you with His blessings, both apparent and hidden? Yet from the people are some who argue about **God** without knowledge, without guidance, and without an enlightening book.

31:21 If they are told: "Follow what **God** has sent down." They say, "No, we will follow what we found our fathers doing." What if the devil had been leading them to the agony of hell?

31:22 Whosoever peacefully surrenders himself completely to **God**, while he is righteous, then he has taken hold of the strongest bond. To **God** all matters will return.

31:23 Whosoever rejects, then do not be saddened by his rejection. To Us is their ultimate return, then We will inform them of what they had done. **God** is fully aware of what is inside the chests.

31:24 We let them enjoy for a while, then We commit them to severe retribution.

31:25 If you ask them: "Who created the heavens and the earth?" They will say, "**God**." Say, "Praise be to **God**." Yet, most of them do not know.

31:26 To **God** belongs everything in the heavens and the earth. **God** is the Rich, the Praiseworthy.

God does not Have Shortage of Words

31:27 If all the trees on earth were made into pens, and the ocean were supplied by seven more oceans, the words of **God** would not run out. **God** is Noble, Wise.

31:28 Your creation and your resurrection are all as one breath. **God** is Hearer, Seer.

31:29 Have you not seen that **God** merges the night into the day and merges the day into the night, and that He has directed the Sun and the Moon, each running to a predetermined term; and that **God** is Ever-aware of everything you do?

The Truth

31:30 That is because **God** is the truth, what they call on besides Him is falsehood, and that **God** is the High, the Great.

31:31 Have you not seen the ships sailing the sea, carrying **God**'s provisions, to show you some of His signs? In that are signs for every one who is patient, thankful.

31:32 When waves surround them like mountains, they call on **God**, sincerely devoting the system to Him. But when He saves them to the shore, some of them revert. None discard Our signs except those who are betrayers, ingrates.

31:33 O people, you shall revere your Lord, and fear a day when a father cannot help his own child, nor can a child help his father. Certainly, **God**'s promise is truth. Therefore, do not be deceived by this worldly life; and do not be deceived from **God** by arrogance.

31:34 With **God** is the knowledge regarding the Moment. He sends down the rain, and He knows what is inside the wombs. No person knows what it may gain tomorrow, nor does any person know in which land it will die. **God** is Knowledgeable, Ever-aware.*

ENDNOTES

031:001 See 2:1; 74:30.

031:014 For an inference regarding human evolution See 46:15; 71:14-17.

031:034 Only two subjects cannot be known by people. Commentaries relying on *hadith and sunna* misinterpret this verse and claim that five subjects cannot be known by people. God's knowledge of something does not necessarily negate another person's knowledge. See 15.87; 20:15.

بسم الله الرحمن الرحيم

32:0 In the name of God, the Gracious, the Compassionate.

32:1 A1L30M40*

32:2 The sending down of this book, without a doubt, is from the Lord of the worlds.

32:3 Or do they say, "He fabricated it!" No, it is the truth from your Lord, so that you may warn a people who never received a warner before you, in order that they may be guided.

32:4 **God** is the One who created the heavens and the earth, and everything between them in six days, then He assumed all authority. You do not have beside Him any Lord, nor intercessor. Will you not then remember?*

The Speed of Light

32:5 He arranges affairs from the heaven to the earth, and then they ascend towards it (heaven) in a day which is equivalent to one thousand of the years which you count.*

32:6 Such is the Knower of the unseen and the seen; the Noble, the Compassionate.

Genesis of Human Being

32:7 The One who perfected everything He created, and He began the creation of the human from clay.

32:8 Then He made his offspring from a structure derived from an insignificant liquid.

32:9 Then He evolved him and blew into him from His Spirit. He made for you the hearing, the eyesight, and the hearts; rarely are you thankful.

32:10 They said, "When we are buried in the ground, will we be created anew?" Indeed, they reject the meeting of their Lord.

32:11 Say, "The controller of death that has been assigned to you will take you, then to your Lord you will be returned."

32:12 If only you could see the criminals when they bow down their heads before their Lord: "Our Lord, we have now seen and we have heard, so send us back and we will promote reforms. Now we have attained certainty!"*

32:13 If We had wished, We could have given every person its guidance, but the word from Me has taken effect, that I will fill hell with Jinn and humans all together.

32:14 So taste the consequences of your ignoring/forgetting this day; for We have now ignored/forgotten you. Taste the eternal retribution in return for what you used to do.*

Appreciating God's Signs

32:15 The only people who affirm Our signs are those whom when they are reminded by them, they fall prostrating, and they glorify by praising their Lord, and they are not arrogant.

32:16 Their sides readily forsake their beds, to call on their Lord out of fear and hope, and from Our provisions to them they give.

32:17 No person knows what is being hidden for them of joy, as a reward for what they used to do.

32:18 Is one who was an affirmer the same as one who was wicked? They are not the same.

32:19 As for those who affirm and promote reforms, for them are eternal Paradises as an abode, in return for their works.

32:20 As for those who were wicked, their abode is the fire. Every time they try to leave it, they will be put back in it, and it will be said to them: "Taste the retribution of the fire in which you used to deny."

32:21 We will let them taste the worldly retribution before the greater retribution, perhaps they will revert.

32:22 Who is more wicked than one who is reminded of his Lord's signs, then he turns away from them? We will exact a punishment from the criminals.

32:23 We have given Moses the book. So do not be in any doubt about his encounter. We made it a guide for the Children of Israel.

32:24 We made from among them leaders who guided in accordance with Our command, for they were patient and had certainty regarding Our signs.

32:25 Your Lord will separate between them on the day of resurrection regarding what they disputed in.

32:26 Is it not a guide for them how many generations We have annihilated before them in whose homes they now walk? In that are signs. Do they not listen?

32:27 Have they not seen that We drive the water to the barren lands, and produce with it crops to feed their livestock, as well as themselves? Do they not see?

32:28 They say, "When is this victory, if you are being truthful?"

32:29 Say, "On the day of the victory, it will not benefit those who rejected if they affirm, nor will they be given respite."

32:30 Therefore, turn away from them and wait, for they too are waiting.

ENDNOTES

032:001 A1L30M40. This combination of letters/numbers plays an important role in the mathematical system of the Quran based on code 19. The function of these letters is stated in the following verse 32:2. See 74:1-56; 1:1; 2:1; 13:38; 27:82; 38:1-8; 40:28-38; 46:10; 72:28.

032:004 For the evolutionary creation of the universe, See 41:9-10.

032:005 Since the first audience of this verse was using the lunar calendar, we can find the velocity of the divine ruling in terms of km/day by calculating the sidereal distance covered by the Moon in one thousand years. Interestingly, when we divide the distance by the number of seconds in a day we end up with the speed of light, which is 299792.5 km/s. In other words, the light is 1000 x 12 x 27.321661 faster than the moon. Time is relative. See 22:47; 70:4.

032:007 See 15:26-29. Also, see 4:119.

032:012 See 6:28.

032:014 It is interesting that the only exception for God's never forgetting nature is in connection with the people who deserved Hell. For a study on verses related to hell see our note for 22:19-22

33:0 In the name of God, the Gracious, the Compassionate.

33:1 O you prophet, you shall be aware of **God**, and do not obey the ingrates and the hypocrites. **God** is Knowledgeable, Wise.

33:2 Follow what is being revealed to you by your Lord. **God** is fully aware of all that you do.

33:3 Put your trust in **God**. **God** suffices as an advocate.

33:4 **God** did not make any man with two hearts in his body. Nor did He make your wives whom you estrange to be as your mothers. Nor did He make your declared associates to be your sons. Such is what you claim with your mouths, but **God** speaks the truth, and He guides to the path.*

33:5 Call them by their father's name. That is more just with **God**. But if you do not know their fathers, then, as your brothers in the system and members of your family. There is no sin upon you if you make a mistake in this respect; but you will be responsible for what your hearts deliberately intend. **God** is Forgiver, Compassionate.

33:6 The prophet is closer to those who affirm than themselves; and his wives are mothers to them. In **God**'s decree, relatives consisting of affirmers and the emigrants have mutual privileges, provided that they take care of their own families first in appropriate way. Such has been decreed in the book.

33:7 When We took from the prophets their covenant. From you, and from Noah, and Abraham, and Moses, and Jesus the son of Mary; We took from them a strong covenant.*

33:8 So that the truthful may be asked about their truthfulness, and He has prepared for the ingrates a painful retribution.

Muslims Defend Themselves Against Aggressors

33:9 O you who affirm, remember **God**'s favor upon you when soldiers attacked you and We sent upon them a wind and invisible soldiers. **God** is Seer of everything you do.

33:10 For they came from above you, and from beneath you; and your eyes were terrified, and the hearts reached to the throat, and you harbored doubts about **God**.

33:11 That is when those who affirm were truly tested; they were severely shaken up.

Hypocrites among Companions of the Prophet

33:12 The hypocrites and those who have a sickness in their hearts said, "**God** and His messenger promised us nothing but illusions!"*

33:13 A group of them said, "O people of Yathrib, you cannot make a stand; therefore, retreat." A small party of them sought permission from the prophet, saying: "Our homes are exposed," while they were not exposed. They just wanted to flee.*

33:14 If the enemy had entered at them from all sides, and then they were asked to betray, they would do so with very little hesitation.

33:15 Indeed they had pledged to **God** before this that they would not turn around and flee. Making a pledge to **God** brings great responsibility.

33:16 Say, "It will not benefit you to flee away from death or from being killed, for you will only have the enjoyment for a short while."

33:17 Say, "Who can protect you from **God** if He intends to harm you, or He intends any blessing for you?" They will never find besides **God** any ally or supporter.

33:18 **God** already knows which of you are the hinderers, and those who say to their brothers: "Come and stay with us." Rarely do they mobilize for battle.

33:19 They are miserable towards you. Then, when fear comes, you see their eyes rolling, as if death had already come to them. But once the fear is gone, they lash out at you with sharp tongues. They are miserable towards doing any good. These have not affirmed, so **God** nullifies their works. This is easy for **God** to do.

33:20 They thought that the opponents had not yet mobilized. If the opponents do appear, they wish that they were out in the desert, seeking out news for you. Even if they were among you, they would not have fought except very little.

Muhammed's Exemplary Courage in the Battlefield

33:21 Indeed, in the messenger of **God** a good example has been set for he who seeks **God** and the Last day and thinks constantly about **God**.*

33:22 When those who affirm saw the opponents, they said, "This is what **God** and His messenger have promised us, and **God** and His messenger are truthful." This only increased their affirmation and their peacefully surrender.*

33:23 From among those who affirm are people who fulfilled their pledge to **God**. Thus, some of them died, while some are still waiting; but they never altered in the least.

33:24 That **God** may recompense the truthful for their truthfulness, and punish the hypocrites if He so wills, or accepts their repentance. **God** is Forgiver, Compassionate.

33:25 **God** drove back those who rejected with their rage; they left empty handed. Thus **God** sufficed the affirmers in the battle. **God** is Powerful, Noble.

33:26 He also brought down those who helped support them among the people of the book from their secure positions. He threw fear into their hearts. Some of them you killed, and some you took captive.

33:27 He inherited you their land, their homes, their wealth, and lands you had never stepped on. **God** is able to do all things.

Public Figures Have More Responsibility

33:28 O prophet, say to your wives: "If you are seeking this worldly life and its vanities, then come, I will make a provision for you and release you in an amicable manner."

33:29 "But if you are seeking **God** and His messenger, and the abode of the Hereafter, then **God** has prepared for the righteous among you a great reward."

33:30 O wives of the prophet, if any of you commits evident lewdness, then the retribution will be doubled for her. This is easy for **God** to do.

33:31 Any of you who is obedient to **God** and His messenger, and does good works, We will grant her double the recompense, and We have prepared for her a generous provision.

33:32 O wives of the prophet, you are not like any other women. If you are righteous, then do not speak too softly, lest those with disease in their hearts will move with desire; you shall speak in a appropriate manner.*

33:33 You shall behave honorably in your homes, and do not act ostentatiously like the olden days of ignorance. You shall maintain the contact prayer, and contribute towards betterment, and obey **God** and His messenger. **God** wishes to remove any affliction from you, O people of the sanctuary, and to purify you completely.*

33:34 Remember **God**'s signs and the wisdom that are being recited in your homes. **God** is Sublime, Ever-aware.

Equal Qualities for Both Genders

33:35 Surely, the peacefully surrendering men, and the peacefully surrendering women, the affirming men, and the affirming women, the obedient men, and the obedient women, the truthful men, and the truthful women, the patient men, and the patient women, the humble men, and the humble women, the charitable men, and the charitable women, the fasting men, and the fasting women, the men who guard their private parts, and the women who similarly guard, and the men who commemorate **God** frequently, and the commemorating women; **God** has prepared for them a forgiveness and a great recompense.

33:36 It is not for an affirming man or affirming woman, if **God** and His messenger issue any command, that they have any choice in their decision. Anyone who disobeys **God** and His messenger, he has gone far astray.

Muhammed Tested regarding his Protectee

33:37 You said to the one who was blessed by **God** and blessed by you: "Keep your wife and revere **God**," and you hid inside yourself what **God** wished to proclaim. You feared the people, while it was **God** you were supposed to fear. So, when Zayd ended his relationship with his wife, We had you marry her, to establish that there is no wrongdoing for those who affirm marrying the wives of their protectees if their relationship is ended. **God**'s command is always done.*

God's Law related to Social Order

33:38 There is no blame on the prophet in doing anything that **God** has decreed upon him. Such was **God**'s *sunna* with the people of old. **God**'s command is a determined duty.*

33:39 Those who deliver **God**'s messages, and revere Him, and they do not revere anyone but **God**. **God** suffices as a Reckoner.

Muhammed the Final Prophet

33:40 Muhammed was not the father of any man among you, but he was the messenger of **God** and the seal of the prophets. **God** is fully aware of all things.

Remember God Frequently

33:41 O you who affirm, you shall remember **God** frequently.

33:42 Glorify Him morning and evening.

33:43 He is the One who supports you, along with His controllers, to lead you out of darkness into the light. He is ever Compassionate towards those who affirm.

33:44 Their greeting the day they meet Him is: "Peace" and He has prepared for them a generous recompense.

33:45 O prophet, We have sent you as a witness, and a bearer of good news, and a warner.

33:46 Inviting to **God**, by His leave, and a guiding beacon.

33:47 Give good news to those who affirm, that they have deserved from **God** a great blessing.

33:48 Do not obey the ingrates and the hypocrites, and ignore their insults, put your trust in **God**; **God** suffices as an advocate.

Marriage Laws and Etiquette Rules

33:49 O you who affirm, if you married affirming women and then divorced them before having intercourse with them, then there is no interim required of them. You shall compensate them and let them go in an amicable manner.

33:50 O prophet, We have made lawful for you the wives to whom you have already given their dowry, and the one who is committed to you by oath, as granted to you by **God**, and the daughters of your father's brothers, and the daughters of your father's sisters, and the daughters of your mother's brothers, and the daughters of your mother's sisters, of whom they have emigrated with you. Also, the affirming woman who had decreed herself to the prophet, the prophet may marry her if he wishes, as a privilege given only to you and not to those who affirm. We have already decreed their rights in regard to their spouses and those who are still dependents. This is to spare you any hardship. **God** is Forgiver, Compassionate.

33:51 You may postpone whom you will of them, and you may receive whom you will. Whomsoever you seek of those whom you have set aside then there is no sin upon you. Such is best that they may be comforted and not grieve; may all be pleased with what you give them. **God** knows what is in your hearts. **God** is Knowledgeable, Compassionate.

33:52 Except those whom you are already committed by contract, no women are lawful to you beyond this, nor that you change them for other wives, even though you may be attracted by their beauty. **God** is watchful over all things.*

33:53 O you who affirm, do not enter the prophet's homes except if you are invited to a meal, without you imposing such an invitation. But if you are invited, you may enter. When you finish eating, you shall leave, without staying to wait for a *hadith*. This used to annoy the prophet, and he was shy to tell you. But **God** does not shy away from the truth. If you ask his wives for something, ask them from behind a barrier. This is purer for your hearts and their hearts. It is not for you to harm **God**'s messenger, nor you should marry his wives after him. This is a gross offence with **God**.*

33:54 If you reveal anything, or hide it, **God** is fully aware of all things.

33:55 There is no sin upon them before their fathers, or their sons, or their brothers, or the sons of their brothers, or the sons of their sisters, or their offspring yet to come, or their women, or those who are still their dependents. Be aware of **God**, for **God** is witness over all things.

God and His Controllers Supported the Prophet

33:56 **God** and His controllers support the prophet. O you who affirm, you shall support him, and yield completely.*

33:57 Surely those who harm **God** and His messenger, **God** will curse them in this life and in the Hereafter; and He has prepared for them a shameful retribution.

33:58 Those who harm the affirming men and the affirming women, with no just reason, they have brought upon themselves a slander and a gross sin.

Advice to Women who are Harassed by Men: Distinguish yourselves from Prostitutes

33:59 O prophet, tell your wives, your daughters, and the wives of those who affirm that they should lengthen upon themselves their outer garments. That is better so that they would be recognized and not harmed. **God** is Forgiver, Compassionate.

33:60 If the hypocrites, and those with disease in their hearts and those who spread lies in the city do not refrain, then We will let you overpower them, then they will not be able to remain as your neighbors except for a short while.

33:61 They are cursed wherever they are found, and they are taken and killed in numbers.

33:62 This is **God**'s way with those who have passed away before, and you will not find any change in **God**'s way.*

The Moment

33:63 The people ask you regarding the Moment. Say, "Its knowledge is with **God**. For all that you know the Moment may be near!"*

33:64 **God** has cursed the ingrates, and He has prepared for them hell.

33:65 Eternally they abide therein. They will find no ally, or helper.

33:66 On the day when their faces will be turned over in the fire, they will say, "Oh, we wish we had obeyed **God**, and obeyed the messenger"

33:67 They will say, "Our Lord, we have obeyed our leaders and our learned ones, but they misled us from the path."

33:68 "Our Lord, double the retribution, and curse them with a mighty curse."

33:69 O you who affirm, do not be like those who harmed Moses. Then **God** cleared him of all they said, and he was honorable before **God**.

33:70 O you who affirm, be aware of **God** and speak only the truth.

33:71 He will then direct your works and forgive your sins. Whosoever obeys **God** and His messenger, has attained a great triumph.

33:72 We have offered the trust to the heavens and the earth, and the mountains, but they refused to bear it, and were fearful of it. But the human accepted it; he became transgressing, ignorant.*

33:73 So that **God** may punish the hypocrite men and the hypocrite women, and the men who set up partners and the women who set up partners. **God** redeems the affirming men and the affirming women. **God** is Forgiver, Compassionate.*

ENDNOTES

033:004 Reflecting on the context of this verse, which reminds us that reality does not change just because people say something, we can deduce that during the time of revelation some people believed that they had two hearts. The Quran rejects these superstitions. From the implication of the verse, we may understand the following too: Heart is a word used for the center of affirming the truth, and love. Hearts should be dedicated to the beautiful attributes of God alone. God is the only Lord and Master.

033:007 See 3:81.

033:012 In Sunni literature, the view of infallible or saint-like companions of Prophet Muhammed is predominant. Sunni literature tries to establish a taboo on discussing or criticizing any companion (*sahaba,* plural *ashab*) whose names are listed in the so-called authentic books as narrators. In a *hadith* falsely attributed to Muhammed, his companions are likened to stars and following any of them would be sufficient for guidance. The Quran, however, gives us a very different picture of Muhammed's companions. The Quran uses the word *sahaba* and its plural *ashab* for all companions of Muhammed, including his opponents, hypocrites, and muslims (34:46; 53:2; 81:22). The Quran informs us about the existence of hypocrites among Muhammed's companions; nevertheless, their names are not exposed to the public. The reason for not exposing the name of the hypocrites is God's will to implement the test freely on earth.

As it is the case in most religious matters, Sunnis prefer fabricated *hadith*s to the Quran. In Sunni teaching there is a hierarchy of righteousness, starting from the stars of the sky to the next generation and the next generation. Through this myth, they craftily established ancestor-worship, one of the versions of polytheism categorically rejected by the Quran. Shiites, on the other hand, have a stricter rule for creating idol stars from dead people. They require Muhammed's or Ali's genes as a pre-requisite for sainthood. Idolizing early generations was a natural consequence of their theology: the authority of *hadith*s solely relies on the credibility and superiority of *sahaba,* and subsequent generations *(tabeen, etba tabeen...*), whose names have been sanctified in the lists of narrators of hearsay. They found it imperative to create a blanket protection for all companions, since allowing one to be scrutinized or questioned would create a slippery slope that would ultimately expose the human nature of the superhuman, ideal, angelic characters. Nevertheless, numerous works by Sunni scholars have challenged the mainstream position.

Even if their history books point at some of those "holy men" killing other "holy men," the Sunnis claim that all will go to heaven, both the murderers and the victims. Sunni sources refuse even to condemn Muawiya and his governor Marwan who committed many atrocities. Though they express their grievances for the brutal murders of Hassan and Hussain, prophet Muhammed's grandchildren, Sunnis again, refuse to criticize, let alone condemn the murderers, who happened to be Muhammed's companions when he was alive, or the companions of his companions! *Sahaba*-worship is one of the main teachings of Sunni sects. Shiites, on the other hand, condemn anyone who somehow participated in the conflict against anyone related to Muhammed through Ali and Fatima's lineage. Shiites, rather than using the universal criteria of justice and truth, use genes as their compass to hell or heaven. They worship only those *ashab* (plural of *sahaba*) who either supported Muhammed's son in-law Ali or his children. In brief, the historical animosity between Sunnis and Shiites originates from the differences in the list of their idols. All other differences are secondary, and plenty can be found among their own subjects.

033:013 Yathrib was the name of the city that Muhammed and his supporters immigrated to and there they established a constitutional federal system accommodating a multi-religious and multi-racial diverse population. But now we cannot find the name of Yathrib on the map. The followers of *hadith* and *sunna* changed the name of the city after Muhammed just as they changed many other things. Since *hadith* is the product of later generations who started calling Yathrib with the word *Medina* (City), Muslims abandoned the original name, the name that is mentioned in the Quran and was used by Muhammed and his companions. This is part of the evidence that proves our assertion that "Muslims" have deserted the Quran and are following hearsay as their guide (25:30).

033:021 Those who idolized Muhammed have taken the verse out of its context and have abused it to justify volumes of stories presented as Muhammed's actions (*sunna*). The verse refers to Muhammed's bravery and constant remembrance of God during the war. The Quran refers to some other good leadership qualities of Muhammed (3:121, 159; 9:40) as well as some of his weaknesses (33:37; 80:1-16). Those who deserted the Quran (25:30), instead of following the progressive teaching of the Quran, extended this good example to irrelevant individual or cultural behavior. They introduced numerous rules to be mimicked, such as Muhammed's clothing, diet, toothbrush, beard, walk, sleep, and even urination. They sanctified growing beards and wearing turbans, ignoring the fact that the Meccan idol worshipers, such as Abu Jahl and Walid b. Mugiyra also had long beards and wore turbans. They adopted a whole concoction of medieval Arabic, Jewish and Byzantine cultures as

God's religion. They regressed to monkeys like previous generations (2:65; 5:60; 7:166).

What is worse, the actions and words ascribed to prophet Muhammed have depicted him with a character that is far from exemplary. The *Hadith* books portray the prophet as a phantasmagoric character with multiple personalities. That character is more fictitious than mythological gods and goddesses, such as Hermes, Pan, Poseidon and Aphrodite. He is a pendulous character, both bouncing up to deity, and down to the lowest degree. He is both wise and a moron. He is sometimes more merciful than God and sometimes a cruel torturer. He is both perfect and criminal, humble and arrogant, chaste and a sex maniac, trustworthy and a cheater, illiterate and an educator, rich and poor, a nepotistic and a democratic leader, caring and a male chauvinist, a believer and a disbeliever, prohibiting *Hadith* and promoting *Hadith*. You can find numerous conflicting personalities presented as an exemplary figure. Choose whichever you like. This peculiar aspect of *Hadith* collection is well-described by the prophetic verses of the Quran (68:35-38).

Furthermore, a similar statement is made about Abraham: "A good example has been set for you by Abraham and those with him" (60:4, 6). If verse 33:21 requires Muhammed's *hadith*, then why would the verses 60:4,6 not also require Abraham's *hadith*? Which books narrate *hadith*s from Abraham? Obviously, the only reliable source for both examples is the Book of God, which narrates the relevant exemplary actions. It also warns us not to repeat the mistakes committed by Muhammed (33:37; 80: 1-10). See 3:18; 3:159; 60:4. 42:21.

033:032 Prophet Muhammed was one of the most successful political and military leaders in history. His revolutionary leadership in theology, philosophy, culture, economy, and politics, put him on top of the list of a prominent Western historian, Michael Hart, in the "History's 100 Most Influential People." Muhammed received a great opposition from the oligarchy of the Meccan theocracy. He also experienced some opposition in his own city-state. Besides violent campaigns to eliminate him and his supporters, he would occasionally become the target of defamation campaigns. Many conspired to embarrass and defame him through his wives. Indeed, this verse distinguishes Muhammed's wives from other Muslim women. Muhammed's wives were asked to be more careful not to provide ammunition for the negative propaganda of regressive forces who wished to stop the progressive movement that started with the light of the Quran.

033:033 One of the verbs in this verse can be understood in two ways, depending on the inclusion or exclusion of the letter W. Traditional commentators read it "*wa QaRna*" (and they should settle), derived from the root *QRR*, meaning to settle, sit down, etc. They translate the phrase as "sit in your homes." This translation followed by generalization of the instruction has been used as one of the reasons for why Muslim women should be confined in her homes. We prefer reading it as *WaQaRna*, which comes from the root *WQR*, which means acting in an honorable, kind, and dignified manner. (See 48:9; 71:13). Our reading would not be accurate if the verb was spelled as *QaRaRna*, from the root *QRR*. (See 7:143; 14:26,29; 22:5; 23:13; 27:61; 38:60; 40:39,64). It is evident that Muhammed's wives had their separate income and wealth.

033:037 In the previous editions of The Quran: a Reformist Translation, I mistranslated the word *ad'iya* as "adopted sons". Though I had rejected Sunni religion in 1986, it took me years to eliminate its influence on me. As it appears I have not yet purified my mind completely from its distortive influence. However, with the passing of time, the moment I become aware of the background noise or distortion, I study the issue and reform my understanding. The word Da'AYa means "to claim, to call, to call upon, to declare". Zayd, most likely a former slave, was declared as an associate, a mentee, or a protectee by Muhammad. It does not mean "adopted sons." For "adopted child" the Quran uses the word *RaByB*, plural *RaBaYiB*" (See 4:23 as adopting wife's child from previous marriage, stepdaughter) or an expression such as *ittakhaza walada* "adopt a son" (See 12:21; 28:9).

Prof David S. Powers of Cornell University has written an academic book on this very verse. In his book titled, *Muhammad is Not the Father of Any of Your Men: The Making of the Last Prophet,* published by University of Pennsylvania Press, Powers contends some controversial ideas, including rejecting the traditional focus on sex and marriage. He shifts the focus to the finality of prophethood, which is stated in the following verse, 33:38. It is indeed interesting that marrying Zayd's divorced wife is related to the finality of the Prophethood. Muhammed's close association to Zayd as his mentor or protector, created expectation of him replacing Muhammed after his death as a prophet. Power's argument and various evidence supporting his argument on this issue is compelling and worth further discussion.

033:038 When the Quran uses the word *hadith* (word; saying; narration) for something other than itself, it usually uses it in negative way or context (7:185; 12:111; 31:6; 33:53; 45:6; 52:34; 66:3; 77:50). Since God knew that that Sunni and Shiite *mushriks* would create other authorities besides God in their religion and call them *hadith* and *sunna* of prophet Muhammed, the Quran uses the word *sunna* (law, ordinance, practice) in conjunction with God, as in Sunnatullah (God's law, ordinance), which depicts the social, political and ecological consequences for rejecting God's message (33:38,62; 35:43; 40:85; 48:23; 8:38; 15:13; 17:77; 18:55; 3:137; 4:26). There is only one valid *sunna* (law, way, practice) and it is God's *sunna*. Those who follow Muhammed's *sunna* or, more accurately, a man-made *sunna* falsely attributed to him, will be convicted to the *sunna* mentioned in 35:43. The singular word *sunna* is never used for other than God to denote a positive path; but its plural, *sunan*, is used once in a positive sense in connection to previous generations (4:26). More interesting, the third fabricated religious authority or idol is called *ijma* (consensus) and whenever this word or its derivatives are used in the Quran for other than God, they are always negative (20:60; 70:18; 104:2; 3:173; 3:157; 10:58; 43:32; 26:38; 12:15; 10:71; 20:64; 17:88; 22:73; 54:45; 28:78; 7:48; 26:39; 26:56; 54:44...). The same is true for the fourth word *sharia* (law) (42:21). The same is true for the word *salaf* (43:56) and for the word *ashab* (53:2). Are all these a coincidence? Since the followers of *hadith* and *sunna* do not translate these words, to expose the nature of their teachings we preferred not to translate some of those words too. This would not only expose their evil scheme but would also highlight one of the prophetic features of the Quran. Faced with these prophetic Quranic challenges, later *Sunni* and *Shiite* scholars tried to hijack one of the attributes of the Quran, *hikma* (wisdom) or *hakim* (wise), for their *hadiths*; but they failed. The abundance of nonsense in *hadith* books did not accept or deserve such a description; they could not market donkey's braying as music, except to the deaf; they could not portray camels as penguins, except to the blind. See 6:112-116; 17:39; 36:1; 39:18; 53:2; 66:3. Also, see 49:1.

033:052 Those who traded the progressive precepts of the Quran with the misogynistic culture of the medieval Pagan, Jewish and Christian communities covered women from head to toe while stripping them of their identities. Why do these people not reflect on this verse and affirm it?

The verse refers to physical attraction. What is it? Is it subjective or is there a universal or measurable factor in our perception of beauty? Beauty, specifically facial attractiveness, has been a subject attracting the curiosity of philosophers and scientists. The issue received a scientific explanation with the computerized studies in the 1990s. In one of those studies, the scientists took the pictures of 96 college students and made their computer-generated composites by mathematically averaging their pixels in key points. They later asked 300 judges to rate them on a 5-point scale, 1 being very unattractive and 5 being very attractive. The composite faces consisting of higher component faces received the highest scores. In other words, our brain considers the average face of the population in which we live as the most attractive. There are interesting genetic, social and cognitive implications of this study. See 12:31; 30:21.

033:053 See 66:3; 33:38, and 33:32. Verse 4:22 prohibits marrying the ex-wives of fathers. To keep the privacy of Muhammed's personal life and prevent a possible competition among would-be candidates, the prophet's wives were legally considered to be the mothers of those who affirm (33:6).

033:056 This verse is one of the most distorted and abused verses. We translated the word *salli ala* as "support/encourage." The same word occurs at 33:43 and 9:99,103. When these verses are studied comparatively, the traditional abuse and distortion becomes evident. Through this distortion, Muslim masses are led to commemorate and praise Muhammed's name day and night, rather than his Lord's and Master's name (33:41-42). Also, see 2:157; 2:136; and 3:84.

033:062 The only valid *sunna* is God's *sunna*. See 33:38.

033:063 We may learn some of the things that God knows. See 20:15.

033:072 See 3:83; 13:15; 41:11.

033:073 The fabricators of *hadith* stating that Muhammed's followers would split into 73 factions (*ahzab*) were most likely inspired by the number of verses in this chapter, which is traditionally named Factions. Since that *hadith*, the number of sects, cults, and orders is in the thousands. See: 6:153-159; 30:32; 98:4; 21:92; 23:52; 2:285.

34:0 In the name of God, the Gracious, the Compassionate.

34:1 Praise be to **God**, to whom belongs everything in the heavens and the earth; and to Him is the praise in the Hereafter. He is the Wise, the Ever-aware.

34:2 He knows what goes into the earth, and what comes out of it, and what comes down from the sky, and what climbs into it. He is the Compassionate, Forgiver.

34:3 Those who rejected said, "the Moment will not come upon us!" Say, "Yes, by my Lord, it will come to you." He is the Knower of the unseen, not even an atom's weight or less than that or greater can be hidden from Him, be it in the heavens or the earth. All are in a clear record.

34:4 That He may reward those who affirm and promote reforms. To them will be forgiveness and a generous provision.

34:5 As for those who sought against Our signs to frustrate them, they will have retribution of painful affliction.

34:6 Those that have been given knowledge will see that what has been sent down to you from your Lord is the truth, and that it guides to the path of the Noble, the Praiseworthy.

34:7 Those who rejected said, "Shall we lead you to a man who will tell you that after you are dismembered you will be created anew?"

34:8 "Has he invented a lie against **God**, or is there madness in him?" Indeed, those who do not affirm the Hereafter will be in retribution and far straying.

34:9 Have they not seen all that is before them and behind them, in the heaven and the earth? If We wished, We could cause the earth to swallow them, or cause pieces of the sky to fall on them. In this is a sign for every obedient servant.

David and Solomon

34:10 We granted David blessings from Us: "O mountains, glorify with him, as well as the birds." We softened the iron for him.

34:11 That you may make armors that fit perfectly, and work righteousness. For I am Seer of what you do.

Transportation and Petroleum

34:12 For Solomon the wind was committed, covering one month in morning and one month in evening. We also caused a spring of petroleum to flow for him. From among the Jinn are those that worked for him by his Lord's leave; and any one of them who turns from Our commands, We shall cause him to taste the retribution of the fire.*

34:13 They made for him what he desired of enclosures, and statues, and pools of deep reservoirs, and heavy pots. "O family of David, work to show thanks." Only a few of My servants are appreciative.*

34:14 Then, when We decreed death for him, nothing informed them of his death until a worm kept eating from his staff, so when he fell down, the Jinn realized that if they had known the unseen, they would not have remained in the humiliating retribution.

People of Sheba

34:15 There was for Sheba a sign in their homeland, with two paradises, on the right and the left. "Eat from the provisions of your Lord, and be thankful to Him." A good land, and a forgiving Lord.

34:16 But they turned away, so We sent them a destructive flash flood, and We substituted their two gardens into two gardens of rotten fruits, thorny plants, and a skimpy harvest.

34:17 We thus requited them for what they rejected. We do not requite except the ingrate.

34:18 We placed between them and between the towns that We blessed, towns that were easy to see; and We measured the journey between them: "Travel in them by night and day in complete security."

34:19 But they said, "Our Lord, make the measure between our journeys longer," and they wronged themselves. So We made them a thing of the past, and We scattered them into small groups. In this are signs for every person who is patient, thankful.*

34:20 Satan has been successful in his suggestions to them, for they followed him, except for a group of those who affirm.

34:21 He did not have any authority over them except that We might know who affirmed the Hereafter from those who are doubtful about it. Your Lord is Keeper over all things.

On the Day of Judgment: Intercession is Denied

34:22 Say, "Call on those whom you have claimed besides **God**. They do not possess even a single atom's weight in the heavens, or the earth. They possess no partnership therein, nor is there for Him any assistant among them."

34:23 Nor will intercession be of any help with Him, except for whom He has already given leave. Until when the terror has subsided from their hearts, they ask: "What did your Lord say?" They will say, "The truth!" He is the Most High, the Great.*

34:24 Say, "Who provides for you from the heavens and the earth?" Say, "**God**! Either we or you are guided, or are clearly astray."

34:25 Say, "You will not be asked about our crimes, nor will we be asked for what you do."

34:26 Say, "Our Lord will gather us together then He will judge between us with truth. He is the Judge, the Knowledgeable."

34:27 Say, "Show me those whom you have set up as partners with Him! No; He is but **God**, the Noble, and the Wise."

34:28 We have sent you to all people to be a bearer of good news, as well as a warner; but most people do not know.

34:29 They say, "When is this promise, if you are truthful?"

34:30 Say, "You have an appointed day, which you cannot delay by one moment, nor advance."

34:31 Those who rejected have said, "We will not affirm this Quran, nor in what is already with him." If you could but see these transgressors when they stand before their Lord, how they will accuse one another back and forth. Those who were weak will say to those who were mighty: "If it were not for you, we would have been those who affirm!"

34:32 Those who were mighty will say to those who were weak: "Did we turn you away from the guidance after it had come to you? No, it was you who were criminal."

34:33 Those who were weak will say to those who were mighty: "No, it was your scheming night and day, when you commanded us to reject **God**, and to set up equals to Him." They will be filled with regret, when they see the retribution. We will place shackles around the necks of those who rejected. Are they not being requited for what they used to do?

34:34 We do not send a warner to any town, except its privileged hedonists would say, "We reject what you have been sent with."

34:35 They said, "We have more wealth and more children, and we will not be punished."

34:36 Say, "My Lord gives provisions to whomever He wishes, or He restricts them, but most people do not know."

34:37 It is not your money or your children that will bring you closer to Us, but only those who affirm and promote reforms, they will receive double the reward for their works, and they will reside in the high dwellings in peace.

34:38 As for those who sought against Our signs, they will be brought to the retribution.

34:39 Say, "My Lord gives provisions to whom He wishes of His servants and He restricts. Anything you spend, He will replace it; and He is the Best of providers."

34:40 On the day when We gather them all, then We will say to the controllers: "Was it you that these people used to serve?"

34:41 They will say, "Be You glorified. You are our Lord, not them. No, most of them were serving the Jinn; most were those who affirm them."*

34:42 So today, none of you can help or harm one another. We will say to the transgressors: "Taste the retribution of the fire that you used to deny."

Signs are not Appreciated by the Majority

34:43 When Our clear signs were recited to them, they said, "This is but a man who wants to turn you away from what your parents were serving." They also said, "This is nothing but a fabricated lie." Those who rejected say of the truth when it has come to them: "This is nothing but evident magic!"

34:44 We had not given to them any book to study, nor did We send to them before you any warner.

34:45 Those before them had also denied, while they did not reach one tenth of what We have given to this generation, so they denied My messengers, how severe was My retribution!

34:46 Say, "I advise you to do one thing: that you stand to **God**, in pairs or as individuals, then reflect. There is no madness in your friend, he is only a warner to you in the face of a severe retribution."

34:47 Say, "I have not asked you for any wage; you can keep it. My wage is from **God**, and He is witness over all things."

34:48 Say, "My Lord rebuts with the truth. He is the Knower of all secrets."

34:49 Say, "The truth has come; while falsehood can neither initiate anything, nor resurrect."

34:50 Say, "If I stray, then I stray to my own loss. If I am guided, then it is because of what my Lord revealed to me. He is Hearer, Near."

34:51 If you could but see, when they will be terrified with no escape, and they will be taken from a place that is near.

34:52 They will say, "We affirm it," but it will be far too late.

34:53 They had rejected it in the past; and they conjectured about the unseen from a place far off.

34:54 They will be separated from what they had desired, as was done for their counterparts before. They have been in grave doubt.

ENDNOTES

034:012 Accepting the average speed of that day to be 6 km/h, then Solomon was reaching 360 km/h.

034:013 Solomon, besides being a leader, was also an art enthusiast. It is ironic to see that those who worship the black stone in Mecca, or turn the tombs of dead people into shrines, or hope for the intercession of the dead prophets and saints, those who commit polytheism in a myriad of ways, have such an opposition to statues made for artistic purposes.

034:019 With the discovery of new water sources and the formation of new towns, competing markets emerged. It seems that the competition did not please Sheba's traders. Instead of adapting to the new conditions of demand and supply, trying to preserve the status quo might lead to loss and extinction.

034:023 See 2:48.

034:041 Following the teaching of Satan is the equivalent of serving him. Even if someone claims to serve God, if he is following other teachings for salvation, he is serving Satan. The only way for protecting oneself is to dedicate one's purpose in life to God alone (15:39-42). Those who hope for the intercession of prophets, saints, and controllers, and consider them to be partners with God in designing the system, in fact do not serve them. Service requires mutual acceptance. Considering someone who rejects any other lord besides God to be one's lord is a contradiction and pleases Satan. In the hereafter, righteous people who were idolized against their will, will reject those who "worship them." (19:82; 46:6).

بسم الله الرحمن الرحيم

35:0 In the name of God, the Gracious, the Compassionate.

35:1 Praise be to **God**, Initiator of the heavens and the earth; maker of the controllers as messengers with powers/dimensions in twos, threes, and fours. He increases in the creation as He wishes. **God** can do all things.*

35:2 Whatever mercy **God** opens for the people none can stop it. What He holds back none can send it other than He. He is the Noble, the Wise.

35:3 O people, remember **God**'s blessing upon you. Is there any creator other than **God** who provides for you from the sky and the earth? There is no god besides Him, so why do you deviate?

35:4 If they deny you, then also messengers before you have been denied. To **God** will all matters be returned.

Beware of Being Deceived by Religious Leaders

35:5 O people, **God**'s promise is the truth; therefore, do not be deceived by this worldly life, and do not let the deceiver deceive you by invoking **God**.

35:6 The devil is an enemy to you, so treat him as an enemy. He only invites his faction to be the dwellers of hell.

35:7 Those who reject will have a painful retribution, and those who affirm and promote reforms, they will have forgiveness and a great reward.

35:8 The one whose evil work is adorned, and he sees it as being good. **God** thus misguides whom/whomever wills, and He guides whom/whomever wills. So do not let yourself grieve over them. **God** is fully aware of what they are doing.

35:9 **God** is the One who sends the winds so it stirs up a cloud, then We drive it to a town that is dead, and We revive with it the land after it had died. Such is the resurrection.

Dignity Belongs to God

35:10 Whosoever is seeking dignity, then know that to **God** belongs all dignity. To Him ascend the good words, and they are exalted by the good deeds. As for those who scheme evil, they will have a painful retribution, and their scheming will fail.*

35:11 **God** created you from dust, then from a droplet, then He made you into pairs. No female becomes pregnant nor gives birth without His knowledge. Nor does anyone have his life extended, or his life is shortened, except in a record. This is easy for **God**.*

35:12 Nor are the two seas the same. One is fresh and palatable, good to drink, while the other is salty and bitter. From each of them you eat tender meat, and you extract jewelry to wear. You see the ships sailing through them, seeking His provisions; and perhaps you may be appreciative.

35:13 He merges the night into the day and merges the day into the night. He has directed the Sun and the Moon; each run for a predetermined period. Such is **God** your Lord; to Him is the sovereignty. As for those whom you call on besides Him, they do not possess a seed's shell.

Ignorance of Asking Help from Shrines

35:14 If you call on them, they cannot hear you. Even if they hear you, they cannot respond to you. On the day of resurrection, they will reject your idolatry. None can inform you like an Ever-aware.*

35:15 O people, you are the poor when it comes to **God**, while **God** is the Rich, the Praiseworthy.

35:16 If He wishes, He could do away with you and He would bring a new creation.

35:17 This is not difficult for **God**.

35:18 None can carry the load of another, and even if it calls on another to bear part of its load, no other can carry any part of it, even if they were related. You will only be able to warn those who fear their Lord while unseen, and they maintain the contact prayer. Whoever contributes, is contributing to himself. To **God** is the final destiny.

35:19 The blind and the seer are not equal.

35:20 Nor are the darkness and the light.

35:21 Nor are the shade and the heat.

35:22 Nor equal are the living and the dead; **God** causes whomever He wills to hear. You cannot make those who are in the graves hear.*

35:23 You are but a warner.

35:24 We have sent you with the truth, a bearer of good news, and a warner. There is not a nation but a warner came to it.

35:25 If they deny you, then those before them have also denied. Their messengers went to them with proofs, and the Psalms, and the enlightening book.

35:26 But then I seized those who rejected; how terrible was My retribution!

Colorful and Diverse Creation

35:27 Do you not see that **God** sends down water from the sky, thus We produce with it fruits of various colors? Of the mountains are peaks that are white, red, or raven black.*

35:28 Among the people, and the animals, and the livestock, are various colors. As such, only the knowledgeable among **God**'s servants revere Him. **God** is Noble, Forgiving.

35:29 Surely, those who recite **God**'s book, and maintain the contact prayer, and spend secretly and openly from what We have provided them, they are seeking a trade that can never lose.

35:30 He will give them their recompense, and He will increase them from His blessings. He is Forgiving, Appreciative.

35:31 What We revealed to you from the book, it is the truth, authenticating what is present. **God** is Ever-aware of His servants, Seer.

35:32 Then We inherited the book to those whom We selected from Our servants. Subsequently, some of them wronged themselves, and others upheld it partly, while others were eager to work righteousness in accordance with **God**'s will; this is the greatest triumph.

35:33 They will enter the gardens of Eden, where they will be adorned with bracelets of gold and pearls, and their garments in it will be of silk.

35:34 They will say, "Praise be to **God** who took away our sorrow. Our Lord is Forgiving, Appreciative."

35:35 "The One who admitted us into the abode of eternal bliss, out of His grace. In it, no boredom touches us, nor does any fatigue."

35:36 As for those who have rejected, for them is the fire of hell, where they do not terminate and die, nor is its retribution ever reduced for them. It is such that We requite every ingrate.

35:37 They will cry out there: "Our Lord, let us out, and we will work righteousness instead of the works we used to do." Did We not give you a long life so that he who would remember would take heed, and a warner came to you? Therefore, taste, for the transgressors will have no supporter.

35:38 **God** knows unseen of the heavens and the earth. He is knowledgeable of what is even inside the chests.

35:39 He is the One who made you successors on the earth. Subsequently, whoever rejects, then to him is his rejection. The rejection of those who do not appreciate only increases their Lord's abhorrence towards them. The rejection of those who do not appreciate only increases their loss.

35:40 Say, "Do you see the partners whom you have called on besides **God**? Show me what they have created on earth. Or do they possess any partnership in the heavens? Or have We given them a book wherein they are taking knowledge from it? Indeed, what the transgressors promise one another is no more than arrogance."

35:41 **God** is the One who holds the heavens and the earth, lest they cease to exist. They would certainly cease to exist if anyone else were to hold them after Him. He is Forbearing, Forgiving.

35:42 They swore by **God** in their strongest oaths, that if a warner came to them, they would be the most guided of all nations, but when a warner came to them, it only increased their aversion!

God's Law Never Changes

35:43 Arrogance on earth, and evil scheming… The evil schemes only backfire on those who scheme them. Were they expecting anything different from the *sunna* used on the people of the past? You will not find any change in **God**'s *sunna*.*

35:44 Did they not roam the earth and note what the consequence was for those who were before them? They were even more powerful than them in strength. But nothing can deter **God** be it in the heavens, or in the earth. He is Omniscient, Omnipotent.

35:45 If **God** were to punish the people for what they have earned, He would not leave a single creature standing. But He delays them to a predetermined time. Then, when their time comes, then **God** is Seer of His servants.

ENDNOTES

035:001 *Malaika* (singular: *Malak*) are powers, managers, or controllers employed by God to carry out God's specific commands and control the physical universe according to God's will. Controllers (malaika) are contrasted with Satan (Satan). Controllers are commonly called angels, and through stories they are depicted females with wings. The word *janaha* (plural *ajniha*) is used 9 times in the Quran to mean "incline" (8:61) or wing of compassion and protection (15:88; 18:24; 20:22; 26:215; 28:32), bird's wing (6:38) and here, which can be understood as a wing of compassion and protection. However, considering another meaning of the word, "side" we may understand it as" dimension." Angels are not like birds and the odd numbers of "wings" turn it into a metaphor, rather than bird's wings. Besides, what is the function of wings? To make a mass afloat in the air. The vast space however has no air. The Christian concept of angels which was later adopted by Muslims contradict the concept of *Malak* mentioned in the Quran, since *Malak*'s space is not limited to that of airplanes and birds. See 37:150; 66:6; 69:17; 89:22; 97:4.

035:010 The principles advocated by the Quran build excellent qualities for individuals as well as communities. For the word *izza* (dignity, prestige) see, 3:26; 19:81; 4:139; 35:10; 38:2; 63:8.

035:011 Could the "book" or "record" mentioned here be the DNA, which consists of various combinations of four molecules in a double helix? We learned recently that our lifespan is determined by some genes in our DNA.

035:014 Leading rational monotheist peacemakers such as Jesus, Mary, Muhammed, Ali, all will reject those who worship them.

035:022 Polytheists are likened to dead people. 6:122; 27:80; 30:50-52.

035:027 The audience of the Quran is everyone. For a farmer, the color of the soil provides valuable information about its use. For a scientist, the colorful stripes indicate different geological layers or archeological data. For a painter, such a view is excellent artwork worth copying onto a canvas. For a miner, the color of the earth is as important as the color of gold and rock. For another one, such as me, the wisdom in the appearance of a raven sitting on the edge of a hill is as important as the others are (20:114).

035:043 Those who follow Muhammed's *sunna* (law, practice) rather than God's *sunna* are addressed by this verse. See 33:38.

36:0 In the name of God, the Gracious, the Compassionate.
36:1 Y10S60*
36:2 By the wise Quran.
36:3 You are one of the messengers.
36:4 Upon a straight path.
36:5 The revelation of the Honorable, the Compassionate.
36:6 To warn a people whose fathers were not warned, for they are unaware.
36:7 The retribution has been deserved by most of them, for they do not affirm.
36:8 We have placed shackles around their necks, up to their chins, so that their head tilted upward.*
36:9 We have placed a barrier in front of them, and a barrier behind them, thus We shielded them so they cannot see.
36:10 Whether you warn them or do not warn them, they will not affirm.*
36:11 You can only warn him who follows the reminder and reveres the Gracious while unseen. Give him good news of forgiveness and a generous reward.
36:12 It is indeed Us who resurrect the dead, and We record what they have done and left behind. Everything We have counted in a clear ledger.

Messengers were Rejected by their People

36:13 Cite for them the example of the people of the town, when the messengers came to it.
36:14 Where We sent two to them, but they rejected them, so We supported them with a third one, thus they said, "We are messengers to you."
36:15 They replied: "You are but human beings like us, and the Gracious did not send down anything, you are only telling lies."
36:16 They said, "Our Lord knows that we have been sent to you."
36:17 "We are only required to give a clear delivery."

36:18 They replied: "We have welcomed you better than you deserve. If you do not cease, we will stone/excommunicate you, and you will receive a painful retribution from us!"
36:19 They said, "Keep your welcome with you, for you have been reminded. Indeed, you are transgressing people."
36:20 A man came running from the farthest part of the city, saying: "O my people, follow the messengers."
36:21 "Follow those who do not ask you for any wage and they are guided."
36:22 "Why should I not serve the One who initiated me, and to Him is your ultimate return?"
36:23 "Shall I take gods besides Him? If the Gracious intends any harm for me, their intercession cannot help me in the least, nor can they save me."
36:24 "Then I would be clearly astray."
36:25 "I have affirmed your Lord, so listen to me!"
36:26 It was said, "Enter Paradise." He said "Oh, how I wish my people only knew!"
36:27 "Of what my Lord has forgiven me and made me of the honored ones."
36:28 We did not send down upon his people after him soldiers from the sky; for there was no need to send them down.
36:29 For all it took was one blast, whereupon they were stilled.
36:30 Alas for the servants. For every time a messenger went to them, they would ridicule him.
36:31 Did they not see how many generations We destroyed before them; do they not go back to them?
36:32 How every one of them will be summoned before Us.

Signs in Nature

36:33 A sign for them is the dead land, We revive it and produce from it seeds from which they eat.
36:34 We made in it gardens of date palms, and grapes, and We cause springs to gush forth therein.
36:35 So that they may eat from its fruits, and what they manufacture with their own hands. Would they be thankful?
36:36 Praise be to the One who created all pairs from what the earth sprouts out and from themselves, and from what they do not know.
36:37 A sign for them is the night, We remove the daylight from it, whereupon they are in darkness.
36:38 The Sun runs to a specific destination, such is the design of the Noble, the Knowledgeable.
36:39 The moon: We have measured it to appear in stages, until it returns to being like an old, curved sheath.
36:40 The Sun is not required to overtake the Moon, nor will the night precede the day; each of them is swimming in its own orbit.*
36:41 A sign for them is that We carried their ancestors on the charged Ship.
36:42 We created for them of its similarity, to ride in.
36:43 If We wished, We could drown them, so that their screaming would not be heard, nor could they be saved.
36:44 Except through a mercy from Us, and as an enjoyment for a while.
36:45 When they are told: "Be aware of your present and your past, that you may attain mercy."
36:46 No matter what sign comes to them from the signs of their Lord, they turn away from it.
36:47 When they are told: "Spend from what **God** has provisioned you." Those who reject say to those who affirm: "Shall we feed those whom **God** could feed, if He so willed? You are clearly misguided!"
36:48 They say, "When is this promise to come, if you are truthful?"
36:49 They will not realize it when one blast overwhelms them, while they dispute.
36:50 They will not even be able to leave a will, nor will they be able to return to their people.
36:51 The horn will be blown, whereupon they will rise from the graves rushing towards their Lord.
36:52 They will say, "Woe to us. Who has resurrected us from our resting

place? This is what the Gracious had promised. The messengers were truthful!"

36:53 It only took one blast, whereupon they are summoned before Us.

36:54 On this day, no person will be wronged in the least. You will be recompensed precisely for whatever you did.

The Dwellers of Paradise and Hell

36:55 The dwellers of Paradise will be, on that day, joyfully busy.

36:56 They and their spouses, both will be shaded, reclining on high furnishings.

36:57 They will have fruits therein; they will have in it whatever they ask for.

36:58 "Peace," a word from a Compassionate Lord.

36:59 "As for you, O criminals, you are singled out."

36:60 "Did I not pledge to you, O Children of Adam, that you should not serve the devil for he is your most ardent enemy?"

36:61 "That you should serve Me? That is a straight path."*

36:62 "He has misled mountain loads of you. Did you not possess any understanding?"

36:63 "This is hell that you have been promised!"

36:64 "Burn in it today, as a consequence of your rejection."

36:65 Today, We shall seal their mouths, and their hands will speak to Us, and their feet will bear witness to everything they had done.

36:66 If We wished, We could blind their eyes, and they would race towards the path, but how would they see?

36:67 If We wished, We could freeze them in their place; thus, they can neither move forward, nor go back.

36:68 For whomever We grant a long life, We weaken him in body. Do they not understand?

The Quran is a Warning for the Living Beings

36:69 We did not teach him poetry, nor does he need it. This is a reminder and a clear Quran.

36:70 To warn those who are alive, and so that the retribution will be deserved by the ingrates.*

36:71 Did they not see that We created for them with Our own hands livestock which they own?

36:72 They were subdued by Us for them. So, some they ride, and some they eat.

36:73 They have in them other benefits, as well as drinks. Would they not be appreciative?

36:74 They have taken besides **God** other gods, perhaps they will help them!

36:75 They cannot help them. They become to them as foot-soldiers.

36:76 So do not be saddened by what they say. We are fully aware of what they conceal and what they declare.

36:77 Has the human being not seen that We have created him from a seed, and yet he would become a clear enemy?

36:78 He cites an example for Us, while forgetting his own creation! He says: "Who can resurrect the bones while they are dust?"

36:79 Say, "The One who made them in the first place will resurrect them. He is fully aware of every creation."*

Burning: A Daily Example of Resurrection of Chemical Energy Stored by Photosynthesis

36:80 The One who produces for you a fire from the green tree, which you burn.

36:81 Is not the One who created the heavens and the earth able to create the like of them? Yes indeed; He is the Creator, the Knowledgeable.

36:82 His command, when He wants anything, is to say to it: "Be" and it is!*

36:83 Therefore, praise be to the One in whose Hand is the sovereignty of all things, and to Him you will be returned.

ENDNOTES

036:001 Y10S60. The combination of these letters/numbers plays an important role in the mathematical system of the Quran based on code 19. For the meaning of these letters, see 1:1; 2:1; 40:1; 13:38; 46:10; 74:1-56.

036:008 The description of arrogant evildoers in hereafter is a parody of their choices, attitudes, and actions in this world. The way they will be shackled mimics their arrogance. The way walls limit their vision is a reminder for their myopic choices in this world.

036:010 See 57:22-23.

036:040 The expression *KuLlun FiY FaLaK* (each in an orbit) describing the motion of the Sun, Moon, and Earth is palindromic. The letters of the expression (K, L, F, Y, F, L, K), with its cyclic nature, symbolizes the circular/ellipsoid orbits. Similar expression is found at 21:33. The Quran is an interesting book (72:1). Note that the Arabic word *kul* (all) refers to more than two bodies and it should include the Sun, the Moon, and the reference planet from the expression "night and day." It is also noteworthy that the Quran frequently uses the expression, "alternation of night and day" rather than "rotation of the Sun," drawing our attention to the rotation of the planet around itself. See 21:33; 27:88; 39:5; 68:1; 79:30.

036:061-62 Compare it to the New Testament verse: ""Get thee behind me, Satan: for it is written, 'Thou shalt worship the Lord thy God, and him only shalt thou serve'" (Luke 4:8).

036:070 It is an implicit and ironic prophecy of the Quran that the chapter containing the only verse stating that the Quran was revealed to remind the living people, has been dedicated to dead people. Those who follow *hadith* and sectarian teachings recite this chapter in funerals. In opposition to the Quran, "living" people recite these verses for the dead, who do not hear. According to the Quran, those people are no different from the dead people. A class of people getting paid for reciting a book that they do not appreciate nor understand over dead people who also do neither, is one of the strangest affairs.

036:079-80 After reminding us that He knows all kinds of creation, God gives us the recreation of energy stored in plants through photosynthesis as fossil fuels such as coal or petroleum. Indeed, conservation of energy and its transformation from physical into chemical form and vice versa, is a good example of resurrection. See 4:82.

036:082 The creation of universe is the result of such an instant order. See 21:30.

بسم الله الرحمن الرحيم

37:01 In the name of God, the Gracious, the Compassionate.
37:1 By the arrangers in ranks.
37:2 By the forces that repel.
37:3 By the memory which follows.
37:4 Your god is indeed One.
37:5 The Lord of the heavens and the earth, and what lies between them, and Lord of the easts.
37:6 We have adorned the lowest heaven with the decoration of planets.
37:7 To guard against every rebellious devil.
37:8 They cannot listen to the command up high; and they are bombarded from every side.
37:9 Outcasts; they will have an eternal retribution.
37:10 Any of them who snatches something away, he is pursued by a piercing flame.
37:11 So ask them: "Are they the more powerful creation, or the others We created?" We have created them out of sticky clay.
37:12 While you were awed, they simply mocked.
37:13 When they are reminded, they do not care.
37:14 When they see a sign, they make fun of it.
37:15 They said, "This is nothing but evident magic!"
37:16 "Can it be that after we die and become dust and bones, that we are resurrected?"
37:17 "What about our fathers of old?"
37:18 Say, "Yes, and then you will be humiliated."
37:19 All it takes is one rebuke, then they will be staring.
37:20 They said, "Woe to us, this is the day of Judgment!"
37:21 This is the day of decisiveness that you used to deny.

Mutual Accusation of the Leaders and their Followers

37:22 Gather the transgressors, their associates, and all they served,
37:23 … beside **God**; and guide them to the path of hell.
37:24 Let them stand and be questioned:
37:25 "Why do you not support one another?"
37:26 No, for today they have totally given up.
37:27 Some of them came to each other, questioning.
37:28 They said, "You used to entice us from the right side."*
37:29 They replied: "No, it was you who were not affirmers."
37:30 "We never had any power over you, but you were a wicked people."
37:31 "So our Lord's decree is now upon us, that we will suffer."
37:32 "We misled you, because we were astray."
37:33 Then, on that day they will all share in the retribution.
37:34 This is how We deal with the criminals.
37:35 When it was said to them: "There is no god except **God**," they would be arrogant.
37:36 They would say, "Shall we leave our gods because of a crazy poet?"
37:37 No, he has come with the truth, and he has confirmed the messengers.
37:38 You will taste the greatest of retribution.
37:39 You are only recompensed for what you have done.
37:40 Except for **God**'s servants who are dedicated.
37:41 For them will be known provisions.
37:42 Fruits, and they will be honored.
37:43 In the gardens of bliss.
37:44 On furnishings which are opposite one another.
37:45 They will be served with cups of pure drinks.
37:46 Clear and tasty for the drinkers.
37:47 There is no bitterness therein, nor will they tire from it.
37:48 With them are attendants with wide eyes and a splendid look.
37:49 They are like fragile eggs.

Conversation between the People of Paradise and Hell

37:50 So then they approached one another, questioning.

37:51 One of them said, "I used to have a friend."

37:52 "Who used to say, "Are you among those who affirm this?"

37:53 "That if we die and turn into dust and bones, that we would be called to account?""

37:54 He said, "Can anyone find him?"

37:55 So when he looked, he saw him amid hell.

37:56 He said, "By **God**, you nearly ruined me!"

37:57 "Had it not been for my Lord's blessing, I would have been with you."

37:58 "Are we then not going to die,"

37:59 "Except for our first death, and we will not be punished?"

37:60 Such is the greatest triumph.

37:61 For this let those who will work endeavor.

The Tree Growing Inside Hell

37:62 Is that a better destination, or the tree of bitterness?

37:63 We have made it a punishment for the transgressors.

37:64 It is a tree that grows amid hell.*

37:65 Its shoots are like the devils' heads.

37:66 They will eat from it, so that their bellies are filled up.

37:67 Then they will have with it a drink of boiling liquid.

37:68 Then they will be returned to hell.

37:69 They had found their parents astray.

37:70 So they too have hastened in their footsteps.

37:71 Most of the previous generations have strayed before them.

37:72 We had sent to them warners.

37:73 Thus note the consequences for those who were warned.

37:74 Except for **God**'s servants who are dedicated.

Noah

37:75 Noah had called upon Us, for We are the best to respond.

37:76 We saved him and his family from the great disaster.

37:77 We made his progeny the one that remained.

37:78 We kept his history for those who came later.

37:79 Peace be upon Noah among the worlds.

37:80 It is such that We reward the righteous.

37:81 He is of Our affirming servants.

37:82 Then We drowned the others.

Abraham and His Dream

37:83 From among his descendants was Abraham.

37:84 For he came to his Lord with a pure heart.

37:85 When he said to his father and his people: "What are you serving?"

37:86 "Is it fabricated gods that you want instead of **God**?"

37:87 "What do you say of the Lord of the worlds?"

37:88 Then he looked towards the stars.*

37:89 He said, "I am sick of this!"

37:90 So they turned away from him and departed.

37:91 He then went to their idols and said, "Can you not eat?"

37:92 "What is the matter, you do not speak?"

37:93 So he then turned on them, striking with his hand.

37:94 Then they approached him outraged.

37:95 He said, "Do you serve what you carve?"

37:96 "While **God** has created you, and all that you make!"

37:97 They said, "Build for him a structure, and then throw him into its fire."

37:98 So they wanted to harm him, but We made them the losers.

37:99 He said, "I am going to my Lord; He will guide me."

37:100 "My Lord, grant me from among the righteous."

37:101 So We gave him good news of a compassionate child.

37:102 When he grew enough to work with him, he said, "My son, I am seeing in a dream that I am sacrificing you. What do you think?" He said, "O my father, do what you are commanded to do. You will find me, **God** willing, patient."

37:103 So when they both had peacefully surrendered, and he put his forehead down.

37:104 We called him: "O Abraham,"

37:105 "You have affirmed the vision." It was such that We rewarded the righteous.*

37:106 Surely, this was an exacting test.

37:107 We ransomed him with a great animal sacrifice.

37:108 We kept his history for those who came later.

37:109 Peace be upon Abraham.

37:110 It is thus that We reward the righteous.

37:111 He was of Our affirming servants.

37:112 We gave him the good news of the coming of Isaac, a prophet from among the righteous.

37:113 We blessed him and Isaac. From among their progeny, some are righteous, and some are clearly wicked.

Moses and Aaron

37:114 Indeed, We have given Our grace to Moses and Aaron.

37:115 We saved them and their people from the great disaster.

37:116 We supported them, so that they became the winners.

37:117 We gave both the clear book.

37:118 We guided them to the straight path.

37:119 We kept their history for those who came later.

37:120 Peace be upon Moses and Aaron.

37:121 We thus reward the righteous.

37:122 Both of them were among Our affirming servants.

Elias

37:123 And Elias was one of the messengers.

37:124 When he said to his people: "Would you not be righteous?"

37:125 "Would you call on Baal and forsake the best Creator?"

37:126 "**God** is your Lord, and the Lord of your fathers of old!"

37:127 But they denied him. Thus, they were called to account.

37:128 Except for **God**'s devoted servants.

37:129 We kept his history for those who came later.

37:130 Peace be upon the family of Elias.

37:131 We thus reward the righteous.

37:132 He was one of our affirming servants.

Lot

37:133 And Lot was one of the messengers.

37:134 When We saved him and all his family.

37:135 Except an old woman who remained.

37:136 Then, We destroyed the rest.

37:137 You pass by their ruins in the morning;

37:138 In the night. Do you not understand?

Jonah and the Fish

37:139 And Jonah was one of the messengers.

37:140 When he escaped to the charged ship.

37:141 He was guilty, so he became one of the losers.

37:142 Thus a whale swallowed him, and he was the one to blame.

37:143 Had it not been that he was one of those who implored,

37:144 He would have stayed in its belly until the day of resurrection.

37:145 So We threw him on the shore while he was sick.

37:146 We caused seaweed to grow on him.

37:147 We sent him to a hundred thousand, or more.

37:148 They affirmed, so We gave them enjoyment for a time.

The Questions

37:149 So ask them: "Are the daughters for your Lord, while the sons are for them?"

37:150 Or did We create the controllers as females while they witnessed?

37:151 Indeed, it is out of their falsehood that they say,

37:152 "The son of **God**." They are liars.

37:153 Has He chosen the daughters over the sons?
37:154 What is wrong with you, how do you judge?
37:155 Will you not remember?
37:156 Or do you have a clear proof?
37:157 Bring forth your book, if you are truthful.
37:158 They invented a kinship between Him and the Jinn. But the Jinn know that they will be gathered.
37:159 **God** be glorified; far from what they describe.
37:160 Except **God**'s servants who are dedicated.
37:161 As for you and what you serve.
37:162 You cannot lead away from Him.
37:163 Except those who are destined to hell.
37:164 Every one of us has a destined place.
37:165 We are the ones who are in columns.
37:166 We are the ones that glorify.
37:167 They used to say,
37:168 "If only we had received a reminder from the previous generations,"
37:169 "We would have surely been **God**'s loyal servants."
37:170 But they rejected it. They will come to know.
37:171 Our word had been decreed to Our servants who were sent.
37:172 That they would be made victorious.
37:173 That Our soldiers are the winners.
37:174 So turn away from them for a while.
37:175 Observe them; for they will see.
37:176 Do they seek to hasten Our retribution?
37:177 Then, when it descends into their courtyard, evil will be the morning, for they have already been warned.
37:178 Turn away from them for a while.
37:179 Observe; for they will see.
37:180 Glory be to your Lord, the Lord of greatness, for what they have described.
37:181 Peace be upon the messengers.
37:182 Praise be to **God**, Lord of the worlds.

ENDNOTES

037:028 The worst deception is the one done under the name of God and good causes. Religious leaders and politicians usually deceive people by ápproaching them through the "right" side. They sell lies in the name of God, they promote aggression, pre-emptive wars, and corruption in the name of venerated terms such as "our god," "patriotism," "liberty," "our flag," "saints," "our lord," or "our way of life." See 20:63.

037:064 See 17:60.

037:088-89 Abraham's people were worshiping stars and idols. (6:76-78)

037:105-107 Killing a human being without justification is a sin for humans, not for God. God creates, kills, and revives without justification. Controllers of God are not subject to the test like us; and they carry out God commandments, including taking the lives of children; teleporting them from this world to a higher universe.

In verses 37:101-113, Abraham told his son that he saw in his sleep (*manam*) that he was sacrificing him. The Quran does not state that God ordered him to do that. He fulfills his dream (not his literal interpretation of it though) by intending and attempting to sacrifice his son. So, God stops him from going further all the way. Though it was not God who ordered him to sacrifice his son, Abraham's perfect submission to his literal interpretation of his vision proved his sincerity. God saved Abraham from completing his misinterpretation of his dream. Abraham was wrong in literally interpreting his dream since the dream was a metaphor, a dramatic metaphor reminding him to make God his only priority by mentally giving up from being preoccupied with his dearest possession. Though his literal understanding was wrong, he demonstrated his perfect submission to God. He fulfilled the real meaning of the dream. God rewarded his intention, not his knowledge of interpretation. (For sleep and dream, see: See 25:47; 8:43; 12:6; 30:23; 39:42; 78:9).

The ethical issue involving this dream has been extensively discussed since Socrates. In Plato's *Euthyphro*, Socrates asked a profound question, "is it right because God says so, or does God say so because it is right?" Because vs Therefore... Though through this question Socrates exposes the contradiction of polytheism, I think the question commits either/or fallacy. The question assumes that God is different from the concept of goodness, while according to the Quran, God is the Truth, the Wise, the Noble, and the Just. The correct answer should be affirmative for both parts of the question. It is good because God says so, AND it is good therefore God says so, since God is Good, the source of goodness.

Jean-Paul Sartre pulled our attention to a different epistemological problem with the story. He questioned the authenticity or reliability of the source. How could Abraham be sure that the voice was of divine origin, rather than of the devil? That is a legitimate question. If there is an Omnipotent and Omniscient God, and if that God wishes to communicate His will to humans without any doubt, He should be able to do so. Society will and should judge a person based on the principle of justice. Had Abraham lived today and had he managed to sacrifice his son, according to God's law, as a society we should hold him responsible for the murder. However, we should remember that the ultimate judge is God on the Day of Judgment. See 7:28.

Genesis 21:1-24; 22:1-19 tell the story of the sacrifice. Though the Quran does not give the identity of the child, Muslim scholars, perhaps because of a bias for the ancestor of the Arabs, to that coveted position have assigned Ishmael, Abraham's son from Hagar. The Old Testament, though, spells out Isaac, Abraham's son from Sara, the ancestor of the Children of Israel. However, the story contains some problems. According to the Bible, when Abraham attempted to sacrifice his son, his only son was Ishmael who was 13, while his second son Isaac was not even born yet (Genesis 16:16; 17:24; 21:5). This information contradicts the anonymous author of the Hebrews 11:17. How can Abraham sacrifice his "only son" Isaac, when the younger Isaac was never Abraham's "only son"? Biblical scholars try to get around this problem by claiming that Ishmael was an illegitimate child. Though Genesis 16:3 specifically considers Hagar to be Abraham's wife, the following verses refer to her as a mistress (Genesis 16:4-9). While Genesis 22:2 does not even consider the older Ishmael to be Abraham's son, New Testament's Galatians 4:22 writes about Abraham's two sons. Regardless of the relationship between Abraham and Hagar, blaming or demeaning a child because of his being born from a slave is unjust discrimination. Even if the Biblical slander regarding Abraham's extramarital affair was true, holding an innocent child responsible for the sins of his parents is unjust both according to the Old Testament, Ezekiel 18:20, and the Quran 6:164.

بسم الله الرحمن الرحيم

38:0 In the name of God, the Gracious, the Compassionate.

38:1 S90, and the Quran that contains the Reminder.*

38:2 Indeed, those who have rejected are in false pride and defiance.

38:3 How many a generation have We destroyed before them. They called out when it was far too late.

38:4 They were surprised that a warner has come to them from among themselves. The ingrates said, "This is a magician, a liar."

38:5 "Has he made the gods into One god? This is indeed a strange thing!"

38:6 The leaders among them went out: "Walk away and remain steadfast for your gods. This thing can be turned back."

38:7 "We never heard of this from the people before us. This is but an innovation."

38:8 "Has the remembrance been sent down to him, from between all of us!" Indeed, they are doubtful of My reminder. They have not yet tasted My retribution.*

38:9 Or do they have the treasures of mercy of your Lord, the Noble, the Benefactor.

38:10 Or do they possess the dominion of the heavens and the earth, and all that is between them? Then let them bring their own solutions.

38:11 The opposing troops they have gathered will be defeated.

38:12 Rejecting before them were the people of Noah, Aad, and Pharaoh with the planks.

38:13 And Thamud, and the people of Lot, the dwellers of the Woods; such were the opponents.

38:14 Each of them rejected the messengers, therefore My retribution came to be.

38:15 What these people are waiting for is a single blast, from which they will not recover.

38:16 They said, "Our Lord, hasten for us our punishment, before the day of Reckoning."

David's Sensitivity Regarding Justice

38:17 Be patient to what they say, and recall Our servant David, the resourceful. He was obedient.

38:18 We directed the mountains to glorify with him, during dusk and dawn.

38:19 The birds were gathered; all were obedient to him.

38:20 We strengthened his kingship, and We gave him the wisdom and the ability to make sound judgment.

38:21 Did the news come to you of the disputing party who came over into the temple enclosure?

38:22 When they entered upon David, he was startled by them. They said, "Have no fear. We are two who have disputed, and one has wronged the other, so judge between us with truth, and do not wrong us, and guide us to the right path."

38:23 "This is my brother, and he owns ninety-nine lambs, while I own one lamb; so, he said to me: 'Let me take care of it' and he pressured me."*

38:24 He said, "He has wronged you by asking to combine your lamb with his lambs. Many who mix their properties take advantage of one another, except those who affirm and promote reforms, and these are very few." David guessed that We had tested him, so he sought forgiveness from his Lord, and fell down kneeling, and repented.*

38:25 So We forgave him in this matter. For him with Us is a near position, and a beautiful abode.

38:26 O David, We have made you a successor on earth. Therefore, you shall judge among the people with truth, and do not follow desire, lest it diverts you from the path of **God**. Indeed, those who stray off the path of **God** will have a severe retribution for forgetting the day of Reckoning.

38:27 We did not create the heaven and the earth, and everything between them, in vain. Such is the thinking of those who rejected. Therefore, because of the fire woe to those who have rejected.
38:28 Or shall We treat those who affirm and promote reforms as We treat those who make corruption on earth? Or shall We treat the righteous as the wicked?
38:29 A book that We have sent down to you, that is blessed, so that they may reflect upon its signs, and so that those with intelligence will take heed.

Solomon is Tested by Wealth

38:30 To David We granted Solomon. What an excellent and obedient servant.
38:31 When, during dusk, well trained horses were displayed before him.
38:32 He said, "I enjoyed the good of materials more than I enjoyed remembering my Lord, until it became totally dark!"*
38:33 "Send them back." He then rubbed their legs and necks.
38:34 We tested Solomon and placed a corpse on his throne, but he then repented.
38:35 He said, "My Lord, forgive me, and grant me a kingship that will never be attained by anyone after me. You are the Grantor."
38:36 So, We directed the wind to run by his command, raining gently where he directed it.
38:37 The devils, builders, and divers.
38:38 Others, held by restraints.
38:39 "This is Our gift, so you may spend or withhold, without any repercussions."
38:40 He has deserved a near position with Us, and a wonderful abode.

Job is Tested by Health and Family Problems

38:41 Recall Our servant Job, when he called upon his Lord: "The devil has afflicted me with an illness and pain."
38:42 "Strike the ground with your foot. Here is a cold spring to wash with and to drink."
38:43 We restored his family to him along with a group like them, as a mercy from Us; and a reminder for those who possess intelligence.
38:44 "Take in your hand a blend of herbs and travel with it. Do not break your oath." We found him steadfast. What a good servant! He was obedient.

The Elite

38:45 Recall Our servants Abraham, Isaac, and Jacob. They were resourceful, and with vision.
38:46 We purified them; with the purity of awareness of the (last) home.
38:47 They are with Us of the elite, the best.
38:48 Recall Ishmael, Elisha, and Isaiah; all are among the best.
38:49 This is a reminder, and the righteous will have a wonderful abode.
38:50 The gardens of Eden, whose gates will be open for them.
38:51 Reclining therein, they will be invited to many fruits and drinks.
38:52 With them are attendants with a splendid look and of equal age.*
38:53 This is what you have been promised for the day of Reckoning.
38:54 Such is Our provisions, it does not run out.

The Transgressors are Disappointed

38:55 This is so, and for the transgressors is a miserable destiny.
38:56 Hell is where they burn. What a miserable abode!
38:57 This is so, and let them taste a boiling drink and bitter food.
38:58 Other multitudes that are similar to that.
38:59 Here is another group to be thrown into hell with you. "We have no welcome for them, for they shall burn in the fire."
38:60 They said, "No, you are the ones without welcome. It was you who misled us, so here is the result!"
38:61 They said, "Our Lord, whoever brought this upon us, then double their retribution in the fire!"
38:62 They said, "Why do we not see some men whom we used to count among the wicked?"

38:63 "Did we mock them erroneously, or have our eyes failed to find them?"
38:64 Surely, this is in truth the feuding of the people of hell.
38:65 Say, "I am but a warner; and there is no god besides **God**, the One, the Supreme."
38:66 "The Lord of the heavens and the earth, and everything between them; the Noble, the Forgiving."
38:67 Say, "It is awesome news."
38:68 "From which you turn away."

The First Racist Disobeys God

38:69 "I had no knowledge of the command up high that they had quarreled."
38:70 "It is only revealed to me that I am a clear warner."
38:71 For your Lord said to the controllers: "I am creating a human being from clay."
38:72 "So when I have evolved him, and breathed into him from My Spirit, then you shall submit to him."
38:73 The controllers submitted, all of them,
38:74 Except Satan; he turned arrogant, and became one of the ingrates.
38:75 He said, "O Satan, what prevented you from submitting to what I have created by My hands? Are you too arrogant? Or are you one of those exalted?"
38:76 He said, "I am better than he; You created me from fire, and created him from clay."
38:77 He said, "Therefore exit from it, you are outcast."
38:78 "My curse will be upon you until the day of judgment."
38:79 He said, "My Lord, respite me till the day they are resurrected."
38:80 He said, "Then, you are granted respite,"
38:81 "Until the appointed day."
38:82 He said, "By Your majesty, I will mislead them all."
38:83 "Except for Your servants who are loyal."
38:84 He said, "The truth, and the truth is what I say.
38:85 That I will fill hell with you and all those who follow you."
38:86 Say, "I do not ask you for any wage, nor am I a fraud."
38:87 "It is but a reminder for the worlds."
38:88 "You will come to know its news after awhile."

ENDNOTES:

038:001 This letter/number plays an important role in the mathematical system of the Quran based on code 19. See 74:1-56; 1:1; 2:1; 13:38; 27:82; 38:1-8; 40:28-38; 46:10; 72:28.

038:001-15 S90. This letter/number plays an important role in the mathematical system of the Quran based on code 19. For the meaning of this letter, see 1:1; 2:1; 40:1; 13:38; 46:10; 74:1-56. Also see 6:124; 26:1-6; 40:28-38,78; 72:24-28.

038:008 *Zikr* (message) is another attribute used for the Quran or for a feature of the Quran. The word message in this verse might be referring to a particular message contained in the Quran. The Quran is message upon message, light upon light (24:35). Could the Quran contain another book coded within its text? For another aspect of the message, see 74:30-35. Also see 15:9; 21:2; 21:24,105; 26:5; 29:51; 38:1,8; 41:41; 44:13; 72:17; 74:49,54.

038:023 See 74:30.

038:024 We are not given the exact reason why David repented. Perhaps we are expected to ponder on our own. For instance, I find at least two reasons for what might have caused David to feel guilty. The complaint and his reaction to it might have reminded him of his own greed and injustice against others. Or perhaps, after making the judgment, he realized that he reached a verdict without listening to the other party. Justice cannot be adjudicated in an ex parte hearing. The plaintiff might be distorting or hiding some facts pertinent to the case. Even if we approach the case in terms of social justice, hearing the defendant is still important. For instance, the plaintiff in this case could have been the owner of vast lands, vineyards and farms, while the entire wealth of the other party could have been no more than a hundred sheep. (See also: 28:15)

The Quran mentions David's name 16 times and describes him in very positive words. He was blessed by God (17:55; 21:79; 34:10), knowledgeable (27:15), resourceful and obedient (38:17), caring and just (38:22-26).

The Old Testament in 2 Samuel 12:1-14 tells a similar story about David with variable details. The Bible describes David as a good-looking young man (1 Samuel 17:25) and praises his intelligence and courage (1 Samuel 17:1-58), at age thirty he becomes king by divine direction (2 Samuel 2:1-4), and after some internal fights, finally becomes king over all of Israel (2 Samuel 4:1-12; 5:1-5; 1 Chronicles 11:1-3), brings the ark of the covenant to Jerusalem (Psalms 24:1-10), and extends his kingdom after fighting a series of wars and committing numerous massacres in a vast territory (2 Samuel 8:1-18; 8:3-13; 10:1-19; 18:7-27; 19:31; 21:11; 29:5). However, the Bible that praises him for wars and bloody conquests, contains lengthy stories blaming him for being a sinful manipulator (2 Samuel 11:2-27), and a sinful proud man who angered God (2 Samuel 23:39), and having affairs with married women (1 Samuel 25:31-44). The Bible then tells us about the reformation of David by depicting him in his last moments as one with abiding appreciation and trust in God after his reign of forty years (2 Samuel 23:1-7; 2 Samuel 5:5; 1 Corinthians 3:4).

038:052 It is interesting that this verse could also be understood as the continuation of the verse before as description of fruits almost touching the earth.

39:0 In the name of God, the Gracious, the Compassionate.

39:1 The sending down of the book is from **God**, the Noble, the Wise.

39:2 We have sent down to you the book with truth, so serve **God** while devoting the system to Him.*

Excuse for Intercession

39:3 Absolutely, to **God** is the true system. Those who set up allies besides Him: "We only serve them so that they may bring us closer to **God**." **God** will judge between them in what they dispute. Surely, **God** does not guide the one who is a liar, an ingrate.*

39:4 If **God** wished to take a son, He could have exalted from among His creation what He pleases. Be He glorified; He is **God**, the One, the Supreme.

39:5 He created the heavens and the earth with truth. He rolls the night over the day, and He rolls the day over the night. He directed the Sun and the Moon, each running for an appointed term. Absolutely, He is the Noble, the Forgiving.*

Triple Darkness

39:6 He created you from one person, then He made from it its mate. He sent down to you eight pairs of the livestock. He creates you in the wombs of your mothers, a creation after a creation in triple darkness. Such is **God** your Lord. To Him belongs the sovereignty. There is no god besides Him. How is it then you deviate?

39:7 If you reject, then know that **God** is in no need of you, and He dislikes rejection for His servants. If you are appreciative, He is pleased for you. None shall bear the burdens of another. Then to your Lord is your return, and He will inform you of everything you had done. He is fully aware of what is inside the chests.

39:8 When the human being is afflicted with adversity, he implores his Lord, turning in repentance to Him. But then, when He grants him a blessing from Him, he forgets his previous imploring, and sets up equals with **God**, in order to mislead others from His path. Say, "Enjoy your rejection for a while; for you are of the dwellers of the fire."

The Value of Knowledge

39:9 As for one who is meditating in the night, prostrating and standing, fearing the Hereafter, and seeking the mercy of his Lord. Say, "Are those who know equal to those who do not know?" Only those who possess understanding will remember.

39:10 Say, "O servants who affirmed, be aware of your Lord." For those who worked righteousness in this world will be a good reward, and **God**'s earth is spacious. Those who steadfastly persevere will receive their recompense fully, without reckoning.

Devoting the System to God Alone

39:11 Say, "I have been commanded to serve **God**, devoting the system to Him.*

39:12 I was commanded to be the frontrunner of those who peacefully surrender."

39:13 Say, "I fear, if I disobeyed my Lord, the retribution of a great day."

39:14 Say, "**God** is the One I serve, devoting my system to Him.

39:15 Therefore, serve whatever you wish besides Him." Say, "The losers are those who lose themselves, and their families, on the day of resurrection. Indeed, such is the real loss."

39:16 They will have coverings of fire from above them and below them. It is as such that **God** makes His servants fearful: "O My servants, you shall revere Me."

39:17 For those who avoid serving evil, and turn to **God** in repentance, for them are glad tidings. So give the good news to My servants.

Listen and Critically Evaluate

39:18 The ones who listen to the word, and then follow the best of it. These are the ones whom **God** has guided, and these are the ones who possess intelligence.*

39:19 As for those who have deserved the retribution; can you save those who are in the fire?

39:20 But those who revere their Lord, they will have dwellings constructed upon dwellings, with rivers flowing beneath them. **God**'s promise; **God** does not break the promise.

39:21 Do you not see that **God** sends down water from the sky, and turns it into streams through the land, then He produces with it plants of various colors, then they grow until they turn yellow, then He makes them dry and broken? In this is a reminder for those of understanding.

39:22 If **God** opens one's chest to peacefully surrender, then he will be on a light from his Lord. So woe to those whose hearts are hardened against remembering **God**. They have gone far astray.

Quran: The Best Word/Narration

39:23 **God** has sent down the best *hadith*, a book that is consistent, relating/contrasting/repeating. The skins of those who revere their Lord shiver from it, then their skins and their hearts soften up to the remembrance of **God**. Such is **God**'s guidance; He guides with it whoever He wills. For whomever **God** misguides, then none can guide him.*

39:24 As for he who saves his face from the terrible retribution on the day of resurrection; and it will be said to the transgressors: "Taste for what you have earned."

39:25 Those before them have denied, and thus the retribution came to them from where they did not perceive.

39:26 So **God** made them taste the humiliation in this worldly life, but the retribution in the Hereafter is far greater, if only they knew.

The Quran Alone or Quran Plus Contradictory Partners?

39:27 We have cited for the people in this Quran from every example, that they may take heed.

39:28 A compilation in Arabic, without any distortion, that they may be righteous.*

39:29 **God** cites the example of a man who has partners that dispute with each other, and a man who has given to dealing with only one man. Are they the same? Praise be to **God**; most of them do not know.*

39:30 You will die, and they will die.

39:31 Then, on the day of resurrection, you will quarrel at your Lord.

The Fabricators of Religious Teachings

39:32 Who then is more wicked than one who lies about **God**, and denies the truth when it comes to him? Is there not in hell an abode for the ingrates?

39:33 Those who came with the truth, and affirmed it, these are the righteous.

39:34 They shall have what they wish at their Lord. Such is the reward for the good doers.

39:35 So that **God** may cancel for them the worst that they did, and He may recompense them their reward for the best of what they used to do.

Isn't God Enough for His Servant?

39:36 Is **God** not sufficient for His servant? They frighten you with others beside Him. Whomever **God** sends astray, then for him there will be no guide.

39:37 Whomever **God** guides, then there will be none that can mislead him. Is **God** not Noble, with Vengeance?

39:38 If you ask them: "Who created the heavens and the earth?" they will say, "**God**." Say, "Do you see what you call on besides **God**, If **God** wanted any harm for me, can they alleviate His harm? Or if He wanted a mercy for me, can they hold back his mercy?" Say, "**God** is sufficient for me; in Him those who trust shall put their trust."

39:39 Say, "O my people, work according to your way, and I will work.

39:40 You will come to know to whom the humiliating retribution will come, and on whom descends the eternal retribution."

39:41 We have sent down to you the book for the people with truth. Then, whoever is guided is guided for himself, and whoever goes astray goes astray to his own loss. You are not a keeper over them.

Person (nafs) = Consciousness

39:42 **God** takes the person when it dies, and during their sleep. He then keeps those that have been overtaken by death, and He sends the others back until a predetermined time. In that are signs for a people who will think.

Polytheists are Disturbed when God is Mentioned Alone

39:43 Or have they taken intercessors besides **God**? Say, "What if they do not possess any power, nor understanding?"

39:44 Say, "To **God** belong all intercessions." To Him belongs the sovereignty of the heavens and the earth, then to Him you will be returned.*

39:45 When **God** Alone is mentioned, the hearts of those who do not affirm the Hereafter are filled with aversion; and when others are mentioned beside Him, they rejoice!*

39:46 Say, "Our God, Initiator of the heavens and the earth, Knower of the unseen and the seen, You will judge between Your servants regarding what they disputed in."

39:47 If those who transgressed owned everything on earth, and its equivalent again with it, they would ransom it to avoid the terrible retribution on the day of resurrection. They will be shown by **God** what they did not expect.

39:48 The sinful works they had earned will be shown to them, and they will be surrounded by what they used to mock!

39:49 So when the human is touched by adversity, he implores Us, then when We bestow a blessing upon him, he says: "I attained this because of knowledge I had!" Indeed, it is a test, but most of them do not know.

39:50 Those before them have said the same thing, yet what they earned did not help them in the least.

39:51 So, they suffered the evil of what they had earned. Those who transgressed from among these will suffer the evil of what they earned; they cannot escape.

39:52 Do they not realize that **God** spreads the provision for whomever He chooses, and withholds? In that are signs for a people who affirm.

Never Despair; Always be an Optimist

39:53 Say, "O My servants who transgressed against themselves, do not despair of **God**'s mercy. For **God** forgives all sins. He is the Forgiver, the Compassionate."*

39:54 Repent to your Lord, and peacefully surrender to Him, before the retribution comes to you. Then you cannot be helped.

39:55 Follow the best of what has been sent down to you from your Lord, before the retribution comes to you suddenly when you least expect it.*

39:56 Lest a person say, "How sorry I am for disregarding **God**'s path; and I was certainly one of those who mocked."

39:57 Or say, "Had **God** guided me, I would have been among the righteous."

39:58 Or say, when it sees the retribution: "If only I could have another chance, I would be among the good doers."

39:59 Yes indeed, My signs came to you, but you denied them and turned arrogant, and became one of the ingrates.

The Arrogant Religious Charlatans

39:60 On the day of resurrection you will see those who lied about **God** their faces will be blackened. Is there not an abode in hell for the arrogant ones?

39:61 **God** will save those who were righteous as their reward. No harm will touch them, nor will they grieve.

39:62 **God** is the Creator of all things, and He is Guardian over all things.

39:63 To Him belongs the keys of the heavens and the earth. Those who rejected **God**'s signs are the losers.

39:64 Say, "Do you order me to serve other than **God**, o you ignorant ones?"

39:65 He has revealed to you and to those before you, that if you set up partners, He will nullify all your work and you will be of the losers.

39:66 Therefore, you shall serve **God**, and be among the appreciative.

Polytheists do not Appreciate God's Greatness

39:67 They have not given **God** His true worth; and the whole earth is within His fist on the day of resurrection, and the heavens will be folded in His right hand. Be He glorified; He is much too high above what they set up.

39:68 The horn will be blown, whereupon everyone in the heavens and the earth will be struck unconscious, except those spared by **God**. Then it will be blown another time, whereupon they will all rise up, looking.

39:69 The earth will shine with the light of its Lord, the record will be placed, and the prophets and the witnesses will be brought forth; it will then be judged between them with truth, and they will not be wronged.

39:70 Every person will be paid for whatever it did, for He is aware of everything they have done.

39:71 Those who rejected will be ushered to hell in groups. When they reach it, and its gates are opened, its guards will say to them: "Did you not receive messengers from among you, who recited to you the signs of your Lord, and warned you about meeting this day?" They said,

"Yes, but the promise of retribution was destined to be upon the ingrates."

39:72 It was said, "Enter the gates of hell, wherein you will abide. What a miserable destiny for the arrogant."

39:73 Those who revered their Lord will be ushered to Paradise in groups. When they reach it, and its gates are opened, its guards will say to them: "Peace be upon you. You have done well, so enter to abide herein."

39:74 They said, "Praise be to **God** who has fulfilled His promise to us, and He made us inherit the earth, enjoying Paradise as we please. What a beautiful recompense for the workers!"

39:75 You will see the controllers surrounding the center of control, glorifying the praise of their Lord. It will be judged between them with truth, and it will be said, "Praise be to **God**, Lord of the worlds"

ENDNOTES

039:002 See 39:11.

039:003 There is not much difference between the Meccan polytheists and modern polytheists who turned monotheistic peacemaking into a limited corporation or limited liability partnership. Those who expect their prophets and saints to one day intercede on their behalf are setting up partners with God, even though they do not admit it (6:23). Also, see 2:48.

039:005 The verb *yuKaWiRu* (rotates, rolls around) used in the verse clearly indicates the roundness of the earth. See 21:33; 27:88; 36:40; 68:1; 79:30.

039:011-14 Decades and centuries after Muhammed, monotheist peacemakers deserted the Quran and gave up dedicating themselves to God alone. They started following man-made religions and sects concocted by the clerics and scholars, a partnership comprising God + prophet + his companions + the companions of his companions + sect imams + imams in a particular set + former scholars + later scholars and/or saints. See 29; 16:52; 39:2,14; 40:14,65; 98:5.

039:018 Knowing that those who abandoned the Quran would follow teachings called *hadith*, *sunna*, and *ijma*, God uses these words almost invariably with a negative connotation when they do not relate to the Quran. It is noteworthy, that the verse asking us to listen to different words does not use the word "*hadith*" but the word "*qawl.*" God does not give even a single positive usage of *hadith* when it is used to denote human words. Isn't there a sign in this? See 33:38; 66:3 and 6:110.

039:023 *Mutashabih* means "similar" or "multi-meaning" (3:7). However, when it is used for the entire book, which also contains single-meaning words, *mutashabih* should be understood as 'similar' or 'consistent.' Some statements of the Quran have multiple meanings (*mutashabih*) and all the statements of the Quran are consistent (*mutashabih*). Like its references, the word *mutashabih* itself has multiple meanings.

039:028 See 43:3.

039:029 Can a person who peacefully surrenders only to the Truth be the same as the one who surrenders to many contradictory authorities such as saints, imams, mujtahids, scholars, sheiks, *hadith* books, sectarian jurisprudence, fatwas?

039:044 See 2:48.

039:045 The majority of the so-called Muslims, despite the testimony in 3:18, insist on adding Muhammed's name. This criterion exposes the fact that those who are not happy with uttering God's name alone, those who enjoy adding the names of Muhammed, Jesus, or any other creature, do not really affirm the hereafter (3:18; 17:46; 63:1).

039:053 Associating partners with God (*shirk*), is a different category than sin (*zanb*); it is an unforgivable crime. See 4:48,116; Also, see 12:87; 15:56; 30:36; 41:49, and 39:10. The pronoun in "my servant" refers to God, since according to numerous verses only God can be the master. This sudden shift in pronouns is called *iltifat* (conversion, turning one's face to someone) in Arabic literature, which is defined as "the change of speech from one mode to another, for the sake of freshness and variety for the listener, to renew his interest, and to keep his mind from boredom and frustration, through having the one mode continuously at his ear." On this and other literary aspects of the Quran, I recommend Neal Robinson's book: Discovering The Qur'ān: A Contemporary Approach to a Veiled Text (1996, SCM Press Ltd.). For other examples of *iltifat,* see 35:9; 69:41-47; 16:1-2; 39:15-16; 89:25-30; 20:113-114; 67:18-19; 5:44; 20:124; 75:1-3; 2:38; 11:37; 50:45; 1:2-5; 19:88; 52:17-19; 16:72; 45:35; 30:38; and 70:39-4; 17:1-3; 17:95; 39:53; 10:22.

039:055 See 39:23.

40:0 In the name of God, the Gracious, the Compassionate.

40:1 **H**8M40*

40:2 The revelation of the book is from **God**, the Noble, the Knowledgeable.

40:3 Forgiver of sins, and acceptor of repentance, severe in retribution, with ability to reach. There is no god other than Him, to Him is the ultimate destiny.

40:4 None dispute **God**'s signs except those who have rejected. So do not be impressed by their influence through the land.

40:5 Before they, the people of Noah denied and the opponents after them. Every nation plotted against their messenger to seize him, and they disputed by means of falsehood to defeat with it the truth. So, I seized them; how then was My punishment!

40:6 As such, the word of your Lord has come to pass upon those who rejected, that they are the dwellers of the fire.

God's Compassion and Knowledge

40:7 Those who carry the throne and all those around it glorify the praise of their Lord, and affirm Him, and they seek forgiveness for the affirmers: "Our Lord, You encompass all things with compassion and knowledge, so forgive those who repented and followed Your path, and spare them the agony of hell."

40:8 "Our Lord, admit them into the gardens of Eden which You had promised for them and for those who did good of their fathers, their spouses, and their progeny. You are the Noble, the Wise."

40:9 "Save them from the sins. Whomever You save from sins, on that day, You have treated him with compassion. That is the greatest triumph."

40:10 Those who had rejected will be told: "**God**'s abhorrence towards you is greater than your abhorrence towards yourselves, for you were invited to affirm, but you chose to reject."

40:11 They will say, "Our Lord, You have made us die twice, and You have given us life twice. Now we have confessed our sins. Is there any way out of this path?"

There is no god but God. Period.

40:12 This is because when **God** Alone was mentioned, you rejected, but when partners were associated with Him, you affirmed. Therefore, the judgment is for **God**, the Most High, the Most Great.

40:13 He is the One who shows you His signs, and He sends down to you provisions from the sky. But none do remember except those who repent.

40:14 Therefore, call on **God** while devoting the system solely to Him, even if the ingrates dislike it.*

40:15 Possessor of the highest ranks, the One with the Throne. He sends the Spirit with His command upon whom He wills from His servants, so that they may warn of the day of Summoning.

40:16 The day when they will be exposed. None of them will hide anything from **God**. To whom is the sovereignty on this day? To **God**, the One, the Supreme.

40:17 Today, every person will be recompensed for what he/she had earned. There will be no injustice today. Truly, **God** is swift in reckoning.

40:18 Warn them of the imminent day, when the hearts will reach the throats, and many will be remorseful. For the transgressors there will be no friend, nor intercessor to be obeyed.

40:19 He knows what the eyes have seen, and what the chests conceal.

40:20 **God** judges with the truth, while those they call on besides Him do not judge with anything. Certainly, **God** is the Hearer, the Seer.

40:21 Have they not roamed the earth and seen what was the consequence of those who were before them? They used to be stronger than them and had built more in the land. But **God** seized them for their sins, and they had no protector against **God**.

40:22 That is because their messengers used to come to them with proofs, but they rejected. Thus, **God** seized them; for He is Mighty, severe in punishment.

Moses versus the Coalition of Arrogance, Might and Greed

40:23 We had sent Moses with Our signs, and a clear authority.

40:24 To Pharaoh, Haman, and Qarun. But they said, "A lying magician!"*

40:25 Then, when the truth came to them from Us, they said, "Kill the children of those who affirmed with him, and shame their women." But the scheming of the ingrates is always in error.

40:26 Pharaoh said, "Leave me to kill Moses, and let him call upon his Lord. I fear that he may change your system, or that he will cause evil to spread throughout the land."

40:27 Moses said, "I seek refuge in my Lord and your Lord from every arrogant one who does not affirm the day of Reckoning."

The Brave Monotheist and the Prophecy

40:28 An affirming man from among Pharaoh's people, who had concealed his affirmation, said, "Will you kill a man simply for saying: 'My Lord is **God**', and he has come to you with proofs from your Lord? If he is a liar, then his lie will be upon him, and if he is truthful, then some of what he is promising you will afflict you. Surely, **God** does not guide any transgressor, liar."*

40:29 "O my people, you have the kingship today throughout the land. But then who will save us against **God**'s torment, should it come to us?" Pharaoh said, "I am only showing you what I see, and I am only guiding you to the right path."

40:30 The one who affirmed said, "O my people, I fear for you the same fate as the day of factions."

40:31 "Like the fate of the people of Noah, Aad, and Thamud, and those after them. **God** does not wish any injustice for the servants."

40:32 "And O my people, I fear for you the day of mutual blaming."

40:33 "A day when you will turn around and flee, you will have no protector besides **God**, and whomever **God** sends astray, then there is none who can guide him."

Opposing a Messenger in the name of a "Final Messenger"

40:34 "Joseph had come to you before with proofs, but you remained in doubt regarding what he came to you with, until when he died, you said, "**God** will not send any messenger after him." It is such that **God** sends astray he who is a transgressor, doubter."*

40:35 Those who dispute about **God**'s signs without any authority that has come to them, it is greatly abhorred by **God** and by those who affirm. **God** thus seals the hearts of every arrogant tyrant.

40:36 Pharaoh said, "O Haman, build for me a high platform that I may uncover the secrets."

40:37 "The secrets of the heavens, and that I can take a look at the god of Moses, although I think he is a liar." Thus, the evil works of Pharaoh were made to appear correct to him, and he was blocked from the path. Pharaoh's scheming brought nothing but regret.

40:38 The one who affirmed said, "O my people, follow me, and I will guide you to the right path."

40:39 "O my people, this worldly life is but an enjoyment, while the Hereafter is the permanent abode."

40:40 "Whosoever does an evil deed, he will not be requited except for its equivalent, and whosoever does good, whether male or female and is an affirmer, so those will be admitted to Paradise, where they will receive provision without limit."

40:41 "And O my people, why is it that I invite you to salvation, while you invite me to the fire!"

40:42 "You invite me to reject **God**, and place partners beside Him that I have no knowledge of, and I am inviting you to the Noble, the Forgiver."

40:43 "There is no doubt that what you invite me to has no basis in this world, nor in the Hereafter. Our ultimate return will be to **God**, and that the transgressors will be the dwellers of the fire."

These Statements are not Limited to Egyptians

40:44 "You will come to remember what I am telling you, and I leave my affair in this matter to **God**; **God** is the Seer of the servants."

40:45 So **God** protected him from the evil of what they schemed, while the people of Pharaoh have incurred the worst retribution.

40:46 The fire, which they will be exposed to morning and evening, and on the day when the Moment is established: "Admit the people of Pharaoh into the most severe of the retribution."

40:47 When they argue in hell, the weak will say to those who were arrogant: "We used to be your followers, can you take from us any portion of the fire?"

40:48 Those who were arrogant will say, "We are all in it together, for **God** has passed judgment upon the servants."

40:49 Those who are in fire will say to the guardians of hell: "Call upon your Lord to reduce for us the retribution, by just one day!"

40:50 They will say, "Did not your messengers come to you with proofs?" They will reply: "Yes." They will say, "Then call out, for the call of the ingrates is nothing but in vain."

40:51 We will indeed grant victory to Our messengers and to those who affirmed the worldly life, and on the day when the witnesses will rise.

40:52 A day when excuses will be of no help to the transgressors, and they will be cursed, and they will have the worst abode.

40:53 We have given Moses the guidance, and We made the Children of Israel inherit the book.

40:54 A guide and a reminder for those who possess intelligence.

40:55 So be patient, for the promise of **God** is true, and seek forgiveness for your sin, and glorify and praise your Lord at dusk and dawn.

Ignoring God's Signs in the Scripture and Nature

40:56 Surely, those who dispute about **God**'s signs without any authority given to them, there is nothing but arrogance in their chests, which they do not perceive. Therefore, seek refuge in **God**; He is the Hearer, the Seer.

40:57 The creation of the heavens and the earth is greater than the creation of people, but most people do not know.

40:58 Not equal are the blind and the seer; nor those who affirm and promote reforms, and those who do evil. Little do you remember.

40:59 Surely the Moment is coming, there is no doubt in it, but most people do not affirm.

40:60 Your Lord said, "Call on Me and I will respond to you." Surely, those who are too arrogant to serve Me, they will enter hell, forcibly.*

40:61 **God**, it is He Who has made the night for your rest and the day to see in. Surely, **God** provides many blessings upon the people, but most people are not thankful.

40:62 That is **God**, your Lord, Creator of all things. There is no god except He, so why do you deviate?

40:63 Thus, those who used to deny **God**'s Signs were deviated.

40:64 **God** is the One who made the earth a habitat for you, and the sky as a structure, and He designed you; how beautifully he designed you! He provided you with good provisions. Such is **God** your Lord. Most Exalted is **God**, Lord of the worlds.*

40:65 He is the Living; there is no god except Him. So, call on Him while devoting the system for Him. Praise be to **God**, Lord of the worlds.*

Before Receiving the Quran, Muhammed was a Polytheist

40:66 Say, "I have been forbidden from serving those whom you are calling upon besides **God**, since the proofs have come to me from my Lord. I have been commanded to peacefully surrender to the Lord of the worlds."*

40:67 He is the One who created you from dust, then from a seed, then from an embryo, then He brings you out as a child, then He lets you reach your maturity, then you become old, and some of you may pass away before this, and so that you may reach an appointed term, in order that you may understand.

40:68 He is the One who gives life and causes death. When He decides upon anything, He simply says to it: "Be" and it is.

The Ingrates of Divine Signs

40:69 Did you not see those who dispute about **God**'s signs, how they have deviated?

40:70 Those who deny the book, and what We have sent Our messengers with. Therefore, they will come to know.

40:71 When the collars will be around their necks and in chains they will be dragged off.

40:72 To the boiling water, then in the fire they will be burned.*

40:73 Then it will be said to them: "Where are those that you have set up as partners—

40:74 beside **God**?" They will say, "They have abandoned us. Nay, we were not calling on anything before!" Thus, **God** leads the ingrates astray.

40:75 That was because you used to gloat on earth without any right, and for what you used to rejoice.

40:76 Enter the gates of hell, abiding therein. What a miserable abode for the arrogant ones.

40:77 So be patient, for the promise of **God** is true. Either We will show you some of what We have promised them, or We will let you die, then it is to Us that they will be returned.

The Discovery of the Mathematical Structure of the Quran was no Coincidence

40:78 We have sent messengers prior to you. Some of them We mentioned to you, and some We did not mention to you. It was not given to any messenger that he should bring a sign except by **God**'s leave. So, when **God**'s judgment is issued, the matter is decided with truth, and the followers of falsehood will be lost.*

40:79 **God** is the One who made the livestock for you that you may ride on some of them, and some of them you eat.

40:80 You have other benefits in them. That you may reach by them what is desired in your chests. On them and on the ships, you are carried.

40:81 He shows you His signs. So which of **God**'s signs do you deny?

40:82 Have they not roamed the earth and noted the consequences for those who were before them? They used to be greater in number than they are and mightier in strength, and they had built more in the land. Yet, all that they had earned could not avail them.

40:83 Then, when their messengers came to them with clear proofs, they were content with what they already had of the knowledge. What they ridiculed became their doom.

God's Law for His Servants

40:84 So when they saw Our might, they said, "We affirm **God** Alone, and we reject all the partners we used to set up!"

40:85 But their affirmation could not help them once they saw Our might. Such is **God**'s *sunna* that has been established with His servants. The ingrates were then totally in loss.

ENDNOTES

040:001 H8M40. This two-letter/number combination initializes seven chapters from chapter 40 to chapter 46 and participates in the mathematical system of the Quran. Furthermore, the frequencies of these letters in these seven chapters create a special pattern built according to a specific formula. Twenty-nine chapters of the Quran start with various combinations of letters/numbers and their frequencies in those chapters demonstrate an amazing interlocking mathematical pattern based on the number 19. The fact that the book does not lose its literal quality while also being embedded with a mathematical structure based on a network of letters, words, verses, and chapters, is great evidence for the authenticity of the Quran. See 2:1; 74:1-56.

040:014 See 39:11.

040:024 Pharaoh, Haman, and Qarun, representing the political, military, capitalist powers, respectively, united to fight against God's message. This oligarchy used magicians (religious leaders) to continue their corrupt and exploitative system (7:112-116).

040:028-38 These eleven verses have prophetic fulfillment in our times. The Arabic text contains supportive clues for those who witnessed the fulfillment of the prophecy. Unappreciative people, who will try to cover the clear prophecies, will resort to Pharaoh's statement. The prophetic nature of these verses is highlighted by the statement of the monotheist in verse 40:44. Verses 40:78-85 also have multiple references. See 3:81; 40:44,78-85; 72:24-28; 74:1-56.

040:034 Similarly, Jews rejected John the Baptist and Jesus, and Christians rejected Muhammed.

040:060 God is our Lord, the Creator, Designer, and the Programmer of our body and mind. Seeking God's help and approval, including for material things, creates a connection between us.

Continuous contact with the source of life and wisdom, updates our system program, thereby protecting us from the harms of diabolic viruses.

040:064 See 15:20; 20:54 and 35:12-13. Also, see 4:119.

040:065 See 39:11.

040:066 See 93:7.

040:072 Those who do not improve their initial program through good thoughts and righteous acts will not be able to sustain God's presence. See 89:22-23.

040:078 Verse 40:34 informs us about people's tendency to consider their idolized messenger to be the last messenger. Those who close their minds and follow convention rather than reason and facts, react negatively to a "new" message delivered by a messenger (38:1-15; 23:44). This verse informs us that there were even many more messengers before Muhammed that the Quran did not mention by name. For instance, after a careful study of Plato's work about Socrates, one may infer that Socrates was a messenger of God to the Athenian people, who criticized their polytheistic religion and superstitions. Since God sent messengers to every nation, there must be many messengers from among the Chinese, Japanese, Hindi, Malay, etc. This verse also informs us that no messenger can come up with a miracle without God's permission. Unveiling the specifically hidden miracle of the Quran also cannot be done without God's special permission. (10:47; 16:36; 35:24; 40:28-44; and 74:30-35).

بسم الله الرحمن الرحيم

41:0 In the name of God, the Gracious, the Compassionate.

41:1 **H**8M40*

41:2 A revelation from the Gracious, Compassionate.

41:3 A book whose signs are detailed, a compilation in Arabic for a people who know.*

41:4 A bearer of good news, and a warner. But most of them turn away; they do not hear.

41:5 They said, "Our hearts are sealed from what you invite us to, in our ears is deafness, and there is a barrier between us and you. So do what you will, and so will we."

Muhammed was a Human Like Us

41:6 Say, "I am no more than a human being like you. It is revealed to me that your god is One God, therefore you shall be upright towards Him and seek His forgiveness. Woe to those who set up partners."

41:7 "The ones who do not contribute towards betterment, and with regards to the Hereafter, they are those who do not affirm."

41:8 Surely, those who affirm and promote reforms, they will receive recompense without limit.

Cosmology and Ecology

41:9 Say, "You are rejecting the One who created the earth in two days, and you set up equals with Him. This is the Lord of the worlds."*

41:10 He placed in it stabilizers from above it, and He blessed it and established its provisions in four equal days, to satisfy those who ask.

41:11 Furthermore, He settled to the heaven, while it was still gas, and He said to it, and to the earth: "Come willingly or unwillingly." They said, "We come willingly."*

41:12 He, moreover, made them into seven heavens in two days, and He revealed to every universe its affair. We adorned the lowest universe with lamps, and for protection.

Such is the design of the Noble, the Knowledgeable.

Arrogance; a Prime Cause of Decline

41:13 But if they turn away, then say, "I have warned you of destruction like the destruction of Aad and Thamud."

41:14 When the messengers came to them, publicly and privately: "You shall not serve except **God**." They said, "Had our Lord willed, He would have sent controllers. We are rejecting what you have been sent with."

41:15 As for Aad, they turned arrogant on earth, without any right, and they said, "Who is mightier than us in strength?" Did they not see that **God**, who created them, was mightier than them in strength? They were denying Our signs.

41:16 Consequently, We sent upon them violent wind, for a few miserable days, that We may let them taste the humiliating retribution in this life, and the retribution of the Hereafter is more humiliating; they can never win.

41:17 As for Thamud, We provided them with guidance, but they preferred blindness over guidance. Consequently, the blast of humiliating retribution annihilated them, because of what they earned.

41:18 We saved those who affirmed and were righteous.

41:19 The day when the enemies of **God** will be gathered to the fire, forcibly.

The Record

41:20 When they come to it, their own hearing, eyes, and skins will bear witness to everything they had done.

41:21 They will say to their skins: "Why did you bear witness against us?" They will reply: "**God** made us speak; He is the One who causes everything to speak. He is the One who created you the first time, and to Him you return."

41:22 There was no way you could hide from the testimony of your own hearing, or your eyes, or your skins. In fact, you thought that **God** was unaware of much of what you do.

41:23 This is the kind of thinking about your Lord that has caused you to fail, and thus you became of the losers.

41:24 If they wait, then the fire will be their destiny, and if they beg to be excused, they will not be excused.

41:25 We assigned to them companions who adorned their present and past actions. Thus, the retribution has been deserved by them the same as previous nations of Jinn and humans; all of them were losers.

Ingrates Try to Obstruct the Message

41:26 Those who rejected said, "Do not listen to this Quran and talk over it that you may succeed."

41:27 We will let those who have rejected taste a severe retribution. We will recompense them for the evil that they used to do.

41:28 Such is the recompense for **God**'s enemies; the fire shall be their eternal abode, as a recompense for their discarding Our signs.

41:29 Those who have rejected will say, "Our Lord, show us those who have misled us from among the Jinn and humans so we can trample them under our feet, and render them the lowliest."

41:30 Surely, those who have said, "Our Lord is **God**," then they did right, the controllers will descend upon them: "You shall not fear, nor shall you grieve. Rejoice in the good news of Paradise that you have been promised."

41:31 "We are your allies in this worldly life and in the Hereafter. There you will have anything your person desires, and in it you will have anything you ask for."

41:32 "A dwelling, from a Forgiver, Compassionate."

The Muslim Activist

41:33 Who is better in saying than one who invites to **God**, and does good works, and says: "I am one of those who have peacefully surrendered."

41:34 Not equal are the good and the bad response. You shall resort to the one which is better. Thus, the one who used to be your enemy may become your best friend.

41:35 None can attain this except those who are patient. None can attain this except those who are extremely fortunate.

God's Signs

41:36 If the devil misleads you in anything, then you shall seek refuge with **God**. He is the Hearer, the Knowledgeable.

41:37 From among His signs are the night and the day, and the Sun and the moon. Do not prostrate to the Sun, nor the moon; you shall prostrate to **God** who created them, if it is truly Him you serve.*

41:38 So, if they become arrogant, then those who are with your Lord glorify Him night and day, and they never despair.

41:39 From among His signs is that you see the land barren, then, as soon as We send down the water upon it, it shakes and grows. Surely, the One who revived it can revive the dead. He is capable of all things.

Quran is Embedded with a Protection Program

41:40 Surely, those who distort Our signs are not hidden from Us. Is one who is thrown into hell better, or one who comes secure on the day of resurrection? Do whatever you wish; He is Seer of everything you do.*

41:41 Surely, those who have rejected the Reminder when it came to them; and it is an Honorable book.*

41:42 No falsehood could enter it, presently or afterwards; a revelation from a Most Wise, Praiseworthy.*

41:43 What is being said to you is the same that was said to the messengers before you. Your Lord has forgiveness, and a painful retribution.

The Primary Language of the Quran is Universal

41:44 Had We made it a non-Arabic compilation, they would have said, "If only its signs were made clear!" Non-Arabic and Arabic, say, "For those who affirm, it is a guide and healing. As for those who reject, there is deafness in their ears, and they are blind to it. These will be called from a place far away."*

41:45 We have given Moses the book, but it was disputed. Had it not been for your Lord's predetermined decision, they would have been judged immediately. Indeed, they harbor many doubts about it.

41:46 Whoever does good works does so for his own person, and whoever works evil shall have the same. Your Lord does not wrong the servants.

Ignorance and Arrogance of Human Beings

41:47 To Him belongs the knowledge regarding the Moment. No fruit emerges from its sheath, nor does any female conceive or give birth, except by His knowledge. On the day He asks them: "Where are My partners?" They will say, "By your leave, none of us will testify to that."*

41:48 They were abandoned by what they used to call on before, and they realized that there will be no escape.

41:49 The human being does not tire in imploring for good things. But if adversity touches him, he is disheartened, desperate!*

41:50 When We let him taste a mercy from Us after adversity had touched him, he will say, "This was by my actions, and I do not think that the Moment will come to pass. Even if I am returned to my Lord, I will find at Him good things for me." Surely, We will inform the ingrates of all they had done, and We will let them taste the severe retribution.

41:51 When We bless the human being, he withdraws and turns away, and when he suffers any adversity, he implores in long prayers!

Prophecy Fulfilled

41:52 Say, "Do you see if this was from **God** and then you rejected it? Who is further astray than those who are in opposition?"

41:53 We will show them Our signs in the horizons, and within themselves, until it becomes clear to them that this is the truth. Is it not enough that your Lord is witness over all things?*

41:54 Indeed, they are in doubt about meeting their Lord; but He is Encompassing over all things.

ENDNOTES

041:003 H8M40. This combination of two letters/numbers plays an important role in the mathematical system of the Quran based on code 19. For the meaning of these letters, see 1:1; 2:1; 40:1; 13:38; 46:10; 74:1-56. Also, see 43:3 and 11:1.

041:009-10 Each "day" of creation represents a period. According to the Quran, time is relative (32:5; 70:4). While the creation of earth took two units of time, the creation and evolution of all conditions for and stages of life took four units. This temporal comparison highlights the importance of the ecological system on planet earth. The conjunction between fragments in this verse is *Wa* (and), thus implying not a sequence of separate events, but overlapping series of events. See 7:69; 25:59; 32:4; 50:38; 57:4.

Though the Bible contains some of the original revelations, Ezra and other Jewish scholars inserted into it many of their misunderstandings, comments, desires, and ignorance (2:59). The Biblical account of the creation of the universe does not match scientific evidence. For instance, Genesis 1:1-2 informs us that the first stage of the universe started with the creation of water. Another example of external inconsistency is the creation of plants on earth before the creation of the Sun (Genesis 1:11-14).

The order of creation in the Bible is given in detail in its first book, which partially contradicts modern scientific findings: First day, light and darkness (Genesis 1:3-5), Second day, atmosphere, and waters (Genesis 1:6-8), Third day, land and making it fruitful (Genesis 1:9-13), Fourth day, Sun, Moon, and stars (Genesis 1:14-19), Fifth day, birds, insects, and fishes (Genesis 1:20-23), Sixth day, land animals, and man (Genesis 1:24,28).

The Bible also contains internal inconsistencies. For instance, a comparative reading of chapter 1 and chapter 2 of Genesis will show that, most likely, the two chapters were authored by at least two different people who had different ideas about how the universe was created. Only through scientific evidence, which supports the Quranic version of creation, can we sort out the discrepancies in the Bible. Trees were created before man was (Genesis 1:11-12,26-27) versus man was created before trees were (Genesis 2:4-9). Birds were created before man was (Genesis 1:11-12, 26-27) versus man was created before birds were (Genesis 2:7,19). Animals were created before man was (Genesis 1:24-27) versus man was created before animals were (Genesis 2:7,19).

041:011 Time, space, matter, and everything in the universe, are muslims following the laws issued by their Creator. This verse also informs us about the early stage of the universe after the Big Bang (21:30).

041:037 The verse does not say "before Sun and Moon," but "for/to Sun and moon." We are commanded not to prostrate to the Sun and Moon, but only to God. The verse prohibits physical and mental prostration to others than God. See 16:49.

041:040 Knowing that polytheists would attempt to distort the message of the Quran, God warns them beforehand mentioning the perfect preservation of His message.

041:041 *Zikr* means, "message" and it is used in connection with the mathematical code of the Quran. See 38:1-8 and 74:30-31.

041:041-42 The Quran is not ink and paper, it is in the hearts of those who have knowledge (29:49). The message is specifically protected by God, not by humans (15:9). The extraordinary mathematical composition of the Quran testifies that any addition, deletion or distortion will be exposed. The code 19 shows that early attempts to distort the Quran have failed. Like a gold ring with a pattern of diamonds on it, the Quran as it was designed by God, has always existed and will always exist. With its built-in error-correction program, the Quran even helps us to identify scribal errors. Those who know the chemical and physical properties of gold and diamonds and those who witness the unique pattern of diamonds can recognize the precious ring and identify any fraudulent addition or even subtraction from its diamonds. None, especially those who had no clue about the unique physical and chemical properties of the ring, could have tampered with the Quran. See 15:9; 9:127.

041:044 The Quran should be translated to all world languages (26:198-200). Monotheists will get the clear divine message despite errors in translations since they study comparatively by using their God-given critical thinking skills. The teacher of the Quran is God Himself (55:1-2). See 43:3; 11:1.

041:047 See 20:15; 15:87; 31:34. Also see 3:18; 39:45; 63:1 and 6:23, 148.

041:049 See 12:87; 15:56; 30:36; 39:53.

041:053 Falsifiable and verifiable physical evidence and sound logical inferences provide scientific or objective confirmation that the Quran is indeed the word of God. Our personal experiences parallel those objective facts and remove all doubts. 4:82; 74:1-56. (Also, see 3:41).

42:0 In the name of God, the Gracious, the Compassionate.

42:1 **H**8M40*

42:2 **A**70S60Q100*

42:3 Similarly, revealing to you and those before you, is **God**, the Noble, the Wise.

42:4 To Him belongs all that is in the heavens and all that is in the earth, and He is the Most High, the Great.

42:5 The heavens would nearly shatter from above them, and the controllers praise the glory of their Lord, and they ask forgiveness for those on earth. Surely, **God** is the Forgiver, the Compassionate.

42:6 Those who take allies besides Him, **God** is watching them; and you are not a guardian over them.

42:7 Thus We have revealed to you an Arabic compilation, so that you may warn the capital town and all around it, and to warn about the day of Gathering that is inevitable. A group will be in Paradise, and a group in hell.*

42:8 Had **God** willed, He could have made them one nation. But He admits whom He wills into His mercy. The transgressors will have neither an ally, nor helper.

42:9 Or have they taken allies besides Him? But **God** is the ally, and He is the One who resurrects the dead, and He is able to do all things.*

42:10 Anything you dispute in, then its judgment shall be with **God**. Such is **God** my Lord. In Him I put my trust, and to Him I repent.

God is Unique

42:11 Creator of the heavens and the earth. He created for you from among yourselves mates, and also mates for the livestock so they may multiply. There is nothing like unto Him. He is the Hearer, the Seer.*

42:12 To Him belongs the possessions of the heavens and the earth. He spreads out the provision for whomever He wills, and He measures it. He is fully aware of all things.

The Same System; Peaceful and Tolerant

42:13 He has decreed for you the same system He ordained for Noah, and what We revealed to you, and what We ordained for Abraham, Moses, and Jesus: "You shall uphold this system, and do not divide in it." Intolerable for those who have set up partners is what you invite them towards. **God** chooses to Himself whoever/whomever He wills; He guides to Himself whoever repents.

42:14 They only divided after the knowledge had come to them, due to resentment among themselves. Had it not been for a predetermined decision from your Lord, they would have been judged immediately. Indeed, those who inherited the book after them are full of doubts.

42:15 For that, you shall preach and be upright, as you have been commanded, and do not follow their wishes. Say, "I affirm all that **God** has sent down from book, and I was commanded to treat you equally. **God** is our Lord and your Lord. We have our deeds and you have your deeds. There is no argument between us and you. **God** will gather us all together, and to Him is the ultimate destiny."

42:16 Those who debate about **God**, after they had been answered, their argument is nullified at their Lord. They have incurred a wrath and will have a severe retribution.

42:17 **God** is the One who sent down the book with truth, and the balance. For all that you know, the Moment may be very near.

42:18 Those who do not affirm it seek to hasten it, while those who affirm are concerned about it, and they know that it is the truth. Certainly, those who dispute the Moment have gone far astray.

42:19 **God** is Gracious to His servants; He gives provisions for whomever He wills, and He is the Powerful, the Noble.

42:20 Whoever desires the harvest of the Hereafter, We will increase for him his harvest. Whoever seeks the harvest of this world, We will give it to him, and he will have no share in the Hereafter.

Accepting Religious Authorities and Laws beside God's Word is Polytheism

42:21 Or do they have partners who decree for them a *sharia* which has not been authorized by **God**? If it were not for the word already given, they would have been judged immediately. Indeed, the transgressors will have a painful retribution.*

42:22 You see the transgressors worried because of what they had done; and it will come back at them. As for those who affirmed and promote reforms, they will be in the paradises of bliss. They will have what they wish from their Lord. This is the great blessing.

42:23 Such is the good news from **God** to His servants who affirm and promote reforms. Say, "I do not ask you for any wage, except that you show compassion to your relatives." Whosoever earns a good deed, We shall increase it for him in goodness. Surely, **God** is Forgiving, Appreciative.*

42:24 Or do they say, "He has fabricated lies about **God**!" If **God** willed, He could have sealed your heart. **God** erases the falsehood and affirms the truth with His words. He is fully aware of what is inside the chests.

42:25 He is the One who accepts the repentance from His servants, and He forgives the sins. He is fully aware of what you do.

42:26 Those who affirm and promote reforms respond to Him, and He increases for them His blessings. As for the ingrates, they have incurred a severe retribution.

42:27 If **God** were to increase the provision for His servants, they would transgress on earth; but He sends down what He wills in a measure. He is Ever-aware and Seer of His servants.

42:28 He is the One who sends down the rain after they had despaired, and spreads His mercy. He is the Supporter, the Praiseworthy.

God's Signs in Nature

42:29 From among His signs is the creation of the heavens and the earth, and the creatures He spreads in them. He is able to gather them, if He wills.

42:30 Any misfortune that happens to you is a consequence of what your hands have earned. He overlooks much.

42:31 You can never escape, and you have none besides **God** as an ally or helper.

42:32 From His signs are the vessels that sail the sea like flags.

42:33 If He willed, He could still the winds, leaving them motionless on top of it. In that are signs for everyone who is patient, thankful.

42:34 Or He may drown them, for what they have earned. He overlooks much.

42:35 Those who dispute Our signs may know that they have no place to hide.

Traits of those who affirm and Trust God

42:36 So whatever you are given is simply an enjoyment of the worldly life, and what is with **God** is far better and more lasting for those who affirm and put their trust in their Lord.

42:37 Those who avoid gross sins and lewdness, and when they are angered, they forgive.

42:38 Those who have responded to their Lord, and they maintain the contact prayer, and their affairs are conducted by mutual consultation among themselves, and from Our provisions to them they give.*

42:39 They are those who seek justice when gross injustice befalls them.

42:40 The recompense for a crime shall be its equivalence, but whoever forgives and makes right, then his reward is upon **God**. He does not like the wrongdoers.

42:41 For any who demand action after being wronged, those are not committing any error.

42:42 The error is upon those who oppress the people, and they aggress in the land without cause. For these will be a painful retribution.

42:43 As for the patient and forgiving, that is an indication of strength.

The Transgressor Opponents

42:44 Whomever **God** sends astray will not have any ally after Him. You will see the transgressors, when they see the retribution, saying: "Is there any way we can go back?"

42:45 You will see them being displayed to it, in fearful humiliation, and looking, while trying to avoid looking. Those who affirmed will say, "The losers are those who lost themselves and their families on the day of resurrection. The transgressors will be in a lasting retribution."

42:46 They had no allies to help them against **God**. Whomever **God** misguides will never find the way.

42:47 Respond to your Lord before a day comes from **God** which cannot be averted. You will have no refuge for you on that day, nor an advocate.

42:48 But if they turn away, then We did not send you as their guardian. You are only required to deliver. When We let the human being taste compassion from Us, he becomes happy with it, and when adversity afflicts them because of what their hands have done, the human being becomes rejecting.

42:49 To **God** is the sovereignty of the heavens and the earth. He creates whatever He wills. He bestows daughters to whomever He wills and bestows sons to whomever He wills.*

42:50 Or, He may bestow them with both daughters and sons, and He makes whom He wills sterile. He is Knowledgeable, Omnipotent.

Methods of Divine Communication

42:51 It is not for any human being that **God** would speak to him, except through revelation, or from behind a barrier, or by sending a messenger to reveal to whom He wills by His leave. He is the Most High, Most Wise.

42:52 Thus, We revealed to you a revelation of Our command. You did not know what the book was, nor the affirmation. Yet, We made this a light to guide whomever We wish from among Our servants. Surely, you guide to a straight path.*

42:53 The path of **God**; to Him belongs what is in the heavens and what is in the earth. Ultimately, all matters revert to **God**.

ENDNOTES

042:001-2 H10M40. **A**70S60Q100. These two combinations of letters/numbers play an important role in the mathematical system of the Quran based on code 19. For the meaning of these letters, see 1:1; 2:1; 40:1; 13:38; 46:10; 74:1-56.

042:007 See 43:3.

042:009 Those who worship their clerics and scholars (9:31; 42:21), especially in Pakistan and India call their religious leaders *Mawlana,* that is "our lord, our master, our patron." By assuming and using such a title, those religious leaders participate and encourage in this polytheistic practice. As long as Sunni and Shiite *mushriks* do not submit themselves to God alone, and as long as they consider their religious scholars and clerics to be their masters and lords, they will remain in the darkness of ignorance. See 2:286.

042:011 The Biblical verse, Genesis 1:26, stating that humans were created in God's image could be understood in a way that might contradict this verse. For detailed information about the nature of initiation or creation of the universe and living organisms, see 6:79; 82:1.

042:021 The religion that the so-called Muslims inherited from their parents and try hard to practice today, has little to do with the system of peacefully surrendering to God alone, which was delivered by Muhammed through the Quran. These clergymen who arrogated themselves and falsely claimed to be the "*ulama*" (people of knowledge), polluted the message of islam with ignorance. They fabricated numerous *sharias* (laws), prohibitions, veils, beards, turbans, rules on how to clean one's bottom, rules on how to pee in the bathroom, toothbrushes, right hands, left hands, right feet, left feet, *hadiths*, *sunnas*, intercession, holy hair, holy clothes, holy teeth, holy feet traces, *hazrats*, lords, saints, *mawlas*, *mahdies*, innocent *emams*, orders, sects, rosaries, amulets, dreams, holy loopholes, prayer caps, circumcisions, shrines, extra prayers, extra prohibitions, and numerous Arabic jargon such as *mandup*, *mustahap*, *makruh*, *sharif*, *sayyid* and more nonsense. Thus, the religion of Sunnis and Shiites contradicts the divine laws in nature and scripture and condemns its sincere followers to misery and backwardness. The religious leaders and their political allies contribute greatly to the backwardness of the Muslim world. God Almighty now wants to reform us and open the path of progress with the message described as "one of the greatest" (74:30-37).

Similar distortion and corruption were inflicted upon the system of islam (submission to God in peace and/or peacemaking) by professional religious leaders. For instance, soon after Jesus, a Pharisee-son-of-Pharisee who claimed to have seen Jesus in his vision, started preaching in the name of Jesus. His passionate, yet diabolic doctrine was rejected by the monotheists and muslims, but a majority of people were duped with his delusional passion and clever salesman skills. As a result, he made major changes, including transforming monotheism into polytheism to the extent of coining the name *Christian* (Acts 11:26). Jesus never silenced women and put them down with xenophobic teachings, but St. Paul asked women to submit to men and hush: (1 Timothy 2:7-15; 1 Corinthians 14:34-35; 1 Peter 3:7). Jesus never asked for money for preaching but St. Paul asked for money shamelessly and likened his audience to a flock of sheep to be milked by the holy shepherd! (1 Corinthians 9:7). He was a master of deceit as opposed to Jesus who did not twist the truth to gain followers; Paul made up anything he deemed helpful to increase the number of his milk-giving flock (1 Corinthians 9:22).

See 9:31; 33:67. Also, see 2:59; 3:45,51-52-52,55; 4:11,157,171; 5:13-15,72-79; 7:162; 19:36.

042:023 This verse is mostly abused by Shiite clergymen. They claim that prophet Muhammed was ordered to ask help for HIS relatives. The special status given to the descendants of Muhammed through Fatima and Ali, supported by a distorted meaning of this verse has created a privileged and "sacred" religious class. Hundreds of thousands of bearded leeches in Iran, Iraq and other Middle Eastern countries claim that they are descendants of Prophet Muhammed (*Sayyeed* or *Shareef*) and that they are thereby entitled to obligatory financial help. They abuse the verse mentioned above to exploit people economically. However, the verse does not say "my relatives." The context of the verse is plain enough to state that Muhammed does not need a wage from the affirmers and if they can help somebody, they should help their own relatives. Indeed, helping the relatives is a divine command repeated in 2:83; 4:36; 8:41; 16:90. Also, see 33:33.

042:038 The most important public task needing consultation is the election of leaders. About thirty years after the departure of Muhammed, polytheists abandoned the election system and established a monarchy or sultanate. Their puppet clerics tried to distort the application of this verse by claiming that consultation to elect a leader was not recommended but the leader, that is, the self-appointed tyrant, should consult. Unfortunately, this travesty of logic became the mainstream view of the followers of *hadith* and *sunna*. Ironically, neither the prophet nor his closest companions imposed themselves by self-appointment or by resorting to a glorious or royal genealogy. Those who split hairs to mimic every aspect of life attributed to Muhammed and his companions, from bathroom etiquette to how to groom one's beard, ignored this most important Quranic instruction that was practiced by Muhammed, and his companions for more than 40 years, starting from the immigration to Yathrib and continuing with Abu Bakr, Omar, Uthman and Ali, an era called the *Khulafa-i Rashideen*, that is, Guided Caliphs.

The later satanic Caliphs, assuming the blasphemous titles of "God's Caliph (successor)" or "God's Shadow on Earth" for centuries committed numerous atrocities in the name of God until 1924, when the caliphate was abolished by a Young Turk, Mustafa Kemal Atatürk. Democracy practiced according to the separation of powers is not perfect, but it is the best system we have discovered so far. As long as the democratic procedure is not corrupted by the money from interest groups or corporations, democracy helps the majority to pursue its self-interest, which ultimately is forced to include individual rights and the interests of minorities.

Ottoman Caliphs prohibited the importation and use of the printing machine based on the religious *fatwa* of a *sheikh al-islam* (highest cleric within the Ottoman Empire) who prohibited the use of the printing machine from 1455 to 1727, for 272 years, for 100,000 precious days, in a vast land stretching from North Africa to Iran, from today's Turkey to the Arabian Peninsula. While Europe indulged in learning God's signs in nature, shared the knowledge via printing machines, and was rewarded by God with a renaissance, reform, technology, and prosperity, Ottomans devolved and sunk further in ignorance. While Europeans engaged in philosophical arguments, the Ottomans recited the holy book no better than a parrot, the book that highlighted the importance of learning, questioning, discovery, and the pursuit of knowledge. They marveled at handwritten books of hearsay and superstition, at the lousy arguments developed by *Ghazzali*, who with the full support of a king, aimed to banish philosophy. While Europe sought a better system to save themselves from the tyranny of kings and the Church, the Ottomans recited handwritten poems to praise their corrupt kings and idols. No wonder why the land, the name, the face, and the religion of the children of this population is now associated with backwardness, ignorance, oppression, violence, and poverty.

Federal democracies facilitate cooperation among diverse groups with different religions, cultures, and legal systems and provide common ground for a productive, peaceful, and freely co-existing society.

042:049 Of course, this should not be understood that none other than God could choose the sex of babies. See 13:8; 31:34.

042:052 For more information about *ruh,* see 17:85; 15:29. None can guide anyone to salvation, including Jesus and Muhammed. Verse 28:56 states that Muhammed could not guide *those he loved.* But it was up to God to choose those who deserve to be guided. Muhammed could guide people in accordance with God's pre-established laws, but he could not guide a particular person of his choice. Every monotheist can guide to the "right path" defined by God. In other words, it is God who defines the system and conditions of guidance, and we are expected to deliver it to others. Accepting Muhammed as independent source of guidance is setting him up as a partner with God. See 6:112-114; 39:11. Also, see 7:30; 28:56.

بسم الله الرحمن الرحيم

43:0 In the name of God, the Gracious, the Compassionate.

43:1 **H8M40***

43:2 The evident book.

43:3 We have made it into an Arabic compilation, perhaps you may understand.*

43:4 It is held honorable and wise in the master record with Us.

43:5 Shall We take away the reminder from you, because you are a transgressing people?

43:6 How many a prophet did We send to the previous generations!

43:7 Every time a prophet went to them, they ridiculed him.

43:8 We have destroyed those who were even more powerful than these, and the example of the previous generations has already been given.

Appreciating God's Power and Blessings

43:9 If you asked them: "Who created the heavens and the earth?" They will say, "They were created by the One who is the Noble, the Knowledgeable."

43:10 He is the One who made the earth a habitat for you, and He made pathways in it that you may be guided.

43:11 He is the One who sends down water from the sky, in exact measure. We then revive with it a dead land. Similarly, you will be brought out.

43:12 He is the One who created all the pairs. He made for you ships and the livestock to ride.

43:13 So that you may settle on their backs; and then when you have settled on them you may recall your Lord's blessing, by saying: "Glory be to the One who commits this for us, and we could not have done so by ourselves."

43:14 "We will ultimately return to our Lord."

Misogynistic and Militaristic Culture Condemned

43:15 They assigned a share to Him from His own servants! The human being is clearly denying.

43:16 Or has He selected daughters from among His creation, while He has left you with the sons?

43:17 When one of them is given news of what he cites as an example for the Gracious, his face becomes dark, and he is miserable!

43:18 "What good is an offspring that is brought up to be beautiful, and cannot help in a fight?"

43:19 They have claimed the controllers who are with the Gracious are females! Have they been made witness to their creation? We will record their testimony, and they will be asked.

43:20 They said, "If the Gracious willed, we would not have served them." They have no knowledge of this; they only conjecture.

43:21 Or have We given them a book before this which they are upholding?

The Trouble with Following Ancestors Blindly

43:22 The fact is, they have said, "We found our fathers following a certain way, and we are following in their footsteps."

43:23 Similarly, We did not send a warner to a town, except its privileged hedonists said, "We found our fathers following a certain way, and we are being guided in their footsteps."

43:24 He said, "What if I brought to you better guidance than what you found your fathers upon?" They said, "We reject the message you deliver."

43:25 Consequently, We took revenge upon them. So see what the consequence of the deniers was!

43:26 When Abraham said to his father and his people: "I am innocent of what you serve."

43:27 "Except for the One who initiated me, He will guide me."

43:28 He made it a word to last in his subsequent generations; perhaps they may turn back.

Ignorance and Arrogance

43:29 Indeed, I have given these people and their fathers enjoyment, until the truth came to them, and a clarifying messenger.

43:30 When the truth came to them, they said, "This is magic, and we reject it."

43:31 They said, "If only this Quran was sent down to a great man from the two towns!"

43:32 Is it they who assign your Lord's mercy? We have assigned their shares in this worldly life, and We raised some of them above others in ranks, so that they would take one another in service. The mercy from your Lord is far better than what they amass.

43:33 If it were not that all the people would become one cluster, We would have provided for those who reject the Gracious silver roofs for their homes, and stairs upon which they could ascend.

43:34 For their homes, gates, and furnishings on which they could recline.

43:35 Many ornaments. All these are the pleasures of this worldly life. The Hereafter with your Lord is for the righteous.

43:36 Whosoever turns away from the remembrance of the Gracious, We appoint a devil to be his constant companion.

43:37 They hinder from the path, but they think they are guided!

43:38 Until he comes to Us, he will say, "Oh, I wish that between you and me were the distance of the two easts. What a miserable companion!"

43:39 It would not benefit you this day, for you have transgressed; you are partners in the retribution.

43:40 Can you make the deaf hear, or can you guide the blind and those who are far astray?

43:41 For when We take you away, We may seek revenge on them.

43:42 Or We may show you what We promised for them; We are in full control over them.

43:43 You shall hold on to what is revealed to you; you are on a straight path.

43:44 This is indeed a reminder for you and your people; and you will all be questioned.

Moses versus the Arrogant, the Oppressor

43:45 Ask those of Our messengers whom We sent before you: "Did We ever appoint gods besides the Gracious to be served?"

43:46 We sent Moses with Our signs to Pharaoh and his entourage, saying: "I am a messenger of the Lord of the worlds."

43:47 But when he came to them with Our signs, they laughed at them.

43:48 Every sign We showed them was greater than the one before it, and We seized them with the torment, perhaps they would revert.

43:49 They said, "O you magician, implore your Lord for us according to what He has agreed with you; we will then be guided."

43:50 But when We removed the torment from them, they broke their word.

43:51 Pharaoh proclaimed among his people: "O my people, do I not possess the kingship of Egypt, and these rivers that flow below me? Do you not see?"

43:52 "Am I not better than this one who is despised and can barely be understood?"

43:53 "Why then are not golden bracelets bestowed on him, or the controllers accompanying him?"

43:54 He thus convinced his people, and they obeyed him; they were a wicked people.

43:55 So when they persisted in opposing Us, We sought revenge from them, and drowned them all.

43:56 We thus made them a thing of the past (*salaf*), and an example for the others.*

Reaction of Arab Nationalists to Jesus

43:57 When the son of Mary was cited as an example, your people turned away from it.

43:58 They said, "Are our gods better or is he?" They only cited this to argue with you. Indeed, they are a quarrelsome people.

43:59 He was no more than a servant whom We blessed, and We made him an example for the Children of Israel.

43:60 If We willed, We could have made some of you controllers to be successors on earth.

Jesus' Birth as a Sign for the End of the World

43:61 He was a sign for knowing the Moment. You should have no doubt about it. Follow Me; this is a straight path.*

43:62 Let not the devil repel you; he is to you a clear enemy.

43:63 When Jesus came with the proofs, he said, "I have come to you with the wisdom, and to clarify some of the matters in which you dispute. So be aware of **God** and obey me."

43:64 "**God** is my Lord and your Lord. So serve Him. This is a straight path."

43:65 The groups disputed among themselves. Woe to those who have been wicked from the retribution of a painful day.

43:66 Do they only wait for the Moment to come to them suddenly, while they do not perceive?

43:67 Friends on that day will become enemies of one another, except for the righteous.

43:68 "O My servants, you will have no fear on this day, nor will you grieve."

43:69 They are the ones who affirmed Our signs and had peacefully surrendered.

Paradise and Hell

43:70 "Enter Paradise, together with your spouses, in happiness."

43:71 They will be served with golden trays and cups, and they will find everything the self desires and the eyes wish for, and you will abide therein forever.

43:72 This is the Paradise that you have inherited, in return for your works.

43:73 In it you will have all kinds of fruits, from which you eat.

43:74 Surely, the criminals will abide in the retribution of hell forever.

43:75 It will not be removed from them; they will be confined therein.

43:76 We did not wrong them, but it was they who were the wrongdoers.

43:77 They called out: "O Malek/angel, please let your Lord terminate us!" He will say, "No, you are remaining."

They Hate the Truth

43:78 We have come to you with the truth, but most of you hate the truth.

43:79 Or have they devised some scheme? We will also devise.

43:80 Or do they think that We do not hear their secrets and private counsel? Yes indeed; and Our messengers are with them, recording.

43:81 Say, "If the Gracious had a son, I would be the first to serve!"

43:82 Glorified be the Lord of the heavens and the earth, the Lord of the Throne, from what they attribute.

43:83 So leave them to speak nonsense and play until they meet their day, which they have been promised.

43:84 He is the One who is a god in the heaven and a god on earth. He is the Wise, the Knowledgeable.

43:85 Blessed is the One who possesses the sovereignty of the heavens and the earth, and everything between them; and with Him is the knowledge of the Moment, and to Him you will be returned.

Those Who Expect to be Saved by the Intercession of Prophets and Saints will be Surprised

43:86 Those whom they call on beside Him do not possess any intercession; except those who bear witness to the truth, and they fully know.*

43:87 If you asked them who created them, they would say, "**God**." Why then do they deviate?

43:88 It will be said, "O my Lord, these are a people who do not affirm."

43:89 So disregard them and say, "Peace." For they will come to know.

ENDNOTES

043:001 **H**10M40. The combination of these two letters/numbers plays an important role in the mathematical system of the Quran based on code 19. For the meaning of these letters see 1:1; 2:1; 40:1; 13:38; 46:10; 74:1-56.

043:003 The root of *ARaBy* (Arabic) is *ARB*, which also means "excellent" or "perfect" (See 56:37). Since the message of the Quran was sent to all humanity and jinns, the eloquence of the language of the Arabic Quran does not necessarily come from its being in the Arabic language, but its Arabic being *ARB*, that is, excellent or perfect. In other words, the message of this verse is not about the proper name Arabic, but about its literary meaning. Every Justin might not be just; every Mr. Smart may not be smart. Similarly, each Arabic text is not necessarily *arabic* (excellent). The language of the Quran is both Arabic and *arabic*, that is, flawless, and perfect. Though the Quran was revealed to an Arabic speaking prophet in Arabic, since it contains universal truth and is taught by the Gracious (55:2), it is a message to all humanity regardless of their language.

043:056 The word *salaf* (past, ancestors, predecessors) is currently used as a label by the radical Sunnis who reject sectarian jurisprudence yet rely directly on *hadith* and *sunna*; they claim to be the followers of the prophet and his companions, whom they call *salaf*. The Salafi movement started with Ibn Taymiyya, a charismatic scholar who rejected some major tenets of orthodox teachings. Orthodox Sunni sects promote blindly following the fatwas or jurisprudence of a particular sect and discourage from interpreting and inferring directly from *hadith*. Ibn Taymiyya argued that every Muslim must have direct access to *hadith* books and learn their religion through studying them, rather than being bound by the views of earlier scholars. Though Salafis justifiably rejected many polytheistic teachings and rituals, such as sectarian jurisprudence, intercession, and imploring shrines, they unfortunately indulged in fabricated *hadith*, and advocated a narrow (mis)understanding of the Quran, such as rejecting the use of metaphors. Since they set *hadith* sources as partners with God in practicing Islam, they ended up with a religion that reflected a concoction of Arabic, Jewish and Christian culture from medieval times. The majority of *salafis* live in Saudi Arabia and they promote a Taliban-style tyrannical system that produces an ignorant, misogynistic, and backward crowd who subject themselves to being herded like sheep by the moral police. As we noted in relation to the Quranic usage of *hadith*, *sunna*, and *ijma*, the word *salaf* too has negative connotations in the Quran. This is a miraculous prophecy of the Quran; major sects who abandoned the Quran by setting up various lists of "holy" idols and teachings as partners with God in His system, use labels and names that are exposed by the Quran. The word *salaf,* in all its derivatives, occurs eight times (2:285; 4:22,23; 5:95; 8:38; 10:30; 69:24; 43:56), and only in one occurrence, here, is it used in a reference to a community of people; and they were ignorant, aggressive, and oppressive. For the prophetic reference of the Quran regarding other sectarian labels, see: 33:38.

043:061 The Quran contains some hints regarding the time of the end of the world (20:15; 15:87). See 47:18

043:086 See 25:30 and 2:48.

بسم الله الرحمن الرحيم

44:0 In the name of God, the Gracious, the Compassionate.
44:1 **H**8M40*
44:2 The clarifying book
44:3 We have sent it down in a blessed night. Surely, We were to warn.
44:4 In it, is a decree for every matter of wisdom.
44:5 A decree from Us. Surely, We were to send messengers.
44:6 A mercy from your Lord. He is the Hearer, the Knowledgeable.
44:7 The Lord of the heavens and the earth, and everything between them. If you were certain!
44:8 There is no god except Him. He gives life and causes death; your Lord and the Lord of your ancestors.
44:9 No; they are in doubt, playing.

Environmental Calamity Prophesied

44:10 Therefore, watch for the day when the sky will bring a visible smoke/gas.
44:11 It will envelop the people: "This is a painful retribution!"
44:12 "Our Lord, remove the retribution from us; we are those who affirm."
44:13 How is it that now they remember, while a clarifying messenger had come to them?
44:14 But they turned away from him and said, "Clearly educated, but crazy!"
44:15 We will remove the retribution in a while; you will then revert back.
44:16 On the day We strike the great strike, We will avenge.

An Example of Previous Tests

44:17 We had tested before them the people of Pharaoh, and an honorable messenger came to them.
44:18 "Restore to me the servants of **God**. I am a trustworthy messenger to you."
44:19 "And, do not transgress against **God**. I come to you with clear authority."
44:20 "I seek refuge in my Lord and your Lord, should you reject me."

44:21 "If you do not wish to affirm, then leave me alone."

44:22 Subsequently, he called on his Lord: "These are a criminal people."

44:23 You shall travel with My servants during the night; you will be pursued.

44:24 Cross the sea quickly; their troops will be drowned.

44:25 How many paradises and springs did they leave behind?

44:26 Crops and luxurious dwellings?

44:27 Blessings that they enjoyed?

44:28 Thus it was; and We caused another people to inherit it.

44:29 Neither the heaven, nor the earth wept over them, and they were not reprieved.

44:30 We saved the Children of Israel from the humiliating agony.

44:31 From Pharaoh; he was a transgressing tyrant.

44:32 We have chosen them, out of knowledge, over the worlds.

44:33 We granted them signs, which constituted a great test.

44:34 These people now are saying:

44:35 "There is nothing but our first death; and we will never be resurrected!"

44:36 "So bring back our forefathers, if you are truthful!"

44:37 Are they better or the people of Tubba and those before them? We destroyed them, they were criminals.

44:38 We have not created the heavens and the earth, and everything between them, for mere play.

44:39 We did not create them except with the truth, but most of them do not know.

A Hellish Metaphor

44:40 Surely, the day of Separation is the appointment for them all.

44:41 That is the day when no friend can help his friend in any way; nor will they be helped.*

44:42 Except the one whom **God** treated with compassion. He is the Noble, the Compassionate.

44:43 Surely, the tree of Bitterness

44:44 Will be the food for the sinful.

44:45 Like hot oil, it will boil in the stomachs.

44:46 Like the boiling of liquid.

44:47 "Take him and throw him into the midst of hell."

44:48 "Then pour upon his head the retribution of boiling liquid."

44:49 "Taste this; surely you are the noble, the generous!"

44:50 Surely, this is what you used to doubt!

A Heavenly Metaphor

44:51 The righteous will be in a place of security.

44:52 Among paradises and springs.

44:53 Wearing silk and satin; facing each other.

44:54 So it is, and We coupled them with wonderful companions.*

44:55 They enjoy in it all kinds of fruits, in perfect peace.

44:56 They do not taste death therein except for the first death, and He has spared them the retribution of hell.

44:57 As a blessing from your Lord. Such is the great triumph.

44:58 We have thus made it easy in your language, perhaps they may take heed.

44:59 Therefore, keep watch; for they too will keep watch.

ENDNOTES

044:001 H10M40. The combination of these two letters/numbers plays an important role in the mathematical system of the Quran based on code 19. For the meaning of these letters, see 1:1; 2:1; 40:1; 13:38; 46:10; 74:1-56.

044:041 See 2:286.

044-054 The word *hur* means intelligent, pure, dazzling friend; male or female. Words derived from the same root are used to mean back and forth conversation (*hiwar, muhawara*) (18:34,37; 58:1), disciple or true friend (*hawary*) (3:52; 5:111-112; 61:14), and to return (84:14). See 52:20; 55:72; 56:22.

044:056 Because of our rebellion in paradise we are sent to this world via a transition between universes, which is called death. See 40:11.

45:0 In the name of God, the Gracious, the Compassionate.

45:1 **H**8M40*

45:2 The revelation of the book is from **God**, the Noble, the Wise.

45:3 In the heavens and the earth are signs for those who affirm.

45:4 In your creation, and what creatures He puts forth are signs for people who are certain.

45:5 The alternation of the night and the day, and what **God** sends down from the sky of provisions to revive the land after its death, and the changing of the winds, are signs for a people who understand.

Which Hadith Besides the Quran?

45:6 These are **God**'s signs that We recite to you with truth. So, in which *hadith*, after **God** and His signs, do they affirm?*

45:7 Woe to every sinful fabricator.

45:8 He hears **God**'s signs being recited to him, then he persists arrogantly, as if he never heard them. Give him news of a painful retribution.

45:9 If he learns anything from Our signs, he makes fun of them. For these will be a humiliating retribution.

45:10 Waiting for them is hell. What they earned will not help them, nor those whom they have taken as allies besides **God**, and for them is a terrible retribution.

Divine Signs

45:11 This is guidance. Those who reject the signs of their Lord, for them is an affliction of a painful retribution.

45:12 **God** is the One who committed the sea in your service, so that the ships can sail in it by His command, and that you may seek of His provisions, and that you may be appreciative.

45:13 He committed in your service all that is in the heavens and in the earth; all from Him. In that are signs for a people who reflect.

45:14 Say to those who affirmed: they should forgive those who do not look forward to the days of **God**. He will fully recompense people for whatever they have earned.

45:15 Whoever works good does so for himself, and whoever works evil will suffer it. Then to your Lord you will be returned.

45:16 We had given the Children of Israel the book and the judgment, and the prophethood, and We provided them with good provisions; and We preferred them over the worlds.

45:17 We gave them proof in the matter, but then they disputed after the knowledge had come to them, out of jealousy amongst themselves. Your Lord will judge them on the day of resurrection regarding everything that they have disputed.

45:18 Then We have established you on the correct path; so follow it and do not follow the desires of those who do not know.

45:19 They cannot help you against **God** in the least. The transgressors are allies to one another, while **God** is the Protector of the righteous.

45:20 This is a physical evidence for the people, and guidance and a mercy for a people who are certain.

45:21 Or do those who work evil expect that We would treat them the same as those who affirm and promote reforms, in their present life and their death? Miserable is how they judge.

45:22 **God** created the heavens and the earth with truth, and so that every person may be recompensed for whatever he/she earned, and they will not be wronged.

The Narrow-Minded Skeptics

45:23 Have you seen the one who took his fancy as his god, and **God** led him astray, despite his knowledge, and He sealed his hearing and his heart, and He made a veil on his eyes? Who then can guide him after **God**? Will you not remember?

45:24 They said, "There is nothing but this worldly life; we die and we live and nothing destroys us except the passing of time!" They have no knowledge about this; they only conjecture.

45:25 When Our clear signs are recited to them, their only argument is to say, "Then bring back our forefathers, if you are truthful."

45:26 Say, "**God** gives you life, then He puts you to death, then He will gather you to the day of resurrection, in which there is no doubt. But most people do not know."

45:27 To **God** belongs the sovereignty of the heavens and the earth. On the day the Moment comes to pass, on that day the falsifiers will lose.

45:28 You will see every nation on their kneels. Every nation will be called to its record. "Today, you are recompensed for everything you have done."

45:29 "This is Our record; it utters the truth about you. We have been recording everything you did."

45:30 As for those who affirm and promoted reforms, their Lord will admit them into His mercy. Such is the clear triumph.

45:31 As for those who reject: "Were not My signs recited to you, but you turned arrogant and were a criminal people?"

45:32 When it was said, "Surely **God**'s promise is the truth, and there is no doubt about the coming of the Moment," you said, "We do not know what the Moment is! We are full of conjecture about it; we are not certain."

45:33 The evils of their works will become evident to them, and the very thing they ridiculed will be their doom.

45:34 It will be said to them: "Today We will forget you, just as you forgot the meeting of this day. Your abode is the fire, and you will have no helpers."

45:35 "This is because you took **God**'s signs in mockery, and you were deceived by the worldly life." So from this day, they will never exit therefrom, nor will they be excused.

45:36 All praise belongs to **God**; the Lord of the heavens, and the Lord of the earth; the Lord of the worlds.

45:37 To Him belongs all majesty in the heavens and the earth. He is the Noble, the Wise.

ENDNOTES

045:001 H10M40. The combination of these two letters/numbers plays an important role in the mathematical system of the Quran based on code 19. For the meaning of these letters, see 1:1; 2:1; 40:1; 13:38; 46:10; 74:1-56.

045:006-9 Those who consider God and His signs to be insufficient, answer this challenge in the affirmative. They list a number of *hadith* books as an answer. See 6:112-114; 12:111; 33:38. Many of those who follow *hadith* make fun of "one of the greatest signs" prophesied in Chapter 74.

بسم الله الرحمن الرحيم

46:0 In the name of God, the Gracious, the Compassionate.

46:1 **H**8M40*

46:2 The revelation of the book from **God**, the Noble, the Wise.

46:3 We did not create the heavens and the earth, and everything between them except with truth, and for an appointed time. Those who reject turn away from what they are being warned with.

46:4 Say, "Do you see those that you call on besides **God**? Show me what they have created on the earth, or do they have a share in the heavens? Bring me a book before this, or any trace of knowledge, if you are truthful."

46:5 Who is more astray than one who calls on others besides **God** that do not respond to him even till the day of resurrection? They are totally unaware of the calls to them!

46:6 At the time when people are gathered, they will be enemies for them, and they will reject their service.*

46:7 When Our clear signs are recited to them, those who rejected said of the truth that came to them: "This is evidently magic!"

46:8 Or do they say, "He fabricated this!" Say, "If I fabricated this, then you cannot protect me at all from **God**. He is fully aware of what you say. He suffices as a witness between me and you. He is the Forgiver, the Compassionate."

46:9 Say, "I am no different from the other messengers, nor do I know what will happen to me or to you. I only follow what is revealed to me. I am no more than a clear warner"

Rabi Juda: A Witness from the Children of Israel

46:10 Say, "Do you see that if it were from **God**, and you rejected it, and a witness from the Children of Israel testified to its similarity, and he has affirmed, while you have turned arrogant? Surely, **God** does not guide the wicked people."*

46:11 Those who had rejected said regarding those who had affirmed: "If it were any good, they would not have beaten us to it." When they are not able to be guided by it, they will say, "This is an old fabrication!"*

46:12 Before this was the book of Moses, as a role model and a mercy. This is an authenticating book in an Arabic tongue, so that you may warn those who have transgressed, and to give good news to the righteous.

46:13 Surely, those who said, "Our Lord is **God**," then they lead a righteous life, there is no fear for them, nor will they grieve.

46:14 These are the dwellers of Paradise, abiding therein, a reward for what they used to do.

The Age of Intellectual and Emotional Maturity

46:15 We instructed the human being to honor his parents. His mother bore him with hardship, gave birth to him in hardship, and his weaning lasts thirty months. Until he has attained his maturity, and reaches forty years, he says: "My Lord, direct me to appreciate the blessings You have bestowed upon me and upon my parents, and to do righteousness that pleases You. Let my progeny be righteous. I have repented to You; I am of those who have peacefully surrendered."*

46:16 It is from these that We accept the best of their deeds, and We shall overlook their sins, among the dwellers of Paradise. This is the promise of truth that they had been promised.

46:17 The one who says to his parents: "Enough of you! Are you promising me that I will be resurrected, when the generations who died before me never came back?" While they both will implore **God**: "Woe to you; affirm! For **God**'s promise is the truth." He would say, "This is nothing but tales from the past!"

46:18 These are the ones against whom the retribution has been deserved in nations who had come before them of humans and Jinn; they are the losers.

46:19 To each will be degrees according to what they did; He will recompense them, and they will not be wronged.

46:20 On the day when those who rejected will be displayed to the fire: "You have wasted the good things given to you during your worldly life, and you took pleasure in it. Consequently, today you will be recompensed with a shameful retribution for your arrogance in the land without any right, and for your evil works."

Hood

46:21 Recall that the brother of Aad warned his people at the dunes, while numerous warnings were also delivered before him and with him: "You shall not serve except **God**. For I fear for you the retribution of a great day."

46:22 They said, "Have you come to us to divert us away from our gods? Then bring us what you are promising us, if you are truthful!"

46:23 He said, "The knowledge of it is with **God**; and I only convey to you what I was sent with. However, I see that you are a people who are ignorant."

46:24 Then when they saw the dense cloud heading towards their valley, they said, "This is a dense cloud that will bring to us much needed rain!" No, this is what you had asked to be hastened; a violent wind with a painful retribution.

46:25 It destroys everything by the command of its Lord. Thus, they became such that nothing could be seen except their homes. We thus requite the criminal people.

46:26 We had established them in the same way as We established you, and provided them with the hearing, and the eyesight, and the heart. But their hearing, eyesight, and hearts did not help them at all. This is because they used to disregard **God**'s signs, and they were stricken by what they used to mock!

46:27 We have destroyed the towns around you, after We had explained the proofs, perhaps they would repent.

46:28 Why then did the idols they set up to bring them closer to **God** fail to help them? Instead, they abandoned them. Such was their lie, and what they fabricated.

The Message of the Quran Transcends Humans

46:29 We sent to you a small number of Jinn, in order to let them listen to the Quran. So when they arrived there, they said, "Pay attention." Then, when it was finished, they returned to their people, to warn them.

46:30 They said, "O our people, we have heard a book that was sent down after Moses, authenticating what is present with him. It guides to the truth; and to a Straight Path."

46:31 "O our people, respond to **God**'s caller, and affirm Him. He will then forgive your sins, and spare you a painful retribution."

46:32 Whosoever does not respond to **God**'s caller, then he cannot escape on the earth, and he will not have besides Him any allies. These are the ones who have gone far astray.

46:33 Did they not see that **God**, Who created the heavens and the earth, and was not tired by their creation, is able to revive the dead? Yes indeed; He is able to do all things.

46:34 The day those who do not appreciate are displayed to the fire: "Is this not the truth?" They will answer: "Yes indeed, by our Lord." He will say, "Then taste the retribution for what you had rejected."

46:35 Therefore, be patient like the messengers of strong-will did before you and do not be in haste regarding them. On the day they will see what they have been promised; it will be as if they had not remained except for one moment of a single day. A proclamation: "Isn't true that none other than the wicked people will be destroyed?"

ENDNOTES

046:001 **H**10M40. The combination of these two letters/numbers plays an important role in the mathematical system of the Quran based on code 19. For the meaning of these letters, see 1:1; 2:1; 40:1; 13:38; 46:10; 74:1-56.

046:006 Compare it to the New Testament, Matthew 7:21-23. Jesus (Esau) warns those who call him "Lord" with deserving the hellfire. Though this verse is distorted by some translations, by comparing different translations, one can infer the original message and distortions made afterwards. We have discussed this issue in "19 Questions For Christian Scholars." See 41:47-48.

046:010 In the 1980's, we learned that Rabi Juda, a French Jewish scholar who lived in eleventh century, had discovered a mathematical system based on the number 19, in the original parts of the Old Testament. The similarity and clarity of the discovery is impressive and fulfills the prophecy of this verse. Rabi Judah's discovery of 19 in the original parts of the Old Testament, and his correction of the modern text by using the code, together with a list of examples, was unveiled in an article authored by Joseph Dan and was published by the University of California in 1978 under the title "*Studies in Jewish Mysticism*." According to Dan's article, Rabi Juda claimed to have written 8 volumes on the code 19-based mathematical structure in the Hebrew Old Testament. Hopefully, one day some scholars will search and discover that important work.

046:011 A group of scholars, including my father and Tayyar Altıkulaç, the former head of Turkey's Department of Religious Affairs, published a book criticizing code 19. The title they picked for their book was no coincidence: 19 Efsanesi (The Myth of 19). Like their choice of key words, such as *hadith*, *sunna*, *ijma*, this too is another prophecy of the Quran. See 6:25.

046:015 This verse informs us that the sum of pregnancy and breastfeeding is 30 months. When evaluated together with verse 31:14 suggesting the time of breastfeeding as 24 months, it takes simple arithmetic to learn that the pregnancy of the person (*nafs*) to be 6 months. It is common knowledge that pregnancy is longer than that. We know from modern embryology that the exact time for a normal pregnancy is 266 days (see 77:23). There is extra information to be extracted from these natural and scriptural signs. If we subtract the 6 months, that is 180 days, from 266 days, we get 86 days. We can easily infer that the Quran does not accept the creature inside the womb to be a person (*nafs*) from the time of conception until the 86th day of pregnancy. Other verses support our inference. For instance, verse 22:5 and 23:14, which explain the evolution of the human being in four stages, does not use the word person (*nafs*) or human (*insan*) to describe the early stages of pregnancy. The four stages of pregnancy are (1) sperm; (2) embryo; (3) about four-inch fetus (4) a new creature. A new creation comes into existence after the 86th day. In other words, the emergence of consciousness/personhood (*nafs*) starts taking place in the brain of the fetus approximately three months after conception. Do we need to add further that reflecting on the *ayat* (signs) of the scripture together with the *ayat* (signs) of nature sheds light on the controversial issue of abortion? Also, see 16:58-59; 17:31

47:0 In the name of God, the Gracious, the Compassionate.

47:1 Those who rejected and repelled from the path of **God**, He will mislead their works.

47:2 Those who affirm and promote reforms, and affirm what was sent down to Muhammed, for it is the truth from their Lord, He cancels for them their sins, and relieves their concern.

47:3 That is because those who reject followed falsehood, while those who affirm followed the truth from their Lord. **God** thus cites for the people their examples.

During the War, Aim to Capture the Enemy

47:4 So, if you encounter those who have rejected, then strike the control center until you overcome them. Then bind them securely. You may either set them free or ransom them, until the war ends. That, and had **God** willed, He alone could have beaten them, but He thus tests you by one another. As for those who get killed in the cause of **God**, He will never let their deeds go to waste.*

47:5 He will guide them and relieve their concerns.

47:6 He will admit them into Paradise, which He has described to them.

47:7 O you who affirm, if you support **God**, He will support you, and make your foothold firm.

47:8 Those who rejected, for them is destruction; and He has misled their works.

47:9 That is because they hated what **God** sent down, thus He nullifies their works.

47:10 Did they not roam the land and note what the end was for those before them? **God** destroyed them and the ingrates in a similar fate.

47:11 This is because **God** is the Master for those who affirm, while those who do not appreciate have no master.*

47:12 **God** admits those who affirm and promote reforms to paradises with rivers flowing beneath them. As for those who reject, they are enjoying and eating as the cattle eat, and the fire will be their abode.

47:13 Many a town was stronger than your own town, which drove you out. We destroyed them, and there was none who could help them.

47:14 Is one who is based on proof from his Lord, as one for whom his evil works have been adorned for him and they followed their desires?

Allegorical Description of Paradise and Hell

47:15 The example of Paradise; that the righteous have been promised with rivers of pure water, rivers of milk whose taste does not change, rivers of intoxicants that are delicious for the drinkers, and rivers of strained honey. For them in it are all kinds of fruits, and forgiveness from their Lord. Compare it to that of those who abide in the fire and are given to drink boiling water that cuts up their intestines.*

47:16 Some of them listen to you, until when they go out from you, they say to those who have been given the knowledge: "What did he say?" Those are the ones whom **God** has sealed upon their hearts, and they followed their desires.

47:17 Those who are guided, He increases their guidance, and grants them their righteousness.

47:18 So are they waiting until the Moment comes to them suddenly? For its conditions have already been met. But once it comes to them, how will they benefit from their message?*

47:19 So know that there is no god besides **God;** ask forgiveness of your sins and also for the affirming men and affirming women. **God** knows your movements and your place of rest.

Difficult Times Expose the Hypocrites

47:20 Those who affirm say, "If only a chapter is sent down!" But when a resolute chapter is sent down, and fighting is mentioned in it, you see those who have a disease in their hearts look at you, as if death had already come to them. It thus revealed them.

47:21 Obedience and to speak righteousness until the matter is decided, then if they trust **God,** it would be better for them.

47:22 So do you plan that when you turn away, that you will corrupt the land and sever your family ties?

47:23 These are the ones whom **God** has cursed and thus He made them deaf and blinded their sight.

47:24 Do they not reflect on the Quran? Or are there locks on their hearts?

47:25 Surely, those who reverted back, after the guidance has been made clear to them, the devil has enticed them and led them on.

47:26 That is because they said to those who hated what **God** had sent down: "We will obey you in certain matters." **God** knows their secrets.

47:27 So how will it be when the controllers put them to death, striking their faces and their backs?

47:28 That is because they followed what angered **God**, and they hated what pleased Him. Thus, He nullified their works.

47:29 Or did those who harbor a disease in their hearts think that **God** would not bring out their evil thoughts?

47:30 If We wished, We would show them to you so you would recognize them by their looks. However, you can recognize them by their speech. **God** is fully aware of your works.

47:31 We will test you until We know those who strive among you and those who are patient. We will bring out your qualities.

47:32 Those who have rejected and repelled from the path of **God** and stood against the messenger after the guidance has been made clear for them, they will not harm **God** in the least, and He will nullify their works.

47:33 O you who affirm, obey **God**, and

obey the messenger. Do not render your work in vain.

47:34 Surely, those who rejected and repelled from the path of **God**, then they died while still rejecting, **God** will never forgive them.

47:35 Therefore, do not be weak in calling for peace. You are on a higher moral ground, and **God** is with you. He will not waste your efforts.

47:36 This worldly life is no more than play and vanity. But if you affirm and lead a righteous life, He will reward you, and He will not ask you for your wealth.

47:37 If He were to ask you for it, to the extent of creating a hardship for you, you would become stingy, and your hidden evil might be exposed.

47:38 Here you are being invited to spend in the cause of **God**, but some among you turn stingy. Whoever is stingy is only being stingy on himself. **God** is the Rich, while you are the poor. If you turn away, He will substitute another people instead of you, then they will not be like you.

ENDNOTES

047:004 The expression "*darb al riqab*" is traditionally translated as "smite their necks." We preferred to translate it as "strike the control center." The Quran uses the word "*unuq*" for neck (17:13,29; 8:12; 34:33; 38:33; 13:5; 26:4; 36:8; 40:71). The root *RaQaBa* means observe, guard, control, respect, wait for, tie by the neck, warn, fear. "*Riqab*" means slave, prisoner of war, since they are controlled or guarded. Even if one of the meanings of the word *riqab* were neck, we would still reject the traditional translation, for the obvious reason: The verse continues by instructing peacemakers regarding the capture of the enemies and the treatment of prisoners of war. If they were supposed to be beheaded, there would not be a need for an instruction regarding captives, which is a very humanitarian instruction. Unfortunately, the Sunni and Shiite terrorists have used the traditional mistranslation and abused it further by beheading hostages in their fight against their counterpart terrorists, Crusaders and their allied coalition, who torture and kill innocent people in even bigger numbers, yet in a baptized fashion that is somehow depicted as non-barbaric by their culture and media. The Quran gives two options regarding the hostages or prisoners of war before the war ends: (1) set them free; or (2) release them to get a fee for their unjustified aggression. Considering the context of the verse and emphasis on capturing the enemy, we could have translated the segment under discussion as, "aim to take captives."

The Old Testament contains many scenes of beheadings and grotesque massacres. For instance, see: 2 Samuel 4:7-12; 2 Kings 10:7, and 2 Chronicles 25:12.

047:011 See 2:286.

047:015 The only verse that uses intoxicants (*KHAMR*) in a positive context is 47:15, and interestingly it is about paradise, that is, the hereafter. A quick reflection on the reason for prohibition of intoxicants will explain the apparent contradiction. The harm of intoxicants, such as drunk driving, domestic violence or alcoholism, is not an issue in the other universe, where the laws and rules are different. In other words, a person who is rewarded by eternal paradise will not hurt himself, herself, or anyone else by getting intoxicated. See 52:23; 76:21. Also see 7:43; 15:47; 21:102; 41:31; 43:71; 2:112; 5:69.

047:018 Since the Quran is the last book, it gives us clues about the end of the world. See 20:13; 15:87.

بسم الله الرحمن الرحيم

48:0 In the name of God, the Gracious, the Compassionate.

48:1 We have given you a clear opening. *

48:2 So that **God** may forgive your present sins, as well as those past, and so that He may complete His blessings upon you, and guide you on a straight path.

48:3 **God** will grant you a conquest which is mighty.

48:4 He is the One who sends down tranquility into the hearts of those who affirm, so that they may increase in affirmation along with their present affirmation. To **God** belong the soldiers of the heavens and the earth, and **God** is Knowledgeable, Wise.

48:5 That He may admit the affirming men and affirming women into Gardens with rivers flowing beneath them, abiding eternally therein, and He will remit their sins from them. With **God** this is a great triumph.

48:6 He will punish the hypocrite men and the hypocrite women, and the idolater men and the idolater women, who think evil thoughts about **God**. Their evil will turn to them, and **God** was angry with them, and He has cursed them and prepared for them hell. What a miserable destiny!

48:7 To **God** belongs the soldiers of the heavens and the earth. **God** is Noble, Wise.

48:8 We have sent you as a witness, a bearer of good news, and a warner.

48:9 So that you may affirm **God** and His messenger, and that you may support Him, honor Him, and glorify Him, morning and evening.*

48:10 Those who exchanging pledge with you, are in-fact exchanging pledge with **God**; **God**'s hand is above their hands. Those of them who violate such a pledge, are violating it only upon themselves. Whosoever fulfills what he has pledged to **God**, then He will grant him a great reward.*

48:11 The Arabs who lagged behind will say to you: "We were preoccupied with our money and our family, so ask forgiveness for us." They say with their tongues what is not in their hearts. Say, "Who then would possess any power for you against **God** if He wanted harm to afflict you or if He wanted benefit for you?" No, **God** is fully Aware of everything you do.

48:12 Alas, you thought that the messenger and those who affirm would not return to their families, and this was deemed pleasant in your hearts, and you thought the worst thoughts; you were a wicked people.

48:13 Anyone who does not affirm **God** and His messenger, then We have prepared for the ingrates a hellfire.

48:14 To **God** is the sovereignty of the heavens and the earth. He forgives whomever He wills and punishes whomever He wills. **God** is Forgiver, Compassionate.

48:15 Those who stayed behind will say, when you venture out to collect the spoils: "Let us follow you!" They want to change **God**'s words. Say, "You will not follow us; this is what **God** has decreed beforehand." They will then say, "No, you are envious of us." Alas, they rarely understood anything.

48:16 Say to those Arabs who stayed behind: "You will be called on to fight a people who are very powerful in warfare, unless they peacefully surrender. Then if you obey, **God** will grant you a good reward, but if you turn away as you turned away before, He will punish you with a painful retribution."

48:17 There is no burden on the blind, nor is there any burden on the cripple, nor is there on the sick any burden. Whosoever obeys **God** and His messenger, He will admit them into paradises with rivers flowing beneath; and whosoever turns away, He will punish him with a painful retribution.

48:18 **God** is pleased with those who affirm who exchanged pledges with to you under the tree. He thus knew what was in their hearts, so He sent down tranquility upon them, and rewarded them with a near victory.*

48:19 Abundant spoils that they will take. **God** is Noble, Wise.

48:20 **God** has promised you abundant spoils that you will take. Thus He has hastened this for you, and He has withheld the people's hands against you; that it may be a sign for those who affirm, and that He may guide you to a straight path.

48:21 The other group which you could not vanquish, **God** took care of them. **God** was capable of all things.

48:22 If the ingrates had fought you, they would have turned and ran, then they would have found neither an ally nor a helper.

God's Law

48:23 Such is **God**'s *sunna* with those who have passed away before, and you will not find any change in **God**'s *sunna.*

48:24 He is the One who withheld their hands against you, and your hands against them in the interior of Mecca, after He had made you victorious over them. **God** is Seer of what you do.

48:25 They are the ones who rejected and barred you from the Regulated Temple and barred your donations from reaching their destination. There had been affirming men and women whom you did not know, and you may have hurt them, and on whose account, you would have committed a sin unknowingly. **God** will admit into His mercy whomever He wills. Had they become separated, We would then have punished those of them who rejected with a painful retribution.

Fury of Ignorance versus the Tranquility of Submission

48:26 Those who rejected had put in their hearts the rage of the days of ignorance, then **God** sent down tranquility upon His messenger and those who affirm, and directed them to uphold the word of righteousness, and they were well entitled to it and worthy of it. **God** is fully aware of all things.

48:27 **God** has fulfilled with truth His messenger's vision: "You will enter the Regulated Temple, **God** willing, secure, with your heads shaven and shortened, having no fear." Thus, He knew what you did not know, and He has coupled with this a near victory.

48:28 He is the One who sent His messenger with the guidance and the system of truth, so that it would expose all other systems. **God** is sufficient as a witness.*

48:29 Muhammed, the messenger of **God**, and those who are with him, are severe against the ingrates, but merciful between themselves. You see them kneeling and prostrating, they seek **God**'s blessings and approval. Their distinction is in their faces, as a result of prostrating. Such is their example in the Torah. Their example in the Injeel is like a plant which shoots out and becomes strong and thick and it stands straight on its trunk, pleasing to the farmers. That He may enrage the ingrates with them. **God** promises those among them who affirm and promote reforms forgiveness and a great reward.

ENDNOTES

048:001 The word F80T400H8 is mistranslated as "conquest" while it simply means "opening" or "break" or "new start." Sunni *mushriks* have abused this chapter and used it to support their aggression and bloody conquest of other people's land. Until 2020, we too unwittingly repeated that distortion in this translation. The first verse refers to the opening of the city

Mecca after the peace treaty with Meccan theocratic oligarchs. Muhammad and his friends returned to their own hometown, without shedding blood. They were unjustifiably forced to emigrate from Mecca and were attacked numerous times and lost many friends and relatives to the aggressive armies of Meccan polytheists. But soon the message of rational monotheism which led to major socio-economic and political revolutions was embraced by nearby towns, even in towns under the control of Byzantine and Persian empire. Within several years, Mecca lost its early power and allies. Thus, Mecca's doors opened to peacemakers without bloodshed. There was total amnesty. Muhammad was one of the greatest peacemakers in history. His success in social, political, cultural, scientific, religious reforms and his resolve against all sorts of attacks, plots, and betrayals, was extraordinary. Muhammad as the founder and elected leader of the federal-secular social democracy in Medina was also the military commander in wars of defense. His remarkable accomplishments in many areas led Dr. Michael Hart to pick him as the number 1 for his controversial book, "The 100: A Ranking of the Most Influential Persons in History," first published in 1978.

048:010-018 Due to the influence of Sunni religion, which I was raised with, in previous editions I mistranslated the verb "*yuBaYiUnaka*" (from B2Y10A'70) as "they pledged allegiance to you." In fact, this is a distorted translation, which is traditionally known *Bay'a*, that is allegiance. The correct translation should be "exchanging pledge with you" or "making social contract with you" since the form (*Mufa'ala*) of the verb is interactive or mutual. The Sunni distortion consider the election process as one-way allegiance to the elected person or the government. So, the government does not have obligations towards people. People and Muhammad gave promises to each other. Muhammad's promise was to be honest, truthful, just, consulting people, trustworthy, avoid nepotism and tribalism as all emphasized in numerous verses.

The verse 48:18 is also very important in terms of human rights and democracy. Almost everyone, including the followers of Sunni and Shiite religions think that women's right to vote is a modern phenomenon.

048:028 We are promised that God's system will prevail over falsehood (9:43; 41:53 and 61:9). The discovery of the mathematical system of the Quran is the start of the new era. See 27:82; 72:28; 74:1-56.

بسم الله الرحمن الرحيم

49:0 In the name of God, the Gracious, the Compassionate.

49:1 O you who affirm, do not advance before **God** and His messenger. Be aware of **God**. **God** is Hearer, Knowledgeable.*

49:2 O you who affirm, do not raise your voices above the voice of the prophet, nor shall you speak loudly at him as you would speak loudly to each other, lest your works become nullified while you do not perceive.

49:3 Surely, those who lower their voices in the presence of the messenger of **God**, they are the ones whose hearts have been tested by **God** for righteousness. They have deserved forgiveness and a great recompense.

49:4 Surely, those who call out to you from behind the private apartments, most of them do not reason.

49:5 If they had only been patient till you came out to them, it would have been better for them. **God** is Forgiver, Compassionate.

Investigate Rumors

49:6 O you who affirm, if a wicked person comes to you with any news, then you shall investigate it. Lest you harm a people out of ignorance, then you will become regretful over what you have done.

49:7 Know that among you is the messenger of **God**. If he were to obey you in many things, you would be in difficulty. But **God** made you love affirmation and He adorned it in your hearts; He made you hate denial, wickedness, and disobedience. These are the rightly guided ones.

49:8 Such is the grace from **God** and a blessing. **God** is Knowledgeable, Wise.

Be a Peacemaker; Observe Justice and Respect Diversity

49:9 If two parties of those who affirm battle with each other, you shall reconcile them; but if one of them aggresses against the other, then you shall fight the one aggressing until it complies with **God**'s command. Once it complies, then you shall reconcile the two groups with justice, and be equitable; for **God** loves those who are equitable.

49:10 Those who affirm are but brothers; so reconcile between your brothers, and be aware of **God**, that you may receive mercy.*

49:11 O you who affirm, let not a people ridicule other people, for they may be better than them. Nor shall any women ridicule other women, for they may be better than them. Nor shall you mock one another or call each other names. Evil indeed is the reversion to wickedness after attaining affirmation. Anyone who does not repent, then these are the transgressors.

49:12 O you who affirm, you shall avoid much suspicion, for some suspicion is sinful. Do not spy on one another, nor shall you gossip one another. Would one of you enjoy eating the flesh of his dead brother? You certainly would hate this. You shall observe **God**. **God** is Redeemer, Compassionate.

There is no Superiority of one Sex over another or One Nation over another; the Only Criterion for Superiority is Righteousness

49:13 O people, We created you from a male and female, and We made you into nations and tribes, that you may know one another. Surely, the most honorable among you in the sight of **God** is the most righteous. **God** is Knowledgeable, Ever-aware.*

49:14 The Arabs said, "We affirm." Say, "You have not affirmed; but you should say, 'We have peacefully surrendered', for affirmation has not yet entered into your hearts." If you obey **God** and His messenger, He will not put any of your works to waste. **God** is Forgiver, Compassionate.

49:15 Those who affirm are those who affirm **God** and His messenger, then they do not doubt; they strive with their money and their lives in the cause of **God**. These are the truthful ones.

49:16 Say, "Are you teaching **God** about your system while **God** knows everything in the heavens and the earth? **God** is knowledgeable of all things."

49:17 They think they are doing you a favor by having peacefully surrendered. Say, "Do not think you are doing me any favors by your peacefully surrender. For it is **God** who is doing you a favor that He has guided you to the affirmation, if you are honest."

49:18 Surely, **God** knows all the unseen in the heavens and the earth; **God** is Seer of everything you do.

ENDNOTES

049:001 Followers of *hadith* and *sunna* claim that God is represented by the Quran, and the messenger is represented by his opinions on the Quran. Thus, they claim that the Quran is not enough for salvation. Some people may not utter this claim straightforwardly. They may even claim that the Quran is complete and enough for our guidance. However, further questioning will reveal that their quran is not "the Quran." The Quran is the one that consists of 114 Chapters and 6346 verses. It is a mathematically coded book. However, their minds are confused, and their quran is contaminated with human speculations and limited by a snapshot interpretation. They try to scare the affirmers by saying "you do not like the messenger." The belief that God is represented by the Quran, and the messenger by his teaching is a satanic claim. There are several points to remember:

- Quran represents God and His messenger.
- Obeying the Quran is obeying God and the messenger.
- Quran never says: "Obey God and Moses," or "Obey God and Muhammed." Rather, Quran consistently states: "Obey God and the messenger." This is because the word messenger (*rasul*) comes from the "message" (*risala*). The message is entirely from God; messengers cannot exist without the message.
- Messengers, as humans, make mistakes. Thus, when affirmers made a covenant with prophet Muhammed, they promised to obey him conditionally, i.e., his righteous orders (60:12). Moreover, God specifically orders Muhammed to consult the affirmers around him (3:159). If nobody can object to the opinion of the messenger, then consultation is meaningless. However, whenever the final decision after consultation is made, it should be followed.
- During their lifetimes, messengers are community leaders. In this regard, messengers are not different than the affirmers who are in charge (4:59); both should be obeyed. But this obedience is not absolute. It is open to consultation and discussion.
- The position of messengers is different during their lives; they are interactive teachers and curious students as well. We have the chance to ask them further questions, discuss issues, learn their intentions, and even correct their mistakes. On the other hand, they have the opportunity to correct our misunderstandings. However, when they pass away, their teaching becomes frozen and loses its advancing three-dimensional character. The frozen, snap-shot fragments of knowledge are a dangerous weapon in the hands of ignorant people; they use it to stop God's perpetual teaching. They defend every plain error in the name of the messenger. They insult every sincere student of God's revelation.

We reject *Hadith* because we respect Muhammed. No sound person would like to have people born several centuries after him roam the earth and collect hearsay attributed to him. Besides, if Muhammed and his supporters really believed that the Quran was not sufficient for guidance, or an ambiguous book, or lacked details, then, surely, they would have been the first ones to write the *hadith* down and collect them in books. After all, their numbers were in the hundreds of thousands, and they had plenty of wealth. They could afford some ink, papyrus paper or leather, and some brain cells, for such an important task. They would not leave it to a guy from faraway Bukhara or his ilk to collect hearsay reports called *hadiths* more than two hundred years later in a land soaked with blood because of sectarian wars. Additionally, Muhammed had many unemployed and handicapped people around who would gladly volunteer for such a mission. To explain away this 200-year gap in history, the traditionalists created an excuse for prophet Muhammed and his supporters. Supposedly, Muhammed and his followers feared that people would mix the Quran with *hadith*s. This is nonsense. They were smart enough to distinguish both, and there were enough people to keep track of them.

Besides, it is the followers of *hadith* and *sunna* themselves who claim that the Quran was a "literary miracle." If their claim of "literary miracle" were true, then it would be much easier to separate the verses of the Quran from *hadith*. Let's assume that they could not really distinguish the text of the Quran from Muhammed's words, then couldn't they simply mark the pages of the scripture with the letter Q for the Quran and letter H for *Hadith*, or let some record only the Quran, or simply color code their covers? Or allocate leather for the Quran and paper for *hadith*, or vice versa? They could find many ways to keep different books separate from each other. They did not need to study rocket science or have computer technology to accomplish that primitive task. The collectors of *hadiths* wished that people would accept their assertion that Muhammed and his supporters did not have ink, paper or leather, minds, and willingness to collect *hadith* before them. It is no wonder they even fabricated a few *hadiths* claiming that Muhammed's companions were competing with dogs for bones to write the verses of the Quran on!

Most likely, Muhammed feared that people would mix his words with the Quran. Not in the primitive way that is claimed by the Sunnis and Shiites, since as we pointed out, there were many ways to eliminate that concern. But the real concern of Muhammed was different. Because of the warnings of the Quran, he feared that muslims would follow the footsteps of Jews and would create their own Mishna, Gomorra, and Talmud: *hadith* would be an authority, as another source besides the Quran, setting him up as a partner with God!

Ironically, the followers of *hadith* and *sunna* accomplished exactly that. They did not need to publish the text of *hadith* together with the Quran--though they have done that in many commentaries—since they have been doing worse. Though they usually have kept *hadith* separate from the Quran physically, when it comes to purposes of guidance and religious authority, they mix them with the verses of the Quran with passion. Even worse, they consider the understanding of the Quran to be dependent on the understanding of *hadith*, thereby elevating *hadith* to a position of authority over the Quran.

Thus, if indeed Muhammed was worried about people mixing his words with the Quran, the followers of *hadith* proved his worries right: centuries after him, not only did they mix his words with the Quran, but also mixed thousands of fabrications and nonsense attributed to him. See 25:30; 59:7.

049:010 Some readers commit the logical fallacy called Accent in understanding of this verse, "Those who affirm are but brethren…" They read the beginning of the Arabic statement, *innama al-muminuna ikhwah,* with an emphasis on the first two words rather than on the last word, leading to a mistranslation, "Only affirmers are brethren," thereby creating contradiction with the verses of the Quran depicting all humanity as brethren from Adam.

049:013 This verse unequivocally rejects sexism and racism and reminds us that neither male nor female, neither this race nor that race, is superior to the other.

Recognition and appreciation of diversity is an important point in our test. The only measure of superiority is righteousness; being a humble, moral, and socially conscientious person who strives to help others. See 4:34; 7:19; 24:61; 30:21-22; 33:5; 48:17; 60:12.

بسم الله الرحمن الرحيم

50:0 In the name of God, the Gracious, the Compassionate.
50:1 Q100, and the glorious Quran.*
50:2 Are they surprised that a warner has come to them from amongst them, so the ingrates said, "This is something strange!"
50:3 "When we are dead and we become dust... This is a far return!"
50:4 We know which of them has become consumed by the earth; and We have with Us a record which keeps track.
50:5 But they denied the truth when it came to them, so they are in a confused state.
50:6 Do they not look at the sky above them, how We built it, and adorned it, and how it has no gaps?
50:7 The land We extended it, and placed in it stabilizers, and We gave growth in it to every kind of healthy pair.
50:8 Something to see and a reminder for every pious servant.

Example of Resurrection

50:9 We sent down from the sky blessed water, and We produced with it gardens and grain to be harvested.
50:10 The palm trees, emerging with clustering fruit.
50:11 A provision for the servants, and We gave life with it to the land which was dead. Such is the resurrection.
50:12 Before their denial was that of the people of Noah, and the dwellers of Al-Raas, and Thamud.
50:13 And Aad, and Pharaoh, and the brethren of Lot.
50:14 The people of the forest, and the people of Tubba. All of them rejected the messengers, and thus the promise came to pass.
50:15 Did We have any difficulty in making the first creation? No, they are in confusion about the next creation.
50:16 We have created the human being and We know what his person

whispers to him, and We are closer to him than his jugular vein.

50:17 When the two receivers meet on the right and on the left.

50:18 He does not utter a word except with having a constant recorder.

50:19 The moment of death came in truth: "This is what you have been trying to avoid!"

The Promised Day

50:20 The horn is blown on the promised day.

50:21 Every person is brought, being driven, and with a witness.

50:22 "You were heedless of this, so now We have removed your veil, and your sight today is iron/sharp!"

50:23 His constant companion said, "Here is one who is a transgressor."

50:24 Cast in hell every stubborn ingrate.

50:25 Denier of good, transgressor, doubter.

50:26 The one who sets up another god beside **God**. So cast him into the severe retribution.

50:27 His constant companion said, "Our Lord, I did not corrupt him, but he was already far astray."

50:28 He said, "Do not argue with each other before Me, I have already presented you with My promise."

50:29 "The sentence will not be changed with Me, and I do not wrong the servants."

50:30 The day We say to hell: "Are you full?", and it says: "Are there anymore?"

50:31 Paradise is brought near to the righteous, not far off.

50:32 This is what you have been promised, for every obedient, steadfast

50:33 The one who feared the Gracious while unseen, and came with a repenting heart

50:34 Enter it in peace. This is the day of eternal life.

50:35 In it they will have what they wish, and We have even more.

50:36 How many a generation before them have We destroyed? They were stronger in power, and they had dominated the land. Did they find any sanctuary?

50:37 In this is a reminder for whoever has a heart, or cares to listen while he is heedful.

50:38 We have created the heavens, the earth and what is between them in six days, and no fatigue touched Us.*

50:39 So be patient to what they are saying; glorify the grace of your Lord before the rising of the Sun and before the setting.

50:40 From the night glorify Him, and after prostrating.

50:41 Listen to the day when the caller will call from a near place.

50:42 The day they hear the scream with truth. That is the day of coming out.

50:43 We are the Ones who give life and bring death, and to Us is the destiny.

50:44 The day when the earth will rapidly crumble away from them; that will be a gathering which is easy for Us.

50:45 We are totally aware of what they say, and you are not to be a tyrant over them. So remind with the Quran those who fear My promise.

ENDNOTES

050:001 Q100. This letter/number plays an important role in the mathematical system of the Quran based on code 19. For the meaning of this letter, see 74:1-56; 1:1; 2:1; 13:38; 27:82; 38:1-8; 40:28-38; 46:10; 72:28.

050:038 This verse refers to the distortion in the Old Testament, Genesis 2:2, and Exodus 31:17. However, in Deuteronomy 5:6-18, God tells the Israelites to remember the Sabbath because He liberated them from slavery. In Deuteronomy, there is no mention of God "resting," and thus it is in harmony with the Quran.

For the evolutionary creation of the universe, see 41:9-10.

بســـــــم الله الرحمن الرحيـــــــم

51:0 In the name of God, the Gracious, the Compassionate.
51:1 By the atoms that scatter.*
51:2 Carrying a charge.
51:3 Flowing smoothly.
51:4 Distributing a command.
51:5 What you are being promised is true.
51:6 The Judgment will come to pass.
51:7 By the sky with orbits.
51:8 You are in verbal discord.
51:9 Misled by it is the deviant.
51:10 Woe to the deceitful.
51:11 Who are in mischief, unaware.
51:12 They ask: "When is the day of Judgment?"
51:13 The day they are tested upon the fire.
51:14 "Taste this trial of yours; this is what you asked to be hastened."
51:15 The righteous are in paradises and springs.
51:16 Receiving what their Lord has bestowed to them, for they were before that pious.
51:17 They used to rarely sleep the whole night.
51:18 Before dawn, they would seek forgiveness.
51:19 And in their money was a portion for the needy and the deprived.
51:20 In the earth are signs for those who reason.
51:21 Within yourselves; do you not see?
51:22 In the heaven is your provision, and what you are promised.
51:23 By the Lord of the heaven and the earth, this is truth just as the fact that you speak.

Abraham's Guests

51:24 Has the story of Abraham's noble guests come to you?
51:25 When they entered upon him, they said, "Peace." He said, "Peace to a people unknown!"
51:26 Then he went to his family and brought a fat roasted calf.
51:27 He offered it to them, he said, "Do you not eat?"
51:28 He then became fearful of them. They said, "Do not fear," and they gave him good news of a knowledgeable son.
51:29 His wife then approached in amazement. She slapped upon her face, and said, "I am a sterile old woman!"
51:30 They said, "It was such that your Lord has said. He is the Wise, the Knowledgeable."
51:31 He said, "What is your undertaking, O messengers?"
51:32 They said, "We have been sent to a criminal people."
51:33 "To send down upon them rocks of clay."
51:34 "Prepared by your Lord for the transgressors."
51:35 We then vacated from it all those who affirmed.
51:36 But We only found in it one house of those who had peacefully surrendered.
51:37 We have left in it a sign for those who fear the painful retribution.

Previous Generations

51:38 Also Moses, for We sent him to Pharaoh with a clear authority.
51:39 But he turned away, in arrogance, and he said, "A magician, or crazy."
51:40 So We took him and his troops; We cast them into the sea, and he was to blame.
51:41 Also Aad, for We sent upon them the hurricane wind.
51:42 Anything that it came upon was utterly destroyed.
51:43 Also Thamud, for it was said to them: "Enjoy for a while."
51:44 But they rebelled against the command of their Lord. So the lightning struck them while they were looking.
51:45 They were unable to rise up, nor could they win.
51:46 The people of Noah before; they were a wicked people.

Expanding Universe

51:47 We constructed the universe with might, and We are expanding it.*
51:48 The earth We furnished; We are fine Providers.

51:49 From everything We created a pair, perhaps you may remember.
51:50 So escape to **God**. I am to you from Him a clear warner.
51:51 Do not make any other god with **God**. I am to you from Him a clear warner.
51:52 Likewise, when a messenger went to those before them, they said, "A magician, or crazy."
51:53 Have they passed down this saying to each other? Indeed, they are a wicked people.
51:54 So turn away from them; you will not be blamed.*
51:55 Remind, for the reminder benefits those who affirm.
51:56 I did not create the Jinn and the humans except to serve Me.*
51:57 I need no provisions from them, nor do I need them to give food.
51:58 **God** is the Provider, the One with Power, the Supreme.
51:59 The transgressors will have the same fate as their previous friends; so let them not be hasty.
51:60 So woe to those who rejected from the day that is waiting for them.

ENDNOTES

051:001-4 These verses are traditionally translated as a description of wind, rain and sailing ships. However, the audience of the Quran transcends time. These words can also be understood as a description of the movement of electrons and their ability to carry information.

051:047 God who created the universe billions of years ago by His command of "Be" (21:30), is continuously expanding the universe in time and space. Remarkably, this 1400-year-old proclamation has just recently in the 20th century been shown by modern science to be a fact. See 4:82.

051:054 Jonah was criticized for abandoning his mission before completing it (37:142). For the Old Testament's account, see the four short chapters of the book of Jonah.

051:056 See 2:35

بسم الله الرحمن الرحيم

52:0 In the name of God, the Gracious, the Compassionate.
52:1 By the mount,
52:2 A recorded book.
52:3 In paper published.
52:4 The frequented/built sanctuary.
52:5 The ceiling which is elevated.
52:6 The sea that is set aflame.
52:7 Your Lord's retribution is unavoidable.
52:8 Nothing can stop it.
52:9 The day when the sky will violently thunder.
52:10 The mountains will be wiped out.
52:11 Woes on that day to the deniers.
52:12 Who are in their recklessness, playing.
52:13 The day they will be called into hell, forcibly:
52:14 "This is the fire which you used to deny!"
52:15 "Is this magic, or do you not see?"
52:16 "Enter it, whether you are patient or impatient, it will be the same for you. You are only being requited for what you used to do."
52:17 The righteous are in paradises and bliss.
52:18 Delighted by what their Lord has granted them; and their Lord has spared them the retribution of hell.
52:19 Eat and drink happily because of what you used to do.
52:20 They recline on arranged furnishings, and We coupled them with wonderful companions.*
52:21 Those who affirmed, and their progeny also followed them in affirmation; We will have their progeny join them. We never fail to reward them for any work. Every person is paid for what he did.
52:22 We will supply them with fruit and meat such as they desire.
52:23 They will exchange one with another a cup in which there is no harm, nor regret.
52:24 Around them will be children like protected pearls.
52:25 They came to one another asking.

52:26 They said, "We used to be compassionate among our people."
52:27 "**God** thus blessed us, and has spared us the agony of the fiery winds."
52:28 "We used to implore Him before; He is the Kind, the Compassionate."
52:29 Therefore, you shall remind. For by the grace of your Lord, you are neither a soothsayer, nor mad.
52:30 Or do they say, "He is a poet; so let us just wait until a disaster befalls him."
52:31 Say, "Continue waiting; for I will wait along with you."
52:32 Or do their dreams dictate this to them, or are they a transgressing people?
52:33 Or do they say, "He made it all up"? No, they simply do not affirm.

The Quran, Unlike Human Words

52:34 Let them produce a *hadith* like this, if they are truthful.*
52:35 Or were they created from nothing? Or was it they who created?
52:36 Or did they create the heavens and the earth? No, they do not reason.
52:37 Or do they possess the treasures of your Lord? Are they in control?
52:38 Or do they have a stairway that enables them to listen? Then let their Hearers show their proof.
52:39 Or to Him belong the daughters, and to you belong the sons?
52:40 Or do you ask them for a wage, thus they are burdened by the fine?
52:41 Or do they know the future, thus they have it recorded?
52:42 Or do they intend to scheme? Indeed, it is the ingrates who are schemed against.
52:43 Or do they have another god besides **God**? **God** be glorified above what they set up.
52:44 If they see an object falling from the sky, they will say, "Piled clouds!"
52:45 So disregard them until they meet the day in which they are struck.
52:46 The day when their schemes will not protect them, nor will they be helped.
52:47 For those who transgressed will be retribution beyond this, but most of them do not know.
52:48 You shall be patient for the judgment of your Lord, for you are in Our sights, and glorify the praise of your Lord when you get up.
52:49 Also during the night glorify Him, and at the setting of the stars.

ENDNOTES

052:020 See 44:54.

052:034 Abu Dawud, a *hadith* book set up as a partner with the Quran by Sunnis, reports a *hadith* alleging Muhammed saying, "I am given the Quran and similar to it." This *hadith* is obviously a polytheist response to the Quran's challenge. See 6:112-116; 12:111; 33:38; 35:43; 45:6.

بســــم الله الرحمن الرحيــــم

53:0 In the name of God, the Gracious, the Compassionate.
53:1 As the star collapsed.*
53:2 Your friend was not astray, nor was he deceived.*
53:3 Nor does he speak from personal desire.*
53:4 It is a divine inspiration.
53:5 He has been taught by One mighty in power.
53:6 Possessor of ultimate wisdom; he became stable.
53:7 While he was at the highest horizon.
53:8 Then he drew nearer by moving down.
53:9 Until he became as near as two bow-lengths or nearer.
53:10 He then conveyed the inspiration to His servant what was to be revealed.
53:11 The heart did not invent what it saw.
53:12 Do you doubt him in what he saw?
53:13 Indeed, he saw him in another descent.
53:14 At the ultimate point.
53:15 Next to the eternal Paradise.
53:16 The whole place was overwhelmed.
53:17 The eyes did not waver, nor go blind.
53:18 He has seen from the great signs of his Lord.

The Meccan Idols Were False Attributes

53:19 Have you considered Allaat and Aluzzah?*
53:20 And Manaat, the third one?
53:21 Do you have the males, while He has the females?
53:22 What a strange distribution!
53:23 These are but names/attributes that you made up, you and your forefathers. **God** never authorized such. They only follow conjecture, and wishful thinking, while the guidance has come to them from their Lord.
53:24 Or shall the human being have what he wishes?
53:25 To **God** belongs the Hereafter, and the world.
53:26 There are many controllers in heaven, who have no power to intercede, except after **God** gives permission for whom He wishes and is satisfied with him.*

The Irony in the Religious Doctrine of Misogynists

53:27 Those who reject the Hereafter have given the controllers feminine names.
53:28 While they had no knowledge about this; they only followed conjecture. Conjecture is no substitute for the truth.
53:29 So disregard he who turns away from Our reminder, and only desires this worldly life.
53:30 This is the extent of their knowledge. Your Lord is fully aware of those who strayed away from His path, and He is fully aware of those who are guided.
53:31 To **God** belongs everything in the heavens and everything on earth. He will requite those who commit evil for their works, and will reward the righteous for their righteousness.
53:32 They avoid major sins and lewdness, except for minor offences. Your Lord is with vast forgiveness. He has been fully aware of you since He initiated you from the earth, and while you were fetuses in your mothers' wombs. Therefore, do not acclaim yourselves; He is fully aware of the righteous.
53:33 Have you noted the one who turned away?
53:34 He gave very little, and then he stopped.
53:35 Did he possess knowledge of the future? Could he see it?
53:36 Or was he not informed of the teachings in the book of Moses?
53:37 Of Abraham who fulfilled?
53:38 None can carry the burdens of another.
53:39 The human being will have what he sought.
53:40 His works will be shown.

53:41 Then he will be paid fully for such works.

53:42 To your Lord is the final destiny.

God: The Ultimate Cause

53:43 He is the One who makes laughter and tears.*

53:44 He is the One who takes life and gives it.

53:45 He is the One who created the pair, male and female.*

53:46 From a seed that is put forth.

53:47 He will affect the recreation.

53:48 He is the One who makes you rich or poor.

53:49 He is the Lord of Sirius.

53:50 He is the One who destroyed Aad the first.

53:51 Of Thamud He left nothing.

53:52 The people of Noah before that; they were evil transgressors.

53:53 The evil communities were the lowliest.

53:54 Consequently, they utterly vanished.

53:55 So which of your Lord's marvels can you deny?

53:56 This is a warning like the former ones.

53:57 The final judgment draws near.

53:58 None beside **God** can relieve it.

53:59 Are you surprised by this *hadith*?

53:60 You are laughing, instead of crying?

53:61 You are insisting on your ways?

53:62 You shall prostrate to **God** and serve.

ENDNOTES

053:001-18 The extraordinary event told in these verses, as it appears, is a miraculous experience. Is the falling of stars an apparent perception because of traveling among galaxies faster than speed of light? What were the signs Muhammed witnessed? Did Muhammed meet God or Gabriel? 20:114.

053:002 Centuries after Muhammed, he and his companions became idols. Muslims started considering the prophet's companions (*sahaba*) or anyone who met him as a Muslim, to be almost infallible. The word *Saheb*, *Sahaba*, and its plural *Ashab* are usually used in a negative context. For instance, out of the 77 occurrences of *Ashab*, and one occurrence of "*Ashabahum*" (their comrades), only 27 are used positively, such as, "*Ashab ul-Jannah*" (People of the Paradise) or "*Ashab ul-Yameen*" (People of the Right). Excluding the few neutral usages of the word, the word "*Ashab*" is usually used to denote ingrates and hypocrites. None of these *Ashab*, the PLURAL of *Sahaba*, refers to peacemakers who lived during time of prophet Muhammed. In only one case, the plural *Ashab* refers to people with Moses (26:61), and we learn from the Quran that most of them were not true affirmers or peacemakers (7:138-178; 20:83-87).

The Quran informs us about the quality of Muhammed's comrades in numerous verses. What we see is not a depiction of perfect holy people, but ordinary people, with weaknesses and shortcomings.

Among the 12 occurrences of the singular and dual form, *sahib*, only five describe a relationship between a prophet and his friends. And, out of these five occurrences, only one of them has a positive connotation. Before quoting the verses, I want to remind you that the word *sahib* (companion, friend) is about a mutual relationship; if someone is your companion, you are their companion too, and vice versa. In four occurrences, the addressees are ingrates or polytheists (34:46, 53:2, 81:22; 12:39). The only positive usage of the word *sahib*, as a companion of a prophet is in 9:40.

In sum, according to the Quranic usage, the words *sahib* (companion, friend), or *ashab*, (companions, friends) by themselves do not denote any positive meaning. In three verses, Muhammed is described as the *sahib* (companion) of ingrates, and in only one verse, he was the *sahib* of an affirmer. Thus, statistically speaking, the word *ashab* refers to hypocrites and polytheists 75% of the time.

According to the books of *hadith*, Abdullah Ibn Masood was one of the top companions of the prophet Muhammed. His *hadith* narrations are among Sunni Muslims' most cherished sources of jurisprudence. Many *hadith* and narration books, including Bukhary and Ibn Hanbal, report that Ibn Masood had a personal copy of the Quran and that he did not put the last two chapters in it. According to those books, he claimed that those two chapters did not belong in the Quran.

Apparently, another companion of the prophet, Ubayy Ibn Kaab, also had a different personal Quran. He added two chapters called "Sura Al-Hafd" and "Sura Al-Khal," and claimed that these were from the Quran (These "chapters" are still being recited by Hanafis in the "*sala al-witr,*" after night prayers.)

For the prophecy of the Quran regarding the words *hadith*, *sunna*, *ijma*, *sharia*, and *salaf*, see 33:38.

053:003-04 Meccan idol worshipers claimed that Muhammed was the author of the Quran (25:5; 68:15). The beginning of Chapter 53 is about the revelation of the Quran. It states that "the Quran is from Him." It is not Muhammed's personal claim; it is a divine statement. Therefore, claiming that the pronoun "it" in the verse 53:4 refers to the words of Muhammed, not of God, is an obvious distortion. According to the above verse, "it" is revelation, without exception. This can be valid only for the Quran. It is nonsense to claim that Muhammed's daily conversation was entirely revelation. For example, God firmly criticizes Muhammed's words to Zayd (33:37). Obviously, the criticism was not about revelation. The beginning phrase of Chapter 97 informs us about the revelation of the Quran: "We revealed it in the Night of Destiny." The "it" in this verse is the same as the "it" in 53:4. . .

053:019-23 Meccan polytheists had an abstract concept of idols. Embracing Abraham's legendary rejection of statues, they discovered new ways of associating partners with God. Meccan polytheists who considered themselves to be the followers of Abraham were commemorating the names of their dead saints for intercession (39:3) and still considered themselves to be monotheists (6:23; 16:35). Those who idolized Muhammed, his companions and numerous "saints" (*awliya*) and commemorate their names for intercession, wanted to cover the similarities between their religion and the religion of Meccan polytheists. Thus, they ventured into fabricating tangible differences; they therefore made up many statues and ascribed them to the Meccan polytheists. Expectedly, they narrated contradictory accounts of their physical shapes (See *Al-Kalbi's Kitab ul-Asnam*). According to the Quran, Meccan *mushriks* (polytheists) were praying in the Sacred Masjid, organizing Debate Conference, and were also fasting (8:35; 9:19,45,54; 2:199).

053:026 See 2:48.

053:032 See 6:100

053:043 For similar remarks, see the Old Testament, 1 Samuel 2:6-7.

053:045-46 The sex of the baby is determined by the sperm in the male fluid. See 4:82.

بسم الله الرحمن الرحيم

54:0 In the name of God, the Gracious, the Compassionate.
54:1 The moment drew near, and the Moon was split.*
54:2 If they see a sign, they turn away and say, "Continuous magic!"
54:3 They rejected, and followed their desires, and every old tradition.*
54:4 While the news had come to them in which there was sufficient warning.
54:5 A perfect wisdom; but the warnings are of no benefit.
54:6 So turn away from them; the day will come when the caller announces a terrible disaster.

Like Cicadas: in Big Numbers, Simultaneously and After a Long Period

54:7 With their eyes humiliated, they come out of the graves like scattered cicadas.*
54:8 Hastening towards the caller, the ingrates will say, "This is a difficult day."
54:9 The people of Noah rejected before them. They rejected Our servant and said, "He is crazy!" and he was oppressed.
54:10 So he called on his Lord: "I am beaten, so grant me victory."
54:11 So We opened the gates of the sky with pouring water.
54:12 We caused springs to gush out of the earth. Thus the waters met to a command which had been measured.
54:13 We carried him on a craft made of slabs and mortar.*
54:14 It ran under Our watchful eyes; a reward for one who was rejected.
54:15 We have left it as a sign. Do any of you wish to learn?
54:16 So how was My retribution after the warnings!

The Quran is Easy to Learn for the Appreciative

54:17 We made the Quran easy to learn. Do any of you wish to learn?*
54:18 Aad rejected. So how was My retribution after the warnings!
54:19 We sent upon them a violent wind, on a day of continuous misery.
54:20 It snatched people away as if they were palm tree trunks; uprooted.
54:21 So how was My retribution after the warnings!
54:22 We made the Quran easy to learn. Do any of you wish to learn?
54:23 Thamud rejected the warnings.
54:24 They said, "Shall we follow one of us; a human being? We will then go astray and be in hell."
54:25 "Has the message come down to him, instead of us? He is an evil liar."
54:26 They will find out tomorrow who the evil liar is.
54:27 We are sending the camel as a test for them. So observe them and be patient.
54:28 Inform them that the water shall be divided between them; each shall be allowed to drink in the specified time.
54:29 But they called on their friend, and he was paid to slaughter.
54:30 So how was My retribution after the warnings!
54:31 We sent upon them one scream, whereupon they became like harvested hay.
54:32 We made the Quran easy to learn. Do any of you wish to learn?
54:33 The people of Lot rejected the warners.
54:34 We sent upon them projectiles. Except for the family of Lot, We saved them at dawn.
54:35 A blessing from Us; it is thus that We reward those who are appreciative.
54:36 He warned them about Our punishment, but they ridiculed the warnings.
54:37 They sought to remove him from his guests; so We diverted their eyes. "Taste My retribution; you have been warned."
54:38 In the early morning, a devastating retribution struck them.
54:39 "Taste My retribution; you have been warned."

54:40 We made the Quran easy to learn. Do any of you wish to learn?

54:41 The warnings had come to the people of Pharaoh.

54:42 They rejected all Our signs. So We took them, the taking of the Noble, the Able.

54:43 Are your ingrates better than those? Or have you been absolved by the book?

54:44 Or do they say, "We are a large group, and we will win."

54:45 The large group will be defeated; they will turn around and flee.

54:46 The moment is their appointed time, and the Moment is far worse and more painful.

54:47 The criminals are in error and will burn.

The Agony of Rejecting a Manifest Sign

54:48 The day when they will be dragged upon their faces into the fire: "Taste the agony of *Saqar*!"*

54:49 Everything We have created in measure.

54:50 Our commands are done at once with the blink of an eye.

54:51 We have destroyed your counterparts. Do any of you wish to learn?

54:52 Everything they did is in the records.

54:53 Everything, small or large, is written down.

54:54 The righteous will be in paradises and rivers.

54:55 In a position of honor, at an Omnipotent King.

ENDNOTES

054:001 *Hadith* narrators and collectors misunderstood this verse and tried to support their misunderstanding by the fabrication of a *hadith* reporting that Muhammed split the Moon by pointing at it with his finger. Some narrators did not stop there and claimed that half of the Moon fell in the backyard of Ali, Muhammed's son-in-law. The affirmers of this *hadith* are silent to the critical questions about the event. Why didn't populations of other countries witness this incredible event? If there were such an astronomical event, many people from China to Africa would have noticed and recorded it. Besides, this traditional interpretation contradicts the Quran, which specifically limits the signs given to Muhammed to be the signs contained in the Quran alone (29:50-51). The splitting of the Moon was fulfilled when Apollo 11 landed on the Moon and astronauts took some rocks from the Moon back to earth in July 1969. The word *shaqqa*, which we translated as "split," does not necessarily mean splitting in half (see 80:26; 50:44). When this event was happening, Dr. Rashad Khalifa was uploading the Quran to his computer in St. Louis. Splitting the surface of the Moon was the start of his work, which led to the historical discovery of the miraculous mathematical system based on code 19, prophesied in chapter 74, which happened in 1974. Thus, the following verse relates to this discovery. See 74:1-56; 27:82; 38:1-8; 40:28-38; 46:10; 72:28.

054:003 The verse also may be read differently and understood as, "They denied and followed their wishful thinking. Indeed, every affair will find its right place."

054:007 The allegory of locusts for resurrection is interesting, since some locusts lay their eggs in the forests and appear years later simultaneously. For instance, in the North-eastern region of the US, Periodical Cicadas emerge in big populations after years of dormancy underground sustaining themselves on tree roots. For some broods, the period is 13 years and for others, 17. After years of disappearance, they emerge all together with a cacophony of songs, testifying a similitude of resurrection for those who appreciate God's signs in nature. I discussed this and many other scientific signs and prophecies of the Quran, in my first book, "*Kuran En Büyük Mucize*" (*Quran, the Greatest Miracle*), co-authored with Ahmad Deedat, a very popular book in Turkey in 1983-1987. "In sum, locusts resurface from under the soil, after a long period, all together, in big crowds. Similarly, humans too will resurface from under the soil, after a long period, all together, in big crowds. Of course, in similes the comparandum and comparatum will have some differences. For instance, locusts are dormant under the soil, while human body disintegrates." (Kuran En Büyük Mucize, Ahmet Deedat and Edip Yuksel, Inkılab, Istanbul, 12th Edition, 1986, p.146). Though this book is now out of print and was not translated to English, a recently published book by the Quran Studies Group led by my friend Caner Taslaman, PhD, presents the scientific aspect of the Quran under the title, Kuran Hiç Tükenmeyen Mucize, which is translated into English under the title, "The Quran: Unchallengeable Miracle" See: 4:82.

054:013 The flood was limited to the region of Dead Sea, with Noah's community, and the ark was a small watercraft carrying Noah's animals from his farm. Storytellers exaggerated the event to a worldwide flood. Many Christians who believe those stories have been looking for the remnants of Noah's ark on top of mountains, wrong mountains, in vain. See 11:44.

054:017-40 Despite the repetitious assertion of the Quran that it is easy to understand, those who traded it with the teachings of *Hadith* and *Sunna* claim otherwise, which turns it into a self-fulfilling prophecy (6:25; 17:46; 18:57). Also, see 25:30.

054:043 See 6:110.

054:048 For the Quranic meaning and prophetic implication of the word *saqar*, see 74:26-56.

بسم الله الرحمن الرحيم

55:0 In the name of God, the Gracious, the Compassionate.
55:1 The Gracious:
55:2 Has taught the Quran.
55:3 He created the human being,
55:4 Taught him how to distinguish.
55:5 The Sun and the Moon are perfectly calculated.*
55:6 The stars and the trees submit.
55:7 He raised the sky and He established the balance.
55:8 Do not transgress in the balance.
55:9 Observe the weight with equity, and do not fall short in the balance.

Lord's Innumerable Favors

55:10 The earth He has made for all creatures.
55:11 In it are fruits, and date palms with their hanging fruit.*
55:12 Grains and the spices.
55:13 So which of your Lord's favors will you two deny?
55:14 He created the human from clay, like pottery.
55:15 The Jinn He created from a smokeless fire.
55:16 So which of your Lord's favors will you two deny?
55:17 The Lord of the two easts and the two wests.
55:18 So which of your Lord's favors will you two deny?
55:19 He merges the two seas where they meet.
55:20 A barrier is placed between them, which they do not cross.*
55:21 So which of your Lord's favors will you two deny?
55:22 Out of both of them come pearls and coral.
55:23 So which of your Lord's favors will you two deny?
55:24 To Him are the ships that roam the sea like flags.
55:25 So which of your Lord's favors will you two deny?
55:26 Everyone upon it will fade away.
55:27 The presence of your Lord will remain, the One with Majesty and Honor.*

55:28 So which of your Lord's favors will you two deny?
55:29 Those in the heavens and earth ask Him, everyday He is in some matter.
55:30 So which of your Lord's favors will you two deny?
55:31 We will call you to account, both of you beings.
55:32 So which of your Lord's favors will you two deny?
55:33 O tribes of Jinn and humans, if you can penetrate the diameters of the heavens and the earth, then go ahead and penetrate. You will not penetrate without authority.
55:34 So which of your Lord's favors will you two deny?
55:35 He sends against both of you projectiles of fire and copper; so you will not succeed.
55:36 So which of your Lord's favors will you two deny?
55:37 Then, when the universe is torn, and turns like a rose-colored paint.
55:38 So which of your Lord's favors will you two deny?
55:39 On that day, no more questioning is asked of human or Jinn as to his sins.
55:40 So which of your Lord's favors will you two deny?
55:41 The guilty will be recognized by their features; they will be taken by the forelocks and the feet.
55:42 So which of your Lord's favors will you two deny?
55:43 This is hell that the criminals used to deny.
55:44 They will move between it and a boiling liquid.
55:45 So which of your Lord's favors will you two deny?
55:46 For he who revered the majesty of his Lord, will be two paradises.
55:47 So which of your Lord's favors will you two deny?
55:48 Full of provisions.
55:49 So which of your Lord's favors will you two deny?
55:50 In them are two springs, flowing.
55:51 So which of your Lord's favors will you two deny?
55:52 In them will be of every fruit, in pairs.
55:53 So which of your Lord's favors will you two deny?
55:54 While reclining upon furnishings lined with satin, the fruits of the two paradises are within reach.
55:55 So which of your Lord's favors will you two deny?
55:56 In them are the best of the fruits hanging low, untouched before by any human or Jinn.
55:57 So which of your Lord's favors will you two deny?
55:58 They look like rubies and coral.
55:59 So which of your Lord's favors will you two deny?
55:60 Is there any reward for goodness except goodness?
55:61 So which of your Lord's favors will you two deny?
55:62 Besides these will be two other paradises.
55:63 So which of your Lord's favors will you two deny?
55:64 Dark green in color.
55:65 So which of your Lord's favors will you two deny?
55:66 In them, two springs which gush forth.
55:67 So which of your Lord's favors will you two deny?
55:68 In them are fruits, date palms, and pomegranate.
55:69 So which of your Lord's favors will you two deny?
55:70 In them is what is good and beautiful.
55:71 So which of your Lord's favors will you two deny?
55:72 Companions, inside grand pavilions.*
55:73 So which of your Lord's favors will you two deny?
55:74 No human has ever touched them, nor Jinn.
55:75 So which of your Lord's favors will you two deny?
55:76 Reclining on green carpets, in beautiful surroundings.
55:77 So which of your Lord's favors will you two deny?
55:78 Glorified be the name of your Lord, Possessor of Majesty and Honor.

ENDNOTES

055:005 The motion of the Sun and Moon follows a very accurate pattern. We can predict their precise locations years ahead of time. Instead of following the *hadith* that recommends eyewitness testimony of the Moon, if people follow God's signs in nature, they will not argue every year about the timing of Ramadan and other holy-days. The eyewitness testimony regarding the new Moon has been proven many times to be false. Ironically, the same people who consider astronomic computation to be a Satanic innovation, rely on astronomic computation regarding the times of *Sala* prayers. Their minds are so distorted with contradictory teachings that they live in a universe of falsehood and contradiction.

055:011 Planet earth is a big spaceship, and we are astronauts beamed here through conception.

055:020 See 25:53.

055:027 See 55:78.

055:072 See 44:54.

بسم الله الرحمن الرحيم

56:0 In the name of God, the Gracious, the Compassionate.
56:1 When the inevitable comes to pass.
56:2 Nothing can stop it from happening.
56:3 Abasing, exalting.
56:4 When the earth will be shaken a terrible shake.
56:5 The mountains will be wiped out.
56:6 So that they become only scattered dust.
56:7 You will be in three groupings.
56:8 So those on the right; who will be from those on the right?
56:9 Those on the left; who will be from those on the left?

The Higher Heaven

56:10 Those foremost; who will be from those foremost?
56:11 They will be the ones who are brought near.
56:12 In the paradises of bliss.
56:13 Many from the first generations.*
56:14 A few from the later generations.
56:15 Upon luxurious furnishings.
56:16 Residing in them, neighboring each other.
56:17 Moving amongst them will be immortal children.
56:18 With cups and pitchers, drinks which are pure.
56:19 They do not run out, nor do they get bored.
56:20 Fruits of their choice.
56:21 Meat of birds that they desire.*
56:22 Wonderful companions,
56:23 Like pearls which are sheltered.
56:24 A reward for their works.
56:25 They never hear any nonsense therein, nor sinful utterances.
56:26 Only the utterances of: "Peace. Peace."

The Lower Heaven

56:27 Those on the right; who will be with those on the right?
56:28 In lush orchards.
56:29 Fruit-trees which are ripe.
56:30 Extended shade.
56:31 Flowing water.
56:32 Many fruits.

56:33 Neither ending; nor forbidden.
56:34 Raised furnishings.
56:35 We have produced them in a special way.
56:36 Made them young.
56:37 A perfect match.
56:38 For those on the right side.
56:39 Many from the early generations.
56:40 Many from the later generations.

Hell

56:41 Those on the left, who will be from those on the left?
56:42 In fierce hot winds and boiling water.
56:43 A shade that is unpleasant.
56:44 Neither cool, nor helpful.
56:45 They used to be indulged in luxury before this.
56:46 They persisted in the great blasphemy.
56:47 They used to say, "After we die and turn to dust and bones, shall we then get resurrected?"
56:48 "As well as our forefathers?"
56:49 Say, "The people of old and the later generations,"
56:50 "Will be summoned to a meeting on a predetermined day."
56:51 "Then you, O rejecting strayers,"
56:52 "Will eat from the tree of Bitterness,"
56:53 "Filling your bellies therewith,"
56:54 "Then drinking on top of it boiling water,"
56:55 "So you will drink like thirsty camels!"
56:56 Such is their share on the day of Judgment.

Observe, Question, Reflect and Appreciate

56:57 We have created you, if you could only affirm!
56:58 Have you noted the semen that you produce?
56:59 Did you create it, or were We the Ones who created it?
56:60 We have predetermined death for you. Nothing can stop Us,
56:61 From transforming your forms, and establishing you in what you do not know.
56:62 You have come to know about the first creation. If only you would remember.
56:63 Have you noted the crops you reap?
56:64 Did you grow them, or were We the Ones who grew them?
56:65 If We wished, We could turn them into hay. Then you would be left in wonderment:
56:66 "We are lost."
56:67 "No, we are deprived!"
56:68 Have you noted the water you drink?
56:69 Did you send it down from the clouds, or was it We who sent it down?
56:70 If We wished, We could make it salty. If only you would give thanks.
56:71 Have you noted the fire you kindle?
56:72 Did you initiate its tree, or was it We who initiated it?
56:73 We rendered it a reminder, and a useful tool for the users.
56:74 You shall glorify the name of your Lord, the Great.

Quran Accessible Only by the Pure

56:75 I do swear by the positions of the stars.*
56:76 This is an oath, if you only knew, that is great.
56:77 It is an honorable Quran.*
56:78 In a protected record.
56:79 None can grasp it except those pure.
56:80 A revelation from the Lord of the worlds.
56:81 Are you disregarding this *hadith*?*
56:82 You made your denial your livelihood?
56:83 So when the time comes and it reaches your throat.
56:84 At that moment you look around.
56:85 We are closer to it than you are, but you do not see.
56:86 If it is true that you do not owe any account.
56:87 Then return it, if you are truthful?
56:88 So, if he is one of those who are made near.
56:89 Then joy, and rose buds, and paradises of bliss.
56:90 If he is one of the people of the right.
56:91 Then: "Peace" from the people of the right.
56:92 But if he is one of the deniers, the strayers.

56:93 Then an abode of boiling water.
56:94 A burning in hell.
56:95 This is the absolute truth.
56:96 You shall glorify the name of your Lord, the Great.*

ENDNOTES

056:013-40 Early supporters of the message are rewarded with higher ranks since they take more risks by accepting the truth when it has greater opposition.

056:022 See 44:54

056:075-76 Our universe consists of billions of galaxies, each containing billions of stars. Each of these heavenly bodies that we are even unable to count, move in orbits determined by the same explosion and by the same universal laws of gravity. The more our knowledge of astronomy and cosmology increases, the more we appreciate the meaning of this oath. Can the expression "the place of stars" refer to black holes? For the function of oaths in the Quran see 89:5.

056:077-79 You will find on the cover of almost all published Quranic manuscripts a few verses. If you check, you will probably find 56:77-79 written in Arabic calligraphy. Why among hundreds of verses describing the Quran, has the convention decided on these verses? Out of more than 50 descriptive noun-adjectives used for the Quran, why would they pick "*Karym*" (Honorable)? *Al-Quran il-Karym*? Why not more frequently used words such as *Zikr* (Message), *Hakym* (Wise), *Mubyn* (Clear), *Nur* (Light), instead of *Karym*? Why is this verse highlighted in connection to the Quran? Why not, for instance, the verses repeatedly reminding us of the easy-to-understand language of the Quran (54:17,22,32,40)? Or, why not one of these verses: 12:111; 15:1; 17:9; 17:88; 17:89; 30:58; 41:3; 55:2 . . .?

We have all the reason to be suspicious of the intention of those who dedicated Chapter 36 (Ya Sin) to be recited in funerals for the DEAD, the chapter that contains the only verse declaring that the Quran is sent to remind the LIVING beings (36:70)! We have all the reason to be suspicious of the intention of those who picked the name *hadith*, a negative word when used for hearsay narrations and teachings other than the Quran, to depict another source besides the Quran! So, why did they pick the word "*Karym*" as the most common adjective for the Quran and verses 56:77-79 as the most common subtitle for the covers of manuscripts?

Those who appreciate the Quran know the answer to the question very well: The polytheistic clergymen and scholars who betrayed the Quran, reached a consensus in not understanding or misunderstanding 56:77-79, and they thought they could repel others from the Quran according to their misunderstanding. They distorted the meaning of these verses by claiming that those who do not have ablution, including the women who they considered "dirty" because of menstruation, should not touch the Quran. Now we may infer why the verse whose meaning has been distorted by consensus is picked for the cover of the Quran manuscripts. Now we may infer why the adjective mentioned in that verse was made the most popular adjective of the Quran. We believe that this is a part of a diabolic conspiracy to prevent the Quran from becoming a pocketbook, a book of quick reference; they keep the Quran on high shelves or nail it on high walls far away from people's reach… Unfortunately, this plan has worked successfully. The Quran has been transformed from being a guide, a reference book, a map, a compass, into a dangerous object, like a train, like a high-voltage transformer station! When the Quran becomes a book too-difficult-to-understand, making it impossible to reach its "high" meanings, and dangerous to touch, then rush in volumes of *hadiths*, loads of *sunna*, barrels of hearsay, mishmash heaps of sectarian teachings, piles of nonsense, tons of superstitions, hordes of holy men, and troops of holy merchants. This explains the misery, backwardness, oppression, repression, division, and corruption rampant in the so-called "Muslím countries."

The unappreciative people who deem the Quran insufficient, because of their prejudice, are not permitted to understand the Quran (6:110; 17:45; 18:57). It is interesting that those who do not accept the Quran as the sole authority do not understand the very verses about understanding the Quran. Thus, 7:3; 17:46; 41:44; 56:79 are delicious art work that function both as the "thesis and the evidence."

056:081 For the meaning and prophetic implication of the word *hadith*, see 7:185; 12:111; 31:6; 33:38; 35:43; 45:6; 52;34; 77:50.

056:096 Or "Exalt your Lord with His great name/attribute."

57:0 In the name of God, the Gracious, the Compassionate.

57:1 Glorifying **God** is everything in the heavens and the earth. He is the Noble, the Wise.

57:2 To Him is the kingship of the heavens and the earth. He brings life and death. He is capable of all things.

57:3 He is the First and the Last, the Evident and the Innermost. He is fully aware of all things.

57:4 He is the One who created the heavens and the earth in six days, and then He settled over the dominion. He knows everything that enters into the land, everything that comes out of it, everything that comes down from the sky, and everything that climbs into it. He is with you wherever you may be. **God** is Seer of everything you do.*

57:5 To Him is the kingship of the heavens and the earth, and to **God** all matters are returned.

57:6 He merges the night into the day, and merges the day into the night. He is fully aware of what is inside the chests.

57:7 affirm **God** and His messenger, and spend from what He has made you successors to. Those among you who affirm and spend have deserved a great recompense.

57:8 Why should you not affirm **God** when the messenger is inviting you to affirm your Lord? He has already taken a pledge from you, if you are those who affirm.

57:9 He is the One who sends down to His servant clear signs, to bring you out of the darkness into the light. **God** is Kind towards you, Compassionate.

57:10 Why do you not spend in the cause of **God**, when **God** possesses all wealth in the heavens and the earth? Not equal among you are those who spent before the victory and fought. They attained a greater rank than those who spent after the victory and fought. For each, **God** promises goodness. **God** is Ever-aware in everything you do.

57:11 Who will lend **God** a loan of goodness, to have it multiplied for him manifold, and he will have a generous recompense?

Light and Truth

57:12 The day will come when you see the affirming men and the affirming women with their light radiating around them and to their right. "Good news is yours today, you will have paradises with flowing rivers, abiding therein. This is the great triumph!"

57:13 On that day, the hypocrite men and the hypocrite women will say to those who affirmed: "Wait for us! Let us absorb some of your light." It will be said, "Go back behind you, and seek light." So a barrier will be set up between them, whose gate separates compassion on the inner side, from retribution on the outer side.

57:14 They will call upon them: "Were we not with you?" They will reply: "Yes, but you led yourselves into temptation, and you hesitated, and doubted, and became misled by wishful thinking, until **God**'s judgment came. You were diverted from **God** by arrogance."

57:15 "This day, no ransom can be taken from you, nor from those who rejected. Your abode is the fire, it is your patron; what a miserable abode."

57:16 Has not the time come for those who affirmed to open up their hearts for the remembrance of **God**, and the truth that is revealed herein? They should not be like those who were given the book before, and the waiting became long for them, so their hearts became hardened, and many of them were wicked.

57:17 Know that **God** revives the land after it had died. We have made clear to you the signs, perhaps you may understand.

57:18 Surely, the charitable men and charitable women, who have loaned **God** a loan of goodness, He will multiply their reward, and they will have a generous recompense.

57:19 Those who affirmed **God** and His messengers are the truthful ones. The martyrs at their Lord will have their reward and their light. As for those who rejected and rejected Our signs, they are the dwellers of hell.

The Purpose of Life on Earth

57:20 Know that this worldly life is no more than play and amusement, and boasting among you, and an increase in wealth and children. It is like plants that are supplied by an abundant rain, which appear pleasing to those who do not appreciate. But then they dry up and turn yellow, and become useless hay. In the Hereafter there is severe retribution, and forgiveness from **God** and approval. This worldly life is no more than a deceiving enjoyment.

57:21 You shall race towards forgiveness from your Lord, and a Paradise whose width is as the width of the heaven and the earth, prepared for those who affirmed **God** and His messengers. Such is **God**'s grace that He bestows upon whomever He wills. **God** is the Possessor of Infinite Grace.

The Key of Happiness: Stoic Submission to God

57:22 No misfortune can happen on earth, or in yourselves, except it is decreed in a record, before We bring it about. This is easy for **God** to do.*

57:23 In order that you do not despair over anything that has passed you by, nor be exultant of anything He has bestowed upon you. **God** does not like those who are boastful, proud.

57:24 They are stingy, and order the people to be stingy. For anyone who turns away, then know that **God** is the Rich, the Praiseworthy.

Iron: an Important Metal

57:25 We have sent Our messengers with clear proofs, and We sent down with them the book and the balance, that the people may uphold justice. We sent down the iron, wherein there is great strength, and many benefits for the people. So that **God** would distinguish those who would support Him and His messengers on affirmation. **God** is Powerful, Noble.*

Prophethood

57:26 We had sent Noah and Abraham, and We placed in their progeny the prophethood and the book. Some of them were guided, while many were wicked.

57:27 Then We sent in their tracks Our messengers. We sent Jesus the son of Mary, and We gave him the Injeel, and We ordained in the hearts of his followers kindness and compassion. But they invented Monasticism which We never decreed for them. They wanted to please **God**, but they did not observe it the way it should have been observed. Consequently, We gave those who affirmed among them their recompense, while many of them were wicked.*

57:28 O you who affirm, you shall revere **God** and affirm His messenger. He will then grant you double the reward from His mercy, and He will make for you a light by which you shall walk, and He will forgive you. **God** is Forgiving, Compassionate.

The People of the Book Should Know

57:29 So that the followers of the book should know that they have no power over **God**'s grace, and that all grace is in **God**'s hand. He bestows it upon whoever/whomever He wills. **God** is Possessor of Infinite Grace.

ENDNOTES

057:004 For the evolutionary creation of the universe, See 41:9-10.

057:022-23 Proving the existence of free will is philosophically and scientifically impossible. Considering Gödel's incompleteness theorem, it is impossible to eliminate axioms in axiomatic deductions. The assertion, "I have freedom of will" is not only non-falsifiable, but also contradicts the deterministic laws governing matter and energy. As Nietzsche and Blanchard argued, it might be impossible to predict the interactive motion of hundreds of billiard balls, yet the motion of each ball depends on the mass and velocity of the ball colliding it. The motion of all balls depends on the motion of the first ball, and its motion depends on the angle and kinetic energy of the cue, which passes it to the balls during the impact. Sartre and other atheistic existentialists assumed the existence of freedom of will by asserting that "existence preceded essence" yet they never explained why one egg ended up with a snake while another with a bird; they ignored the fact that there could not be existence without essence.

Nevertheless, if there is freedom of will, it cannot be possible without an omnipotent, the greatest paradox-solver God. Without God, there is no justification for believing in one's freedom in a world where every atom and molecule works in a deterministic way. Does the human brain function according to the quantum laws governing quarks? So far, there is no scientific evidence showing such a relationship. Muslims cannot be accused of being fatalistic; to the contrary, freedom of will is only possible with the existence of God, and muslims can only be accused of believing in freedom of will. See 6:110; 7:15,30; 13:11; 18:29; 42:13,48; 46:15; 57:22.

However, acceptance of freedom of will based on divine communication creates other logical problems or paradoxes. Leaving aside the uncertainty principle and speculations related to quantum physics, it seems impossible to assert the freedom of an agent who did not even choose his or her genetic makeup, the time, the space, and the conditions where he or she would be born. Obviously, this is a paradox since it is impossible to choose existence or the qualities of existence without having an existence or the qualities for choosing qualities! In other words, the paradox of freedom of will is triggered at the moment of creation. Did God ask us whether to give us freedom of will and subject us to a test? If yes, then how could we have the freedom to choose or reject freedom, if we did not have it in the first place? If no, then, how can we have freedom of will, while we were forced to have it in the first place?

Though believing that we humans have freedom of will is one of the paradoxes most difficult to digest, I affirm it because of the proven book (18:29; 57:22).

God, as a demonstration of His creative powers, chose to test the results of creating a being with the ability to freely choose its own destiny (18:29; 6:110; 13:11). God downloaded His *rouh* (revelation/ information/logic) to the prototypical human that would provide him with innate rules of reasoning to distinguish falsehood from truth, bad from good (15:29; 32:9; 38:72;). Messengers and books containing *rouh* were only a bonus mercy, mere reminders of the facts that could be discovered by reason (2:37-38; 10:57; 11:17; 16:89; 21:107; 29:51; 16:2; 36:69; 37:87; 39:21; 42:52; 58:22). After the two human prototypes failed the first test (2:11-27), God created life and death on this planet to give another chance, to reform ourselves, this time subjecting billions of individuals inheriting and operating the same basic program to the same test (67:2). After letting the program run for a period, an individual is deemed accountable to God (46:15). God decided to punish those who freely choose a path contradictory to its original program as they corrupt it through false ideas and actions (2:57; 4:107; 6:12,20,26,110; 7:9,30,53,177; 59:19). The programs that are infected with viruses will experience a regretful stage in a quarantine called Hell (Hell and Paradise are allegories: 13:35; 17:60; 37:62-64; 7:44-50). In this stage the corrupt programs and their chief infector (Satan) will be penalized (7:11-27; 38:71-88), and then they will altogether be annihilated there (87:13). The only virus that will not be healed on the day of Judgment is the virus that creates a schizophrenic personality, a personality that submits itself to others besides God, a personality that does not free itself from false gods thereby alienating itself from its origin, that is God (4:48,116).

057:025 This chapter, which mentions the qualities of iron provides some interesting information regarding the physical structure of the element. Since it is the only chapter elaborating on the benefits and harms of iron, the chapter is called *Sura al-Hadid*, that is, Chapter Iron. Decades ago, when I was a young Sunni scholar, my research on the numerical structure of the chapter led me to interesting discoveries which was published as a booklet. (Kuran'da Demirin Kimyasal Esrarı, Edip Yuksel, Timaş, Istanbul, 1984).

We know that during the era of Quranic revelation, Arabs were not using the so-called Arabic numerals. They adopted those numerals about two centuries after Muhammed. During the time of Muhammed, Arabs were using their Alphabet letters as numbers. The 28 letters of the Arabic alphabet then were arranged in a different order that started with the letters ABJD, thus known as *Abjad*, or Gematria. Since Arabic and Hebrew are closely related languages, their numerical system too resembled each other. Each letter corresponded to a number, such as A for 1, B for 2, J for 3, D for 4. When it reached 10, the corresponding numerical values would continue as multiples of tens and when reaching 100, they would continue as multiples of hundred ending with 1000. Archeological evidence shows that then they were distinguishing letters from numbers by using different color ink or simply by putting a line over the letter, resembling the Roman numerals. However, a literary text might use letters/numbers for multiple purposes; that is for both their semantic or lexicon function and at the same time for their numerical value. Such a multi-use requires an extreme command and knowledge of language, and the author is limited by the language. When Arabs abandoned the *Abjad* system, it was still used by poets and charlatan healers who wished to take advantage of an antiquated numerical system that became a curiosity for the gullible. Poets wrote verses about events or epithets for important figures, with the date of the event or the death embedded in the numerical values of the letters of a word or a phrase in their verses. The usage of the *Abjad* in the Quran is extensive, impressive, and prophetic (For another example see 74:1-2).

When we add up the numerical values of letters comprising the word ***Ha**DYD* (iron), it gives the atomic number of <u>any</u> iron element: 8+4+10+4=26.

If we add up the numerical values of the letters comprising the word *AL-**Ha**DYD* (the iron, or a particular version of iron), it gives the atomic weight of a particular isotope of the iron element: 1+30+8+4+10+4=57.

Of course, this interesting coincidence has nothing to do with the Quran, since the word ***Ha**DYD* already existed in the language of common people, including the opponents of the Quran. However, when we study the details of this chapter, we will learn that the author of the book was aware of the unique relationship between this word, its numerical values, and the related physical or chemical properties.

This chapter is listed as the 57th chapter of the Quran, corresponding to the numerical value of its popular name, *AL-HaDYD*, and at the same time corresponding to the atomic weight of a particular isotope of iron. Iron has five isotopes, with the atomic number of 54, 55, 56, 57, and 58, containing 29, 30, 31, and 32 neutrons, respectively. We also observe that the only verse mentioning the physical properties of iron is listed as the 26th verse, including the unnumbered verse, Bismillah, in the beginning of chapter. (Bismillah, as the first and the most repeated verse of the Quran has a unique role in the mathematical structure of the Quran). The chapter contains 29 numbered verses or 30 verses if Bismillah included; each number giving us the number of neutrons of different isotopes of iron. Iron with 30 neutrons has the most stable nucleus possible, and thus cannot be subjected to fission or fusion. Formation of iron marks the formation of planets and death of stars. Iron with 31 neutrons has a nuclear spin and is used in industry. The word God occurs 32 times in the chapter, two of them in the verse mentioning iron. Thus, the verse about the iron contains the 26th occurrence of the word God, with or without including the unnumbered Bismillah. The verb "we sent down" is interesting, since it describes the formation of planets.

057:027 According to a law attributed to Moses, eunuchs were not respected. "No one who has been emasculated by crushing or cutting may enter the assembly of the Lord" (Deuteronomy 23:1). However, Christians who later fabricated *hadith* attributing them to Jesus, turned castration or hermitic life into a righteous act: "Jesus replied, 'Moses permitted you to divorce your wives because your hearts were hard. But it was not this way from the beginning. I tell you that anyone who divorces his wife, except for marital unfaithfulness, and marries another woman commits adultery.' The disciples said to him, 'If this is the situation between a husband and wife, it is better not to marry.' Jesus replied, 'Not everyone can accept this word, but only those to whom it has been given. For some are eunuchs because they were born that way; others were made that way by men; and others have renounced marriage--because of the kingdom of heaven. The one who can accept this should accept it'" (Matthew 19:8-12). "These are they which were not defiled with women; for they are virgins..." (Revelation 14:4). St. Paul, the voice behind the Gospels of the Nicene Conference, praised hermitic life: "Now for the matters you wrote about: It is good for a man not to marry" (1 Corinthians 7:1).

Today, only the Catholic Church and some small sects are struggling to keep this innovation alive. However, as the Quran states, they do not follow their own fabricated practice of celibacy. Worse, their churches have become notorious for pedophilia and child abuse, and many rich churches have recently declared bankruptcy because of a series of litigations by abused children. The unnatural religious innovation in the name of God, led many priests to engage in child abuse and sodomy, perhaps attracting homosexuals to the ranks of priesthood for various reasons. Some religious men with homosexual tendencies who thought celibacy to be a cure for their desires ended up getting more opportunity to spend time with altar boys in their churches. Perhaps some homosexuals picked the profession just to indulge in their perverted act behind holy curtains. In 2004, the Conference of Catholic Bishops found 11,000 cases of sexual abuse by about 4,000 priests and deacons since 1950. The real numbers are most likely much higher. The sex-abuse lawsuits cost the Church billions of dollars in compensation, leading some dioceses to financial bankruptcy.

The moral bankruptcy and perversion, however, was going on for centuries and was kept a "holy secret" by the powerful and secretive Church. It is only in the last decades that children have more of a voice, and the Church's financial and social power to keep the evil deeds of its priests as secrets behind the confession session is weakening. The Catholic Church, which is also having problems in recruiting priests, is in a dilemma. If they adhere to their centuries-old, celebrated rule, then they will continue suffering from the perverted behavior of priests and a shortage of new recruits. If they amend their rules and abandon the practice, then their credibility and claim of infallibility will get hurt, again. If history is a guide, the Church will resist until reality will force them to abandon it. They will always find a way to justify their blunder to their faithful, who consider reason and religion to be like ice and fire, with their eternally "infallible" teachings of their Popes! Those charlatans did not only steal the power of forgiveness, infallibility, the keys of heaven from God, but also stole the Biblical title of God, "holy father." According to the Quran, the Pope is no different than Pharaoh since he is claiming divine powers. According to the Quran, Catholics are mentally and spiritually enslaved by their Popes. Isn't it time to accept the truth so that truth will set us all free?

The Quran encourages man and woman to marry. Sexual intercourse, if it is done within the limit of God's law, is considered one of God's great blessings. See 30:21. Also see 3:71; 4:161; 9:24; 2:42; 2:188.

بسم الله الرحمن الرحيم

58:0 In the name of God, the Gracious, the Compassionate.

58:1 **God** has heard the woman who argues with you regarding her husband, and she complains to **God**. **God** hears the argument between you. **God** is Hearer, Seer.*

58:2 Those among you who estrange their wives by saying to them: "You are as my mother." They can never be as their mothers, for their mothers are the women who gave birth to them. Indeed, they are uttering what is strange and a falsehood. **God** is Pardoner, Forgiver.

58:3 Those who had estranged their wives in this manner, then they again repeat it, they shall free a slave before they have sexual contact between them. This is to enlighten you. **God** is well aware of everything you do.*

58:4 If he cannot find anyone, then he shall fast two consecutive months before any sexual contact between them. If he cannot, then he shall feed sixty poor people. That is so you would affirm **God** and His messenger. These are the limits set by **God**. Those who do not appreciate have incurred a painful retribution.*

58:5 Those who oppose **God** and His messenger will be disgraced, like their previous counterparts were disgraced. We have sent down clear proofs, and the ingrates have incurred a shameful retribution.

58:6 The day when **God** resurrects them all, then He informs them of what they had done. **God** has recorded it, while they have forgotten it. **God** witnesses all things.

God is Omnipresent

58:7 Do you not realize that **God** knows everything in the heavens and everything on earth? No three people can meet secretly without Him being their fourth, nor five without Him being the sixth, nor less than that, nor more, without Him being there with them wherever they may be. Then, on the day of resurrection, He will inform them of everything they had done. **God** is fully aware of all things.

58:8 Have you noted those who were forbidden from holding meetings in secret, but then insist on it? They meet secretly to commit sin, transgression, and disobedience of the messenger. When they come to you, they greet you with a greeting other than what **God** greets you with. They say inside themselves: "Why does **God** not punish us for our utterances?" Hell will be sufficient for them, wherein they burn; what a miserable destiny.

58:9 O you who affirm, if you must meet secretly, then you shall not meet to commit sin, transgression, and to disobey the messenger. You shall meet to work righteousness and piety. Be aware of **God**, before whom you will be summoned.

58:10 The secret meetings are from the devil, to cause grief to those who affirmed. However, he cannot harm them except by **God**'s will. In **God** those who affirm shall put their trust.

During Conventions, Assemblies, Conferences

58:11 O you who affirm, if you are told to make room in the councils, then you shall make room. **God** will then make room for you. If you are asked to step down, then step down. **God** will raise those among you who affirm, and those who acquire knowledge to higher ranks. **God** is fully aware of everything you do.

Lobbying

58:12 O you who affirm, if you wish to hold a private meeting with the messenger, you shall offer a charity before you do so. This is better for you, and purer. If you cannot afford it, then **God** is Forgiver, Compassionate.*

58:13 Are you reluctant to offer a charity before such a meeting? If you do not do it; and **God** had forgiven you; then you shall maintain the contact prayer, contribute towards betterment, and obey **God** and His messenger. **God** is fully Aware of everything you do.

58:14 Have you noted those who befriended people with whom **God** is angry? They are now neither from you nor from them. They deliberately swear to lies while they know!

58:15 **God** has prepared for them a severe retribution. Miserable indeed is what they used to do.

58:16 They used their treaties/oaths as a means of repelling from the path of **God**. Consequently, they have incurred a shameful retribution.

58:17 Neither their money, nor their children will help them against **God** in the least. They have incurred the hellfire; in it they will abide.

58:18 The day will come when **God** resurrects them all. Then they will swear to Him, just as they swear to you, thinking that they actually are right! Indeed, they are the liars.

The Party of Devil versus the Party of God

58:19 The devil has overtaken them and has caused them to forget the remembrance of **God**. These are the party of the devil. Absolutely, the party of the devil are the losers.

58:20 Surely, those who oppose **God** and His messenger, they will be with the lowliest.

58:21 **God** has decreed: "I and My messengers will be the victors." **God** is Powerful, Noble.

58:22 You will not find any people who affirm **God** and the Last day befriending those who oppose **God** and His messenger, even if they were their parents, or their children, or their siblings, or their tribe. For these, He decrees trust into their hearts, and supports them with a Spirit from Him, and He admits them into paradises with rivers flowing beneath, wherein they will abide. **God** is pleased with them, and they are pleased with Him. These are the party of **God**. Most assuredly, the party of **God** are the winners.

ENDNOTES

058:001 This is the only chapter that contains the word Allah (God) in each verse.

058:003 The Quran abolished slavery. See 4:25; 90:1-20.

058:004 The beginning of the verse, "*faman lam yajid*" (literally: as for the one who does not find), clearly indicates that the slaves instructed to be freed in the previous verse are not expected to belong to peacemakers. Otherwise, an expression such as "if one does not possess" should have been used. Slavery is prohibited by the Quran as one of the greatest sins. See: 3:79; 4:3,25,92; 5:89; 8:67; 24:32-33; 58:3-4; 90:13; 2:286; 12:39-42; 79:24.

058:012 Muhammed was appointed by God as a messenger and was accepted by affirmers, with their free will, as a messenger of God. Muhammed was also elected as a leader of a constitutional city-state government by a diverse community comprising of Christians, Jews, and Pagans, beside affirmers of the Quran. Muhammed is no longer alive, yet the rule applies to our leaders and governments. Democracies suffer greatly from the corrupting influence of lobbies that promote special interests, usually of the most powerful segments of the population. Financial contributions and favors provided by lobbies create a *quid pro quo* relationship between the lawmakers and the wealthy. Big governments allied with big corporations create one of the worst governments: the governments of the wealthy, by the wealthy and for the wealthy, which we are asked to avoid (59:7). Such governments may drag an entire nation to unjustified wars of aggression to benefit big corporations, especially the weapons industry and the contractors who profit from such adventures, adventures that usually cost the lower class in terms of blood and higher cost of living. Thus, this verse recommends those who wish to lobby to pay some charity to the needy before asking for a conference with their elected leaders. This financial contribution for public good, not the wallets of the elected officials, is a reminder to the parties that their conference should not betray the publicly declared interest of the people.

59:0 In the name of God, the Gracious, the Compassionate.

59:1 Glorifying **God** is everything in the heavens and the earth, and He is the Noble, the Wise.

The Mass Exile

59:2 He is the One who drove out those who rejected among the people of the book from their homes at the very first mass exile. You never thought that they would leave, and they thought that their fortresses would protect them from **God**. But then **God** came to them from where they did not expect, and He cast fear into their hearts. They destroyed their homes with their own hands and the hands of those who affirm. So take a lesson, O you who possess vision.*

59:3 Had **God** not decreed to banish them, He would have punished them in this life. In the Hereafter they will face the retribution of the fire.

59:4 This is because they challenged **God** and His messenger. Whosoever challenges **God**, then **God** is severe in punishment.

59:5 Whether you cut down a tree or left it standing on its root, it was by **God**'s leave. He will surely humiliate the wicked.

Distribute the Wealth; Avoid Creating Monopolies

59:6 What **God** provided to His messenger, without you having to battle for it on horses or on foot, was because **God** sends His messengers against whomever He wills. **God** is capable of all things.

59:7 Whatever **God** provided to His messenger from the people of the townships, then it shall be to **God** and His messenger; for the relatives, the orphans, the poor, and the wayfarer. Thus, it will not remain monopolized by the rich among you. You may take what the messenger gives you, but do not take what he withholds you from taking. Be aware of **God**, for **God** is mighty in retribution.*

59:8 For the immigrants who are poor and were driven out of their homes and deprived of their properties, they sought **God**'s grace and pleasure, and they supported **God** and His messenger. They are the truthful ones.

59:9 Those who provided them with shelter, and were those who affirm before them; they love those who immigrated to them, and they find no hesitation in their hearts in helping them, and they readily give them priority over themselves, even when they themselves need what they give away. Indeed, those who overcome their hedonism are the successful ones.

59:10 Those who came after them saying: "Our Lord, forgive us and our brothers who preceded us to the affirmation, and do not place in our hearts any animosity towards those who affirmed. Our Lord, You are Kind, Compassionate."

The Conspiracy of Hypocrite Warmongers

59:11 Have you noted those who are hypocrites, they say to their companions in denial among the people of the book: "If you are driven out we will go out with you, and we will never obey anyone who opposes you. If anyone fights you, we will support you." **God** bears witness that they are liars.

59:12 If they were driven out, they would not have gone out with them, and if anyone fought them, they would not have supported them. Even if they supported them, they would have turned around and fled. They could never win.

59:13 Indeed, you strike more trepidation in their hearts than **God**. This is because they are people who do not reason.

59:14 They will not fight you all unless they are in fortified towns, or from behind walls. Their might appears formidable among themselves. You would think that they are united, but their hearts are divided. This is because they are people who do not understand.

59:15 Like the example of those who preceded them. They suffered the consequences of their decisions. They have incurred a painful retribution.

59:16 Like the example of the devil, when he says to the human being: "Reject," then as soon as he rejects, he says: "I disown you. I fear **God**, the Lord of the worlds."

59:17 So the destiny for both of them is the fire, abiding therein. Such is the recompense for the wicked.

Be God-conscious

59:18 O you who affirm, be aware of **God**, and let every person examine what it has put forth for tomorrow. Be aware of **God**; **God** is fully aware of everything you do.

59:19 Do not be like those who forgot **God**, so He made them forget themselves. These are the wicked.

59:20 Not equal are the dwellers of the fire and the dwellers of Paradise; the dwellers of Paradise are the winners.

59:21 Had We sent down this Quran to a mountain, you would have seen it trembling, crumbling, out of reverence for **God**. Such are the examples We cite for the people, that they may reflect.

59:22 He is **God**; there is no other god beside Him. Knower of all secrets and declarations. He is the Gracious, the Compassionate.*

59:23 He is **God**; there is no other god beside Him. The King, the Holy, the Peace, the Trusted, the Supreme, the Noble, the Powerful, the Dignified. **God** be glorified; far above what they set up.

59:24 He is **God**; the Creator, the Initiator, the Designer. To Him belong the most beautiful names. Glorifying Him is everything in the heavens and the earth. He is the Noble, the Wise.

ENDNOTES

059:002-6 According to *hadith* books, there were three Jewish tribes in Yathrib (Medina): Banu Qaynuqa, Banu al-Nadir and Banu Qurayza. They provoked muslims and the first two tribes were forced to leave the city with their transportable possessions. However, prophet Muhammed allegedly did not forgive Banu Qurayza; their necks were struck, and their children were made slaves. Estimates of those killed vary from 400 to 900. The Quran refers to the event and contrary to the claims of hearsay collections, never mentions killing or enslaving them, which is in direct contradiction to many verses of the Quran. The Quran, in these verses, informs us that a group from "The People of the Book" was forced to leave the territory because of their violation of the constitution and secretly organizing war together with their enemies against peacemakers (59:1-4). Verse 59:3 clearly states that they were not penalized further in this world.

The credibility of the story of Muhammed massacring Bani Qurayza Jews has been a controversial subject since the time it was published by Ibn Ishaq. Ibn Ishaq, who died in 151 A.H., that is 145 years after the event in question, was severely criticized by his peers for relying on highly exaggerated Jewish stories. He was also harshly criticized for presenting forged poetry attributed to famous poets. Some of his contemporary scholars, such as Malik, called him "a liar." However, his work was later copied by others without critical examination. This is an example of hearsay used by dubious Jewish reporters for propaganda purposes.

Modern scholars found astonishing similarities between Ibn Ishaq and the account of historian Josephus regarding King Alexander, who ruled in Jerusalem before Herod the Great, hung upon crosses 800 Jewish captives, and slaughtered their wives and children before their eyes. Many other similarities in the details of the story of Banu Qurayza and the event reported by Josephus are compelling.

Besides, the lack of reference or justification in the Quran for such a massacre of great magnitude, and the verses revealing principles for muslims to abide by removes all credibility from this story (35:18: 61:4). The Quran gives utmost importance to human life (5:32) and considers racism and anti-Semitism evil (49:11-13).

059:007 The Quran clearly warns us against the power of a few wealthy people and asks us to distribute the wealth of a nation among its population in a more equitable way. See 58:12.

Contrary to the political alliance of many clergymen with the wealthy and the privileged throughout history, the Bible too contains many verses protective of the weak and poor. "He upholds the cause of the oppressed and gives food to the hungry. The Lord sets prisoners free, the Lord gives sight to the blind, the Lord lifts up those who are bowed down, the Lord loves the righteous" (Psalm 146:7-8). "Is not this the sort of fast that pleases me: to break unjust fetters, to undo the thongs of the yoke, to let the oppressed go free, and to break all yokes? Is it not sharing your food with the hungry, and sheltering the homeless poor…?" (Isaiah 58:6-7). "The Spirit of the Lord GOD is upon me; because the LORD hath anointed me to preach good tidings unto the meek; he hath sent me to bind up the brokenhearted, to proclaim liberty to the captives, and the opening of the prison to them that are bound" (Isaiah 61:1).

The following verses from the Bible are very instructive today since it rejects unequivocally the deity of Jesus and the baptizing of Capitalism. "As Jesus started on his way, a man ran up to him and fell on his knees before him. 'Good teacher,' he asked, 'what must I do to inherit eternal life?' 'Why do you call me good?' Jesus answered. 'No one is good—except God alone. You know the commandments: 'Do not murder, do not commit adultery, do not steal, do not give false testimony, do not defraud, honor your father and mother." 'Teacher,' he declared, 'all these I have kept since I was a boy.' Jesus looked at him and loved him. 'One thing you lack,' he said. 'Go, sell everything you have and give to the poor, and you will have treasure in heaven. Then come, follow me.' At this the man's face fell. He went away sad, because he had great wealth. Jesus looked around and said to his disciples, 'How hard it is for the rich to enter the kingdom of God!' The disciples were amazed at his words. But Jesus said again, 'Children, how hard it is to enter the kingdom of God! It is easier for a camel to go through the eye of a needle than for a rich man to enter the kingdom of God.' " (Mark 10:17-25).

Muhammed was both the messenger of God and elected leader of a diverse group in the city-state of Yathrib. As a leader and commander, he had the right to allocate the spoils of war. The verse does not mean that Muhammed gave us another book, or a television set, or silly stories, or numerous prohibitions; it is about a specific case, that is, the distribution of spoils of war. The followers of *hadith* and *sunna*, abuse this verse and take it out of context hoping to justify volumes of books containing numerous contradictory and silly hearsay reports. They want us to believe that to "take whatever Muhammed has given" means to take whatever story Bukhari and his ilk told several centuries after Muhammed's departure! See 49:1.

059:022-24 The Quran contains more than 100 attributes of God (most likely, 114). The Bible lists many similar attributes for God:

- None beside Him: Deuteronomy 4:35; Isaiah 44:6;
- None before Him: Isaiah 43:10;
- None like to Him: Exodus 9:14; Deuteronomy 33:26; 2 Samuel 7:22; Isaiah 46:5,9;
- None good but He: Matthew 19:17;
- Fills heaven and earth: 1 Kings 8:27; Jeremiah 23:24;
- God Is light: Isaiah 60:19; James 1:17; 1 John 1:5;
- Love: 1 John 4:8,16; Invisible: Job 23:8-9;
- Unsearchable: Job 11:7; 37:23; Psalms 145:3; Isaiah 40:28;
- Eternal: Deuteronomy 33:27; Psalms 90:2;
- Omnipotent: Genesis 17:1; Exodus 6:3;
- Omniscient: Psalms 139:1-6; Proverbs 5:21;
- Omnipresent: Psalms 139:7; Jeremiah 23:23;
- Immutable: Psalms 102:26-27;
- Glorious: Exodus 15:11; Psalms 145:5;
- Most High: Psalms 83:18; Acts 7:48;
- Perfect: Matthew 5:48;
- Holy: Psalms 99:9; Isaiah 5:16;
- Just: Deuteronomy 32:4; Isaiah 45:21;
- True: Jeremiah 10:10; John 17:3;
- Upright: Ps 25:8; 92:15;
- Righteous: Psalms 145:17;
- Good: Psalms 25:8; 119:68;
- Great: 2 Chronicles 2:5; Psalms 86:10;
- Gracious: Exodus 34:6; Psalm 116:5;
- Merciful: Exodus 34:6-7;
- Compassionate: 2 Kings 13:23;
- Forgiving: Numbers 14:18; Micah 7:1;
- Should be worshipped in spirit and in truth: John 4:24.

However, the Bible also contains a few attributes that do not fit the overall depiction of God in the Quran. For instance, according to the Old Testament, God is Jealous (Joshua 24:19; Nahum 1:2), or a Consuming Fire (Hebrews 12:29). The God of the Old Testament demonstrates weak qualities, since He regrets creating Adam (Genesis 6:6-7); worries about his reputation (Ezekiel 20:8-10); cannot see Adam (3:8-10); rests, eats, sleeps (Exodus 31:17; Genesis 18:1-8; Psalm 78:65); can be seen and cannot be seen (Genesis 32:30, compare it to Exodus 33:17-20; John 1:18); loses a wrestling match with Jacob (Genesis 32:24-30); and is forgetful (Lamentations 5:20; Psalms 13:1).

For God's attributes in the Quran, see 7:180.

بســـــم الله الرحمن الرحيـــــم

60:0 In the name of God, the Gracious, the Compassionate.

60:1 O you who affirm, do not take My enemy and your enemy as allies, you extend love to them, even though they have rejected the truth that has come to you. They drive you and the messenger out, simply because you affirm **God**, your Lord. If you are mobilizing to strive in My cause, seeking My blessings, then how can you secretly love them? I am fully aware of everything you conceal and what you declare. Whosoever of you does this, then he has gone astray from the right path.

60:2 If they encounter you, they treat you as enemies, and they extend their hands and tongues against you to hurt you. They desire you to reject.

60:3 Neither your relatives nor your children will benefit you; on the day of resurrection He will separate between you. **God** is Seer of everything you do.

Role Models

60:4 There has been a good example set for you by Abraham and those with him, when they said to their people: "We are innocent from you and what you serve besides **God**. We have rejected you, and it appears that there shall be animosity and hatred between us and you until you affirm **God** alone." Except for the saying of Abraham to his father: "I will ask forgiveness for you, but I do not possess any power to protect you from **God**." "Our Lord, we have put our trust in You, and we turn to You, and to You is the final destiny."*

60:5 "Our Lord, do not let us become a test for those who rejected, and forgive us. Our Lord, You are the Noble, the Wise."

60:6 Certainly, a good example has been set by them for those who seek **God** and the Last day. Whosoever turns away, then **God** is the Rich, the Praiseworthy.

Ultimate Social and Political Goal: Compassion, Peace, and Justice among People

60:7 Perhaps **God** will grant compassion between you and those you consider enemies; and **God** is Omnipotent. **God** is Forgiving, Compassionate.*

60:8 **God** does not forbid you from those who have not fought you because of your system, nor drove you out of your homes, that you deal kindly and equitably with them. For **God** loves the equitable.

60:9 But **God** forbids you befriending only those who fought you because of your system, and drove you out of your homes, and helped to drive you out. You shall not ally with them. Those who ally with them, then such are the transgressors.

Regarding Women Seeking Asylum during War

60:10 O you who affirm, if the affirming women come emigrating to you, then you shall test them. **God** is fully aware of their affirmation. Thus, if you establish that they are those who affirm, then you shall not return them to those who do not appreciate. They are no longer lawful for them, nor are those who do not appreciate lawful for them. Return the dowries that were paid. There is no sin upon you to marry them, if you have paid their dowries to them. Do not keep the wives who do not affirm, and ask back what dowries you paid. Let them ask back what dowries they had paid. Such is **God**'s judgment; He judges between you. **God** is Knowledgeable, Wise.*

60:11 If any of your wives have gone over to the camp of those who do not appreciate, and you are granted victory over them, then you shall compensate those whose wives have gone over, to the equivalent of what dowry they spent. Be aware of **God**, whom you affirm.

A Group of Women Vote for Muhammed's Leadership by Entering into a Social Contract

60:12 O you prophet, if the affirming women come to make allegiance to you that they will not set up anything beside **God**, nor steal, nor commit adultery, nor kill their born children, nor fabricate any falsehood, nor disobey you in any matter which is righteous, then you shall accept their allegiance, and ask **God** to forgive them. **God** is Forgiver, Compassionate.*

60:13 O you who affirm, do not ally a people with whom **God** is angry; for they have given up regarding the Hereafter, just like the ingrates have given up on the people who are already in the graves.

ENDNOTES

060:004 See 33:21. The Arabic word *wahdahu* (alone) occurs in the Quran six times and one of them is used to instruct us to follow the Quran alone (17:46). The expression *Allahu wahdahu* (God alone) occurs in 7:70; 39:45; 40:12,84 and 60:4. These numbers add up to 361, or 19x19 and thus it numerically emphasizes that the main message of the Quran is serving God ALONE. For the same reason, the numerical value of God's attribute WA**H**iD (One; 6+1+8+4) equals 19, the code of the Quran's numerical structure discovered in 1974. For the important role of the number 19, see chapter 74.

We may pray for polytheists to attain salvation; but we cannot ask God to forgive them (4:48, 116).

060:007-9 These verses make it clear that war is allowed only in self-defense. See 9:5,29; 8:19; 47:35.

060:010 We should take a lesson from the meticulous care for the rights of individuals that have participated in war. Returning the dowries paid to the women who sought refuge among muslims to those who live with the aggressive enemies demonstrates the highest standard for a peaceful society that puts the value of justice and peace far above financial gain. It is no wonder Muhammed and his followers soon conquered the hearts of the entire population and were able to return peacefully to their hometown as victors. This shows that Muhammed's mission was not to gain more wealth, as some biased critic claim, since women would be encouraged to smuggle as much as they can from enemy ranks.

060:012 Verse 60:12 informs us about the rights and privileges enjoyed by women in the early muslim community during the life of prophet Muhammed. In this verse, the prophet affirms women's right to vote, by taking the pledge of believing women to peacefully surrender themselves to God alone and lead a righteous life. The word "mu*BaYA**a***" used in the verse implies the political nature of the pledge; they accepted the leadership of the prophet individually, with their free choice. Note that it is not *BaYA* as Sunni and Shiite mushriks have distorted, but it is *muBaYaA*. It is not a contract with only one party given promises and assuming responsibilities; but it is a two-way contract where the elected leader too, here Muhammad, has to follow the terms such as be just, do not waste their taxes, consult them in their affairs, etc. This verse is not about some pagan or *mushrik* women embracing monotheism but is about a group of monotheist women publicly announcing their allegiance to Muhammed, who became a founder of a federally secular constitutional government in central Arabia. This is a historical document that shows that monotheist women were not considered default appendices of their decision-making husbands, brothers, fathers, or male guardians, but were rather treated as independent political entities who could vote and enter into social contracts with their leaders. Unfortunately, many of the human rights recognized by islam were later one by one taken away from individuals, especially from women, by the leaders of Sunni and Shiite religions. They replaced the progressive and liberating teaching of the Quran and practice of the early muslims with reactionary and enslaving teachings, thereby resurrecting the dogmas and practices of "the days of ignorance." It took humanity centuries to recognize for women their God-given rights. For instance, the US recognized the right of women to vote in 1919 by passing the 19th Amendment. Exactly, 13 centuries after it was recognized by the Quran. As for the region that once led the world in human rights and freedom, it is now more than 13 centuries behind! After women, the men too lost their dignity to elect their leaders. What a regression!

61:0 In the name of God, the Gracious, the Compassionate.

61:1 Glorifying **God** is everything in the heavens and everything on the earth. He is the Noble, the Wise.

61:2 O you who affirm, why do you say what you do not do?

61:3 It is most abhorent with **God** that you would say what you do not do.

Organized Fighters

61:4 **God** loves those who fight in His cause as one column; they are like bricks in a wall.

61:5 When Moses said to his people: "O my people, why do you harm me, while you know that I am **God**'s messenger to you?" But when they deviated, **God** diverted their hearts. **God** does not guide the wicked people.

The Messenger after Jesus

61:6 When Jesus, son of Mary, said, "O children of Israel, I am **God**'s messenger to you, authenticating what is present with me of the Torah and bringing good news of a messenger to come after me whose name will be 'most acclaimed.'" But when he showed them the clear proofs, they said, 'This is clearly magic.'*

61:7 Who is more evil than one who fabricates lies about **God**, while he is being invited to peacefully surrender? **God** does not guide the evil people.

61:8 They wish to put out **God**'s light with their mouths. But **God** will continue with His light, even if the ingrates dislike it.

61:9 He is the One who sent His messenger with the guidance and the system of truth, so that it will manifest it above all other systems, even if those who set up partners dislike it.

61:10 "O you who affirm, shall I lead you to a trade that will save you from painful retribution?"

61:11 "That you affirm **God** and His messenger and strive in the cause of **God** with your money and your lives. This is best for you, if only you knew."

61:12 He will then forgive your sins, and admit you into paradises with rivers flowing beneath, and beautiful mansions in the gardens of Eden. This is the greatest triumph.

61:13 Also you will receive what you love: a triumph from **God** and a victory that is close at hand. Give good news to those who affirm.

The Supporters (Ansar/Nasara/Nazarenes)

61:14 O you who affirm, be **God**'s supporters, as Jesus the son of Mary said to his disciples: "Who are my supporters towards **God**?" The disciples said, "We are **God**'s supporters." Thus, a group from the Children of Israel affirmed, and another group rejected. So, We supported those who affirmed against their enemy, and they were successful.

ENDNOTES

061:006 The word *ahmad* is an adjective meaning "most acclaimed" or "most celebrated." Traditional sources consider it a proper name for Muhammed. This contradicts historical facts. The name of the prophet that came after Jesus was Muhammed, which is used in the Quran four times. Centuries after the departure of Muhammed, Muhammed-worshipers fabricated 99 names, including Ahmad, for Muhammed, in order to compete with the attributes of God. They could not accept one God having so many beautiful attributes, with their second god having only one attribute, Muhammed. They included many divine attributes, such as, "The First, The Last, The Judge.." in their list for Muhammed. Furthermore, we do not find the word Ahmad in the Bible. Rather we see the translation of the Greek adjective, "*paracletos*".

"And I will pray the Father, and he shall give you another Comforter/Counselor (*paracletus*), that he may abide with you forever" (John 14:16). Also, see John 14:26; 15:26 16:7.

Jesus predicted the coming of another prophet. The one whose coming was foretold by Jesus is mentioned as "*Paracletos*" or "*Periclytos*" in Greek manuscripts. *Paracletos* means advocate, comforter, or counselor. *Periclytos*, on the other hand, means "admirable one" (in Arabic *ahmad*). The "spirit" here, does not mean other than human. There are cases where the word "spirit" is used for humans (2 Thessalonians 2:2; 1 John 4:1-3).

If indeed Jesus had prophesied the proper name of the prophet after him, this prophecy would have gotten the attention of almost every one of his supporters. Furthermore, we would see many people among his supporters and later Christians giving the name Ahmad to their sons, hoping that their sons would fulfill the prophecy. But we do not even see a single Christian named Ahmad. Therefore, the Aramaic or Hebrew equivalent of this word did not become a name. However, the name Muhammed sharing the same root and similar meaning with the word *ahmad* is instructive.

Bible-thumpers assert that there are more than two thousand prophecies foretelling the coming of the Messiah in the Old Testament. Evidently, this is a preposterous claim. But an impartial reader will find several clear prophecies. Then, the reader may ask, "What is wrong with Jews? Why don't they see those plain prophecies in their book?"

Obviously, this problem is not peculiar to Jews. Whenever a new messenger comes, the bigoted religious people reject them in the name of previous messengers. For instance, Egyptians rejected Moses by claiming that Joseph was the last messenger (40:28-38). Jews rejected (J)esu(s) when he started his mission. Those who started worshiping Muhammed too claimed Muhammed to be the last messenger, in contradiction to the Quran (3:81). Christians were and are not different.

The Old Testament in Deuteronomy 18:18 prophesies the coming of Muhammed, since he was more like Moses and came from among the Arabs, the brethren of the Children of Israel.

"I will raise them up a Prophet from among their brethren, like unto thee, and will put my words in his mouth; and he shall speak unto them all that I shall command him." (Deuteronomy 18:18).

Christians try hard to use 18:18 as a prophecy about Jesus, but their fictional "son of God" who "died for the sins of the world" is not like Moses; neither is his miraculous birth, his marital status, nor his political power. However, when we compare, we see that Muhammed is much more like Moses than Jesus was. Besides, the Biblical prophecy does not say "among themselves" or "among Children of Israel," but "among their brethren." The word brethren is used by the Old Testament to describe the relationship between the family of Isaac and the family of Ishmael (See Genesis 16:12; 25:18).

The New Testament too contains some clues regarding the coming prophet, other than the Messiah. John 1:19-25 lists the expected three persons: Christ, Elijah, and That Prophet. These verses, if they are not a fabrication of John, provide us with clues that Jewish people were expecting a "prophet" besides Christ, a prophet that had been prophesied in Deuteronomy 18:18. It is not a surprise that some Christian scholars are using their famous formula, in order to hide this obvious Biblical fact. "Christ + Elijah + That Prophet = Jesus." This is Trinity in Trinity or Pandora's Trinity.

62:0 In the name of God, the Gracious, the Compassionate.

62:1 Glorifying **God** is everything in the heavens and everything on the earth; the King, the Holy, the Noble, the Wise.

62:2 He is the One who sent to the Gentiles a messenger from among themselves, to recite to them His signs, to purify them, to teach them the book and the wisdom. Before this, they were clearly astray.

62:3 To other generations subsequent to them. He is the Noble, the Wise.

62:4 Such is **God**'s grace, which He bestows upon whoever/whomever He wills. **God** is Possessor of Infinite Grace.

62:5 The example of those who were given the Torah, but then failed to uphold it, is like the donkey that is carrying a cargo of books. Miserable indeed is the example of people who denied **God**'s signs. **God** does not guide the wicked people.

The Arrogance of Racism

62:6 Say, "O you who are Jewish, if you claim that you are **God**'s chosen, to the exclusion of all other people, then you should long for death if you are truthful!"

62:7 But they will never long for it, because of what their hands have brought forth. **God** is fully aware of the wicked.

62:8 Say, "The death that you are fleeing from, it will come to find you. Then you will be returned to the Knower of all secrets and declarations, then He will inform you of everything you had done."

The Congregational Prayer

62:9 O you who affirm, if the contact prayer is called to on the day of congregation, then you shall hasten towards the remembrance of **God**, and cease all selling. This is better for you, if you only knew.*

62:10 Then, once the contact prayer is complete, you shall disperse through the land and seek **God**'s provisions, and remember **God** frequently, so that you may succeed.

62:11 If they come across any trade, or some entertainment, they rush to it and leave you standing, say, "What **God** possesses is far better than entertainment or trade. **God** is the best Provider."

ENDNOTES

062:009 See 16:97. There is a question regarding the implication of the *Yawm al Jumua* (the day of congregation). The traditional understanding considers it a reference to the 5th day of the week, that is, Friday. However, the Arabic word could be a description of any day picked by a group of people for assembly prayer, for a public event with political, social and spiritual purposes. If Arabs had another word for Friday (such as, *Aruba*) during the era of Quranic revelation, then we should understand this day to be any day chosen by a group of people in a particular location to meet and pray together. This understanding might be supported by the condition of calling or inviting people to the prayer. For an already known day and time, such as Friday noon, such an invitation or announcement loses its meaning. 20:114.We should also note that the day set for congregational prayer is not the day of complete rest, such as Sabbath. For researchers, I recommend *Studies in Islamic History and Institutions* by Shelomo Dov Goitein, 1966.

بسم الله الرحمن الرحيم

63:0 In the name of God, the Gracious, the Compassionate.

63:1 When the hypocrites come to you they say, "We bear witness that you are the messenger of **God**." **God** knows that you are His messenger, and **God** bears witness that the hypocrites are liars.*

63:2 They have chosen their oath as a deceit, thus they repel from the path of **God**. Miserable indeed is what they do.

63:3 That is because they affirmed, then rejected. Hence, their hearts are sealed; they do not understand.

Paranoid Show-offs

63:4 When you see them, you are impressed by their physical stature; and when they speak, you listen to their eloquence. They are like blocks of wood propped up. They think that every call is against them. These are the enemies, so beware of them. May **God** condemn them; they have deviated.

63:5 If they are told: "Come and let the messenger of **God** ask for your forgiveness," they turn aside their heads, and you see them walking away in pride.

63:6 It is the same for them, whether you ask for their forgiveness, or do not ask for their forgiveness; **God** will not forgive them. For **God** does not guide the wicked people.*

63:7 They are the ones who say, "Do not spend on those who are with the messenger of **God**, unless they abandon him!" To **God** belongs the treasures of the heavens and the earth, but the hypocrites do not reason.

63:8 They say, "When we go back to the city, the noble therein will evict the lowly." But all nobility belongs to **God** and His messenger, and those who affirm. However, the hypocrites do not know.

63:9 O you who affirm, do not be distracted by your money and your children from the remembrance of

God. Those who do this, will be the losers.

63:10 Give from what We have provided to you, before death comes to one of you, then he says: "My Lord, if only You could delay this for a short while, I would then be charitable and join the righteous!"

63:11 **God** will not delay any person if its time has come. **God** is Ever-aware of all that you do.

ENDNOTES

063:001 Islam, which is peacemaking or peacefully surrendering to God, is the system followed by all righteous monotheists since Adam, and its basic principle is testimony to the absolute and sole authority of God. This testimony is mentioned and advocated in 3:18. As if God forgot a crucial part of the testimony, Sunnis have the audacity to add in the statement *Muhammedun Rasulullah* (Muhammed is God's Messenger). The Shiite faction went even a step further and added *Ali Hujjatullah* (Ali is God's Argument). This mentality of adding the names of idolized creatures next to God has been exposed by 39:45. Verse 63:1, is the only verse in the Quran that contains the second part of the *shahada* (testimony). As with the word *hadith*, *sunna*, *ijma*, and *sharia* (33:38; 35:43; 42:21; 45:6), God knew in advance that the polytheists would add the name of their idols next to His name. Thus, God mentioned the second part as the utterance of hypocrites. Modern hypocrites do not really believe that Muhammed was merely a messenger since they claim that he was a partner with God in authoring God's religion. They think that Muhammed completed where God had forgotten and that he even abrogated some of the Quranic verses. A messenger of God would never do such a thing, since their only mission is to deliver God's message. Thus, those who add a testimony about Muhammed's messengership in the *shahada* in fact are lying in that testimony; the Muhammed in their mind is no messenger, but rather their imaginary god besides God.

063:006 So much for the power of intercession! See 2:48.

بسم الله الرحمن الرحيم

64:0 In the name of God, the Gracious, the Compassionate.

64:1 Glorifying **God** is everything in the heavens and everything on the earth. To Him is all kingships, and to Him is all praise, and He is capable of all things.

64:2 He is the One who created you, then among you there is the ingrate, and among you is the affirmer. **God** is Seer over what you do.

64:3 He created the heavens and the earth with truth. He designed you and made your design the best, and to Him is the final destiny.

64:4 He knows what is in the heavens and the earth, and He knows what you conceal and what you declare. **God** is aware of what is inside the chests.

64:5 Did the news not come to you of those who had rejected before? They had tasted the consequences of their decision and they incurred a painful retribution.

64:6 That was because their messengers came to them with clear proofs, but they said, "Shall humans guide us?" So, they rejected and turned away. Yet, **God** has no need. **God** is Rich, Praiseworthy.

64:7 Those who reject claim that they will not be resurrected. Say, "Yes, by my Lord, you will be resurrected, then you will be informed of everything you have done, and this is easy for **God** to do."

64:8 Therefore, you shall affirm **God** and His messenger, and the light that We have sent down. **God** is Ever-aware of all that you do.

64:9 The day when He will gather you, the day of Gathering. That is the day of mutual blaming. Whosoever affirms **God** and does good works, He will remit his sins, and will admit him into paradises with rivers flowing beneath to abide therein forever. Such is the greatest triumph.

64:10 As for those who reject and deny Our signs, they are the dwellers of the fire; they will abide therein. What a miserable destiny!

64:11 No misfortune strikes except with **God**'s leave. Whosoever affirms **God**, He will guide his heart. **God** is fully aware of all things.

64:12 Obey **God** and obey the messenger. If you turn away, then it is only required of Our messenger to deliver clearly.

In God You Shall Trust

64:13 **God**, there is no god besides Him. In **God** those who affirm shall put their trust.

64:14 O you who affirm, from among your spouses and your children are enemies to you; so beware of them. If you pardon, overlook, and forgive, then **God** is Forgiver, Compassionate.

64:15 Your money and children are a test, and with **God** is a great recompense.

64:16 Therefore, be aware of **God** as much as you can, and listen, and obey, and give for charity for your own good. Whosoever is protected from his own hedonistic desires, then these are the successful ones.

64:17 If you lend **God** a loan of righteousness, He will multiply it for you many fold, and forgive you. **God** is Appreciative, Compassionate.

64:18 The Knower of all secrets and declarations; the Noble, the Wise.

بسم الله الرحمن الرحيم

65:0 In the name of God, the Gracious, the Compassionate.

65:1 O you prophet, when you divorce the women, then they should be divorced while ensuring that their required interim is fulfilled and keep count of the interim. You shall revere **God** your Lord, and do not evict the women from their homes, nor should they leave, unless they have committed a proven adultery. These are **God**'s limits. Anyone who transgresses **God**'s limits has wronged his person. You never know; perhaps **God** will make something come out of this.*

65:2 Then, once the interim is fulfilled, either you remain together equitably, or part ways equitably and have it witnessed by two just people from among you; and give the testimony for **God**. This is to enlighten those who affirm **God** and the Last day. Whosoever reveres **God**, He will create a solution for him.

65:3 He will provide for him whence he never expected. Anyone who puts his trust in **God**, then He suffices him. **God**'s commands will be done. **God** has decreed for everything its fate.

65:4 As for the women who have reached menopause, if you have any doubts, their interim shall be three months. As for those whose menstruation has ceased, and already pregnant, their interim is until they give birth. Anyone who reveres **God**, He makes his matters easy for him.

65:5 This is **God**'s command that He sends down to you. Anyone who is aware of **God**, He will remit his sins, and will improve his reward.

Duties Towards Divorced Women

65:6 You shall let them reside in the home you were in when you were together, and do not coerce them to make them leave. If they are pregnant, you shall spend on them until they give birth. Then, if they nurse the infant for you, you shall pay them their due for such. You shall maintain the amicable relations between you. If you disagree, then another woman may nurse the child.

65:7 The rich shall provide support in accordance with his means, and the poor shall provide according to the means that **God** bestowed upon him. **God** does not burden any person more than He has given it. **God** will provide ease after difficulty.

The Protected Message, a Messenger

65:8 Many a community rebelled against the commands of its Lord and His messengers. Thus, We called them to account, and requited them a terrible requital.

65:9 They tasted the result of their actions, and the consequence of their actions was a total loss.

65:10 **God** has prepared for them a severe retribution. Therefore, O the intelligent people who have affirmed, you shall be aware of **God**. **God** has sent down to you a remembrance.*

65:11 A messenger who recites to you **God**'s signs, which are clear, to lead those who affirm and work righteousness out of the darkness into the light. Anyone who affirms **God** and does good works, He will admit him into paradises with rivers flowing beneath; they abide therein forever. **God** has granted for him an excellent reward.

65:12 **God** who created seven heavens and the same number of earths. His command descends between them, that you may know that **God** is capable of all things, and that **God** has encompassed all things with His knowledge.*

ENDNOTES

065:001 The divorced women cannot be evicted from their homes. This is one of the many verses that protects women from unjust traditions of a misogynistic society. A divorced woman must wait three menstruation periods until she can marry again (2:228). See 2:226-230 and 4:19. Compare this verse to the Old Testament, Deuteronomy 24:1-3.

065:010-11 The Quran is a universal messenger; it is good news and a warner. See 41:4

065:012 What it is the implication of this verse? Seven planets like earth or the atomic structure of earthly materials? The word *ard* means both planet earth and ground. According to modern chemistry, atoms have a maximum of seven main energy levels. According to our current atomic model, theoretically the maximum number of protons in a stable atom is 114. In other words, the number of natural and artificially created stable elements equals the number of Chapters of the Quran. See 4:82.

بسم الله الرحمن الرحيم

66:0 In the name of God, the Gracious, the Compassionate.

66:1 O you prophet, why do you prohibit what **God** has made lawful for you, seeking to please your wives? **God** is Forgiver, Compassionate.

66:2 **God** has already given the law, regarding the cancellation of oaths. **God** is your Lord, and He is the Knowledgeable, the Wise.*

Spreading the Words of the Prophet

66:3 When the prophet disclosed a *hadith* to some of his wives, then one of them spread it, and **God** revealed it to him, he recognized part of it and denied part. So when he informed her, she said, "Who informed you of this?" He said, "I was informed by the Knowledgeable, the Ever-aware."*

66:4 If the two of you repent to **God**, then your hearts have listened. But if you band together against him, then **God** is his master. Gabriel, the righteous of those who affirm, and the controllers are his supporters.*

66:5 It may be that he would divorce you, then his Lord will substitute other wives in your place who are better than you; peacefully surrendering, affirming, devout, repentant, serving, active in their societies, responsive, and foremost ones.*

66:6 O you who affirm, protect yourselves and your families from a fire whose fuel is people and rocks. Guarding it are stern and powerful controllers who do not disobey **God** in what He commanded them; and they carry out what they are commanded to.

66:7 O you who have rejected, do not apologize today. You are being requited only for what you did.

66:8 O you who affirm, you shall repent to **God** a sincere repentance. It may be that your Lord will remit your sins and admit you into paradises with rivers flowing beneath. On that day, **God** will not disappoint the prophet and those who affirmed with him. Their light will radiate around them and to their right. They will say, "Our Lord, keep perfect our light for us, and forgive us; You are able to do all things."

66:9 O prophet, strive against the ingrates and the hypocrites and be stern with them. Their abode is hell, and a miserable destiny.

Bad and Good Examples of Women

66:10 **God** cites as examples of those who have rejected, the wife of Noah and the wife of Lot. They were married to two of Our righteous servants, but they betrayed them and, consequently, they could not help them at all against **God**. It was said, "Enter the fire, both of you, with those who will enter it."*

66:11 **God** cites as an example of those who affirmed, the wife of Pharaoh. She said, "My Lord, build a home for me near You in Paradise, and save me from Pharaoh and his works; and save me from the transgressing people."

66:12 Also Mary, the daughter of Imran, who maintained her chastity. So We blew into her from Our Spirit, and she affirmed the words of her Lord and His books; and she was obedient.

ENDNOTES

066:002 God is the only *mawla* (lord and master) of people. See 2:286.

066:003 The word *hadith* (word, utterance) occurs twice in the Quran in connection with Muhammed. One is here; the other is at 33:53. Here the sharing or reporting of the *hadith* heard from Muhammed is criticized and in verse 33:53, listening to the *hadith* of Muhammed is criticized. Knowing that polytheists would create other authorities besides His word and call them *hadith*, *sunna* and *ijma*, God convicted those words. See 33:38; 45:6.

066:004 This verse has been generally mistranslated in traditional translations since they include controllers and affirmers to be Muhammed's *mawla* (lord/patron/master). There is only one *mawla*, and it is none other than God. See 2:286.

066:005 Traditional translations mistranslate the last three adjectives used here to describe Muslim women. They distort their meaning as "fasters, widows and virgins." When the issue is about women, somehow, the meaning of the Quranic words passes through rapid mutations.

For instance, we know that the Sunni and Shiite scholars who could not beat cows and examples found it convenient and fair to beat women (see 4:34). Those of us who have rejected other religious sources besides the Quran are still struggling to clean our minds from these innovations that even have sneaked into the Arabic language long after the revelation of the Quran. There is, in fact, nothing whatsoever about fasting, widows and virgins in this verse. We are rediscovering and relearning the Quran.

The third word from the end of the verse, *SaYiHat,* which we have translated as "active in their societies" simply means to travel or move around for a cause. About two centuries after the revelation of the Quran, when the rights of women were one by one were taken through all-male enterprises called *hadith*, *ijtihad* and *tafseer*, Muslim communities found themselves thinking and living like the enemies of Islam in the Days of Ignorance. The misogynistic mind of orthodox commentators and translators simply could not fathom the notion of a Muslim woman traveling around alone to do anything – and so they pretended that the word in question was not *SaYaHa,* but *SsaWM* – fasting! Socially active women were indeed more difficult to control than the women who would fast in their homes; they were even less costly, since they would eat less. For the usage of the verb form of the root, see 9:2. The word *SaYaHa* has nothing to do with fasting; the Quran consistently uses the word *SaWaMa* for fasting (2:183-196; 4:92; 5:89,90; 19:26; 33:35; 58:4).

The second word from the end is *THaYiBat, which* means “those who return, or those who are responsive”. Various derivatives of the same root are used to mean “reward” or “refuge” or “cloths”. For instance, see 2:125; 3:195. The Arabic words for widow are *ARMiLa* or *AYaMa*. The Quran uses *AYaMa* for widow or single; see: 24:32.

The last word of this verse, *aBKaR*, which means those who are "young," "early risers" or "foremost," has traditionally, and implausibly, been interpreted as "virgins" in this passage. The resulting distorted meaning of the verse supports a sectarian teaching that justifies a man marrying more than one virgin. The Arabic word for virgin is *BaTuL* or *ADRa*.

This false interpretation has become so popular that it is apparently now considered beyond any challenge. Excluding Edip Yuksel's Turkish translation, Mesaj; published in 1999, we have not seen any published translation that does not duplicate this centuries-old error. For a comparative discussion of this verse, see the Sample Comparisons section in the Introduction.

066:010 See 2:48.

67:0 In the name of God, the Gracious, the Compassionate.

67:1 Most exalted is the One in whose hands is all sovereignty, and He is capable of all things.

67:2 The One who created death and life, that He may test you, which of you will do better works? He is the Noble, the Forgiving.

67:3 He created seven heavens in harmony. You do not see any disorder in the creation by the Gracious. Keep looking; do you see any flaw?

67:4 Then look again twice; your eyes will come back humiliated and tired.

67:5 We have adorned the lower heaven with lamps, and We made it with projectiles against the devils; and We prepared for them the retribution of the blazing fire.

67:6 For those who rejected their Lord will be the retribution of hell. What a miserable destiny.

67:7 When they are thrown therein, they hear its furor as it boils.*

67:8 It almost explodes from rage. Whenever a group is thrown therein, its keepers would ask them: "Did you not receive a warner?"

67:9 They would say, "Yes indeed; a warner did come to us, but we rejected and said, **God** did not reveal anything, you are being led astray."

67:10 They would say, "If we had listened or understood, we would not be among the dwellers of the blazing fire!"

67:11 Thus, they confessed their sins. So away with the dwellers of the blazing fire.

67:12 As for those who revere their Lord unseen, they have attained forgiveness and a great reward.

67:13 Whether you keep your utterances secret, or declare them, He is fully aware of what is inside the chests.

67:14 Should He not know what He created? He is the Sublime, the Ever-aware.*

67:15 He is the One who made the earth subservient to you. So roam its paths, and eat from His provisions; and to Him is the final summoning.

67:16 Are you secure that the One in heaven will not cause the earth to rupture, thus causing it to shake?

67:17 Or are you secure that the One in heaven will not send upon you a violent storm? Then you will know the value of the warning.

67:18 Those before them have rejected; so how terrible was My requital!

67:19 Have they not looked to the birds above them, flapping their wings? The Gracious is the One who holds them in the air. He is Seer of all things.

67:20 Where is this army of yours to grant you victory without the Gracious? Indeed, ingrates are deceived.

67:21 Where is this one who can give you provisions if He holds back His provisions? Indeed, they have plunged deep into transgression and aversion.

67:22 Is one who walks with his face groveling better guided, or one who walks straight on the right path?

67:23 Say, "He is the One who initiated you, and granted you the hearing, the eyes, and the hearts. Little do you give thanks."

67:24 Say, "He is the One who placed you on earth, and to Him you will be gathered."

67:25 They say, "When will this promise come to pass, if you are being truthful?"

67:26 Say, "The knowledge is only with **God**, and I am but a clear warner."

67:27 So when they see it near, the faces of those who rejected will turn miserable, and it will be proclaimed: "This is what you had called for!"

67:28 Say, "Do you see? If **God** destroys me and those with me, or He bestows mercy upon us, who is there to protect ingrates from a painful retribution?"

67:29 Say, "He is the Gracious; we affirm Him, and we put our trust in Him. You will come to find out who is clearly astray."

67:30 Say, "What if your water becomes deep underground, who then can provide you with pure water?"

ENDNOTES

067:007-11 Many future events are told in past tense, indicating the certainty of their coming into being.

067:014 It could also be translated as "Doesn't the One who created know?"

بسم الله الرحمن الرحيم

68:0 In the name of God, the Gracious, the Compassionate.
68:1 N50, the pen, and what they write.*
68:2 You are not, by the blessing of your Lord, crazy.
68:3 You will have a reward that is well deserved.
68:4 You are on a high moral standard.
68:5 So you will see, and they will see.
68:6 Which of you are tormented.
68:7 Your Lord is fully aware of those who strayed off His path, and He is fully aware of those who are guided.
68:8 So do not obey those who deny.
68:9 They wish that you compromise, so they too can compromise.

Bad Character

68:10 Do not obey every lowly swearer.
68:11 A slanderer, a backbiter.
68:12 Forbidder of charity, a transgressor, a sinner.
68:13 Ignoble, and additionally, mischievous.
68:14 Because he possessed money and children.
68:15 When Our signs are recited to him, he says: "Tales from the past!"
68:16 We will mark him on the path.
68:17 We have tested them like We tested those who owned the farms, when they swore that they will harvest it in the morning.
68:18 They were without doubt.
68:19 So a passing sent from your Lord came to it while they all were asleep.
68:20 Thus, it became barren.
68:21 They called on one another when they awoke.
68:22 "Let us go this morning to harvest the crop."
68:23 So they went, while conversing.
68:24 "Let not a poor person come to your presence today."
68:25 They went, ready to harvest.
68:26 But when they saw it, they said, "We have gone astray!"
68:27 "Now, we have nothing!"
68:28 The best among them said, "If only you had glorified!"
68:29 They said, "Glory be to our Lord. We have transgressed."
68:30 Then they started to blame each other.
68:31 They said, "Woe to us. We sinned."
68:32 "Perhaps our Lord will grant us better than it. We repent to our Lord."
68:33 Such was the punishment. But the retribution of the Hereafter is far worse, if they only knew.
68:34 The righteous have deserved, at their Lord, gardens of bliss.

A Prophetic Description of the Followers of Hadith

68:35 Should We treat the ones who peacefully surrendered the same as those who are criminals?
68:36 What is wrong with you, how do you judge?
68:37 Or do you have another book which you study?
68:38 In it, you can find what you wish?
68:39 Or do you have an oath from Us, extending until the day of resurrection, that you can judge as you please?
68:40 Ask them: "Who of them will make such a claim?"
68:41 Or do they have partners? Then let them bring their partners, if they are truthful.*
68:42 The day will come when they will be exposed, and they will be required to prostrate, but they will be unable to.*
68:43 With their eyes subdued, humiliation will cover them. They were invited to prostrate when they were whole and able.
68:44 Therefore, let Me deal with those who reject this *hadith*; We will entice them from where they do not perceive.
68:45 I will lead them on; for My scheming is formidable.
68:46 Or did you ask them for a wage, so they are burdened by the fine?
68:47 Or do they know the future? So they have it recorded?

The Companion of the Fish

68:48 You shall be patient for the judgment of your Lord. Do not be like the companion of the fish who called out while he was in sorrow.*

68:49 Had it not been for his Lord's grace, he would be ejected to the shore, in disgrace.

68:50 But his Lord blessed him and made him righteous.

68:51 Those who have rejected almost attack you with their eyes when they hear the reminder, and they say, "He is crazy!"

68:52 It is but a reminder for the worlds.

ENDNOTES

068:001 N50. This letter/number plays an important role in the mathematical system of the Quran based on code 19. For the meaning of this letter, see 74:1-56; 1:1; 2:1; 13:38; 27:82; 38:1-8; 40:28-38; 46:10; 72:28.

Ibn Kathir is a popular commentary of the Quran, which is respected because of its reliance on *hadith* to "explain" verses of the Quran. Ibn Kathir (d. 1372), in the classic commentary carrying his name, makes the following remarks on verses 2:29 and 68:1. For this commentary, he relies mainly on a *hadith* from Abu Dawud (d. 888), one of the so-called authentic Sunni holy *hadith* books:

> "Ibn Abbas told all of you by Wasil b. Abd al-Ala al-Asadi- Muhammed b. Fudayl- al-Amash- abu Zabyan- ibn Abbas: the first thing God created is the pen. God then said to it: write! Whereupon the pen asked: what shall I write, my lord! God replied: write what is predestined! He continued: and the pen proceeded to (write) whatever is predestined and going to be to the coming of the hour. God then lifted up the water vapor and split the heavens off from it. Then God created the fish (*nun*), and the earth was spread out upon its back. The fish became agitated, with the result that the earth was shaken up. It was steadied by means of the mountains, for they indeed proudly (tower) over the earth."

After learning the intellectual level of the "affirmers," collectors, narrators, and commentators of the above *hadith*, such as Abu Dawud (d. 888), al-Tabari (d. 1516), and Ibn al-Baz (1995), it becomes clear why Muhammed would utter the words in 25:30. See 21:33; 27:88; 36:40; 39:5; 79:30.

068:041 See 42:21; 9:31; 6:112-116,145-150.

068:042 Commentaries relying heavily on *hadith* have distorted the meaning of this verse. According to a *hadith* reported by Bukhari thrice, to prove His identity, God will uncover his leg showing it to the prophets!

068:048 Instead of using the proper name of the prophet, Jonah (Yunus), this verse uses *sahib ul-hut*, that is, the Companion of the Fish. Thus, our attention is pulled to the letter N. For other examples, see 21:87; 50::13; 3:96; 7:69.

بسم الله الرحمن الرحيم

69:0 In the name of God, the Gracious, the Compassionate.
69:1 Reality!
69:2 What is the reality?
69:3 Do you know what the reality is?
69:4 Thamud and Aad rejected the Shocker.
69:5 As for Thamud, they were annihilated by the devastation.
69:6 As for Aad, they were annihilated by a furious violent wind.
69:7 He unleashed it upon them for seven nights and eight days, in succession. You could see the people destroyed in it, as if they were decayed palm trunks.*
69:8 Do you see any legacy for them?
69:9 And Pharaoh, and those before him, and the sinners, came with wickedness.
69:10 They disobeyed the messenger of their Lord. So He took them with a devastating requital.
69:11 When the water flooded, We carried you on the vessel.
69:12 That We would make it as a reminder for you, and so that any listening ear may understand.

The End

69:13 When the horn is blown once.
69:14 The earth and the mountains will be removed from their place and crushed with a single crush.
69:15 On that day the unavoidable event will come to pass.
69:16 The heavens will be torn, and on that day it will be flimsy.
69:17 The controllers will be on its borders. That day, eight will carry the dominion of your Lord above them.*
69:18 On that day, you will be exposed, nothing from you can be hidden.
69:19 As for the one who is given his record in his right, he will say, "Here, come and read my record!"
69:20 "I knew that I was going to be held accountable."
69:21 So he shall be in a life, well-pleasing.
69:22 In a lofty Paradise.
69:23 Its fruits are within reach.
69:24 "Eat and drink merrily in return for your works in days past."
69:25 As for he who is given his record in his left, he will say, "Oh, I wish I never received my record,"
69:26 "That I never knew my account,"
69:27 "I wish the end had been final,"
69:28 "My money cannot help me,"
69:29 "All my power is gone."
69:30 Take him and shackle him.
69:31 Then to hell cast him.
69:32 Then, in a chain that is the length of seventy hands, tie him up.
69:33 For he did not affirm **God**, the Great.
69:34 Nor did he advocate the feeding of the poor.
69:35 Consequently, he has no friend here today.
69:36 Nor any food, except for pollutants.
69:37 Food for the sinners.
69:38 I swear by what you see.*
69:39 What you do not see.
69:40 This is the utterance of an honorable messenger.
69:41 It is not the utterance of a poet; rarely do you affirm.
69:42 Nor the utterance of a soothsayer; rarely do you take heed.
69:43 A revelation from the Lord of the worlds.

A Reminder

69:44 Had he attributed anything falsely to Us.
69:45 We would have seized him by the right.
69:46 Then, We would have severed his life-line.
69:47 None of you would be able to prevent it.
69:48 This is a reminder for the righteous.
69:49 We know that some of you are deniers.
69:50 That it is a distress for the ingrates.
69:51 It is the absolute truth.
69:52 Therefore, you shall glorify the name of your Lord, the Great.*

ENDNOTES

069:007 The expression "seven nights and eight days" denotes a continuous period. From this expression, we might infer that the Quranic day starts from daylight rather than night, as is the case in Jewish tradition. Those who contend otherwise rely on the common Quranic usage of the word night before the word day, but they must also account for 91:3-4, where the day is mentioned before the night. The statement in verse 36:40, that night cannot pass the day, might implicitly support our position. Verse 2:187 describes the starting points of day and night. Verses 3:27 and 17:12, talk about the dawn and evening, the two overlapping periods between night and day. According to these verses, the night starts with the sunset and day ends with the disappearance of Sun light from the horizon. Similarly, day starts at down and night starts with the sunset. Also, see 10:67; 25:62; 36:37.

069:017 See 7:143; 89:22 and 74:30.

069:038 Perhaps, the vast portion of our universe is made of invisible matter. For the function of oaths, see 89:5.

069:052 Or, "Then praise your Lord by His greatest attribute/name"

70:0 In the name of God, the Gracious, the Compassionate.
70:1 Someone asked about the inevitable retribution.
70:2 For the ingrates, there is nothing that will stop it.
70:3 From **God**, Possessor of the ascending portals.
70:4 The controllers and the Spirit ascend to Him in a day which is equivalent to fifty thousand years.*
70:5 So be patient with a good patience.
70:6 They see it as far away.
70:7 We see it as near.
70:8 On the day the sky is like molten copper.
70:9 The mountains are like wool.
70:10 No friend will ask about his friend.
70:11 When they see it, the criminal will wish he can ransom his children against the retribution.
70:12 As well as his mate and his brother.
70:13 As well as his whole clan that protected him.
70:14 All who are on earth, so that he can be saved!
70:15 No, it is a flame.
70:16 Eager to roast.
70:17 It calls on those who turned away.
70:18 Who hoarded and counted.
70:19 Indeed, human being is created anxious.
70:20 When adversity touches him he is miserable.
70:21 When good touches him he is stingy.

Attributes of Muslims

70:22 Except for those who are supportive.
70:23 Who are always maintaining their support/contact prayers.
70:24 Those who set aside part of their wealth.
70:25 For the seeker and the deprived.
70:26 Those who affirm the day of Judgment.
70:27 Those who are fearful of their Lord's retribution.
70:28 The retribution of their Lord is not to be taken lightly.

70:29 Those who conceal their private parts.
70:30 Except around their spouses or those they have committed to through unofficial contract, there is no blame.*
70:31 Then, whoever seeks anything beyond this, they are the transgressors.
70:32 Those who are trustworthy and keep their pledges.
70:33 Those who uphold their testimonies.
70:34 Those who are dedicated to their support/contact prayer.
70:35 They will be honored in gardens.

The Ingrates

70:36 So what is wrong with the ingrates staring at you?
70:37 From the right and the left, in crowds?
70:38 Does every one of them hope to enter a paradise of bliss?
70:39 No, We have created them from what they know.
70:40 So I swear by the Lord of the east and the west that We are able
70:41 To replace them with better people, We can never be defeated.
70:42 So let them talk in vain and play, until they meet their day which they are promised.
70:43 When they will come out of the graves in a rush, as if they are racing towards a goal.
70:44 Their eyes are cast down, with shame covering them. This is the day which they were promised.

ENDNOTES

070:004 Time is relative. See 22:47; 32:5.

070:030 See 4:25 and 23:6.

71:0 In the name of God, the Gracious, the Compassionate.
71:1 We have sent Noah to his people: "Warn your people before a painful retribution comes to them."
71:2 He said, "My people, I am to you a clear warner."
71:3 "That you shall serve **God** and be aware of Him and obey."
71:4 "He shall forgive your sins and delay you to a predetermined time. When **God**'s time comes, then it cannot be delayed, if you know."
71:5 He said, "My Lord, I have called on my people night and day."
71:6 "But my calling only drove them away!"
71:7 "Every time I called on them so that You may forgive them, they put their fingers in their ears and they covered their heads with their outer garments and they insisted, and they became greatly arrogant."
71:8 "Then I called to them openly."
71:9 "Then I announced to them, and I spoke to them in secret."
71:10 "I said, Seek forgiveness from your Lord, for He was forgiving."
71:11 "He sends the clouds upon you constantly."*
71:12 "He provides you with money and children, and He makes for you gardens, and He makes for you rivers."

Our Evolution

71:13 "Why do you not seek **God** humbly."
71:14 "While He created you in stages?"*
71:15 "Did you not see how **God** created seven heavens in harmony?"
71:16 "He made the Moon to illuminate in them, and He made the Sun to be a lit flame?"
71:17 "**God** made you grow from the earth as plants."
71:18 "Then He returns you to it, and He brings you out totally?"
71:19 "**God** made the land for you as a plain."
71:20 "So that you may seek in it ways and paths?"
71:21 Noah said, "My Lord, they have disobeyed me and have followed one whose money and children only increased him in loss."
71:22 They plotted a great plotting.
71:23 They said, "Do not abandon your gods. Do not abandon Destroyer, neither Night Group, Helper, Deterrer, nor Eagle."
71:24 They have misguided many, but We only increase the wicked in misguidance.
71:25 Because of their sins they were drowned, then they were admitted to the fire, and they could not find beside **God** any victor.
71:26 Noah said, "My Lord, do not leave on the earth any of the ingrates at all."
71:27 "If you are to leave them, then they will misguide Your servants and they will only give birth to a wicked ingrate."
71:28 "My Lord, forgive me and my parents and whoever enters my home as an affirmer, and the affirming men and the affirming women; and do not increase the wicked except in destruction."

ENDNOTES

071:011 See 11:40-44.

071:014-17 Evolution is a marvelous assembly line of creation designed by God. See 29:18-20; 41:9-10; 7:69; 15:26; 24:45; 32:7-9. Evolution of species through mutation and cumulative selection, as subscribed by the modern scientific community, provides sufficient evidence for the existence of immanent intelligent design in nature. The theory of evolution provides more evidence for an intelligent designer than a fingerprint on a canvas could provide clues about the identity of its human painter. Inferring the existence and some attributes of an intelligent designer from nature is as equally scientific as inferring the existence and some attributes of an unknown creature from its footprints left in the sand. For a debate on Evolution versus Intelligent Design, see the Appendix titled "*The Blind Watch-Watchers or Smell the Cheese: An Intelligent and Delicious Argument for Intelligent Design in Evolution*" by Edip Yuksel.

72:0 In the name of God, the Gracious, the Compassionate.

72:1 Say, "It has been revealed to me that a group of Jinn were listening." They said, "We have heard an interesting recitation/Quran!"*

72:2 "It guides to what is correct, so we affirmed it, and we will not set up anyone with our Lord."

72:3 "Exalted is the majesty of our Lord, He has not taken a wife nor a son."

72:4 "It was the foolish one amongst us who used to tell lies about **God**."

72:5 "We had thought that neither people nor the Jinn would ever utter a lie against **God**."

72:6 "There were men from among people who used to seek help from the men among the Jinn, but they only helped increase them in sin."

72:7 "They thought as you thought that **God** would not send anyone."

72:8 "We touched the heavens but found it full of powerful guards and projectiles."

72:9 "We used to sit in it in places of listening, but anyone who sits now finds a projectile seeking him."

72:10 "We do not know, is it bad that is intended for those on earth, or does their Lord want them to be guided?"

72:11 "Among us are those who are good doers, and some of us are opposite to that, we are in many paths."

72:12 "We affirm that we cannot escape **God** on earth, nor can we escape Him if we run."

72:13 "When we heard the guidance, we affirmed it. So whoever affirms his Lord, then he will not fear a decrease in reward, nor a burdensome punishment."

72:14 "Among us are those who peacefully surrendered, and among us are the compromisers. As for those who have peacefully surrendered, they have sought what is correct."

72:15 As for the compromisers, they are firewood for hell.

72:16 Had they walked on the right path, We would have provided them with abundant water.

72:17 To test them with it. Whosoever turns away from the remembrance of his Lord, He will enter him a severe retribution.*

72:18 The temples are for **God**, so do not call on anyone with **God**.

Prophecy Regarding the Numerical System

72:19 When **God**'s servant stood up to call on Him, they nearly banded to oppose him.*

72:20 Say, "I only call on my Lord, and I do not associate anyone with Him."

72:21 Say, "I have no power to cause you harm nor to show you what is right."

72:22 Say, "No one can protect me from **God**, and I will not find any refuge except with Him."

72:23 It is but an announcement from **God**, and His messages. Whosoever disobeys **God** and His messenger, then he will have the fire of hell to dwell eternally therein.

72:24 Until they see what they are promised, then they will know who has the weakest ally and is least in number.

72:25 Say, "I do not know if what you are promised is near, or if my Lord will make it distant."

72:26 The knower of the unseen, He does not reveal His knowledge to anyone.

72:27 Except to whom He has accepted as a messenger; then leads way in its present and future as a secret observer.

72:28 So that He knows that they have delivered the messages of their Lord, and He encompasses all that is with them, and He has counted everything in numbers.*

ENDNOTES

072:001 Jinns are descendants of Satan or Satan (7:27), invisible (6:86; 72:1), made of energy (10:27), swift (27:10, 39), righteous and evildoers (72:11), possess power/ability to exit the atmosphere of Earth (72:8), and have physical powers not common to humans (27:39). Also see 35:1

072:017 For the word *ZKR* (reminder) see 15:9; 21:2; 21:24,105; 26:5; 29:51; 38:1,8; 41:41; 44:13; 74:31,49,54.

072:019-28 The prophecy involving numbers was fulfilled in 1974. The person's name who was chosen to unveil the prophecy in Chapter 74, which is summarized with the number 19, is mentioned in the Quran, together with all its derivatives exactly 19 times and 4 of them occur in this chapter (72:2; 72:10; 72:14; 72:21). Those who notice statistical anomalies and reflect on the relationship of statistical anomalies and their semantic context will have no doubt about this divine revelation. Furthermore, the name of the person with its exact format is mentioned twice in the Quran (40:29 and 40:38) and they sandwich a quote from unappreciative people who rejected a new messenger by arguing, "God will not send any messenger after him." Interestingly, a polytheist terrorist group affiliated with Bin Laden's al-Qaida, al-Fuqra or al-Fuqara, killed that messenger in early 1990, as another prophetic remark in verses 72:19 and 40:28. Considered to be al-Qaida's first terrorist act in the USA, the events following the murder brought the Islamic reformation movement started by the messenger to the attention of the entire world. In this and many other events, there are signs for those who reflect. See 3:81; 40:28-38; 74:1-56.

بسم الله الرحمن الرحيم

73:0 In the name of God, the Gracious, the Compassionate.
73:1 O you burdened with heavy responsibility,
73:2 Stand the night except for a little.
73:3 Half of it, or a little less than that.
73:4 Or a little more and enunciate the Quran thoughtfully and distinctly.
73:5 We will place upon you a saying which is heavy.
73:6 The night-time production is more efficient and better for study.
73:7 For you have many duties during the day.
73:8 Remember the name of your Lord and devote to Him completely.
73:9 The Lord of the east and the west, there is no god but He; so take Him as a protector.
73:10 Be patient over what they say and depart from them in a good manner.
73:11 Leave Me to deal with the deniers who have been given the good things and give them time for a while.
73:12 We have with Us chains and a raging fire.
73:13 Food that chokes, and a painful retribution.
73:14 The day the earth and the mountains shake, and the mountains become a crumbling pile.
73:15 We have sent to you a messenger as a witness over you, as We have sent to Pharaoh a messenger.
73:16 But Pharaoh disobeyed the messenger, so We took him in a severe manner.
73:17 So how can you be righteous if you have rejected, on a day when the children become gray-haired?
73:18 The heavens will shatter with it. His promise is always delivered.
73:19 This is a reminder, so let him who wishes to take a path to his Lord.
73:20 Your Lord knows that you rise a little less than two thirds of the night, and half of it, and one third of it, as well as a group of those who are with you. **God** measures the night and the day. He knows that you will not be able to keep up, so He pardons you. So study what is made easy of the Quran. He knows that there will be sick among you, and others that venture out in the land seeking from **God**'s bounty, and others who are fighting in the cause of **God**, so study what you can of it. Maintain the contact prayer and contribute towards betterment and give **God** a loan of righteousness. Whatever you put forth yourselves, you will find it with **God**, for it is better and a greater reward. Seek **God**'s forgiveness, for **God** is Forgiving, Compassionate.

بسم الله الرحمن الرحيم

74:0 In the name of God, the Gracious, the Compassionate.

74:1 O you hidden one.*

74:2 Stand and warn.

74:3 Your Lord glorify.

74:4 Your garments purify.*

74:5 Abandon all that is vile.

74:6 Do not be greedy.

74:7 To your Lord be patient.

74:8 So when the holes are punched.*

74:9 That will be a very difficult day.

74:10 Upon the ingrates it will not be easy.

74:11 So leave Me alone with the one I have created.

74:12 I gave him abundant wealth.

74:13 Children to bear witness.

74:14 I made everything comfortable for him.

74:15 Then he wishes that I give more.

74:16 No. He was stubborn to Our signs.

74:17 I will exhaust him in climbing.

74:18 He thought and he analyzed.

74:19 So woes to him for how he analyzed.

74:20 Then woe to him for how he analyzed.

74:21 Then he looked.

74:22 Then he frowned and scowled.

74:23 Then he turned away in arrogance.

Prophecy Fulfilled

74:24 He said, "This is nothing but an impressive magic."

74:25 "This is nothing but the words of a human."

74:26 I will cast him in the *Saqar*.

74:27 Do you know what *Saqar* is?

74:28 It does not spare nor leave anything.*

74:29 Manifest to all the people.*

74:30 On it is nineteen.*

74:31 We have made the guardians of the fire to be angels/controllers; and We did not make their number except as a test for those who have rejected, to convince those who were given the book, to strengthen the affirmation of the affirmers, so that those who have been given the book and those who affirm do not have doubt, and so that those who have a sickness in their hearts and the ingrates would say, "What did **God** mean by this example?" Thus **God** misguides whoever/whomever He wishes, and He guides whoever/whomever He wishes. None knows your Lord's soldiers except Him. It is but a reminder for people.

74:32 No, by the moon.*

74:33 By the night when it passes.

74:34 By the morning when it shines.

74:35 It is one of the great ones.

74:36 A warning to people.

74:37 For any among you who wishes to progress or regress.

74:38 Every person is held by what it earned;

74:39 Except for the people of the right.

74:40 In gardens, they will be asking

74:41 About the criminals.

74:42 "What has caused you to be in *Saqar*?"

74:43 They said, "We were not of those who maintained solidarity with community (or observed contact prayer)."*

74:44 "We did not feed the poor."

74:45 "We used to participate with those who spoke falsehood."

74:46 "We used to deny the day of Judgment."*

74:47 "Until the certainty came to us."

74:48 Thus, no intercession of intercessors could help them.

74:49 Why did they turn away from this reminder?*

74:50 Like fleeing zebras,

74:51 Running from the lion?

74:52 Alas, every one of them wants to be given separate manuscripts.

74:53 No, they do not fear the Hereafter.

74:54 No, it is a reminder.

74:55 Whosoever wishes will take heed.

74:56 None will take heed except if **God** wills. He is the source of righteousness and the source of forgiveness.

ENDNOTES

074:001-2 The All-Wise God, though He revealed the Quran to Muhammed about 14 centuries ago, kept its mathematical miracle as a gift to our times.

Rashad Khalifa, an Egyptian biochemist living in the USA, did not have any knowledge that his curiosity regarding the meaning of the alphabet letters that initialize 29 chapters of the Quran would end up with the discovery of its mathematical system. His computerized study that started in 1969 gave its fruits in early 1974 by the discovery of the 14-century-old SECRET. The discovery of this code opened a new era that has been changing the paradigm of those who do not turn off the circuits of their brains in matters related to God. This discovery provided very powerful verifiable and falsifiable evidence for God's existence, His communication with us, and the fulfillment of a great prophecy. Besides, it saved monotheists from the contradiction of trusting past generations regarding the Quran, knowing that they have fabricated, narrated and followed volumes of man-made teachings, demonstrated ignorance, distorted the meaning of the Quranic words, and glorified gullibility and sheepish adherence to the fatwas of clerics.

If the Code 19 was going to provide strong evidence for the existence of God and for the authenticity of the Quran, then it is reasonable to expect that the identity of the discoverer and the time of the discovery would not be coincidental. Indeed, the events have demonstrated a prophetic design in the timing of this miraculous mathematical design.

The number 19 is mentioned only in a chapter known as "The Hidden," the 74th chapter of the Quran. Juxtaposing these two numbers yields 1974, exactly the year in which the code was deciphered. (Calendar based on the birth of Jesus and the solar year are accepted by the Quran as units of calculating time. See, 19:33; 43:61; 18:24. Besides, this is the most commonly used calendar in the world.) If we multiply these two numbers, 19x74, we end up with 1406, the exact number of lunar years between the revelation of the Quran and the discovery of the code.

In January 31, 1990, Rashad Khalifa was assassinated in Tucson, Arizona, by a terrorist group affiliated with al-Qaida. Ironically, soon after his departure, ignorant people started idolizing him and created a cult distorting his message of strict monotheism. Also See 27:82-85; 40:28-38; 72:19-28. For further information on this historical discovery, see the Appendix *On it is Nineteen*.

074:004 What could the metaphorical message of this verse be?

074:008 NaQaRa covers the following meanings: Pecking a thing with a pointed instrument; engraving stone; making sound by snapping fingers; making smacking sound through tongue and palate; perforating a hole in a thing; pecking hole with beak; making a groove in datestone; searching, inquiring, investigating, scrutinizing, or examining an issue; opening a little hole; and more. I think it is a prophetic description of programming or entering data via punching holes in cards of mainframe computers in early years of computers. Dr Rashad Khalifa entered the Quran into an IBM Mainframe Computer in 1969 exactly as it is described by these two words derived from the same root, NQR. (also see 4:53,124)

074:028 The Quran does not want us to rush into defining the meaning of the word *saqar*, since it introduced it with a question followed up by an explanation. Past generations understood it as another word for Hell. After the discovery of the role of 19 in the Quran and fulfillment of the prophecies in this chapter and in other verses of the Quran, now we know that *Saqar* refers to an intellectual punishment. The extreme allergy and aversion demonstrated by Sunni and Shiite *mushriks* against the number 19 is part of the prophetic fulfillment of these verses.

074:029 This verse, like many of the verses and words in this chapter, before 1974 had multiple meanings (*mutashabih*) and commentators were aware of them. Though all commentators knew the primary meaning of this verse to be "manifest to all the people," few of them mentioned it in their commentaries, and most of them chose the alternative meaning, "scorching the skin." Some, picked a mixture and preferred "scorching humans." Commentators who lived in pre-1974 had an excuse to use 'scorch' and 'burn,' but after the discovery of the secret, after the prophecy of this chapter was unveiled, translators and commentators have no excuse for repeating the misunderstanding of previous generations.

We translate the word *bashar* as "humans" while many classic commentaries rendered it as "skin." The Quran uses it invariably for humans. For instance, it used for humans just two verses down, at the end of verse 31. The word we render as "manifest" or "succeeding tablets/screens," is *lawaha* and the Quran never use it to mean "scorching," "burning" or "shriveling." See 7:145,150,154; 54:13; 75:22.

After learning about the discovery of the mathematical system based on the number 19, it takes a blindness and bigotry. Denying such an obvious mathematical pattern in the Quran, I think, is another miracle. Nothing short of miracle can accomplish this. Those who assert that the only response the Lord of the Universe could give to a skeptic who challenged the authenticity of His word is to tell such a skeptic, "I will throw you in fire and scorch you!", do not appreciate God as He should be appreciated. These people can accept a God who can produce the best literary work, but they cannot accept Him as the ultimate mathematician and advocate. They see hell, fire and smoke in verses where an amazing intellectual piece of evidence is prophesied. See 6:91; 22:74; 39:67. For a comparative discussion of this verse, see the Sample Comparisons section in the Introduction.

074:030-37 I have presented and discussed the many details of this physical, verifiable and falsifiable miracle in a Turkish book titled Üzerinde 19 Var (On it is Nineteen) and its improved English version is published under the title, NINETEEN: *God's Signature in Nature and Scripture*. In fact, you do not need to read that book to witness the great miracle for yourself. All you need is to hear that the literal and numeral units of the Quran are designed according to a mathematical system based on code 19. If you do your research objectively, with a critical mind, you will discover many features that already have been discovered and witnessed by many. You should also be informed about two extreme reactions and groups. The majority of the followers of *hadith* and *sunna* have a great allergy and aversion against this number. Their blood pressure goes up when they hear the number 19. They are quick in repeating excuses, or false claims made by their scholars. Another group, however, entails the manipulators and innumerates who are too gullible and eager to discover even more examples. They are somehow impressed by the pattern of 19 in the Quran, but because of their lack of comprehension of the laws of probability and solid base in mathematics, they juggle with numbers in an arbitrary and anecdotal fashion and come up with many so-called miracles.

A careful and objective reader will notice that short verses build and prepare the reader to verse 74:30 and the verses afterwards describe the function of the number 19 and the reaction of people to it. Verse 74:30 is the longest verse in the Quran in proportion to the average length of verses in the chapters they belong. The reader, at the start, encounters "the hidden one" and is informed about a typical profile of an arrogant and ignorant antagonist with a fast paced series of short descriptions that lead to a question and an enigmatic answer in 74:30. The enigma is described in cryptic, yet prophetic words of in 74:31. Later, following a powerful emphasis on the importance of the prophecy, verse 74:37 describes those who would witness the prophetic fulfillment of these verses as "progressive" and depicts those who deprive themselves from such a blessing as "regressive" people. The rest of the chapter elaborates on the characteristics of two groups and connects them to their reaction to the prophetic message (ZiKRa).

Many verses of the Quran implicitly or explicitly refer to this prophetic event that will be witnessed only by those who are appreciative. For instance, see 1:1; 2:2; 2:118; 6:4-5; 6:25-26; 6:67; 6:104; 6:158; 7:146; 10:1; 10:20; 12:1; 13:1; 13:38-40; 15:1; 15:9; 17:88-96; 20:133-135; 25:4-6; 26:1-6; 27:1; 27:82-85; 27:93; 28:1-2; 29:1-2; 29:50-52; 31:1-2; 38:1-8; 38:29; 38:87-88; 40:44; 40:78-83; 41:53; 46:10; 54:1-5; 59:21; 72:28; 78:27-30; 83:7-21; 98:4. Also, see 4:82 and 13:38. We highly recommend Rashad's book, Quran: Visual Presentation of the Miracle, which is available at amazon.com. For the details of this extraordinary and intricate mathematical design, new discoveries, and refutations of major criticism see: NINETEEN: God's Signature in Nature and Scripture, Edip Yuksel, Brainbow Press, 2009.

Also, see 2:1; 2:6-7 and 7:180.

074:032 For the meaning and function of the Quranic oaths, see 89:5.

074:043 Statements starting from this verse describe both the religious and nonreligious unappreciative opponents. The word we translated as "we were not of those who offered support (or observe contact prayer)" reflects both meanings within the context. See 2:157; 9:99,103; 33:43,56; 75:31.

074:046 The majority of Jews, Christians, Sunnis and Shiites do not affirm the Day of Judgment. According to the definition of the Quran, the ruler and judge of that day is God alone (1:4) and none can help or harm the other (82:19). Those who believe in intercession, by their very belief, have denied the Day of Judgment as defined by God. Atheists, on the other hand, deny the Day of Judgment explicitly.

074:049 For the prophetic use of the word *ZKR* (reminder), see 15:9; 21:2; 21:24,105; 26:5; 29:51; 38:1,8; 41:41; 44:13; 72:17; 74:31,54.

بسم الله الرحمن الرحيم

75:0 In the name of God, the Gracious, the Compassionate.
75:1 I swear by the day of Judgment.*
75:2 I swear by the person which is self-blaming.
75:3 Does the human being think that We will not gather his bones?
75:4 Indeed, We were able to make his fingertips.*
75:5 No, the human being desires to exceed the limits.
75:6 He asks: "When is the day of resurrection?"
75:7 So, when the sight is dazzled.
75:8 The Moon is eclipsed.
75:9 The Sun and Moon are joined together.
75:10 Man will say on that day: "Where can I escape!"
75:11 No. There is no refuge.
75:12 To your Lord on that day is the abode.
75:13 Man will be told on that day what he has put forward, and what he has done.
75:14 Indeed, the human being will testify against himself.
75:15 Even though he puts forth his excuses.

God is the One Who Explains the Quran

75:16 Do not move your tongue with it to make haste.*
75:17 It is for Us to collect it and relate it.
75:18 So when We relate it, you shall follow its revelation.
75:19 Then it is for Us to clarify it.
75:20 Alas, you all like this world.
75:21 Neglecting the Hereafter.
75:22 Faces on that day will be shining.
75:23 Looking at their Lord.
75:24 Faces on that day will be gloomy.
75:25 Thinking that a punishment is coming to them.
75:26 Alas, when it reaches the throat.
75:27 It will be said, "Who can save him?"
75:28 He assumes it is the time of passing.
75:29 The leg is buckled around the other leg.

75:30 To your Lord on that day he will be driven.
75:31 For he did not affirm nor support.*
75:32 But he denied and turned away.
75:33 Then he went to his family admiring himself.
75:34 Woe to you and woe to you.
75:35 Then woe to you and woe to you.
75:36 Did the human being think that he will be left neglected?
75:37 Was he not a seed from sperm put forth?
75:38 Then he was an embryo, so he was created and developed.
75:39 Then He made the two pairs, male and female.
75:40 Is One as such then not able to resurrect the dead?

ENDNOTES

075:001-2 For the meaning and function of the Quranic oaths, see 89:5.

075:004 Is this verse also implying that our fingerprints, with their unique designs, are like our identity cards?

075:016 See 20:114

075:031 See 74:43.

76:0 In the name of God, the Gracious, the Compassionate.
76:1 Was there not a time in the past when the human being was nothing to even be mentioned?
76:2 We have created the human from a seed that is composite, We test him, so We made him hear and see.
76:3 We have guided him to the path, either to be thankful or to reject.
76:4 We have prepared for the ingrates chains and collars and a blazing fire.
76:5 As for the pious, they will drink from a cup which has the scent of musk.
76:6 A spring from which the servants of **God** drink, it gushes forth abundantly.
76:7 They fulfill their vows, and they fear a day whose consequences are widespread.
76:8 They give food out of love to the poor and the orphan and the captive.
76:9 "We only feed you for the sake of **God**; we do not desire from you any reward or thanks."
76:10 "We fear from our Lord a day, which will be horrible and difficult."
76:11 So **God** shielded them from the evil of that day, and He cast towards them a look and a smile.
76:12 He rewarded them for their patience with paradise and silk.
76:13 They are reclining in it on raised couches, they do not have in it excessive Sun nor bitter cold.
76:14 The shade is close upon them, and the fruit is hanging low within reach.
76:15 They are served upon with bowls of silver and glasses of crystal.
76:16 Crystal laced with silver, measured accordingly.*
76:17 They are given to drink in it from a cup which has the scent of ginger.
76:18 A spring therein which is called 'Salsabeel'.

76:19 They are encircled with eternal children. If you see them you will think they are pearls which have been scattered about.

76:20 If you look, then you will see a blessing and a great dominion.

76:21 They will have garments of fine green silk, necklaces and bracelets from silver, and their Lord will give them a cleansing drink.

76:22 "This is the reward for you, and your struggle is appreciated."

76:23 We have sent down to you the Quran at once.

76:24 So be patient for the judgment of your Lord, and do not obey from them any sinner or ingrate.

76:25 Remember the name of your Lord morning and evening.

76:26 From the night you shall prostrate to Him and praise Him throughout.

76:27 These people like the current life, and they ignore a heavy day.

76:28 We have created them, and We have made them resolute. When we wished, We replaced their kind completely.

76:29 This is a reminder, so let whoever wills take a path to his Lord.

76:30 You cannot will, except if **God** wills. **God** is Knowledgeable, Wise.*

76:31 He admits whoever/whomever He wills to His mercy. As for the wicked, He has prepared for them a painful retribution.

ENDNOTES

076:016 Translucent cups made of silver and trees growing in the middle of fire are parables. See 2:26; 17:60 (For the meaning of the word we translated as "translucent," see 27:44.

076:030 See 57:22-23.

77:0 In the name of God, the Gracious, the Compassionate.
77:1 By those which are sent to benefit.*
77:2 By the blows that explode.
77:3 By the distributors that dispense.
77:4 By the separators that divide.
77:5 By the transmitters of the message.
77:6 As an excuse or a warning.
77:7 What you are being promised will come to pass.
77:8 So when the stars are dimmed.
77:9 When the sky is opened.
77:10 When the mountains are destroyed.
77:11 When the messengers are gathered.
77:12 For what day has it been delayed?
77:13 For the day of Separation.
77:14 Do you know what the day of Separation is?
77:15 Woes on that day to the deniers!
77:16 Did We not destroy the ancient people
77:17 Then We made others succeed them?
77:18 It is such that We do to the criminals.
77:19 Woes on that day to the deniers!
77:20 Did We not create you from a simple water/liquid,*
77:21 Then We made it in a place of protection,
77:22 Until a time that is predetermined?
77:23 So We measured, and We are the best to measure.
77:24 Woes on that day to the deniers.
77:25 Did We not make the earth an abode,
77:26 Living and dead.
77:27 We made massive stabilizers in it, and We gave you to drink fresh water?
77:28 Woes on that day to the deniers!
77:29 Depart unto what you have denied.
77:30 Depart unto a shadow with three columns.
77:31 Neither does it give shade, nor does it avail from the flames.
77:32 It throws sparks as huge as logs.
77:33 As if they were yellow camels/ropes.
77:34 Woes on that day to the deniers!
77:35 This is a day when they shall not speak.
77:36 Nor will it be permitted for them so they can make excuses.
77:37 Woes on that day to the deniers!
77:38 This is the day of Separation where We have gathered you with the ancient people.
77:39 So if you have a scheme, then make use of it.
77:40 Woes on that day to the deniers!
77:41 The righteous are among shades and springs.
77:42 Fruit from what they desire.
77:43 "Eat and drink comfortably for what you used to do."
77:44 It is such that We reward the good doers.
77:45 Woes on that day to the deniers!
77:46 "Eat and enjoy for a little while, for you are criminals."
77:47 Woes on that day to the deniers!
77:48 When they are told to kneel, they do not kneel.*
77:49 Woes on that day to the deniers!
77:50 So in what *hadith* after it will they affirm?

ENDNOTES

077:001-6 For the meaning and function of the Quranic oaths, see 89:5.

077:020-23 According to the findings of modern embryology, the exact pregnancy period is 266 days or 38 weeks.

077:048 For a discussion of the meaning of bowing and prostrating, see 16:49.

بسم الله الرحمن الرحيم

78:0 In the name of God, the Gracious, the Compassionate.
78:1 What are they inquiring about?
78:2 About the grand news.*
78:3 The one which they are in disagreement about.
78:4 No, they will come to know.
78:5 No, then again, they will come to know.
78:6 Did We not make the earth a resting ground?
78:7 The mountains as pegs?
78:8 We created you in pairs?
78:9 We made your sleep for resting?
78:10 We made the night as a covering?
78:11 We made the day to work in?
78:12 We constructed above you seven mighty ones?
78:13 We made a flaming light?
78:14 We sent down abundant water from the rain clouds.
78:15 To bring out with it seeds and plants.
78:16 Gardens of thick growth?
78:17 The day of Separation is an appointed time.
78:18 The day when the horn is blown and you come in crowds.
78:19 The heaven is opened, so it becomes gates.
78:20 The mountains will be moved as if they were a mirage.
78:21 For hell is in wait.
78:22 For the transgressors it is a dwelling place.
78:23 They will abide in it for eons.
78:24 They will not taste anything cold in it nor drink.
78:25 Except for boiling water and filthy discharge.
78:26 A precise recompense.

The Surprise

78:27 They did not expect a reckoning/computation.*
78:28 They denied Our signs greatly.
78:29 Everything We have counted in a record.
78:30 So taste it, for no increase will come to you from Us except in retribution.
78:31 As for the righteous, they will have success.
78:32 Gardens and vineyards.
78:33 Scrumptious and ripe.*
78:34 A cup that is full.
78:35 They do not hear in it any vile talk or lies.
78:36 A reward from your Lord, in recognition for what is done.

Not Intercession; Only Testimony to the Truth

78:37 The Lord of heavens and earth and what is between them, the Gracious. They do not possess any authority beside Him.
78:38 The day when the Spirit and the controllers stand in line, none will speak unless the Gracious permits him and he speaks what is true.
78:39 That is the day of truth, so let whoever wills seek a way to his Lord.
78:40 I have warned you of a retribution which is close, the day when man will look at what he has brought forth and the ingrate will say, "I wish I were dust!"

ENDNOTES

078:002 The revelation of the Quran created a big controversy when it was first revealed to Muhammed. The discovery of its secret too might be considered another controversial big event. See 38:67

078:027 Maybe the word *hesab* has another implication besides the reckoning or computation of one's work on the Day of Judgment?

078:033 The verse is translated by many male translators as description of women, but in the context, it makes more sense to consider it as the description of the fruits of the vineyards mentioned in the previous verse.

79:0 In the name of God, the Gracious, the Compassionate.
79:1 By those that pull forcibly.*
79:2 Those that release vigorously.
79:3 Those that float along.
79:4 Those that press forward in a race.
79:5 So as to carry out a command.
79:6 On the day the ground shakes.
79:7 It will be followed by the second blow.
79:8 Hearts on that day will be terrified.
79:9 Their eyes cast down.
79:10 They will say, "Shall we be returned to live our lives."
79:11 "Even after we were crumbled bones?"
79:12 They said, "This is an impossible recurrence."
79:13 But all it takes is one blow.
79:14 Whereupon they will rise up.

The Story of Moses

79:15 Did the *hadith* of Moses come to you?
79:16 His Lord called him at the holy valley of Tuwa.
79:17 "Go to Pharaoh, for he has transgressed."
79:18 Tell him: "Would you not be purified?"
79:19 "I will guide you to your Lord, that you may turn reverent."
79:20 He then showed him the great sign.
79:21 But he rejected and rebelled.
79:22 Then he turned away in a hurry.
79:23 So he gathered and proclaimed.
79:24 He said, "I am your lord, the most high."*
79:25 So **God** seized him with retribution in the Hereafter, as well as in the first life.
79:26 In that is a lesson for those who are aware.
79:27 Are you a more powerful creation than the sky which He built?
79:28 He raised its height, and perfected it.
79:29 He covered its night and brought out its morning.
79:30 The land after that, He made it like an egg.*
79:31 He brought forth from it its water and pasture.
79:32 The mountains He fixed firmly.
79:33 All this to be a provision for you and your livestock.
79:34 Then, when the great blow comes.
79:35 The day when the human remembers all that he strove for.
79:36 Hell will be apparent to all who can see.
79:37 As for the one who transgressed.
79:38 He was preoccupied with the worldly life.
79:39 Then hell will be the abode.
79:40 As for the one who revered the majesty of his Lord and restrained himself from desire.
79:41 The Paradise will be the abode.
79:42 They ask you about the Moment: "When is its appointed time?"
79:43 You have no knowledge of it.
79:44 To your Lord is its term.
79:45 You are simply to warn those who fear it.
79:46 For the day they see it, it will be as if they had remained an evening or half a day.

ENDNOTES

079:001-5 For the meaning and function of the Quranic oaths, see 89:5.

079:024 Slave owners are called *rab* (lord, master) in Arabic (12:41). Pharaohs were claiming to be the lord/master of those humans whom they have subjugated. Enslaving humans and thus claiming to be their *rab* (lord, master) is polytheism, or associating one's ego with God as His partner See 12:39-42; 90:13; 4:25.

079:030 The Arabic word *dahaha* comes from root *dahy*, which means egg. The earth, with its physical shape and geologic layers resembles an egg. Though the roundness of the earth was known by philosophers before the revelation of the Quran, ordinary people did not find it a reasonable or common-sense fact. Therefore, almost all the commentators of the Quran tried to interpret the word so that they could save their flat earth from being curved. They reasoned that the verse must be an allegorical description, that God must have meant the nest or the place of the egg from the word egg. So, they translated this verse as follows: "and after that He spread the earth"!

Abd al-Aziz bin Baz was the chief cleric of Saudi Arabia, the head of the Council of Senior Religious Scholars for three decades. He had a great impact on regressive and oppressive laws in Saudi Arabia, including the ban on women's driving. A book authored by Bin Baz was published in 1975 carried the following title: "*Scientific and Narrative Evidence for that the Earth is Fixed, the Sun is Moving and it is Possible to Go to the Planets.*" The book was not published by any publishing house; it was published by none other than the Islamic University of Medina. In that book, Bin Baz complains about a new heresy; he is saddened to see, well more accurately, hear, people talking about the motion of the earth, and he wants to put a stop to that heresy. In page 23, Bin Baz after listing some *hadiths*, issues a fatwa, asserting that those who believe that the earth is rotating are *kafirs* (ingrates), and if they were Muslims before, they became apostates. The Saudi Sunni leader does not stop there and explains the ramification of the fatwa: any Muslim believing in the rotation of the earth loses his or her right to life and property; they should be killed! This same cleric was the head of an international conference of Sunni scholars representing 38 countries discussing the Salman Rushdi affair. Then, Saudi Arabia was competing with Iran regarding leadership in the Islamic world and this was the hot issue. The conference issued a unanimous fatwa on March 19, 1989, condemning Rushdi and Rashad Khalifa, as apostates. The Western world by then knew Rushdi, but not many westerners were familiar with the second name in the fatwa. Dr. Rashad Khalifa, the late leader of the modern reformist movement and the discoverer of the mathematical code of the Quran, in less than a year after this fatwa, would be assassinated by a terrorist group affiliated with the Saudi terrorist Osama bin Laden on January 31, 1990, in his mosque in Tucson, Arizona.

Bin Baz, in his book, quoted some verses and many *hadith* to support his position that the earth is fixed. After his expressing his religious verdict of the death penalty for the apostates who believe in a moving earth, he included the following reasoning as his scientific evidence:

> "If the world was rotating as they assert, countries, mountains, trees, rivers, seas, nothing would be stable; humans would see the countries in the west in the east, the ones in the east in the west. The position of the *qibla* would change continuously. In sum, as you see, this claim is false in many respects. But, I do not wish to prolong my words."

Towards the end of the 20th century, a "university" of a Sunni country publishes such nonsense authored by the highest-ranking religious leader of that country! Considering the Ottoman chief clerics banned the import and use of the printing machine for about 300 years, it becomes clear why the so-called Muslim countries are so trailing behind in civilization; socially, politically, and in science and technology.

For a comparative discussion of this verse, see the Sample Comparisons section in the Introduction.

The Bible, because of the tampering of many human hands, contains questionable information about the shape of earth: Isaiah 11:12; Matthew 4:8; Revelation 7:1.

See 4:82; 68:1 and 39:5.

80:0 In the name of God, the Gracious, the Compassionate.
80:1 He frowned and turned away.
80:2 When the blind one came to him.
80:3 What makes you know, perhaps he is seeking to purify?
80:4 Or to remember, so the remembrance will benefit him?
80:5 As for the one who was conceited;
80:6 To him you attended.
80:7 Why does it concern you that he does not want to purify?
80:8 As for the one who came to you seeking.
80:9 While he was fearful.
80:10 You were too preoccupied for him.
80:11 No, this is but a reminder.
80:12 For whoever wills to remember.
80:13 In records which are honorable.
80:14 Exalted and pure.
80:15 By the hands of scribes.
80:16 Honorable and righteous.
80:17 Woe to human being; how unappreciative!
80:18 From what did He create him?
80:19 From a seed He created him and designed him.
80:20 Then the path He made easy for him.
80:21 Then He made him die and buried him.
80:22 Then if He wishes He resurrects him.
80:23 Alas, he did not fulfill what He commanded him.
80:24 Let the human being look to his provisions.
80:25 We have poured the water abundantly.
80:26 Then We cracked the land with cracks.
80:27 We made grow in it seeds.
80:28 Grapes and pasture.
80:29 Olives and palm trees.
80:30 Gardens in variety.
80:31 Fruits and vegetables.
80:32 An enjoyment for you and your livestock.
80:33 So when the screaming shout comes.
80:34 The day when a person will run from his brother.
80:35 His mother and father.
80:36 His mate and children.
80:37 For every person on that day is a matter that concerns him.
80:38 On that day are faces which are openly displayed.
80:39 Laughing and seeking good news.
80:40 Faces on that day with dust on them.
80:41 Being burdened by remorse.
80:42 Those are the ingrates, the wicked.

بســـــم الله الرحمن الرحيـــــم

81:0 In the name of God, the Gracious, the Compassionate.
81:1 When the Sun is rolled.
81:2 When the stars fade away.
81:3 When the mountains are moved.
81:4 When the reproduction is ended.
81:5 When the beasts are gathered.
81:6 When the seas are made to boil.
81:7 When the persons are paired.*
81:8 When the girl killed in infancy is asked,
81:9 "For what crime was she killed?"
81:10 When the records are displayed.
81:11 When the sky is removed.
81:12 When hell is ignited.
81:13 When Paradise is made near.
81:14 Every person will know what it had done!
81:15 So, I swear by the collapsing/elusive stars.*
81:16 That run and hide.
81:17 The night when it passes.
81:18 The morning when it breathes.
81:19 It is the saying of an honorable messenger.
81:20 Authorized by the Possessor of the throne.
81:21 Obeyed, and trustworthy.
81:22 Your friend is not crazy.
81:23 He saw him by the clear horizon.*
81:24 He has no knowledge of the future.
81:25 It is not the saying of an outcast devil.
81:26 So where will you go?
81:27 It is but a reminder for the worlds.
81:28 For whomever of you wishes to be straight.
81:29 You cannot will anything except if **God** also wills, the Lord of the worlds.*

ENDNOTES

081:007 See 15:29 and 17:85.

081:008 See 46:15; 16:58-59; 17:31.

081:015-19 For the meaning and function of the Quranic oaths, see 89:5.

081:023 See 53:1-18

081:029 See 57:22-23.

بسم الله الرحمن الرحيم

82:0 In the name of God, the Gracious, the Compassionate.
82:1 When the sky is cracked.
82:2 When the planets are scattered.
82:3 When the seas burst.
82:4 When the graves are laid open.
82:5 Then the person will know what it has brought forth and what it has left behind.
82:6 O mankind, what has turned you arrogant against your Lord, the Most Generous?
82:7 The One who created you, then designed you, then proportioned you?*
82:8 In any which form He chose, He constructed you.
82:9 No, you are but deniers of the system.
82:10 Over you are those who watch.
82:11 Honorable scribes.
82:12 They know what you do.
82:13 The pious are in Paradise.
82:14 The wicked are in hell.
82:15 They will enter it on the day of *Deen*.
82:16 They will not be absent from it.
82:17 Do you know what the day of *Deen* is?
82:18 Then again, do you know what the day of *Deen* is?
82:19 The day when no person possesses anything for any other person, and the decision on that day is to **God**.*

ENDNOTES

082:007 See 4:119.

082:019 This is the only chapter that ends with the word *Allah* (God) and it occurs in the 19th verse as the 19th occurrence from the end of the Quran. For the context of the 19th occurrence of the word *Allah* from the beginning of the Quran see 2:55. For the Day of Judgment, see 1:4 and 74:46.

بسم الله الرحمن الرحيم

83:0 In the name of God, the Gracious, the Compassionate.
83:1 Woe to those who cheat.*
83:2 Those who when they are receiving any measure from the people, they take it in full.
83:3 When they are the ones giving measure or weight, they give less than due.
83:4 Do these not assume that they will be resurrected?
83:5 To a great day?
83:6 The day people will stand before the Lord of the worlds.

Functions of the Numerically Coded Book

83:7 No, the record of the wicked is in *Sijjeen.*
83:8 Do you know what *Sijjeen* is?
83:9 A numerical book.*
83:10 Woes on that day to the deniers.
83:11 Those who denied the day of recompense.*
83:12 None will deny it except every transgressor who knows no bounds.
83:13 When Our signs are recited to him, he says: "Tales of the ancients!"
83:14 No, a covering has been placed on their hearts for what they have earned.
83:15 No, they will be blocked from their Lord on that day.
83:16 Then they will be entered into the hell.
83:17 Then it will be said, "This is what you used to deny!"
83:18 No, the record of the pious is in *Elliyeen.*
83:19 Do you know what *Elliyeen* is?
83:20 A numerical book.
83:21 To be witnessed by those brought closer.
83:22 The pious are in Paradise.
83:23 Upon luxurious furnishings, observing.
83:24 You know in their faces the look of Paradise.
83:25 They are given drink from a pure sealed vial.
83:26 Its seal will be of musk, so in that let those who are in competition compete.
83:27 Its taste will be special.
83:28 A spring from which those who are brought near will drink.
83:29 Those who were criminals used to laugh at those who had affirmed.
83:30 When they passed by them, they used to wink to each other.
83:31 When they returned to their people, they would return jesting.
83:32 If they see them, they say, "These are indeed misguided!"
83:33 But they were not sent over them as caretakers.
83:34 Today, those who had affirmed are laughing at the ingrates!
83:35 Upon the luxurious furnishings they are observing.
83:36 Have the ingrates not been reprised for what they used to do?

ENDNOTES

083:001-4 For similar ethical reminders see the Old Testament, Deuteronomy 25:13-16.

083:009-20 The book that confounds the wicked like a prison (*sijjeen*) and serves the do-gooders as an elevator (*illiyin*), is described as *kitabun marqum*, that is, numbered book or numerical book. The wicked rejects it, while the good people witness its marvels.

083:011 The *yawm ul-din* is described as the day of Judgment. Those who believe that they will be saved by their idol's intercession are in denial of that day. See 74:46-48 and 1:4

بسم الله الرحمن الرحيم

84:0 In the name of God, the Gracious, the Compassionate.
84:1 When the sky is ruptured.
84:2 Attends to its Lord and is ready.
84:3 When the earth is stretched.
84:4 It spits out what is in it and becomes empty.
84:5 It attends to its Lord and is ready.
84:6 O mankind, you will be returning to your Lord with your actions and meeting Him.
84:7 So whoever is given his record in his right.
84:8 He will then receive an easy reckoning.
84:9 He will return to his family in joy!
84:10 As for he who is given his record behind his back.
84:11 He will ask for his destruction.
84:12 He will enter a blazing fire.
84:13 He used to be joyful amongst his people!
84:14 He thought he would not be returned.
84:15 No, He is ever seeing of him.
84:16 So, I swear by the redness of dusk.
84:17 The night and what it is driven on.
84:18 The Moon when it is full.
84:19 You will ride a stage/layer upon a stage/layer.
84:20 So what is the matter with them that they do not affirm?
84:21 When the Quran is being related to them, they do not prostrate.
84:22 No, those who have rejected are in denial.
84:23 **God** is more aware of what they gather.
84:24 So inform them of a painful retribution.
84:25 Except for those who affirm and promote reforms, they will have a reward that will not end.

بسم الله الرحمن الرحيم

85:0 In the name of God, the Gracious, the Compassionate.
85:1 By the universe laden with galaxies.*
85:2 The appointed day.
85:3 A witness and a witnessed.
85:4 Woes to people of the canyon.
85:5 The fire supplied with fuel.
85:6 Then they sat around it.
85:7 They were witness to what they did to those who affirm.
85:8 They hated them because they affirmed **God**, the Noble, the Praiseworthy!
85:9 The One to whom belongs the kingship of heavens and earth, and **God** is witness over everything.
85:10 Those who have put the affirming men and women under ordeal, and then did not repent, they will have the retribution of hell, and they will have the retribution of burning.
85:11 Those who affirm and promote reforms, they will have gardens with rivers flowing beneath. Such is the great reward.
85:12 The punishment of your Lord is severe.
85:13 It is He who initiates and then returns.
85:14 He is the Forgiver, the Most Kind.
85:15 Possessor of the throne, the Glorious.
85:16 Doer of what He wills.
85:17 Has news come to you of the soldiers?
85:18 Pharaoh and Thamud?
85:19 No, those who rejected are in denial.
85:20 **God,** after them, is Encompassing.
85:21 No, it is a glorious Quran.
85:22 In a tablet, preserved.*

ENDNOTES

085:001 The word *buruj* means constellations or galaxies. Millions of gullible people still believe in the superstition that Zodiac constellations or one's birthday has a great impact on the character and destiny of a person. For the function of oaths in the Quran, see 89:5.

085:022 The mathematical system shows that every element of it is protected by God. The word *lawh* here and the word *lawaha* in 74:29 share the same root. See 74:30.

86:0 In the name of God, the Gracious, the Compassionate.
86:1 The sky and the herald.*
86:2 Do you know what the herald is?
86:3 The piercing star.
86:4 Every person has a recorder over it.
86:5 So let the human being see from what he was created.*
86:6 He was created from water that spurts forth.
86:7 It comes out from between the spine and the testicles.
86:8 For He is able to bring him back.
86:9 The day when all is revealed.
86:10 Then he will not have any power or victor.
86:11 The sky which gives rain.
86:12 The land with small cracks.
86:13 This is the word that separates matters.
86:14 It is not a thing for amusement.
86:15 They are plotting a scheme.
86:16 And I plot a scheme.
86:17 So respite the ingrates, respite them for a while.

ENDNOTES

086:001 For the function of oaths in the Quran, see 89:5.

086:005 Those who are not familiar with the function of oaths question the wisdom of God "swearing" to convince His audience. The Quran uses oaths to pull our attention to a particular object, concept, or relationship. We could translate all oaths by starting with "I pull your attention to:..." When we utter an oath, we make God our witness. This testimony could work for us or against us depending on how accurate our position is. Thus, our use of God's name in our oaths has no bearing in proving our assertion or negation. We refer to oaths to persuade our audience by using our apparent relationship with God as consideration for contracts. If we are lying, our oath only increases the magnitude of our lie and thus increases our responsibility. The Quranic oaths, however, are not used to convince us; but to make us think. When God uses His creatures in His oaths, He pulls our attention to the signs in nature and scripture to encourage us and invite us to be rational, consistent, and logical.

The Quran reminds us consistently that the God who authored the *ayat* (signs) of the book is the same God who created the *ayat* (signs) of the physical universe; both are products of the same God and share the same characteristics. Besides this important reminder, the objects, events, and concepts used in oaths prepare our mind for the upcoming issues or theses; they provide aesthetic and semantic mental pictures and reference points. For instance, do the above oaths pull our attention to the ten-based and digital number systems? Is there a relationship between the fourth verse and 74:33? See 56:75; 69:38; 70:40; 75:1,2; 77:1-7; 79:1-5; 81:15-19; 84:16; 85:1-3; 86:1,11-13; 90:1; 91:1-8; 92:1-4; 93:1-3; 95:1-4; 100:1-6; 103:2.

87:0 In the name of God, the Gracious, the Compassionate.
87:1 Glorify the name of your Lord, the Most High.
87:2 The One who created and designed.
87:3 The One who measured and then guided.
87:4 The One who brought out the pasture.
87:5 So He made it dry up into hay.
87:6 We will make you study, so do not forget.
87:7 Except for what **God** wills, He knows what is declared and what is hidden.
87:8 We will make easy for you the way.
87:9 So remind, perhaps the reminder will help.
87:10 The reverent will remember.
87:11 The wicked will avoid it.
87:12 He will enter the great fire.
87:13 Then he will neither die in it nor live.*
87:14 Whosoever purifies will succeed.
87:15 Remembers the name of his Lord and reaches out.
87:16 No, you desire the worldly life.
87:17 But the Hereafter is better and more lasting.
87:18 This has been revealed in the previous scripts.
87:19 The scripts of Abraham and Moses.

ENDNOTES

087:013 Unlike Paradise, Hell itself is not eternal, while the people of hell will end up there for eternity. (See, the Appendix titled *Eternal Hell and Merciful God?*, which is also available at www.19.org or www.islamicreform.org).

بسم الله الرحمن الرحيم

88:0 In the name of God, the Gracious, the Compassionate.
88:1 Has the news come to you of the Overwhelming?
88:2 Faces on that day will be desolate.
88:3 Laboring and weary.
88:4 They will enter a blazing fire.
88:5 Be given to drink from a boiling spring.
88:6 They will have no food except from a thorny plant.
88:7 It does not nourish nor avail against hunger.
88:8 Faces on that day are joyful.
88:9 For their pursuit they are content.
88:10 In a high paradise.
88:11 You will not hear in it any nonsense.
88:12 In it is a running spring.
88:13 In it are raised beds.
88:14 Cups that are set.
88:15 Cushions arranged in rows.
88:16 Rich carpets spread out.
88:17 Will they not look at the camels/clouds, how are they created?
88:18 To the sky, how is it raised?
88:19 To the mountains, how were they set?
88:20 To the land, how was it flattened?
88:21 So remind, for you are but a reminder.
88:22 You have no power over them.
88:23 Except for he who turns away and rejects.
88:24 Then **God** will punish him with the great retribution.
88:25 Indeed, to Us is their return.
88:26 Then to Us is their judgment.

ENDNOTES

088:017 As God's marvelous creation, we are given a limited power to mimic God's creation and create. Mimicking the marvelous organs and functions of plants and animals has always been an inspiration for inventors. Recently, engineers turned more seriously to biology for inspiration which led to the establishment of a new scientific field called Biomimicry. We all know the example of Velcro, which was inspired by barbs on weed seeds. According to the information provided by the Biomimicry Institute, the application of this new science in technology is limitless. For instance, the hydrodynamic design of fish is now inspiring car designers to increase cars' aerodynamics, the morpho-butterfly is teaching fabric manufacturers to create structural colors without dyes, and the gecko lizard with tiny split hairs under its feet is inspiring chemists to come up with perfect adhesives. The examples are many. For instance, Orb-weaver spider silk for manufacturing fiber (used in parachute wires, suspension bridge cables, protective clothing, etc.) without using heat, high pressure, or toxic chemicals. Porcupine quills are a source of inspiration for agronomists to breed better wind resistance in wheat and barley, Rhesus monkeys for new sources of nutrient minerals… Sharks and other marine creatures that live with all kinds of pathogens for new antibiotics. Abalone mussel nacre for manufacturing lightweight but fracture-resistant windshields and bodies of solar cars, and airplanes. Antler's teeth, bones, and shells with their natural biomineralization for building 3-D objects layer by layer using CAD, and ink-jet technology. Blue mussel adhesive for making paints and coatings that do not need primer or catalysts to work. Elastin, the elastic protein in the heart muscle for materials and fibers that can stretch and contract in response to heat, light, and chemical changes. Fish antifreeze to freeze human transplant organs without injury… For more information on the subject, you may visit www.biomimicry.org.

89:0 In the name of God, the Gracious, the Compassionate.
89:1 By the dawn.
89:2 The ten nights.
89:3 The even and the odd.
89:4 The night when it passes.
89:5 In this is an oath for the one with intelligence.
89:6 Did you not see what your Lord did to Aad?
89:7 Iram, with the great columns?*
89:8 The one which was like no other in the land?
89:9 Thamud who carved the rocks in the valley?
89:10 And Pharaoh with the pyramids?
89:11 They all transgressed in the land.
89:12 Made much corruption therein.
89:13 So your Lord poured upon them a measure of retribution.*
89:14 Your Lord is ever watchful.
89:15 As for man, if his Lord tests him and grants him much, then he says: "My Lord has blessed me!"
89:16 If his Lord tests him and gives him little wealth, then he says: "My Lord has humiliated me!"
89:17 No, you are not generous to the orphan.
89:18 You do not look to feeding the poor.
89:19 You consume others inheritance, all with greed.
89:20 You love money, a love that is excessive.
89:21 No, when the earth is pounded into rubble.
89:22 Your Lord comes with the controllers row after row.
89:23 Hell on that day is brought. On that day the human being will remember, but how will the remembrance now help him?
89:24 He says: "I wish I had worked towards my life!"
89:25 On that day, no other will bear his punishment.
89:26 Nor will anyone be able to free his bonds.
89:27 "As for you, O the content person."
89:28 "Return to your Lord pleasing and pleased."
89:29 "Enter in amongst My servants."
89:30 "Enter My Paradise."

ENDNOTES

089:002 See 7:142.

089:007 The rock towers of Iram were discovered 4800 years after the destruction and disappearance of the city of Iram in Umman. The Challenger spaceship equipped with SIR-B radars that could see under the sand made this discovery possible in 1992.

089:013 Individual or societal violations of God's law governing nature cause many disasters. For instance, wasting energy and reckless consumerism contribute to air, water and ground pollution; social and political aggression such as dictatorship and racism contribute to civil wars, poverty and famine; accepting no boundaries in sexual conduct contribute to epidemics in sexually transmitted diseases, abandoned children, and increase in crime; disregard for the physical laws, for instance gravity, causes increase in loss of lives during earthquakes; use of alcoholic beverages, heroin, and other drugs cause many problems and disasters for both the individual and society; nonproductive industries such as gambling, fraud, and usury cause social and economic problems; not recognizing the freedom of expression of individuals and groups may cause and invite many problems as well.

بسم الله الرحمن الرحيم

90:0 In the name of God, the Gracious, the Compassionate.

90:1 I swear by this land.*

90:2 While you are a legal resident at this land.

90:3 A father and what he begets.

90:4 We have created the human being to struggle.

90:5 Does he think that no one is able to best him?

90:6 He says: "I spent so much money!"

90:7 Does he think that no one saw him?

90:8 Did We not make for him two eyes?

90:9 A tongue and two lips?

90:10 We guided him to both paths?

90:11 He should choose the better path.

90:12 Do you know which the better path is?

90:13 The freeing of slaves.*

90:14 Or the feeding on a day of great hardship.

90:15 An orphan of relation.

90:16 Or a poor person in need.

90:17 Then he has become one of the affirmers, and exhort one another to patience, and exhort one another to kindness.

90:18 Those are the people of happiness.

90:19 As for those who rejected Our signs, they are the people of misery.

90:20 Upon them is a fire closed over.

ENDNOTES

090:001-4 For the function of oaths in the Quran, see 89:5.

090:013 Appreciative people, monotheists, cannot have slaves. To be a slave owner is equivalent to claiming to be a "lord" and thus tantamount to polytheism. See 4:3,25,92; 2:286; 3:79; 5:89; 8:67; 24:32-33; 58:3-4; 90:13; 2:286; 12:39-42; 79:24.

بسم الله الرحمن الرحيم

91:0 In the name of God, the Gracious, the Compassionate.
91:1 By the Sun and its brightness.*
91:2 The Moon that comes after it.
91:3 The day which reveals.
91:4 The night which covers.
91:5 The sky and what He built.
91:6 The earth and what He sustains.
91:7 A person and what He made.
91:8 So He gave it its evil and good.
91:9 Successful is the one who betters it.
91:10 Failing is the one who hides it.
91:11 Thamud denied their transgression.
91:12 They followed the worst amongst them.
91:13 **God**'s messenger said to them: "This is **God**'s camel, let her drink."
91:14 They rejected him, and they killed her. So their Lord repaid them for their sin and leveled it.*
91:15 Yet, those who came after remain heedless.

ENDNOTES

091:001-9 For the function of oaths in the Quran, see 89:5.

091:014 See 26:155; 54:27-29.

بسم الله الرحمن الرحيم

92:0 In the name of God, the Gracious, the Compassionate.
92:1 By the night when it covers.*
92:2 The day when it appears.
92:3 He created the male and female.
92:4 Your works are various.
92:5 As for he who gives and is righteous.
92:6 Trusts in goodness.
92:7 We will make the easy path for him.
92:8 As for he who is stingy and holds back.
92:9 Denies goodness.
92:10 We will make the difficult path for him.
92:11 His wealth will not avail him when he demises.
92:12 It is upon Us to guide.
92:13 To Us is the end and the beginning.
92:14 I have warned you of a fire that blazes.
92:15 None shall have it but the wicked.
92:16 The one who denies and turns away.
92:17 As for the righteous, he will be spared it.
92:18 The one who gives his money to develop with.
92:19 Seeking nothing in return.
92:20 Except the face of His Lord, the Most High.
92:21 He will be satisfied.

ENDNOTES

092:001 See 89:5.

93:0 In the name of God, the Gracious, the Compassionate.
93:1 By the late morning.
93:2 The night when it falls.
93:3 Your Lord has not left you, nor did He forget.
93:4 The Hereafter is better for you than the first.
93:5 Your Lord will give you and you will be pleased.
93:6 Did he not find you an orphan and He sheltered you?
93:7 He found you lost, and He guided you?*
93:8 He found you in need, so He gave you riches?
93:9 As for the orphan, you shall not make him sad.
93:10 As for the beggar, you shall not reprimand.
93:11 You shall proclaim the blessings from your Lord.

ENDNOTES

093:007 See 42:52; 47:19; 48:2.

94:0 In the name of God, the Gracious, the Compassionate.
94:1 Did We not relieve your chest,*
94:2 Take from you your load,
94:3 Which had put strain on your back?
94:4 We have raised your remembrance,
94:5 So with hardship comes ease.
94:6 With hardship comes ease.
94:7 So when you are done, then stand.
94:8 To your Lord you shall seek.

ENDNOTES

094:001 The verse that mentions relieving the chest has been distorted by fabricated *hadith*, which report a literal surgery on Muhammed's chest by Gabriel. The audience of this verse is not Muhammed alone, but all monotheists. See 6:125; 20:25; 39:22.

بسم الله الرحمن الرحيم

95:0 In the name of God, the Gracious, the Compassionate.
95:1 By the fig and the olive.*
95:2 The mount of ages.
95:3 This secure land.
95:4 We have created the human being in the best form.
95:5 Then We returned him to the lowest of the low.
95:6 Except the affirmers and carry out reforms; they will have a well-deserved reward.
95:7 So what would make you deny the system after that?
95:8 Is **God** not the wisest of the wise?

ENDNOTES

095:001-3 If the words Sina and "this secure land" are symbols referring to Moses and Muhammed, then could the fig and olive represent the location of other messengers? See 89:5.

بسم الله الرحمن الرحيم

96:0 In the name of God, the Gracious, the Compassionate.
96:1 Read in the name of your Lord who has created.*
96:2 He created the human being from an embryo.*
96:3 Read, and your Lord is the Generous One.
96:4 The One who taught by the pen.
96:5 He taught the human being what he did not know.
96:6 Alas, the human being is bound to transgress.
96:7 When he achieves, he no longer has need!
96:8 To your Lord is the return.
96:9 Have you seen the one who deters,
96:10 A servant from being supportive?
96:11 Have you seen if he was being guided,
96:12 Or he ordered righteousness?
96:13 Have you seen if he lied and turned away?
96:14 Did he not know that **God** can see?
96:15 Alas, if he does not cease, we will strike the frontal lobe.
96:16 A frontal lobe which lies and errs.
96:17 So let him call on his assembly.
96:18 We will call on the guardians.
96:19 So do not obey him, prostrate and come near.

ENDNOTES

096:001-19 This chapter is traditionally known to be the first revelation of the Quran, chronologically. Those who know Arabic will notice the different spellings of two homophone words, which we transliterate as *bismi* (in the name). A reflection on these two differently spelled homophones, together with the first word, is another proof that Muhammed was a literate man. See 2:78; 7:157-158. This chapter is the 19th chapter from the end of the Quran and has 19 verses.

096:002 The Arabic word *alaq* is a multi-meaning word: (1) blood clot; (2) hanging thing; (3) leech. Medieval commentators preferred the first meaning because of its organic nature. However, Maurice Bucaille, a French medical doctor who converted to islam while serving as a medical doctor for the Saudi royal family, had a problem with this traditional rendering. As someone who had deep knowledge in human embryology (see 3:7), he became the first person who noticed the problem with the traditional understanding and he rightly reminded us of the right meaning of the word, which perfectly describes the embryo, the hanging fertilized cells on the wall of the uterus, like a leech. It is true that in the stages of creation of humans, starting from semen, fertilization of the egg until birth, there is no "clot" stage. The Quran locates *alaq* (hanging thing) as the stage after the semen (22:5; 23:14). In his landmark book, The Bible, the Quran, and Science, Dr. Bucaille discusses the scientific accuracy of the Quran and compares it with the Bible and archeological and scientific evidences. Though his book is celebrated by Sunni and Shiite *mushriks*, many are disappointed by his criticism against *hadith*; he argued that the *hadith* fails the tests of science. This brave voice irritated the illusions of the followers of *hadith* and *sunna*. Thus, some publishing houses in the so-called Muslim countries, published the book after expunging the section containing his criticism of *hadith*, without the permission of the author.

97:0 In the name of God, the Gracious, the Compassionate.

97:1 We have sent it down in the Night of Decree.*

97:2 Do you know what the Night of Decree is?

97:3 The Night of Decree is better than one thousand months.

97:4 The controllers and the Spirit come down in it by their Lord's leave to carry out every matter.

97:5 It is peaceful until the coming of dawn.

ENDNOTES

097:001 For every appreciative person there is a night of *qadr* (determination; power) night, in which they decide to dedicate themselves to God alone by fully using their intellectual faculties. My night of power is the night of July 1986, in which I decided to to dedicate myself to my Lord alone.

98:0 In the name of God, the Gracious, the Compassionate.

98:1 Those who rejected amongst the people of the book and those who set up partners would not leave until proof came to them.

98:2 A messenger from **God** reciting purified scripts.

98:3 In them are valuable books.

98:4 Those who had previously received the book did not divide except after the proof came to them.

98:5 They were not commanded except to serve **God** and be loyal to His system, monotheism, and maintain the contact prayer and contribute towards betterment. Such is the valuable system.*

98:6 Those who rejected from the people of the book and those who set up partners are in the fires of hell abiding therein, those are the worst of creation.

98:7 As for those who affirm and promote reforms, they are the best of creation.

98:8 Their reward with their Lord is the gardens of Eden with rivers flowing beneath them; they abide eternally therein. **God** is satisfied with them, and they are satisfied with Him. That is for whoever feared His Lord.

ENDNOTES

098:005 See 39:11.

99:0 In the name of God, the Gracious, the Compassionate.

99:1 When the earth rumbles and shakes.

99:2 The earth brings out its mass.

99:3 The human being will say, "What is wrong with her?"

99:4 On that day it will inform its news.

99:5 That your Lord had revealed to her to do so.

99:6 On that day, the people will be brought out in throngs to be shown their works.

99:7 So whoever does an atom's weight of good will see it.

99:8 Whoever does an atom's weight of evil will see it.

100:0 In the name of God, the Gracious, the Compassionate.
100:1 By the fast gallopers.*
100:2 Striking sparks.
100:3 Charging in the morning.
100:4 Forming clouds of dust.
100:5 Penetrating into the midst together.
100:6 Surely, the human being is ungrateful to his Lord.
100:7 He will indeed bear witness to this.
100:8 He loves wealth tenaciously.
100:9 Does he not realize that when what is in the graves are scattered.
100:10 What is in the chests is gathered.
100:11 That their Lord has been fully cognizant of them?

ENDNOTES

100:001-5 Past generations saw the description of horses in these verses. We can understand it as the description of jet airplanes which intake oxygen on one end and spew fire on the other end. We are not required to understand every verse in the same way the previous generations did. For the function of oaths, see 89:5.

101:0 In the name of God, the Gracious, the Compassionate.
101:1 The Shocker.
101:2 What is the Shocker?
101:3 How would you know what the Shocker is?
101:4 The day when people come out like swarms of butterflies.
101:5 The mountains will be like fluffed up wool.
101:6 As for him whose weights are heavy.
101:7 He will be in a happy life.
101:8 As for him whose weights are light.
101:9 His destiny is the lowest.
101:10 How would you know what it is?
101:11 A blazing fire.

102:0 In the name of God, the Gracious, the Compassionate.
102:1 Hoarding has distracted you.*
102:2 Until you visit the graves.
102:3 No, you will find out.
102:4 Then again, you will find out.
102:5 No, if only you knew with certainty.
102:6 You would then see the hell.
102:7 Then you would see it with the eye of certainty.
102:8 Then you will be questioned, on that day, about the blessings.

ENDNOTES

102:001 Many forget the purpose of this life while getting preoccupied with competing to increase their share of more money, real estate, children, votes, awards, fame, etc.

Compare it to the New Testament: "Woe unto you that are rich! For you have received your consolation. Woe unto you who are full! You shall be hungry. Woe unto you who laugh now! You shall weep and mourn" (Luke 6:24).

103:0 In the name of God, the Gracious, the Compassionate.
103:1 By time,
103:2 The human is indeed in loss.
103:3 Except those who affirm, carry out the reforms, exhort one another with the truth and exhort one another with perseverance.

104:0 In the name of God, the Gracious, the Compassionate.
104:1 Woe unto every backbiter, slanderer.*
104:2 Who gathered his wealth and counted it.
104:3 He thinks that his wealth will make him immortal.
104:4 Never! He will be thrown into the *Hutama*.
104:5 Do you know what the *Hutama* is?
104:6 **God**'s kindled fire.
104:7 Which reaches the inside of the hearts.
104:8 It will confine them therein.
104:9 In extended columns.

ENDNOTES

104:001 For the function of oaths, see 89:5.

105:0 In the name of God, the Gracious, the Compassionate.
105:1 Have you not noticed what your Lord did to the people of the elephant?
105:2 Did He not cause their schemes to go astray?
105:3 He sent upon them swarms of flying creatures.
105:4 Striking them with fiery projectiles.
105:5 Until He turned them into like devoured hay.

106:0 In the name of God, the Gracious, the Compassionate.
106:1 For the unity and security of Quraysh.
106:2 Their unity and security during their journey of the winter and summer.
106:3 So let them serve the Lord of this sanctuary.
106:4 The One who fed them from hunger and protected them from fear.

107:0 In the name of God, the Gracious, the Compassionate.
107:1 Do you notice who rejects the system?
107:2 It is the one who mistreats the orphan.
107:3 Does not encourage the feeding of the poor.
107:4 So woe to those who offer support (or observe contact-prayer).*
107:5 Who are totally heedless of their support (or of observing their contact-prayer).
107:6 They only want to be seen,
107:7 They refuse to aid.

ENDNOTES

107:004 *Sala* prayers and other rituals were being practiced since Abraham and were a common knowledge among Meccan polytheists. See 21:73

108:0 In the name of God, the Gracious, the Compassionate.
108:1 We have given you plenty.
108:2 Therefore, you shall reach out to your Lord, and devote yourself.
108:3 Indeed your rival will be the loser.

109:0 In the name of God, the Gracious, the Compassionate.
109:1 Say, "O ingrates,"
109:2 "I do not serve what you serve,"
109:3 "Nor do you serve what I serve,"
109:4 "Nor will I serve what you serve,"
109:5 "Nor will you serve what I serve,"
109:6 "To you is your system, and to me is mine."

110:0 In the name of God, the Gracious, the Compassionate.
110:1 When **God**'s victory and conquest comes.*
110:2 You will see people entering into **God**'s system in groups.
110:3 You shall glorify your Lord's grace, and seek His forgiveness. He is the Redeemer.

ENDNOTES

110:001-3 This chapter consists of 19 words, and its first verse has 19 letters. See 48:28.

111:0 In the name of God, the Gracious, the Compassionate.
111:1 Condemned is the power of the flaming provocateur; condemned indeed.*
111:2 His money will not avail him, nor what he has earned.
111:3 To a flaming fire he will be cast.
111:4 His wife carrying the logs.
111:5 With a twisted rope on her neck.

ENDNOTES

111:001 The expression *Abu Lahab* means the "father of flame" or provocateur.

Traditional commentaries tie this description to Muhammed's uncle Abd al-Uzza bin Ab al-Muttalib. Even if the first person who was implied by this verse were Muhammed's uncle, the chapter by using a metaphor rather than a proper name, refers to all warmongers and their allies who provokes people to violence. In this chapter, the wife has two different roles: either she is supplying more fuel for her husband in support of his bigoted campaign against peacemakers, or she is supplying fuel for her husband who is burning himself with flames of hatred.

Some of the followers of the *hadith* and *sunna*, who consider the name to be only a proper name, present this chapter as evidence for the divinity of the Quran, by arguing that Abu Lahab could have falsified the Quran simply by professing Islam after hearing these verses about him. This assertion is the product of poor thinking. If the Quran was the product of Muhammed, Muhammed would never accept his conversion to Islam, and would continue condemning him with additional accusations, such as, him being a lying hypocrite. And Muhammed would be right (not necessarily in his claim of the origin of the Quran) regarding Abu Lahab, since he could never honestly affirm the truthfulness of a book condemning him to be a misguided loser; his conversion would create a contradiction. In other words, such a claim cannot be falsified, and thus cannot be used as an example of prophecies.

112:0 In the name of God, the Gracious, the Compassionate.
112:1 Say, "He is **God**, the One,"*
112:2 "**God**, the Absolute/First-cause,"
112:3 "Never did He beget, nor was He begotten,"
112:4 "None is equal to Him."

ENDNOTES

112:001 This chapter is called *ikhlas* (devotion), since all its verses are dedicated to God. In contradiction to many Biblical verses, such as John 14:28, polytheistic Christians wish to equate Jesus to God through the Trinity, a fiction written by the Nicene Conference in 325 AC. Their fictional character erratically oscillates between humanity and divinity, between 1/3 of godhead and the entire godhead. Nothing can be compared to God, and nothing can be symbolically represented in proportion to God. From absurd premises, one can never get a sound conclusion. That is why, like other polytheists, Trinitarian Christians hate reason, and that is why they promote wishful thinking under the glorified word "faith."

See 2:59; 3:18,45,51,55; 4:11,171,157; 5:72-79; 7:162; 17:36; 19:36.

بسم الله الرحمن الرحيم

113:0 In the name of God, the Gracious, the Compassionate.
113:1 Say, "I seek refuge in the Lord of the dawn,"
113:2 "From the evil of what He has created,"
113:3 "From the evil of darkness as it falls,"
113:4 "From the evil of those who whisper into the contracts and affairs,"
113:5 "From the evil of the envious when they envy."

114:0 In the name of God, the Gracious, the Compassionate.
114:1 Say, "I seek refuge in the Lord of the people,"
114:2 "The King of the people,"
114:3 "The God of people,"
114:4 "From the evil of the sneaky whisperer,"*
114:5 "Who whispers into the chests of the people,"
114:6 "From among the Jinn and people."

ENDNOTES

114:004 See 7:116-117.

1

Some Key Words and Concepts

There are certain key words or concepts found in the Quran that are important to the understanding of the scripture. In many cases these terms have been misunderstood. A list of these terms is provided below along with their Quranic meanings and implications.

Allah	Allah is the First Cause, the Eternal, the Benevolent, the Omniscient, and the Omnipotent Creator of the universe. Allah is the most repeated word in the Quran (2698 times in numbered verses). It is not a proper name, but rather a contraction of the definite article *al* (the) and *Elah* (god), meaning "the god" or simply God. It is etymologically related to *Eli*, the Aramaic word for God, still quoted in English translations of the Bible: "*Eli, Eli, lama sabachthani*?" (Matthew 27:46; Mark 15:34). The Quran contains about 114 attributes for God. The most frequently used attributes of God are, the All-Knowing (*Alim*), the Lord (*Rabb*), the Loving/Caring (*Rahim*), god (*Elah*), the Wise (*Hakim*), the Forgiving (*Ghafur*), the Honorable (*Aziz*), the Compassionate (*Rahman*), the Hearer (*Sami*), the Planner (*Qadir*), the Knower (*Khabir*), and the Seer (*Basir*).
Islam	Islam is not a proper name; it simply means submission and peacemaking. There are many aspects of being a Muslim or submitter to God. There are at least three spheres of divine law. Submitting oneself to those laws is the only way to attain peace and eternal happiness. These laws are: **Natural Realm**. Studying, discovering, affirming and submitting to the divine laws imposed in nature create worldly progress and prosperity. Ignoring and defying those laws bring disasters and discomfort. **Social Realm**. There are universal laws inherent in social structure and are learned through experience and through logical rules that dictate avoiding contradictions. The last six articles of the Ten Commandment can be learned from human experience. Societies that submit to the universal laws that justify communal life prosper socially and politically. **Intellectual realm**. This is only possible through an intellectual or philosophical decision to submit to the Creator, to attain knowledge and conviction about resurrection and the day of judgment. Intellectual submission to the truth completes submission and brings eternal happiness (For a detailed discussion on this topic, see the article titled "How Much Muslim Are You?" available at 19.org). **Islam in the Quran** As we stated above, Islam is not a proper name, nor it is the religion founded by Muhammed as it is falsely claimed. Islam is the paradigm of all those who submit themselves peacefully to God alone. The Quran

teaches us that all people affirming the truth were named by God with the word Submitters or Peacemakers (Muslims) (22:78). Islam is a descriptive name meaning submission and peace, or within the context of the Quran, peacefully surrendering self to God alone. Islam is a paradigm and a way of life that emphasizes personal freedom from all powers by accepting God alone as the only absolute authority. Thus, the only way of life acceptable by God is Islam (3:19-85). According to the Quran, all messengers, together with their supporters, were Muslims; they were peacemakers who submitted themselves to the will of God alone. Noah, Abraham, Moses, David, Solomon, Jesus and their followers were all described with the same word, Peaceful Submitters or Peaceful Surrenderers (Muslims) (10:72; 2:128; 10:84; 27:31; 5:111; 72:14). Since all messengers delivered the message in the language of their people, they described their submission to God, not using the Arabic word, "Islam" or "Muslim," but rather through their own languages (14:4). In fact, according to the Quran, with the exception of the minds of ingrates and *mushriks*, the entire universe, including the material bodies of rejecters, is Muslim, since every particle, atom, molecule, planet, star, light, galaxy in the universe submits perfectly to God's law (41:11). The shadows submit to God's law, even the shadows of the opponent's body (13:15). Everything in the universe has submitted to the God's system; submission to God in peace and peacemaking (Islam) is the law of the universe (3:83). İslam:

- is not a proper name, but a descriptive noun coming from the Arabic root of surrendering/submission/peace, used by God to describe the system delivered by all His messengers and prophets (5:111; 10:72; 98:5), which reached another stage with Abraham (4:125; 22:78).
- is peacefully surrendering to God alone (2:112,131; 4:125; 6:71; 22:34; 40:66).
- is a system with universal principles, which are in harmony with nature (3:83; 33:30; 35:43).
- requires objective evidence in addition to personal experience (3:86; 2:111; 21:24; 74:30).
- demands conviction not based on wishful thinking or feelings but based on reason and evidence (17:36; 4:174; 8:42; 10:100; 11:17; 74:30-31).
- esteems knowledge, education, and learning (35:28; 4:162; 9:122; 22:54; 27:40; 29:44,49).
- promotes scientific inquiry regarding the evolution of humankind on earth (29:20).
- rejects clergymen and intermediaries between God and people (2:48; 9:31-34).
- condemns profiteering from religion (9:34; 2:41,79,174; 5:44; 9:9).
- stands for liberty, accountability, and defiance of false authorities. (6:164).
- stands for freedom of expression (2:256; 18:29; 10:99; 88:21-22).
- requires consultation and representation in public affairs (42:38; 5:12).

	• promotes a democratic system where participation of all citizens is encouraged and facilitated (58:11). • prohibits bribery and requires strict rules against the influence of interest groups and corporations in government (2:188). • requires election of officials based on qualifications and principles of justice (4:58). • promises justice to everyone, regardless of their creed or ethnicity (5:8). • affirms the rights of citizens to publicly petition against injustices committed by individuals or government (4:148). • encourages the distribution of wealth, economic freedom, and social welfare (2:215, 59:7). • promotes utmost respect to individuals (5:32). • relates the quality of a society to the quality of individuals comprising it (13:11). • recognizes and protects an individual's right to privacy (49:12). • recognizes the right to the presumption of innocence and right to confront the accuser (49:12). • provides protection for witnesses (2:282). • does not hold innocent people responsible for the crimes of others (53:38). • protects the right to personal property (2:85,188; 4:29; exception 24:29; 59:6-7). • discourages a non-productive economy (2:275; 5:90; 3:130). • encourages charity and caring for the poor (6:141; 7:156). • unifies humanity by promoting gender and race equality (49:13). • values women (3:195; 4:124; 16:97). • values intellect (5:90). • offers peace among nations (2:62; 2:135-136, 208). • considers the entire world as belonging to God and supports immigration (4:97-98). • promotes peace, while deterring the aggressive parties (60:8,9; 8:60). • pursues the golden-plated brazen rule of equivalence, that is, retaliation with occasional forgiveness (42:20; 17:33). • stands up for human rights and the oppressed (4:75). • encourages competition in righteousness and morality (16:90). • stands for peace, honesty, kindness, and deterring from wrongdoing (3:110). • expects high moral standards (25:63-76; 31:12-20; 23:1-11). • asks us to be in harmony with nature and the environment (30:41). • No wonder that the only system/law approved by God is Islam (3:19,85).
Muslim	A person who peacefully submits or surrenders herself or himself to

	God alone is a Muslim. In other words, it is a person who has accepted God as the only authority for eternal salvation and tries his or her best to follow and respect divine laws elucidated in scripture and nature. With the exception of thoughts, prejudices and choices of ingrates and of those who associate partners to God, everything else in the universe is, in a sense, by definition a Muslim or submitter to God. There is no ritual, ceremony, or magical word to become a Muslim. Anyone who seeks the truth with their God-given reasoning faculties and senses is in essence Muslim; Muslims are those who guard against the influence of peers, parents, clergymen, ancestors, crowds, dogmas, ideology, and wishful thinking in the quest to find answers to various philosophical or theological questions. A Muslim, by not submitting her or his mind to any authority besides God, is a model free person. Since God is the truth, a Muslim is a perpetual seeker of truth and facts, and in this quest none other than truth and facts can have authority over him or her. Thus, a Muslim is a critical thinker, a brave free man standing on his or her own feet. He or she demonstrates a reasonable dose of skepticism in regard to accepting assertions, especially those that are related to his or her eternal salvation.
Belief, Faith	Before going further, we should also define the meaning of "belief" or "faith" in the context of religion. The word "belief" or "faith" is one of the most abused and misused words among people who subscribe to a certain religion or creed. Organized religions use it as a euphemism for conjecture, wishful thinking, a wild guess, or, in most cases, for blindly joining the closest, the most crowded, or the loudest bandwagon. This euphemism enables those who join the crowd for the sole purpose of comfort, or those who pick a religion based on psychological and purely subjective reasons to hide the problematic character of their reasons for their beliefs, even from their own cognition. The euphemistic statement "I believe" gives them the illusion of a self-righteous comfort that they are blessed by God by their faith, that they somehow reached a particular faith not by their psychological need or petty self-interest of going along with the group, but by a mysterious divine grace. Packaging wishful thinking and introducing it as absolute or eternal truth is one of the most revolting and harmful frauds in human history. The price of this delusion is a sacrifice of the truth, and an inability to communicate based on reason. Masses have been manipulated, exploited politically, economically, and socially, and mobilized against each other by the gate-keepers of faith. History is full of examples of atrocities, unholy wars, and injustices done in the name of faith. It may surprise many that Martin Luther, the founder of Protestant movement who was considered a progressive clergyman compared to the Pope, was a bigot. He invited his followers to give up their reasoning faculties and discard their brains: > "Reason is the greatest enemy that faith has; it never comes to the aid of spiritual things, but more frequently than not struggles against the divine Word, treating with contempt all

that emanates from God." (Martin Luther, Last Sermon in Wittenberg, 17 January 1546.)

If you are the supplier or the client of absurd stories in the name of God, then of course you will have problem with God's greatest gift to you: your reasoning faculties. Marx hit the nail in the head by rejecting religions with his famous description: "opium of the masses."

Perhaps we have a hate-and-love relationship with truth. We are fed lies all the time, from fairy tales in cradles to religious stories in places of worship. Fiction books are the best-selling books. Actors, whose entire profession is based on faking other people, or nonexistent people, are treated like gods. A belief based on intellect, reason and empirical evidence is preferable, even if it is not reliable, to a belief based mainly on human emotions! An individual may be excused for an inability to find the best answer, but Spiritual reasons, more accurately "spiritual feelings" or "spiritual experiences," are leading millions of gullible people to justify numerous silly or absurd stories about God. Worse yet, these spiritual affirmers are deprived of their ability to discard their absurd beliefs when they are shown good reason or evidence to the contrary. On the other hand, the group that relies on logical, mathematical, or scientific justification has more chance to see problems in their belief system and come closer to both truth and Truth.

So, as reformed Muslims, when we say, "we believe," we believe as the concept is defined by the Quran, which can be translated to English more accurately as "affirmation." The Quran does not accept conjecture or the testimony of the majority, or hearsay as the basis of philosophical and theological quests for truth. *Mumin* is the one who affirms the truth and continuously seeks knowledge for a better understanding of reality. Here are some characteristics of those who affirm. They:

- Do not accept information on faith; critically evaluate it with their reason and senses (17:36)
- Ask the experts if they do not know (16:43)
- Use intelligence, reason, and historical precedents to understand and carry out God's commands (7:179; 8:22; 10:100; 12:111; 3:137)
- Do not dogmatically follow the status-quo and tradition; are open to new ideas (22:1; 26:5; 38:7)
- Are open-minded and promote freedom of expression; listen to all views and follow the best (39:18)
- Do not follow conjecture (10:36,66; 53:28)
- Study God's creation in the heavens and land; explore the beginning of creation (2:164; 3:190; 29:20)
- Attain knowledge, since it is the most valuable possession in their appreciation of God (3:18; 13:13; 29:43,49)
- Are free individuals, do not follow crowds, and are not afraid of crowds (2:112; 5:54,69; 10:62; 39:36; 46:13)
- Do not follow the religion of your parents or your nation blindly (6:116; 12:103,112)
- Do not make profit from sharing God's Message with others (6:90; 36:21; 26:109-180)

- Read in order to know, and they read critically (96:1-5; 55:1-4)
- Do not ignore divine revelation and signs (25:73)
- Do not miss the main point by indulging in small and inconsequential details (2:67-71; 5:101-102; 22:67)
- Hold their judgment if they do not have sufficient information; do not rush into siding with a position (20:114)
- …
- Speak the truth; do not lie, although stratagem is allowed against adversaries (8:7-8; 25:72; 33:70; 12:70-81)
- Are kind and forgiving (42:40,43)
- Are active, dynamic, creative and courageous people (2:30-34; 4:75-77; 15:28-30)
- Are not egoistic and proud (25:43; 17:37)
- Are steadfast and humble (31:17-18)
- Are brave (33:23)
- Do not lose hope; are optimistic (12:87; 39:53)
- Walk humbly on earth and when harassed by ignorant people; they ignore the harassers with dignity and respond to them by saying "peace" (25:72: 29:63)
- Hold firmly to principles, but are flexible in methods (2:67-71, 142; 3:103; 5:54; 22:67)
- Are not proud of their accomplishments, and are not saddened by their losses (57:23)
- Seek unity not division; do not divide themselves due to jealousy (3:103; 6:159; 61:4; 42:14)
- Put moral considerations uppermost, but do not disregard their due material interests (28:77)
- Fulfill promises (17:34)
- Eat and drink moderately, and avoid intoxicants and gambling (7:31; 2:219)
- Dress decently (24:30-31)
- Do not ridicule or mock one another (49:11)
- Abstain from vain talk (23:4)
- Do not discriminate based on gender and race; they know that superiority is only through God-consciousness (49:13)
- Are loyal to what they have been entrusted and keep their pledges (23:9)
- Treat everyone with civility and give greetings to all (2:83; 28:55; 43:89; 4:86)
- Do not escape from problems, but rather actively focus on problems to solve them equitably, even if the solution requires a fight (49:9)
- Avoid suspicion, spying and backbiting among the monotheists; seek peace, since they are only brothers and sisters (49:9-10,12)
- Do not follow monarchs, princes, emirs, sultans, except when they are forced; obey those who are in charge among themselves, and when they dispute in any matter, they refer it to God and the messenger (4:59)
- Are the party of God (58:22)

	• Do not aggress, but defend themselves against aggression (7:33; 42:39) • … • Serve God alone and do not associate partners in His authority (17:22-23) • Dedicate themselves to God alone; they do not kill unjustly, and do not commit adultery (29:68) • Believe in God and live righteously (2:62, 112) • Love God, the Truth, more than anything else (5:54; 9:23) • Do not fear mankind, but fear God (5:44; 33:37, 39) • Maintain the contact prayers with God (23:10) • Repent for their sins (25:70-71) • … • Respect and honor their parents (17:23-24) • Love their spouses and treat them with care and compassion (30:21) • Are charitable to relatives, the poor and destitute and contribute towards public welfare (9:60; 17:26) • … • Practice consultations to solve social and political problems (42:36) • Act justly, do not commit evil and rule according to God's laws, i.e., justice, truth, and mercy (4:58, 135; 5:8; 7:28-29; 5:44) • Do not turn their cheeks arrogantly from people, nor roam the earth insolently, since they know that God does not like the arrogant showoffs (31:18; 57:23) • Perform prayers and other rites of worship, without quarrelling over methodology, and they share their blessings with those who have less than they have (98:5; 22:67) • Obey just leaders, respect, honor and support them, but they do not idolize them (4:59; 33:56; 9:30-31) • Do not practice bribery and corruption (2:188) • Do not practice usury, but they practice charity (2:275-80) • Give charity in moderation (29:67) • Are honest and fair in financial and economic dealings (6:152) • Do not bear false witness (25:72) • Are not extravagant and wasteful, nor are they stingy (17:26-29) • Save lives and do not kill except in the cause of justice (17:33) • Respond equally, but they know that forgiveness is the best route (2:179; 13:14; 45:14; 64:14) • Do not devour the properties of orphans (17:34) • Enter marriage with those who affirm, do not marry polytheists, and they do not commit adultery (5:5; 23:6-7; 30:21; 17:32) • … • Are not apathetic (5:79) • Enjoin good and forbid evil (3:104)

<table>
<tr>
<td></td>
<td>
<ul>
<li>Hasten to do righteous work (3:114)</li>
<li>Are active towards betterment (23:5)</li>
<li>Support each other in the cause of God (8:74)</li>
<li>Raise knowledgeable people in the society in order to learn the laws (9:122)</li>
<li>Cooperate and help each other in good works; do not cooperate in evil works (5:2)</li>
<li>Fight in the cause of justice and truth with their wealth and their lives (4:75; 9:111)</li>
<li>Promote legal education (9:122)</li>
<li>Persevere in any good effort and do not fear to face difficulties and hardships; success comes only after hardships (2:45, 177; 94:5-8)</li>
<li>Fight for the rights of those who are oppressed (4:75)</li>
<li>Seek peace (2:208; 4:90; 8:61; 2:208)</li>
<li>Are progressive (15:24; 74:37)</li>
<li>Reform themselves and are reformers (6:48,54; 7:35,56,142; 11:117; 12:101; 21:72,105)</li>
</ul>
(Thanks to Kassim Ahmad of Malaysia for his contribution in the compilation of this list.)
</td>
</tr>
<tr>
<td>Reasoning,
Rationalism</td>
<td>
The Quran criticizes its opponents for not using their God-given reasoning faculties (Aql) (For instance, see 10:100). Belief in one God can be rational, and it should be rational. Rational Monotheism is not an oxymoron like "Square Circle" nor is it tasteless like "Garlic Strawberry Jam" as some might say but is rather an idea that liberates the person from illusions and satisfies both the intellect and emotions.
We "know" something when we have direct experience of something. We "believe" something when we have circumstantial evidence testifying to its existence or its attributes. A deep analysis of these two concepts might demonstrate their merger or overlap. For instance, modern man claims that he knows that the earth is revolving around the sun. Most educated people express their belief in the Copernican model in terms of absolute truth. They "know" that the earth revolves around the sun. However, a little inquiry will show that many people are affirmers of scientific claims just as they are affirmers of religious claims. In his revolutionary book, The Copernican Revolution, Thomas Kuhn writes:
<blockquote>"The idea that the earth moves seems initially equally absurd. Our senses tell us all we know of motion, and they indicate no motion for the earth. Until it is reeducated, common sense tells us that, if the earth is in motion, then the air, clouds, birds, and other objects not attached to the earth must be left behind. A man jumping would descend to the earth far from the point where his leap began, for the earth would move beneath him while he was in the air...." (Thomas S. Kuhn, The Copernican Revolution; Planetary Astronomy in the Development of Western Thought, Harvard University Press,</blockquote>
</td>
</tr>
</table>

Cambridge, Mass., 1985, pp. 43-44)

It is true that without the concept of inertia and related experiments, we would rationally object to the idea of a moving earth. How could the cities, mountains, and valleys move, and we not even feel their movement? Prof. Daniel Kolak, in his deliciously written book, Lovers of Wisdom, criticizes our tendency to accept scientific theories or claims without a thorough examination. We, it seems, have developed too much trust in science and scientists.

> "Well, but let's think about it. Could we, if we wanted to and had all our technology at our disposal, go and see whether the Sun goes around the earth, or the earth goes around the sun? Of course, it would be silly, and we'd be laughed at if we actually went to anyone, doubting what everybody obviously knows, and said we just want to see for ourselves. It would be like going to pull on Santa's beard in the department store to try and verify if that is really Santa. You'd have to be a little crazy to do it. So, let's be a little crazy—even though we're all certain that the earth has been seen by somebody to move around the sun. Let's doubt the obvious; let's pretend we really are crazy enough to doubt what is so obviously true. How could we go to see the truth for ourselves? Science, after all, is based on experience. The empirical method will allow us to see that the earth goes around the Sun, surely.
>
> "We step outside and look up. What do we see? The Sun rises, then sets. The Moon rises, then sets. The stars move across the heaven. . . . so far, everything seems to be moving around the earth. Not very convincing for the sun-centered view! We had better go up in a spaceship . . .
>
> "Now, inside our spaceship, looking down from our synchronous orbit, what do we see? Below us is the earth, with the Florida coastline visible beneath a layer of clouds. The earth is perfectly motionless. We look up. The Sun, the Moon, all planets, and all the stars are seen to move around the earth. . . We fire our thrusters and retrorockets and fly away until we are in a synchronous orbit around the Sun (the air conditioning is very good). What do we see?
>
> "Lo and behold, we now see the Sun perfectly, still beneath us and all the planets and stars moving about the sun. Finally! Copernicus and Galileo vindicated! But hold on. We now fly to Mars. From our orbit around the Mars, what do we see? Earth, the other planets, the Sun, and all the stars revolving around Mars. So Mars is at the center. No, wait . . what is going on?" (Lovers of Wisdom, Daniel Kolak, Wadsworth, 2001, pp. 210-211)

Well, motion depends on one's reference point and there is no absolute vantage point in the universe to assess the "true" motions of all heavenly bodies. Then, why have we all come to an agreement on the sun-centered model? Is it because the Copernican model is mathematically artistic? Or is it because technology, the popular product of the scientific enterprise, has increased our faith in scientists and their theories and "facts"? Discussing this issue is beyond the scope of this article, but this example shows that the line between faith and knowledge is not always as clear as it seems.

Another way to differentiate "belief" or "faith" from "knowledge" is to check whether the subject of faith or knowledge is examinable by others. A belief that is not subject to verification or falsification is personal and the holder of such a belief has no reasonable grounds to invite others to share his or her belief. Furthermore, "belief" might have many levels of strength. In our daily conversation, we occasionally use this word to mean "suppose" or "assume" or "suspect." Conversely, we might use the word "believe" to assert that something is true.

Any belief or action held based on a consistent reason, any consistent reason, is rational. According to this contextually self-evident definition, we can conceive of two types of rationality. (This categorization is not done after hours of reflection or research; it is the product of a quick analysis. Accordingly, treat it as a starting point, perhaps a not-well-articulated one, for a new approach to the topic).

Subjective rationalism:

Any idea or act justified by personal reasons or causes, which are not obvious or are not communicated to other rational beings, is subjectively rational. For instance, someone might hold his nose with his hand while talking. One might wonder about the sanity of that person. But if that person just had a stubborn nosebleed before his lecture, his action might seem pretty rational by him and by those who are aware of his condition. A person who joins a cult that worships extraterrestrials (or a bloody cross that killed the hero) might also have personal reasons for this conventionally irrational belief. For instance, his girlfriend might be a member of the ET-worshipers cult, or membership to the cult will bring attention, even though a negative one, that he craves. Worshiping ET is subjectively rational if it is consistent with the purpose of the worshiper. Most believers of religious dogmas have undeclared rationalizations for their beliefs and practices. The more their beliefs and practices are consistent with their goals, the more subjectively rational they are. As long as a particular belief and practice serves the goal of a person, it is subjectively rational. A politician frequenting a church might be performing a rational act depending on the brand of church and its consistency with the politician's goal of getting more votes.

A particular belief or action will not be considered "subjectively rational" if the believer or the actor is either unaware of the reasons for those beliefs/actions at all, or if aware of such reasons, their beliefs and actions still yet do not serve their purpose. Imagine a person who is in love with a feminist girl trying to get her attention by boasting about his macho relationship with his former girlfriend. Or think of a "madman" who, after trashing his clothes and shoes, walks on the street in the middle of winter, nude and barefoot. If we have sufficient reasons to believe that the nude man has no justifiable personal reasons for such an act, then we may conclude that such an act is irrational.

However, we would be reaching this conclusion based on ignorance, which is not rational, that is, not consistent with our goal of learning the truth of the matter. There is most likely an external or internal reason triggering such an unconventional or "abnormal" behavior. Most likely, since we learn that the more, we learn about the history and condition of apparently irrational people, we find compelling reasons for why they behave that way. Based on inductive reasoning, therefore, we hesitate to consider beliefs and actions irrational in an absolute sense, even if they appear to be extremely bizarre. Perhaps, we are justified to describe the action of the nude man as an irrational act if we see him sometimes nude and sometimes dressed and cannot find any reason for this oscillation. In other words, beliefs and actions that are random deserve to be called irrational. Again, someone who is a determinist like me cannot fathom the existence of randomness (Ironically, I hold two contradictory beliefs regarding free will. While on one hand, I cannot provide a rational explanation for freedom of will, I believe in the existence of freedom of will based on my rational belief in God and His proven word. This is another murky issue to ponder.).

As for an adult who is afraid of cockroaches (such as I), he might not know, let alone explain to others, any reason for such a fear. An objective observer too might not be able to justify the fear by a giant human being of a poor little insect. However, the reasons or causes of such an objectively irrational behavior might be buried in the hardware of his brain that was shaped during childhood or as a recessive gene through millions of years of evolution. A child's dramatic experience with insects might have planted or triggered such a permanent alarm button in his brain.

In sum, our beliefs and acts have reasons. We call the products of unknown, but knowable reasons as subjectively rational. We may call the products of unknowable reasons as irrational.

Objective Rationalism:

Any belief or act that is consistent with objective reality or truth is objectively rational. For instance, not putting my hand in fire is usually objectively rational. However, to save a vital document from burning, I might still have a good reason to put my hand in the fire for a few seconds to save the document. The rationality of a belief or act depends on its consistency with our goal. If our goal is to build a house to raise a family, then stealing from the cement of the house is irrational, since it will contradict our purpose. If a criminal trying to escape at a speed of 90 miles per hour from a chasing police officer enters a turn that requires a maximum of 20 miles per hour speed limit, then, his escape might be considered objectively irrational. His purpose would contradict the laws of gravity and motion.

What about deism or theism as opposed to atheism or agnosticism? What about monotheism as opposed to polytheism? Before I venture to discuss this issue, you should know that I assume that we agree that happiness is the ultimate goal of rational beings. Furthermore, I assert that this is a self-evident fact.

	In the case of belief or disbelief in God, the same principle of consistency or non-contradiction applies. If there is no God, an intelligent and eternal first cause, then atheism or more accurately agnosticism is an objectively rational position, provided that it makes the person happier. Similarly, belief in a fictional God may make a person happier, therefore making this belief rational. However, if there is a God, then rejection of such an entity might be rationally problematic, if such a God cares about the beliefs and actions of his creatures. Let's reflect on those who claim to have faith in God. If their purpose is mere happiness in this world, they might attain happiness by that belief regardless of the merit in their reasons for such a belief. They might believe in God to conform to their family and peers or use such a belief like opium to escape from the harsh realities of life. Rationality of such a belief is evaluated according to its economic, social or psychological benefits to the person or society. If the purpose of those who believe in the existence of God is to accept reality and attain eternal salvation, then the rationality of this position is contingent upon their method of reasoning and the consistency of their conclusion with objectively verifiable or falsifiable facts. I agree that a person who believes in a non-existing God with the hope of being resurrected by such an entity with eternal happiness, still might be considered rational, since he increased his hope and happiness in his only life. However, if there is a God who cares about the choice of his creatures, attaining unjustified beliefs about God may pose eternal risks. As for Pascal's Wager, it is a rational justification to believe in God, if only God were a Christian God (which denomination?) and rewarded gamblers.
Rational Monotheism	Now, after this lengthy introduction, let me briefly explain why monotheism is a rational conviction. My belief in God started as a subjectively rational position, a mere imitation of the religious conviction of my parents; however, after a period of skepticism, questioning and reflection, it is now grounded on both subjective and objective rationalism. Though I rejected many of the religious beliefs of my parents, my trust in God has strengthened with time. I do not find it accurate to use the word "faith" or "belief" to describe my conviction about God. The Quranic word "*iman*" or the verb "*amana*" does not exactly correspond to English "faith" or "to believe." These words in the Quran do not mean conjecture or guesswork; to the contrary, they are used in contrast to conjecture and guesswork. The "*iman*" approved by the Quran is a conviction or conclusion that is substantiated by rational and empirical evidence. Thus, we initially chose acknowledgement, then affirmation for the word *iman*. We are created with innate information that allows us to communicate with God (30:30). This divine information is called rouh (32:9) which can be accessed by using a logical/reasoning program called *Aql*. Our logic comes before the scripture or verbal revelation; in fact, our decision of whether a particular message is divine depends on our logic

and reasoning. No wonder, the Quran never criticizes the act of reasoning or use of logic. To the contrary we are repeatedly encouraged to question assumptions and conventional teachings with our reasoning faculties. We evaluate each external datum and statement with this internal measure. To decide whether a particular person is right in his claim of messengership, or a particular book is indeed a divine book as it claims, we use deductive and inductive reasoning. We can accept, reach and communicate with God without angels, prophets and scripture; but we cannot do so without using our reason. The messengers are merely reminders and mercy from God.

I have many reasons for believing that one God exists. Here are just some of the keywords for such a conviction: singularity, big bang, existence, the exact amount of energy in the universe, structure of an atom, fine-tuned constants in the universe, natural laws governing evolution, consciousness, accuracy of Quranic verses about various sciences, code 19, and profound personal experiences.

Before his messengership, Abraham, as a young philosopher, reached the idea of the "greatest" by a series of hypothetical questions. His method of proving the existence of the creator of all things was both empirical and rational. He invited people to observe the heavenly bodies and afterwards deduce the existence of an absolute creator from their contingent characteristics (6:74-81).

> "Such was our argument, with which we supported Abraham against his people..." (6:83)

Abraham not only supported his monotheistic paradigm through rational arguments, he also falsified the claims of his opponents via rational arguments by breaking the little statues of his pagan people and sparing the biggest one. When the idol-worshipers inquired about the "disbeliever" who committed such a blasphemous act to their idols, Abraham stood up and pointed at the biggest statue:

> "He said, 'It is that big one who did it. Go ask them, if they can speak.' They were taken aback, and said to themselves, 'Indeed, you have been wrong.' Yet, they reverted to their old ideas: 'You know full well that these cannot speak.' He said, 'Do you then worship besides God what possesses no power to benefit you or harm you? Shame on you and on whatever you worship besides God. Do you not understand?'" (21:63-67).

The Quran provides a rational argument for why God cannot have partners or equals. The argument is a logical one called Modus Tollens where the consequent is denied:

> "Have they found gods on earth who can create? If there were in them other gods besides God, there would have been chaos. Glory is to God, the Lord with absolute authority. He is high above their claims." (21:21-22).

Thus, it is no wonder that the Quran invites us not to be gullible. We should not follow anything without sufficient knowledge, including belief in God.

> "You shall not follow any information that you do not have knowledge about it. I have given you the hearing, the eyesight, and the brain, and you are responsible for using them." (17:36).

Trinity

Millions of Christians believe in the "Holy Trinity" on faith. Through this formula they transformed Jesus, the son of Mary, into the "Son of God," and even God himself. However, history, logic, arithmetic, the Old Testament, and the New Testament prove the contrary: Jesus was not Lord; he was a creation of God just like Adam was.

The doctrine of Trinity is found in many pagan religions. Brahma, Shiva, and Vishnu are the Trinitarian godhead in Indian religions. In Egypt there was the triad of Osiris, Isis and Horus; in Babylon, Ishtar, Sin, Shamash; in Arabia, Al-Laat, Al-Uzza, and Manat. The Encyclopedia Britannica (1975) gives a critical piece of information:

> "Trinity, the doctrine of God taught by Christians that asserts that God is one in essence but three in 'person,' Father, Son, and Holy Spirit. Neither the word Trinity, nor the explicit doctrine as such, appears in the New Testament, nor did Jesus and his followers intend to contradict the scheme in the Old Testament: 'Hear, O Israel: The Lord our God is one Lord'" (Deut. 6:4)

This information on the Trinity contradicts the faith of most Christians. They believe that Matthew 28:19 and John 1:1 and some other verses clearly provide a basis for the doctrine of Trinity. However, the New Catholic Encyclopedia (1967 edition, Vol: 14, p. 306) affirms that the Trinity doctrine does not exist in the Old Testament, and that it was formulated three centuries after Jesus. The Athanasian Creed formulated a polytheistic doctrine with the following words: "We worship one God in Trinity, and Trinity in Unity; neither confounding the Persons, nor dividing the Substance (Prayer Book, 1662). The Father is God, the Son is God, and the Holy Spirit is God, and yet there are not three Gods but one God." It is unanimously accepted that the doctrine of Trinity is the product of the Nicene Conference (325 C.E.).

Questions such as, "How could the Father, the Son, and the Holy Spirit be totally different and yet participate in the one undivided nature of God?" have given Christian scholars a hard time for centuries. To explain the nature of the Trinity, they have written volumes of books full of interpretations and speculations ending up with a divine paradox, or a divine mystery, amounting to no more than holy gobbledygook. So, it would not be worthwhile to question the meaning of the Trinity further as the answer, ultimately, will be that it is a divine mystery which cannot be understood. Instead, we will question the compatibility of the doctrine with the Bible.

Trinity is not taught in any of the thirty-nine books of the Old Testament. Neither do Noah, Abraham, Moses, David etc. mention the

Trinity. To the contrary, they emphasized God's oneness (Deuteronomy 4:39; 6:4; 32:39. Exodus 20: 2-3. 1 Samuel 2:2. 1 Kings 8:60, Isaiah 42:8; 45:5-6).

The concept of the Trinity was fabricated within several centuries through gradual distortion and gradual exaggeration of the powers of the hero. Initially, there were many Christian communities rejecting the idea of deity of Jesus, such as the Ebionites, but ultimately, the followers of St. Paul won against the true supporters of Jesus, and they used force, occasionally in a very cruel way, to impose their authority.

The Bible contains many verses rejecting the doctrine of the Trinity. For instance, according to the Bible, we are all children of God, thereby contradicting the idea that Jesus was the only son (Matthew 5:9; 6:14, Luke 20:36; John 8:47, 1 John 5:18,19). According to the Bible, Jesus rebukes someone calling him "Good," by asking him rhetorically, "Why do you call me good?" and then answering the question, "Only God is truly good!" (Mark 10:18-19). Furthermore, in response to those who asked about the time of the end of the world, Jesus rejected the concept of the Trinity, which equates God to Jesus: "No one knows about that day or hour, not even the angels/controllers in heaven, nor I myself, but only the Father" (Mark 13:32). If Jesus were Lord, as St. Paul's followers assert, how could he not know the future? If Jesus were kept out of the loop to play his role, then he could not have been equal to God as the doctrine of Trinity asserts. Many verses in the Gospel reject the concept of deity of Jesus and promote strict monotheism. "Jesus replied, 'The one that says, 'Hear, O Israel! The Lord our God is the one and only God. And you must love him with all your heart and soul and mind and strengths'" (Mark 12:29). Also see, 4:10; 6:24, Mark 10:18, Luke 18:19.

Another verse quoting the Old Testament depicts Jesus as a Servant of God (Matthew 12:17-18). Isn't there a difference between a Servant of God and God? The polytheist Christians will find it very difficult to answer this simple question and they will seek refuge in their dark cave called "mystery." They will do anything to continue their belief in the fabricated doctrine of the "Pharisee son of Pharisee" as their prime holy teaching (Matthew 16:11-12; 23:13-33; Luke 12:1-2; Acts 23:6). Jesus, according to the Gospels, was a "messenger of God" (Matthew 21:11,46; Luke 7:16; 24:19; John 4:19; 6:14); yet to Jesus-worshiping polytheists there is not much difference between the "messenger of God" and "God." If you bring them a dictionary and demand some rational justification, you will hit the wall of "mystery." Biblical words do not have much meaning, or they mutate and transform in many incredible ways in St. Paul's Wonderland. Whether Jesus saw God or not, might be another important question in refuting Trinity, but verse John 1:18 provides two contradictory answers in two different versions.

The Trinity is not a logical or rational theory by its very absurdity (1=1+1+1) and therefore non-falsifiable. One may find hundreds of Biblical verses rejecting the deity of Jesus and still unable to convince a Christian inflicted with this virus. The doctrine is based on revering and consecrating a clear logical contradiction: a being who was a creature, a human being, and at the same time a non-human creator! It ignores the

	fact that nothing can be both a man and a God. According to the very definition of the words "man" and "God," God is not created, but man is created; the non-created cannot be at the same time created. God is eternal, but man is mortal; the eternal cannot be mortal. So on and so forth. Therefore, the doctrine of the Trinity is a virus that attacks and destroys the immunity system of the brain first. A person who received such a virus "on faith," by blindly following the teaching of a particular church or the proximate crowd, will not be healed by rules of logic, mathematics, history, archeological findings, scientific evidence, by nothing. The creator of this virus empowered it with ability to mutate, since he boasted to his "flock" that to win as many adherents as possible, he becomes everything, anything, and all things to all men just to win them over (1 Corinthians 9:20-22): "But even if we, or an angel from heaven, preach any other gospel to you than what we have preached to you, let him be accursed. As we have said before, so now I say again, if anyone preaches any other gospel to you than what you have received, let him be accursed" (Galatians 1:8-9). No wonder, the idol-carver Pharisee-son-of-a-Pharisee has plenty of contempt for those who reason and wrote the best praise for foolishness and packaged them as part of the "gospel" with an immune system against reason, evidence, and even God's angels. Tertullian, the one who gave birth to the doctrine of the Christian Trinity, wrote one of the fanciest defenses for dogmatism, bigotry, and narrow-mindedness. He tried to banish reason by a weak reasoning: "These are human and demonic doctrines, engendered for itching ears by the ingenuity of that worldly wisdom which the Lord called foolishness, choosing the foolish things of the world to put philosophy to shame. For worldly wisdom culminates in philosophy with its rash interpretation of God's nature and purpose. It is philosophy that supplies the heresies with their equipment… After Jesus Christ we have no need of speculation, after the Gospel no need of research. When we come to believe, we have no desire to believe anything else; for we begin by believing that there is nothing else which we have to believe." (The Prescriptions Against the Heretics). Ironically, a handful verses are abused to justify the Trinity, and ALL of them are questionable. For instance, many Christian scholars affirm that the crucial word "begotten" in John 1:14,18 and 3:16,18 does not exist in the original manuscripts. Why was the phrase "son of God" changed into "the only begotten Son of God"? See: 5:72-79; 4:171-176; 4:11.
Islamic Reform	Islamic Reform is not the same as Reform in Islam; they are two different things. The former does not imply there is anything wrong with Islam, but the latter does. Since, we believe/know that the teaching

of Islam is available in its pristine form in the Quran, and since we believe/know that the Quran is God's infallible and protected word (the Final Testament), we cannot suggest reform in Islam. (If we carelessly use the expression “Reform in Islam,” obviously we mean reform in the distorted version of Islam.) Such a suggestion would be ignorance and clear blasphemy. However, a great majority of those who call themselves Muslims by qualifying the word Muslim with Sunni, Shiite or any other name of a sect, order, or cult, are far away from the original message of Muhammed, which is no different than the message of previous messengers. Creation of sects was the consequence of betraying the teaching of the Quran and was originally a heresy (6:159; 30:32). They have doomed themselves to a man-made religion that sanctifies medieval Arab culture and practices in the name of God and prophet. Therefore, we promote Islamic reformation in Muslim societies by inviting them to dedicate their religion to God alone by upholding the Quran alone as the only source of their religion. We will explain this paradigm-changing invitation later.

What do we mean by REFORM? Those who have been Arabized, rather than Islamized, have developed an allergy to other non-Arabic languages, which are considered God's blessing by God's own words (30:21). For instance, they demand converts to change their beautiful native names into Arabic ones, which has nothing to do with the teaching of the Quran. Ironically, the practice was advised neither in their fabricated *hadith* nor in their *sunna* liturgy. Somehow, they are not allergic to the Arabic word for reform, which is *islah*. They have even embraced an Arabic word which has unacceptable theological implications, "*tajdeed* of islam," that is, renewal of Islam! So, if we promote "Islamic *Islah*" or "Islamic *Tajdeed*" instead of "Islamic reform," they would not reject it with prejudice. But the English translation of the Arabic words such as submission (Islam) or reform (*islah*) elicits an immediate negative reaction from them. This might be due to the association of the word in the history with Christianity. But the Arabic word too has many negative connotations in the history of the "Muslim World." Besides, the reform movement against the Catholic church had raised many good points. These were in harmony with the teaching of the Quran, if not inspired by it. Their aversion to the "reform" might be due to its use in Christendom. Well, the reformation movement against Catholicism contained many good ideas promoted by Islam, such as rejection of intercession and selling of indulgences, etc. Regardless, the derivatives of the root SaLaHa (reform) are mentioned in the Quran 180 times, and in only one case is it used in a negative context. Sunni and Shiite *mushriks* have abandoned the Quran thereby brought a range of disasters upon themselves.

The Islamic reform movement that we are actively participating in began in early 1974. It has nothing to do with the political agenda of USA-Inc. and its neocolonialist foreign policies, based on the interests of major corporations. Many monotheist reformers have been persecuted by reactionary governments backed by US-Inc., such as Saudi Arabia, Pakistan, and Egypt. It is only after the 9/11 attacks that the US government began supporting political reform in the Middle

	East. We have yet to see whether this initiative is merely another pretext for planting mutated versions of puppet regimes, or a reflection of an honest realization of how imperative freedom and democracy are for the global community. Only time will tell.

2

The "Holy" Viruses of the Brain

(From *NINETEEN: God's Signature in Nature and Scripture;* By Edip Yüksel)

> Do you have the same religion as your parents? Score 0 points if you do and have never doubted or questioned its teachings. Score 2 for any other answer. This is an example of dogmatism, the blind acceptance of received ideas. Religion itself is not the issue here; rather, its acceptance without question is the important matter. To adhere unflinchingly to childhood beliefs on any subject, to shut your mind to new ideas, or even to other old ideas, is death to the intellect. Besides, religions should have nothing to hide. They ought to encourage doubts and questions so that they can lay them to rest and reinforce faith. (Brain Building, Marilyn vos Savant & Leonore Fleischer, Bantam Books, 1990, p. 38.)

> "As long as the prerequisites for that shining paradise is ignorance, bigotry and hate, I say the hell with it." (Henry Drummond, Inherit the Wind).

Ask the people who are leaving church after the Sunday sermon in a modern neighborhood of San Diego: "Why do you believe that Jesus is God in the flesh and was sacrificed by God for other people's sins?" As an answer, you might hear, "Because the Bible says so." If you then subject them to a follow-up question, "Well, how you know that the Bible is the word of God?" you might hear the following answer while witnessing the smile on the face of your audience fading: "The Bible says that it is the word of God." Should you remind your audience that his/her reasoning is a circular argument; the dialogue is likely to end immediately. If your audience allows you to ask more questions, you might receive the ultimate answer: "Because I believe so; I have faith in the Bible." You might not be able to hear the *reason* behind the faith of many affirmers; moreover, you might never hear the *real* reason. None of the Catholics, Protestants, Baptists, or Mormons will tell you that they believe as they do because their parents and/or their immediate friends believe that way. This is, unfortunately, the reality for most affirmers.

If you ask the same questions of a Hindu who has just purified himself in the waters of the Ganges, you will receive similar answers. The answers of a Muslim praying in the Blue Mosque of Istanbul or a Buddhist chanting in a Tokyo temple will not be any different.

If you were born in India, most likely you are a Hindu, in Saudi Arabia a Muslim, in Israel a Jew. Since you are in the USA, you are most likely a Christian. The dominant religion of your family and your country is more likely to be adopted by you. Why? What is the relationship between religion and geography or ethnicity?

Years ago, I did some psychological experiments to explore certain common human behaviors. The most interesting one was on conformity and compliance. I wanted to find out how we, as individuals, behave under strong group pressure. How does a minority of one react against a unanimous majority? The results were incredible.

The Arrow Test

For the experiment, I gathered five persons in a room and had them sit in a line. These participants would be my confederates. I told them that we would perform an experiment on the next person who would enter the room. He would be the last in the line. In the beginning, I would ask them two warm-up questions, and trained them to give me the correct answers. But, when I would ask them the third question (the real one), I trained my confederates to loudly give me the wrong answer one by one.

When the real participant entered the room, I announced that we would have a test--as if I had never discussed the subject with the group before. Then, I asked two warm-up questions. I drew simple figures on the board and asked them one by one the routine question: Which one is similar to this one? After all the five participants gave the correct answer, the real participant also gave the correct answer. They were easy questions.

Then, it came to the real question, the easiest one. I asked the following question: Which figure on the right side is similar to the figure on the left side?

→ A ➔
B →
C →

One by one, my confederates gave the wrong answer. The first said "C." The second also said "C." The third, fourth and the fifth also followed with "C." The real participant was in shock. He was amazed at the discrepancy between what he saw and what he heard. After hearing five straight C's, when his turn came, he agreed with the majority that the "C" was the right answer.

Later, I learned that I was not the first one who conducted this experiment. Between 1951-56, S. E. Asch performed a series of studies on compliance and conformity. Let me summarize the results of his experiments:

Asch made his experiments with different lengths of lines. He asked the participants to match the standard line with the lines on the left. Out of 123 participants, only 29 did not ever conform to the group's decision. 61 participants went along with their groups on every occasion. However, 33 conformed to their groups numerous times, agreeing on the obviously wrong answer almost every time.

Some participants in the Asch study claimed to have actually seen the wrong line as a correct match. They privately accepted the belief of the majority opinion. About half of the rest of the conformists claimed that they had seen the lines correctly, but that when they heard the majority choice, became convinced they must have been wrong. They then went along with the group. The remaining conformists said they knew that the answer was not correct but that they had gone along with the group anyway. (Small Group Discussion: a theoretical approach, Charles Pavitt & Ellen Curtis, Gorsuch Scarisbric, Scottsdale, AZ., 1992, p 160-165)

Conformity, whether in the form of compliance or private acceptance, occurs in every group. If a gang member steals a car the first time, most likely he will continue to do so. After the first criminal activity, the reluctance and moral anguish that he experienced in the first time will decrease and finally disappear with more involvement. He will probably justify his stealing in order to maintain his internal harmony. The same is true for new members of religious groups. The initial hesitation and questions are replaced by justification after participating the first ritual or baptism ceremony.

Marilyn vos Savant, author of the popular American newspaper column Ask Marilyn, asked her readers whether they laugh more when watching movies in theaters rather than their homes. She then went on to evaluate the impact of a group on an individual, stating:

> This is a good example of the human tendency to put aside one's own thinking and accept the thinking of others. Common to all of us is the pressure to go along with the group, at least to some extent. Also, we feel more comfortable, safer in a group; our opinions aren't attributable to us, and we don't stand out. It's no accident that television sitcoms come complete with laugh tracks; people feel better about laughing out loud if they can hear others laughing too. But sitcoms also come with "gasp" tracks and "awwww" tracks as well; your responses are being subjected to professional manipulation. What we may be timid about doing or saying as individuals, we will do or say in concert with others. However, this type of behavior has a numbing effect upon the intellect. It tends to validate and maintain whatever "groupthink" is current, whether or not it's accurate or true. Worse, it puts the mind out of the habit of thinking. People who let others direct their thinking eventually stop thinking for themselves entirely. (Marilyn vos Savant & Leonore Fleischer, *Brain Building*, Bantam Books, 1990, p. 35.)

The worst place for the brain is not the theaters, since at least there you have certain control over which movie to watch. Further, movies do not control your attitude and decisions regarding issues as crucial as life and death. The worst enemy of the brain, perhaps far ahead of drugs and alcoholic beverages, unfortunately, are those places that are associated with God: churches, mosques, synagogues, and temples of any religion. Usually, those places are picked for you by your parents, and even by your government. When you go there, the "sacred" dogmas and teachings in your brain are reinforced and you are told to close your eyes again in faith and condemn everyone who dares to question them. In time, a large territory of your brain is claimed and operated by religious virtual viruses that manipulate your thought in the interest of clergymen who do so. All in the name of a conventional god. There is no easy cure for this "holy" bug. Jomo Kenyatta, the first prime minister and president of Kenya, once depicted the role of religion in the history of his country as the "opium of masses":

> "When the missionaries came to Africa, they had the Bible and we had the land. They said: 'Let us pray.' We closed our eyes. When we opened them, we had the Bible and they had the land."

Some religious books use an effective psychological trick to gain converts. For instance, The Book of Mormon suggests the following test for skeptics:

> "And when you shall receive these things, I would exhort you that you would ask God, the Eternal Father, in the name of Christ, if these things are not true; and if you shall ask with a sincere heart, with real intent, having faith in Christ, he will manifest the truth of it unto you, by the power of the Holy Ghost. And by the power of the Holy Ghost you may know the truth of all things." (The Book of Mormon, Moroni 10:4-5).

Should it come as a surprise that a good number of people who take this "test" end up experiencing transformation in their lives? The power in this so-called proof of divinity is produced by priming the gullible subject to a self-executing conversion. First, the subject must already have accepted as fact the orthodox dogma regarding the deity of Jesus and all related stories concocted by St. Paul, *the Pharisee, Son of Pharisee*. Second, the subject must believe that the verses of Moroni will lead him to find the truth about the very verses prescribing how to find the truth. Third, the subject is ready to interpret any usual or unusual event occurring in the next days in favor of these tenets! The primed mind will perhaps witness many miracles and "feel" the Holy Ghost inside his or her mind. Fourth, the Church has won another convert who will fill its treasury with money. and a potential volunteer recruiter who would use the same test to attract others to the church.

Many Sufi leaders also use similar psychological tricks. For example, they ask the candidate to utter certain prayers in certain numbers and fashions while thinking about the Sheik before going to bed. Most of those who follow the instructions end up seeing dreams and interpreting them as expected. They become fanatic followers. Beside their night dreams, they start daydreaming. Their minds along with their pockets are intruded and manipulated by their religious leaders.

It may surprise many that Martin Luther, the founder of Protestant movement who was considered a progressive clergyman compared to the Pope, was a bigot. He invited his followers to give up their reasoning faculties and discard their brains:

> "Reason is the greatest enemy that faith has; it never comes to the aid of spiritual things, but more frequently than not struggles against the divine Word, treating with contempt all that emanates from God." (Martin Luther, Last Sermon in Wittenberg, 17 January 1546.)

If you are the supplier or the client of absurd stories in the name of God, then of course you will have problem with God's greatest gift to you: your reasoning faculties. Marx hit the nail in the head by rejecting religions with his famous description: "opium of the masses."

History is filled with tragedies created by those who gave up thinking or questioning those in power, be it of religious leaders or political heroes. No wonder millions of people accept absurd claims on faith and feel self-righteous about promoting nonsense. For instance, millions passionately reject the theory of evolution without even studying it. In the following excerpt from *Inherit the Wind*, Henry Drummond, the defense lawyer, makes a powerful point about the importance of critical thinking:

> **Matthew Harrison Brady**: We must not abandon faith! Faith is the most important thing!
> **Henry Drummond**: Then why did God plague us with the capacity to think? Mr. Brady, why do you deny the one thing that sets above the other animals? What other merit have we? The elephant is larger, the horse stronger and swifter, the butterfly more beautiful, the mosquito more prolific, even the sponge is more durable. Or does a sponge think?
> **Matthew Harrison Brady**: I don't know. I'm a man, not a sponge!
> **Henry Drummond**: Do you think a sponge thinks?
> **Matthew Harrison Brady**: If the Lord wishes a sponge to think, it thinks!
> **Henry Drummond**: Does a man have the same privilege as a sponge?
> **Matthew Harrison Brady**: Of course!
> **Henry Drummond**: [Gesturing towards the defendant, Bertram Cates] Then this man wishes to have the same privilege of a sponge, he wishes to think!

Ironically, those who have reduced their critical thinking abilities to the level of sponge show the audacity to peddle and even impose their silly stories as the ultimate truth. The audacity of arrogant affirmers led the fictional character Henry Drummond to utter one of the most memorable statements on this point: "As long as the prerequisite for that shining paradise is ignorance, bigotry and hate, I say the hell with it."

Before putting anything in our mouths we observe the color, sniff its smell, then we check its taste. If a harmful bit fools all those examinations, our stomach come to rescue and throws them up. There are many other organs that function as stations for testing, examination, and modification of imported material into our bodies. They ultimately meet our smart and vigilant nano-guards: white cells. Then, it is a mystery how we put information and assertions, especially the most bizarre ones, into our brains without subjecting them to rigorous test of critical thinking. We should not turn our brains into trash cans of false ideas, holy viruses, unexamined dogmas, and superstitions! We should be wise!

Carol Tavris, the author of influential books such as *The Mismeasure of Woman* and *Invitation to Psychology*, pulls our attention to the psychological aspect of religious beliefs:

> "One of the problems with the skeptical movement is that it attempts to take important beliefs away from people without replacing them. People believe that skeptics and scientists are forever telling them their ideas are wrong, stupid, and naïve—"No, you cannot talk to Uncle Henry from beyond the grave; that medium is a fraud" or "No, crushed aardvark bones can't cure your cancer." One problem with the critical thinking movement, which came from philosophy, was that it missed the psychological and emotional reasons that people don't think critically and don't want to think critically. Until you understand the forces that make people want to believe something, you can't just expect people to listen rationally to a set of arguments that will skewer their deepest, most cherished ideas." (Michael Shermer, *The Measure of a Woman: An Interview With Social Scientist Carol Tavris*, Skeptic, Vol. 7, No. 1, 1999. p. 71.)

Considering the prevalence of this psychological factor for most religious people, my skeptical approach to religions is not likely to appeal to many. It is very likely that devout members of organized religions will never be able to study the presentation of empirical and rational evidence demonstrating the authenticity of Quran's claims objectively, since their choice of religion is not based on their intellect, but on their emotional reaction to social pressure.

Religion: the best nest for conformists

Organized religions may give a myriad of different answers for a single question. Dogmas attract the highest rate of conformists. Conformity, eventually, causes the private acceptance or justification of the dogma. Some people become fanatics, dedicating themselves to the dogma. The old conformists cause the newcomers to conform. This chain attraction goes on.

Why is the percentage of religious conformists and their private acceptance so high? There are many reasons. Here is my kaleidoscopic, albeit incomplete list:

- We are exposed to dogmas from childhood. It is called "boiled frog syndrome." If one puts a frog in a container and pours hot water over its body, the frog will jump to save its life. But if the temperature of the water is gradually increased, the frog will not notice the heat and will boil to death. Well, we all experience the so-called "boiled frog syndrome" in many aspects of our daily lives. One of the worst examples of this syndrome is very common among religious people. Our early exposure to religion has a great impact on us. For a Hindu, thousands of human gods, the caste system, and holy cows make more sense than anything else does. For a Christian, a God with three personalities sacrificing his innocent son to criminals provides the only answer for the purpose of life. And for a Sunni Muslim, living a life according to medieval Arab culture, and glorifying Muhammed's name is the only password to heaven.
- Recent studies using Implicit Association Test, or IAT show that our unconscious attitudes create strong biases. Those who do not engage in self-examination, critical thinking, and deliberate and constant struggle to be open-minded fall

victim of this stealthy diabolic mental infection.

- Religious answers are not simple. On the contrary, they are mostly complex and vague. One can interpret any dogma and make it acceptable to him. The way is wide open for justification through endless speculations.
- Many answers do not have objective validity or a verifiable/falsifiable thesis. Since we cannot verify them, we can easily accept them.
- Professional priesthood survives on dogmas, so there will always be some well-trained holy salespeople around. They are the most effective pitchmen ever seen in this world, and they are adept at adapting to new ways.
- The common religious norms such as "Have faith without reason" or "Don't question" can close all the circuits for any possible intellectual light. As long as a person has swallowed the Trojan horse of "faith without reason," with its head and tail, even the most absurd religious teachings will have access through a back door to the brain of the victim.
- Religions do not nakedly expose their false dogmas and myths. They exploit the truth and craftily amalgamate it with myths. Phraseology like "Good moral values," attracts many. For the sake of some truth, we may accept the mixture as the whole truth.
- Religious peer pressure is very strong. Because of this, the social and psychological punishment for not complying with the religion of our family and friends usually has a deterrent effect. Therefore, we may employ an intellectual censorship to avoid a possible confrontation.
- Our enigmatic brains can reinforce our private acceptance by playing odd games. Selective cognition and logical fallacies can create spiritual experiences.
- The socio-economic benefits of a religion or cult may force us to rationalize and justify their dogmas.
- The so-called third world countries that suffer from chronic economic and political problems are governed by an elite minority who exploit the resources through authoritarian repression and all that comes with it; bribery, nepotism, monopoly, and usury. In those countries, much of the population is condemned to struggle with unemployment, poverty, and ignorance. In such an environment of corruption and injustice, a religion or a sect that provides the oppressed and deprived with an identity and radical opposition may attract the masses. In this context, the popular religion or sect is a political tool, a courage pill, and a symbol of rebellion. The suppressed hate and rage erupt with slogans colored with the name of God and religious heroes. In such an environment, religion and religious orders do not represent reason and reality, but the complex emotions caused by social and economic frustration.
- Religion, combined with Nationalistic hormones, is used with great success throughout history to send the children of the poor to wars declared by the wealthy elite who enjoy more power and obscene profits during wars. The resources of other countries are plundered in the background of holy hymns and patriotic songs. No wonder clergymen of all religions usually have been the accomplices of corrupt and oppressive kings, slave owners, colonialists, imperialists, invaders, oppressors, and greedy.
- Religion may provide the ultimate feeling of superiority for those who suffer from an inferiority complex.

- Religions promise hope to the poor and sick for eternal bliss after their miserable lives on this earth. In Karl Marx's words, "religion is the opium of masses." Ironically, to a lesser extent, the same concern becomes essential for the rich when they realize that they are aging and cannot control their rapid decline towards the grave, which will separate them from all their luxury and power. A church, a mosque or a temple of any religion may offer them all they want: a clergyman's voice declaring their salvation and entitlement to go to heaven. Now, the poor and sick can bear their pain, and the rich can continue throwing parties and collecting luxury cars in their mansions.
- We might fall in love with the faith that we adhere to. This love affair produces hormones in our brain. Losing an established faith is scary since it threatens the current chemical structure and neurological connections of the brain. A fanatic believer may demonstrate a much stronger obsession or addiction than that of a cigarette smoker (Surely, this is valid for fanatic ingrates too.).

Unfortunately, most believers are ignorant of or uninterested in the intellectual and philosophical aspect and implication of theology. How many religious people do you know who changed their religion because of a rational investigation? How many so-called Muslims do you know who subject their faith to a rational and empirical test, as recommended by their holy book?

> "You shall not accept any information unless you verify it for yourself. The hearing, the eyesight, and the mind are responsible for it." (17:36).

> "Most of them follow nothing but conjecture, and conjecture is no substitute for the truth. GOD is fully aware of everything they do." (10:36).

Indeed, in the Quranic terminology, the words "*mumin*" (acknowledger of truth, *affirmer*) and "*gullible*" denote mutually exclusive characteristics. Unfortunately, in polytheistic religions they are synonymous.

3
"On it is Nineteen"

"No, don't you see? This would be different. This isn't just starting the universe out with some precise mathematical laws that determine physics and chemistry. This is a message. Whoever makes the universe hides messages in transcendental numbers so they'll be read fifteen billion years later when intelligent life finally evolves. I criticized you and Rankin the time we first met for not understanding this. 'If God wanted us to know that he existed, why didn't he send us an unambiguous message?' I asked. Remember?"

"I remember very well. You think God is a mathematician."

"Something like that. If what we're told is true. If this isn't a wild-goose chase. If there's a message hiding in pi and not one of the infinity of other transcendental numbers. That's a lot of ifs."

"You're looking for Revelation in arithmetic. I know a better way."

"Palmer, this is the only way. This is the only thing that would convince a skeptic. Imagine we find something. It doesn't have to be tremendously complicated. Just something more orderly than could accumulate by chance that many digits into pi. That's all we need. Then mathematicians all over the world can find exactly the same pattern or message or whatever it proves to be. Then there are no sectarian divisions. Everybody begins reading the same Scripture. No one could then argue that the key miracle in the religion was some conjurer's trick, or that later historians had falsified the record, or that it's just hysteria or delusion or a substitute parent for when we grow up. Everyone could be a believer." (Sagan, Carl. Contact. Simon and Schuster. New York: 1985, p 418-419)

The above excerpts are quoted from CONTACT, a book by Dr. Carl Sagan the late astronomer who became popular with the TV series, Cosmos. Sagan's CONTACT is a novel expression of philosopher's prime dream: Mathematical evidence for God's existence.

Mathematics is considered as a priori knowledge, gained independently of experience. Most philosophers highly relied on mathematics. Descartes, who employed extreme doubt as a method to reach knowledge (certainty), could not doubt mathematics. The language of mathematics is universal.

The Most Controversial Concept

Hindus believe that he is incarnated in many human beings. Christians pontificate that he has multiple personalities, one of them being sacrificed for humanity. Jews assert that he is Jehovah. Muslims claim that he is Allah. Many question his gender. Millions die for him, millions fight for him, millions cry for him. Clergymen use his name as a trademark for their business, and the very same name motivates many devotees to give away their belongings as charity. Many joyfully sing songs for his love, others outrageously declare dialectic or scientific wars against him. Some even exclaim that he is no longer alive.

Libraries full of books are published for and against him. Big lies are attributed to him while scientific hoaxes are arranged to deny him. He is in the courts, he is on the money, he is in the schools, he is in the mind of saints and in the mouth of hypocrites. Yes, he is everywhere. And yet, philosophers continuously question his existence. In fact, world religions, with numerous versions of odd gods, have not helped philosophers prove his existence. On the contrary, they created further intellectual problems and logical obstacles for questioning minds who try to reach him.

The Prime Evidence

The "prime" evidence comes in the form of a highly sophisticated mathematical code embedded in an ancient document. Computer decoding of this document was originally started by Dr. Rashad Khalifa, a biochemist, in 1969. In 1974, this study unveiled an intricate mathematical pattern based on a prime number. (Being interested in the subject the author, like many others, I examined Dr. Khalifa's findings and assisted him in his further research.)*

For more than 14 centuries it was a hidden secret in the most read yet one of the most ignored books, the Quran (The Book of Recitation). After its discovery in 1974, the discovery of the code not only explained many verses, but it also exposed the diabolic nature of sectarian teachings and the work of clergymen.

With the computer decoding of the Quran, summarized below, the argument for the existence of God gained examinable physical evidence. Although the Quran had been in existence for fourteen centuries, its mathematical code remained a secret until computer decoding became possible. As it turned out, the code ranges from extreme simplicity to complex, interlocking intricacy. Thus, it can be appreciated by persons with limited education, as well as by scholars.

This ancient document is the Quran, revealed to Muhammed of Arabia in early sev-enth century as the Final Testament. The following is a condensed summary of this unique literary code. Please note that one does not need to know Arabic, the original language of the Quran, to examine most of the evidence presented below. For some of them one may only need to recognize the 28 letters of the Arabic alphabet.

The Message For The Computer Generation

Chapter 74 of the Quran is dedicated to the PRIME number 19. This chapter is called "*Al-Muddassir*" (The Hidden Secret). The number 19 is specifically mentioned in that Chapter as a "punishment" for those who state that the scripture is human-made (74:25). This number is also called "One of the greatest portents" (74:35). In 74:31, the purpose of the number 19 is described: to remove all doubts regarding the authenticity of the Quran, to increase the appreciation of the affirmers, and to be a scientific punishment for hypocrites and ingrates. However, the implication of this number as a proof for the authenticity of the Quran remained unknown for centuries. For fourteen centuries, the commentators tried in vain to understand the function and fulfillment of the number 19.

Before The Secret Was Decoded

Before the discovery of the 19-based system, we were aware of a symmetrical mathematical wonder in the Quran. For example:

- The word "month" (*shahr*) occurs 12 times.
- The word "day" (*yawm*) occurs 365 times.
- The word "days" (*ayyam, yawmayn*) occurs 30 times.
- The words "Satan/satan" (*shaytan*) and "controller/angel" (*malak*), each occur 88 times.
- The words "this world" (*dunya*) and "hereafter" (*ahirah*), each occur 115 times.
- The words "they said" (*qalu*) and "you say" (*qul*), each occur 332 times.
- The proper names of 26 messengers mentioned in the Quran and the frequencies of all the derivatives of the root word RaSaLa (message/send messenger), each occur 512 times.

A Great Prophecy is Fulfilled, and the Secret is Unveiled

The miraculous function of the number 19 prophesized in Chapter 74 was unveiled in 1974 through a computerized analysis of the Quran. While in retrospect the implications of 19 in Chapter 74, traditionally called Hidden One, were obvious, they remained a secret for 1406 (19x74) lunar years after the revelation of the Quran. Ironically, the first words of the Chapter 74, The Hidden One, were revealing, yet the code was a divinely guarded gift reserved for the computer generation; they were the ones who would need and appreciate it the most. As we have demonstrated in various books, hundreds of simple and complex algorithms, we witness the depth and breadth of mathematical manipulation of Arabic, an arbitrary human language, to be profound and extraordinary.

This is the fulfillment of a Quranic challenge (17:88). While the meaning of the Quranic text and its literal excellence were kept, all its units, from chapters, verses, words to its letters, were also assigned universally recognizable roles in a creation of mathematical patterns. Since its discovery, the number 19 of the Quran and the Bible increased the appreciation of many believers, removed doubts in the minds of many People of the Book, and caused discord, controversy, and chaos among those who have traded the Quran for man-made sectarian teachings. This is indeed a fulfillment of a Quranic prophecy (74:30-31).

Various verses of the Quran mention an important miracle that will appear after its revelation.

10:20 They say, "If only a sign was sent down to him from His Lord." Say, "The future is with **God**, so wait, and I will wait with you."

21:37 The human being is made of haste. I will show you My signs; do not be in a rush.

38:0 In the name of God, the Gracious, the Compassionate.
38:1 S90, and the Quran that contains the Reminder.
38:2 Indeed, those who have rejected are in false pride and defiance.
38:3 How many a generation have We destroyed before them. They called out when it was far too late.
38:4 They were surprised that a warner has come to them from among themselves. The ingrates said, "This is a magician, a liar."

38:5 "Has he made the gods into One god? This is indeed a strange thing!"

38:6 The leaders among them went out: "Walk away, and remain patient to your gods. This thing can be turned back."

38:7 "We never heard of this from the people before us. This is but an innovation."

38:8 "Has the remembrance been sent down to him, from between all of us!" Indeed, they are doubtful of My reminder. Indeed, they have not yet tasted My retribution.

38:87 "It is but a reminder for the worlds."

38:88 "You will come to know its news after awhile."

41:53 We will show them Our signs in the horizons, and within themselves, until it becomes clear to them that this is the truth. Is it not enough that your Lord is witness over all things?

72:28 ... He has counted everything in numbers.

The beginning of Chapter 25 refers to the arguments of the opponents who denied the divine nature of the Quran:

25:4 Those who rejected said, "This is but a falsehood that he invented and other people have helped him with it; for they have come with what is wrong and fabricated."

25:5 They said, "Tales of the people of old, he wrote them down while they were being dictated to him morning and evening."

The awaited miracle, the hidden mathematical structure

The subsequent verse gives an enigmatic answer to the assertion of those who claimed that the Quran is manmade.

25:6 Say, "It was sent down by the One who knows the secrets in the heavens and the earth. He is always Forgiving, Compassionate."

How can "knowing the secret" constitute an answer for those who assert that the Quran is Muhammed's work? Will the proof or evidence of the divine authorship of the Quran remain a secret known by God alone? Or will the antagonists be rebuffed by that divine mystery? If there should be a relationship between the objection and answer, then we can infer from the above verse that a SECRET will demonstrate the divine nature of the Quran.

The miracle promised throughout the Quran might have been destined to appear after Muhammed's death:

13:38 We have sent messengers before you and We have made for them mates and offspring. It was not for a messenger to come with any sign except by **God**'s leave, but for every time there is a decree.

13:39 **God** erases what He wishes and affirms, and with Him is the source of the book.

13:40 If We show you some of what We promise them or if We let you pass away, for you is only to deliver, while for Us is the reckoning.

The last phrase of verse 13:40 quoted above is interesting, The meaning of this phrase becomes clearer after the discovery of Code 19, since the Arabic word "*HeSaaB*" refers to both the "day of judgment" and "numerical computation."[2] The deliberate use of multi-meaning words is very common in the Quran. For instance, the word "*AaYah*" occurs 84 times in the Quran and in all occurrences, it means "miracle," "sign" or "law." However, its plural form "*AaYAat*" also means "revelation" or "verses of the Quran." This unique usage equates minimum three verses of the Quran (in Arabic, a different form is used for duality) with miracle. It also draws our attention to the parallels between God's signs/laws in the universe and God's revelation in human language: they share the same source and the same truth. In the following verse the plural word "AaYAat" is used not to mean "revelation" but "miracle, sign or physical manifestation":

6:158 Do they wait until the angles/controllers will come to them, or your Lord comes, or some signs from your Lord? The day some signs come from your Lord, it will do no good for any person to affirm if it did not affirm before, or it gained good through its affirmation. Say, "Wait, for we too are waiting."

The verse 30 of Chapter 74, *Al-Muddassir* (The Hidden/The Secret) of the Quran reads exactly, "**Over it Nineteen**." The entire chapter is about the number 19. Let's read the chapter from the beginning.

74:0 In the name of God, the Gracious, the Compassionate.
74:1 O you hidden one.
74:2 Stand and warn.
74:3 Your Lord glorify.
74:4 Your garments purify.
74:5 Abandon all that is vile.
74:6 Do not be greedy.
74:7 To your Lord be patient.
74:8 So when the trumpet is sounded.[3]
74:9 That will be a very difficult day.
74:10 Upon the ingrates it will not be easy.
74:11 So leave Me alone with the one I have created.
74:12 I gave him abundant wealth.
74:13 Children to bear witness.
74:14 I made everything comfortable for him.
74:15 Then he wishes that I give more.
74:16 No. He was stubborn to Our signs.
74:17 I will exhaust him in climbing.
74:18 He thought and he analyzed.
74:19 So woes to him for how he analyzed.
74:20 Then woe to him for how he analyzed.
74:21 Then he looked.
74:22 Then he frowned and scowled.
74:23 Then he turned away in arrogance.

---

[2] The word "*HeSaaB*" is used for the "day of judgment" since on that day our good and bad deeds will be computed.

[3] Traditional commentaries translate it as "When the trumpet is sounded" or "horn is blown," which is an allegorical expression for making a declaration. The root of the word "*NaQuuR*" as a verb means "strike" or "groove," and as a noun it means "trumpet" or "the smallest matter." Based on our contemporary knowledge of the prophecy mentioned in this chapter, we may translate this verse as "when the microchips are grooved."

74:24 He said, "This is nothing but an impressive magic."
74:25 "This is nothing but the words of a human."
74:26 I will cast him in the *Saqar*.
74:27 Do you know what *Saqar* is?
74:28 It does not spare nor leave anything.
74:29 Manifest to all the people.[4]
74:30 **On it is nineteen.**

The punishment issued for the opponent is very interesting: **nineteen**. Almost all numbers mentioned in the Quran serve as adjectives for nouns. Forty nights, seven heavens, four months, twelve leaders. But here the numerical function of nineteen is emphasized. Nineteen does not define or describe anything. The disbeliever will be subjected to the number nineteen itself. Then, what is the mission or function of this nineteen? Those who tended to understand the meaning of *Saqar* as "hell" naturally understood it as the number of guardians of hell. However, the punishment that is described with phrases such as, difficult task, precise, and universal manifestations, was an intellectual punishment, a mathematical challenge. Indeed, the following verse isolates the number nineteen from the number of controllers and lists five goals for it.

74:31 We have made the guardians of the fire to be angels/controllers; and We did not make their number except as a test for those who have rejected, to convince those who were given the book, to strengthen the affirmation of the affirmers, so that those who have been given the book and those who affirm do not have doubt, and so that those who have a sickness in their hearts and the ingrates would say, "What did **God** mean by this example?" Thus, **God** misguides whoever/whomever He wishes, and He guides whoever/whomever He wishes. None knows your Lord's soldiers except Him. It is but a reminder for people.

Traditional commentators of the Quran had justifiably grappled with understanding this verse. They thought that ingrates would be punished by 19 guardians of hell. That was fine. But they could not explain how the number of guardians of hell would increase the appreciation of affirmers and convince the skeptical Christians and Jews regarding the divine nature of the Quran. Finding no answer to this question, they tried some explanations: the Christians and Jews would believe in the Quran since they would see that the number of guardians of hell is also nineteen in their scripture. Witnessing the conversion of Christians and Jews, the appreciation of Muslims would increase.

---

[4] Traditional translations that tend to render Saqar as hellfire, mistranslate the verse "*Lawahatun lil bashar*" as "scorches the skin." Though this rendering might be obtained by using different dialects of Arabic, the Quranic Arabic is very clear regarding the meaning of the two words making up this verse. The first word "*LaWaHa*", if considered a noun, literally means "manifold tablets" or "manifestations" and if considered a verb, means "making it obvious." The second word "*lil*" means "for." And the third word "*BaShaR*" means "human being" or "people." "For the other derivatives of the first word, *LaWaHa*, please look at 7:145, 150, 154; 54:13; and 85:22. In all these verses the word means "tablets." For the other derivatives of the third word, *BaShaR*, please look at the end of the verse 31 of this chapter and 36 other occurrences, such as, 3:79; 5:18; 14:10; 16:103; 19:17; 36:15, etc. The traditional translation of the verse is entirely different than the usage of the Quran. Previous generations who were not aware of the mathematical structure of the Quran perhaps had an excuse to translate it as a description of hell, but contemporary Muslims have no excuse to mistranslate this verse.

This orthodox commentary has three major problems. First, neither the Old nor the New Testament mentions nineteen as the number of the guardians of hell.[5] Second, even if there was such a statement or one similar to it, this would not remove their doubts but, on the contrary, would likely increase their doubts since they would consider it as evidence supporting their claim that the Quran plagiarized many stories from the Bible. Indeed, many Biblical events are reiterated in the Quran, although with a few differences. Third, no one so far converted to Islam because of the guardians of hell.

Some scholars noticed this flaw in traditional commentaries. For instance, Fahraddin el-Razi, in his classic commentary offered many speculations, including that the number nineteen indicates the nineteen intellectual faculties of human beings. Though it is a clever interpretation, it fails to explain the emphasis on the number nineteen itself and also fails to substantiate the speculation.

The following verses emphasized the crucial function of number nineteen:

74:32 No, by the moon.
74:33 By the night when it passes.
74:34 By the morning when it shines.
74:35 It is one of the great ones.[6]
74:36 A warning to people.
74:37 For any among you who wishes to progress or regress.

The purpose of "oaths" in the Quran is different from common usage. The Quran uses oaths to draw our attention to divine signs or lessons regarding a particular subject matter. The Quran does not use oaths to make us *believe*, but to make us *think* (see: 89:5). The passing of the night and the shining of the morning are obviously allegories used to indicate intellectual enlightenment or salvation. But the expression "by the Moon" is literal and draws our attention to the relationship of the Moon and the number nineteen. The year Apollo 11 astronauts dug the Moon's surface and brought a piece of the Moon to the Earth, the same year, a biochemist named Dr. Rashad Khalifa started feeding the Quran into a computer in St. Louis, which would end up in the discovery of Code 19. This might be considered a mere coincidence, but a Quranic verse implies the correlation between the two events.

54:1 The moment drew near, and the Moon was split.[7]

---

[5] The word "nineteen" occurs in the Bible twice: Joshua 19:38 (nineteen cities) and 2 Samuel 2:30 (nineteen men). The word "nineteenth" occurs thrice: 2 King 25:8; 1 Chronicles 24:16, 25:26; and Jeremiah 52:12.

[6] The majority of traditional commentators understand the references in this verse and the ending phrase of verse 31 as "hell fire" instead of "number nineteen." You will find translations of the Quran using parentheses to reflect this traditional exegesis. According to them, "it (the hellfire) is a reminder for the people," and "this (the hellfire) is one of the greatest (troubles)." Their only reason for jumping over two textually nearest candidates, that is, "their number" or "nineteen," was their lack of knowledge about Code 19. They did not understand how a number could be a "reminder" or "one of the greatest miracles." Indeed, their understanding was in conflict with the context; the topic of the previous verse 31 is the number nineteen, not hellfire. Finally, the traditional understanding contradicts verse 49. Those who know Arabic may reflect on the word "ZKR" in the ending phrase of verse 31 with verse 49 and reflect on the fact that hellfire is not something the Quran wants us to accept and enjoy!

54:2	If they see a sign, they turn away and say, "Continuous magic!"
54:3	They rejected, and followed their desires, and every old tradition.
54:4	While the news had come to them in which there was sufficient warning.
54:5	A perfect wisdom; but the warnings are of no benefit.

Let's continue the reading of Chapter 74 (The Hidden):

74:38	Every person is held by what it earned;
74:39	Except for the people of the right.
74:40	In paradises, they will be asking
74:41	About the criminals.
74:42	"What has caused you to be in *Saqar*?"
74:43	They said, "We were not of those who offered support (or observed contact prayer)."
74:44	"We did not feed the poor."
74:45	"We used to participate with those who spoke falsehood."
74:46	"We used to deny the day of Judgment."
74:47	"Until the certainty came to us."
74:48	Thus, no intercession of intercessors could help them.
74:49	Why did they turn away from this reminder?

How can someone who does not believe in the Day of Judgment believe in intercession? Those who believe in the intercession of saints and prophets on their behalf surely believe or at least claim to believe in the Day of Judgment. Nevertheless, according to the Quranic definition of the "Day of Judgment" a person cannot simultaneously believe in both intercession and the Day of Judgment. The Quran defines the "Day of Judgment" as "the day that no person can help another person, and all decisions, on that day, will belong to God." (82:19). That day neither Muhammed nor Ahmad, neither Jesus nor Mary, neither Ali nor Wali can help those who reject the number nineteen, its role and implication in the Quran.[8]

Simple To Understand, Impossible To Imitate

The mathematical structure of the Quran, or The Final Testament, is simple to understand, yet impossible to imitate. You do not need to know Arabic, the original language of the Quran to examine it for yourself. Basically, what you need is to be able to count until 19. It is a challenge for atheists, an invitation for agnostics and guidance for affirmers. It is a

---

[7] The Arabic word "*inShaQQa*" has been traditionally translated is "being divided in the middle." *Hadith* books report a "miracle" as the fulfillment of this verse. According to those narration Muhammad pointed at the Moon and the Moon was split for a while. Some of the reports even provide further details about this fabricated "miracle": half of the Moon fell on Muhammad's cousin, Ali's backyard. The word "*ShaQQa*" has a range of dimensions, from splitting into half to simply breaking or cracking as in 80:25-26: "We pour the water generously. Then we split the soil open." If you wonder the usage of past tense for a prophetic statement, it is a well-known Quranic style to indicate the certainty of the upcoming event and the meta-time nature of God's knowledge. (See 39:68; 75:8-9; 25:30; 7:44-48; 6:128; 20:125-126; 23:112-114). For those who know Arabic and wonder about the meaning of "*infi'al*" form, see 2:60.

[8] For the Quran's position regarding intercession please see 2:48; 6:70,94; 7:5; 9:80; 10:3; 39:44; 43:86; 16:20,21; 78:38.

perpetual miracle for the computer generation. Dr. Rashad Khalifa introduces this supernatural message as follows:

> "The Quran is characterized by a unique phenomenon never found in any human authored book. Every element of the Quran is mathematically composed-the chapters, the verses, the words, the number of certain letters, the number of words from the same root, the number and variety of divine names, the unique spelling of certain words, and many other elements of the Quran besides its content. There are two major facets of the Quran's mathematical system: (1) The mathematical literary composition, and (2) The mathematical structure involving the numbers of chapters and verses. Because of this comprehensive mathematical coding, the slightest distortion of the Quran's text or physical arrangement is immediately exposed" (Rashad Khalifa, Quran The Final Testament, 1989, p. 609).

Physical, Verifiable and Falsifiable Evidence

Here are some examples of this historical message:

- The first verse, i.e., the opening statement "*Bismillahirrahmanirrahim*", shortly "Basmalah," consists of 19 Arabic letters.
- The first word of Basmalah, *Ism* (name), without contraction, occurs in the Quran 19 times.
- The second word of Basmalah, *Allah* (God) occurs 2698 times, or 19x142.
- The third word of Basmalah, *Rahman* (Gracious) occurs 57 times, or 19x3.
- The fourth word of Basmalah, *Rahim* (Compassionate) occurs 114 times, or 19x6.
- The multiplication factors of the words of the Basmalah (1+142+3+6) add up to 152 or 19x8.
- The Quran consists of 114 chapters, which is 19x6.
- The total number of verses in the Quran including all unnumbered Basmalahs is 6346, or 19x334. If you add the digits of that number, 6+3+4+6 equals 19.
- The Basmalah occurs 114 times, (despite its conspicuous absence from chapter 9, it occurs twice in chapter 27) and 114 is 19x6.
- From the missing Basmalah of chapter 9 to the extra Basmalah of chapter 27, there are precisely 19 chapters.
- The occurrence of the extra Basmalah is in 27:30. The number of the chapter and the verse add up to 57, or 19x3.
- Each letter of the Arabic alphabet corresponds to a number according to their original sequence in the alphabet. The Arabs were using this system for calculations. When the Quran was revealed 14 centuries ago, the numbers known today did not exist. A universal system was used where the letters of the Arabic, Hebrew, Aramaic, and Greek alphabets were used as numerals. The number assigned to each letter is its "Gematrical Value." The numerical values of the Arabic alphabet are shown below: [the table is omitted]
- There are exactly 114 (19x6) verses containing all these 14 letters.
- A study on the gematrical values of about 120 attributes of God which are mentioned in the Quran, shows that only four attributes have gematrical values which are multiples of 19. These are "*Wahid*" (One), "*Zul Fadl al Azim*" (Possessor of Infinite Grace), "*Majid*" (Glorious), "*Jaami*" (Summoner). Their gematrical value are 19, 2698, 57, and 114 respectively, which are all divisible by

19 and correspond exactly to the frequencies of occurrence of the Basmalah's four words.

- The total numbers of verses where the word "Allah" (God) occurs, add up to 118123, and is 19x6217.
- The total occurrences of the word Allah (God) in all the verses whose numbers are multiples of 19 is 133, or 19x7.
- The key commandment: "You shall devote your worship to God alone" (in Arabic "*Wahdahu*") occurs in 7:70; 39:45; 40:12,84; and 60:4. The total of these numbers adds up to 361, or 19x19.
- The Quran is characterized by a unique phenomenon that is not found in any other book: 29 chapters are prefixed with "Quranic Initials" which remained mysterious for 1406 years. With the discovery of the code 19, we realized their major role in the Quran's mathematical structure. The initials occur in their respective chapters in multiples of 19. For example, Chapter 19 has five letters/numbers in its beginning, **K**20**H**8**Y**10**A'**70**S**90, and the total occurrence of these letters in this chapter is 798, or 19x42.
- For instance, seven chapters of the Quran starts with two letter/number combinations, **H**8**M**40, and the total occurrence of these letters in those chapters is 2347 (19x113). The details of the numerical patterns among the frequency of these two letters in the seven chapters they initialize follows a precise mathematical formula.
- To witness the details of the miracle of these initials, a short chapter which begins with one initial letter/number, **Q**100, will be a good example. The frequency of "Q" in chapter 50 is 57, or 19x3. The letter "Q" occurs in the other Q-initialed chapter, i.e., chapter 42, exactly the same number of times, 57. The total occurrence of the letter "Q" in the two Q-initialed chapters is 114, which equals the number of chapters in the Quran. The description of the Quran as "*Majid*" (Glorious) is correlated with the frequency of occurrence of the letter "Q" in each of the Q-initialed chapters. The word "*Majid*" has a gematrical value of 57. Chapter 42 consists of 53 verses, and 42+53 is 95, or 19x5. Chapter 50 consists of 45 verses, and 50+45 is 95, or 19x5.
- The Quran mentions 30 different cardinal numbers: 1, 2, 3, 4, 5, 6, 7, 8, 9, 10, 11, 12, 19, 20, 30, 40, 50, 60, 70, 80, 99, 100, 200, 300, 1000, 2000, 3000, 5000, 50000, & 100000. The sum of these numbers is 162146, which equals 19x8534. Interestingly, nineteen is mentioned the 30th verse of chapter 74 and the number 30 is 19th composite number.
- In addition to 30 cardinal numbers, the Quran contains 8 fractions: 1/10, 1/8, 1/6, 1/5, 1/4, 1/3, 1/2, 2/3. Thus, the Quran contains 38 (19x2) different numbers. The total of fractions is approximately 2.
- If we write down the number of each verse in the Quran, one next to the other, preceded by the number of verses in each chapter, the resulting long number consists of 12692 digits (19x668). Additionally, the huge number itself is also a multiple of 19.

Code 19: The Real Bible Code

It is significant that the same 19-based mathematical composition was discovered by Rabbi Judah in the 12th century AD in a preserved part of the Old Testament. Below is a quote from Studies in Jewish Mysticism.

"The people (Jews) in France made it a custom to add (in the morning prayer) the words: " 'Ashrei temimei derekh (blessed are those who walk the righteous way)," and our Rabbi, the Pious, of blessed memory, wrote that they were completely and utterly wrong. It is all gross falsehood, because there are only nineteen times that the Holy Name is mentioned (in that portion of the morning prayer), . . . and similarly you find the word Elohim nineteen times in the pericope of Ve-'elleh shemot . . . .

"Similarly, you find that Israel were called "sons" nineteen times, and there are many other examples. All these sets of nineteen are intricately intertwined, and they contain many secrets and esoteric meanings, which are contained in more than eight volumes. Therefore, anyone who has the fear of God in him will not listen to the words of the Frenchmen who add the verse " 'Ashrei temimei derekh (blessed are those who walk in the paths of God's Torah, for according to their additions the Holy Name is mentioned twenty times . . . and this is a great mistake. Furthermore, in this section there are 152 words, but if you add " 'Ashrei temimei derekh" there are 158 words. This is nonsense, for it is a great and hidden secret why there should be 152 words . . ." (Studies In Jewish Mysticism, Joseph Dan, Association for Jewish Studies. Cambridge, Massachusetts: 1978, p 88.)

Running like Zebras

The last section of Chapter 74 (The Hidden) likens those who turn away from the message of nineteen to zebras running away from a lion.

74:49 Why did they turn away from this reminder?
74:50 Like fleeing zebras,
74:51 Running from the lion?
74:52 Alas, every one of them wants to be given separate manuscripts.
74:53 No, they do not fear the Hereafter.
74:54 No, it is a reminder.
74:55 Whosoever wishes will take heed.
74:56 None will take heed except if **God** wills. He is the source of righteousness and the source of forgiveness.

Numerous books and articles rejecting the importance of the number nineteen in the Quran have been published in many languages worldwide. Some of the publications were freely distributed by the support of petrol-rich countries, such as Kuwait and Saudi Arabia.

The Quran is the only miracle given to Muhammed (29:51). Muhammed's mushrik companions could not comprehend that a book could be a miracle, and they wanted miracles "similar" to the ones given to the previous prophets (11:12; 17:90-95; 25:7,8; 37:7-8). Modern *mushriks* also demonstrated a similar reaction when God unveiled the prophesied miracle in 1974. When the miracle demanded from them the dedication of the system to God alone, and the rejection of all other "holy" teachings they have associated with the Quran, they objected, "How can there be mathematics in the Quran; the Quran is not a book of mathematics" or, "How can there be such a miracle; no previous messenger came up with such a miracle!" When a monotheist who was selected to fulfill the prophecy and discover the code started inviting his people to give up polytheism and the worship of Muhammed and clerics, he was officially declared an apostate by Muslim scholars gathered in Saudi Arabia from 38 different countries in March 19, 1989. Within less than a year he was assassinated by a group linked to al-Qaida in early 1990 in Tucson, Arizona. See 3:81; 40:28-38; 72:24-28; 74:1-56. For the prophetic use of the word reminder (ZKR), see 15:9; 21:2-3; 21:24,105; 26:5; 29:51; 38:1,8; 41:41; 44:13; 72:17; 74:31,49,54.

Was the Discovery of the Code 19 a Coincidence?

Dr. Rashad Khalifa did not have any knowledge that his curiosity regarding the meaning of the alphabet letters that initialize 29 chapters of the Quran would end up with the discovery of its mathematical system. His computerized study that started in 1969 gave its fruits in 1974 by the discovery of the 14-century old SECRET.

If the Code 19 was going to provide strong evidence for the existence of God and for the authenticity of the Quran, then it is reasonable to expect that the identity of the discoverer and the time of the discovery would not be coincidental. Indeed, the events have demonstrated a prophetic design in the timing of this miraculous mathematical design.

The number 19 is mentioned only in a chapter known "The Hidden," the 74th chapter of the Quran. Juxtaposing these two numbers yields 1974, exactly the year in which the code was deciphered. (Calendar based on the birth of Jesus and the solar year is accepted by the Quran as units of calculating time. See, 19:33; 43:61; 18:24. Besides, this is the most commonly used calendar in the world.) If we multiply these two numbers, 19x74, we end up with 1406, the exact number of lunar years between the revelation of the Quran and the discovery of the code.

Furthermore, the first statement expressed in the first two verses of Chapter 74 is about the unveiling of the secret. It is interesting that if we consider one version of spelling the first word, which contains three *Alifs* instead of two, there are 19 letters in the first statement of chapter 74. More interestingly, when we add the numerical values of each letter in these two verses the sum is a very familiar number. Here is the value of each letter in "*Ya ayyuhal Muddassir; qum fa anzir*" (O you Hidden one, stand and warn).

Y	10
A	1
A	1
Y	10
H	5
A	1
A	1
L	30
M	40
D	4
TH	500
R	200
Q	100
M	40
F	80
A	1
N	50
Z	700
R	200
TOTAL	**1974**

And **1974**[9] is the year when the hidden secret came out and warned us!

Adding to this prophetic mathematical design is the fact that the derivatives of the name of the discoverer, RS̲hD (guidance), occurs in the Quran exactly 19 times. (See 2:186; 2:256; 4:6; 7:146; 11:78; 11:87; 11:97; 18:10; 18:17; 18:24; 18:66; 21:51; 40:29; 40:38; 49:7; 72:2; 72:10; 72:14; 72:21.). The exact form, Rashad, occurs twice and they sandwich the claim of unappreciative people who wish to end the messengership (4:28-38).

In sum, the relationship between the following seven elements is more than interesting:

- The mathematical code (19).
- The number of the chapter mentioning the code (74).

---

[9] The absolute values of these 19 numbers (the numerical values of letters) add up to 57 (19x3).

- The year of the discovery of the code (1974).
- The number of lunar years between the revelation of the Quran and the year of the discovery (19x74).
- The numerical value of the 19 letters comprising the first statement of the chapter 74 (1974).
- The frequency of derivatives of the discoverer's name in the Quran (19).
- The context of the verses where the exact name of the discoverer is mentioned as an adjective (40:28-38).

In January 31, 1990, Rashad Khalifa was assassinated in Tucson, Arizona, by a terrorist group affiliated with al-Qaida. Ironically, soon after his departure, ignorant people started idolizing him and created a cult distorting his message of strict monotheism.[10] Also See 27:82-85; 40:28-38; 72:19-28.

27:82 When the punishment has been deserved by them, We will bring out for them a creature made of earthly material, it will speak to them that the people have been unaware regarding Our signs.[11]

27:83 The day We gather from every nation a party that denied Our signs, then they will be driven.

27:84 Until they have come, He will say, "Have you denied My signs while you had no explicit knowledge of them? What were you doing?"

27:85 The punishment was deserved by them for what they transgressed, for they did not speak.

---

10 For those who claim the infallibility of Dr. Khalifa, I would like to give a sample of verses that I think carry some minor or important translational errors: 2:114; 2:275*; 2:282*; 4:34&*; 4:79*; 4:127; 8:64; 9:29; 10:34!; 14:4; 16:75*; 18:16*; 19:26!; 20:96&*; 20:114; 21:90* x 21:73; 21:96*; 29:12 x 29:13; 32:5!; 34:41; 35:24 x 25:51; 43:11 x 41:12*; 43:36*; 47:11 x 42:15; 49:1 x 38:26&7:3; 65:12* x 42:29; 66:5; 73:15! (Asterisks are for footnotes and/or subtitles, exclamation marks for missing phrases, and "x" for contradictions.)

11 Unlike the creatures made of water (24:45) this one is made from earthly elements. After the discovery of the code and from the context of this verse, this creature (DaBBah) can be understood as reference to computer. In traditional books there are many bizarre descriptions of this prophesized creature. Contrary to the Quran's positive depiction of the earth-based creature, *hadith* books contain negative descriptions.

Some may object to my interpretation by saying that the word "DaBBah" implies something alive that moves. I think this objection is reasonable. But, I prefer the computer to an animal since I can understand the phrase "dabbatan min al -ardi" not a creature geographically from earth, but a creature made of earth. Thus, I can accept some differences in mobility between the "dabbeh from water and "dabbeh from earth." Since a computer has many moving parts, from its hard disk to the information carried by trillions of electrons, I do not see it a far-fetched understanding of the implication of "DaBBah".

How Can We Explain This Phenomenon?

There are basically four possible explanations:

1. Manipulation: One may be skeptical about our data regarding the mathematical structure of the Quran. However, one can eliminate this option by spending several hours of checking the data at random. (We recommend *Quran: The Visual Presentation of Miracle* by Rashad Khalifa and the upcoming book, *Nineteen: God's Signature in Nature and Scripture* by Edip Yuksel). Muslim scholars and clerics who have traded the Quran with primitive mediaeval fabrications, that is, *Hadith* and *Sunna*, strongly reject this mathematical system, since the mathematical system exposes the corruption of religions by clergymen.

2. Coincidence: This possibility is eliminated by the statistical probability laws. The consistency and frequency of the 19-based pattern is much too overwhelming to occur coincidentally.

3. Human fabrication: While fabricating a literary work that meets the criteria of the document summarized here is a stunning challenge for our computer generation, it is certainly even more improbable during the time of initiation of the document, namely, 610 AD. One more fact augments the improbability of human fabrication. If a certain person or persons had fabricated this literary work, they would want to reap the fruits of their efforts; they would have shown it to people to prove their cause. In view of the originality, complexity, and mathematical sophistication of this work, one has to admit that it is ingenious. However, no one has ever claimed credit for this unique literary code; the code was never known prior to the computer decoding accomplished by Dr. Khalifa. Therefore, it is reasonable to exclude the possibility of human fabrication.

The timing of the discovery may be considered additional evidence for the existence and full control of the Supreme Being: The mystery of the number 19 which is mentioned as "one of the greatest events" in the chapter 74 (The Hidden Secret) was discovered by Dr. Khalifa in1974, exactly 1406 (19x74) lunar years after the revelation of the Quran. The connection between 19 (the code) and 74 (the number of the chapter which this code is mentioned) is significant in the timing of the discovery.

4. Super Intelligent Source: The only remaining reasonable possibility is that a super intelligent source is responsible for this document; one who designed the work in this extraordinary manner, then managed to keep it a well-guarded secret for 14 centuries, for a predetermined time. The mathematical code ensures that the source is super intelligent and that the document is perfectly intact.

▼

The discoverer of the Code 19, Dr. Rashad Khalifa, was assassinated in Tucson, Arizona, by an international terrorist group al-Fuqra or al-Fuqara, which was affiliated to Usama Ben Laden, in 1990.[12] However, the power of this message is promising a new era of reformation

---

[12] My mentor and friend, Dr. Rashad Khalifa, was assassinated in 1990 by a terrorist organization organized in Salt Lake City. The members of al-Fuqra, which was later claimed to be affiliated with Ben Laden's newly founded Al-Qaida, stabbed him to death in our Tucson Mosque. The assassination of my mentor and its aftermath was widely covered by Arizona Daily Star and Tucson Weekly and local radio and TV stations. After September 11, the national media picked up the story. For instance, Newsweek and Dan Rather at CBS Evening News declared this incident to be Al-Qaida's first terrorist act in the USA. See: CBS Evening News with Dan Rather, on October 26, 2001; cover story of Newsweek, January 14, 2002, p.44. On March 19, 2002, KPHO-TV at Phoenix, a CBS affiliate, in its evening news, broadcast an interview with me under

in religions, particularly Islam and Christianity. Ironically, some of Khalifa's "followers" reverted to the days of ignorance by trying to turn the community of Submitters into a cult after the departure of that brave iconoclastic scientist. Though there are still some monotheists in the group, as it seems, a gang is actively busy in repeating the history of ignorant people by carving a new idol for the group. The gang claims infallibility of Rashad Khalifa and considers all his writings, appendices, footnotes, fingernotes, and articles in the newsletter and his speeches in video recordings to be divine revelation. The gang members consider Rashad's re-re-revised translation to be error-free, and even justify the obvious errors such as the spelling ones. They have replaced their former idol with a new one. The Submitters community has been fatally infected by the idol-carvers and with the rate of infection; the group may excommunicate all the monotheists within a decade. I have exposed the distortion of this gang in a booklet titled, United but Disoriented. Ten years ago, many considered some of my predictions to be "unreal" or "paranoid", but my recent debate with the gang members proved my predictions. They are now making claims that none expected to hear, including themselves. Like most polytheists, they have mastered their arguments for their version of polytheism while pretending to be monotheists. The book's first edition is out of print, but it is available on the net. God willing, an updated edition of the book tracking the group's devolution since departure of Rashad will be published by Brainbow Press, most likely under the title: Idolizing Iconoclasts: A Cult after Dr. Rashad Khalifa

Edip Yuksel's comprehensive demonstration of the mathematical structure of the Quran, and answers to major criticisms, has been published by BrainbowPress under titles:

- NINETEEN: God's Signature in Nature and Scripture
- Running Like Zebras

---

the headline: *Traces of Al Qaeda Cell in Tucson*. However, despite its importance in revealing the theological vulnerability of Al-Qaida, this first terrorist event did not receive the attention it deserved. Curiously, the *9/11 Commission Report* (2002) left out this important first act of terrorism in the US from their report.

4

Which One Do You See: HELL OR MIRACLE?

(From 19 Questions for Muslim Scholars by Edip Yuksel)

From the smoke of hellfire in their minds, hypocrites and unappreciative people are not able to witness one of the greatest miracles.

I know that the title and the subtitle of this article are quite challenging. If you have developed an attitude against witnessing the mathematical miracle of the Quran, you are justified to get upset with these words and perhaps get little angry. You might have already blinded your eyes and closed your ears with one or more false, contradictory, trivial, or irrelevant excuses, such as:

1. God uses literary styles to prove the authenticity of His word; but not mathematics. God is not a mathematician, and His word has nothing to do with math.
2. If Muhammed was not aware of a mathematical miracle in the Quran, then it cannot be true. Muhammed knew everything in the Quran. The knowledge contained in the Quran is limited to Muhammed's knowledge and understanding more than 14 centuries ago.
3. If I accept a mathematical structure in the Quran based on number 19, then I will be denying two Quranic verses in the end of Chapter 9, and thus would contradict the majority of Muslims and their manuscripts. Besides, I would be contradicting the verses guaranteeing the perfect preservation of the Quran.
4. Some verses of the Quran are Mutashabih; none can understand their meaning except God. The proponents of 19-based mathematical system in the Quran are indulging in those Mutashabih verses.
5. The number 19 is heresy, since it is the holy number of Bahai sect.
6. Though I can balance my check and shop in the stores, I am innumerate when it comes to the 19.
7. We can find many mathematical phenomena if we spend enough time on any book. For instance, the Bible Code claimed that the distorted translations of the Bible contain impressive, coded prophecies.
8. The Quran cannot contain secrets, mysteries, or prophecies; it is a clear book.
9. What about the previous generations; will they all go to hell?
10. The claim of a 19-based mathematical code is a myth.
11. The claim of a 19-based mathematical code is a magic.
12. There are many other numbers in the Quran and the number 19 has no special place.
13. There are problems in the count of letters or words. For instance, the first verse of the Quran does not have 19 letters; it has 18, 20, 21, or 22 letters, but not 19. For instance, the counts of A.L.M. letters have errors.
14. Part of the numerical claim is based on the Gematria system (ABJAD), a Jewish fabrication and a deceptive tool used by numerologists and astrologists.

15. The pronoun "it" in verse 74:30 is feminine, and it refers to Hell not the Quran. Thus verse 74:30-31 is about guardians of hell, not a mathematical proof of authenticity.
16. Even if there is such a code in the Quran it is not important. We need to follow the instruction of the book. We need to focus on how to fight against infidels.
17. The discoverer of Code 19 claimed to be the messenger of God. He deserved divine retribution. Thus, he was killed in early 1990 by an Al-Qaida affiliate American cult, al-Fuqra, as Al-Qaida's first act in the USA.
18. If we believe in code 19, then we would end up rejecting holy teachings of *Hadith* and *Sunna*.
19. Code 19 is a Zionist trick.
20. I already believe in the Quran, and I do not need miracles.
21. I believe in the Quran because its message is true.
22. I believe in the Quran because it does not contain contradictions.
23. I believe in the Quran on blind faith, and I do not need any reason for my belief.
24. And many more excuses, reasons, or lack of them...

I will deal with all these excuses, and God willing, I will respond each of them one by one; but first thing first:

Whatever are your EXCUSES or REASONS to ignore, reject, or ridicule the number 19, this article will expose your true intentions, which you might be trying to hide even from yourself (11:5). If you are claiming to believe in the Quran, and after reading this article still continue ignoring or ridiculing the mathematical code of the Quran, you will be TORMENTED all your life by repeatedly witnessing an exciting fact and losing its sight afterwards, like a person who witnesses succeeding events of lightning in the darkness (2:17-20). You will neither appreciate nor comprehend the miracle, nor will you be satisfied with your denial. You will perpetually oscillate between momentary belief end prolonged disbelief, between private doubt and public denial. You will be doomed to SAQAR here and in the hereafter.

But, if you have affirmed the truth and gained some goodness in your affirmation, then this article, by God's will, may wake you up from ignorance, and may change your paradigm (6:158). You will never be the same; you will be among the progressives (74:30-37). You will be one of the few who are blessed by God to witness one of the greatest miracles. You will be sure about your conviction, not based on self-deceptive claims, on conformity with a religious group, or on blind faith, or wishful thinking; but a faith based on knowledge gathered from empirical and rational evidence corroborated by spiritual experience (41:53; 74:31). You will experience God's presence in your life and attain happiness promised to affirmers and submitters (9:124; 10:64; 16:89). You will dedicate all your life to serve God alone without remorse or fear (2:62; 3:170; 10:62; 46:13). You will understand many Quranic verses that had not much meaning for you (2:1,108). Every time you see an unappreciative disbeliever denying or mocking the 19-based system, your knowledge-based faith will be justified. And the more people do not see what you and few others have witnessed will increase the importance, power and the wonder of the prophetic description of 74:31 in your mind. You will attain certainty (74:31; 27:1-3; 2:260; 13:2; 2:118; 45:20). However, witnessing this miracle will also put some responsibility on you (5:115; 47:25).

Let's start from the translation of verses 21-37 of chapter 74, The Hidden One:

74:21	He looked.
74:22	He frowned and scowled.
74:23	Then he turned away arrogantly.
74:24	He said, "This is but an impressive illusion/magic!"

74:25 "This is human made."
74:26 I will cast him into SAQAR!
74:27 And what will explain to you what the SAQAR is?
74:28 It leaves nothing, it lets nothing (NOT MORE, NOT LESS; PRECISE; PERFECT);
74:29 VISIBLE (LAWAHATUN) to PEOPLE (BASHAR) (universal)!
74:30 On it is Nineteen.
74:31 As guardians of fire we appointed none except angels/controllers, and we assigned their number (1) to torment the unappreciative ingrates, (2) to convince People of the Book, (3) to strengthen the affirmation of the affirmer, (4) to remove all traces of doubt from the hearts of People of the Book, as well as the affirmers, and (5) to expose those who have disease in their hearts, and the unappreciative ingrates; they will say, "What did God mean by this allegory?" God thus sends astray whomever He wills (or, whoever wills), and guides whomever He wills (or, whoever wills). None knows the soldiers of your Lord except He. IT (HIYA) is a reminder for the people.
74:32 Indeed, by the Moon.
74:33 By the night as it passes.
74:34 And the morning as it shines.
74:35 This (NUMBER) is one of the GREATEST (KUBRA).
74:36 A WARNING (NAZEER) for the PEOPLE (BASHAR).
74:37 For those who want to progress or regress.

Though the function of 19 is listed in Chapter 74, conventionally known *Muddathir* (The Hidden One), its implication and fulfillment were kept hidden according to God's will as the Quran's secret for 19x74 lunar years after its revelation to Muhammed. However, the All-wise God, unveiled this secret via a monotheist scientist in 1974, according to the most commonly used calendar, a calendar allegedly based on the birth of Jesus, which is considered a holy day and the sign of the End (19:15). As a result, the number 19, as the miraculous code of the Quran and the Bible strengthened and continues to strengthen the affirmation of affirmers, removed and continues to remove doubts in the hearts of people of the book, and intellectually tormented and continues to torment hypocrites and unappreciative ingrates.

Now let's first discuss the words written in CAPITAL letters in the translation of the verses above. Then, we will discuss each of the popular excuses used by the unappreciative ingrates of the miracle 19.

We should understand the meaning of these words by first referring to their usage in the Quran. Their immediate context as well as their use in other verses is usually sufficient to illuminate their meanings. For affirmers who have been lucky to witness this great miracle, the meaning of the verses above is good news; it gives them hope, enlightens them, informs them, and turns their position from wishful thinking or conjecture to knowledge-based paradigm. Thus, the rhetorical value of these verses is very high for affirmers:

74:27 ---> SAQAR (saqar, to be defined by the following verses)
74:28 ---> (NOT MORE, NOT LESS; PRECISE; PERFECT);
74:29 ---> LAWAHA (obvious; visible; tablet; screen)
74:29 ---> BASHAR (humans; people)
74:31 ---> HIYA (it; reference to the number 19)
74:31 ---> ZIKRA (reminder; message)
74:35 ---> HA (it; referring to the number 19)
74:35 ---> KUBRA (great miracles)

74:36 ---> NEZEER (warner to be embraced and supported)

On the other hand, those who have deprived themselves from witnessing the miracle 19 because of their ill intentions or their dogmatic rejection try hard to render these key words incompatible with the semantic context of the Quran. They conceive God of the Quran as an angry and despotic God who is not able provide any reasonable argument against those who question the Quran's authenticity, but only resorts to intimidation: "I will burn you in hell!!!" The God they depict has double standard: He asks the ingrates to bring their evidence for their argument (2:111; 11:17; 21:24; 27:64; 28:75; 35:40) but for His argument He only wants to scare them! The opponents of the 19-based miracle, by distorting the meaning of the words in these verses, manage to blind themselves to one of the most profound philosophical and theological arguments and evidences in history. Not only they divert themselves from the right path they try to divert others too (6:25-26; 22:3).

Thus, the understanding (more accurately, the misunderstanding) of those who cannot appreciate God as He should be (6:90-91), the argument for Quran's authenticity is scorching, burning, dark, hellish, misfortune, disastrous, and scary. Thus, in the minds of opponents of miracle 19, the rhetorical value of these verses is simply a threat to burn and torture:

74:27 ---> SAQAR (hellfire)
74:28 ---> (NEITHER LEAVES THE FLESH NOR THE BONES; DESTROYS CONTINUALLY);
74:29 ---> LAWAHA (scorching; burning; shriveling)
74:29 ---> BASHAR (skin)
74:31 ---> HIYA (it; referring to hell-fire)
74:31 ---> ZIKRA (news of disaster)
74:35 ---> HA (it; referring to hell-fire
74:35 ---> KUBRA (great punishment; gravest misfortune; dire scourge)
74:36 ---> NAZEER (warning to be escaped)

Now let's one by one discuss each of these words, which were widely and perhaps JUSTIFIABLY misunderstood by pre-1974 generations, and yet are intentionally distorted by the post-1974 opponents of one of the greatest miracles.

Does "Saqar" In 74:27 Mean Hell OR Something Else?

Though prominent Arabic dictionaries such as Lisan-ul Arab and specialized dictionaries such as Mufradat Fi Gharib-il Quran affirm that the word might be of foreign origin with no Arabic derivatives, these and other dictionaries and commentaries of the Quran do not hesitate to define it as hell or heat radiating from Sun. Lisan-ul Arab refers to a *Hadith* which uses a bizarre derivative of the word Saqar to mean "liars". The word SaQaR is mentioned four times in the Quran, three times in Chapter 74 and once in 54:48. In the later one, the word SaQaR is used in a statement warning that when criminals will be dragged to the fire they will be told: "taste the touch of Saqar". From this verse one might infer that Saqar is another word for fire; but a better inference is that Saqar is a negative feeling or state of mind one tastes after being committed to the divine punishment. In this case, there is no reason to think that this negative state of mind could not be obtained from experiences other than fire.

The Arabic word for hell is GaHyM or GaHaNnaM. The word NAR (Fire), though not a specific name for hell, is also frequently used to denote the same phenomenon. However, the word NAR (fire) is also used in its literal meaning, which is simply fire. For instance, verses 20:10-14 describe God's communication with Moses through fire. Obviously, this fire in the holy land which God spoke through cannot be the Hell. Similarly, the word SaQaR does not necessarily mean hell. In fact, the semantic connection of SaQaR with hell is allegorical, since SaQaR is a descriptive word derived from the verb SaQaRa rather than a noun like GaHYM, GaHaNnaM.

In any case, the verse 74:27 does not ask nor expect us to rush into defining the meaning of this word, which it appears to be its first usage in the Quran. We are warned against rushing to define Quranic words or attempting to preempt the Quranic definition by prematurely assigning a meaning to a verse before considering its immediate or Quranic context or fulfillment (20:114; 75:16-19). Doing so is a sign of pretension and arrogance. Since the word is rarely used in the Quran (total four and three of them are in this chapter) and this verse is most likely the first usage of this rare word, it is more appropriate to wait the Quran explain the word. In brief, rushing to limit/define the meaning of the word Saqar in the verse "Do you know what SaQaR is?" is a disrespectful act against the Quran. When the Quran asks, "Do you know what X means?" it does not want us to try to understand the meaning of X in the very question about its meaning! It is a rhetorical question. God wants to draw your attention to its modified or new meaning. The Quran uses this question 13 times either to modify the meaning of an already used word or to add another nuance (See: 69:3; 74:27; 77:14; 82:17-18; 83:8, 19; 86:2; 90:12; 97:2; 101:3, 10; 104:5).

Nevertheless, we see that almost all commentators or translators of the Quran have rushed to translate SaQaR as Hell or Fire:

> **Yusuf Ali**: "Soon I will cast him into Hell-Fire!"
> **Marmaduke Pickthall**: "Him shall I fling unto the burning."
> **T.B. Irving**: "I'll roast him by scorching!"
> **M. H. Shakir**: "I will cast him into Hell."
> **MaulaHUM Muhammed Ali**: "I will cast him into hell."
> **N. J. Dawood**: "I will surely cast him to the fire of Hell. "
> **Muhammed Asad**: "[Hence,] I shall cause him to endure hell-fire [in the life to come]!"
> **Rashad Khalifa:** "I will commit him to retribution"

As you see, none leave the word SaQaR as it is. However, Rashad Khalifa, as the one who was chosen to fulfill this great prophecy, renders the word SaQar accurately by translating it with a general word, "retribution." Yasar Nuri Ozturk, a Turkish theology professor, in his post-1974 translation of the Quran, translated the word SaQaR under the light of the descriptive verses and related discoveries as the COMPUTER with succeeding screens manifesting the mathematical miracle of the Quran to all people.

So, we should not prime our minds or blind ourselves with prejudice by assuming SaQaR as HELL-FIRE before reading the following verses (75:16-19; 20:114).

Does "It Leaves Nothing; It Lets Nothing" in 74:28 Mean "Exact, Precise" OR "Destroys Flesh And Bones"?

Though the short verse is not generally mistranslated, but its meaning and implication are distorted. In the light of the context and post-1974 discoveries we should understand it is a description of the 19-based mathematical structure that is exact. It does neither leave extra (BaQaYa) nor it let anything necessary go away (WaZaRa); in other words, it is perfect and precise; it is not one more not one less!

However, parroting the pre-1974 commentaries of the Quran, many translations and commentaries still convey the same misunderstanding. Since English translations usually do not comment on this particular verse and leave it to be understood within the context of CORPORAL PUNISHMENT, HELL and FIRE rather than the crucial role of the NUMBER NINETEEN, I will give you a sample of some popular Arabic commentaries of the Quran (which I have easy access to in my personal library) in ascending chronological order. If the name of the commentary different than the name of the author his name will be indicated in the parenthesis. The number in parenthesis is the year of the authors' death:

Tabari (922): Neither kills nor leaves alive; or devours them all and when they are recreated do not leaves them until eating them.
Tha'alibi (1035): Does not leave the dweller of hell alone and burns them.
Bagawi (1122): Neither kills nor leaves alive; eats everything thrown in; does not leave their flesh nor take their bones.
Nasafi (1143): Does not leave the flesh nor does let the bones of its dwellers; destroys everything in it and then restores them back to their starting point.
Baydawi (1292): It does not leave anything thrown in and does not leave it until destroys it.
Qurtubi (1272): Does not leave for them neither flesh, nor bone, nor blood, but burns them. Or does not leave anything from them and then does not let them go when they are recreated.
Zad-ul Masir (Ibn Cawzi, 1200): Does not leave but destroys their flesh and does not let them go when they are recreated.
Ibn Kathir (1372): Eat their flesh, sweat, bones, and skin. They do not die in this condition or they live.
Jalalayn (Celaleddin, 1459): Destroys flesh and bones, then starts it again.
Ruh-ul Maani (Alusi 1853): Destroys flesh, bones of everything thrown in.

Interestingly, one can understand the verse as pre-1974 commentators understood. Though they all heavily relied on *Hadith* for understanding of this verse and knowing that there were literally thousands competing to fabricate *hadith* for various reasons we cannot prove whether Muhammed and his believing friends too understood that way. In fact, since we know that Muhammed and his friends had no clue about the mathematical structure of the Quran but had affirmed its promise of QURANIC MIRACLES being manifested in the future (10:20; 25:4-6; 29:50-51; 41:53), it is highly conceivable that they understood the fact that these verses were about the importance of the number 19 and that their meaning would be fulfilled in the future. I further assert that Muhammed and those who dedicated their religion to God alone by upholding the Quran as the sole authority in their religion, stopped speculating about these verses as soon as they received the divine instruction in 75:16-19. As we know, after Muslims started following the fabricated *hadiths*, *sunna*, and man-made teachings of various sects (6:112-145, 6.159, 7.29, 9:31; 16:52; 18:57; 39:2,11;14; 39:29-37; 39:43-45; 40:14,65; 42:21; 98:5), they lost their capacity for understanding of the Quran (6:23-25; 17:46). (See ENDNOTE 1)

However, it is important to remember that the verses 74:26-37 contain many words that can be understood both a description of Hell and a description of a great miracle, though the latter is a much better fit. This linguistically marvelous aspect is well appreciated by those who witness the mathematical miracle and understand the original language of the Quran.

Does "Lawaha" In 74:29 Mean Scorching/Burning OR Obvious/Visible?

The derivatives of the word LWH are used in the Quran to mean a surface used for recording information, board, and flat wood; and nowhere is it used to mean scorch or burn. Before the fulfillment of the prophecy, translators and commentators of the Quran had difficulty in understanding the simple meaning of this word and thus, they resorted to external sources and often odd meanings, such as scorch, or burn. In fact, the drive to justify a particular meaning for some "difficult" Quranic words is one of the many reasons for fabricating *Hadith*. (See: ENDNOTE 2).

Verse 74:29 is very interesting and crucial in understanding the rest of the chapter. Though it consists of only 2 words, this verse is translated in several different ways. Here are some examples from English Translations:

> **Yusuf Ali**: "darkening and changing the color of man"
> **Marmaduke Pickthall**: "It shrivelleth the man"
> **T.B. Irving**: "as it shrivels human (flesh)."
> **M. H. Shakir**: "It scorches the mortal."
> **MaulaHUM Muhammed Ali**: "It scorches the mortal"
> **N. J. Dawood**: "it burns the skins of men."
> **Muhammed Asad**: "making (all truth) visible to mortal man."
> **Rashad Khalifa**: "obvious to all the people."

Those who do not know Arabic might think that the words are really difficult to understand and translate. In fact, the meaning of these two words, LaWwaHa and BaSHaR is very clear in the Quranic context. The word LaWwaHa, which comes from the root LWH, is the sister of the word LaWH (85:22) and its plural aLWaH. The plural form aLWaH is used in verses 7:145, 150, 154 for the "tablets" given to Moses, and in verse 54:13 for broad planks used by Noah to build his ark. The medieval commentators, not knowing the mathematical implication of the verses, mostly chose an unusual meaning for the word: scorching, burning, shriveling, etc. Ironically, most of them did affirm the obvious meaning of the word as "open board, tablet" (See Baydawi, Fakhruddin Er-Razi, etc.) Few preferred the "obvious" to the obscure. For instance, Muhammed Asad, who had no idea of the mathematical code, preferred the most obvious meaning. Rashad Khalifa who fulfilled the prophecy and discovered the implication of the entire chapter reflected the same obvious meaning. That "obvious" meaning, however, was obscured by the smoke of "scorching fire" burning in the imaginations of generations before him.

In 7:145; 7:150; 7:154, the word alwah, the plural of lawha is used to depict the tablets on which the Ten Commandments were inscribed. In 54:13 it is used to describe the structure of the Noah's ship that made of wood panes. In 85:22 the same word is used for the mathematically protected record of the original version of the Quran. As for the lawaha of 74:29, it is the amplified noun-adjective derived from the root of verb LWH, meaning open tablets, succeeding screens, obvious, manifesto, or clearly and perpetually visible.

Ironically, the Quran uses different words to describe burning or scorching. For instance, for burning the derivatives of *haraqa* (2:266; 3:181; 7:5; 20:97; 21:68; 22:9; 22:22; 29:24; 75:10), or for scorching the derivatives of *salaya* (4:10; 4:30; 4:56; 4:115; 14:29; 17:18; 19:70; 27:7; 28:29; 29:31; 36:64; 38:56; 38:59; 38:163; 52:16; 56:94; 58:8; 69:31; 74: 26; 82:15; 83:16; 84:12; 87:12; 88:12; 92:15), or *nadaja* are used (4:56).

Again, we should note that the understanding of pre-1974 commentators was not without basis. Though their understanding did not rely on the Quranic usage of the words and created some problems (such as explaining the verse 74:31), they had some justifiable excuses to understand words the way they did. The word *lawaha* also meant burn and *bashara* was another word for skin in Arabic language. As I mentioned above, the multiple meaning of these verses allowed the impatient pre-1974 generations to have an understanding, though a temporary and not primarily intended one. In fact, it was better for them to have patient and not rush to speculate on these verses without knowledge (20:114; 75:16-19). It was the computer generation destined to understand their real meaning (10:37-46).

Does "Bashar" in 74:29 Mean Skin OR Human?

The translation of the second word, BaSHaR is also among the distorted one. Many old commentaries translated it as "skin" rather than "human being" or "people" or "humans." For instance, N. J. Dawood parrots such a translation. The meaning of BaSHaR is also obvious in the context of the Quran. The word BASHAR occurs in the Quran 36 times. It is also mentioned as BASHARAYN (two bashars). If we exclude the BASHAR of 74:29 for the sake of the argument, we see that the word BASHAR is used to mean human beings in all 36 verses: 3:47; 3:79; 5:18; 6:91; 11:27; 12:31; 14:10; 14:11; 15:28; 15:33; 16:103; 17:93; 17:94; 18:110; 19:17; 19:20; 19:26; 21:3; 21:34; 23:24; 23:33; 23:34; 25:54; 26:104; 26:186; 30:20; 36:15; 38:71; 41:6; 42:51; 54:24; 64:6; 74:25; 74:29; 74:31; 74:36.

Then, why those who have allergy against the code 19 still insist to translate the word BaSHaR as skin while in its all 36 occurrences the word BaSHaR there is not a single instance where the word BaSHaR is used to mean skin; but always used to mean human beings? Especially, despite the fact that the word BASHAR occurs thrice in the same Chapter: after 74:29, in 74:31 and 74:36, and the fact that the verse 74:36 witnesses for the 36th times for "human beings," and that they were obliged to translate the last two occurrences as "human beings," how could they still insist on translating the BaSHaR of 74:29 as "skin." Since the prophetic verses did not make any sense for them before its fulfillment in 1974, translators and commentators of the Quran had an excuse to translate it differently before the fulfillment of the prophecy. However, after learning the discovery, none has an excuse to continue parroting the overly stretched meaning.

Furthermore, the Quran uses "GeLD" for skin. This organic wrap that protects our body and provides tactile sense is referred in 13 verses with GLD and its derivatives: 4:56; 4:56; 16:80; 22:20; 24:2; 24:2; 24:4; 24:4; 39:23; 39:23; 41:20; 41:21; 41:22.

In short, while the Quran consistently uses GLD and its derivatives for skin and related words and while it consistently uses BShR for human beings, a convincing reason must be provided for ignoring all these examples and translating the BShR of 74:39 as "skin." The pre-1974 generations can be excused for trying to understand these verses even if it meant accepting some uncommon usages or dialects. As for post-1974 generation, if they are still stuck with hell and fire, then they share the same disease with the ingrates of the past (2:75; 4:46; 5:13; 5:41).

After the fulfillment of the prophecy and divine clarification of its context, it is distortion to render this two-word verse 74:29 as "shrivels/scorches the skin" or "shrivels/scorches the man" rather than "it is clearly visible for human beings" or "it is a manifesto/successive-screens for human beings." However, the repetitive stretch and distortion committed by the post-1974 translators/commentaries on the many verses of Chapter 74 are bizarre and extraordinary. This pattern is prophetically described by verses 2:18; 3:7; 11:28; 41:44; 17:72; 25:73; 27:81.

Does The Pronoun "Hiya" (She) in the End of 74:31 Refer to the Hell OR to the Number 19?

We know that verse 74:30 does not qualify the reference of 19. It draws our attention to a NUMBER ALONE. The verse does not say "on it nineteen angles" or "on it nineteen guards" or "on it nineteen this or that." The verse says, "on it nineteen." Period. Verse 74:31, after informing us that the number of guardians of hell is 19, isolates this number from hell again describes its role.

Yet, those whose minds are stuck in Hell cannot even notice this evident emphasis to the number. Thus, they violate the general grammar rules and universal linguistic logic and identify the reference of HIYA (SHE) in the end of 74:31 as HELL FIRE.

According to the grammar rules and common linguistic logic, the reference of a pronoun should be sought in proximity first. Sure, if there is no compelling empirical or rational reason for skipping a closer noun. For instance, "Yesterday, at Mary's home I saw Lisa. She was very thoughtful." In this sentence grammar and logic leads us to think that the thoughtful person was Lisa not Mary. But "Yesterday, in Delhi's streets on the back of an elephant I saw Lisa. She was swinging her trunk/hose left and right and walking majestically." (I had given the original example in Turkish and in Turkish the same word is used for both elephant's trunk and hose. For didactic purposes please assume that it is the same in English.) Though grammatically the best candidate for reference of SHE is Lisa, we have an empirical reason to ignore the general rule since the one who would swing her trunk/hose is the elephant. But if these sentences are in a story and if we are told that Lisa was walking in the street with a hose in her hand, then the SHE in the sentence may equally refer to elephant OR Lisa. There will be ambiguity and if it is intentionally done then we wonder the purpose of the ambiguity. But, if the sentences continue, "When she hit the trunk/hose in her hand to a store sign…," then we become sure that it was a hose and not a trunk. "Your mother had given you a walnut and a book. Have you eaten and finished it?" A sound mind can easily deduce that that "it" refers not to the book, but to the walnut. But if the question were "Your mother had given you a walnut and a book. Have you finished it?" then, an ambiguity will arise. However, if the context of the question is known then we can deduce whether the query is about finishing the book or the walnut or both.

The pronoun HIYA (SHE), based on the rules of grammar, the emphasis of the verse in its context, and based on the rhetorical superiority of its meaning, must be referred to the grammatically feminine closest noun IDDATAHUM (their number, that is Nineteen). How can we explain the act of skipping IDDATAHUM (their number, Nineteen) by ignoring the grammar, context, emphasis, and rhetoric, and insisting to reach the FIRE?

We will see in the miraculously elegant and prophetic expression of the Quran that hypocrites and unappreciative people have deserved the SaQaR penalty both in this world and in the hereafter. In this world as the 19-based mathematical code of the Quran; they will be intellectually tormented by its powerful evidences. And in the hereafter as the 19 guards of symbolic hellfire; they will be convicted to eternally face eternelly the number 19 in the presence of guarding angles. (See ENDNOTE 3)

Does The "Zikra" (Reminder) In 74:31 Refer To Hell OR The Evidence?

This brings us to the last phrase of verse 74:31 "it/this is a warning/reminder for mankind." Old commentaries and their parroting contemporaries refer "it/this" to the FIRE or HELL.

Those who have difficulty in accepting the possibility of an intellectual argument in The Hidden One (Muddathir), insert a HELL in the translation or the understanding of the last phrase of 74:31:

> "This (HELL) is a warning for people."

This way, they transform the ZiKRa to a warning to be scared, to a penalty to avoid. However, all the derivatives of ZKR, including ZiKRa, are used in the Quran 273 times; and if we include DKR which is mentioned 7 times; this word is mentioned 280 times in the Quran and NOWHERE is it used to describe HELL or FIRE. You may check for yourself all the 21 verses where the word ZiKRa is mentioned: 6:68; 6:69; 6090; 7:2; 11:114; 11:120;

21:84; 26:209; 29:51; 38:43; 38:46; 39:21; 40:54; 44:13; 50:8; 50:37; 51:55; 74:31; 80:4; 87:9; 89:23. You may also check the two verses where the word is suffixed with pronouns ZiKRaha and ZiKRahum: 47:18; 79:43.

More interestingly, however, you will find another derivative of the same word "ZiKRa" (reminder) is used in verse 74:49 as "taZKiRa."

> "Why do they turn away from this message/reminder (taZKiRa)?"

The ingrates had ignored the reminder! Obviously, hell cannot be that reminder since you are not supposed to embrace and face hell while you are alive on this earth. You are expected to turn away from Hell and turn to the message (taZKiRa). Therefore, taZKiRa and ZiKRa cannot be both HELL and the MESSAGE of the Quran. All the derivatives of ZKR are consistently used in the Quran for the divine message that revives and bestows eternal happiness.

Because of the self-ignited and self-inflicted hell fire burning in their minds, the ingrates and hypocrites ignore the PLEASANT meaning associated to the word 280 times, and fight with us to keep it as HELL. Well, this was exactly what they were promised and what they deserve. Their initial decision, their prejudice, ignorance and arrogance, have led them to bet their faith on hell, a questionable one chance against 280 chances!

How does the Number Nineteen Increase the Faith of Affirmers and also Remove Doubts From the Minds of People of the Book as Promised in 74:31?

With the exception of Rashad Khalifa, all translations and commentaries listed above have had big problem with this question. Since they did not accept it as a prophecy left to be fulfilled in the future, and since they rushed to speculate on them without knowledge, they dug a hole for themselves. Surely, they always could deceive themselves and others by claiming that THEIR FAITH was increased because of the association of the number 19 with the guardians of hell. This claim obviously would be a non-falsifiable claim. One can say anything about his or her own faith. However, this claim could not explain why specifically 19 and not another number or another thing. Let's assume they could pretend that their faith was really increased by the nineteen guardians of hell, but how could they explain the other functions of the number nineteen? How could this number nineteen remove the doubts of people of the book? The proponents of HELL-19 just could not make up an answer for this, since the issue involved an OBJECTIVE fact, not a subjective claim.

Interestingly, those who saw the Hellfire in these prophetic verses claimed that the people of the book would believe in the Quran because they would see that the number of guardians of hell was also nineteen in their own books! What? Please read it again. Yes, this is the explanation of all the prominent Sunni and Shiite scholars of the past and unfortunately it is exactly parroted by the contemporary ones!

First, there is no such a statement in the Bible. Second, even if there was such a statement there would not be anything special about nineteen, since there are many principles, stories of prophets, and instructions mentioned in the Quran that are also in the Bible. In fact, the very existence of similarity had an opposite effect on many people of the book; they claimed and still claim that the Quran was plagiarized from the Bible. Third, none has been persuaded regarding the authenticity of the Quran because the number of guardians of hell is nineteen! I personally communicated and met dozens of Christians or Jews whose doubts about the authenticity of the Quran was removed because of their witnessing the mathematical miracle of the Quran based on the number nineteen. You can find some of these people on Internet forums. This is happening despite the aggressive misinformation and disinformation campaign carried out by numerous Sunni and Shiite groups. Compare the prophetic role of MIRACLE-19 to the role of HELL-19 in their powers of removing doubts from the minds of People of the Book. Two DECADES versus more than 14 CENTURIES, and several THOUSAND people versus more than a BILLION people.

The HELL-19 advocates cannot show a single Christian or Jew who converted to their version of Islam because their doubts were removed after they noticed that the number of guardians of hell is also 19 in their books! This failure alone should be sufficient to wake up those who are still eager to see Hell in these verses. Yet, they still shamelessly repeat this lie to defend their hellish version.

Some opponents of the CODE-19 even tried to use a phrase in verse 74:31 against understanding of the very verse. They quote the fifth function of the number nineteen, "to expose those who have disease in their hearts, and the ingrates; they will say, 'What did God mean by this allegory?'" and then charge us: "You 19ers are asking this question; therefore, you have disease in your hearts." First, none of us have asked this question, including Rashad Khalifa. We happened to learn the meaning of number nineteen even before asking such a question. We know the meaning of this allegory and we do not challenge that meaning. Furthermore, this question is not a question of a curious person who sincerely wants to learn the meaning of the Quran. This question is the challenge of arrogance and ignorance. It is a question that many still millions of Sunni and Shiite people are asking themselves. They will keep asking this question as long as they follow their ancestors blindly and worship Prophet Muhammed and religious scholars.

Does the "Ha" In 74:35 Refer to Hell OR the Number 19? Does the Word "Kubra" in the Same Verse Refer to "Great Disasters and Calamities" OR to "Great Signs and Miracles"?

When we reflect on the context and the usage of the words of verse "This is one of the greatest" we can easily deduce that this refers to the number 19 and its prophetic function as conclusive evidence for the authenticity of the Quran and a great test for people. Unfortunately, many of those who did not witness the incredible mathematical structure based on the number 19, again related this verse to Hell.

Let's one by one see several different translations of this verse. (Again, the pre-1974 translations have an excuse in their mistranslations since they did not witness the fulfillment of the prophecy of the mathematical miracle):

> **Yusuf Ali**: "this is but one of the mighty (portents)"
> **Marmaduke Pickthall**: "Lo! This is one of the greatest (portents)"
> **T.B. Irving**: "Surely it is one of the gravest (misfortunes)."
> **M.H. Shakir**: "Surely it (hell) is one of the gravest (misfortunes)."
> **MaulaHUM Muhammed Ali**: "Surely it is one of the gravest (misfortunes)"
> **N. J. Dawood**: "it is a dire scourge."
> **Muhammed Asad**: "Verily, that [hell-fire) is Indeed one of the great [forewarnings]"
> **Rashad Khalifa**: "This is one of the greatest miracles."

I leave for you to assess the contextual reference of the pronoun in 74:35. Here, I will draw your attention to the descriptive word KuBeR, which is the plural of *KaBeeR* derived from KBR (great). The derivatives of KBR are used in the Quran to describe positive or negative events, people, or things. For instance, in verses 20:23; 44:16; 53:18; 79:20 it is used to describe God's *ayaats* or miracles. The derivatives of the same word are used for disaster in 79:34 and hell in 87:12.

In other words, if we do not consider the context of the group of verses, we may understand the reference of THIS IS ONE OF THE GREATEST as either HELL FIRE or one of God's greatest MIRACLES. Through this verse alone, one may see flames of fire or signs of a miracle. Nevertheless, considering the context of this verse it becomes clear that the first vision is a man-made hallucination and the later a divine gift.

Does the Word Nazer in 74:36 Refer to the Hell Fire OR to the Devine Evidences?

NAZER is an adjective derived from the root NZR and means "WARNER" Throughout the Quran the various derivatives of this word occur 130 times. Here are the verse numbers where the form NaZeR (warner) occurs 44 times: 2:119; 5:19; 5:19; 7:184; 7:188; 11:2; 11:12; 11:25; 15:89; 17:105; 22:49; 25:1; 25:7; 25:51; 25:56; 26:115; 28:46; 29:50; 32:3; 33:45; 34:28; 34:34; 34:44; 34:46; 35:23; 35:24; 35:24; 35:37; 35:42; 35:42; 38:70; 41:4; 43:23; 46:9; 48:8; 51:50; 51:51; 53:56; 67:8; 67:9; 67:17; 67:26; 71:2; 74:36.

In NONE of these verses the word NaZeR is used for HELL or FIRE. In these verses NaZeR describe God's messengers, books, revelation, and signs.

Then Why Those Who Claim to be Muslims do not Understand or Appreciate the Prophetic Fulfillment of These Ayaat (Revelations/Signs/Miracles)? Why They See and Work Hard to See HellFire, Instead of the Precise and Universal Great Miracle?

Surely, previous generations who were not aware the discovery of Code 19 via computer had an excuse to twist the meaning of obvious verses in order to make sense out of them. They could not imagine that the 19 was the code of an elaborate mathematical system. Nevertheless, some early commentators of the Quran who did not limit the understanding of the Quran with *Hadith* and *Sunna*h sensed something beyond HELL from these verses and discussed them as alternative understanding. For instance, Fakhruddin er-Razi in his famous Tafsir al Kabir, in his 23rd comment on Basmalah, speculates on the 19 letters of Basmalah and lists numerous implications, such as the difference between the number of 5 prayers and the number of hours in a day; the number of physical, intellectual, and emotional faculties that humans are blessed with, etc. Among the modern commentators who lived before 1974, Muhammed Asad is the only one that subscribes to Razi's interpretation of the number nineteen of 74:30. (See ENDNOTE 4). However, Asad too could not avoid but see Hell everywhere.

The opponents of the miracle 19 are no different than those who rejected the entire Quran in the past. The description of these verses fits both groups, which is another miraculous aspect of the Quran:

> 74:21 He looked.
> 74:22 He frowned and scowled.
> 74:23 Then he turned away arrogantly.
> 74:24 He said, "This is but an impressive illusion/magic!" (See ENDNOTE 5)
> 74:25 "This is human made."

In conclusion, those who reject to witness the miracle 19 are insulting the wisdom of God. According to them God's only answer to someone who challenges the authenticity of the Quran is: "Get lost! I will burn you in the hellfire" Depicting God as someone who cannot engage in an intellectual argument with His opponent but can only employ the cheapest method of persuasion (threat) cannot be the path of affirmers who are warned by the incredible prophecies of this wonderful chapter, The Secret.

Turning back to the question: Then why those who claim to be Muslims do not understand or appreciate the prophetic fulfillment of these ayaat (revelations/signs/miracles)? Why they see and work hard to see hellfire, instead of the precise and universal great miracle?

The Quran provides prophetic answers to this question in numerous verses. Please read and reflect on them:

29:1 A1L30M40
29:2 Did the people think that they will be left to say, "We affirm" without being put to the test?
29:3 While We had tested those before them, so that **God** would know those who are truthful and so that He would know the liars.
29:4 Or did those who sinned think that they would be ahead of Us? Miserable indeed is their judgment!
2:17 Their example is like one who lights a fire, so when it illuminates what is around him, **God** takes away his light and leaves him in the darkness not seeing.
2:18 Deaf, dumb, and blind, they will not revert.

7:146 I will divert from My signs those who are arrogant on earth unjustly, and if they see every sign, they do not affirm it, and if they see the path of guidance, they do not take it as a path; and if they see the path of straying, they take it as a path. That is because they have denied Our signs and were heedless of them. (See, ENDNOTE 6)
29:54 They hasten you for the retribution; while hell surrounds the ingrates. (Please read from 29:48)
3:7 He is the One who sent down to you the book, from which there are definite signs; they are the essence of the book; and others, which are multi-meaning. As for those who have disease in their hearts, eager to cause confusion and eager to derive their interpretation, they will follow what is multi-meaning from it. But none knows their meaning except God and those who are well founded in knowledge; they say, "We affirm it, all is from our Lord." None will remember except the people of intellect. (See, ENDNOTE 7)
25:73 Those who when they are reminded of their Lord's signs, they do not fall on them deaf and blind.
6:158 Do they wait until the controllers will come to them, or your Lord comes, or some signs from your Lord? The day some signs come from your Lord, it will do no good for any person to affirm if s/he did not affirm before, or s/he gained good through his/her affirmation. Say, "Wait, for we too are waiting."
6:25 Among them are those who listen to you; and We have made covers over their hearts to prevent them from understanding it, and deafness in their ears; and if they see every sign they will not affirm; even when they come to you, they argue, those who reject say, "This is nothing but the tales from the past!"
6:26 They are deterring others from it and keeping away themselves; but they will only destroy themselves, yet they do not notice.
27:84 Until they have come, He will say, "Have you denied My signs while you had no explicit knowledge of them? What were you doing?"
78:27 They did not expect a reckoning/computation.
78:28 They denied Our signs greatly.
78:29 Everything We have counted in a record.

74:56 None will take heed except if **God** wills. He is the source of righteousness and the source of forgiveness.

ENDNOTES:

1. To see the examples of misunderstanding and distortion of Quranic verses by the followers of *Hadith* and *Sunna*h, please read the comparative section in the beginning of the Reformist Translation of the Quran.

2. Soon after Muhammed's death thousands of *hadith*s (words attributed to Muhammed) were fabricated and two centuries later compiled in the so-called "authentic" *hadith* books, to support the teaching of a particular sect against another (for instance, what nullifies ablution; which sea food is prohibited); to flatter or justify the authority and practice of a particular king against dissidents (such as, Mahdy and Dajjal); to promote the interest of a particular tribe or family (such as, favoring Quraysh tribe or Muhammed's family); to justify sexual abuse and misogyny (such as, Aisha's age; barring women from leading *Sala* prayers); to justify violence, oppression and tyranny (such as, torturing members of Urayna and Uqayla tribes, massacring Jewish population in Madina, assassinating a female poet for her critical poems); to exhort more rituals and righteousness (such as, nawafil prayers); to validate superstitions (such as, magic; worshiping the black stone near Kaba); to prohibit certain things and actions (such as, prohibiting drawing animal and human figures, playing musical instruments, chess); to import Jewish and Christian beliefs and practices (such as, circumcision, head scarf, hermitism, using rosary); to resurrect pre-islamic beliefs and practices common among Meccans (such as, intercession; slavery;); to please crowds with stories (such as the story of Mirage and bargaining for prayers); to idolize Muhammed and claim his superiority to other messengers (such as, numerous miracles, including splitting the moon); to defend *hadith* fabrication against monotheists (such as, condemning those who find the Quran alone sufficient); and even to advertise a particular fruit or vegetables (such as, the benefits of date grown in Ajwa farm). In addition to the abovementioned reasons, many *hadith*s were fabricated to explain the meaning of the "difficult" Quranic words or phrases, or to distort the meaning of verses that contradicted to fabricated *hadith*s, or to provide trivial information not mentioned in the Quran (such as, Saqar, 2:187; 8:35…)

3. For an article titled "Eternal Hell and Merciful God" visit www.yuksel.org OR www.19.org.

4. After translating the verse 74:30 as "Over it are nineteen (powers)" Muhammed Asad explains his parenthetical comment with the following footnote:

   "Whereas most of the classical commentators are of the opinion that the "nineteen"' are the controllers that act as keepers or guardians of hell, Razi advances the view that we may have here a reference to the physical, intellectual and emotional powers *within man himself*: powers which raise man potentially far above any other creature, but which, if used wrongly, bring about a deterioration of his whole personality and, hence, intense suffering in the life to come. According to Razi, the philosophers *(arbab al-hikmah)* identify these powers or faculties with, firstly, the seven organic functions of the animal - and therefore also human - body (gravitation, cohesion, repulsion of noxious foreign matter, absorption of beneficent external matter, assimilation of nutrients, growth, and reproduction); secondly, the five "external" or physical senses (sight, hearing, touch, smell and taste); thirdly, the five "internal" or intellectual senses., defined by Ibn Sina - on whom Razi apparently relies - as (1) perception of isolated sense-images, (2) conscious apperception of ideas, (3) memory of sense-images, (4) memory of conscious apperceptions, and (5) the ability to correlate sense-images and higher apperceptions; and, lastly, the emotions of desire or aversion (resp. fear or anger), which have their roots in both the "external" and "internal" sense-categories - thus bringing the total of the powers and faculties which preside over man's spiritual fate to nineteen. In their aggregate, it is these powers that confer upon man the ability to think *conceptually,* and place him, in this respect, even above the controllers (cf. 2:30 ff. and the corresponding notes; see also the following note)."

5. Magic is not considered an extraordinary, paranormal event by the Quran. Its influence on people is described by its two aspects: illusion and suggestion/bluffs/hypnosis (7:103-120). Magic is an art of influencing gullible people via illusion and hypnosis (7:11; 20:66; 2:102). So, when the unappreciative disbeliever describes the Quran OR the mathematical miracle of the Quran as "influencing magic," what should be understood is that it is just a manipulation and hyperbole.

6. The plural word AYAAT means revelation, sign, evidence, or miracle. However, its singular form AYAT is mentioned in the Quran 84 times and in all its occurrences it means sings, evidence or miracles, not literary revelation of the Quran.

7. This verse is a crucial verse in understanding the Quran and ironically this very verse is one of the most commonly misunderstood verses in the Quran. For a detailed argument on this verse please visit: www.yuksel.org or www.19.org.

5
Manifesto for Islamic Reform

There is virtually no difference between the teaching of Sunni or Shiite Religions and of Al-Qaeda, ISIS, or Taliban. They all use the same men-made religious sources fabricated by Ibn Filan and Abu Falan after the Quran, such as Hadith, Sunna, Ijma, Syrah, sectarian jurisprudence… Sunni and Shiite scholars distorted the meaning of many Quranic verses, which I have discussed in *Quran: a Reformist Translation.* I have exposed the Sunni/Shiite religions in books such as *Manifesto for Islamic Reform, Peacemaker's Guide to Warmongers*, various articles, speeches, debates & interviews such as: "ISIS follows Sunni Religion".

For a detailed discussion of the table below, see: MANIFESTO for ISLAMIC REFORM

	Quran	Hadith	ISIS
Religious dogmas, stories should be accepted on faith	No	Yes	Yes
Adulterers are killed by stoning to death	No	Yes	Yes
Slavery & having sex slaves are permitted	No	Yes	Yes
The heads of prisoners of war are chopped	No	Yes	Yes
Apostates and heretics are killed	No	Yes	Yes
Insulting prophet Muhammed is punished	No	Yes	Yes
A 54 y/o man marrying with a 9 y/o girl is ok	No	Yes	Yes
Muslims who do not pray may be beaten and even killed	No	Yes	Yes
Sculpture/picture of animals and humans is prohibited	No	Yes	Yes
Muhammad was illiterate	No	Yes	Yes
Testimony is: La ilaha illa Allah Muhammad Rasulullah	No	Yes	Yes
Muhammad was Better than Jesus	No	Yes	Yes
Shrimp, Lobster… are prohibited	No	Yes	Yes
Christians/Jews should pay extra tax (jizya)	No	Yes	Yes
Women cannot work with men	No	Yes	Yes
Women must cover their hair	No	Yes	Yes
Women are buried alive in sacks; must cover their faces	No	Yes	Yes
Some Quranic verses contradict and abrogate each other	No	Yes	Yes
There was a verse about stoning adulterers to death, but after Muhammad's death, a hungry goat ate it	No	Yes	Yes
The Quran is not detailed; it needs hadith, sunna, fatwas and stories of Ibni Filan and Abu Falan	No	Yes	Yes
Expecting intercession, asking help from dead people, praying at shrines is good	No	Yes	No

The following is from the last pages of Manifesto for Islamic Reform and was first published in 2008

"O people, a proof has come to you from your Lord, and We have sent down to you a guiding light." (4:174)

"...and do not make corruption on the Earth after it has been reformed..." (7:85)

"It is one of the great ones. A warning to humanity. For any among you who wishes to progress or regress." (74:36-37)

An Invitation to Jews, Christians, Muslims, and all Humanity

In this foreword, we focused on the incredible number of distortions made in the message delivered by Muhammed. Christianity and Judaism are no different. Today's Christianity, with its dogmas and practices, is far from the monotheistic teachings of Jesus, the son of Mary.

If Moses, Jesus, and Muhammed were back today, Jews would condemn the first as Anti-Semite, Christians would denounce the second as Anti-Christ, and Muslims would revile the third as the Dajjal (The imposter).

Imagine a religion that its members worship the murder weapon, perform rituals to pretend that they are drinking the blood and flesh of their heroic victim, claim that 1+1+1 equals to 1, adopt a word as their name which was used by none of the early followers, misspell and mispronounce the name of their hero, follow someone's teaching who was prophetically condemned by their hero, accept a formula that was coined by a self-appointed commission 325 years after the founder, sing love and peace yet be responsible for most of the blood-shed and weaponry in the world, mobilize even children for centuries of barbarism called Crusades, sell parcels of heaven, excommunicate scientists, burn the first translator of their holy book, burn women in witch-hunt craze, invent ingenious torture devices and torture many in their holy courts, declare the earth as the flat center of the world for more than a millennium, lead and pray for colonialists, defend and practice slavery and racism until the cause was lost, mostly side with kings and the wealthy, deny women from many of their rights, condemn the theory of evolution, support occupations and wars with jingoistic slogans... Yes, how can such a religion, with a fake name, with a fabricated doctrine, with bizarre pagan practices, and with such a miserable historical record and bitter fruits belong to God? How can it be attributed to a philosopher, to a peacemaker, to an advocate of the rights of the weak, to a human messenger of God? (For a more detailed critical evaluation of modern Christianity, see 19 Questions For Christians, by Edip Yüksel)

Idolization of human beings is the epidemic of all religions, and it is the most common tragedy of human history. According to the original teachings of all God's messengers, idol worship or setting up partners to God, is the biggest offense against God. Besides, the idolization of prophets, messengers, saints and the faith of human intercession creates religious abuse, oppression, conflict and fighting between children of Adam, who are servants of God.

When affirmers start idolizing their previous religious leaders, they develop the tendency to idolize their living religious leaders too. Instead of seeking the truth, they are attracted to names and titles. The clergymen, in order to take advantage of that weakness and gain more power over their subjects, focus their preaching on praising the departed heroes, instead of God.

These clergymen and their fanatic followers killed many people, destroyed many homes in the name of their incarnated gods. They fabricated many rules and prohibitions in the name of God, and with such a complicated religion, they secured their jobs as professional holy men. They made money and fame in the name of those human gods. And they claimed to have the power of intercession in their names--so much so that they sold keys to the heavens, turned temples and churches to big businesses.

If we want to follow the basic principles common among the Old, the New and the Final Testaments, if we want to stop religious exploitations, if we don't want to use our God-given reasoning faculties to its maximum capacity, if we want the unity of all the affirmers of all religions, freedom for everyone, including for non-religious people, and if we want to attain eternal salvation, we must start a "Copernican revolution" in theology. Instead of Krishna-centered, Jesus-centered, Mohammed-centered religions, we must turn to the original center, to the God-centered model. To achieve this revolution, each of us must start questioning the formulas and teachings that have created gods out of humans like us.

In a time where religious fanatics are pushing the world for another Crusade or Holy War, in a time where the words Messiah, Rapture, Armageddon, Mahdi invite hostile masses to shed more and more of each other's blood, in a time when those in power and in positions of making profit from curtailing civil liberties, in a time when wars and occupations are playing on jingoistic and religious emotions, yes in such a time, people of intelligence and good intentions should come together and plant the seeds of tolerance, peace, reason, human rights, and unity of humanity.

On Israel, Palestine, Suicide Bombers, and Terror

Compared to their small population, the Jewish influence is immense in the global arena, financially, politically, and culturally. Disproportionate to their population, Jews have exhibited astonishing examples in both good and bad, in both success and blunder, and they have enjoyed vivid presence in world politics for millenniums. This explains why the Quran mentions them so frequently. Well, maybe it is also true the other way around.

After being subjected to genocide and atrocious tortures by fascist forces, Jews were scattered around the world as immigrants. Yet, they did not disappear from the global scene or take centuries to recover, as many other nations would do. Not surprisingly, with the help of major powers of the time they were able to establish their own independent state in 1948, soon after their almost utter annihilation; a state not in Germany, but in their historical land, which has once again become the focal point of a global conflict; stirring the world by showcasing human aggression, greed, hatred, cruelty, racism, and terror.

As it seems, victim nations too might repeat the crimes of their predators. One would expect Israel to be the first against racism and colonialism, yet Israel was the last government to cut its relationship with the racist apartheid regime in South Africa, reflecting the depth of its racist policy against Palestinians. One would expect Israel to be the first nation against the weapons industry, yet Israel is one of top weapon manufacturers and exporters in the world. The racist and colonial policy of Israel by no means should be generalized to all Jews. There are more Jews in the world who condemn this policy then those who perpetrate it, and many are ashamed of what is being done in their name. While we should condemn terrorism as a method to get back one's land and independence, we should also mention that there are many Arabs who are hoping for a just solution and peaceful co-existence with their Jewish cousins.

Jews and Muslims lived together in peace for centuries, and their current conflict is partially due to the early terrorist tactics used by Zionist guerillas, and partially due to a myriad of external forces who are trying to keep the fires burning. These external forces include the ambitions of UK-Inc and USA-Inc, the racist Zionist zealots, corrupt Mullahs, racist Sunni and Shiite zealots, Evangelical Crusaders, Weapon and Oil industries, who make massive amounts of money from the tension in the region. Unfortunately, superpowers who mediated the negotiations have not honestly sought justice in this conflict. Perhaps, they deliberately wanted a continuous, yet controlled conflict in the region so that they could exploit its rich resources through puppet regimes.

In their pre-emptive war in 1967, the Israeli soldiers carried the verse 2:249 of the Quran over their tanks when they entered Sinai after defeating the Arabs, and their misguided Arab nationalism. Ironically, the evildoers among them pushed for further land-grab in the East, thereby subjecting Palestinian natives to racial discrimination, dislocation, humiliation, massacres, destruction of property/infrastructure, legalized torture, and assassinations. Israel deliberately did not set a border, rather it kept its borders flexible seeking excuses and occasionally provoking its dehumanized subjects so that it could invade new territories and create more settlements. Decades of suffering under the brutal and humiliating fascist occupation destroyed the hopes and aspiration of a Palestinian population and it gave birth to suicide bombers, which in turn provided more excuses for the occupational force to continue its invasion and barbarism. The West's propaganda machine distorts the real picture of the conflict and deceives Christendom by depicting the victimized Palestinians as the aggressor. The numbers speak clearly. The number of Palestinian civilians and children killed by Israeli occupying forces far greater than the number of Israeli civilians and children killed by suicide bombers. Palestinians gave up continuing a hopeless fight with slings and rocks against tanks. The world's indifference against injustice in the region, and on top of that, the support of the superpowers of the brutal racist occupation, gave birth to global resentment and hatred among Muslims, triggering a global gang-terrorism challenging the legalized and glorified state-terrorism.

Islam (more accurately, Hislam) has been around for centuries, and compared to other religious groups Muslims do not appear more violent. An objective study of suicide terrorism will inform us that it has more to do with brutal occupations than religions or ideologies. Religion and ideologies are mostly used for justification and propaganda of the political cause. Robert Pape, of the University of Chicago, in his book Dying to Win: The Logic of Suicide Terrorism, rightly argues that suicide terrorism is not driven by religion but by occupations. He provides many examples, such as the suicide attacks of Marxist Tamil Tigers organization in Sri Lanka in 1990's that inspired Palestinians who were using slings, rocks and rifles against occupying Israeli soldiers and tanks before the Intifada of 2000. In fact, a great majority of suicide-terrorist campaigns carried out in Lebanon, Sri Lanka, Chechnya, Kashmir, and Palestine aimed to compel occupation forces to withdraw. No wonder, Ayman al-Zawahiri and his terrorist organization Islamic Jihad was born after Israel's occupation during the 1967 pre-emptive war. No wonder, Russian invasion and occupation of Afghanistan, together with the legalized US occupation of Middle East through puppet and oppressive kings and emirs gave birth to Osama bin Laden and al-Qaida. No wonder, Russian brutal occupation of Chechnya gave birth to Shamil Basayev and his terrorist organization. Again, it is no wonder that US occupation of Iraq gave birth to Abu Musab al-Zarqawi and hundreds of other suicide bombers. Though compared to rebels or insurgents, occupiers commit much worse acts of barbarism and terrorism on the population of lands they occupy, that state terrorism is cleverly hidden from the world. Ironically, occupiers who create these terrorist insurgents or contribute substantially to their growth, use the terrorist attacks to justify and continue their occupation. Occupying forces cleverly use fear, xenophobia, and patriotic emotions of the taxpayers and take advantage of their ignorance about foreign affairs. Government agencies work cleverly to depict their brutal and bloody occupation as a justified act against evil and barbarism. Secret agencies are showered with money to stage covert operations, flood the world with misinformation and disinformation campaigns. Talking heads in media and academics are secretly hired to promote the policy of occupations. No wonder, despite all the obvious fraud, deception, and lies, the mainstream American media gave green card to the Neocon-Zionist-Crusader coalition to justify their pre-emptive war against Iraq. The pictures of Rumsfeld shaking hands with Saddam at the time Saddam was committing his horrendous atrocities against Kurds and Iran as a puppet of US-Inc, somehow became a footnote, rather than an incriminating headline, demonstrating the hypocrisy of warmongers.

The cycle of violence has since been accelerated by religious fanatics on all sides. The Zionist-Crusader-Capitalist coalition on one side, and the Salafi-Mullah-Taliban coalition on the other side, each adding more fuel to the fire. Each with their own agendas. Zionists hope to grab more land, Crusaders pray for a bloody Armageddon followed by Rapture, the capitalist salivates for more profit from wars; and the other gang weep for the Mahdi to come with its sword to seek out Jews hiding behind rocks. Another aspect of recent conflict between Christendom and Muslims is the empty shoes of "evil" after the demise of communism. Global oligarchs, who strengthen their political and financial capital during conflicts and mass paranoia, were looking for a substitute to communism. With a mixture of covert operations, provocations, unjustified wars, tyrant puppets, the lesser-of-two-evil policy, and training future terrorists, the mission is almost accomplished.

Now, Muslims in general, and Arabs in particular will be christened as the new face of evil. Knowing history, we should not be surprised to witness another genocide and another use of nuclear weapons, followed by tears of regret, confession sessions, and cries of "Never again!" So long as people do not use their God-given reasoning and follow their clergymen and politicians blindly, Satan will use every tool at his disposal to create artificial divisions, hostility, and hatred among the children of Adam. And Satan, who has a successful record of enticing since Cain & Abel, has always found religious clergymen and jingoistic politicians to be his best allies in his acts of corruption, destruction, and bloodshed on earth.

To the East, Muslims, and the Middle East:

The following words are not from an enemy of yours, but from someone who shares the same book and the same history. These are the words of someone who cares a great deal about you. Someone who cries at night for your plight, for the tragedies which have befallen you. This is someone who knows your generosity, your sincerity, your unfulfilled dreams, your aspirations, your tragedies, your fears, your follies, and delusions. You should listen, at least once. Enough prejudice and bigotry. Enough paranoia and hatred.

We must affirm the truth so that the truth will set us free.

Before looking around to point fingers at the cause of your problems, first look at the mirror. I do not mean that you should ignore the imperialistic ambitions of other nations and their open or clandestine interferences with the politics, economy, and culture of your people. But you cannot change your condition unless you change yourself. You cannot glorify the invasions, aggressions, massacres, and imperialistic policy of corrupt Umayyad, Abbasid, and Ottoman caliphs in your history and at the same time morally be critical of others for doing the same. Had God given you the same superiority, perhaps you would inflict the earth with more corruption and destruction than your current powerful enemies. You cannot kick them out from your home unless you reform yourself and your home. You cannot demand mercy from others if you do not have mercy on yourself.

We must affirm the truth so that the truth will set us free from self-righteousness.

Go check the list of patents issued last year. Check and see how many of them belong to a group, nation, religion you identify with. It should tell you a lot you a lot about your position in a world where information and technological progress are so crucial. Go check the list of prosperous countries. Check and see how many of them belong to a group, nation, religion you identify with. Centuries ago, you were a role model for civilization, justice, democracy, and freedom; once you were a pioneer in mathematics, astronomy, medicine, and philosophy. Now look around and look at the mirror; who are you? You followed the religious *fatwa* of a *sheikh ul-islam* (highest cleric within the Ottoman Empire) who prohibited the use of printing machine from 1455 to 1727 for 272 years, for 100,000 precious days, in a vast land stretching from North Africa to Iran, from today's Turkey to Arabian Peninsula. While Europe indulged in learning God's signs in nature, shared the knowledge via printing machines, and was rewarded by God with renaissance, reform, technology, and prosperity; you devolved and sunk further in your ignorance. While Europeans engaged in philosophical arguments, you recited the holy book no better than a parrot, the book that highlighted the importance of learning, questioning, discovery, and pursuit of knowledge. You marveled at handwritten books of hearsay and superstition, at the lousy arguments developed by Ghazali who with the full support of a king aimed to banish philosophy. While Europe sought for a better system to save themselves from the tyranny of kings and church, you recited handwritten poems to praise your corrupt kings and idols. No wonder why, your land, your name, your face, your religion is now associated with backwardness, ignorance, oppression, violence, and poverty. You have become the bum of the world.

We must affirm the truth so that the truth will set us free from our ignorance.

Once the religious among you hoped that the theocracy of mullahs would fulfill your dream, would bring back the glorious days of your past. They promised "*istiqlal, azadi, hukumat-i islami*" (independence, freedom, Islamic government); yet what you ended up with a swarm of leaches with turbans, repression, and a satanic government. Some of you hoped that a Sunni Taliban in Afghanistan would bring dignity and glory to you. What they brought was worse than the Saudi regime: they put women in black sacks, revived the barbaric stoning practice, regressed to the times of tribalism, denied women education, exponentially increased ignorance, and turned Afghanistan into an international farm for opium. You did not question the religion and sect you inherited from your parents or the teachings of the mullah, the sheikh, or the imam. You little examined the nightmare sold to you as dreams.

We must affirm the truth so that the truth will set us free from our own transgression.

God blessed you with crucial natural resources, so that you could utilize it for your prosperity. Yet, their proceeds are wasted by corrupt, hedonistic, shortsighted, backward and oppressive kings, emirs, tribal leaders, and mullahs. Instead of gaining your freedom, instead of establishing the democratic system instructed by the holy book you claim allegiance, you are wasting your time in cafeterias, on the streets, and in rotten offices of antiquity, which produces nothing but zeros.

We must affirm the truth so that the truth will set us free from apathy and slavery.

Look at half of your population, your wives, mothers, sisters, daughters. What have you done to them? How can you hope to progress and attain peace, prosperity and God's mercy, while you have buried many of them alive? You cannot expect happiness, while you despise half of God's creation, your wives, mothers, sisters, and daughters; while you deprive them from their human rights given by their Creator, turn them to fractionally humans. You cannot tell God that you did all those evil things to please the idols called Bukhari, Muslim, Tirmizi, Ibn Hanbal, Ibn Maja, Abu Dawud, Malik, Kafi, and a herd of other imams, mullahs and clergymen. None of those idols will save you from God's justice. You are already paying dearly for your misogynistic beliefs and practices. You must apologize to your mothers, wives, sisters, and daughters for treating them like your slaves; you must repent for acting like Pharaohs against them.

We must affirm the truth so that the truth will set us free from the dark holes of our deception.

The world knows that Israel has transformed from a victim nation to a racist colonial power. Many progressive Jews too are painfully accepting this fact and they are fighting against it. The world sees and most people affirm the fascist policy, occupation, atrocities, massacres, and humiliation committed against the Palestinian people since 1948. The world knows that Israel has killed many more Palestinian children than the Palestinian suicide bombers have done. The numbers and events are out there recorded to prove that Israel has used state terrorism against Palestinian people. The world knows that a coalition of Crusaders, Zionists, and weapon/oil and other interest groups, nested in towers of power are using American tax money, military, and political power to perpetuate this tragedy, hoping for the Armageddon, more land, or bloody profits from wars. Nevertheless, again you must look in the mirror. What have you done, what have you become? You have become as racist as the Zionist you condemn. You condemn Jews without discrimination, Jews that raised many great prophets, philosophers, scientists, and inventors whom you revere and admire. You have become a suicidal nation. Though there were more than mere pacifism into Gandhi's resistance against British colonialism, Gandhi's struggle provided a great example for you. Instead, you followed ignorant leaders, racist and manipulative politicians, terror organizations, misguided religious clerics, and your hormones. If you had taken lessons from modern history and you had used your mind more than your animalistic instincts, if you had followed the Quran rather than the religious teachings that promote violence and racism, by now you would be living next to Israel sharing Jerusalem peacefully as brothers and sisters. You cannot have God's mercy if you respond to hatred with hatred, racism with racism, atrocities with atrocities. You cannot attain freedom and peace without sincerely asking the same thing for your enemies. How can you claim to be muSLiMs, while you have taken SiLM (peace), out of it?

We must affirm the truth so that the truth will set us free from violence that has surrounded us.

By continuing along the path of denial and sectarianism, you are risking more than just happiness and dignity in this world, but you also risk shame and retribution in the Hereafter…

> "Those who had rejected will be told: 'God's abhorrence towards you is greater than your abhorrence towards yourselves, for you were invited to affirm, but you chose to reject.' They will Say, 'Our Lord, You have made us die twice, and You have given us life twice. Now we have confessed our sins. Is there any way out of this path?' This is because when God Alone was mentioned, you rejected, but when partners were associated with Him, you believed. Therefore, the judgment is for God, the Most High, the Most Great." (40:10-12)

Unless you are willing to take the necessary and painful steps of reform through self-examination and research, you will be led by the mold of complacency and blind followings into the abyss that is becoming your fate. You must turn to the true system of Islam, as revealed by God through His messenger, and stop blindly following your scholars and leaders into distortions and unauthorized teachings. You have been losing continuously because you have abandoned the word of God and replaced it with other religious laws and teachings which in-turn has caused God to abandon you and leave you to your folly.

This life is not just about fun and games…it is about fulfilling our part of the pledge with God and proving that we can serve Him Alone.

> "And when God Alone is mentioned, the hearts of those who do not believe in the Hereafter are filled with aversion; and when others are mentioned beside Him, they rejoice!" (39:45)

Are you ready to embrace the path of God Alone and abandon all your idolatry? Or will you continue to lose? We must affirm the truth so that the truth will set us free.

To the West, Christians, and Americans:

The following words are not from an enemy of yours, but from someone who is a member of your society and cares about your interest as much as you care. These are not the words of a politician either, who is ready to break a world record in somersault to appease you; neither the words of a religious leader who lives in a parallel universe of deception and hallucination. These are the words of a common man who left his country behind to seek peace, justice, and liberty. These are the words of a grateful person who found such a refuge in your midst. So, do not treat these words of advice with prejudice, but with care. Do not be scared to hear the truth about your "way of life" which always highlighted the freedom of expression and justice for all. Do not expect me to count the list of the many good things you have accomplished; you hear them frequently from speeches and news in your media, and you celebrate them in your holidays. Sure, you should remember the good things in your past, present, and remember them, so that you can continue repeating those good things. However, you need to hear the other voice too; the voice that you have not yet allocated a holiday to hear. You should open your ears to what you do not hear from those who have invested interest in caressing your ego, nationalism, patriotism, and feelings. I think you do not wish to be aloof from the facts around you and repeat the pattern of all fallen civilizations in history. Do not be arrogant, aloof, self-righteous, and selfish, since this will only inflict further harm on you.

We must affirm the truth so that the truth will set us free.

Since you have separated the church from state, since you have appreciated the importance of freedom, God has blessed you with progress, abundance, and prosperity. Though your history is tainted with wars, oppression, superstitions, and injustices, such as crusades, inquisition, indulgences, sectarian wars, witch-hunt, holocaust, slavery, racism, colonialism, misogynistic practices, sexual abuse, you seem to have learned from the past mistakes and have come up with a better functioning society that tolerates diversity and respects science. Though your society suffers from a myriad of problems such as promiscuous lifestyle, sexually transmitted diseases, high divorce rate, high crimes, videogames teaching violence, addiction to drugs and alcohol, gambling, greed, big gap between rich and poor, children abused by priests, high number of prisoners per capita, homelessness, waste, pollution, jingoism, apathy, etc., your constitutions, courts, congresses, and academic institutions are still functioning. Freedom has its own side effects and having the freedom of living one's life according to one's own choice, without the fear of government repression is by far the greatest blessing. The greatest danger to your society is the corruption of the democratic process through the influence of money and lobbies. When big corporations control your finance, media, and congress, your democracy and freedom will be only an illusion. However, there is hope since you frequently demonstrate the confidence to be self-critical and you are able to acknowledge your weaknesses and shortcomings. You have also demonstrated times and again that you have the ability to find novel solutions for social, economic, and political problems. You have shown grace and generosity against your former enemies.

We must affirm the truth so that the truth will set us free from self-righteousness.

The world has shrunk due to increase in population, pollution, economic interdependence, mass transportation, and speed and ease in communication; thus, you can no longer have a world with half of it eating themselves to obesity while the other half starving to death. You can no longer spend billions of dollars on pets, millions of dollars for cosmetic surgeries (including on your pets), gulp world's limited resources to feed your ever-increasing appetite for consuming, and yet expect love and admiration from the rest of the world for your capitalism, the system you adhere to like a religion. How can you convince the world that you are the bastion of liberty while your prison industry is booming, and you have the highest number of prisoners per capita in the world? You can no more support cruel, corrupt, regressive puppet regimes and occupying military forces, and expect not being hurt by those who you have deprived directly or indirectly of freedom, education, progress, prosperity, and hope. You can no more self-righteously claim to be a free and civilized nation while spending a great portion of your national production on conventional and unconventional weapons, which transforms you into arrogant beasts running from one war to another, from one occupation to another. You can no more fool yourselves to be a peaceful nation while you have been shedding the blood of millions of people around the globe in more than a hundred wars, covert operations, and occupations in less than a century! You cannot condemn terrorism without apologizing to humanity for destroying not one, but two cities in their entirety in retaliation for an attack by your enemy on your military base. If terrorism means to intimidate the enemy by aiming at civilians, then you should look in the mirror without trying to find justification for your own aggression and acts of terrorism. You cannot talk about a free and better world while you reject banning landmines that kill and maim so many innocent people every day.

We must affirm the truth so that the truth will set us free from our own transgression.

Watch out for the right (wrong) wing religious organizations; when they are passionate about a social or political issue think thrice. If their historical record is a measure, they occasionally get it right, but usually they are wrong, very wrong. You cannot let the left-behind fiction fans lead your global policy. However, pay more attention to the other wing, to the other groups, such as the Quakers. Their record, their conscience, their heart, their stand for peace and justice, is what you need. We are not telling you to turn your left cheek when you are slapped; but beware of getting intoxicated with power. While you might be pretending to be David, without knowing, your arrogance and transgression might transform you into a Goliath. The change might be slow, so you might not be able to notice it by looking at yourself through the mirror; especially when there are some politician magicians and their entourage whose job is to distort and contort the mirror so that you cannot see yourself as you should.

We must affirm the truth so that the truth will set us free from our sins.

You can no longer give lip service to the Biblical advice regarding the speck and plank in the eyes. You can no longer ignore the fact that those who live by sword are destined to die by the sword. You can no longer preach, "Love your enemy" while you are out there trying every means possible to hurt your friends, half-friends, and potential friends. You can no longer talk about "the golden rule" while you are working hard to justify the "iron rule" under the euphemistic expression, "preemptory strike." You can no longer talk about human rights and freedom while at the same time, you have turned little islands and navy ships into torture centers, and you have become the inventors of a diabolic scheme called "offshore interrogation." How come America that once led the establishment of the United Nations and promotion of Human Rights, now has turned torture into an international enterprise and high-tech affair? How can you allow the gulags such as, Abu Gharib and Gitmo happen? Yes, you have not broken the records of Stalin, Mao, Hitler, Pol Pot; but you should not be competing with them. Your founding fathers did not fight for independence and did not draft one of the best legal documents in human history so that you could become a super war machine and be the cowboy of the world. You carpet bombed dozens of countries in your short history, destroying hundreds of cities and killing millions of people. You destroyed two big cities with its civilian population as retaliation to losing less than three thousands of your soldiers in Pearl Harbor. As retaliation to losing less than three thousand civilians by a terrorist organization, which once you trained and financed, you started two wars, killed hundreds of thousands, destroyed many cities, and are still looking for more countries to destroy. How can you label your revenge, your aggression, your disrespect to the lives of other people, as "freedom" or "civilization"? You cannot change the reality by fabricating fancy names in your PR rooms and spinning them as the corporate media your accomplice. You cannot fool the world by replacing one puppet regime with another, by supporting oppressive and cruel tyrants in Saudi Arabia, Egypt, Pakistan, Israel, and then congratulate yourselves for being the champion of "freedom" and "democracy." You cannot preach about morality, rights, and God, as long as you do not value the lives of each innocent human being equally, regardless of their religion, nationality, and color. You have been taking wars, destruction, death, horror, and terror to many nations around the world without even changing your fancy lifestyle at home. Now, you are enraged, and you demand justice from the world because you have tasted a small fraction of what others have tasted.

We must affirm the truth so that the truth will set us free from the dark holes of our deception.

Why should we treat terrorizing an entire nation, destroying their cities, killing, torturing, and humiliating their children and youth in the name of democracy and liberty lightly? Why should killing tens of thousands of civilians should be forgiven if the murderers, who are also proven congenial liars, use the magic word "collateral damage?" Why should smashing the brains of children with bombs or severing their legs and arms should be considered civilized and treated differently than beheadings? Why should destroying an entire neighborhood or city and massacring its population by the push of a button from the sky should not be considered equally or more evil than the individual suicide bomber blowing himself or herself up among his powerful enemies who snuffed out all their hope? Why should surviving to push another button to kill more people should be considered a civilized action not the action of those who gave their own lives while doing the killing? Why should the smile of a well-fed and well-armed mass murderer be deemed more sympathetic than the pain and anger of a poor person? How can one honestly call an occupying foreign military force to be freedom fighters? How can one call the native population to be terrorists just because they are fighting against an arrogant and lethal occupation army, which was mobilized against them through lies and deception? Why are the children of poor Americans used to kill the children in poor countries?

We must affirm the truth so that the truth will set us free from violence that has surrounded us.

You should not favor one criminal over another because of their religion or nationality. Your media did not depict the Serbian rapists and murderers as Christian Murderers, nor they labeled IRA terrorists who engaged in a long sectarian terror campaign that took the lives of thousands, as Christian terrorists. The right-wing Christian militia that massacred thousands of Palestinian refugees in Sabra and Shatilla camps somehow lost their religion when they became news on your media. The same with terrorist groups who claimed the cause of Zionism. Furthermore, you should know that state terrorism, regardless of the nationality and religion of the population, is much more cruel, dangerous, and sinister than the group or individual terrorism. In your stand against war, violence, and terrorism, you must be consistent and fair. Peacemakers and promoters must protest and condemn the atrocities regardless as to whether those engaged in atrocities have a uniform on them or not. If military uniform justifies the acts of terror, destruction, or genocide, then Nazi soldiers should receive your sympathy.

We must affirm the truth so that the truth will set us free.

One World and Shared Destiny

We must eliminate the nationalistic virus that alienates the children of Adam and turns them into monsters against each other. This does not mean that we should eliminate national borders or abolish the social contracts among groups of people. We must consider the entire world as one community and work accordingly. This is not only morally right but is the only way we can survive on this little planet. We can no longer be reckless in treating this planet, this precious earth, and can no longer be myopically selfish in our dealings with other nations. Otherwise, we will inhale and poison ourselves with each other's pollution, we will suffer calamities caused by global warming, we will spend a great portion of our national production, we will overpopulate the land, we will shed the blood of many innocent people, and we will lose our individual freedoms for security because of the economic and political problems in other parts of the world. The world has become smaller, and troubles are shared more than ever before. We must act now as a world and revive the spirit of the United Nations with a new vision. We can no longer afford jingoism, macho attitudes, another world war, always looking for an "evil" outside us, retaliating against violence and terror with our own version of violence and terror. We should not let terrorists or warmongers define our vision, our destiny, since they will only bring more disasters for humanity. We should not allow evil whisperers to dupe us into inflicting another holocaust against another race; we should have learned our lesson. We should not tolerate authoritarian regimes, corrupt leaders, kings, and emirs in our countries; we must be braver than the corrupt bullies. We should be vigilant against the myopic and greedy interest groups that have grown like cancerous tumors in our democracies, infecting legislation, judiciary, executive branches, and the mass media.

Hopefully, this century will be the century of unity under the banner of "God Alone," so that the children of Adam will greet each other with peace by saying, "your system/religion is for you and my system/religion is for me." All humanity, including atheists and polytheists, should share this planet in peace and justice.

> "The Lord your God is One God. You shall worship the Lord your God with all your heart, with all your soul, with all your mind, with all your strength." (Old Testament, Deuteronomy 6:4; New Testament, Mark 12:29-30)

> "God bears witness that there is no god but He, as do the controllers, and those with knowledge, He is standing with justice. There is no god but Him, the Noble, the Wise." (The Quran, 3:18)

> "Say, 'O followers of the scripture, let us come to a logical agreement between us and you: that we do not worship except God, that we never set up any idols besides Him, and never set up each other as gods beside God.' If they reject such an agreement, then say, 'Bear witness that we are Submitters.'" (The Quran 3:64)

Therefore:

- Let's follow the Quran alone and dedicate the system to God alone.
- Let's stand against marginal elements among us, oppressive puppet regimes, brutal wars, occupations, and clandestine operations.
- Let's topple the oppressive monarchs and elect our own leaders so that we can have peace, liberty and justice on our own volition.
- Let's fight not with bullets or bombs, but with intelligence and wisdom.
- Let's give up superstitions and medieval culture and start engaging in scientific enterprise.
- Let's stop subjugating our mothers, sisters, daughters, and wives; let's give them back their dignity, equal rights, liberty, and identity.
- Let's unite our voices and prayers with genuine Christians, Buddhists, Jews, agnostics, anyone who seeks justice and peace, rather than injustice and war.

- Let's organize local and international conferences to discuss this issue. We may invite religious scholars of every sect or cult, but we should not let them run them, since our experience shows that they have not done a good job in leading.
- Let's affirm the truth so that the truth will set us free.

6
Why Trash All the Hadiths as Secondary Authority Besides the Quran?

After witnessing the comparison between the *hadith* and the Quran, how can a sound mind still insist on *hadith*? How can people still call those books *sharif* (honorable), or *sahih* (authentic)? How can they forgive the *hadith* narrators and collectors who sold them all kinds of lies and stories, containing so much ignorance and distortion? How can they get mad at Salman Rushdie, while much worse insults charged against Muhammed by *hadith* narrators and collectors?

When followers of *hadith* and *sunna* cannot defend the nonsensical and contradictory *hadith*s (narrations) abundant in their so-called authentic *hadith* books, they suggest picking and choosing those *hadith*s that are not contradictory to the Quran. The following brief argument with a Sunnite shows how deceptive and meaningless this apparently innocent suggestion. We call these people compromisers, or Selective Sunnis. Let's now follow a debate between a Selective Sunni and a Monotheist Muslim:

Sunni:

1. How can you claim that several thousand *sahih hadith*s are necessarily false while you cite only a few *sahih hadith*s which have debatable contents? Is this not generalization from scanty data?

2. Why do you assume that either all *sahih hadiths* should be rejected or all of them should be accepted? Why not judge each *hadith* based on its individual merit according to all the available data about its *isnad*, its transmitters, and so on?

3. Suppose we cease to use *hadith* as a source of information about the Prophet, his life, and his career. Then we notice that the Quran itself says very little about the Prophet's life. It also says nothing about how it was complied.

4. The historicity of the Quran is based on *hadith*s. It is from *hadith*s that we know how the Quran was compiled. It is also from *hadith* that we know about the life of the Prophet.

Muslim:

1. If any book contains a few lies (and we have more than just "a few"), then, the endorsement of that book is not reliable. If you see dozens of repeated fabrications introduced as trustworthy (*sahih*) *hadith*, then how can you still rely on other narrations of the same book? How can you trust Bukhari, Muslim, and Ibn Hanbal who narrate the LAST *HADITH* of the prophet Muhammed in his death bed, rejecting the recording of any *hadith* through a declaration from the mouth of Omar Bin Khattab and the acquiescence of all prominent muslims that "*Hasbuna kitabullah*" (God's book is enough for us)? (Bukhari: Itisam 26, Ilm 39,49, Janaiz 32, Jihad 176, Jizya 6, Marza 17, Magazi 83; Muslim, Janaiz 23, Vassiya 20-22; Hanbal 1/222,324,336,355).

2. Judging each *hadith* on its individual merit may seem attractive for those who are not satisfied with God's book, but it is a waste of time and a deceptive method. If the signature of narrators (*sanad*) cannot provide authenticity about the source of *hadith*, then our only guide to decide on the content of *hadith*s (*matn*) will be our personal wish or our current inclinations. How can we decide which *hadith* has merit? How can we decide which *hadith* is accurate? We may say "by comparing them with the Quran!" But what does this really mean? If it is "me" who will compare a *hadith* to the Quran, if it is again "me" who will ultimately judge whether it contradicts the Quran or not, then, I will end up with "*hadith*" which supports "my" personal understanding of the Quran. In this case a *hadith* cannot function as an explanation of the Quran. It will be confirmation or justification of my or someone else's understanding of the Quran; with literally tasteless, grammatically lame language.... Furthermore, what about *hadith*s that bring extra duties and prohibitions?

3. Again, there are many *hadith*s about the prophet's life, which you cannot accept with a sober mind. They are narrated repeatedly in many so-called authentic books. We cannot create a history out of a mishmash of narration by a subjective method of pick and choose. We can create many conflicting portraits of Muhammed out of those *hadith*s. As for pure historical events that are isolated from their moral and religious implications, they are not part of the religion, and we don't need them for our salvation. I never said "we should not read *hadith*." In fact, we may study *hadith* books to get an approximate idea about the people, culture and events of those times. We can even construct a "conjecture" about the history, without attributing them to God or his prophet. Please don't forget that "history" is not immune to filtration, censorship and distortion by the ruling class. You can see many different versions of histories (!) regarding the era of early Islam. Just read Sunni and Shiite histories.

4. We cannot disregard God's frequent assertion that the Quran is sufficiently detailed, complete, clear, and easy to understand. What do you think about the verse 12:111? "In their stories is a lesson for the people of intelligence. It is not a *hadith* that was invented, but an authentication of what is already present, a detailing of all things, and a guidance and mercy to a people who affirm." Or, what about 17:46? "When you preach your Lord, in the Quran ALONE, they run away with aversion."

5. *Hadith* books are full of contradictory teachings. They eventually lead us to a sanctified and justified sectarian division in the name of the Prophet. Their very nature is another proof that *hadith* collections cannot be divine, since God, characterizes his word and system as not having contradiction: "Why do they not study the Quran careful? If it were from other than God, they would have found in it numerous contradictions." (4:82). This verse clearly refutes the traditional argument that *hadith* books contain other revelations besides the Quran, since the followers of *Hadith* and *Sunna* confuse some of the reference to Quran with *hadith*, as in: "Your friend (Muhammed) is not astray, nor is he deceived. Nor is he speaking out of a personal desire. It is a divine inspiration." (53:2-4). Furthermore, verses 39:27-28 describe the Quran and the following verse distinguishes the divine teaching from other teachings. "God cites the example of a man who deals with disputing partners, compared to a man who deals with only one man. Are they the same? Praise be to God; most of them do not know." (39:29). Obviously, *hadith* narrators and collections are "disputing partners," while the Quran is a consistent source.

6. Our conviction regarding the divinity of the Quran and even its protection does not come from our trust in the number of people, but from the evidence contained in the book, which is another number, a number that is not appreciated by those who determine the truth based on the number of heads with turbans. (Wonder about that number? See 74:30).

7. We reject *Hadith* because we respect Muhammed. No sound person would like to have people born several centuries after him roam the earth and collect a bunch of hearsay attributed to him. Besides, if Muhammed and his supporters really believed that the Quran was not sufficient for guidance, an ambiguous book, or lacked details, then, surely, they would be the first ones who would write them down and collect them in books. After all, their numbers were in tens of thousands, and they had plenty of wealth. They could afford some ink, papyrus paper or leather, and some brain cells, for such an important task. They would not leave it for a guy from far Bukhara or his ilk who would come more than two hundred years to collect *hadith*s in a land soaked with blood because of sectarian wars. Besides, Muhammed had many unemployed or handicapped people around who could gladly volunteer for such mission. The traditional excuse fabricated for Prophet Muhammed and his supporters is absurd. Supposedly, Muhammed and his followers feared that people would mix the Quran with *hadith*s. This is nonsense. They were smart enough to distinguish both, and there were enough people to keep track of them. Besides, what is the use of separating both, if we will need the second for our salvation as much as we need the first? In practice, the followers of *Hadith* have perfectly mixed both. Worse, in most instances they have preferred *hadith* over the Quran.

Furthermore, it is the followers of *hadith* and *sunna* themselves who claim that the Quran was a "literary miracle". If their claim of "literary miracle" were true, then it would be much easier to separate the verses of the Quran from *hadith*. Let's assume that they could not really distinguish the text of the Quran from Muhammed's words, then couldn't they simply mark the pages of the scripture with the letter Q for the Quran and letter H for *Hadith*, or let some record only the Quran, or simply color code their covers? Or allocate leather for the Quran and paper for *hadith*, or vice versa? They could find many ways to keep different books separate from each other. They did not need to study rocket science or have computer technology to accomplish that primitive task. The collectors of *hadith*s wished that people would accept their assertion that Muhammed and his supporters did not have ink, paper or leather, mind, and care to collect *hadith* before them. No wonder, they even fabricated a few *hadith*s claiming that Muhammed's companions were competing with dogs for bones to write on the verses of the Quran!

Well, most likely, Muhammed feared that people would mix his words with the Quran. Not the primitive way that is depicted by the Sunnis and Shiites, since as we pointed out, there were many ways to eliminate that concern. But the real concern was different. Because of the warnings of the Quran, he had all the reasons to fear that Muslims would follow the footsteps of Jews and would create their own Mishna, Gomorra, and Talmud: *hadith* would be considered as an authority, as another source besides the Quran, setting him as partner with God! Ironically, the followers of *hadith* and *sunna* accomplished exactly that. They did not need to publish the text of *hadith* together with the Quran--though they have done that in many commentaries--they have been doing worse. Though they usually have kept *hadith* separate from the Quran physically, as far as for the purpose of guidance and religious authority, they mix it with the Quran. Even worse, they make the understanding of the Quran dependent on the understanding of *hadith*, thereby elevating *hadith* to position of authority over the Quran. Thus, if indeed Muhammed was worried about people mixing his words with the Quran, the followers of *hadith* proved his worries right: centuries after him, they did not only mix his words with the Quran, but they also mixed thousands of fabrication and nonsense attributed to him. See 25:30; 59:7.

8. Give me one, only one "*hadith*" that you think is necessary for my salvation besides the Quran. If you are not ready to discuss the necessity and accuracy of a single *hadith*, then please give up inviting people to *hadith* and *Sunna*.

Further Discussion

Sunni: The bound collection of testimony from any court is certain to contain some lies and some errors. The reliability of any piece of evidence remains debatable. Where the narrators agree, where there is no irreconcilable conflict with the Quran, where the *hadith* is not offensive to *tawhid*, etc., we may well be justified in accepting it as reliable. And if a collector collects a thousand *hadith* and makes a few errors, neither is he to be condemned as unreliable.

Muslim: Not a single court will accept the testimony of Bukhari who collected contradictory *hadith*s about the Prophet

Muhammed, narrated from generation to generation 200 years after his departure. You try to minimize the number and size of errors. There are hundreds of lies, not "a few errors." And they are grave ones. They attribute silly and contradictory laws and words to God. They create a manmade religion in the name of God! They are full of insult to God and his messenger. They are not trivial, since God Almighty does not accept those "few errors" as trivial:

> " . . . Who is more evil than the one who fabricates lies and attributes them to God?" (29:68)

Sunni: If the *hadith* are not *mutawatir*, the monotheist Muslim should know by now that most scholars would say that one is free to disregard it, though not necessarily without peril. The issue the Muslim raises about the difficulties of decision regarding *hadith* also apply to personal interpretation of the Quran. No, the Quran makes it clear, we cannot disregard any evidence out of hand, not even the evidence of an unrighteous man; how much less the evidence of those against whom we have no evidence of unrighteousness or lack of caution?

Muslim: First, can you please tell us how many *mutawatir* (accepted with consensus) *Hadith* are there. What are they and where are they? Second, can you give me a few names of those "most scholars" who would say that I am free to disregard non-*mutawatir hadith*s? As far as for evidences.... Sure, we cannot disregard evidences for our daily affairs, even for an unrighteous man. But God's religion is not left to the mercy of that evidence. God explained and sufficiently detailed his religion in his book, which is described as complete, detailed, and perfect. It does not contain any doubt. Furthermore, God promised to preserve it. And He did it with a unique mathematical system which hypocrites and ingrates are unable to see.

Sunni: I have answered The Muslim about of a number of these *hadith*. Certainly, I personally have trouble with certain *hadith*; however, I must always ask myself whether or not it is my own view, which is in error, rather than the *hadith*. Perhaps there is something I have not thought of.

For example, there is a *hadith*, which The Muslim loves to cite mentioning the drinking of camel's urine, which he seems to believe, is particularly ridiculous. Does he base this on a scientific study of the virtues of drinking camel's urine? I think not. Nor does he ever mention that nomadic peoples, not just Arabs but including them, often consume the waste products of their animals. So "cannot accept" is definitely culturally conditioned. But no one has claimed that drinking camels' urine is required of any Muslim.

Muslim: Well, prescribing camels' urine is a minor problem with that *hadith*. You can even find some Sunni doctors who pontificate that camel's urine is a panacea for every disease. The big problem was about gouging their eyes after pruning all their legs and hands, etc. You craftily skipped that part.

Sunni: The Muslim confuses *Hadith* and *Sunna*. *Hadith* is only one of a number of major sources of *Sunna*, other major sources being the Qur'an and the practice of the community. The latter is how we generally learn to pray, by the way. To answer the question about necessity of *hadith* without going deeply into the whole concept of necessity is impossible.

But I will answer this way: if a *hadith* transmits a wisdom required in a particular situation, and one turns away from that wisdom merely because it was a *hadith* (and not some other preferred modality), then one becomes culpable for failure to act correctly in the situation. This could, indeed lead to hellfire. Of course, the same is true of the Qur'an, or even the preaching of a Christian.

Muslim: If you think that someone is wrong and even misguided because of his rejection of *hadith* and that person challenges you with that question you don't answer like you did above. You did not or could not answer my challenge. Answering questions is not an act of writing irrelevant lines after the question. Please come to the point.

7
A Forsaken God?

(From *19 Questions for Christians* by Edip Yuksel)

The last words of Jesus are one of the few words kept untranslated. Jesus calls God Eli, which has the same root as the Arabic *Elah* or *Allah*, and he complains about his fate.

> "And about the ninth hour Jesus cried with a loud voice, saying, E'li, E'li, la'ma sa-bach'-tha-ni? that is to say, My God, my God, why hast thou forsaken me?" (Matthew 27:46, Mark 15:34)

Obviously, the followers of St. Paul use sensational language to dramatize the scenario of Crucifixion. They are like a plastic surgeon who gouges out the eye while trying to make an eyebrow! This verse is not only at cross purposes with the fabricated doctrine of Christianity, but also reflects the confusion on the part of the authors of the Gospels. (Compare the wavering hero of Mark 14:36 and Luke 22:42 to the brave hero of John 12:27).

Evangelists are fond of using the Crucifixion of Jesus by Jews as a proof of his deity. Their "strongest evidences" about the deity of Jesus is based on "the deep understanding capacity of Jews." Evangelists pontificate: "Look, if Jesus did not claim that he was God or literally Son of God, monotheistic Jews would not have stoned him for blasphemy". To support their evidence, they feel the obligation to add that "Jews definitely understood his teaching." But the Bible says the contrary:

> "Why is my language not clear to you? Because you are unable to hear what I say" (John 8:43 and Mark 4:13).

Not only the disbeliever Jews, but even his disciples, sometimes had difficulty understanding him. See Mark 9:32 Luke 18:34; 9:45 & John 8:27; 12:6.

Unfortunately, the evangelists and clergy do not have a better understanding of his teaching than the Jews who stoned him. Josh McDowell, in his "one-million-in-print-book" tries to persuade us of the Jews' deep understanding capacity:

> "Jesus is threatened with stoning for 'blasphemy.' The Jews definitely understood his teaching. . ." (More Than a Carpenter, Josh McDowell, Tyndale, Illinois, 1989, p. 17)

E. Calvin Beisner, a professional evangelist, advocates another common logical fallacy related to the fictitious divine sacrifice:

> "Think what kind of act gets the highest praise among men: isn't it when someone voluntarily sacrifices his life in order to save the lives or others? Such self-sacrifice is a tremendous good. The greatest such sacrifice was when God sacrificed His life in the Person of Jesus Christ to save the lives of all who believe in Jesus." (Answers for Atheists, p. 10).

Before listing my questions, let me share a brief argument:

I asked an evangelist: "Why should God sacrifice 'His son' in order to show His love and forgive us?" He replied with a counter question, "If you love a girl too much what do you do?" I said, "I'll try to help her." "Wouldn't you die for her?" he suggested. "Why?" I questioned, "To the contrary, I would try to keep myself healthy and handsome. When we feel desperately helpless and not able to find a solution, then we may sacrifice ourselves, in another word, commit suicide. But God does not run out of solutions, and He is never helpless." The evangelist friend did not have an answer; his attempt to justify his faith by reason had failed. So, it was time to resort back to his first and last, perhaps the only refuge: "faith." He was not yet prepared to get out of the dark and windy tunnel he was inviting me to it with a little candle that could not even stand the shortest breath of reason.

Now, here are my questions:

- Do fanatic religious people kill others only because they claim that they are God? How many prophets were killed by "monotheistic" Jews before Jesus? Should we infer that all those prophets claimed to be God? Why did the "monotheistic" Church condemn Galileo? Did he claim to be God? Why did the "monotheistic" Christians burn Tyndale? Did he also claim to be God?

- According to that verse (Matthew 27:46, Mark 15:34), Jesus was not even a human hero. History (even this author) has witnessed so many brave warriors who did not cry under torture, but they roared their slogans into the ears of their murderers. How can Jesus deserve to be God, while he, according to your record, demonstrates a weakness which is considered shameful for a human warrior? Who is more courageous according to your own record: the first Christian martyr, Stephen (Acts 7:59-60), or Jesus (Mat 26:38, 39)?

- What was the main mission of Jesus' birth according to Paul's disciples, that is, the Gospel writers? Wasn't it to sacrifice himself for the salvation of the human race? Then, why did Jesus try to escape from that mission? And why did he start to wail and whimper even before his mission was carried out (Mat 26:38,39)?

- According to the story of the "crucifiction", did Jesus not accept his divine mission? Did God promise to save him or to let him 'die for the sins of humankind'? Then, how can he accuse God of forsaking him? Isn't this a slander to God? How can God be a betrayer?

- If the purpose of Jesus in this life was to die for our sins, as it is claimed in the Pauline New Testament, would he not have said on the cross: "My God, my God, thank you for fulfilling my mission!"?

- You believe that Jesus, your god, prayed as "My God, my God, why have you forsaken me." Do you really believe that God prayed to himself and asked help from himself? Does God call himself out as "My God, my God"? Did your god fight against himself to save himself?

- When Jesus was dead for three days, was God also dead? If "yes," then who controlled the universe during those 3 days? If your answer is No, then Jesus is not equal to God.

- "E'li, E'li, la'ma sa-bach'tha-ni" is one of the few Aramaic phrases in the New Testament. Ironically, it is one of the most obvious fabrications in the Bible. Did the authors of the Gospel decide to keep that part in Aramaic specially to make us believe that those words were accurately transmitted?

- How can killing an innocent person be considered the method for salvation? Do we have to kill somebody unjustly to deserve salvation? Is this divine wisdom?

- Is it just and wise to punish your son because of the misbehavior of your neighbors' children?

- What kind of love is it to kill your own innocent son in order to be *able* to forgive your neighbors' children?

- We can forgive our friends, even our enemies without feeling an obligation to kill our loved ones. Couldn't God forgive us without killing His "own son"?
- According to your dramatized teaching, a "son" is a greater hero than his "father", since he volunteered to sacrifice his life, and his father behaved selfishly.
- Your "divine sacrifice" story does not deserve to be entitled a "sacrifice," since according to your own doctrine, God has sacrificed only one-third of his personalities. When a human sacrifices his life, he sacrifices whatever he has. Doesn't your "divine sacrifice" have less importance than "human sacrifice" in terms of "sacrifice"?
- If "all his disciples forsook him and fled" (Mark 14:50), then, who are the eye-witness narrators of the events following Christ's arrest? How can we trust the hearsay that came through his murderers and bystanders?
- Don’t the contradictory reports of this most important event, even in the carefully selected four Gospels, impeach the narrators? (Mk 15:23 v Mt 27:34; Mk 14:36 & Lk 22:42 v Joh 12:27; Mt 27:11-14 v Joh 18:33-40; Mt 27:32 & Lk 23:26 & Mk 15:21 v Joh 19:17; Lk 23:46 v Joh 19:30; Mt 27:50-51 & Mk 15:37-38 v Lk 23:45-46; Mt 27:54 & Mk 15:39 v Lk 23:47; Mt 12:40 v Mk 15:42-43 & Joh 20:1; Mk 16:1 v Joh 19:39-40; Mt 28:9 v Joh 20:11-17; Mk 16:12 v Lk 24:34). Which court of justice would accept the testimony of the witnesses, riddled with so many contradictions?
- Some Christian scholars claim that the Greek word "stauros" means a piece of timber, not cross. If this is correct, then why do you transform the stake into a cross?
- If timber or cross was the murder weapon and torture device, then how can you glorify and worship such a thing?
- If the salvation of humanity could be accomplished only by the Crucifixion, why did God spend thousands of years preparing a chosen people for this task?

Also, see 2:59; 3:45,51-52,55; 4:11,171; 5:72-79; 7:162; 19:36.

8
Eternal Hell and the Merciful God?

As a monotheist, I have compelling scientific, philosophic and spiritual reasons to believe in the Quran, yet I have to admit that I have not digested all the verses of the Quran. Some verses challenge my cultural norms or the mainstream ideology, and a few also appear to contradict other clear verses of the scripture and/or the laws of nature. Knowing that my culture is relative, I usually handle well the first category, but those that create contradiction among God's signs (ayat) of the scripture or of nature act like viruses infecting my faith. Those who have no intellectual problem with any verses of the Quran, in my opinion, are either gullible people who happened to inherit/acquire their faith because of peer pressure, geographic proximity, or any other extraneous reason; or they are hiding their intellectual problems from others and perhaps from even their own cognition. Neither type, however, can set a good example.

My doubt is not about the veracity of the Quran, but about the veracity of my understanding of some verses of the Quran. Since I rejected Sunni precepts that require blindly following the opinion of orthodox scholars and clerics, and since I accepted the Quran as the only source of my religion, whenever I encounter a problem with my understanding of a verse that puts it at odds with 4:82, I follow the divine advice to act patiently in seeking knowledge (20:114), ask the experts (21:7) without following them blindly (17:36), avoid wishful thinking and hearsay (53:28), and know that God is the one who will ultimately provide explanation (75:16-19). Sometimes, I attain a coherent understanding within months, but sometimes it takes years and even decades. Each of my intellectual and spiritual experience is a testimony to the following facts:

7:52 We have brought them a book that we have detailed with knowledge to be a guidance and mercy for the people who believe.

55:1-2 The Gracious Teaches the Quran

Before sharing with you my observation about the duration of Hell, I would like to share with you several examples of my intellectual struggle and their results. (If you do not have patience with this lengthy introduction or preparation, please jump to the subtitle Is Hell Eternal.)

For instance, my inquiry on 5:38 evolving about fifteen years resulted in three understanding: (a) cutting or marking a thief's hand as a means of public humiliation and identification, (b) physically cutting off a thief's hand, or (c) cutting off a thief's means to steal and burglarize (presumably through rehabilitation or imprisonment). Depending on the economic and social circumstances, frequency of theft, its risk to the society, and the economic, social and psychological cost of punishment, a society may pick any of the suggested punishments. In other words, I am now convinced that the deliberate use of a semantically flexible key word, QaTTa'A, is to accommodate time, mood, culture, and circumstances of diverse populations. Freezing the message of the Quran with the understanding and practice of the first generation (including Prophet Muhammed and his companions) is stripping the Quran from its prophetic divine nature that makes its message universal.

My inquiry on the apparent contradiction between 2:233 and 46:15, within several months led me to a conclusion that I never wished to reach: taking the normal length of pregnancy as 266 days (or 38 weeks), abortion within 86 days of pregnancy would not be considered murder. In the first trimester, the fetus was not considered as a person. Becoming a person is with the emergence of consciousness (that is, Nafs, which is usually wrongly translated as Soul, because of Plato's influence on later Muslim scholars). "Personhood" is described as the stage of "new creation" that follows the stages of being a sperm, embryo, bite-size fetus forming bones and flesh, and finally a new creature (23:14; 22:5). My problem with the traditional understanding of 4:34 and reconciling it with 30:21 was solved within a year through research and "accidental" events. I learned that men were not rulers over women, but providers for them; women were described not as devotees of their husbands, but of their Creator; the issue was not disobedience to husband but disloyalty to the marriage contract; and husbands were not advised to beat their wives but to separate from them before deciding on divorce.

Similarly, my problem with reconciling traditional understanding of 4:3 and 4:129 was solved decades later when I noticed a universal mistranslation of a phrase in 4:127. Though the Quran permits polygamy (4:3), it discourages and restricts its actual practice by requiring significant preconditions: men may marry more than one wife only if the latter ones are widows with orphans, and they should treat each wife equally and fairly. (See 4:19-20, 127-129.) Unfortunately, verse 4:127 has been traditionally misinterpreted and mistranslated in such a way as to suggest that God permits marriage with juvenile orphans. This was clearly not the case.

Let me give you one more example. I had a problem with the traditional mistranslation of a key word in verse 2:106, since it implied contradiction in the Quran and made any verse in the Quran a vulnerable subject to the claim of abrogation. The word "ayat", the plural of "ayah," is used in the Quran to mean both (a) signs/miracles, and (b) verses/revelations of the Quran itself. Since verses of the Quran are miracles/signs, the plural form occasionally conveys both meanings simultaneously. A single verse of the Quran is not deemed to be a miracle since some short verses of the Quran (for instance: 55:3; 69:1; 74:4; 75:8; 80:28; 81:26) are not unique and can be found in daily conversation of Arabic-speaking people. In fact, the Quran determines the minimum unit of miraculous nature as a chapter (10:38), and the shortest chapters consist of 3 verses (103; 108; 110). Therefore, only the plural form of "ayah", that is, "ayat", can be used as reference to the verses/revelation of the Quran. However, the singular form, AYAH, in all its 84 occurrences in the Quran is always used to mean sign or miracle. Therefore, I choose to translate the singular form "ayah" in verse 2:106 as "sign."

Is Hell Eternal?

God, as demonstration of ultimate creation, chose to test the results of creating a being with the ability to freely choose its own destiny (18:29; 6:110; 13:11). God downloaded His revelation/commands/logic (rouh) to the prototype human that would provide him with innate rules of reasoning to distinguish falsehood from truth, bad from good (15:29; 32:9; 38:72;). Messengers and books containing rouh were only a bonus mercy, mere reminders of the facts that could be discovered by reason (2:37-38; 10:57; 11:17; 16:89; 21:107; 29:51; 16:2; 36:69; 37:87; 39:21; 42:52; 58:22). Though believing that we humans have freedom of will is one of the paradoxes most difficult to digest, I accept it on faith (18:29; 57:22). God created life and death on this planet to test His ultimate creature (67:2). After a certain age, an individual is deemed accountable by God (46:15). God decided to punish those who freely choose a path contradictory to its original program as they corrupt it through false ideas and actions (2:57; 4:107;

6:12,20,26; 7:9,53177; 59:19). The programs that are infected with viruses will experience and a regretful stage called Hell (Hell and Paradise are allegories: 13:35; 17:60; 37:62-64; 7:44-50). In this stage the corrupt programs and their chief infector (Satan) will be penalized (7:11-27; 38:71-88), and then altogether they will be annihilated. The only virus that will not be healed on the day of Judgment is the virus that creates a schizophrenic personality, a personality that submits itself to others besides God, a personality that does not free itself from false gods thereby alienating itself from its origin, that is God (4:48,116).

The popular belief that Hell will burn eternally bothered me for decades, but I suppressed my problem by saying "God is Merciful and Just; He knows something that I do not know." Of course, God knows many things that we do not know. But what if we are protecting our superstitions and false beliefs through such an excuse? What if we are stopping ourselves from using God's greatest gift: reason, which distinguishes a believer from a disbeliever, a human from an animal? (2:73, 170, 171, 242, 269; 3:118, 190; 7:169; 8:22; 10:42, 100; 11:51; 12:2, 111; 13:4, 19; 16:67; 21:10, 67; 23:80; 24:61; 29:63; 30:28; 38:29; 39:9, 18, 21; 40:54; 59:14). Sure, there was a danger in confusing "reason" with my personal wishes, ignorance, and cultural biases. I could distort the meaning of God's Word to appease my wishes or to conform to my limited knowledge. There was a fine line. Should I use my reason to question an understanding that I inherited from a particular sect, or should I follow everything without using my mind? Knowing that the Quran strongly admonishes us against following the crowd, the footsteps of our parents, or religious scholars blindly, (6:116; 2:170; 9:31; etc.). I rejected blind faith and chose faith based on knowledge and reason (17.36). To prefer an unorthodox understanding, I have adopted a two-pronged rule: I should be able to support it by the original language of the scripture, AND the unorthodox understanding should not create a contradiction either among the divine laws and precepts in the scripture or between scripture and divine laws in nature.

About six years ago, I read a Turkish translation of a booklet, The Universal Salvation, written by Musa Jarullah Bigiyev (1874-1949). In that booklet, Bigiyev argued that according to the Quran and *Hadith*, Hell was not eternal. When I finished the booklet my excitement and hope faded as the author had not dealt with the many pertinent verses that led hundreds of millions of Muslims to believe that hell was eternal. He was making a radical assertion, but he had weak arguments to support it. He was utilizing more emotional appeal more than scholarly evaluation of related verses. Disappointed, I continued in my belief in eternal Hell, albeit as a contradictory concept which continuously irritated my faith and intellect lurking in the background. I could not ignore numerous Quranic verses/signs that were threatening ingrates or *mushriks* with suffering in hell for eternity. But I also could not ignore the other fact that God's most repeated attribute in the Quran was God's mercy (RaHYM 114 times, RaHMaN 57 times, etc.). God had decreed mercy as His attribute (6:12,54) and His mercy was immense (6:147; 40:7). I frequently took solace in the implication of the following dialogue between Jesus and God that will take place on the day of judgment:

5:118-119 'If you punish them, they are your creatures. If you forgive them, you are almighty, wise.' God will say: 'This is a day when their truth will benefit the truthful ones.' They have deserved gardens with flowing streams. They abide there forever. God is pleased with them and they are pleased with Him. This is the greatest achievement.

His justice was also reminded frequently (3:182; 4:40; 8:51; 11:101; 16:33,118; 22:10; 41:46; 43:76; 50:29; 99:7-8). How can a merciful and just God torture his creatures in eternal hell for their crimes committed in a very short time, a period that is almost zero compared to eternity?! How could divine mercy and justice be challenged by my limited mercy and justice? If I had a part of God's revelation/knowledge/logic (rouh) in my genetic program, then I should be able to find a way to embrace, not necessarily comprehend, God's mercy and justice without reservation.

Mushriks And Ardent Ingrates Are Condemned To Stay In Hell Eternally!

Please note that the subtitle does not assert that "*Mushriks* and Ardent Ingrates Are Condemned To Stay In Eternal Hell!" And, according to the language used in the Quran there is a difference. Let me explain:

All forms derived from the root of *KHaLaDa* (to be eternal, live forever, to remain forever in place, or to stay for long time, or lifetime) occur 87 times in the Quran. If my preliminary count is correct, in 40 occurrences it describes the duration of reward in Paradise (2:25; 2:82; 3:15; 3:107; 3:136; 3:198; 4:13; 4:57; 4:122; 5:85; 5:119; 8:42; 9:22; 9:82; 9:89; 9:100; 10:26; 11; 23; 11:108; 14:23; 18:108; 20:86; 21:102; 23:11; 25:15; 25:16; 25:76; 29:58; 31:9; 39:73; 43:71; 46:14; 48:5; 50:34; 57:12; 58:22; 64:9; 65:11; 98:8; 50:34), and in 40 occurrences it describes the duration of punishment in Hell-fire (2:39; 2:81; 2:162; 2:217; 2:257; 2:285; 3:88; 3:116; 4:14; 4:93; 4:169; 5:80; 6:128; 7:36; 9:17; 9:63; 9:68; 10:26; 10:27; 10:52; 11:107; 13:5; 16:29; 20:101; 21:99; 23:103; 25:29; 25:69; 32:14; 33:65; 39:72; 40:76; 41:28; 43:74; 47:15; 59:17; 64:10; 72:23; 85:17; 98:6).

The word *KHaLaDa* also conveys the meaning of long duration. For instance, the classic Arabic dictionary Lisanul Arab lists the plural form *KHawaLiD* to mean mountains and rocks since they last very long. The Quran, at least in one instance, uses the past tense of the word to describe an act lasting lifetime (7:176). The verse describes a fanatic disbeliever and tells us that he "*Akhlada ilal ardi*." that is, stuck to the ground, stuck to lowly ideas! In this article I do not argue that the meaning of the word KHaLaDa and all its derivatives convey only the idea of a very long time or a period of a particular lifetime. Though there might be some evidence for such an argument, and the lack of usage of this word for God might be considered supporting evidence, at present I am not convinced.

As for the adverb *ABaDa* (eternally, ever, forever), it occurs 28 times in the Quran, and out of these, in nine occurrences it is used to describe the duration spent in paradise (4:57; 4:122; 5:119; 9:22; 9:100; 18:3; 64:9; 65:11; 98:8) and in three verses this word is used for the duration spent in hell (4:169; 33:65; 72:23). Verses 5:37; 22:22; 32:20 state that the ingrates will want to exit Hell, but they will never be able to do so. The arguments of those who reject eternal punishment is rejected (3:24).

ABaDa is used in 9:84; 9:108; 24:4; 33:53; 59:11; 62:7 to mean eternity contingent with the life of the subject:

"You shall never (*La.... ABaDa*) pray for any of them when he dies... " (9:84) "You shall never (*La ... ABaDa*) pray in such a masjid..." (9:108) "... and never (*La ... ABaDa*) accept t any testimony from them..." (24:4) "... You shall never (*La ... ABaDa*) marry his wives after him." (33:53) "... We will never (*La ... ABaDa*) obey anyone who against you..." (59:11) "... They will never (*La ... ABaDa*) long for it... " (62:7)

All these negative statements use *ABaDa* to express a prohibition that will last forever. More accurately, as long as the conditions exist. For instance, when the person prohibited from a funeral prayer himself dies, the prohibition too ends naturally. A dead person cannot pray at the grave of another dead person and therefore, this prohibition does not practically last for eternity. Similarly, when the wives of the Prophet all died, the prohibition to marry them ceased to exist. Therefore, the eternity of prohibition was, in fact, limited by the condition or lifetime of the subject. In other words, in the above examples, the word *ABaDa* indicates the entirety of a particular period.

Eternal Punishment In Hell Does Not Necessarily Mean That The Hell Or Its Inhabitants Are Eternal

Remember Jonah. When he escaped from his duty he was swallowed by a whale.

37: 143-144 But had he not been of those who glorify, he would have stayed in its belly to the day of resurrection.

It is a fact that neither Jonah nor the whale was immortals that could live to the day of resurrection. God knows this. Thus, this Quranic expression simply informs us that Jonah would die or end up in the belly of the whale. Although both would perish in a short time, the whale would be the ultimate destiny of Jonah until the day of resurrection. Had Jonah not intended to be resurrected (together with the whale, 6:38), using the same logic, the verse would state: "he would have stayed in its belly eternally.'"

Similarly, verses informing us about ingrates or idol worshipers staying in Hell eternally, does not necessarily mean that the Hell is eternal, unless we are informed that Hell itself is eternal. It simply means that ingrates and idol worshipers would end up in Hell and nothing else. If Hell, together with its inhabitants, one day ceased its existence, then the ingrates or idol worshipers would still stay in Hell eternally. Their punishment would be the entire life of Hell.

In fact, the Quran informs us that both the eternal punishment in Hell and reward in Paradise is conditioned with their life span:

11:107 Eternally they abide therein (Hell), for as long as the heavens and the earth endure, in accordance with the will of your Lord. Your Lord is doer of whatever He wills. As for fortunate ones, they will be in Paradise. Eternally they abide therein, for as long as the heavens and the earth endure, in accordance with the will of your Lord-an everlasting reward.

The Quran informs us that the Earth and Heavens will be changed to different Earth and Heavens (14:48). If the re-creation referred to in this verse is the one before the Day of Judgment, then there is possibility of another re-creation. There are indications that Paradise will be preserved or created again. For instance, the word KHuLD (eternal, everlasting) is not used for Hell, but it is used as an adjective to describe Paradise (25:15). On the other hand, the same adjective is not used to describe Hell, but to describe the punishment IN HELL (10:52; 41:28).

Hell, together with its Inhabitants, will be annihilated

Our language contains synonyms, antonyms, complements, counterparts, etc. The Quran frequently uses pairs of words/concepts to contrast, compare, or complement each other. Usually, the frequencies of semantically related words also demonstrate mathematical harmony, which is another subject I extensively demonstrated in my books. For instance, Akhirah (Hereafter) and Dunya (Lowly World) , Malak (Angel) and Shaytaan (Satan), Rahmah (Mercy) and Huda (Guidance), Qul (Say) and Qalu (they said), Khalq (Creation) and Helaak (Annihilation), Hayat (Life) and Mawt (Death) are semantically and mathematically related words.

Let's reflect on verse 40:11 and 67:2. The first verse refers to two creations and two deaths and the second refers to creation of death and life. These two verses can be understood better if we know that in the Quranic language death cannot exist without life and vice versa. They exist together, since Death is permanent halt of the brain's conscious activity (39:42; 16:21), but a temporal stage to be followed by resurrection (29:57; 10:56; 22:6). Death is a process leading to life. A living creature will die and a dead creature will get a new life (22:66). Vegetation experiences successive lives and deaths through seasons (2:64; 3:27;

6:95; 16:65; 22:5-6; 30:19-50; 35:9). With the first creation there was neither death nor life; we just existed. But God decided to create death, and life (67:2). Creation, death, life (current), death, life (resurrection). In other words, two deaths and two lives (40:11). The word HaLaKa, on the other hand, is occasionally used to describe the death of an individual (40:34), but it usually means irreversible destruction and annihilation, or total existential extinction of an entity (5:17; 6:6; 6:47; 8:42; 20:128; 21:95; 22:45; 28:59; 36:31; 69:5; 77:16).

8:42 . . . whereby those destined to be perished/annihilated were perished/annihilated for an obvious reason, and those destined to be saved/revived were saved/revived for an obvious reason (or clear argument)

The above verse not only refers to the loss of lives and surviving in a particular battle, but also to a higher cosmic event: ingrates will perish forever since they relied on falsehood while affirmers will last forever since they relied on a clear argument No wonder life and death are used as metaphors for attaining truth or falsehood (6:122). In fact, witnessing and affirming the truth leads to life, both metaphorically and literally:

8:24 O you who believe, respond to God and the messenger when he calls you to that which grants you life. You should know that God in between a man and his heart, and that before Him you will be gathered

From the above verse, it is fair to infer that those who reject the divine message will lose eternal life. Will they die in Hell? The answer is No:

20:74 Anyone who comes to His Lord guilty will deserve Hell, wherein he never dies, nor stays alive.

Then, the alternative is obvious: total annihilation, ontological extinction, together with Hell. Those programs with free choice that chose to corrupt themselves with the worst diabolic "viruses" (such as associating partners with God or killing an innocent program) will be sentenced to an eternal punishment: after resurrection they will experience a period of diagnosis, justice, regret and then, with the creation of a new earth and heaven, they will be hurled to non-existence together with Hell. Perhaps, their memories as well will be erased from the minds of their relatives who chose eternal life by dedicating their religion to God alone and by leading a righteous life with the Day of Judgment in mind. It is interesting that the only exception for God's never forgetting is in connection with the people who deserved Hell:

32:14 So taste the consequences of your ignoring/forgetting this day; for We have now ignored/forgotten you. Taste the eternal retribution in return for what you used to do.

GOD IS JUST AND MERCIFUL

The Quran repeatedly comforts us by reminding us of God's perfect justice. The following verse provides us with a precise idea of God's justice and mercy:

10:27 For those who did good work there will be the best and more. . . As for those who earned evil, they will receive equivalent evil.

Suffering in an eternal Hell creates a contradiction between this divine justice, since eternity cannot be equal to an evil committed during the limited life span of a human. However, eternal punishment in a lifetime of a non-eternal Hell avoids such a contradiction. After receiving an equivalent punishment, the chief evildoers like those referred in verse 4:48, will be eliminated from existence. They will end up in Hell and Hell will end up in oblivion, eternally.

No wonder the first and the most repeated verse of the Quran reminds us over and over that God is Gracious, Merciful (1:1).

9
No Contradiction in the Quran

Verse 4:82 of the Quran claims that it is free of contradictions. Any internal contradiction or contradiction between the Quran and God's laws in the nature will falsify the claim. I found the following claims of contradiction posted at an evangelical site disguised as "humanist." Here are the 10 charges, followed by my answers.

QUESTION 1: What was man created from, blood, clay, dust, or nothing?

1. "Created man, out of a (mere) clot of congealed blood," (96:2).

2. "We created man from sounding clay, from mud molded into shape, (15:26).

3. "The similitude of Jesus before Allah is as that of Adam; He created him from dust, then said to him: "Be". And he was," (3:59).

4. "But does not man call to mind that We created him before out of nothing?" (19:67, Yusuf Ali). Also, 52:35).

5. "He has created man from a sperm-drop; and behold this same (man) becomes an open disputer! (16:4).

ANSWER 1: Human being was created from earthly materials and water according to a divinely guided evolution

The criticism presented above is a classic example of EITHER-OR fallacy, or the product of a frozen mind that does not consider or perceive time and evolution as reality. If he uses the same standard, the critic of these verses will find contradiction in almost every book. If he looks into biology books, he will similarly get confused. In one page he will learn that he is made of atoms, in another that he is made of cells, of DNA, sperm, egg, embryo, earthly materials, etc. He would express his disbelief and confusion with a similar question. A careful and educated reading of the Quran will reveal the following facts about creation:

1. There were times when man did not exist. Billion years after the creation of the universe humans were created. In other words, we were nothing before we were created:

> "Did the human being forget that we created him already, and he was nothing?" (19:67).

2. Humans were created according to a divinely guided evolution:

> "Have they not seen how GOD initiates the creation, and then repeats it? This is easy for GOD to do. Say, 'Roam the earth and find out the origin of life.' For GOD will thus initiate the creation in the Hereafter. GOD is Omnipotent." (29:19-20).

"He is the One who created you in stages. Do you not realize that GOD created seven universes in layers? He designed the Moon therein to be a light and placed the Sun to be a lamp And GOD germinated you from the earth like plants." (71:14-17).

3. Creation of man started from clay:

> "We created the human being from aged mud, like the potter's clay." (15:26).

Our Creator started the biological evolution of microscopic organisms within the layers of clay. Donald E. Ingber, professor at Harvard University, in an article titled "The Architecture of Life" published as the cover story of Scientific American stated the following:

> "Researchers now think biological evolution began in layers of clay, rather than in the

> primordial sea. Interestingly, clay is itself a porous network of atoms arranged geodesically within octahedral and tetrahedral forms. But because these octahedra and tetrahedra are not closely packed, they retain the ability to move and slide relative to one another. This flexibility apparently allows clay to catalyze many chemical reactions, including ones that may have produced the first molecular building blocks of organic life."

Humans are the most advanced fruits of organic life that started millions of years ago from the layers of clay.

4. Human being is made of water:

> "Do the ingrates not realize that the heaven and the earth used to be one solid mass that we exploded into existence? And from water we made all living things. Would they believe?" (21:30).

The verse above not only emphasizes the importance of water as an essential ingredient for organic life, it also clearly refers to the beginning of the universe, what we now call the Big-Bang. The Quran's information regarding cosmology is centuries ahead of its time. For instance, verse 51:47 informs us that the universe is continuously expanding. "We constructed the sky with our hands, and we will continue to expand it." Furthermore, the Quran informs us that the universe will collapse back to its origin, confirming the closed universe model: "On that day, we will fold the heaven, like the folding of a book. Just as we initiated the first creation, we will repeat it. This is our promise; we will certainly carry it out." (21:104).

> "And GOD created every living creature from water. Some of them walk on their bellies, some walk on two legs, and some walk on four. GOD creates whatever He wills. GOD is Omnipotent." (24:45).

Walking on two legs is a crucial point in evolution of humanoids. Walking on two feet may initially appear to be insignificant in evolutionary process, many scientists believe that walking on two feet made a significant contribution to human evolution by enabling the Homo Erectus to use tools and gain consciousness, thereby leading to Homo Sapiens.

5. Human being is made of dust, or earth, that contains essential elements for life:

> "The example of Jesus, as far as GOD is concerned, is the same as that of Adam; He created him from dust, then said to him, "Be," and he was." (3:59).

6. Human being is the product of long-term evolution and when human sperm and egg, which consist of water and earthly elements, meet each other in right condition, they evolve to embryo, fetus, and finally, after 266 days to a human being:

> "Was he not a drop of ejected semen?" (75:37).

> "He created the human from a tiny drop, and then he turns into an ardent opponent." (16:4).

> "He created man from an embryo." (96:2).

> " O people, if you have any doubt about resurrection, (remember that) we created you from dust, and subsequently from a tiny drop, which turns into a hanging (embryo), then it becomes a fetus that is given life or deemed lifeless. We thus clarify things for you. We settle in the wombs whatever we will for a predetermined period. We then bring you out as infants, then you reach maturity. While some of you die young, others live to the worst age, only to find out that no more

> knowledge can be attained beyond a certain limit. Also, you look at a land that is dead, then as soon as we shower it with water, it vibrates with life and grows all kinds of beautiful plants." (22:5).

As you noticed we do not translate the Arabic word *ALaQ* as clot; since neither in interspecies evolution nor intra-species evolution, there is no a stage where human is clot. This is a traditional mistranslation of the word, and the error was first noticed by medical doctor Maurice Bucaille. Any decent Arabic dictionary will give you three definitions for the word *ALaQ*: (1) clot; (2) hanging thing; (3) leach. Early commentators of the Quran, lacking the knowledge of embryology, justifiably picked the "clot" as the meaning of the word. However, the author of the Quran, by this multi-meaning word was referring to embryo, which hangs to the wall of uterus and nourishes itself like a leach. In modern times, we do not have excuse for picking the wrong meaning. This is one of the many examples of the Quran's language in verses related to science and mathematics: while its words provide a kind of understanding to former generations, its real meaning shines with knowledge of God's creation and natural laws.

QUESTION 2: Is there or is there not compulsion in religion according to the Qur'an?

1. "Let there be no compulsion in religion: Truth stands out clear from Error: whoever rejects evil and believes in Allah hath grasped the most trustworthy hand-hold, that never breaks. And Allah heareth and knoweth all things," (2:256).

2. "And an announcement from Allah and His Messenger to the people (assembled) on the day of the Great Pilgrimage,- that Allah and His Messenger dissolve (treaty) obligations with the Pagans. If then, ye repent, it were best for you; but if ye turn away, know ye that ye cannot frustrate Allah. And proclaim a grievous penalty to those who reject Faith," (9:3).

3. "But when the forbidden months are past, then fight and slay the Pagans wherever ye find them, and seize them, beleaguer them, and lie in wait for them in every stratagem (of war); but if they repent, and establish regular prayers and practice regular charity, then open the way for them: for Allah is Oft-forgiving, Most Merciful," (9:5). "Fight those who believe not in Allah nor the Last Day, nor hold that forbidden which hath been forbidden by Allah and His Messenger, nor affirm the religion of Truth, (even if they are) of the People of the Book, until they pay the Jizya with willing submission, and feel themselves subdued," (9:29).

ANSWER 2: Yes, there is no compulsion in religion according to the Quran, and Muslims are permitted to defend themselves against aggressors and murderers.

The Quran promotes freedom of opinion, religion, and expression. The critic is taking the verses from Chapter 9 out of its context and present it as a contradiction with the principle expressed in 2:256 and other verses. The Chapter 9 starts with an ultimatum Meccan *mushriks* who not only tortured, killed, and evicted muslims from their homes, they mobilized several major war campaigns against them while they established a peaceful multinational and multi-religious community. The beginning of the Chapter refers to their violation of the peace treaty and gives them an ultimatum and four months to stop aggression. Thus, the verses quoted from Chapter 9 have nothing to do with freedom of religion; it is a warning against aggressor murderer religious fanatics.

I discussed this subject extensively in my first debate and argued that Sunni Tyrants have distorted the meaning of the word JIZYA as a taxation of non-muslims, while the word means "compensation" or more accurately, "war reparation," which was levied against the aggressing party who initiated the war. My argument on Quran's position on war and peace is posted at Articles section of 19.org under the title "To the Factor of 666."

QUESTION 3: The first Muslim was Muhammed? Abraham? Jacob? Moses?

1. "And I [Muhammed] am commanded to be the first of those who bow to Allah in Islam," (39:12).

2. "When Moses came to the place appointed by Us, and his Lord addressed him, He said: "O my Lord! show (Thyself) to me, that I may look upon thee." Allah said: "By no means canst thou see Me (direct); But look upon the mount; if it abides in its place, then shalt thou see Me." When his Lord manifested His glory on the Mount, He made it as dust. And Moses fell down in a swoon. When he recovered his senses he said: "Glory be to Thee! to Thee I turn in repentance, and I am the first to believe." (7:143).

3. "And this was the legacy that Abraham left to his sons, and so did Jacob; "Oh my sons! Allah hath chosen the Faith for you; then die not except in the Faith of Islam," (2:132).

ANSWER 3: Many Prophets and messengers were the first muslims in their time and location.

If we check google.com with the following words: Olympic first place 100-meters and running, we will find many names of athletes who got the first place. If we use the logic of the critic, we would think that there is a great confusion and contradictory claims regarding the first place winner of 100-meters. What is wrong with that logic? Obviously, we need to consider the two important elements called time and space! Abraham was first muslim (submitter and promoter of peace) in his time and location. Similarly, Moses and Muhammed were too pioneer muslims of their times.

QUESTION 4: Does Allah forgive or not forgive those who worship false gods?

1 "Allah forgiveth not that partners should be set up with Him; but He forgiveth anything else, to whom He pleaseth; to set up partners with Allah is to devise a sin Most heinous indeed," (4:48 ; Also 4:116).

2 "The people of the Book ask thee to cause a book to descend to them from heaven: Indeed they asked Moses for an even greater (miracle), for they said: "Show us Allah in public," but they were dazed for their presumption, with thunder and lightning. Yet they worshipped the calf even after clear signs had come to them; even so we forgave them; and gave Moses manifest proofs of authority," (4:153).

ANSWER 4: God does not forgive those who associate other powers or gods to Him, if they do not repent on time.

The Quran contains numerous verses regarding idol-worshipers or *mushriks* accepting the message of islam.

> "He is the One who accepts the repentance from His servants, and remits the sins. He is fully aware of everything you do." (42:25).

Most of the supporters of and companion of messengers and prophets were associating partners to God before they repented and accepted the message. For instance, the Quran informs us that even Muhammed was an idol-worshiper before he received revelation, and obviously, after his affirmation of the truth he repented from his ignorance and God forgave him:

> "Say, 'I have been enjoined from worshiping the idols you worship beside GOD, when the clear revelations came to me from my Lord. I was commanded to submit to the Lord of the universe.'" (40:66).

> "Thus, we revealed to you a revelation proclaiming our commandments. You had no idea about the scripture, or

faith. Yet, we made this a beacon to guide whomever we choose from among our servants. Surely, you guide in a straight path." (42:52).

"He found you astray, and guided you." (93:7).

"Whereby GOD forgives your past sins, as well as future sins, and perfects His blessings upon you, and guides you in a straight path." (48:2).

QUESTION 5: Are Allah's decrees changed or not?

1. "Rejected were the messengers before thee: with patience and constancy they bore their rejection and their wrongs, until Our aid did reach them: there is none that can alter the words (and decrees) of Allah. Already hast thou received some account of those messengers," (6:34).

2. "The word of thy Lord doth find its fulfillment in truth and in justice: None can change His words: for He is the one who heareth and knoweth all, (6:115).

3. "None of Our revelations do We abrogate or cause to be forgotten, but We substitute something better or similar: Knowest thou not that Allah Hath power over all things?" (2:106).

4. "When We substitute one revelation for another,- and Allah knows best what He reveals (in stages),- they say, "Thou art but a forger": but most of them understand not," (16:101)

ANSWER 5: God's decrees do not change.

This is a valid criticism against those who do not follow the Quran alone, since because of fabricated *hadith*s, they have distorted the meaning of 2:106 and 16:101. What Quran describes as does not change are *Sunnatullah* (God's law) and *Kalimatullah* (God's word) as in 6:34 and 6:115. Verses 2:106 and 16:101 contain neither of these words; they are about God's AYAT (Sign; miracle), which were given to prophets and messengers. The translation the critic is quoting contains a translation error with grave theological ramifications.

According to the official faith of Hislam, some verses of the Quran abrogate other verses and even some *hadith* have abrogated some verses. This outrageous lie has been supported by distorting the meaning of this verse. The Quran has a peculiar language. The singular word "Ayah" occurs 84 times in the Quran and nowhere is it used for the verses of the Quran; rather it is always used to mean "sign, evidence, or miracle." However, the plural form of this word, "Ayat" besides being used to mean its singular form, additionally it is used for the verses of the Quran. The fact that a verse of the Quran does not demonstrate the miraculous characteristics of the Quran supports this peculiar usage or vice versa. For instance, there are short verses that comprised of only one or two words, and they were most likely frequently used in daily conversation, letters and poetry. For example, see: 55:3; 69:1; 74;4; 75:8; 80:28; 81:26. Furthermore, we are informed that the minimum unit that demonstrates Quran's miraculous nature is a chapter (10:38) and the shortest chapter consists of 3 verses (103; 108, 110). The first verse of the Quran, commonly known as Basmalah, cannot be a miracle on its own, but it gains a miraculous nature with its numerical network with other letters, words, verses and chapters of the Quran. By not using the singular form "Ayah" for the verses of the Quran, God made it possible to distinguish the miracles shown in the text and prophecies of the scripture from the miracles shown in nature. See 4:82 for further evidence that the Quranic verses do not abrogate each other.

QUESTION 6: Was Pharaoh killed or not killed by drowning?

1. "We took the Children of Israel across the sea: Pharaoh and his hosts followed them in insolence and spite. At length, when overwhelmed with the flood, he said: 'I believe that there is no god except Him Whom the Children of Israel believe in: I am of those who submit (to Allah in

Islam).' (It was said to him): 'Ah now!- But a little while before, wast thou in rebellion!- and thou didst mischief (and violence)! This day shall We save thee in the body, that thou mayest be a sign to those who come after thee!' But verily, many among mankind are heedless of Our Signs!" (10:90-92).

2. "Moses said, 'Thou knowest well that these things have been sent down by none but the Lord of the heavens and the earth as eye-opening evidence: and I consider thee indeed, O Pharaoh, to be one doomed to destruction!' So he resolved to remove them from the face of the earth: but We did drown him and all who were with him," (17:102-103).

ANSWER 6: Pharaoh was killed by drowning and his body was saved via mummification.

Verse 10:92 does not say that God will keep Pharaoh alive; it informs us that God will preserve His body after he was drowned.

QUESTION 7 Is wine consumption good or bad?

1. "O ye who believe! Intoxicants and gambling, (dedication of) stones, and (divination by) arrows, are an abomination,- of Satan's handwork: eschew such (abomination), that ye may prosper," (5:90).

2. "(Here is) a Parable of the Garden which the righteous are promised: in it are rivers of water incorruptible; rivers of milk of which the taste never changes; rivers of wine, a joy to those who drink; and rivers of honey pure and clear. In it there are for them all kinds of fruits; and Grace from their Lord. (Can those in such Bliss) be compared to such as shall dwell forever in the Fire, and be given, to drink, boiling water, so that it cuts up their bowels (to pieces)?" (47:15).

3. "Truly the Righteous will be in Bliss: On Thrones (of Dignity) will they command a sight (of all things): Thou wilt recognize in their faces the beaming brightness of Bliss. Their thirst will be slaked with Pure Wine sealed," (83:22-25).

ANSWER 7: Consumption of wine is bad in this world.

The Quran strongly rebukes the consumption of intoxicants for affirmers. This is not an enforced legal prohibition; but left individuals to decide. The reason for this prohibition is obvious: intoxicants, though may provide some social or psychological benefits to the consumer, they impair their judgment and intelligence and causes too many problems for both the individual and the society. The Quran prohibits intoxicants to individuals because of various reasons such as: moral (the designer and creator of your body and mind asks you not to intentionally harm the body lent to you for a lifetime), intellectual (the greatest gift you have is your brain and its power to make good judgment, do not choose to be stupid or stupider than you are already!") and pragmatic (you and your society will suffer grave loss of health, wealth, happiness, and many lives; do not contribute to the production and acceleration of such a destructive boomerang).

This said, let me suggest a correction. The verses 83:22-25 does not mention wine; thus, the translation is erroneous. The only verse that uses intoxicants (KHAMR) in a positive context is 47:15, and interestingly it is about paradise, that is, hereafter. A quick reflection on the reason for prohibition of intoxicants will explain the apparent contradiction. The harm of intoxicants, such as drunk driving, domestic violence or alcoholism, is not an issue in other universe, where the laws and rules are different. In other words, a person who is rewarded by eternal paradise will not hurt himself or herself or anyone else by getting intoxicated (See 7:43; 15:47; 21:102; 41:31; 43:71; 2:112; 5:69).

QUESTION 8: Has the Quran been abrogated?

No, the Quran is perfect and can never be

abrogated. Yes, some verses have been abrogated.

"There is none to alter the decisions of Allah." (6:34).

"Perfected is the Word of thy Lord in truth and justice. There is naught that can change His words." (6:115).

"There is no changing the Words of Allah." (10:64).

"And recite that which hath been revealed unto thee of the Scripture of thy Lord. There is none who can change His words." (18:27).

Yes, some verses have been abrogated.

"And when We put a revelation in place of (another) revelation, - and Allah knoweth best what He revealeth - they say: Lo! thou art but inventing. Most of them know not." (16:101).

"Nothing of our revelation (even a single verse) do we abrogate or cause be forgotten, but we bring (in place) one better or the like thereof." (2:106).

ANSWER 8: No, there is no abrogation in the Quran.

This question received its answer when I answered the Question 5, above.

QUESTION 9: Who chooses the devils to be the friends of ingrates?

Allah:

"We have made the devils protecting friends for those who believe not." (7:27).

The ingrates:

"A party hath He led aright, while error hath just hold over (another) party, for lo! they choose the devils for protecting supporters instead of Allah and deem that they are rightly guided." (7:30).

ANSWER 9: Ingrates choose evil and devils in accordance with God's law established to test us on this planet.

While the Quran states that every event happens in accordance with God's design and permission (8:17; 57:22-25), the Quran also informs us regarding our freedom to choose our path (6:110; 13:11; 18:29 42:13,48; 46:15).

QUESTION 10: Will all Jews and Christians go to hell?

Yes, all Christians will go to hell.

"Whoso seeketh as religion other than the Islam it will not be accepted from him, and he will be a loser in the Hereafter." (3:85).

"They surely disbelieve who say: Lo! Allah is the Messiah, son of Mary. ... Lo! whoso ascribeth partners unto Allah, for him Allah hath forbidden paradise. His abode is the Fire. For evil-doers there will be no helpers." (5:72).

No, some will not.

"Those who are Jews, and Christians, and Sabaeans - whoever believeth in Allah and the Last Day and doeth right - surely their reward is with their Lord, and there shall no fear come upon them neither shall they grieve. (2:62).

"Lo! those who believe, and those who are Jews, and Sabaeans, and Christians - Whosoever believeth in Allah and the Last Day and doeth right - there shall no fear come upon them neither shall they grieve." (5:69).

ANSWER 10: Some Jews and Christians will go to hell.

First, the Quran calls the followers of Jesus with the word Nazarenes, rather than Christians. Second, the word Sabaean is not a proper name referring to a particular religion, rather than it is a verb meaning "those who are from other religions."

The critic assumes that peacefully surrendering to God is only possible if someone utters an Arabic magic word. Islam is not a proper name, neither did it start with Muhammed, nor it ended with Muhammed. Any person, regardless of the name of their religion, who dedicate himself to God alone, believe in the day of judgment and live a righteous life, that person is considered muslim (who surrender themselves to God alone and promote peace). There are many people among Christians and Jews who fit this description.

10
"How Can we Observe the *Sala* Prayers by Following the Quran Alone?"

"How can we observe the *Sala* prayers by following the Quran alone?" is a favorite question among the adherents of Sunni and Shiite sects who follow derivative texts, religious instruction, teachings, and laws, all of which were authored by men. With this question, both sects try to justify the necessity and proliferation of contradictory sectarian teachings, medieval Arab culture, oppressive laws filled with numerous prohibitions and regulations--all falsely attributed to God and His prophet. Perhaps, the best answer for those who direct that question is the following:

> If you are expecting the intercession of Muhammed and many other saints, if you are associating your religious leaders as partners to God in authoring your religion, justifying authoritarian regimes, violating women's rights and putting them in black sacks, showing no tolerance for the expression of opposing ideas and cultures, justifying the punishment of stoning to death for adulterers, adhering to numerous superstitions, entering the restroom with left foot, forcing your child to eat with right hand, prohibiting music and visual arts….. In short, if you are condemning yourself and your society to a backward and miserable life just because you learn from those sectarian sources how to bend your belly or where to put your hands, then it is better for you not to pray at all. Such a prayer (more accurately, physical exercise) is not worth exchanging monotheism with polytheism, reason with ignorance, light with darkness, peace with conflicts, universalism with tribalism, progress with regression.

But this answer may not be able to wake them up and save them from the cobwebs of clergymen. They will still challenge us to produce a manual for prayer. How and how many times to bow, where and how to put their hands, what to do with their fingers, where and how to stand on their feet, etc. For many adherents of Sunni or Shiite sects, the question is just an excuse to avoid the pain of reforming themselves. They know that the Quran does not provide extensive explanation on every itsy-bitsy spider crawling in their minds. So, the question is usually asked in a rhetorical sense: "You cannot find in the Quran how many times to bow and prostrate, or how to hold my index or pinky finger in *Sala* prayer, so I will continue following all those volumes of contradictory books filled with silly stories and outrageous instructions!"

As a result, Monotheism is redefined as a Limited Partnership, in which the recognition and submission to God alone becomes an oxymoron; a contradiction in terms in which other 'partners' are submitted to and accepted by these 'affirmers.' The most common set-up for Sunni *shirk* is: the Quran (God) + *hadith*s and *sunna* (messenger) + the practice of the Prophet's companions + the practice of the companions of the Prophet's companions + the opinions of emams (ijtihad) + consensus of *ulama* in a particular sect (*ijma'*) + the comments and opinions of their students + the comments and opinions of early *ulama* + the comments and opinions of later *ulama* + the fatwas of living *ulama.*

In Shiite version of *shirk*, in addition to the aforementioned partners, the 12 Infallible Emams (all relatives and descendants of the Prophet Muhammed starting with Ali) and the living substitute emam is added to the board of directors of the Holy Limited Partnership. The Quran is usually considered an ambiguous book and is basically distorted and abused for their justification of this *shirk*, that is, setting partners with God.

Islam, which means Peacemaking and Submission to God in peace, is the state of mind of all people who submit to God alone. All messengers, Noah, Abraham, Moses, Solomon, Jesus, and all previous monotheists were Peacemakers and Submitters (2:131; 5:111; 7:126; 10:72,84; 22:78; 27:31,42,91; 28:53; 72:14). Thus, the only religion approved by God is Submission to God (3:19). It is God Almighty who uses this attribute to describe those who submit to His law (22:78). Islam is referred to as the "Religion of Abraham" in many verses since Meccan idol worshipers were claiming that they were following their father Abraham (2:130,135; 3:95; 4:125; 6:161; 12:37-38; 16:123; 21:73; 22:78). And Muhammed was a follower of Abraham (16:123).

Ignorant of the fact that Abraham observed the contact prayers (21:73), many contemporary muslims challenge God by asking where we can find the number of units in each contact prayer. Ignorant of the fact that God claimed Quran to be complete (6:11-116), they do not "see" that ALL religious practices of Submission/Monotheism were established and practiced before the Quranic revelation (8:35; 9:54; 16:123; 21:73; 22:27; 28:27). Messengers after Abraham practiced *Sala* prayers, obligatory charity, and fasting (2:43; 3:43; 11:87; 19:31,59; 20:14; 28:27; 31:17).

The Meccan *mushriks* used to believe that they were followers of Abraham. They were not worshipping statutes or icons as claimed by *hadith* fabricators, but they were praying for *shafaat* (intercession) from some holy names, such as al-Lat, al-Uzza, and al-Manaat (53:19-23). So, contrary to their false assertion of being monotheists (6:23), they were accused of being *mushrik* or associating partners with God (39:3).

Sunni and Shiite scholars subsequently fabricated stories in an attempt to erase any similarities between themselves and the *mushriks*, but in doing so exposed their own inherent lies in contradictory descriptions of those statutes (For instance, see Al-Kalbi's classic book on statutes: Kitab-ul Asnam). Meccan *mushriks* who were proud of Abraham's legend could not practice the literal observance of idol-worship; they settled for a more metaphysical satanic trap by accepting intercession and man-made religious prohibitions (6:145-150; 39:3). They were metaphysical or spiritual idol worshipers.

Meccan *Mushriks*, during the era of Prophet Muhammed were respecting the Sacred *Masjid* built by Abraham (9:19). They were practicing the contact prayers, fasting, and Debate Conference (2:183,99; 8:35–the meaning of this verse is deliberately distorted in traditional translations–; 9:54; 107:4-6). Although they knew *Zaka* (obligatory charity) they were not fulfilling their obligation (53:34). During the time of the Prophet Muhammed people knew the meaning of *Sala*, *Zaka, Sawm*, and *Hajj*. They were not foreign words.

God sent the Quran in their language. As with each proceeding Book, the revelation was given in its time, in the language of the people receiving the revelation. God commands and reveals in a manner which can be both understood and observed, and then He details the requirements for his people throughout His Book (16:103; 26:195). Moreover, if God wants to add a new meaning to a known word, He informs us. For instance, the Arabic word *al-din* in 1:4 is explained in 82:15-19.

Verse 16:123 is a direct proof that all religious practices in Islam were intact when Muhammed was born. Thus, he was ordered to "follow the religion of Abraham." If I ask you to ride a bicycle, it is assumed that you know what a bicycle is, and you know or would learn from others how to ride it. Similarly, when God enjoined Muhammed to follow the practices of Abraham (16:123), such practices must have been well known.

Nevertheless, contrary to the popular belief, the Quran details contact prayers. While neither Quran nor *hadith* books contain illustration for *Sala* or video clips showing how the prophets observed their *Sala*, Quran does describe prayer. The Quranic description of *Sala* prayer is much more superior for the following reasons:

- The language of the Quran is superior to the language of hadiths. Hadiths are collections of narration containing numerous different dialects and are afflicted with chronic and endemic linguistic problems. The language of the Quran is much simpler as witnessed by those who study both the Quran and hadith. The eloquence of the Quranic language is emphasized in the Quran with a repeated rhetorical question (54:17, 22, 32, 40).

- Hadith books may contain more details. But are those details helpful and consistent with the Quran? How does a believer decide between conflicting details? Does he just pick the word of his favorite Imam? If we follow the words of a particular, favorite imam, does that mean that we are really following the practice of the prophet? For instance, you may find dozens of hadiths in Sahih Muslim narrating that the Prophet Muhammed read the first chapter of the Quran, al-Fatiha, and bowed down, without reading any additional verses from the Quran. You will find many other hadiths claiming that the Prophet read this or that chapter after al-Fatiha. There are also many conflicting hadiths regarding ablution which is the source of different rituals among sects. Hadith, more or less of them compounding God's Word with contradictory details, cannot guide to the truth. It has become a necessary evil for ignorant affirmers and community leaders who manipulate them.

- Hadith books narrate a silly story regarding the times of Sala prayer and its ordinance. The story of Mirage is one of the longest hadiths in the Bukhari. Reportedly, after getting frequent advice from Moses by going up and down between the sixth and seventh heaven, Muhammed negotiated with God to reduce the number of prayers from 50 times a day (one prayer for every 28 minutes) to 5 times a day. This hadith portrays Muhammed as a compassionate union leader saving his people from God's unmerciful and impossible demand.

Sala Prayer According to the Quran

- Observing *Sala* prayer is frequently mentioned together with giving charity and thus emphasizing the social consciousness and communal responsibility of those who observe the prayer (2:43,83,110; 4:77; 22:78; 107:1-7).
- *Sala* prayer is observed to commemorate and remember God alone (6:162; 20:14).
- *Sala* remembrance protects us from sins and harming others (29:45).
- *Sala* prayer should be observed continuously until death (19:31; 70:23,34).
- *Sala* is for God's remembrance (20:14).
- *Sala* is conducted regularly at three times each day (24:58, 11:114, 2:238, 17:78).
- *Sala* is for men and women (9:71).
- *Sala* requires a sober state of mind (4:43).
- *Sala* requires cleanliness (5:6).
- *Sala* is done while facing one unifying point (2:144).

- *Sala* is done in a stationary standing position (2:238-239).
- *Sala* involves the oath being made to God (1:1-7) followed by the Scripture being recited (29:45).
- *Sala* requires a moderate voice (17:110).
- *Sala* involves kneeling and prostration (3:43; 4:102).
- *Sala* is ended by prostrating and saying specific words (4:102, 17:111).
- *Sala* can be shortened in case of war (4:101).
- *Sala* can be done on the move or sitting in case of worry (2:239).
- *Sala* is to be prescribed to your own family (20:133).
- *Sala* is also performed with congregation (62: 9-10).

Ablution

To observe prayer, one must make ablution (4:43; 5:6). Ablution is nullified only by sexual intercourse or passing urine or defecation. Ablution remains valid even if one has passed gas, shaking hands with the opposite sex, or a woman is menstruating. A menstruating woman may observe contact prayers, contrary to superstitious cultural beliefs (5:6; 2:222; 6:114-115).

Dress Code

There is not a particular dress code for prayer, in fact, if you wish you may pray nude in your privacy. Covering our bodies is a social and cultural necessity aimed to protect ourselves from harassment, misunderstanding and undesired consequences (7:26; 24:31; 33:59).

Times for Prayer

Quran mentions three periods of time in conjunction with *Sala* prayer. In other words, the Quran qualifies the word "*Sala*" by three different temporal words: (1) *Sala*-al Fajr (Morning Prayer), (2) *Sala*-al Esha (Evening Prayer), (3) *Sala*-al Wusta (Middle Prayer). The Morning Prayer (24:58) and Night Prayer (24:58) should be observed at both ends of the day and part of the night (11:114). (We will discuss the times of *Sala* prayers later in detail at the end of this article).

Direction for Prayer

For the prayer one should face the Sacred Masjid built by Abraham, the Ka'ba (2:125, 143-150; 22:26). To find the correct qibla a person should keep in mind that the world is a globe, far different from Mercator's flat map. Since the prayer during emergency and fear is reduced to one unit, in normal conditions the prayer should be at least two units and during the prayer one must dramatically reduce his/her contact with the external world (4:101-103). Prayers, unlike fasting, cannot be performed later after they are missed; they must be observed on time (4:103).

Congregational Prayer

Affirmers, men and women, once a week are invited to a particular location to pray together every Juma (Congregational) Day. They go back to their work and normal daily schedule after the Congregational Prayer which could be led by either man or a woman (62:9-11). The mosques or masjids should be dedicated to God alone, thus, the invitation should be restricted to worship God alone, and no other name should be inscribed on the walls of masjids and none other than God should be commemorated (72:18-20). Those who go to masjids should dress nicely since masjids are for public worship and meetings (7:31).

Position for Prayer

One should start the *Sala* prayer in standing position (2:238; 3:39; 4:102) and should not change his/her place except during unusual circumstances, such as while riding or driving (2:239). Submission to God should be declared physically and symbolically by first bowing down and prostrating (4:102; 22:26; 38:24; 48:29). This physical ritual is not required at the times of emergencies, fear, and unusual circumstances (2:239).

Comprehension, Purpose of Prayer

We must comprehend the meaning of our prayers, as these are the moments in which we communicate directly with God (4:43). We must be reverent during our contact prayers (23:2). Along with understanding what we say, we can recall one of God's attributes, depending on our need and condition during the time of our prayer (17:111). Prayer is to commemorate God, and God Alone (6:162; 20:14; 29:45). Prayer is to praise, exalt and remember His greatness, His Mercy and ultimately our dependence on each of these attributes (1:1-7; 20:14; 17:111; 2:45). So that even mentioning other names besides God's contradicts our love and dependence on Him (72:18; 29:45).

Recitation during the *Sala* Prayer

Preferring the Quran for recitation has practical benefits since affirmers from all around the world can pray together without arguing on which language to choose or which translation to use. The chapter al-Fatiha (The Opening) is the only chapter which addresses God in its entirety and is an appropriate prayer for *Sala*. For non-Arabs it should not be too difficult to learn the meaning of words in al-Fatiha since it consists of seven short verses. Those who are unable to learn the meaning of al-Fatiha should pray in the language that he or she understands. I see no practical reason for reciting in Arabic during individually observed prayers.

We should recite *Sala* prayers in moderate tone, and we should neither try to hide our prayers nor try to pray it in public for political or religious demonstration (17:111). If it is observed with congregation, we should listen to the recitation of the men or women who leads the prayer (7:204; 17:111). After completing the *Sala* prayer, we should continue remembering God (4:103).

Units of Prayer

The Quran does not specify any number of units for prayers. It leaves it to our discretion. We may deduce some ideas regarding the length of the prayers from verses 4:101 and 102. Verse 101 allows us to shorten our prayers because of the fear of being ambushed by enemy during wars. The following verse explains how to pray with turn; it mentions only one prostration, thus implying one unit. If shortening the prayers is considered as reducing the number of their units, then one may infer that prayers at normal times should consist at least of two units. The units of the Congregational Prayer being 2 is revealing since it is more likely to be accurately preserved. Again, the units of prayer are not fixed by the Quran; it is up to the individual and groups.

Funeral Prayer

There is no funeral *Sala* prayer. However, remembering those who died as monotheists and providing community support for their relatives is a civic duty.

Sectarian Innovations

There are many sectarian innovations that differ from sect to sect. Some of the innovations are: combining the times of prayers, performing the prayers omitted at their proper times, shortening the prayers during normal trips, adding extra prayers such as "*sunna*" and "nawafil," innovating a paid cleric occupation to lead the prayers, prohibiting women from leading the prayers, while sitting reciting a prayer "at-Tahiyat" which addresses to the Prophet Muhammed as he is alive and omnipresent, adding Muhammed's name to the Shahada, reciting zamm-us Sura (extra chapters) after the al-Fatiha, indulging in sectarian arguments on details of how to hold your hands and fingers, washing mouth and nose as elements of taking ablution, brushing the teeth with "misvak" (a dry branch of a three beaten into fibers at one end as a toothbrush) just before starting the prayers, wearing turbans or scarves to receive more credits...

HOW MANY PRAYERS A DAY?

Only three Contact Prayers are mentioned by name in the Quran. In other words, the word "*Sala*" is qualified with descriptive words in three instances. These are:

- *Sala*-al Fajr-DAWN PRAYER (24:58; 11:114).
- *Sala*-al Esha-EVENING PRAYER (24:58; 17:78; 11:114)
- *Sala*-al Wusta- MIDDLE PRAYER (2:238; 17:78)

All of the verses that define the times of the prayers are attributable to one of these three prayers. Now let's see the related verses:

DAWN & EVENING PRAYERS by their names:

> ". . . This is to be done in three instances: before the DAWN PRAYER, at noon when you change your clothes to rest, and after the EVENING PRAYER. . ." (24:58).

For other usage of the word "esha" (evening) see: 12:16; 79:46

The times of DAWN & EVENING PRAYERS defined:

> "You shall observe the contact prayers at both ends of the daylight, that is, during the adjacent hours of the night. . . " (11:114)

Traditional translators and commentators consider the last clause "zulfan minal layl" of this verse as a separate prayer indicating to the "night" prayer. However, we consider that clause not as an addition but as an explanation of the previous ambiguous clause; it explains the temporal direction of the ends of the day. The limits of "Nahar" (daylight) are marked by two distinct points: sunrise and sunset. In other words, two prayers should be observed not just after sunrise and before sunset, but before sunrise and after sunset.

Furthermore, the traditional understanding runs into the problem of contradicting the practice of the very tradition it intends to promote. Traditionally, both morning and evening prayers are observed in a time period that Quranically is considered "LaYL" (night) since Layl starts from sunset and ends at sunrise. The word "Layl" in Arabic is more comprehensive than the word "night" used in English.

If the expression "tarafayin nahar" (both ends of the day) refers to morning and evening prayers which are part of "Layl" (night), then, the last clause cannot be describing another prayer time.

The time of NOON and EVENING PRAYER defined.

> "You shall observe the contact prayer when the Sun goes down until the darkness of the night. You shall also observe the Quran at dawn. Reading the Quran at dawn is witnessed." (17:78).

The decline of the Sun can be understood either its decline from the zenith marking the start of the Noon prayer or its decline behind the horizon marking the start of the Evening prayer. There are two opposing theories regarding the purpose of the usage of "duluk" (rub) in the verse; nevertheless, either understanding will not contradict the idea of 3 times a day since both Noon and Evening prayers are accepted.

MIDDLE PRAYER (Noon)

> "You shall consistently observe the contact prayers, especially the MIDDLE PRAYER, and devote yourselves totally to God." (2:238).

verse 38:32 implies that the evening prayer is from dusk until the full darkness of the night. We can easily understand the MIDDLE prayer as a prayer between the two other prayers mentioned by name (Dawn and Evening).

The Old Testament has at least three verses referring to Contact Prayers (*Sala*) and they confirm this understanding. Though we may not trust the Biblical translations verbatim, we may not consider them as errors since both internal and external consistency of the Biblical passages regarding the Contact Prayers are striking.

> "And as soon as the lad was gone, David arose out of a place toward the south, and fell on his face to the ground, and bowed himself three times: and they kissed one another, and wept one with another, until David exceeded." (1 Samuel 20:41)

> "As for me, I will call upon God; and the Lord shall save me. Evening, and morning, and at noon, will I pray, and cry aloud: and he shall hear my voice." (Psalms 55:16-17) (PS: crying aloud apparently means praying with passion).

> "Now when Daniel knew that the writing was signed, he went into his house; and his windows being open in his chamber toward Jerusalem, he kneeled upon his knees three times a day, and prayed, and gave thanks before his God, as he did aforetime." (Daniel 6:10)

The followers of Shiite sect observe 5 prayers in 3 times: morning, noon and evening. This strange practice perhaps was the result of a historical compromise with the dominant Sunni 5-times-a-day practice.

ADDENDUM (2009)

Though the 3 times of prayer is my current understanding, I do not find fault in praying five times a day. Since we have decided to devote ourselves to God alone and uphold the Quran alone as a sufficient authority for guidance, we have exposed and rejected numerous false teachings. During this process of purification, we have been passing through internal debate, both individually and as groups. The debate on Sala, prayer has been one of the longest lasting and more contentious. Sometimes, we tend to ignore our agreement on so many issues yet focus on our differences. I have received accusations of being a "divider" from some of the followers of the Quran alone and engaged in numerous debates on this issue.

With bigotry and expectation that everyone must understand everything exactly like we do has been one of the main causes of division. Furthermore, the great majority of those who follow the Quran have so much in common... But the dividers focus on the marginal ones and focus on the few issues that are still being discussed: they scream division! Their obsession with the military-style lockstep and uniform walk is perhaps the very cause of the division they complain about...

To reject numerous verses informing us that the Quran is easy to understand, fully detailed, complete, and sufficient for guidance; and to add volumes of fabricated religious rules made up by Ibn Filan and Abu Falan, Sunni and Shiite mushriks ask us: "If the Quran is complete, fully detailed and clear as God says, can you tell us how we can pray by using Quran alone?"

They are so confident as if their hadith books contain video clips of their idol, Muhammad while praying. Furthermore, their prayer is not based on description of hadiths either, since hadith books contain numerous contradictory descriptions of prayer, and the number of *rak'as* (units) which they are so obsessed with.

Ironically, the information regarding the details of prayer is not their only problem. They have bigger problem, much bigger. Without their fabricated *hadiths, sunna* and *sharia,* they cannot eat, they cannot sleep, and they cannot even go to bathroom!

We are told many things ranging from, which order you should clip your fingernails in, how long you should grow your beard and moustache, on which side you should sleep, to which foot you should use to first step into the bathroom, and how many rocks you should use to clean your buttock. You will find long chapters in hadith books dedicated to details of potty training. (9:31; 42:21).

If we make different understandings/ practices of a detail that is not even mentioned in the Quran (such as *rakaas*, that is units of the *sala* prayers, or the position of our hands during standing in *sala*), as the cause of hostility and division among ourselves then we are doomed to be divided into many sects and sub-sects. Interestingly, we are following a book that contains verse 2:62 among others.

May God guide us, teach us, and help us to show the courage and wisdom to accept the truth and correct ourselves. 20.114.

Bacca, Mecca, Jerusalem, Qibla

EDIP: If after Muhammed there was such a major geographical distortion there would be great division over this OBVIOUS change. Thus, the conspiracy theory is not realistic.

LAYTH: I may be being repetitive on this issue, but the last time I checked the written history it said there was a "civil war" that broke out in which the Muslims slaughtered one another shortly after the prophet died...The same history also states that the Kaaba in Mecca was attacked and destroyed by the Muslim Khalifa (Abdul-Malik Bin Marwan) who brought it to the ground and, at the same time, enhanced the structure of the Dome at Jerusalem (not an act any pious Muslim leader would do if the Kaaba were indeed the undisputed Regulated Temple).

Now, add to all this the fact that the 1st Qibla was Jerusalem, that Jesus preached in Jerusalem, that the Jews only know Jerusalem, that Mohammed allegedly made the night journey from Jerusalem to heaven, that Mohammed struggled to fight the Romans who were based north and controlled Jerusalem (a mighty and powerful force) and insisted to try again until he won, that old mosques are centered towards Jerusalem, that Bakka is the name of Jerusalem, that the fig and the olive are features of Jerusalem, that the Maqam of Abraham lies at the heart of the temple (while in Mecca it is outside the temple), that Bani Israel take the Regulated Temple as part of the prophecy of sura 17 when they are all gathered in one land....The list goes on and on, but I think the facts are overwhelming beyond any 'conspiracy'.

EDIP: Interesting and plausible points, yet I need to do a thorough research on this issue. I discussed the issue with Dan Gibson

11
The Blind Watch-Watchers or Smell the Cheese

An Intelligent and Delicious Argument for Intelligent Design in Evolution

(From upcoming, *19 Questions For Atheists*, by Edip Yuksel)

Let's do it backwards. I will start with quoting a sample of reactions I received from people, mostly my close friends, to the draft version of this article. I do not hope they would influence you like those "two-thumbs-up" movie reviews, but I hope that they will confuse you regarding the merits of this article before engaging you in a philosophical and scientific argument. The mixed reaction I received so far taught me this: a great many of my readers will close their eyes and touch the tail, the trunk or the ear of this elephantine article and they will perceive it as they feel. I wrote this article for the lucky few who will not get distracted by its musings or the side arguments; they will see both the watch and the watch-maker as clearly as they see these letters. Here is a sample from that feedback:

> "Very nice and heavily scientific and philosophical as well. You are using simple logic to explain a complex topic and this is a great art." (Ali Bahzadnia, MD., my endocrinologist friend, USA).
>
> "I loved the cheese!" (Mark Sykes, PhD, J.D., my rocket-scientist lawyer friend, Tucson, Mars and Beyond).
>
> "Interesting and thought-provoking." (Megan C. PhD, Biochemist, USA, not my friend)
>
> "Your arguments are against the existence of man and all living, reproducing organisms. Unless we are only God's nightmare without corporal existence, your arguments are foolish. You may want to return to restudy the very simple tenets of evolution. You have a better mind than this paper suggests. Arguing against evolution is not the problem. Your "straw man" argument is... Try again with a little more scholarship... (David Jones, PhD., my psychologist/educator atheist friend, USA).
>
> "I read the article tonight and enjoyed the article very much... The overall feel of the article for me was that it was a different look at the anthropic principle; and in many ways a restatement of it..." (Oben Candemir, MD., my ophthalmologist friend, Australia).
>
> "Very ... " (Kristen Lorenz, OD., my physicist friend, USA, who is still reading it).
>
> "Irrelevant B.S.! Bachelor of Science in philosophy is not the right muscle to dissect or rummage the messy details of fossils, genes, enzymes, and hormones. When lawyers enter a scientific debate, it is time to write its obituary. Irrelevant B.S.! Jurisdiction denied!" (XYZ PhD, my critic from ABC; or my "The Demon-Haunted" skeptic personality).
>
> "This is not a scientific paper. Because many assertions are flat wrong. Evolution IS falsifiable, for example Static fossil records would falsify it or finding a way that would prevent mutation from accumulating. Marvels of Marble is an extremely bad example. Property

of two marbles together is not much different than one, survival of the fittest does not play any part, throwing the marbles down terminates in a finite event of a short period of time. I kind of agree with XYZ." (Fereydoun Taslimi, entrepreneur and philanthropist, a monotheist friend, indeed a good friend, USA)

"I thank and congratulate Edip for taking on." (Mustafa Akyol, a columnist friend expert on evolution versus creation debate, Turkey).

When my older son turned teenager, like others in his age group, his voice and face started mutating. I complimented his evolution from childhood to puberty by jokingly depicting it as devolution. "Yahya, when will you be going to get the kiss that will turn you back into a prince?" He knew well that I was not expecting him to get a kiss from a sweetheart until he graduated from college. Though he did not get that kiss (as far as I know), within a couple of years he started turning into a prince, again.

Please do not spoil your reading of this delicious article by telling yourself, "This guy does not know even the meaning of the words *mutation* and *evolution* in the context of the evolution versus creation." I do not wish to sound arrogant, but I do know this and even more. Though I studied philosophy and received my doctorate degree in law, I took a graduate course entitled "Philosophy of Evolution" just for the fun of it. I am also one of the first people who tried to get some legal inspiration from biology. In the mid 1990's, I wrote articles with bizarre titles, such as, "Biology and Law" or "Biology of Human Rights." (Since they did not possess the characteristics of a "serious" article, such as numerous references, boring language, and lengthy exposition, they were not material for a scholarly journal. Thus, I published them at my personal website: www.yuksel.org). Furthermore, I have read numerous boring and exciting books and articles on this subject matter.

So, I decided to write an essay for the laymen who know that they know very little about the scientific aspect of the debate, yet they feel that they must take sides on this highly controversial issue that has enormous political and theological ramifications. (As for those laymen who do not know that they know very little, even Socrates could not be of any help.) No wonder we see many of those who have no clue about the intricacy of the debate appear to be ready to abort each other on this issue. To them "irreducible complexity" may sound complex, and the sudden appearance of complex life forms in the event called Cambrian Explosion may mean less than Noah's Flood or last year's Emmy's Awards. The argument of one party might be primarily based on the "God of gaps" and of the opposing party on "anything but God." They may not even know more than one or two names besides Darwin. For instance, Empedocles, Cicero, Ibn Sina, Ibn Rushd, Ibn Khaldun, Al-Hazen, Hume, Paley, Mendel, Huxley, Johnson, Dawkins, Gould, Behe, or Dembsky may not spark any ideas in the minds of those who are well versed about fictional characters such as Samson who killed a thousand men with the jaw of an ass and collected foreskins of his enemies as his wedding present. Similarly, those names may not mean much for those who are well versed about fictional characters such as Hamlet who talked with an archaic British accent starting with "Methinks..."

This essay is aimed to reduce the complexity of the debate on the most sensitive point of the controversy. I hope that this will bring the opposing parties in the controversy closer to each other. As the most delicious part of a sandwich is usually its middle, I argue that the truth of this matter is also somewhere in the middle. It is time to start a revolution in the evolution debate and smell the cheese inside the buns.

Let me remind the reader what this article is NOT about. This article is NOT rejecting the theory of evolution; on the contrary it supports the theory of evolution, and its position will not be affected a bit even if we accept that humans are descendants of chimpanzees. I am making this point clear since many pro-evolution zealots tend to demonstrate a knee-jerk reaction to the article without even understanding its argument. My statements regarding some weaknesses of the theory are not used as premises for the conclusion of my arguments, but only to inform the reader about some of the controversial issues between the parties. Even if you smell a bias in my depiction of these issues, even if you think that I am very wrong in those depictions, do not get distracted. Trash those side issues and convict me as "ignorant" or "biased" on them, and get to the main argument, which is:

Evolution of species through mutation and cumulative selection, as subscribed by the modern scientific community, provides sufficient evidences for the existence of immanent intelligent design in nature. The theory of evolution provides evidences about an intelligent designer more than a fingerprint on a canvass could provide clues about the identity of a human painter. Inferring the existence and some attributes of an intelligent designer from nature is as equally scientific as inferring the existence and some attributes of an unknown creature from its footprints left on the sand.

The Genesis

We all started our adventure on this planet as the tiny champions of a vital and brutal competition. Half of all our genetic material was once an individual sperm akin to a tadpole. Hopefully, the events immediately preceding our lives included some laughter and mutually affectionate kisses. After a day-long marathon in a tube not longer than a pen, starting from vagina through the cervix and uterus we finally met our other half and won the award for or condemnation to life. (I am aware that this individual genesis would be told in reverse order if the author of this essay were a woman: "Half of all of our genetic material was once individual eggs waiting…") After reaching the eggs of the chosen female, as the champion sperms, most of us caused the eggs to close their entrances and condemned the other millions of our brothers to death. Whether we like it or not, we started as a selfish gene by causing the demise of millions of viable yet a bit slower or unlucky sperms like us. We are merely the children of murderers who call themselves victors throughout the history. We also started our lives by mass-murdering potential brothers. We are the children of Cane; we are the survivors of ferocious wars, both in macro and micro worlds.

Yes, after our organic rockets hit our organic planets, we became zygotes and we started the 266 daylong evolution, hopefully sans-mutations, in our mother's belly. The approximately six billion bits of DNA program coded in the language of four bases or nucleotide, Adenine, Cytosine, Guanine, and Thiamine create the three-pound jelly, the human brain, whose complexity is beyond our (or, ironically its own) immediate imagination.

There is evolution everywhere: in genes and organs; in stars and planets. Everything, from the smallest organisms to a human... As once a Greek named Heraclitus said, "Everything changes except change itself." You may wish to exclude God, math, or universal laws from this universal statement, but you cannot deny this fact. The mutation of the flu viruses is a well-known fact. The germs are mutating and those that survive antibiotics are now causing a great concern for the health industry. This fact alone is sufficient evidence indicating to at least an intra species evolution.

Though the theory of evolution has produced a brilliant explanation for many questions regarding the origin and diversity of life on this planet, it has also failed in producing explanations for numerous questions. Furthermore, the theory arguably lacks some important characteristics of a good scientific theory since it is not falsifiable. Let's listen to both sides:

- Why did that animal not survive?
- **Because it did not fit the environment.**
- How do you know that it did not fit?
- **Because it did not survive. If they fit, they survive; if they survive they are fit!**
- What? If F then S or if S then F?
- **No, If F then S and if S then F.**
- Wow!

Let's try another one:

- Can you give me an example of a falsifiable claim regarding evolution?
- **Of course! For instance, when populations of bacteria A and B are exposed to low levels of toxic substance X, the fraction of the bacteria resistant to X will increase with time.**
- So what?
- **The experiment is run and the hypothesis correctly predicts the outcome for bacteria A, but not B. Success or failure for evolution?**
- Your hypothesis is not falsifiable as you claim.
- **Why?**
- Because it is circular and the word "low" is too subjective.
- **How?**
- It is circular since it is no more than saying "those who do not die because of their strength will survive." If none survives, you can easily claim that there were no resistant bacteria. Second, the word "low" is not defined before the event in question. If none survives you will call it high, if some survive you will call it low. Furthermore, the predictive power of your statement regarding the bacteria is close to the predictive power of "Dear Nancy, you will give birth to either a boy or a girl."
- **But, what about the Intelligent Design argument? Is it falsifiable?**
- No. For any of the 'not-so-intelligent design' examples you bring, the proponent might reject by saying, "In the past, people claimed similar things for this or that, and with time, when we got more information about their purpose and function we learned that they were indeed very intelligent designs. For instance, once scientists thought that the sharp hairs, awns, or bristles were useless, and they tried to remove them from spikelets. Guess what? After obtaining grains without those pointy hairs, they learned to their dismay that those sharp appendages were protecting the grains from birds. So, we should investigate the reason behind apparent flaws."
- **What about birth defects? Abnormal mutations?**
- The proponent of intelligent design might even accept flaws by saying, "Flaws are there to highlight design through contrast. Without the existence of flaws, we could not know or appreciate design. The existence of a single example of an intelligent design is sufficient to show the existence of an intelligent designer."

It is also argued that the theory of evolution does not have predictive power on specific events:

- With the humans giving up from hunting in the jungles and turning to sedentary office workers, would this ecological change ultimately select the spherical nerds?
- **Spherical nerds?**
- Yes, brains with horizontally grown bodies!
- **It depends...**
- Will humans finally get wings?
- **It depends...**
- Will the thumbs of the descendants of my X-boxed son finally end up with fast and furiously big thumps the size of hot dogs?
- **It depends...**

- Will cats learn how to use remote control?
- **It depends...**
- Wow!

Some proponents of the theory of evolution argue that the theory of evolution demonstrates all the characteristics of a scientific theory. For instance, proving that dinosaurs and humans co-existed would falsify the theory. Even if the critics of the theory were right regarding their assertion on the falsifiability and predictive power of the theory, the theory of evolution is more scientific than the stories of creation believed by billions of people, since it provides a consistent, parsimonious, progressive and verifiable explanation regarding the diversity and complexity of life forms on this planet. My argument in this paper does not rely on this issue. Regardless of the value of the theory of evolution, I argue that the presence of intelligent design is self-evident.

Methinks it is Like a Blind Watch-watcher

To refute the Creationist's argument of the impossibility of a monkey typing the work of Shakespeare, Richard Dawkins provides probability calculations of a random work on a computer using 26 alphabet letters and a space bar, totaling 27 characters. To randomly type Hamlet's 28-character statement, METHINKS IT IS LIKE A WEASEL, it would take 27 to the power of 28 keystrokes, which would be a very small odd, about 1 in 10,000 million million million million million million. Instead of single-step selection of random variation, Dawkins suggests us to program the computer to use *cumulative selection*. The computer generates some random 28 characters and selects the one that most resembles the target phrase, METHINKS…

> "What matters is the difference between the time taken by *cumulative* selection, and the time which the same computer, working flat out at the same rate, would take to reach the target phrase if it were forced to use the other procedure of *single-step selection*: about a million million million million million years. This is more than a million million million times as long as the universe has so far existed. ... Whereas the time taken for a computer working randomly but with the constraint of cumulative selection to perform the same task is of the same order as humans ordinarily can understand, between 11 seconds and the times it takes to have lunch... If evolutionary progress had had to rely on single-step selection, it would have never got anywhere. If, however, there was any way in which the necessary conditions for cumulative selection could have been set up by the blind forces of nature, strange and wonderful might have been the consequences. As a matter of fact that is exactly what happened on this planet, and we ourselves are among the most recent, if not the strangest and most wonderful, of those consequences." (Richard Dawkins, The Blind Watchmaker, Norton, 1987, p.49).

Though he is a bright and articulate scientist, Dawkins takes too many facts and events for granted without even mentioning them: such as the number of characters, their proportion, the computer programmer and program that selects the right characters, the energy that accomplishes the work, the existence of characters, time and space, the continuity of their existence, etc. In the following page, Dawkins distinguishes his METHINKS example from the live evolutionary process.

> "Evolution has no long-term goal. There is no long-distance target, no final perfection to serve as a criterion for selection, although human vanity cherishes the absurd notion that our species is the final goal of evolution. ... The 'watchmaker' that is cumulative natural selection is blind to the future and has no long-term goal." (Id, p.50).

Here Dawkins affirms that he added his intelligence and teleological intention by determining a target, criterion for selection. Thus, Dawkins takes for granted many facts and events, and gives an analogy of a computer program in which he interjects his intelligence, a target, and a selection criterion to explain something that according to him has none of them.

Dawkins who depicts human life as the work of a blind process has a much bigger problem: His theory and its conclusion do not have the light of reason. Let me explain with some analogies. If you now feel an urge to seek an immediate refuge in Hume, please be reminded that they are given to explain inferences to the best explanation. (I recommend Elliott Sober's Philosophy of Biology, containing a brief yet sound criticism of Hume's critique of analogies).

The Assembly Line, the Gullible and the Blind

Assume that we have constructed a completely automated assembly line that manufactures automobiles run on fuel-cells. It receives raw materials such as steel and plastic from one end, and after passing through an assembly line run by computers and robots, it spews out automobiles from the other end.

Now assume that we brought two members of a primitive tribe living in an isolated jungle and placed them in front of the exit door. When a car emerges from the exit door, you enter the car and start driving it. You then stop and watch the reaction of the two tribesmen. You see that the one on the right is awed by the moving beast and is thanking God for showing him a miracle by creating such a complex creature in a few seconds.

Let's assume that the other tribesman on the left side is more curious and adventurous. He wonders about the whereabouts of the room behind the exit door. After some trials, he finds an opening somewhere and able to peek into the room. He sees some robots spraying paint on a car. He touches the paint and notices that it is liquid. After that observation, he comes back and shares what he saw with the believing man on the right. "The shiny stuff on this beast is not too thick. In fact, it was liquid before it was sprayed thinly over its solid skin." But, what about the skin, what about the round circle that determines its direction, and what about the power that moves it? The curious man makes numerous trips, entering some other rooms of the assembly line compound, either by forging a key or luckily discovering a peephole… He learns that the raw materials are spilled in molds upon their arrival and the beast is gradually assembled from simple parts. For instance, the doors are attached by robotic hands through hinges. Though he is not able to access some rooms to explain some stages of the assembly line, he gets a good idea how from simple raw building blocks a complex and powerful beast called automobile could emerge. After getting some ideas about the modus operandi of the assembly line, the curious infers what could have happened in the rooms that he could not access. The believing man outside, who is still intoxicated in spiritual awe, is not impressed by the finding of the curious tribesman. He finds problem in the theory of the curious man since he is not able to explain some events in the assembly line. "You see, you cannot ignore the divine mystery and hand in the creation of this beast!"

The believing man declares that an Omniscient and Omnipotent Creator or an Intelligent Designer created the beast in a second or at worst case scenario in six seconds out of steel and plastic. The believing man goes further and declares his friend to be a heretic disbeliever deserving to burn in Hell forever. The curious man, on the other hand, declares that there is no God of gaps, nor an Intelligent Designer or Engineer, since he had seen none in those rooms. Besides, the curious man brags about his knowledge of most of the events in the evolution of the beast and declares that his friend is a delusional lunatic who deserves to be restricted from expressing his opinion on the evolution of the beasts, especially in public places and in front of children.

Why do most affirmers in God ignore empirical evidences in His creation, while on the other side, most of those who study the empirical evidences ignore intelligent inferences? Parties in the evolution controversy may see each other in these two characters, but perhaps none will identify himself with them. So, let me change my story. Instead of human characters, I will pick some marbles.

Marvels of Marbles

Now let's entertain a thought experiment. We have a gigantic box full of millions of glass marbles. Marbles in different colors, different shapes and sizes… You are an eternal, infinitely patient and curious observer. The box is in a huge empty room and every minute it is tilted by a machine and the marbles are spilled over the clean and smooth surface of the empty floor. Let's assume that you are not interested in the box, machines and the basic laws they follow. You are just interested in the adventure of marbles. Each time, marbles create a particular design randomly and they are filled back to the box to start over the process.

Assume that these events continue for billions of years, trillions of times, without generating anything categorically different. But, in one of the occasions, some of the marbles that were spread over the floor come together and join each other. They then start moving around as a group, slivering through other marbles. Then this gang of marbles start jumping and multiplying. Some even start talking to you. You now may imagine the rest of the story, the marvels of these marbles.

Given an infinite number of trials and years could these happen? If your answer is a "No" then why no? Because they are just made of glasses? What is the difference between glass of marbles and atoms? What is the difference between a cluster of glass marbles and molecules? Well, now you are ready to think about a question and find the answer that somehow eludes some of the brightest scientists. Now, you are ready to see the light of the Intelligent Design in everything, including evolution, including in evolution between species. Do you smell the cheese? Not yet.

The Genius in Hydrogen

Now let's leave the marbles in their box and focus on the simplest atom, Hydrogen. You know that a hydrogen atom has one proton in its nucleus, one electron in its shell, and it does not contain a neutron. Though the structure of each atom is a very complex and precise design, they are somehow seen by the blind watch-watching evolutionists like children see marbles.

The masses of stars are mostly made of Hydrogen atoms. When two hydrogen atoms fuse together they release some energy and particles, and they "mutate" to a Helium atom, a different "species" in periodic table of elements. We know that Hydrogen and Helium atoms have different characteristics, and they behave differently and associate with other atoms differently. When you put two pennies or marbles next to each other or fuse them together they do not act differently; they are still what they are. Their mass and gravitational force may increase, but that is it.

When two Hydrogen atoms fuse together, the information about Helium must have been innate or intrinsic in both of them. Since both Hydrogen atoms are the same, they must contain exactly the same information necessary to create the characteristics of Helium. The information might be triggered by the pressure of fusion. Each Hydrogen atom must contain particular information, since two Hydrogen atoms do not create any characteristics, but the particular characteristics of the atom we call Helium. Thus, Helium must be immanent in Hydrogen. Since Helium and Hydrogen fused together may create Lithium, then the information about Lithium too must be immanent in Hydrogen. In fact, based on the same reasoning, we must expect

Hydrogen to contain all the information regarding the characteristics of each element in periodical table. It is the change in the quantity of protons that leads to qualitative change.

When two Hydrogen atoms associate with one Oxygen atom they create water, the essential ingredient of life as we know. However, when two Hydrogen atoms associate with two Oxygen atoms they create Hydrogen Peroxide, a powerful oxidizer that kills living organisms. Thus, the Hydrogen and Oxygen atoms must contain the information for both molecules. The information inherent in them must lead to Water when they are combined as H2O and must lead to Hydrogen Peroxide when they are combined as H2O2. Since we know that the information of Oxygen must be immanent in the Hydrogen, all this information must be contained in every Hydrogen atom.

I hear the voice of my rocket-scientist friend opposing to my Hydrogen example. So, let's sidetrack a bit to deal with his voice. (If you are a prototype layman who thinks that rocket-scientists are a different species, then you may skip this section and go to the paragraph starting with "In sum, ..."):

> "So all of mathematics is immanent in 1 since the combination of 1 and 1 is 2, therefore the properties of 2 must be immanent in one. But also 3.141592654 is obtained by the spatial ordering of different combinations of 1, therefore 3.141592654 must be immanent in 1. I think there is something of a problem here. From 1 alone, one cannot intuit 2 or any other number except by application of rules which (in this example) can be somewhat arbitrary when applied to 1. Are all verbs immanent in the noun? These things are part of a larger context, perhaps indivisible from that context."

What a wonderful refutation, isn't it? My friend just explained the diversity of elements in the periodic table and their millions of off-springs in the nature, by reducing Hydrogen to our poor and ignorant number 1 which is oblivious of even numbers, prime numbers, perfect numbers, Fermat numbers, and infinite of other numbers begotten as a result of numerical polygamy among the clones of the number One! Interestingly, my scientist friend picked two of his examples from IMAGINARY world of human mind: math and human language. Though the language of nature is written in mathematics, as Galileo once articulated, it does not reflect the "properties" of numbers. Yes, "one odd number plus one odd" becomes an even number, but "one odd chair plus one odd chair" does not become "even chairs." In other words, the properties of numbers are not reflected in the real world. The same is true for our grammar rules. (On this issue, I highly recommend the section in chapter about Pythagoras titled "Where is the Number 2?" in Lovers of Wisdom by Daniel Kolak.).

IN SUM, millions of organic and inorganic compounds, including the ones that yet to be discovered, with their distinct chemical and physical characteristics, must be the materialization of the information immanent in the tiniest building block of the universe, that is, Hydrogen. Going backwards, the same qualities must be imputed for the most fundamental subatomic particle. No wonder Heraclitus had brilliantly inferred that intrinsic law permeating the universe, and called it "logos."

Furthermore, when a particular combination of a particular set of elements in particular proportions generates the function we call *life*, the laws or rules of such an event must have existed before the event occurred. In other words, the laws and rules determining how a particular DNA sequence would behave must have preceded the actual occurrence of the event. Why should a particular configuration of particular molecules made of a particular combination of elements lead to a cell or a living organism? Who or what determined such a magical configuration? None, just chance? No, not a chance! *No, not by a chance!* Chance does not lead to laws. In fact, chance itself is subject to the laws of probability. The laws dominating the universe came into existence with the first moment of Big Bang. If you bet your entire wealth in a casino, you will most likely lose it and you will deserve the title of "another mathematically challenged person" and you may even receive a silver medal in the next Darwin's Award. But you can bet your entire wealth on a scientific prediction based on natural laws and you will most likely win.

It is because of the natural laws of cause and effect that scientists can employ reason and predict events. Mendeleyev knew that elements were not acting haphazardly, so he discovered the periodical table. Thus, it is irrelevant how many millions or billions of years passed before the first organism came into existence among random and chaotic chain of chemical and physical events. Starting from the first seconds of creation of material particles 13.7 billion years ago, the conditions and laws of life must have come into existence too. What scientists do is not inventing, they merely discover. Scientists do not invent laws of physics or chemistry; they learn those laws bit by bit, after tedious experimentation, and based on the information they acquired they put together the pieces of Legos. The characteristics of each newly discovered shape was coded in their nature since the beginning of the universe.

Thus, when a blind watch-watcher refers to the age of the world and its size to explain the marvels of blind cumulative selection, we should not be blindly accepting his argument. The information or laws of life existed billions of years even before the emergence of life. So, we should demand an explanation regarding the a priori information of creating the design of living organisms. Ken Harding, in an article entitled, "Evolution for Beginners," articulates the role of information encoded in genes:

"One of the most common misunderstandings regards "information". The difference between living and non-living things is that living things have information embedded in them which is used to produce themselves. Rocks contain *no* instructions on how to be rocks; a fly contains information on how to be a fly.

"Information is not a thing. It, like an idea, is dimensionless. It's simply a comparison between one thing and another, like a list of differences. Information is not a physical property. Information becomes tangible only when it is encoded in sequences of symbols: zeros and ones, letters and spaces, dots and dashes, musical notes, etc. These sequences must then be decoded in order to be useful. For information to be stored or transmitted, it must be put into some physical form- on paper, computer disk, or in DNA- all processes that take energy.

"Life's information (the instructions on how it works) is encoded in genes, which are decoded by biological mechanisms. Then these mechanisms manufacture parts that work together to make a living organism. Like a computer that builds itself, the process follows a loop: information needs machinery, which needs information, which needs machinery, which needs information. This relationship can start very simply, and then over many generations build into something so complicated that some people can't imagine how it ever could have gotten started in the first place. It is important to recognize that the information encoded in DNA is not like a blueprint, which contains a scale model image of the final product, it is like a recipe-- a set of instructions to be followed in a certain order. Life's complexity arises from remarkable simplicity.

> DNA's message says, "Take this, add this, then add this… stop here. Take this, then add this…" These actions are carried out by a variety of proteins. The result is all the intricacy and diversity of the biological realm.
>
> **http://www.evolution.mbdojo.com/evolution-for-beginners.html**

The issue, however, gets even more interesting. Not only living *organisms*, but their products too must be the consequence of "blind" evolution.

Just take the beginning of the universe and our modern world. Do not let anyone distract you by the events that occurred in between. How can our modern world together with everything in it be the product of a big chaotic explosion? How can such an explosion create the libraries, computer programs and all cars on our streets, in less than 14 billion years? Now, the blind watch-watchers want us to believe that all the books in the Library of Congress, including all the data in our computers, our inventions, and technological marvels, yes all of them are the result of marriage between Mr. Drunk Chaos who is unpredictable and Mrs. Blind Evolution who works according to the principle of *cumulative selection*. If the laws of the universe are deterministic, then the immense amount of information and design permeating our libraries, factories and stores must also be the necessary product of the Big Bang. Not only the initial conditions of the universe had the potential for all the subsequent things at the very start, following deterministic laws they were bound to create human intelligence and leading to landing on the Moon and the I-pod. Even a small fraction of products designed by human intelligence cannot fit in a trillion-year-old universe, let alone one that aged 13.7 billion years-old, via probability calculations. Nor can they be explained by "random (or not random) mutation" and "cumulative selection."

I hear the voice of my rocket scientist friend, again. I cannot ignore that melodious voice. Let's all listen to it:

> "Rather than close the door on the question, wouldn't it be fun to try and figure it out by trying to understand how things work? Could a religious person approach the universe with an open mind and, regardless of the processes they work to slowly identify and better understand, consider the effort a joy and giving of glory? Or does God need to be put into a box where the outcomes of all such investigations are predetermined by those who find a more limited deity more palatable?"

I do not feel compelled to respond to these rhetorical questions, since I do not have a problem with accepting mysteries. I myself am a mystery. But I would like to remind my friend that I have no intention to put God in a box. I saw a box and I said that it must have been created by a box-maker. I never claimed that the box-maker was in the box, nor that he/she/it was limited with only making boxes. In fact, I would expect that the box-maker is capable of making cylinders, spheres and many other shapes and things beyond my poor perception and imagination.

Our blind watch-watchers would like us to accept the emergence of human intelligence and its products as a magical moment, as a miracle. A miracle that terminates the application of deterministic laws and guarantees for all its products the immunity from the probability calculations! Because of that miracle or magic, we are asked not to include the probability of authoring millions of books, articles, computer programs, websites, movies, machines, electronic devices, and everything in the Wal-Mart into our equation. The "anything but God" crowd may even talk in quantum language to de-emphasize the deterministic nature of the universe.

All those "Anything but God" people, in fact, believe in many gods!

Ironically, the blind watch-watchers are proud in declaring their disbelief in God or the irrelevancy of God, while they are fanatic affirmers in infinite number of gods. They are polytheists. Every atom contains all the information necessary for life! Whatever affirmers in God attribute to the Creator, the blind watch-watchers attribute to atoms, matter, or energy. Though they are proud of depicting their gods as "random," "blind" or "stupid," after some interrogations we learn that is not to be the case. Just replace the word God with the word matter, energy or nature and you will have the tenets of faith of blind watch-watchers.

- God is the first cause.
- God is eternal.
- God is the source of information.
- God created everything.
- God created life.

Accepting a God that is not bound by the laws of this universe is much simpler and reasonable than accepting all atoms having all the attributes of a deistic God, and again more coherent than creating our modern world, together with human intelligence and this article, out of their blind and stupid collisions. I prefer believing in the creation of rabbits popping up from a magician's hat, than in a universe coming out of nothing and then blindly creating this planet and the intelligent life on it. So, I assert that if Occam's Razor is sharp for every argument, then it must first shave off the idea of stupid atoms coming into existence out of nothing and billions of years later, several billions of them blindly evolving and transforming into Dawkins' mind.

Some atheists might resort to a false argument by pointing at their "undetectable Purple Cow in the sky." Yes, it is a funny example, but far from being persuasive. They craftily wish to equate the argument for an Intelligent Designer to a Purple Cow. This is a cheap rhetoric, since being Purple or Cow has nothing to do with our argument. However, the intelligence and design are in every atom, in every molecule, in every organism of this universe. Besides they are detectable.

We understand why the majority of religious people tend to have problem with science and philosophical inquiry. But why have many scientists become "anything but God" fanatics? It might be because of the ridiculous claims and arguments of religious zealots who oppose the theory of evolution in the name of God. Atheists have not taken even a small step to answer the fundamental questions related to the issue. What is the cause of the universe or singularity? There is a particular amount of mass in the universe, let's say, N number; why is it N number, not more nor less? Who or what determined the exact amount of mass or the exact number of atoms/particles/energy in the universe? (We would not have this question, of course, if the entire universe was homogenous). How is the probability of the existence of a universe with fine-tuned constants essential to life? Did our universe have infinite time? Are there infinite universes? Is infinity really pregnant to all possibilities? Why is there something rather than nothing? Why is the universe governed by laws? Why do the biological organisms have propensity to mutate? They might believe that answers for these questions are not in the domain of science. Then, how can they claim that the universe and evolution of living beings, from the structure of atoms to the structure of brain and its products, does not need God?

I should again share with you the voice of my scientist friend:

> "Actually, there are many scientists pondering these questions (but the last), and many or some may be atheists. Does it matter? If an atheist drives a car, does that mean the believer should not? With regards to the last question, are affirmers afraid that not needing God in the theories formulated to try to explain observations of life and the universe will prove there is no God? I think that is the fear of many anti-evolutionists. It exposes the weakness of their own faith, that

> they need compelling external evidence that God must exist."

Well said. But I do not think that it applies to me and many other "rational monotheists," since my acceptance of God is not based on "faith," a euphemism for "joining the band wagon" or "wishful thinking." My acceptance or knowledge of God is based on extensive scientific evidences and philosophical inferences, which I am hoping to share with others in a book titled, "19 Questions for Atheists."

We might be able to duplicate or copy life in the bio-world, but we have not yet been able to imitate the full capabilities of biological assembly line in our technology. We have not yet seen any computer giving birth to other computers. Perhaps, with the progress of our production technology, we may witness it in the future. Assume that a scientist discovered a method for evolving computers or gadgets that could multiply by RANDOM MUTATIONS and CUMULATIVE SELECTION. Wouldn't this SIMPLE task be INCREDIBLY INGENUIOUS? What if "nature" had created inorganic materials with such a quality? Would you consider such a "creation" lacking intelligent design? Or would you just say that the "the evolving and multiplying computers by random mutations negate God's intelligence and involvement in the creation process completely"? What about your intelligence? Aren't you a product of nature? How come an intelligent person like you was generated by a dumb and stupid process?

Intelligent design is in every moment and point of evolution (71). There is an intelligent power and wisdom that designs incredibly simple assembly lines that can manufacture incredibly complex organisms and creatures, including the intelligent watch-watchers and blind watch-watchers. The signature of the Intelligent Designer in the book of nature is paradoxically as obvious as the number 19 in 74, and as concealed as the number 19 in 74.

Let me give one more chance to the voice of my scientist friend:

> "Perhaps the signature is found in our perception of beauty of how things work? Don't know about the numerological references - I think most audiences might scratch their heads and wonder what was up with that?"

Yes, indeed. Let those audiences keep scratching their heads. Who knows, if they are curious enough, they will smell the beef after tasting the cheese and learn what was up and down with my numerological references. After all, "On it is nineteen!"

PS:

A delusional cult leader from my country of birth is doing a great disservice to Islam by copying and promoting the works of Evangelical Christians and the Discovery Institute. The theory of evolution is supported by many verses of the Quran as I discussed in the endnotes of the *Quran: a Reformist Translation.* (For instance, see: 15:28-29; 24:45; 32:7-9; 71:14-7)

In fact, the theory was first promoted by Muslim scientists. My colleague Dr. T. O. Shanavas, in his book, *Islamic Theory of Evolution: the Missing Link between Darwin and the Origin of Species*, provides references from the works of major Muslim scientists such as Ibn Sina (Avicenna), Ibn Rushd (Averroes), Muhammed Al-Razi, Ibn Khaldun, Abu Bakr bin Tufayl, Muhammed al-Haytham (Alhazen), Al-Biruni, and provides substantial evidence that Darwin got his inspiration from them through his father Erasmus Darwin. In fact, Darwin's contemporary opponents accused him being influenced by "Barbarian Muhammedans."

It is a travesty that today Muslims have regressed so much they are now peddling pseudo-science against God's system in creation.

12
Sample Comments and Discussions on the Reformist Translation of the Quran

Predictably, reactions to the Reformist Translation have been controversial; from very positive to very negative. The QRT so far helped many Sunnis, Shiites, Christians, Jews and agnostics to embrace islam, that is peacemaking and submitting to God alone in peace. It also sparked passionate discussions in many other Internet forums and groups. Several interesting reviews were also published. A selection from those discussions and reviews will soon be published within the Rainbow Articles series. Below are a few samples from those emails and discussions:

THANK YOU, DEAR GOD THANK YOU, I'm coming back because of your work. Edip, your work has freed me from year of condemnation, cruelty and misinterpretation of islam by my ex husband. Your work has freed me from the pain I've carried for so long and gave me back basic self esteem that was mine from God but was slowly eroded by misogyny. Your work gave me the wings I had lost. Again thank you. ~ **Martina D.**

Your translation is fabulous. Dear Mr. Yuksel, I purchased a copy of your translation of the Qur'an a few months ago. I will admit that at first I approached it a bit warily. But now, I wish I had one in a pocket-size volume so I could use it in a similar way to how Christians use their Bibles. not to pound people over the head with the content and try to convert them, but to show them that there are many misconceptions about what the text says and means. I get at least one comment or question every day that shows me how many people have NO clue what the Qur'an really says or means in places, even Muslims who can recite it in Arabic from Opening to end. As you point out, many Muslims have no clue that there could be a different legitimate meaning to what they read. Many non-Muslims have picked up older translations, such as Mevlana Ali's translation, looked for the only bits and pieces they know about (through the popular press), and confirm in their minds that Islam is a backward, woman-hating, war-mongering religion. The Reformist Translation is an excellent tool in defusing that position in that it clearly shows that many passages have been misunderstood, and that the true meaning is something much more compassionate and sensible. It helps to affirm the concept that "There is a reason the Qur'an begins with "In the Name of God, the Merciful, the Compassionate"." Your translation is a brilliant breath of fresh air, and I am grateful for it. I like that you show not only how a section of the text can be translated, but the REASONING behind it. This is the key to the reader being able to see that there are legitimate "alternative" translations to what we've been given up to now, rather than just once again blindly accept another point of view to replace the one they're coming from. Blessings. ~ **Cait Ramshaw, Ft. Pierce, Florida, USA**

A refreshing and insightful new translation of the Qur'an. The most cogent advice I received when I started reading the Qur'an (several translations/transliterations ago) was to understand and absorb it as a whole, all-at-once message and not through extracted excerpts. Reality is not linear but the written word is necessarily so. Truth-seeking is likewise a somewhat linear experience, if it has been a life-long effort for the individual.

I suggest that the most integrated understanding of the Qur'an can only be realized by synthesizing the full message in one's heart, as a single experience. With this in mind, this

Reformist Version does an unusually fine job in clarifying those elements (such as gender imbalance) which have been perceived as dissonant within the whole message in the "standard" translations. This version, which is not revisionistic, presents an integrated consistency not found in those translations and it elucidates issues not commonly grasped by modern readers (in any language). Those with an open mind and heart, who only understand modern Arabic and not the dialect in which it was originally revealed, have the opportunity to experience comfort and inner peace by absorbing this clean, Reformist translation. With this in mind, this version can only be judged following a thoughtful read of the entire volume.

In addition, additional analyses within the commentaries, shed the light of new understandings, which can be a relief for many open-minded, humble readers, Muslim and non-Muslim alike. This work also offers the possibility of beginning to open the door to the resolution of the conflicts, among the Abrahamic religions, including among the Islamic peoples, so common over past centuries. This can only work through peaceful, thoughtful contemplation and discussion among those with varying views and sincere hearts. To my understanding, there is a consensus within the broad, thoughtful Islamic community, that on the "final day," we are all alone before our Creator, taking personal responsibility for our life. If there are errors within the translation or my understandings, they are owned by whoever created the errors. This important new English version is a well-intended effort and a positive contribution to the ancient and honorable Islamic principle of ijtihad. For those who disagree, at the deepest levels within their "hearts," they should remember that there really can be "no compulsion in religion." ~ **Jeffrey Garrison "Earthyman", Colorado Springs, Colorado, USA.**

Earth-breaking! Truly outstanding approach to understanding Quran. An intellectual revolution, yet called a reform. RTQ brings students of Quran to a whole new level of research complexity, taking it away from medievalist methodology, connecting with modern advances in sciences, thus bridging the artificial gap between West and East. ~ **Daniyar Shekebaev, Moscow, Russian Federation**.

A Searching Christian. I am a Christian and I have a great affinity for Islam, and especially al Quran. Islam has captivated me for the better part of 20 years. During that time I have studied many different translations of al Quran, but your translation is the greatest that I have read. My walk with our Creator has been intense. As such, I know that God has been with me throughout my triumphs and my challenges. I would not be able to manage otherwise. God has numbered our hairs and our breaths; God is intimately involved in sustaining our lives. To not give our Creator the Glory and remembrance that he deserves is a crime. I have been graced by our Creator...I know that he carries me through the day.

And so, in a sense, I consider myself a muslim. Yes, I am well aware of the issues that separate our religions; however there is much that binds us. One of the things that I have wrestled with in Islam is the whole Jesus (pbuh) disagreement. I understand the logical argument of Islam for considering Jesus (pbuh) a Prophet. I solidly cannot wrap my head around the rejection of Jesus' divinity. Christians tend to say it is a matter of faith, and Muslims tend to claim that Christian scriptures are not accurate. This is not a basis for coming to a common ground. One thing is for sure, we all believe in One God. I often point out to my Muslim friends that Allah has 99 names, and infinite qualities. The Creator in Christianity revealed only three distinct qualities. This argument does not take any traction though… God be with you Edip. Peace. ~ **Thomas Lane**

God Bless You Edip. I thank you for the work; you are doing in bringing and supporting much needed reform. I took *shahada* years ago in a Sunni mosque and immediately afterward they started telling me that I could not wear gold or wear silk and a ton of other rules that bogged me down. I did not go back to the mosque after that. That seed of pure monotheism that drew me to Islam became covered up and I could no longer appreciate the Quran because I was brainwashed that to reject their brand of Islam I was rejecting the Quran. But then I came across an article by you online. Then ordered the translation of the Quran you worked on. It is beautiful. As I read and contemplate God's message I am thankful for the work you have done. I have been wrapping my head around a few things and was wondering if you might answer a few questions I had. Feel free to answer if you are willing when you have time. ~ **Jason C.**

I was soon ready to quit studying and never pick up the Quran again! For me, my curiosity about Islam was a sudden thing that started not long ago. Now in my late twenties, I was raised as a Protestant Christian, but early in my teens I came to feel that Christianity did not emphasize the pure message of Jesus (pbuh) because it misunderstood His true role on earth. As time went on, I stayed out of "organized religion". However in the past few years, I began studying Eastern religions and philosophies and gained a deep appreciation for Buddhism because of its emphasis on kindness, good works, and peace. I still appreciate Buddhism, but it failed to address the importance of maintaining a connection with God which I firmly believe is the highest goal of life. Although I was certainly aware of Islam, I think that I ignored it because of my ignorance in assuming that it gave little or no importance to Jesus (pbuh). I now know better! However, at the time, I had no real opinion or bias one way or the other, and I did not know any Muslims who might inform me otherwise.

So as odd as it sounds, one night I went to sleep, and the next morning when I woke, (literally) the very first thought that came to mind was that I needed to learn about Islam. It was surprising how strong an impulse it was, but I could not deny it. It was just something that I suddenly felt was very important.

So began my study of the Quran and the faith in general. Coming to my studies, I already knew that any faith worth accepting should be built on a foundation of encouragement of reason and critical study, the "golden rule", improvement of personal character, loving kindness towards all beings, and equality of rights for all humans.

So you can imagine my confusion and dismay as I studied and discovered the opinions of various scholars and sects, subtly promoting everything from aggressive violence to sexism and racism. I also found the vehement insistence on certain minor details among Muslims to be troubling, not to mention the quickness with which some Muslims would declare others as apostates for simply having a different understanding of things.

Some things that I read in the Quran also deeply troubled me. And when it came to the Hadith, I was dismayed. Where some Hadith showed the Prophet Muhammed (pbuh) as a man of reason and kindness, others were so unreasonable and so cruel, that I hardly knew what to believe! I knew very soon that the contradictions of the Hadith meant I could not hold them in high esteem. My study was starting to be disappointing. I could not believe that the peace which was supposed to be the very essence of Islam could support and encourage such harshness towards others. I was soon ready to quit studying and never pick up the Quran again! But some part of me had come to believe that perhaps just as with Christianity, what I was reading and discovering was full of man-made innovation and sectarian bias.

I knew that there must be Muslims whose understanding captured the true, clear spirit of Islam. So I continued studying. It was finally quite "by accident" that while on-line I discovered a very negative rant about *Quran: A Reformist Translation* (Alhamdulillah, I now own a copy), and this became my introduction to the Truth. I now see that the trouble is not with Islam as I was beginning to think, but with the traditional human interpretations of Islam that are buried under piles of man-made tradition, superstition, sectarian bias, and rigid historical cultural values. It is unfortunate that so many converts like myself tend to never become familiar with the Islam that I now know. But Allah knows best. I am just an infant in my faith. God willing, as I journey through life, may I continue to learn and grow in faith and be a beacon of truth to others (as your work has truly been to me). Thank you. ~ **Palace Le Grand**

This translation needs to be spread. Salam, I realize that I'm probably one of thousands here, but if it is at all possible could you please let the author of this translation know that this is the most fantastic, all encompassing, amazing translation and commentary, detailed and intelligent that I have ever read and to date I have read of 50 translations of the Qu'ran. This is the message in its purest form, much respect to the author. This translation needs to be spread, and translated into all possible languages and distributed world-wide to as many people as humanly possible insha-allah. This could finally get the muslim world out of the horrible slump they are in!! Kind regards. ~ **Adam Abd-Rabbou**

I have been receiving positive feedback from all around the world regarding the *Quran: a Reformist Translation*. However, I also receive hostile letters. Below is one of them written by a Briton. His usage of the word "deen" implies his affiliation with Muslims, yet the language and racist tone he is using to express his religious zeal contradicts the very teachings he is trying to defend. When I was a Sunni fanatic militant, I never used such obscenities and neither any of my comrades did. Perhaps, this is a new trend and culture. To give you an example from a segment of our audience, here is the letter which ironically starts with "dear sir" ending up with expletives:

Greetings from England. Dear Sir, I have just dipped into your Reformist translation of the Quran and I am absolutely appalled. Please remember the clandestine agencies and so called Muslim gay & incestous lobby that has supported your endeavour will fail. All that will happen is that you will receive a Nobel prize and more dollars so you can destroy the deen. All I can do is sincerely curse you from the bottom of my heart and hope that you are savaged by dogs. May your children be saved from your abominable actions. F...k off you Jew loving and gay loving bastard. I hope the Turks wipe out the m...... f...... Kurdish monsters.....you have a filthly blood line any way... .. ~ **Richard Summers, UK**

Dear Richard, you are angry, arrogant, and your mouth smells bad words. Besides, you have hallucinations. Where in my translation have you found my connection with any lobby or my desire to get a prize from anyone? My book is already shunned by major media and it is reviled by bigots from both East and West, by Christianists and Islamists as well. I do not support homosexuality, yet I do not believe that people should be punished for the acts they do in their privacy without imposing on others. I am not for handing clubs to those who arrogate themselves as "moral police." I also condemn Zionist fascism, yet I have no hostility against Jews just because of their race or religious affiliation. Racism is one the biggest crimes against God's law and Satan was the first racist when he rejected serving Adam by claming the superiority of energy over matter. Peace, ~ **Edip**

BRAINBOW? A BLATENT allusion to the LIBERAL COLLUSION. I will be honest here. I have not read this book. Nor do I intend to. Why not, you might ask? Its very simple. Not only is it the korean, the very antithesis to the HOLY BIBLE, but as if that wasn"t enough, it is supposedly a LIBERAL FEMINEST translation (with extensive annotations) to boot!!! I'm not sure which one is even worse!

All I know for sure, is that I certianly wouldn't want a bunch of RUDE WITCHCRAFT WORSHIPING LIBERAL FEMINAZIS running around after a bra burning party and trying to come up with a new translation of the WORD of JESUS CHRIST, which is the BIBLE. Let a bunch of know-nothing LIBERALS, HIGH on PMS, go defile the HOLY BIBLE with their LIBERALY BIASED, MENOPAUSAL WET-FLASHES? NO THANKS!

Fot those wishing to know what all that Koran says, I would have to recomend this book, for its erudition and scholarship, if for nothing else, or, at the very least, for being readable. Apparently, this is a new translation, translated by MOHOMADEAN LIBERAL FEMINISTS. One thinks One wouldn't want a MOHOMADEAN LIBERAL FEMINIST re-translating the HOLY BIBLE Deluxe Miracle Jesus - Action Figure Has Glow in the Dark hands - Comes with 5 Loaves of Bread, 2 Fish, 1 Water into Wine Jug so presumably one woulden't want them doing it to the Koran. To tell you the truth, I dont know WHAT to think about reading the Koran, or even if a CHRISTIAN SHOULD ALLOW HIMSELF TO DO SO. I believe that one should know ones ENEMIES, so thats the only reason why I've started in on this book.

I do wonder though, what it must have been like when WE CHRISTIANS were into useing blood, power thirst, Papal Magic: Occult Practices Within the Catholic Church and violence in our own religion, amongst others as well as between our own kind, like the MOHOMADEANS are doing now, way back in the dark ages before GOD, speaking through the POPE, allowed him to tell women to say that they too, were posessed of SOULS?? GOD BLESS AMERICA!!!

If there is one thing that the Taliban-mohhammedians have gotten right it is their take on the SANCTITY of a woman's place within the HOME. You cant expect LIBERAL FEMINISTS, raised on this SECULAR CULTURE, to respect or to understand concepts of deep SPIRITUAL IMPORT such as this. So how could thier translation of ANY book be trustworthy?

Furthermore, this book is published by some outfit called the BRAINBOW PRESS. BRAINBOW? A BLATENT allusion to the LIBERAL COLLUSION with and AFFINITY for, the GAYHOMOSEXUAL LIFESTYLE. BRAINBOW? RAINBOWS. GET IT? Rainbows used to be for all of us. Now though, DECENT AMERICANS can't lookup into the sky after a spring rain without being reminded of the LOW level to which our CULTURE has sunk since the LSD 1960's. SAD. ~ **Mark Twain** (Commenting at Amazon.com)

Brainbow Press

Quran: A Reformist Translation
Translated and annotated by: Edip Yuksel; Layth Saleh al-Shaiban; Martha Schulte-Nafeh. Brainbow Press, 2007, 600+ pages.

Test Your Quranic Knowledge
Contains six sets of multiple-choice questions and their answers. Edip Yuksel, Brainbow Press, 2007, 80 pages.

Manifesto for Islamic Reform
Edip Yuksel, Brainbow Press, 2008, 128 pages.

The Natural Republic
Layth Saleh al-Shaiban (ProgressiveMuslims.org), Brainbow Press, 2008, 198 pages.

Critical Thinkers for Islamic Reform
Editors: Edip Yuksel, Arnold Mol, Farouk A. Peru, Brainbow Press, 2009, 262 pages.

Peacemaker's Guide to Warmongers
Exposing Robert Spencer, Osama bin Laden, David Horowitz, Mullah Omar, Bill Warner, Ali Sina and other Enemies of Peace. Edip Yuksel, BrainbowPress, 2010, 432 pages.

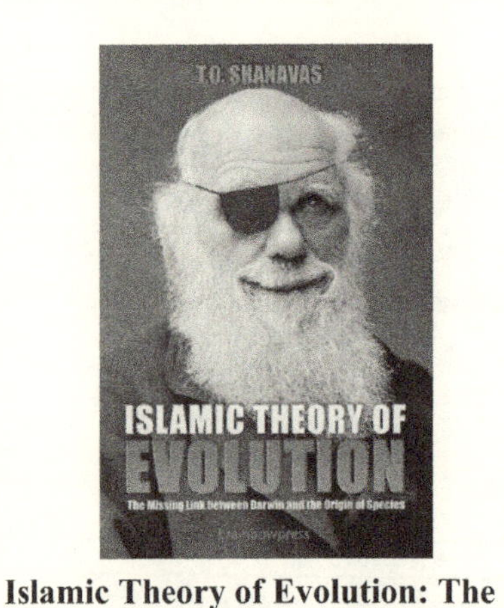

Islamic Theory of Evolution: The Missing Link between Darwin and the Origin of Species
T.O. Shanavas, Brainbow Press, 2010, 240 pages.

Exploring Islam in a New Light: a View from the Quranic Perspective
Abdur Rab, 2010, 460 pages.

<table>
<tr>
<td>

Islamic Renaissance: A New Era Has Started,
Kassim Ahmad
Brainbow Press, 2011, 272 pages,
</td>
<td>

Edip Yüksel
vs
Bilal Philips, Carl Sagan, Daniel Lomax, Ayman, James Randi, Michael Shermer, and other Ingrates
Running Like Zebras
Brainbow Press, 2011, 504 pages,
</td>
</tr>
<tr>
<td>
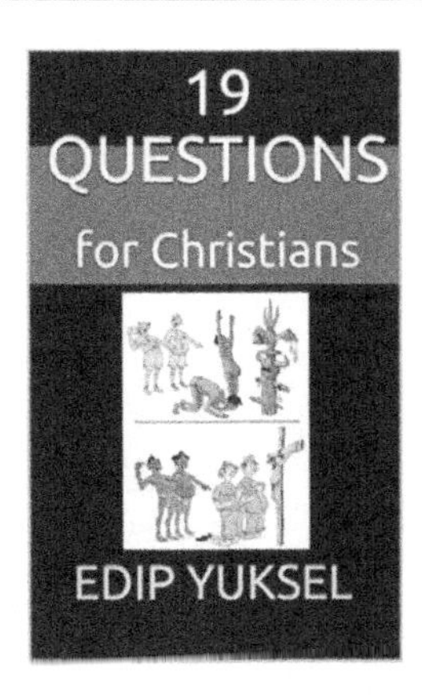

19 Questions for Christians
Edip Yuksel, Brainbow Press, 2019, 197 pages.
</td>
<td>

10 Questions for Atheists
Edip Yuksel, Brainbow Press, 2019, 219 pages.
</td>
</tr>
</table>

NINETEEN: God's Signature in Nature and Scripture

A comprehensive demonstration of the prophetic miracle.

Made in the USA
Las Vegas, NV
31 May 2025